Emerging Contaminants and Associated Treatment Technologies

Series Editors

Muhammad Zaffar Hashmi, Institute of Biology and Biotechnology, The University of Lahore, Lahore, Pakistan

Vladimir Strezov, Department of Environmental Sciences, Macquarie University, Sydney, Australia

Emerging Contaminants and Associated Treatment Technologies focuses on contaminant matrices (air, land, water, soil, sediment), the nature of pollutants (emerging, well-known, persistent, e-waste, nanomaterials, etc.), health effects (e.g., toxicology, occupational health, infectious diseases, cancer), treatment technologies (bioremediation, sustainable waste management, low cost technologies), and issues related to economic development and policy. The book series includes current, comprehensive texts on critical national and regional environmental issues of emerging contaminants useful to scientists in academia, industry, planners, policy makers and governments from diverse disciplines. The knowledge captured in this series will assist in understanding, maintaining and improving the biosphere in which we live. The scope of the series includes monographs, professional books and graduate textbooks, edited volumes and books devoted to supporting education on environmental pollution at the graduate and post-graduate levels.

Manoj Kumar Jindal • Parul Baranwal
Bhumika Das
Editors

Emerging Contaminants in Water

Springer

Editors
Manoj Kumar Jindal
Indian Institute of Science
Bangalore, Karnataka, India

Bhumika Das
Lakhmi Chand Institute of Technology
Bilaspur, Chhattisgarh, India

Parul Baranwal
Oregon Department of
Environmental Quality
Portland, OR, USA

ISSN 2524-6402 ISSN 2524-6410 (electronic)
Emerging Contaminants and Associated Treatment Technologies
ISBN 978-3-031-89590-6 ISBN 978-3-031-89591-3 (eBook)
https://doi.org/10.1007/978-3-031-89591-3

This Springer imprint is published by the registered company Springer Nature Switzerland AG
The registered company address is: Gewerbestrasse 11, 6330 Cham, Switzerland

Preface

Water is the essence of life, yet its quality is increasingly threatened by emerging contaminants—substances that pose significant risks to human health, ecosystems, and water security. These contaminants include pharmaceuticals, personal care products, microplastics, pesticides, per- and polyfluoroalkyl substances (PFAS), radionuclides, rare earth elements, and various industrial chemicals. Despite their often low concentrations, their persistence and bioaccumulative nature make them a growing global concern.

Rapid industrialization, urbanization, and technological advancements have accelerated the release of these contaminants into the environment. Human activities play a crucial role in their proliferation, particularly in regions where regulatory frameworks and treatment technologies are still evolving. The presence of these pollutants in surface water, groundwater, and even drinking water supplies necessitates an urgent and systematic approach to understanding their sources, pathways, environmental impact, and mitigation strategies.

This book, ***Emerging Contaminants in Water***, serves as a comprehensive resource that consolidates scientific knowledge, policy insights, and technological advancements related to this pressing issue. It introduces a structured framework for classifying and assessing contaminants based on their occurrence, distribution, and economic implications. Furthermore, it explores their impact on human health, aquatic ecosystems, and agricultural productivity, emphasizing the need for effective monitoring and management strategies.

One of the key aspects covered in this book is the scientific classification of contaminants and status of these contaminants, which are crucial for understanding their pathways, interactions, and potential risks. Additionally, it highlights innovative solutions for their detection, treatment, and policy implementation.

Emerging contaminants are intricately linked to several United Nations Sustainable Development Goals (SDGs), particularly SDG 6 (Clean Water and Sanitation), which focuses on ensuring safe drinking water by addressing contamination risks. Other relevant SDGs include: *SDG 3 (Good Health and Well-Being):* Mitigating health risks associated with prolonged exposure to contaminants. *SDG 14 (Life Below Water):* Preserving aquatic ecosystems from pollution. *SDG 1* and *2*

(No Poverty and *Zero Hunger):* Safeguarding water resources crucial for food security and economic stability. *SDG 8* and *11 (Economic Growth* and *Sustainable Cities):* Addressing industrial pollution's effects on urban and rural environments. *SDG 12* and *13 (Responsible Consumption and Climate Action):* Promoting sustainable industrial practices to minimize environmental damage.

The importance of this subject cannot be overstated. Access to clean water is a fundamental human right and a pillar of sustainable development. If left unaddressed, emerging contaminants could compromise global water security, disrupt biodiversity, and exacerbate public health crises. This book aims to bridge the gap between research, policy, and practical implementation by providing a comprehensive understanding of national and international challenges and offering solutions from multiple perspectives.

This book is designed for researchers, environmental professionals, policymakers, and students, providing a well-rounded perspective on the critical issue of emerging contaminants in water. Equipped with knowledge, data-driven insights, and innovative approaches, readers will gain a deeper understanding of these challenges. We hope this book inspires action toward safeguarding water resources for future generations.

The world today is grappling with the severe consequences of water pollution. Across the globe, alarming levels of contamination are being reported in rivers, lakes, and groundwater systems. In many countries, the unchecked use of antibiotics and pharmaceuticals has led to the proliferation of antibiotic-resistant bacteria in water bodies. Similarly, radionuclides, PFAS, and especially microplastics—byproducts of modern consumer habits—are now ever-present in both marine and freshwater ecosystems. These issues highlight the urgency of understanding, monitoring, and addressing emerging contaminants before their effects become irreversible.

Emerging Contaminants in Water has been carefully structured into seven parts: *Background Information About Emerging Contaminants*; *Detection Techniques*; *Current Status of Contaminants*; *Red Line for Emerging Contaminants*; *Effects of Emerging Contaminants*; *The Fate of Microplastics, Plasticizers, and PFAS: Emerging Contaminants in Aquatic Ecosystems*; and *Mitigation and Recommendation.* This book examines scientific advancements in analytical methods, the role of policy and regulation, and innovative approaches to wastewater treatment and environmental remediation. With contributions from experts in environmental science, engineering, and public health, it bridges the gap between academia, industry, and policymaking. By shedding light on this critical issue, this book seeks to empower stakeholders with the knowledge needed to develop effective strategies for managing and mitigating emerging contaminants. It is our hope that this work contributes to the global effort to protect water resources and promote sustainable environmental practices for generations to come.

Bangalore, Karnataka, India — Manoj Kumar Jindal
Portland, OR, USA — Parul Baranwal
Bilaspur, Chhattisgarh, India — Bhumika Das

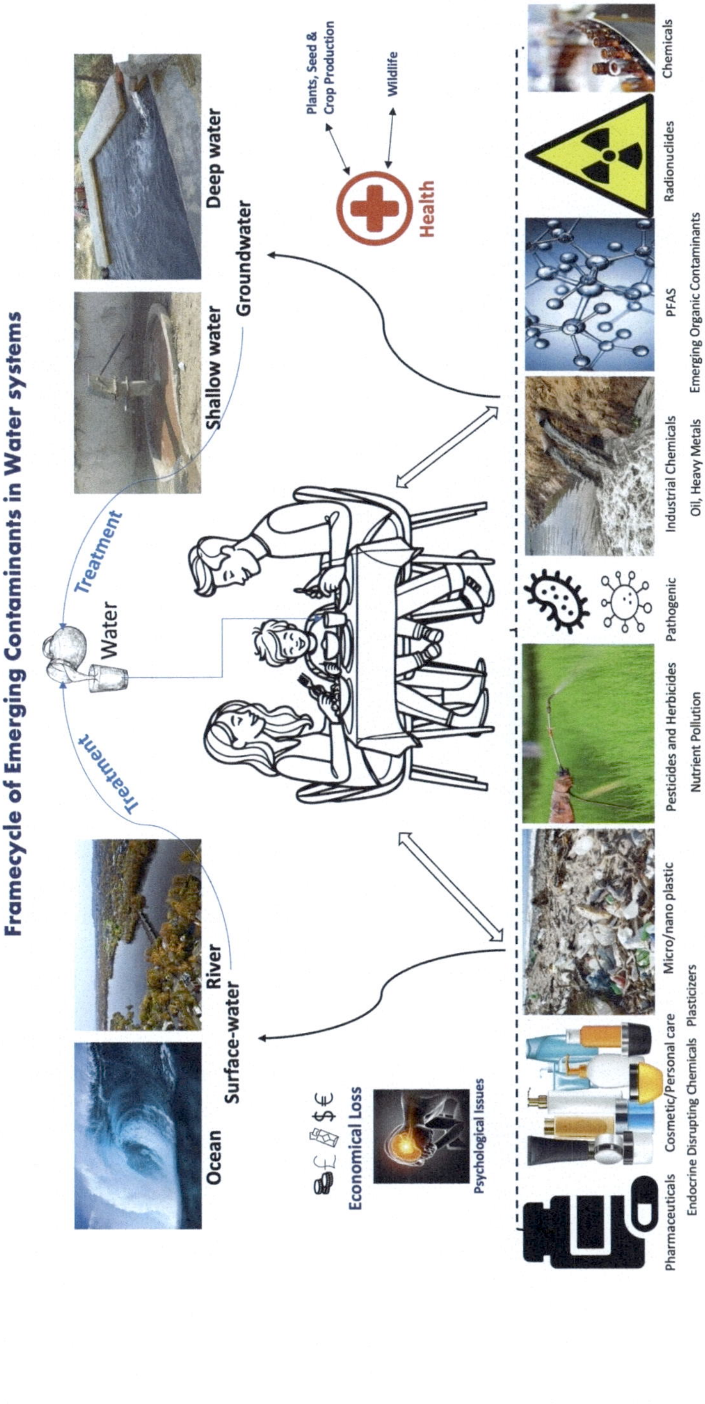
Framecycle of Emerging Contaminants in Water systems
Ocean
River
Surface-water
Shallow water
Deep water
Groundwater
Treatment
Water
Treatment
Economical Loss
Psychological Issues
Health
Plants, Seed & Crop Production
Wildlife
Pharmaceuticals
Cosmetic/Personal care
Endocrine Disrupting Chemicals
Plasticizers
Micro/nano plastic
Pesticides and Herbicides
Nutrient Pollution
Pathogenic
Industrial Chemicals
Oil, Heavy Metals
PFAS
Emerging Organic Contaminants
Radionuclides
Chemicals

Contents

Part I
Background Information About Emerging Contaminants

Chapter 1
Classifying Emerging Contaminants: A Systematic Framework

Sarad Pawar Naik Bukke, Rajasekhar Komarla Kumarachari, Santosh Karajgi, Sangeetha Mani, and Lakshman Kumar Dogiparthi

1 Introduction

In the evolving field of environmental science, "emerging contaminants" (ECs) have become a critical focus because of their potential to disrupt ecosystems and pose risks to human health. Emerging contaminants are broadly defined as synthetic or naturally occurring substances that are infrequently monitored in the environment but have the potential to enter and adversely affect it. According to the World Health Organization and the Environmental Protection Agency (EPA), these substances include pharmaceuticals, personal care products, industrial chemicals, pesticides, microplastics, and specific pathogens. The EPA has identified over 200 emerging contaminants, categorized on the basis of their origins, such as industrial

S. P. N. Bukke (✉)
Department of Pharmaceutics and Pharmaceutical Technology, Kampala International University, Ishaka, Bushenyi, Uganda
e-mail: drsaradpawar@kiu.ac.ug

R. K. Kumarachari
Department of Pharmaceutical Chemistry, Meenakshi Faculty of Pharmacy, MAHER University, Thandalam, Mevalurkuppam, Tamil Nadu, India
e-mail: professorpharma@maher.ac.in

S. Karajgi
BLDEA's SSM College of Pharmacy and Research Centre, Vijayapur, India

S. Mani
Sri Ramachandra Faculty of Pharmacy, Sri Ramachandra Institute of Higher Education and Research, Porur, Chennai, India
e-mail: sangeetha.m@sriramachandra.edu.in

L. K. Dogiparthi
Department of Pharmacognosy, MB School of Pharmaceutical Sciences, Mohan Babu University, Tirupati, Andhra Pradesh, India

M. K. Jindal et al. (eds.), *Emerging Contaminants in Water*, Emerging Contaminants and Associated Treatment Technologies,
https://doi.org/10.1007/978-3-031-89591-3_1

waste, agricultural runoff, or household discharges. For instance, microplastics alone have been found in 93% of bottled water samples worldwide, highlighting their pervasive presence. This classification aids in understanding the sources, behavior, and impacts of ECs, forming the basis for their monitoring and management. Their dynamic nature, complex interactions, and often unforeseen consequences make them pivotal subjects of study [1].

Understanding and addressing the challenges posed by emerging contaminants is crucial because of their significant health risks and ecological impacts. These substances can disrupt hormonal systems [2], contribute to antibiotic resistance [3], and act as carriers for harmful chemicals and pathogens [4]. Additionally, some chemicals can bioaccumulate in food chains, leading to ecosystem-wide consequences, whereas microplastics degrade habitats and threaten species survival [5]. A major obstacle in managing emerging contaminants is the lag in regulatory frameworks, which often do not keep pace with scientific advancements. This gap hinders effective monitoring, assessment, and remediation efforts. Therefore, bridging the divide between scientific understanding and regulatory action is essential for safeguarding environmental quality [6]. The systematic classification of emerging contaminants is a vital tool in this context. Researchers can gain a comprehensive view of their properties and behaviors by grouping these substances based on their shared characteristics, such as their chemical structure or environmental fate [7]. This classification serves multiple purposes: (1) It provides a structured framework for navigating the diversity of contaminants. (2) It acts as a risk assessment toolkit, prioritizing contaminants based on their toxicity and ecological impacts. (3) It uncovers patterns and trends, informing research priorities and targeted mitigation strategies. Moreover, interdisciplinary collaboration is essential for developing comprehensive solutions to mitigate risks and promote sustainable environmental practices. Integrating expertise from chemistry, toxicology, ecology, engineering, and social sciences is crucial for effective management. In summary, systematic classification is not merely an academic exercise but also a practical tool for understanding and managing emerging contaminants. By organizing and categorizing these substances, researchers can inform decision-making and drive effective management strategies, ensuring the protection of environmental health for future generations.

2 Systematic Classification Frameworks

Existing systematic classification frameworks for identifying and categorizing emerging contaminants are essential tools. They organize diverse substances and facilitate research, risk assessment, and regulatory decision-making [8]. These frameworks encompass various dimensions of contaminants, each shedding light on different aspects of their behavior and impact (Fig. 1.1).

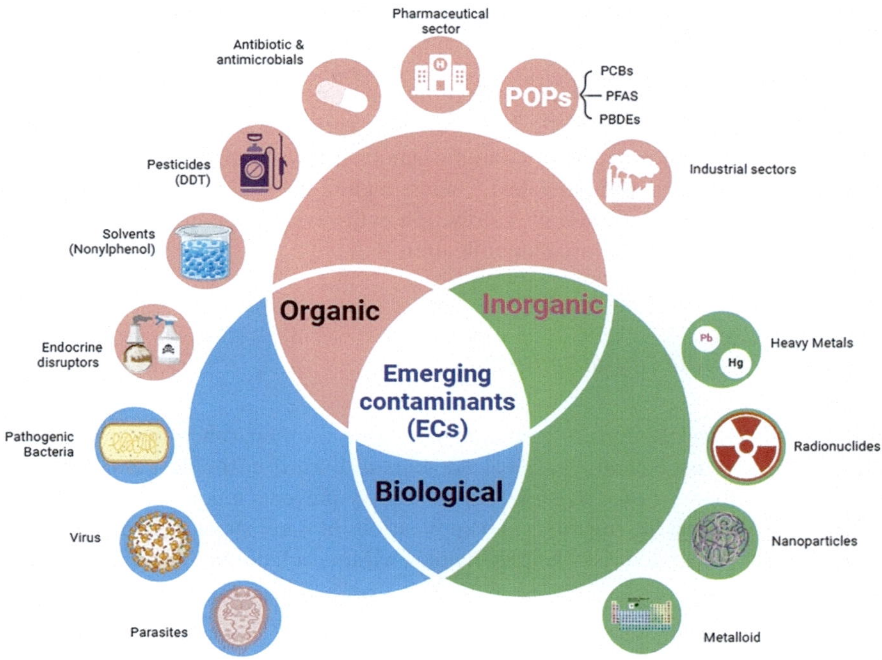

Fig. 1.1 Systematic classification frameworks, used to identify and categorize emerging pollutants

2.1 Chemical Structure-Based Classification

This framework sorts of emerging contaminants on the basis of their chemical structures or functional groups. For example, pharmaceuticals like ibuprofen (a propionic acid derivative) and antibiotics (β-lactams, aminoglycosides, and tetracyclines), personal care products such as triclosan (a chlorinated hydrocarbon), persistent organic pollutants such as polycyclic aromatic hydrocarbons, per- and polyfluorinated alkyl substances, polychlorinated biphenyls (PCBs), flame retardants like polybrominated diphenyl ethers, endocrine disruptors like steroid hormones, radioactive substances such as uranium, toxins such as cyanotoxin, and industrial chemicals such as bisphenol A (a phenol) fall into distinct categories. By discerning parallels in molecular makeup and behavior, researchers have gained insights into the characteristics and potential impacts of these substances [9].

2.2 Source-Based Classification

Contaminants are classified on the basis of their origins or sources, such as pharmaceutical manufacturing (antibiotics or hormones), agricultural activities (pesticides/fertilizers), industrial processes (volatile organics/heavy metals), and microorganisms (bacteria, viruses, and parasites). Understanding the sources of contamination helps unravel the intricate web of distribution pathways, aiding in targeted mitigation efforts and pollution source identification [10].

2.3 Functional Classification

This classification categorizes contaminants based on their functions or intended uses. From pharmaceuticals that heal ailments to toxins, pesticides, and personal care products that enhance hygiene, each category presents unique challenges and risks. By delineating the functions and applications of these substances, researchers have revealed associated risks and challenges within each domain [11, 12].

2.4 Persistence, Bioaccumulation, and Toxicity (PBT) Classification

Contaminants are categorized based on their persistence, ability to bioaccumulate, and toxicity. This includes persistent organic pollutants, bioaccumulative substances such as mercury, and highly toxic chemicals such as arsenic and lead. These properties make them particularly harmful, as they resist degradation, accumulate in living organisms, and pose significant health and environmental risks. Identifying substances with PBT properties allows researchers to prioritize their scrutiny [13].

2.5 Environmental Media-Based Classification

This framework categorizes contaminants based on the environmental media they inhabit—water, soil, air, sediment, or biota. Waterborne contaminants such as pharmaceutical residues, soil contaminants such as pesticides, and airborne pollutants such as volatile organic compounds each occupy distinct stages. Understanding how contaminants interact with different environmental media can shed light on exposure pathways and potential impacts on receptor organisms [14, 15].

2.6 Mode of Action-Based Classification

Contaminants are grouped based on their modes of action or toxicity mechanisms, such as genotoxicity (pesticides), endocrine disruption (bisphenol A), bacterial resistance (antibiotics), and oncogene activation (polybrominated diphenyl ethers). Understanding their modes of action helps researchers decipher the consequences of contamination on organisms and ecosystems [16, 17].

2.7 GHS Classification

The UN, to standardize chemical regulations and legislation across national borders, established the globally harmonized system of classification and labeling of chemicals (GHS). The GHS provides information that labels dangerous substances, and classification standards for environmental, physical, and health risks should be developed. The GHS increases international trade and improves ecological and public health protection. It also reduces the need for recurrent chemical testing and evaluation by offering a harmonized methodology [18]. While each classification framework offers unique insights into emerging contaminants, integrating multiple approaches can enhance our understanding and management of these substances. These frameworks guide researchers, policymakers, and regulators toward effective strategies for safeguarding environmental quality and public health.

3 Contaminant-Specific Classification

Contaminant-specific classification delves into the unique characteristics and implications of individual substances within the broader spectrum of emerging contaminants. Each contaminant possesses distinct attributes, behavior patterns, and environmental impacts, necessitating a focused examination to fully comprehend its significance [19]. Analyzing the chemical structure of pharmaceuticals, pesticide contamination sources, or personal care product functions reveals the complexities of each contaminant [20]. This approach guides targeted mitigation strategies and helps in understanding the intricacies of individual substances that shape the environmental landscape.

3.1 Personal Care Products

Contaminants from this category include a diverse array of substances originating from commonly used hygiene and beauty items. These contaminants are ubiquitous in the environment and can enter water bodies, soils, and even the atmosphere through wastewater discharge, urban runoff, and improper disposal. Personal care items, including shampoos, soaps, cosmetics and scents, pose ecological hazards to aquatic life, terrestrial ecosystems, and human health [21]. Personal care product contaminants can be classified based on several criteria. First, they can be categorized according to their intended functions, such as cleansing agents, moisturizers, fragrances, or preservatives. Second, contaminants can be categorized by their chemical makeup, differentiating between organic substances such as surfactants, parabens, and phthalates and inorganic materials such as heavy metals or nanoparticles. Additionally, contaminants can be classified based on their persistence in the environment, with some exhibiting long-term effects due to bioaccumulation and biomagnification [22, 23].

3.1.1 Triclosan

The aromatic ether and phenol triclosan have chloro- and 2,4-dichlorophenoxy groups at C-2 and C-5, respectively. It functions as a xenobiotic, persistent organic pollutant, fungicide, antimalarial, antibacterial agent, and drug allergen. However, because of its tenacity and capacity to upset ecosystems, it poses significant risks to environmental stability and public health [24]. The molecular formula of triclosan is $C_{12}H_7Cl_3O_2$. At 56–58 °C, it melts, and at 120 °C, it sublimes. While triclosan dissolves more easily in alcohol and bases, it is only marginally soluble in water. Moreover, it does not degrade under alkaline conditions but disintegrates in neutral and acidic conditions [25]. It works well against a variety of fungi, bacteria, and some viruses by inhibiting the fatty acid synthesis pathway in bacteria, specifically by targeting the FabI enzyme, leading to cell membrane disruption [26]. It resists biodegradation and accumulates in aquatic environments, disrupting ecosystems and contributing to antibiotic resistance [27]. Research on its photodegradation and biodegradation has improved our understanding of its persistence and breakdown products [28–30]. By analyzing large datasets, deep learning models can be used to predict triclosan toxicological profiles, such as endocrine disruption potential. Machine learning algorithms model its environmental fate and transport, predicting concentrations in water, soil, and sediment. These models consider adsorption, photolysis, and biotic uptake, aiding in the development of regulatory policies and remediation strategies. Triclosan is linked to several GHS hazard statements, including H315 (skin irritation), H319 (serious eye irritation), H400 (very toxic to aquatic life), and H410 (very toxic to aquatic life with long-lasting effects).

3.1.2 Parabens

Preservatives called parabens are frequently used in medicines, cosmetics, and personal hygiene methods to prevent germs and fungi from growing and to lengthen the shelf-life of products. Despite their usefulness, parabens have sparked controversy due to their harmful effects on human health and the environment [31, 32]. Parabens, derived from para-hydroxybenzoic acid, are commonly used esters. Examples include methylparaben ($C_8H_8O_3$, 152.15 g/mol), ethylparaben ($C_9H_{10}O_3$, 166.17 g/mol), propylparaben ($C_{10}H_{12}O_3$, 180.19 g/mol), and butylparaben ($C_{11}H_{14}O_3$, 194.22 g/mol). These chemicals are made by esterifying parahydroxybenzoic acid with the appropriate alcohol while having an excess of the relevant alcohol and an acid catalyst (such as sulfuric acid). Following the neutralization of the acid with caustic soda, the mixture was cooled, centrifuged, cleaned, vacuum-dried, ground, and blended. Alcohol and acetone make parabens more soluble than they are in water. Normal conditions allow them to remain stable, but high temperatures and extremely high pH can cause them to break down [33–37]. Parabens are effective antimicrobial agents that disrupt microbial cell membranes and interfere with enzymatic activity. This makes them valuable in preventing bacterial and fungal contamination [38]. However, parabens are moderately persistent in the environment, are absorbed into soils and sediments, and are detected in various water bodies. Their degradation can produce p-hydroxybenzoic acid, potentially impacting ecosystems [39].

Parabens are known endocrine disruptors that mimic estrogen and potentially disrupt hormone function. These compounds are associated with reproductive toxicity and carcinogenicity [40]. Research on photodegradation and biodegradation via spectroscopic and chromatographic methods helps researchers understand the persistence and breakdown products of parabens [41]. The chemical properties of methylparaben contribute not only to its antimicrobial activity but also to its environmental impact. The use of advanced analytical techniques and deep learning models is essential for understanding the full scope of paraben effects, informing regulatory policies, and developing effective mitigation strategies. Methylparaben in formulations should not contain more than 0.1% of the ingredients and should adhere to Food Chemicals Codex requirements [42, 43]. Methylparaben is classified under several GHS hazard statements, including H319, H315, H335 (respiratory irritation), and H412 (long-lasting harm to aquatic life). Condensation of 4-hydroxybenzoic acid with ethanol produces ethylparaben [44], an ethyl ester with a GHS hazard statement of H334 (which may induce breathing difficulties or allergy symptoms if inhaled). Propylparaben belongs to the class of benzoate esters and parabens and is derived as the propyl ester of 4-hydroxybenzoic acid. It functions similarly to propan-1-ol and 4-hydroxybenzoic acid. A pH of 4–5 is optimal for stability. When used as a preservative in pesticide formulations for animals, residues of propyl p-hydroxybenzoate are exempt from the need for tolerance when they are employed as inert (or occasionally active) compounds and do not exceed 0.1% in formulations. This exemption applies to residues used in line with good agricultural practices [45–47]. Propylparaben is associated with three GHS hazard statements:

H319, H315, and H335. One standardized chemical allergen is butylparaben. Butylparaben affects the body by promoting cell-mediated immunity and the release of histamine. It is a naturally occurring chemical found in *Strychnos cathayensis* and *Carica papaya* [48, 49]. Butylparaben is classified under the following GHS hazard statements: H315, H318, H319, H335, and H412 (harmful to aquatic life with long-lasting effects).

3.1.3 Phthalates

Phthalates have faced criticism for their widespread use as plasticizers in consumer products, including building materials, plastics, and personal care items. In 2017, 65% of the global plasticizers consumed were esters of phthalic acid, often comprising alkyl chains, including common variants such as dibutyl phthalate (DBP), diethyl phthalate (DEP), and di(2-ethylhexyl) phthalate (DEHP) [50]. Commercial production of DEHP involves the esterification of 2-ethylhexanol and phthalic acid anhydride. At high temperatures, esterification can be finished using a metal or acid catalyst or without one at all [51]. It functions as a plasticizer, an agonist at the androstane receptor, and an inhibitor of apoptosis. It is both a diester and a phthalate ester. The DEHP has several GHS hazard statements, one of which is H360 (might impair fertility or the unborn child). These are the P203, P280, P318, P405, and P501 codes for precautionary statements [52]. Known for their endocrine-disrupting properties, phthalates interfere with hormone signaling pathways in mammals, leading to detrimental effects on reproductive development and fertility [53]. Certain ortho-substituted phthalates, such as DEHP, have been restricted by some countries since the 1990s because of their adverse impact on the male reproductive tract in mammals [54, 55]. Additionally, the International Agency for Research on Cancer (IARC) has categorized DEHP as potentially carcinogenic to humans (Group 2B), although human data are lacking [56]. Consequently, diisononyl phthalate (DiNP) has been employed as a substitute plasticizer for polyvinyl chloride (PVC) in numerous applications [57]. Seventeen phthalates or phthalate combinations have been classified as materials of extreme concern by the EU (European Union) because of their potential to harm the environment, cause endocrine disruption in humans, and be hazardous to reproduction [58, 59]. Phthalates can leach out of products and enter the environment through pathways such as wastewater discharge, landfill leaching, and atmospheric deposition [60]. The human health impacts associated with phthalate exposure include allergies [61, 62], asthma [63], developmental and reproductive abnormalities [64, 65], and some types of cancer [66–68]. Given their ubiquitous presence and potential to impact both human health and ecosystems, ongoing research and monitoring efforts are crucial for mitigating the adverse effects of phthalates. Advanced analytical techniques, coupled with deep learning models and interdisciplinary approaches, are essential for comprehensively understanding their complex interactions in the environment and devising effective management and regulatory strategies.

3.1.4 Microplastics

Microplastics, which are usually smaller than 5 mm, are significant environmental contaminants present in water, soil, and air (Fig. 1.2). They pose potential risks to ecosystems and human health due to their pervasive presence and persistence [69]. Microplastics are composed of polymers such as PVC, polyethylene terephthalate, polypropylene, polyethylene, and polystyrene. They often contain additives such as plasticizers, stabilizers, and colorants [70]. These particles vary in shape (e.g., fragments 31%, fibers 48.5%, beads 6.5%, and films 5.5%) and density, which influences their environmental distribution and behavior [71]. Their polymeric nature makes them resistant to biodegradation, allowing them to persist in the environment and fragment into nanoplastics over time [72].

Microplastics originates from primary sources (manufactured products) and secondary sources (fragmentation of larger plastic debris). They are found in marine and freshwater ecosystems, soils, and air [73, 74]. Microplastics can cause physical harm, such as ingestion and entanglement, and act as vectors for harmful chemicals and pathogens [75]. Ingested microplastics can accumulate in the digestive systems of organisms, reducing their feeding, growth, and reproductive success. In various food products, drinking water, and air, these compounds can cause physical damage to tissues, inflammatory responses, and chemical leaching of harmful additives [76–78]. However, uncertainties remain due to the limited evidence on the toxicological and epidemiological risks of microplastics, especially nanosized particles. Microplastics are pervasive contaminants with significant potential to impact

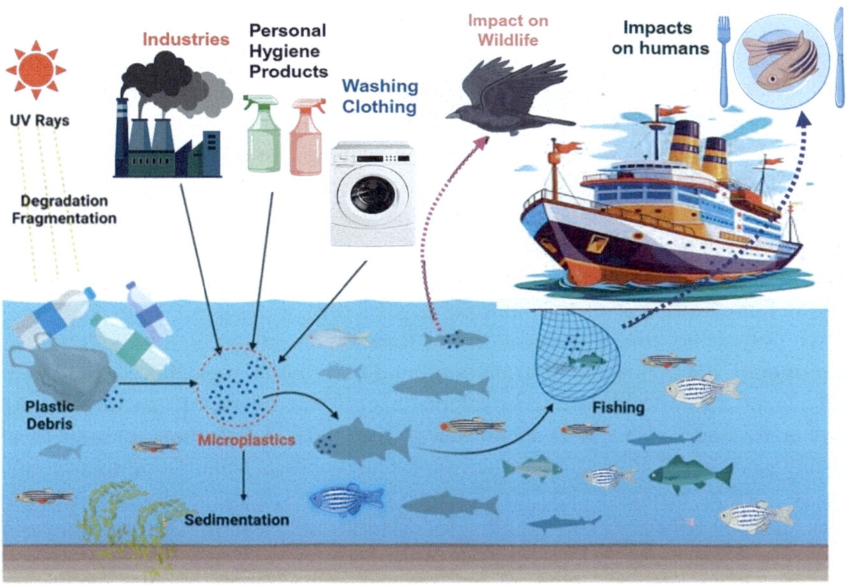

Fig. 1.2 Microplastics in aquatic environments

ecosystems and human health. Research and monitoring, leveraging advanced analytical methods and deep learning models, are essential for unraveling the complex environmental interactions of microplastics and formulating effective management and regulatory strategies.

3.1.5 Synthetic Musk Fragrances

One family of synthetic aromatic chemicals that is frequently employed in personal care products, detergents, and air fresheners. These fragrances, which are designed to mimic natural musks derived from animals and plants, have sparked environmental and health concerns because of their long-lasting nature and potential impacts. Synthetic musks can be divided into several categories, with nitro musks and polycyclic musks being the most concerning emerging contaminants [79, 80]. Two well-known nitro musks are musk xylene and musk ketone. Musk xylene has a melting point of 111–112 °C and a log K_{ow} of 4.9. Its chemical formula is 1-tert-butyl-3,5-dimethyl-2,4,6-trinitrobenzene ($C_{12}H_{15}N_3O_6$). It is made by alkylating m-xylene with isobutene and then nitrating. Musk xylene is known for its reactivity and potential for self-decomposition, self-ignition, or hazardous polymerization under heat, friction, or impact. The typical end-product concentrations are as follows: 0.2% for perfumes, 0.01% for creams and lotions, 0.003% for detergents, and 0.03% for soaps. According to REACH, it is categorized as a very persistent and very bioaccumulative (VPVB) material. Hazard statements for the compound include that it is explosive (H201), very hazardous to aquatic life (H400 and H410), and may cause cancer (H351). Its metabolism involves the reduction of nitro groups, with significant metabolites present as glucuronic acid conjugates [81]. The chemical formula for Musk ketone is 4′-tert-butyl-2′,6′-dimethyl-3′,5′-dinitroacetophenone ($C_{14}H_{16}N_2O_5$), and its melting point is between 135 and 136 °C. Its log K_{ow} value is 4.3. Friedel–Crafts acetylation is the method used for synthesis, followed by nitration. Musk ketone can release heat and combustible fumes when it reacts with a variety of acids and bases. The final product concentrations are typically comparable to those of musk xylene. Musk Ketone's hazard claims that aquatic life is severely harmed by it (H400 and H410) and could result in cancer (H351). Enzymatic nitro-reduction produces the primary metabolite 2-amino-Musk ketone, which is found in human urine as glucuronide conjugates [82].

Two polycyclic musks, galaxolide and tonalide, are also commonly employed. Galaxolide ($C_{18}H_{26}O$) is a chemical compound with the formula 4,6,6,7,8,8-hexam ethyl-1,3,4,6,7,8,8-hexahydro-2H-cyclopenta[g]-2-benzopyran. Its melting point is between 57 and 58 °C. It is synthesized from 1,1,2,3,3-pentamethylindane via hydroxyalkylation followed by ring closure. Galaxolide is found in high concentrations in various products, including soaps, detergents, creams, lotions, and perfumes, and is very toxic to aquatic life (H400 and H410). Despite not being genotoxic in micronucleus tests, high concentrations can affect steroidogenesis in vitro [83].

The melting point of 54.5 °C and log K_{ow} of 5.70 are two characteristics of tonalide, which has the chemical formula 7-acetyl-1,1,3,4,4,6-hexamethyl-1,2,3,4-tetrahydronaphthalene ($C_{18}H_{26}O_2$). Acetylation is used to prepare 1,1,2,4,4,7-hexamethyltetralin. Galaxolide and tonalide are used in comparable products and dosages. Tonalide has several known hazards, including being extremely poisonous to aquatic life (H400 and H410) and dangerous if eaten (H302) [84]. In general, synthetic musks are lipophilic, which means that they dissolve well in organic solvents but poorly in water. They are frequently liquids or solids that are colorless or pale yellow and have a distinct musky smell. The chemical stability and lipophilicity of synthetic musks enable them to accumulate in the environment, particularly in water bodies, sediments, and biota, leading to biomagnification in the food chain [85, 86]. Some synthetic musks exhibit endocrine-disrupting properties, interfering with hormone function in wildlife and potentially in humans [87]. Exposure to synthetic musks has been linked to various health effects, including skin sensitization [88], reproductive toxicity [89], and potential carcinogenicity [90].

3.1.6 Bisphenol A

Bisphenol A (BPA) is becoming an increasingly common contaminant despite its wide use because of its possible impacts on the environment and human health [91]. With the chemical formula $C_{15}H_{16}O_2$, BPA is a diphenylmethane derivative composed of two hydroxyl phenyl functional groups joined to a central carbon atom by a methyl bridge. The IUPAC name for 4,4′-(propane-2,2-diyl) diphenol is 228.29 g/mol. At normal temperature, BPA is a white solid with melting and boiling points of 158–159 °C and 5 mmHg, respectively. It dissolves very well in ethanol and ethyl acetate but only very minimally in water. Under typical circumstances, the chemical stability of BPA is maintained; nevertheless, when strong acids and bases are present, BPA may hydrolyze [92]. The production of BPA involves the condensation of phenol and acetone with the use of an acid catalyst [93]. BPA can be broken down by microbiological activity, but because it is resistant to biodegradation in some situations, it frequently lingers. BPA can accumulate in aquatic habitats and cause creatures to become more biomagnified in the food chain [94–96]. Rodent liver microsomal enzymes metabolize BPA to produce nine metabolites, which include isopropyl-hydroxyphenol (produced by BPA cleavage) and BPA dimers [97, 98]. BPA may have estrogenic effects, increasing the likelihood of developing several malignancies. It functions as an endocrine disruptor, xenobiotic, environmental pollutant, and xenoestrogen [99].

BPA exposure has been linked to reproductive disorders [100, 101], neurological problems [102, 103], cardiovascular diseases [104], and certain cancers [105]. In aquatic life, BPA can cause developmental and reproductive issues [106].

3.1.7 Nonylphenol and Nonylphenol Ethoxylates

Nonylphenol (NP) and nonylphenol ethoxylates (NPEs) are classified as alkylphenols and alkylphenol ethoxylates, respectively, in general chemical terms. The nine-carbon nonyl group can attach to the phenol ring at several points ("ortho," "meta," or "para"). It can also be branched or linear. With trace levels of ortho- and di-substituted nonylphenols, NPs are complex mixtures of highly branched nonylphenols that are primarily mono-substituted at the para-position. There is more than one chemical structure to it. Small quantities of branched C8 and C10 alkyl groups are also present. NP1EO and NP2EO, which are mono- and di-ethoxy NPEs, are breakdown products of polymeric NPEs and can be found in the environment in significant proportions alongside nonionic surfactants, or NPEs, which have one hydrophilic end and one hydrophobic end [107]. The names and numbers (CASRNs) of the Chemical Abstract Service registry that are frequently used to identify nonylphenols or nonylphenol ethoxylates may not be a true representation of those compounds. The most commonly produced nonylphenol, branched 4-nonylphenol, is CASRN 84852-15-3. A significant portion of the literature discusses the specific para regioisomer, or 4-n-nonylphenol (CASRN 104--40--5), which is sometimes referred to as linear (or normal) nonylphenol (CASRN 25152--52--3). Owing to errors in the source material, many references provide false information regarding the identities of the compounds described as nonylphenols. When branched 4-nonylphenol (CASRN 84852-15-3) is delivered, a nonylphenol supplier may use the designation CASRN 104-40-5, which denotes 4-n-nonylphenol [108, 109]. NP is very poisonous to aquatic creatures, persistent in the aquatic environment, and somewhat bioaccumulative, as illustrated in Fig. 1.3. NP is used primarily in the production of NPEs. NPEs are very hazardous to aquatic creatures and break down into more environmentally friendly NPs, although they are less toxic than are NPs [110]. NP is present in human blood, urine, and breast milk and affects rodent development and reproduction [111–115].

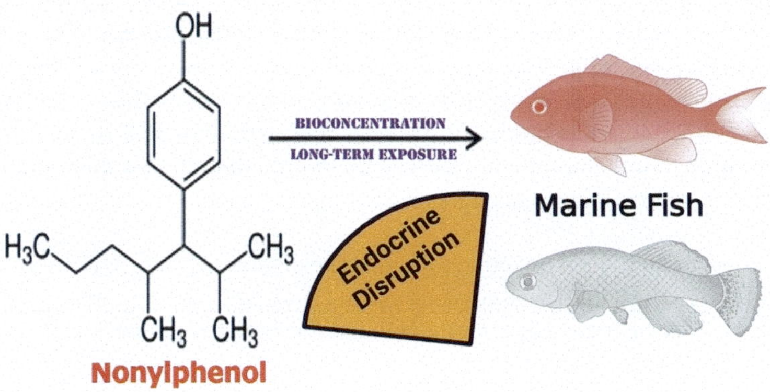

Fig. 1.3 Bioconcentration and endocrine disruption of nonylphenol in aquatic creatures

3.1.8 Per- and Polyfluoroalkyl Substances

Perfluorooctanoic acid (PFOA) and perfluorooctane sulfonate (PFOS) are two of the most widely used and studied chemicals within the per- and polyfluoroalkyl substance (PFAS) families. In the United States, PFOA and PFOS have been replaced in recent years by additional PFAS. One major cause for concern is that many PFAS degrade very slowly and can eventually accumulate in humans and the environment [116–118]. Both PFOA and PFOS are synthetic substances that are not found in the environment by nature. PFASs, or perfluorinated alkyl acids, are a subset of the total fluorinated organic chemicals that include them. Perfluoroalkyl chemicals and polyfluoroalkyl substances are the two subgroups that constitute the PFAS group [119]. Perfluoroalkyl compounds, such as PFOS and PFOA, are those in which the hydrogen atoms of all the carbons are replaced with fluorines, with the exception of the carbons attached to functional groups [120]. The compounds classified as polyfluoroalkyls have some, but not all, of the hydrogen atoms on their carbon atoms, which are swapped out for fluorine. Owing to the exceptionally strong and stable chemical link between the carbon and fluorine atoms, perfluoroalkyl substances (PFASs) are highly persistent in the environment [121].

In addition, a wide range of related PFAS or precursor chemicals can be broken down by the environment or by larger creatures to produce PFOS and PFOA. These stable compounds consist of eight-carbon chains [122, 123]. Because of their resistance to oil and water, they have been applied in surface protection products, and processing aids in the production of fluoropolymers, membranes for waterproof materials, electrical wire casings, and plumbing thread seal tape [124, 125]. A synthetic compound called PFOS is used to confer resistance to water, oil, grease, and stains. A completely fluorinated anion, PFOS, is frequently added to larger polymers or utilized as a salt. PFOSs and their closely related chemicals belong to the broad family of perfluoroalkyl sulfonate substances, which also include compounds that can produce PFOS or impurities in PFOS [126]. Chemical electrochemical fluorination (ECF) is the main commercial method for producing PFOS and PFOS-related chemicals. Perfluorooctanesulfonyl fluoride (PFOSF) is the main intermediate used to create PFOS and compounds linked to PFOS. A mixture of isomers and homologs with approximately 35–40% 8-carbon straight-chain PFOSF are produced via the ECF technique. Approximately 70% of the contaminants in the commercial PFOSF products were linear, and the remaining 30% were branched PFOSF derivatives. PFOSF is hydrolyzed chemically or enzymatically to produce PFOS. Among the additional processes used to produce perfluoroalkylated materials are oligomerization and telomerization. Tetrafluoroethylene reacts with other substances containing fluorine during the telomerization process to produce fluorinated intermediates, which are easily transformed into perfluorooctanoic acid (PFOA) [127–129]. It is anticipated that PFOS will have an atmospheric half-life longer than 2 days. This is because, in every test that has been conducted, PFOS has shown extremely high resilience to deterioration. However, an AOP computer modeling tool v1.91 has been used to derive an atmospheric half-life of 114 days for PFOS. At 25 °C, the indirect photolytic half-life of perfluorocarbons (PFOSs) has been calculated to exceed 3.7 years [130]. PFOA is a carcinogenic agent [131], surfactant [132], endocrine disruptor [133], and environmental contaminant [134]. The chemical formula of 2,2,3,3,4,4,5,6,6,7,7,8,8,8-pentadecafluorooctanoic acid is $C_8HF_{15}O_2$, and its molecular weight is 414.07 g/mol. This is its IUPAC name. The boiling and melting points of PFOA are 192 °C and 54.3 °C, respectively. Its LogP, which measures the partition coefficient between water and octanol, is 6.3. The GHS hazard statements show that PFOA has several risks. It is dangerous if eaten (H302), can burn the skin severely, injure the eyes (H314 and H318), is dangerous if inhaled (H332), and may be carcinogenic (H351). Furthermore, PFOA may harm children who are breastfed (H362), damage unborn children or fertile individuals (H360), and damage organs via prolonged or recurrent exposure (H372) [135].

3.1.9 Polybrominated Diphenyl Ethers

Because they are widely used in consumer products, polybrominated diphenyl ethers (PBDEs) have been a major source of environmental pollution for the past 20 years. Diet, especially meat, fish, and dairy products, and indoor surroundings, especially dust, are major routes of exposure [136, 137]. Because they are most exposed to home dust and breast milk, infants and toddlers typically have the highest body burdens [138]. Several tetra-, penta-, and hexaBDEs, as well as decaBDE, are frequently identified in humans among the 209 potential PBDE congeners. Despite being phased out or outlawed in many cases, PBDEs are still present in the environment [139]. Despite differing in the quantity and arrangement of their bromine atoms, permuted diphenyl ether molecules (PBDEs) have similar structures. The PBDE congeners can be categorized on the basis of their number of bromine atoms, with flame retardant forms having four to ten bromine atoms. Pentabromodiphenyl ether (c-pentaBDE), octabromodiphenyl ether (c-octaBDE), and decabromodiphenyl ether (c-decaBDE) are the three primary PBDE mixtures that have been utilized commercially. Even though c-pentaBDE and c-octaBDE were no longer produced or imported after 2004, humans and the environment still included their related congeners, and some reports even suggest that their numbers are growing. However, data suggest that c-DecaBDE may be converted via debromination processes to more dangerous congeners, which is why it is still in use [140–142].

Decabromodiphenyl ether, a prominent PBDE variant, is used alongside antimony trioxide in polymers, particularly high-impact polystyrene (HIPS). With a molecular weight of 959.2 g/mol and a log K_{ow} of 9.97, its chemical formula is $C_{12}Br_{10}O$ [143]. When decomposed, it emits toxic fumes of hydrogen bromide. The technical composition typically includes decabromodiphenyl oxide, nonabromodiphenyl oxide, and octabromodiphenyl oxide. Production involves reacting bromine with phenyl ether in an ethylene dibromide solvent, facilitated by an aluminum bromide catalyst. The reaction temperature typically approximates the boiling point of the bromine solvent (~59 °C). Decabromodiphenyl ether is dangerous if eaten (H302) and comes into contact with the skin (H312 and H319. It may also harm organs by repeated or prolonged exposure (H373) and is suspected of producing genetic problems (H341). It might also have detrimental long-term impacts on aquatic life (H413). It is essential to take precautions, such as safe handling and disposal, to reduce these risks [144].

3.2 Pesticides

Chemically varied substances known as pesticides are used to eliminate undesired plants (weeds), fungi, insects, and rodents. Because of their intrinsic toxicity, pesticides need to be used and disposed of responsibly because they can affect both humans and other organisms. Pesticides that are older and less costly, including dichlorodiphenyltrichloroethane (DDT) and lindane, can persist in soil and water, causing significant harm to the environment and accumulation in the food chain. Consequently, nations that ratified the Stockholm Convention in 2001 outlaw these compounds [145–147].

3.2.1 Organochlorine Insecticides

The broad class of substances known as organochlorine insecticides is distinguished by the presence of hydrogen, chlorine, and carbon. These pesticides have larger molecular weights than the volatile chlorinated hydrocarbons present in solvents do, which makes them less volatile [148]. Organochlorine insecticides are categorized into four structural groups: diphenyl aliphatics (such as DDT and its derivatives), cyclodienes (such as aldrin, dieldrin, endrin, endosulfan, and heptachlor), cyclohexanes (such as γ-hexachlorocyclohexane or lindane), and polychloroterpenes (such as toxaphene). Dined cage compounds, mirex, and chlordecone are difficult to classify but are occasionally combined with cyclodienes [149]. Interestingly, the toxicity and biological activity of these insecticides do not always match their structural classifications. Substances with comparable structures can indeed have remarkably diverse actions and toxicities. Hexachlorocyclohexane's γ-isomer, for example, exhibits insecticidal qualities and excites the central nervous system, but other isomers have opposite effects and cause depression in the central nervous system [150].

3.2.1.1 DDT (Clofenotane)

The organochlorine insecticide DDT (1,1,1-trichloro-2,2-diphenylethane) is well known for its efficacy. It is a crystalline substance that is white, tasteless, and odorless. It is soluble in the majority of organic solvents, fats, and oils. This characteristic allows it to accumulate in ecosystems, where it functions as an acaricide, carcinogen, persistent organic pollutant, and endocrine disruptor. The chemical formula of DDT is $C_{14}H_9Cl_5$, its molecular weight is 354.5, its melting and boiling points are 108.5 and 260 °C, respectively, and its log K_{ow} value is 6.91. While stable under normal conditions, DDT dehydrochlorinates to DDE at temperatures above its melting point, a reaction catalyzed by iron(III) or aluminum chlorides and UV light. It resists oxidation but may degrade in the presence of alkali or organic bases. Commercial production of DDT yields a mixture of isomeric products. The nervous system is the main organ affected by DDT exposure, which can lead to symptoms, including tremors, nausea, dizziness, incoordination, convulsions, and seizures. Studies on animals have led to the suspicion that it may be carcinogenic. The GHS hazard statements for DDT are as follows: H301+H311 (harmful if ingested or come into contact with the skin), H351 (might lead to cancer), H372 (causes frequent or extended exposure that damages organs), H400 (extremely toxic to aquatic life), and H410 (very harmful and has long-lasting consequences for aquatic life) [151, 152].

3.2.1.2 Endosulfan

Endosulfan is an organochlorine pesticide that belongs to the cyclic sulfite ester family. Its chemical structure is described as 6,7,8,9,10,10-hexachloro-1,5,5a,6,9,9a-hexahydro-6,9-methano-2,4,3-benzodioxathiepine-3-oxide, where positions 6, 7, 8, 9, 10, and 10’ are substituted with chloro groups. With the chemical formula $C_9H_6Cl_6O_3S$ and molecular weight of 406.7, endosulfan has log K_{ow} values of 3.62 (beta) and 3.83 (alpha), with a melting point of 106 °C. Endosulfan efficiently blocks nerve signal transmission in pests by acting as an antagonist of GABA-gated chloride channels. Endosulfan, which is specifically categorized as a cyclodiene organochlorine pesticide, is recognized as a major ecological and health problem because of its persistent nature in the environment [153–155]. Alpha and beta endosulfan are two stereoisomers of the pesticide endosulfan and differ in their molecular structure and physical properties. These structural differences lead to variations in their physical properties. Compared with beta-endosulfan, which has a lower melting point and greater water solubility, alpha-endosulfan has a higher melting point, is more environmentally persistent and is less soluble in water [156].

In terms of toxicity, alpha-endosulfan is typically seen as being more harmful to aquatic life, while beta-endosulfan also poses significant toxicity but can vary depending on the organism and environmental conditions. Their environmental behaviors also differ; alpha-endosulfan is more likely to persist due to its stability and lower solubility, whereas beta-endosulfan may degrade or transform more rapidly because of its higher solubility and lower stability. Both isomers are commonly found in commercial endosulfan formulations, which typically contain a mixture of alpha and beta endosulfan. Regulatory restrictions on endosulfan apply to both isomers owing to their combined environmental and health impacts. Despite their shared use in pest control, the differences between alpha and beta endosulfan influence their handling, environmental impact, and possible dangers to ecosystems and human health. According to the Food and Agricultural Organization (FAO) specifications, technical endosulfan must contain 94% endosulfan as the active ingredient, with alcohol and ether impurities restricted to a maximum of 2% and 1%, respectively. Endosulfan has major toxicity hazards, according to the GHS hazard statements: H300+H330 (32.94%): deadly if consumed or breathed; H300 (96.47%): deadly if consumed; H312 (96.47%): dangerous when skin-in contact; H330 (96.47%): deadly if inhaled; H400 (100%): extremely toxic to aquatic life; and H410 (100%): extremely toxic to aquatic life with persistent effects [157, 158].

3.2.1.3 Lindane

There are eight isomeric variants of the synthetic compound hexachlorocyclohexane (HCH). Lindane, sometimes referred to as gamma-HCH (γ-HCH), is widely used as an insecticide. It is a white solid capable of evaporating into a colorless vapor with a faint musty odor. In addition to its agricultural applications, lindane is also employed in medicinal formulations such as lotions, creams, or shampoos for treating head and body lice, as well as scabies. Despite not being manufactured domestically since 1976, lindane continues to be imported for insecticidal purposes once used as an insecticide in the United States, and technical-grade HCH usually contains 10–15% gamma-HCH in addition to other isomeric forms, including alpha (α), beta (β), delta (δ), and epsilon (ε). Interestingly, gamma-HCH exhibited the most insecticidal activity. However, the United States has neither produced nor utilized technical-grade HHC for more than 20 years [159, 160]. Lindane, specifically the gamma-isomer, can be synthesized from benzene and chlorine through a process called photochlorination. This reaction produces benzene hexachloride, which comprises various isomers, with the desired gamma-isomer being lindane. To isolate the gamma isomer, the reaction mixture was treated with either methanol or acetic acid. These solvents selectively dissolve the alpha and beta isomers, leaving behind the desired gamma-isomer, lindane [161]. The substance presents various hazards, as indicated by the GHS hazard statements. It is deemed toxic if it is swallowed (H301), harmful upon contact with the skin (H312), and suspected of causing cancer (H351). Additionally, it poses a significant risk to aquatic life and is highly toxic, with both acute (H400) and long-lasting effects (H410) on aquatic environments [162].

3.2.1.4 Toxaphene

1,4,5,6,7,7-Hexachloro-2,2-bis(chloromethyl)-3-methylidenebicyclo[2.2.1]heptane is the IUPAC name for toxaphene. It has a molecular weight of 411.8 and the formula $C_{10}H_8C_{18}$. Its LogP is 5.90, and its melting point is between 65 and 90 °C. Convulsive seizures are the primary outcome of stimulation of the central nervous system (CNS). Animals exposed orally for an extended period of time may experience negative effects on their immune system, central nervous system, liver, kidney, spleen, adrenal, and thyroid glands. Prolonged inhalation exposure in humans can result in reversible pulmonary toxicity [163–165]. The GHS hazard statements suggest that there are serious dangers associated with this chemical. It is classified as hazardous when applied topically (H312), poisonous if ingested (H301), irritable when applied topically (H315), or potentially irritating the respiratory system (H335). Moreover, it is categorized as very harmful to aquatic life, with both immediate (H400) and long-term (H410) effects on aquatic habitats, and it is suspected of causing cancer (H351) [166].

3.2.1.5 Chlordecone

Chlordecone is an organochlorine chemical and cyclic ketone with insecticidal properties that is a persistent organic contaminant and insecticide. Unlike naturally occurring chemicals, chemically similar but different, mirex and chlordecone are synthetic insecticides. Chlordecone looks like a tan-white crystalline solid, in contrast to the white crystalline solid appearance of Mirex [167, 168]. Its IUPAC name is 1,2,3,4,6,7,8,9,10,10-decachloropentacyclo [5.3.0.02,6.03,9.04,8] decan-5-one, with a molecular formula of $C_{10}Cl_{10}O$ and a molecular weight of 490.6. Its LogP is 5.41. When sulfur trioxide is present, two molecules of hexachlorocyclopentadiene are condensed, and the resulting product is hydrolyzed to produce chlordecone. Technical-grade chlordecone usually contains 3.5–6.0% water, 0.1% hexachlorocyclopentadiene, and 88.6–99.4% chlordecone [169, 170]. Chlordecone is absorbed by the liver, where it is reduced enzymatically to chlordecone alcohol, conjugated with glucuronic acid, partially converted to unknown polar forms, and excreted mostly as glucuronide conjugates into bile [171]. Toxicity symptoms include nervousness; tremors; visual disturbances such as rapid erratic eye movement; and sporadic ataxia, chest pain, arthralgia, erythematous skin eruptions, and weight loss [172].

The substance poses significant hazards, as reflected in GHS hazard statements, indicating its toxicity upon ingestion (H301), contact with skin (H311), and potential carcinogenicity (H351). It is also deemed very toxic to aquatic life, both in terms of acute (H400) and long-lasting effects (H410) on aquatic environments. Adhering to precautionary statement codes is vital for minimizing the risks associated with chlordecone exposure. These include maintaining awareness of hazards, practicing good hygiene, avoiding environmental release, utilizing protective measures, and ensuring proper disposal. Such precautions are essential for mitigating the adverse impacts of chlordecone exposure [173].

3.2.2 Organophosphorus Pesticides

Esters of phosphoric acid, known as organophosphates, are widely employed as pesticides, herbicides, and insecticides. They cause acetylcholine to build up and become neurotoxic by inhibiting acetylcholinesterase. The common organophosphate pesticide diazinon has a molecular weight of 304.35, a boiling point of 125 °C, a chemical formula of $C_{12}H_{21}N_2O_3PS$, and a log K_{ow} of 3.81. Diethylthiophosphoric acid is applied after isobutyramidine and acetoacetate are condensed to create it. Diazinon poses a risk to one's health if it is swallowed or comes into contact with the skin. It can also cause genetic flaws, cancer, and organ damage. Aquatic life is likewise highly poisonous. H302 (hazardous if consumed), H312 (detrimental to skin touch), H341 (thought to be the source of genetic abnormalities), H350 (might lead to cancer), H370 (damages the organs), and H373 (may result in organ damage from repeated or extended exposure) are among the GHS hazard statements for diazinon [174–176]. Notably, other organophosphates include tribufos and ethoprophos. Ethoprophos are extremely harmful to aquatic life, poisonous if eaten, lethal when used topically, and may trigger an allergic skin reaction [177]. Tribufos kills the skin, is extremely harmful to aquatic life, and is toxic if eaten [178]. Phosmet is poisonous if ingested, damaging if skin contact occurs, lethal if it is breathed, and extremely hazardous to aquatic life. It may also impair fertility [179].

3.2.3 Dioxins

Dioxins are a class of environmentally persistent compounds that are chemically related. These compounds include dibenzofurans, polychlorinated dibenzo-p-dioxins, and certain polychlorinated biphenyls. Particular chlorine locations (2, 3, 7, and 8) on the carbon rings are linked to their toxicity; 2,3,7,8-tetrachlorodibenzo-p-dioxin (TCDD) is particularly dangerous [180, 181]. The molecular weight of TCDD is 322.0, its melting point is 305 °C, its chemical formula is $C_{12}H_4Cl_4O_2$, and its log K_{ow} value is 6.8. TCDD is an unintentional byproduct that arises from the burning of garbage and the production of specific compounds. TCDD does not break down until the temperature exceeds 500 °C; it can break down gradually through photochemical and microbiological activities, but it remains stable in ethanol, acetone, DMSO, and water. Skin rashes, liver damage, weight loss, and immune system dysfunction can all result from TCDD exposure. H300 (fatal if eaten), H319 (major eye irritation), H400 (highly hazardous to aquatic life), and H410 (extremely toxic to aquatic life with long-lasting effects) are the GHS hazard statements for TCDD [182].

3.2.4 Neonicotinoids

Neonicotinoids are neuroactive insecticides similar to nicotine that target insect nicotinic acetylcholine receptors. Owing to their water solubility, they are effective when applied to soil and taken up by plants. Commercially available neonicotinoids include imidacloprid, acetamiprid, clothianidin, thiamethoxam, thiacloprid, nitenpyram, and dinotefuran [183]. Imidacloprid, for example, has the molecular formula $C_9H_{10}ClN_5O_2$, a molecular weight of 255.66, a melting point of 144 °C, and a log K_{ow} of 0.57. Imidacloprid poses long-lasting health risks to aquatic species because of its high toxicity and toxicity if eaten. The GHS hazard statements for imidacloprid are H410 (very harmful and having long-lasting consequences for aquatic life), H301 (toxic if ingested), H302 (dangerous to ingest), and H400 (extremely harmful to aquatic life) [184–186].

3.2.5 Pyrethrin and Pyrethroid Insecticides

A substance from chrysanthemum flowers called pyrethrum has strong insecticidal effects. Six isomers make up pyrethrins, a refined version of pyrethrum that separates its insecticidal components. Pyrethroids, synthetic insecticides structurally similar to pyrethrins, are modified to increase their environmental stability and insecticidal effectiveness. Although pyrethrins are more biodegradable, pyrethroids are photostable and therefore preferable to natural pyrethrins. Unlike pyrethrum, both pyrethrins and pyrethroids do not contain allergenic components. They are less toxic to mammals, making them suitable replacements for more hazardous organophosphate insecticides [187, 188]. Six ester compounds known as "natural pyrethrins" are produced when particular alcohols and acids condense. The alcohol component can be any of three unsaturated hydroxy-cyclo ketones (rethrolones): cinerolone, jasmolone, or pyrethrolone. The acid component can be either pyrethric acid or chrysanthemic acid. Two types of pyrethrins are distinguished: Pyrethrin I, which comprises compounds containing chrysanthemic acid, and Pyrethrin II, which comprises compounds containing pyrethric acid. The compound Pyrethrin I is derived from (+)-trans-chrysanthemic acid, whereas Pyrethrin II is produced when pyrethric acid is produced through the oxidation of a methyl group to a carboxymethyl group [189, 190].

Pyrethrins undergo extensive metabolism through two major pathways: the oxidation of double bonds and methyl groups and the hydrolysis of ester bonds. Pyrethrin I is metabolized primarily oxidatively, whereas Pyrethrin II undergoes both hydrolytic and oxidative processes. Two forms of pyrethroids are distinguished by their structural makeup and poisoning symptoms. Because they do not include the α-cyano group, type I pyrethroids have a lower toxicity level. On the other hand, the inclusion of the α-cyano group makes Type II pyrethroids more hazardous [191]. Among the common physical qualities shared by all pyrethroids are low vapor pressure, low Henry's law constants, high octanol–water partition coefficients (K_{ow}), restricted water solubility, the same physical properties for enantiomers, and distinct physical attributes for diastereomers. Different techniques are used in pyrethroid synthesis to produce new compounds with targeted biological activity. Notable techniques include substructure splicing, which joins fragments of active insecticide, and biomimetic synthesis, which mimics natural reactions. The main synthesis method is esterification, which necessitates the synthesis of alcohols (or aldehydes) and derivatives of chrysanthemic acid [192].

3.3 Pharmaceutical

Unlike traditional pollutants, pharmaceuticals are specifically designed to have biological effects, which can lead to unintended consequences when they enter the environment. These contaminants can adversely affect living organisms even at very low concentrations and persist in the environment for extended periods due to their stable structures [193]. Pharmaceuticals are categorized into many classes according to their chemical structures, modes of action, and intended therapeutic applications. Major therapeutic classes include antibiotics, analgesics and anti-inflammatory agents; antidepressants; antivirals; anticoagulants; hormones; sedatives; and cardiovascular drugs. Antibiotics and hormones are particularly important to study as emerging contaminants because of their medical and therapeutic potential, frequency of use, and impact on microbial infections and estrogenic and androgenic impacts on life [194]. Antibiotics belonging to the macrolide class include clarithromycin, azithromycin, roxithromycin, and erythromycin. A macrocyclic lactone ring with one or more connected deoxy-sugar or amino sugar residues makes up these antibiotics. By attaching to the 50S ribosomal subunit, they obstruct the production of bacterial proteins. Depending on the number and type of bacteria, they can have either bacteriostatic or bactericidal effects. Macrolides come in topical, intravenous, and ocular drop forms and are helpful in treating a variety of diseases, including those caused by atypical mycobacterial infections and *H. pylori*, respiratory, skin, and soft tissue infections. In addition to their antimicrobial action, macrolides have immunomodulatory effects and can contribute to the development of antimicrobial resistance, disrupting microbial functions and structures in the environment [195–197]. Endocrine-disrupting pharmaceuticals include nonsteroidal substances, veterinary growth hormones, glucocorticoids, and sex hormones. Steroid hormones are a significant class of contaminants due to their potential adverse health impacts. Detecting these hormones in aquatic environments is challenging because of their low stability and concentration. Hormones can be classified as natural (produced by the human body) or artificial (produced industrially). Steroids of concern include estrogens and contraceptives, which are potent endocrine disruptors. Both humans and animals depend on estrogens for development and reproduction, and synthetic estrogen derivatives are widely used as alternatives and to treat menopausal symptoms, physiological conditions, and some types of cancer. Both the synthetic estrogen 17α-ethinylestradiol (EE2) and natural estrogens such as estrone (E1), 17β-estradiol (E2), and estriol (E3) are known for their strong estrogenic activity, which can have negative consequences even at low doses [198–201]. Typically, the structure of estrogens consists of a phenolic group and occasionally an aliphatic hydroxyl group. The endocrine disruptors with the highest potency are synthetic EE2 and E2. When hormone replacement therapy (HRT) is used, endogenous estrogen E2 must be administered for the shortest duration possible at the lowest effective dose. EE2 is a semisynthetic alkylated estradiol that is frequently present in HRTs and contraceptive pills. The main estrogenic steroid produced during pregnancy, E3, is helpful in treating menopausal symptoms since it is quickly digested and eliminated from the body. Estrogens are highly metabolized in both humans and animals, mainly in the liver, where E2 is rapidly oxidized to estrone. Estriol, the principal excretion product, can then be produced by additional processing of estrone. Conjugates are the primary type of EE2 elimination; other metabolic changes are unlikely to occur [202, 203]. Despite their significant impact, there is limited information on the environmental dispersion and quantitative effects of pharmaceuticals on other organisms. A number of other medications, in addition to hormones and antibiotics, are dangerous for the environment and public health.

3.4 *Heavy Metals*

Owing to their detrimental impacts on both the environment and human health, heavy metals such as lead, mercury, cadmium, and arsenic are important developing pollutants. These metals frequently contaminate home, industrial, and agro-industrial wastewaters, which contaminate rivers, lakes, and oceans. Even at low quantities, the biosorption and accumulation of heavy metals in aquatic food chains can cause serious health issues for people [204].

3.4.1 Lead

Earth's crust naturally contains bluish-gray lead. Human activities such as mining, manufacturing, and the burning of fossil fuels discharge them into the environment. Despite reductions in lead use in gasoline, paint, and other products, lead exposure remains a critical issue because of its toxicity. Lead can cause nervous system damage, kidney damage, and disrupted vitamin D metabolism, especially in children. The GHS hazard statements draw attention to its hazardous effects on aquatic life, propensity to affect breast-fed children, and detrimental effects if eaten or inhaled [205–207].

3.4.2 Arsenic

It combines with sulfur, chlorine, and oxygen to generate inorganic compounds that are mostly employed to preserve wood. The use of organic arsenic compounds as insecticides carries additional dangers. Exposure to arsenic is linked to cancer and can result in acute poisoning, skin irritation, and significant eye impairment. It has long-lasting effects and is highly harmful to aquatic life [208–211] (Fig. 1.4).

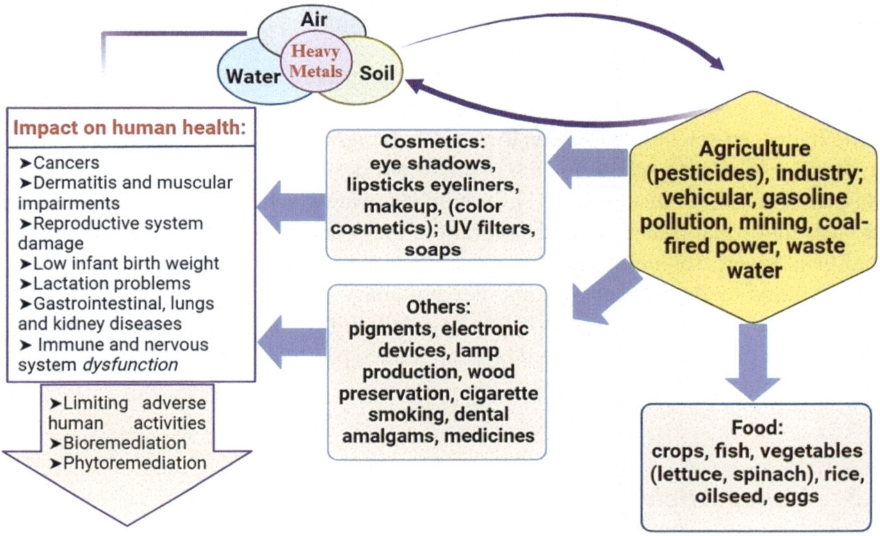

Fig. 1.4 Pathways of exposure, health impacts, and risk mitigation strategies for toxic metals

3.4.3 Mercury

Mercury can be found in many different forms, such as metallic, organic, and inorganic mercury compounds. Methylmercury can bioaccumulate in aquatic species since it is generated primarily by microbes in soil and water. Metallic mercury vaporizes readily at room temperature and is utilized in the manufacture of dental fillings, thermometers, caustic soda, and batteries. Exposure to mercury can result in serious health problems, such as acute toxicity if inhaled or consumed, toxicity to reproduction, and damage to organs. Aquatic life is likewise extremely poisonous [212–215].

3.4.4 Nickel

Nickel compounds are carcinogenic, and prolonged exposure can lead to severe health problems. Nickel is widely used in industrial applications, including in alloys and batteries, and as a catalyst in chemical reactions, posing significant health risks [216, 217].

3.4.5 Cadmium

Long-term cadmium exposure has been linked to bone loss, kidney damage, impaired lung function, and an increased risk of bone fractures. It is used in batteries and coatings and as a stabilizer in plastics. Its persistence in the environment poses severe health risks [218–220].

3.4.6 Selenium

Selenium is metabolized in the body into selenoproteins, nonspecific proteins, or excretory products, with forms such as selenite and selenomethionine converting to toxic hydrogen selenide. Doses as low as 1.5–6 mg/kg can be lethal to animals, and livestock consuming selenium-rich plants often develop conditions such as "blind staggers" or "alkali disease" [221]. In humans, excessive selenium intake can lead to gastrointestinal issues, skin discoloration, tooth decay, hair loss, and nail abnormalities. Severe selenium toxicity, or selenosis, is characterized by hair and nail loss, mottled teeth, skin lesions, and nerve damage. This occurs at blood selenium levels above 1346 μg/L, corresponding to a daily intake of approximately 1260 μg. Even lower intakes, above 750 μg/day, can reduce prothrombin time and blood glutathione levels [222].

3.5 *Nanoparticle Contaminants*

NPs, classified as emerging contaminants, present distinct challenges to aquatic ecosystems and environmental health owing to their small size, expansive surface area–volume ratio, and capacity for interactions with organisms and ecosystems. Many consumer goods, including electronics, cosmetics, textiles, and food packaging, are made from nanoparticles. During manufacturing processes, nanoparticles may be released into wastewater streams or discharged directly into the environment [223]. NPs can interact with aquatic organisms at the cellular and molecular levels, potentially leading to adverse effects on growth, development, reproduction, and behavior. Some nanoparticles, such as silver nanoparticles, have been shown to exhibit antimicrobial properties but may also pose risks to nontarget organisms, including aquatic bacteria, algae, invertebrates, and fish [224]. NPs may alter the physicochemical properties of water, sediment, and biological matrices, affecting nutrient cycling, sediment stability, and ecosystem function. Changes in nanoparticle concentrations and distributions may also influence microbial communities, food webs, and biodiversity in aquatic ecosystems [225].

3.5.1 Metal Nanoparticles

Aluminum, cadmium, cobalt, copper, gold, iron, lead, silver, titanium, and zinc are commonly employed in the synthesis of metal nanoparticles. Their unique properties, such as antibacterial activity, UV protection, and photocatalytic efficiency, make them integral to consumer products such as titanium dioxide, zinc oxide, and silver nanoparticles. However, these particles can leach from textiles, coatings, and personal care items into water systems, potentially harming aquatic ecosystems [226].

3.5.2 Metal Oxide Nanoparticles

These chemically stable nanoparticles consist of metallic cations paired with oxygen anions. Common examples include silicon dioxide, titanium dioxide, zinc oxide, and aluminum oxide. Compared with their metallic counterparts, these nanoparticles offer superior properties, making them highly versatile [227].

3.5.3 Carbon Nanomaterials

Carbon nanomaterials include five key types: nanotubes, fullerenes, graphene, nanofibers, and carbon black. Fullerenes, also known as buckyballs, are spherical or ellipsoidal carbon structures. Graphene, a two-dimensional (2D) hexagonal lattice of carbon atoms, forms a planar sheet approximately 1 nm thick, whereas cylindrical graphene structures are classified as nanotubes. Carbon nanofibers are derived from graphene nanostructures, and carbon black, an amorphous material, consists of spherical particles.

These carbon nanomaterials are widely used in industries such as electronics, composite materials, and environmental remediation. However, they may enter water systems during manufacturing or from the use of products containing these materials [228, 229].

3.5.4 Polymer Nanoparticles

Polymer nanoparticles, including polymeric micelles, dendrimers, and nanoparticle-based drug delivery systems, are widely utilized in pharmaceuticals, cosmetics, and medical devices. Nanocapsule micelles, nontoxic and biodegradable nanoparticles with hollow interiors, are thermolabile and photolabile particles. Dendrimers with 3D structures feature internal cavities capable of encapsulating molecules, making them promising candidates for drug delivery. These nanoparticles can enter aquatic ecosystems through wastewater from manufacturing processes or through the disposal of nanoparticle-containing products [230, 231].

3.5.5 Quantum Dots

Quantum dots are semiconductor nanoparticles with unique optical and electronic properties that are used in applications such as displays, lighting, and biological imaging. Quantum dots may be released into water bodies through manufacturing processes, the use of consumer electronics, and the disposal of electronic waste [232].

3.6 *Microorganisms*

Bacteria and viruses can cause food- or water-borne illnesses that can result in acute poisoning, chronic illnesses, disability, or even death.

3.6.1 Bacterial Pathogens

Salmonella *Salmonella* is a common bacteria that can be found in eggs, chickens, and other animal products. It can cause fever, headache, nausea, vomiting, abdominal pain, and diarrhea. It affects millions annually and can have severe or fatal outcomes [233].

Campylobacter This bacterium is associated primarily with raw milk, undercooked poultry, and contaminated water. It leads to symptoms similar to those caused by *Salmonella* [234].

Escherichia coli (*E. coli*) *E. coli* infections are linked to undercooked meat, unpasteurized milk, and tainted fresh food. They can also result in vomiting, diarrhea, and severe abdominal cramps [235].

Listeria *Listeria* spp. are found in ready-to-eat meals and unpasteurized dairy products, and *Listeria* spp. can lead to serious infections, especially in immunocompromised people, elderly individuals, neonates, and pregnant women, potentially leading to miscarriage or death [236].

Vibrio cholerae This bacterium causes cholera, which is characterized by vomiting, severe watery diarrhea, and abdominal pain. Severe dehydration and death can result from cholera. Common sources include contaminated water and foods such as seafood, vegetables, rice, and millet gruel [237].

3.6.2 Viruses

Norovirus It is a very infectious virus that induces watery diarrhea, vomiting fits, nausea, and cramping in the abdomen. It frequently results in foodborne illnesses [238].

Hepatitis A Spread in contaminated food, especially raw or undercooked fruit and shellfish, it can cause chronic liver damage [239] (Fig. 1.5).

3.7 Miscellaneous

3.7.1 Parasites

3.7.1.1 Trematodes

3.7.1.1.1 Tapeworms (Echinococcus spp. and Taenia spp.)

These parasites can infect humans through food or close contact with animals [239, 240].

3.7.1.1.2 Other Parasites (Ascaris, Cryptosporidium, Entamoeba histolytica, and Giardia)

They can affect fresh produce and enter the food chain through contaminated soil or water [241].

3.7.2 Prions

3.7.2.1 Bovine Spongiform Encephalopathy (BSE)

Known as mad cow disease in cattle, BSE is linked to variant Creutzfeldt–Jakob disease (vCJD) in humans. The consumption of contaminated meat products is the primary route of transmission [242].

3.7.3 Toxins

3.7.3.1 Natural Toxins

These toxins include cyanogenic glycosides, marine biotoxins, mycotoxins (such as ochratoxin and aflatoxin from moldy cereals), and toxins from toxic mushrooms. Prolonged exposure can lead to cancer, impair normal growth, or weaken the immune system [243]. Effective management of food and waterborne illnesses requires stringent food safety practices, monitoring, and public health interventions to mitigate the risks posed by these diverse contaminants.

3.7.4 Radiologicals

Radiologicals, a subset of emerging contaminants, include radioactive substances that pose significant environmental and health risks due to their ionizing radiation. In aquatic environments, radiological contaminants interact with water through dissolution, dispersion, adsorption, and bioaccumulation. Once released, radionuclides

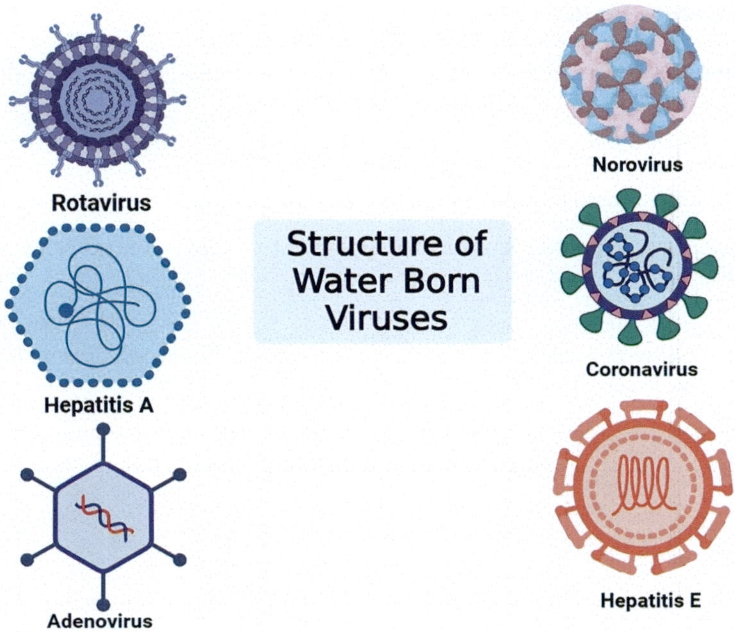

Fig. 1.5 Structures of some of the viruses that have been detected in water sources

may dissolve, form colloidal suspensions, or adsorb onto sediment particles. These processes enable their dispersion and potential bioaccumulation in aquatic organisms, leading to biomagnification through food chains. Their persistence is exacerbated by their long half-lives and radioactive properties [244].

3.7.4.1 Uranium (U)

Found naturally in rocks, soil, and groundwater, uranium contamination often results from mining and ore processing. In aerobic environments, uranium primarily exists as U(VI), forming soluble complexes with ligands such as carbonate and bicarbonate, whereas under reducing conditions, insoluble U(IV) precipitates dominate. Prolonged exposure to contaminated water or crops can harm human health and ecosystems.

3.7.4.2 Radon-222

Radon-222 is a radioactive gas formed from the decay of uranium and thorium, and radon-222 dissolves in groundwater and surface water. Its migration from soil and rock into water sources contributes to increased cancer risk, particularly for lung cancer, when it is consumed or inhaled.

3.7.4.3 Cesium-137 (Cs-137)

As a byproduct of nuclear fission, Cs-137 is water soluble and highly mobile in aquatic environments. It spreads rapidly, forming complexes influenced by water pH and competing ions. Its beta and gamma radiation emissions and ability to bio-accumulate in aquatic organisms present long-term ecological and health hazards.

3.7.4.4 Strontium-90 (Sr-90)

Produced during nuclear reactions and fallout, Sr-90 mimics calcium, integrates into biological systems and accumulates in bones. Although less water soluble than Cs-137 is, Sr-90 can adsorb onto sediments and contaminate surface water and groundwater, posing risks of internal radiation exposure and ecosystem disruption.

The environmental behavior of radiological contaminants is complex and is governed by factors such as pH, redox potential, temperature, and ion interactions. Understanding these processes is essential for assessing their transport, persistence, and impacts on ecosystems and public health [245, 246].

4 Conclusion

In conclusion, systematic classification and comprehensive understanding of emerging contaminants are imperative for addressing the significant environmental and health risks they pose. This chapter delves into the various categories of these contaminants, including heavy metals, pesticides, personal hygiene products, pharmaceuticals, nanoparticles, radiologicals, and microorganisms. The physicochemical properties, hazard statements, key uses, and interactions of these contaminants have been highlighted, underscoring their complex nature and far-reaching impacts. The classification of emerging contaminants serves as a foundational tool that enables researchers and policymakers to navigate the vast diversity of these substances. By grouping contaminants on the basis of their shared characteristics, we can prioritize those that pose the greatest risks and inform targeted mitigation strategies. The systematic approach to classification not only aids in risk assessment but also facilitates the identification of patterns and trends that are crucial for advancing scientific knowledge and regulatory frameworks. Addressing the challenges posed by emerging contaminants requires an interdisciplinary and collaborative approach. The chapter emphasized the importance of integrating expertise from various fields, such as chemistry, toxicology, ecology, engineering, and social sciences. This holistic approach is essential for developing effective management strategies that can keep pace with the rapid industrialization and urban expansion contributing to the proliferation of these contaminants. Furthermore, proactive measures to preserve water quality and protect aquatic ecosystems are urgently needed. This chapter advocates for collective action and collaborative research efforts to address the multifaceted challenges posed by emerging contaminants. By fostering such collaboration, we can ensure the sustainability of our water resources and safeguard the health of current and future generations. This chapter serves as a critical resource for stakeholders in academia, research, industry, and government, urging them to unite in an effort to mitigate the growing crisis of emerging contaminants. By understanding their classification and characteristics, we can develop and implement strategies that are both scientifically sound and practically effective, ensuring a healthier and more sustainable environment. In summary, the classification of emerging contaminants is not just an academic pursuit but also a practical necessity. It equips us with the knowledge and tools needed to address the pressing environmental challenges of our time, promoting a safer and more sustainable future for all.

References

1. Wang, F., Xiang, L., Sze-Yin Leung, K., et al. (2024). Emerging contaminants: A one health perspective. *Innovation, 5*(4), 100612. PMID: 38756954; PMCID: PMC11096751. https://doi.org/10.1016/j.xinn.2024.100612
2. Fei, C., McLaughlin, J. K., Lipworth, L., & Olsen, J. (2009). Maternal levels of perfluorinated chemicals and subfecundity. *Human Reproduction, 24*(5), 1200–1205. https://doi.org/10.1093/humrep/den490
3. Koch, N., Islam, N. F., Sonowal, S., & Prasad, R. (2021). Hemen Sarma, Environmental antibiotics and resistance genes as emerging contaminants: Methods of detection and bioremediation. *Current Research in Microbial Sciences, 2*, 100027. ISSN 2666-5174. https://doi.org/10.1016/j.crmicr.2021.100027
4. Kinigopoulou, V., Pashalidis, I., Kalderis, D., & Anastopoulos, I. (2022). Microplastics as carriers of inorganic and organic contaminants in the environment: A review of recent progress. *Journal of Molecular Liquids, 350*, 118580. ISSN 0167-7322. https://doi.org/10.1016/j.molliq.2022.118580
5. Saeedi, M. (2023). How microplastics interact with food chain: A short overview of fate and impacts. *Journal of Food Science and Technology, 61*(3), 1–11. Epub ahead of print. PMID: 37360257; PMCID: PMC10010214. https://doi.org/10.1007/s13197-023-05720-4
6. Li, X., Shen, X., Jiang, W., Xi, Y., & Li, S. (2024). Comprehensive review of emerging contaminants: Detection technologies, environmental impact, and management strategies. *Ecotoxicology and Environmental Safety, 278*, 116420. ISSN 0147-6513. https://doi.org/10.1016/j.ecoenv.2024.116420
7. Sauvé, S., & Desrosiers, M. (2014). A review of what is an emerging contaminant. *Chemistry Central Journal, 8*(1), 15. PMID: 24572188; PMCID: PMC3938815. https://doi.org/10.1186/1752-153X-8-15
8. Ngwenya, S., Mashau, N. S., Mhlongo, S. E., & Traoré, A. N. (2023). A systematic review of the risk management frameworks for potentially toxic chemical elements. *Toxicology and Industrial Health, 39*(12), 679–686. Epub 2023 Oct 18. PMID: 37853620; PMCID: PMC10655695. https://doi.org/10.1177/07482337231209092
9. Hastings, J., Magka, D., Batchelor, C., Duan, L., Stevens, R., Ennis, M., & Steinbeck, C. (2012). Structure-based classification and ontology in chemistry. *Journal of Cheminformatics, 4*, 8. PMID: 22480202; PMCID: PMC3361486. https://doi.org/10.1186/1758-2946-4-8
10. Annapurna, S. M., Anitha, M., & Kumar, L. S. (2024). Composition and source based aerosol classification using machine learning algorithms. *Advances in Space Research, 73*(1), 474–497. ISSN 0273-1177. https://doi.org/10.1016/j.asr.2023.09.068
11. Ba, F., Peng, P., Zhang, Y., & Zhao, Y. (2023). Classification and identification of contaminants in recyclable containers based on a recursive feature elimination-light gradient boosting

machine algorithm using an electronic nose. *Micromachines (Basel), 14*(11), 2047. PMID: 38004904; PMCID: PMC10673532. https://doi.org/10.3390/mi14112047
12. Gramatica, P., Pozzi, S., Consonni, V., & Di Guardo, A. (2002). Classification of environmental pollutants for global mobility potential. *SAR and QSAR in Environmental Research, 13*(2), 205–217. https://doi.org/10.1080/10629360290002695
13. Escher, B. I., Altenburger, R., Blüher, M., et al. (2023). Modernizing persistence-bioaccumulation-toxicity (PBT) assessment with high throughput animal-free methods. *Archives of Toxicology, 97*(5), 1267–1283. Epub 2023 Mar 23. PMID: 36952002; PMCID: PMC10110678. https://doi.org/10.1007/s00204-023-03485-5
14. Huang, P., Jin, Y., Hou, D., Yu, J., Tu, D., Cao, Y., & Zhang, G. (2017). Online classification of contaminants based on multi-classification support vector machine using conventional water quality sensors. *Sensors, 17*(3), 581. https://doi.org/10.3390/s17030581
15. Xiao, C. H., Meng, X. Z., Li, B. X., & Gao, H. W. (2023). A systematic review and meta-analysis of pollutants in environmental media. *Environmental Science and Pollution Research International, 30*(53), 113205–113217. https://doi.org/10.1007/s11356-023-30347-5
16. Kienzler, A., Barron, M. G., Belanger, S. E., Beasley, A., & Embry, M. R. (2017). Mode of action (MOA) assignment classifications for ecotoxicology: An evaluation of approaches. *Environmental Science & Technology, 51*(17), 10203–10211. https://doi.org/10.1021/acs.est.7b02337
17. Spycher, S., Netzeva, T. I., Worth, A. P., & Escher, B. I. (2008). Mode of action-based classification and prediction of activity of uncouplers for the screening of chemical inventories. *SAR and QSAR in Environmental Research, 19*(5-6), 433–463. https://doi.org/10.1080/10629360802348803
18. Committee on the Design and Evaluation of Safer Chemical Substitutions. (2014). A Framework to Inform Government and Industry Decision; Board on Chemical Sciences and Technology; Board on Environmental Studies and Toxicology; Division on Earth and Life Studies; National Research Council. In *A Framework to Guide Selection of Chemical Alternatives*. National Academies Press (US). Appendix D, Overview of the GHS Classification Scheme in Hazard Classification. Retrieved from https://www.ncbi.nlm.nih.gov/books/NBK253967/
19. Arman, N. Z., Salmiati, S., Aris, A., Salim, M. R., Nazifa, T. H., Muhamad, M. S., & Marpongahtun, M. (2021). A review on emerging pollutants in the water environment: Existences, health effects and treatment processes. *Water, 13*(22), 3258. https://doi.org/10.3390/w13223258
20. Chakraborty, A., Adhikary, S., Bhattacharya, S., Dutta, S., Chatterjee, S., Banerjee, D., Ganguly, A., & Rajak, P. (2023). Pharmaceuticals and personal care products as emerging environmental contaminants: Prevalence, toxicity, and remedial approaches. *ACS Chemical Health and Safety, 30*(6), 362–388. https://doi.org/10.1021/acs.chas.3c00071
21. Zhou, C., Chiang, C., Brehm, E., Warner, G. R., & Flaws, J. A. (2022). Chapter 44. Personal care products and cosmetics. In R. C. Gupta (Ed.), *Reproductive and developmental toxicology* (3rd ed., pp. 867–909, ISBN 9780323897730). Academic Press. https://doi.org/10.1016/B978-0-323-89773-0.00044-8
22. Khalid, M., & Abdollahi, M. (2021). Environmental distribution of personal care products and their effects on human health. *Iranian Journal of Pharmaceutical Research, 20*(1), 216–253. PMID: 34400954; PMCID: PMC8170769. https://doi.org/10.22037/ijpr.2021.114891.15088
23. Government of Canada. (2008). *Guidance document: Classification of products at the cosmetic-drug interface*. Government of Canada
24. Vosátka, R., Krátký, M., & Vinšová, J. (2018). Triclosan and its derivatives as antimycobacterial active agents. *European Journal of Pharmaceutical Sciences, 114*, 318–331. ISSN 0928-0987. https://doi.org/10.1016/j.ejps.2017.12.013
25. National Center for Biotechnology Information. *PubChem compound summary for CID 5564, Triclosan*. Retrieved June 7, 2024, from https://pubchem.ncbi.nlm.nih.gov/compound/Triclosan

26. Sinicropi, M. S., Iacopetta, D., Ceramella, J., Catalano, A., Mariconda, A., Pellegrino, M., Saturnino, C., Longo, P., & Aquaro, S. (2022). Triclosan: A small molecule with controversial roles. *Antibiotics (Basel), 11*(6), 735. PMID: 35740142; PMCID: PMC9220381. https://doi.org/10.3390/antibiotics11060735
27. Carey, D. E., & McNamara, P. J. (2015). The impact of triclosan on the spread of antibiotic resistance in the environment. *Frontiers in Microbiology, 5*, 780. PMID: 25642217; PMCID: PMC4295542. https://doi.org/10.3389/fmicb.2014.00780
28. Yueh, M. F., & Tukey, R. H. (2016). Triclosan: A widespread environmental toxicant with many biological effects. *Annual Review of Pharmacology and Toxicology, 56*, 251–272. PMID: 26738475; PMCID: PMC4774862. https://doi.org/10.1146/annurev-pharmtox-010715-103417
29. Latch, D. E., Packer, J. L., Stender, B. L., VanOverbeke, J., Arnold, W. A., & McNeill, K. (2005). Aqueous photochemistry of triclosan: formation of 2,4-dichlorophenol, 2,8-dichlorodibenzo-p-dioxin, and oligomerization products. *Environmental Toxicology and Chemistry, 24*(3), 517–525. https://doi.org/10.1897/04-243r.1
30. Kanetoshi, A., Katsura, E., Ogawa, H., et al. (1992). Acute toxicity, percutaneous absorption and effects on hepatic mixed function oxidase activities of 2,4,4′-trichloro-2′-hydroxydiphenyl ether (Irgasan® DP300) and its chlorinated derivatives. *Archives of Environmental Contamination and Toxicology, 23*, 91–98. https://doi.org/10.1007/BF00226000
31. United States Food and Drug Administration. (2022). *Parabens in Cosmetics*. Retrieved from https://www.fda.gov/cosmetics/cosmetic-ingredients/parabens-cosmetics
32. Soni, M. G., Carabin, I. G., & Burdock, G. A. (2005). safety assessment of esters of P-hydroxybenzoic acid (Parabens). *Food and Chemical Toxicology, 43*, 985–1015. https://doi.org/10.1016/j.fct.2005.01.020
33. National Center for Biotechnology Information. *PubChem compound summary for CID 7456, Methylparaben*. Retrieved June 4, 2024, from https://pubchem.ncbi.nlm.nih.gov/compound/Methylparaben
34. National Center for Biotechnology Information. *PubChem compound summary for CID 8434, Ethylparaben*. Retrieved June 4, 2024, from https://pubchem.ncbi.nlm.nih.gov/compound/Ethylparaben
35. National Center for Biotechnology Information. *PubChem compound summary for CID 7175, Propylparaben*. Retrieved June 4, 2024, from https://pubchem.ncbi.nlm.nih.gov/compound/Propylparaben
36. National Center for Biotechnology Information. *PubChem compound summary for CID 7184, Butylparaben*. Retrieved June 4, 2024, from https://pubchem.ncbi.nlm.nih.gov/compound/Butylparaben
37. Ambarak, M. F. (2020). Determination of methylparaben in some cosmetics AND pharmaceutics using liquid–liquid extraction and spectrophotometric technique. *Asian Journal of Green Chemistry, 4*(2), 192–201. https://doi.org/10.22034/AJGC/2020.2.7
38. Stroppel, L., Schultz-Fademrecht, T., Cebulla, M., Blech, M., Marhöfer, R. J., Selzer, P. M., & Garidel, P. (2023). Antimicrobial preservatives for protein and peptide formulations: An overview. *Pharmaceutics, 15*(2), 563. PMID: 36839885; PMCID: PMC10217790. https://doi.org/10.3390/pharmaceutics15020563
39. Pereira, A. R., Simões, M., & Gomes, I. B. (2023). Parabens as environmental contaminants of aquatic systems affecting water quality and microbial dynamics. *Science of the Total Environment, 905*, 167332. ISSN 0048-9697. https://doi.org/10.1016/j.scitotenv.2023.167332
40. Nowak, K., Ratajczak-Wrona, W., Górska, M., & Jabłońska, E. (2018). Parabens and their effects on the endocrine system. *Molecular and Cellular Endocrinology, 474*, 238–251. https://doi.org/10.1016/j.mce.2018.03.014
41. Nguyen, V.-H., Thi, L.-A. P., Sri Chandana, P., Do, H.-T., Pham, T.-H., Lee, T., Nguyen, T. D., Le Phuoc, C., & Huong, P. T. (2021). The degradation of paraben preservatives: Recent progress and sustainable approaches toward photocatalysis. *Chemosphere, 276*, 130163. ISSN 0045-6535. https://doi.org/10.1016/j.chemosphere.2021.130163

42. Burdock, G. A. (Ed.). (2010). *Fenaroli's handbook of flavor ingredients* (6th ed., p. 1297). CRC Press.
43. Marks, J. G., & Deleo, V. A. (1997). *Contact and occupational dermatology* (2nd ed., p. 115). Mosby.
44. O'Neil, M. J. (Ed.). (2013). *The Merck Index—An encyclopedia of chemicals, drugs, and biologicals* (p. 708). Royal Society of Chemistry.
45. Sigma–Aldrich. *Safety data sheet for Propyl 4-hydroxybenzoate. Product number: P53357, version 3.6 (Revision Date 01/02/2015)*. Retrieved October 7, 2016, from https://www.sigmaaldrich.com/safety-center.html
46. Burdock, G. A. (Ed.). (2010). *Fenaroli's handbook of flavor ingredients* (6th ed., p. 1757). CRC Press.
47. Dionisio, K. L., Phillips, K., Price, P. S., Grulke, C. M., Williams, A., Biryol, D., Hong, T., & Isaacs, K. K. (2018). The chemical and products database, a resource for exposure-relevant data on chemicals in consumer products. *Scientific Data, 5*, 180125. https://doi.org/10.1038/sdata.2018.125
48. *FDA Pharm Classes. Butylparaben*. Retrieved from https://dailymed.nlm.nih.gov/dailymed/browse-drug-classes.cfm
49. *LOTUS—The natural products occurrence database. Butylparaben*. Retrieved from https://www.wikidata.org/wiki/Q3302873
50. Wang, Y., & Qian, H. (2021). Phthalates and their impacts on human health. *Healthcare (Basel), 9*(5), 603. PMID: 34069956; PMCID: PMC8157593. https://doi.org/10.3390/healthcare9050603
51. Bingham, E., Cohrssen, B., & Powell, C. H. (2001). *Patty's toxicology* (Vol. 1-9, 5th ed., p. V6 841). Wiley.
52. National Center for Biotechnology Information. *PubChem compound summary for CID 8343, Bis(2-ethylhexyl) phthalate*. Retrieved June 1, 2024, from https://pubchem.ncbi.nlm.nih.gov/compound/Bis_2-ethylhexyl_-phthalate
53. Hlisníková, H., Petrovičová, I., Kolena, B., Šidlovská, M., & Sirotkin, A. (2020). Effects and mechanisms of phthalates' action on reproductive processes and reproductive health: A literature review. *International Journal of Environmental Research and Public Health, 17*(18), 6811. PMID: 32961939; PMCID: PMC7559247. https://doi.org/10.3390/ijerph17186811
54. Benjamin, S., Masai, E., Kamimura, N., Takahashi, K., Anderson, R. C., & Faisal, P. A. (2017). Phthalates impact human health: Epidemiological evidences and plausible mechanism of action. *Journal of Hazardous Materials, 34*, 360–383. https://doi.org/10.1016/j.jhazmat.2017.06.036
55. Foster, P. M. D., Mylchreest, E., Gaido, K. W., & Sar, M. (2001). Effects of phthalate esters on the developing reproductive tract of male rats. *Human Reproduction Update, 7*, 231–235. https://doi.org/10.1093/humupd/7.3.231
56. IARC Working Group on the Evaluation of Carcinogenic Risks to Humans. (2013). Some chemicals present in industrial and consumer products, food and drinking-water. In *IARC monographs on the evaluation of carcinogenic risks to humans* (Vol. 101, pp. 9–549). PMID: 24772663; PMCID: PMC4939416.
57. Qadeer, A., Kirsten, K. L., Ajmal, Z., Jiang, X., & Zhao, X. (2022). Alternative plasticizers as emerging global environmental and health threat: Another regrettable substitution? *Environmental Science & Technology, 56*(3), 1482–1488. https://doi.org/10.1021/acs.est.1c08365
58. Ghosh, A., Tripathy, A., & Ghosh, D. (2022). Impact of endocrine disrupting chemicals (EDCs) on reproductive health of human. *Proceedings of the Zoological Society, 75*, 16–30. https://doi.org/10.1007/s12595-021-00412-3
59. Meeker, J. D., Sathyanarayana, S., & Swan, S. H. (2009). Phthalates and other additives in plastics: human exposure and associated health outcomes. *Philosophical Transactions of the Royal Society of London. Series B, Biological Sciences, 364*(1526), 2097–2113. PMID: 19528058; PMCID: PMC2873014. https://doi.org/10.1098/rstb.2008.0268

60. Wojnowska-Baryła, I., Bernat, K., & Zaborowska, M. (2022). Plastic waste degradation in landfill conditions: The problem with microplastics, and their direct and indirect environmental effects. *International Journal of Environmental Research and Public Health, 19*(20), 13223. PMID: 36293805; PMCID: PMC9602440. https://doi.org/10.3390/ijerph192013223
61. Bølling, A. K., Sripada, K., Becher, R., & Bekö, G. (2020). Phthalate exposure and allergic diseases: Review of epidemiological and experimental evidence. *Environment International, 139*, 105706. https://doi.org/10.1016/j.envint.2020.105706
62. Bekö, G., Callesen, M., Weschler, C. J., et al. (2015). Phthalate exposure through different pathways and allergic sensitization in preschool children with asthma, allergic rhinoconjunctivitis and atopic dermatitis. *Environmental Research, 137*, 432–439. https://doi.org/10.1016/j.envres.2015.01.012
63. Bornehag, C. G., & Nanberg, E. (2010). Phthalate exposure and asthma in children. *International Journal of Andrology, 33*(2), 333–345. https://doi.org/10.1111/j.1365-2605.2009.01023.x
64. Radke, E. G., Glenn, B. S., Braun, J. M., & Cooper, G. S. (2019). Phthalate exposure and female reproductive and developmental outcomes: A systematic review of the human epidemiological evidence. *Environment International, 130*, 104580. ISSN 0160-4120. https://doi.org/10.1016/j.envint.2019.02.003
65. Kirti Sharma, A., & Bhatnagar, P. (2022). Comparative reproductive toxicity of phthalate on male and female reproductive potential of rodent when exposure occurs during developmental period. *Materials Today Proceedings, 69*(Part 4), A22–A29. ISSN 2214-7853. https://doi.org/10.1016/j.matpr.2023.04.013
66. Yang, L., Liu, X., Peng, Z., et al. (2024). Exposure to di-2-ethylhexyl phthalate (DEHP) increases the risk of cancer. *BMC Public Health, 24*, 430. https://doi.org/10.1186/s12889-024-17801-w
67. Ahern, T. P., Spector, L. G., Damkier, P., et al. (2022). Medication-associated phthalate exposure and childhood cancer incidence. *Journal of the National Cancer Institute, 114*(6), 885–894. https://doi.org/10.1093/jnci/djac045
68. Ahern, T. P., Broe, A., Lash, T. L., et al. (2019). Phthalate exposure and breast cancer incidence: A Danish Nationwide Cohort Study. *Journal of Clinical Oncology, 37*(21), 1800–1809. https://doi.org/10.1200/JCO.18.02202
69. Ziani, K., Ioniță-Mîndrican, C. B., Mititelu, M., Neacşu, S. M., Negrei, C., Moroșan, E., Drăgănescu, D., & Preda, O. T. (2023). Microplastics: A real global threat for environment and food safety: A state of the art review. *Nutrients, 15*(3), 617. PMID: 36771324; PMCID: PMC9920460. https://doi.org/10.3390/nu15030617
70. Andrady, A. L. (2011). Microplastics in the marine environment. *Marine Pollution Bulletin, 62*, 1596–1605. https://doi.org/10.1016/j.marpolbul.2011.05.030
71. Kooi, M., & Koelmans, A. A. (2019). Simplifying microplastic via continuous probability distributions for size, shape, and density. *Environmental Science & Technology Letters, 6*(9), 551–557. https://doi.org/10.1021/acs.estlett.9b00379
72. Thakur, B., Singh, J., Singh, J., Angmo, D., & Vig, A. P. (2023). Biodegradation of different types of microplastics: Molecular mechanism and degradation efficiency. *Science of the Total Environment, 877*, 162912. ISSN 0048-9697. https://doi.org/10.1016/j.scitotenv.2023.162912
73. Duis, K., & Coors, A. (2016). Microplastics in the aquatic and terrestrial environment: Sources (with a specific focus on personal care products), fate and effects. *Environmental Sciences Europe, 28*(1), 2. . Epub 2016 Jan 6. PMID: 27752437; PMCID: PMC5044952. https://doi.org/10.1186/s12302-015-0069-y
74. Kiran, B. R., Kopperi, H., & Venkata Mohan, S. (2022). Micro/nano-plastics occurrence, identification, risk analysis and mitigation: Challenges and perspectives. *Reviews in Environmental Science and Biotechnology, 21*, 169–203. https://doi.org/10.1007/s11157-021-09609-6
75. Zhao, B., Rehati, P., Yang, Z., Cai, Z., Guo, C., & Li, Y. (2024). The potential toxicity of microplastics on human health. *Science of the Total Environment, 912*, 168946. ISSN 0048-9697. https://doi.org/10.1016/j.scitotenv.2023.168946

76. Smith, M., Love, D. C., Rochman, C. M., & Neff, R. A. (2018). Microplastics in seafood and the implications for human health. *Current Environmental Health Reports, 5*, 375–386. https://doi.org/10.1007/s40572-018-0206-z
77. Yang, D., Shi, H., Li, L., Li, J., Jabeen, K., & Kolandhasamy, P. (2015). Microplastic pollution in table salts from China. *Environmental Science & Technology, 49*, 13622–13627. https://doi.org/10.1021/acs.est.5b03163
78. Campanale, C., Massarelli, C., Savino, I., Locaputo, V., & Uricchio, V. F. (2020). A detailed review study on potential effects of microplastics and additives of concern on human health. *International Journal of Environmental Research and Public Health, 17*(4), 1212. PMID: 32069998; PMCID: PMC7068600. https://doi.org/10.3390/ijerph17041212
79. Homem, V., Silva, J. A., Ratola, N., Santos, L., & Alves, A. (2015). Long lasting perfume—A review of synthetic musks in WWTPs. *Journal of Environmental Management, 149*, 168–192. ISSN 0301-4797. https://doi.org/10.1016/j.jenvman.2014.10.008
80. Wang, T., Zou, H., Li, D., Gao, J., Qingwei, B., & Wang, Z. (2023). Global distribution and ecological risk assessment of synthetic musks in the environment. *Environmental Pollution, 331*(2), 121893. ISSN 0269-7491. https://doi.org/10.1016/j.envpol.2023.121893
81. National Center for Biotechnology Information. *PubChem compound summary for CID 62329, Musk xylene*. Retrieved June 1, 2024, from https://pubchem.ncbi.nlm.nih.gov/compound/Musk-xylene
82. National Center for Biotechnology Information. *PubChem compound summary for CID 6669, Musk ketone*. Retrieved June 1, 2024, from https://pubchem.ncbi.nlm.nih.gov/compound/Musk-ketone
83. National Center for Biotechnology Information. *PubChem compound summary for CID 91497, Galaxolide*. Retrieved June 1, 2024, from https://pubchem.ncbi.nlm.nih.gov/compound/Galaxolide
84. National Center for Biotechnology Information. *PubChem compound summary for CID 89440, Tonalid*. Retrieved June 1, 2024, from https://pubchem.ncbi.nlm.nih.gov/compound/Tonalid
85. Tasselli, S., Rogora, M., Orrù, A., & Guzzella, L. (2023). Behaviour of synthetic musk fragrances in freshwaters: occurrence, relations with environmental parameters, and preliminary risk assessment. *Environmental Science and Pollution Research International, 30*(50), 109643–109658. Epub 2023 Sep 30. PMID: 37777704; PMCID: PMC10622367. https://doi.org/10.1007/s11356-023-30030-9
86. Arruda, V., Simões, M., & Gomes, I. B. (2022). Synthetic musk fragrances in water systems and their impact on microbial communities. *Water, 14*(5), 692. https://doi.org/10.3390/w14050692
87. Li, X., Zhao, Y., Chen, B., Zhu, Z., Kang, Q., Husain, T., & Zhang, B. (2022). Inhalation and ingestion of synthetic musks in pregnant women: In silico spontaneous abortion risk evaluation and control. *Environmental International, 158*, 106911. https://doi.org/10.1016/j.envint.2021.106911
88. Li, X., He, W., Zhao, Y., Chen, B., Zhu, Z., Kang, Q., & Zhang, B. (2022). Dermal exposure to synthetic musks: Human health risk assessment, mechanism, and control strategy. *Ecotoxicology and Environmental Safety, 236*, 113463. https://doi.org/10.1016/j.ecoenv.2022.113463
89. Patel, S., Homaei, A., & Sharifian, S. (2021). Need of the hour: To raise awareness on vicious fragrances and synthetic musks. *Environment, Development and Sustainability, 23*, 4764–4781. https://doi.org/10.1007/s10668-020-00829-4
90. Xu, L., & Cao, Y. (2016). Native musk and synthetic musk ketone strongly induced the growth repression and the apoptosis of cancer cells. *BMC Complementary and Alternative Medicine, 16*(1), 511. PMID: 27931220; PMCID: PMC5146870. https://doi.org/10.1186/s12906-016-1493-2
91. Mora Lagares, L., & Vračko, M. (2023). Ecotoxicological evaluation of bisphenol A and alternatives: A comprehensive in silico modelling approach. *Journal of Xenobiotics, 13*(4), 719–739. https://doi.org/10.3390/jox13040046

92. National Center for Biotechnology Information. *PubChem compound summary for CID 6623, Bisphenol A*. Retrieved June 5, 2024, from https://pubchem.ncbi.nlm.nih.gov/compound/Bisphenol-A
93. Weber, M., et al. (1999–2017). Acetone. In *Ullmann's encyclopedia of industrial chemistry* (7th ed.). Wiley.
94. Tian, K., Yu, Y., Qiu, Q., Sun, X., Meng, F., Bi, Y., Gu, J., Wang, Y., Zhang, F., & Huo, H. (2022). Mechanisms of BPA degradation and toxicity resistance in Rhodococcus equi. *Microorganisms, 11*(1), 67. PMID: 36677360; PMCID: PMC9862853. https://doi.org/10.3390/microorganisms11010067
95. NITE; Chemical Risk Information Platform (CHRIP). *Biodegradation and Bioconcentration*. NITE. Retrieved November 7, 2017, from https://www.safe.nite.go.jp/english/db.html
96. EU; European Union Risk Assessment Report, CAS: 80-05-7 EINECS No: 201-245-8, Environment Addendum of April 2008, 4,4′-ISOPROPYLIDENEDIPHENOL (Bisphenol-A), Part 1 Environment. Retrieved November 7, 2017, from https://publications.jrc.ec.europa.eu/repository/bitstream/111111111/15063/1/lbna24588enn.pdf
97. Ousji, O., Ohlund, L., & Sleno, L. (2020). Comprehensive in vitro metabolism study of bisphenol A using liquid chromatography-high resolution tandem mass spectrometry. *Chemical Research in Toxicology, 33*(6), 1468–1477. https://doi.org/10.1021/acs.chemrestox.0c00042
98. Schmidt, J., Kotnik, P., Trontelj, J., Knez, Ž., & Mašič, L. P. (2013). Bioactivation of bisphenol A and its analogs (BPF, BPAF, BPZ and DMBPA) in human liver microsomes. *Toxicology In Vitro, 27*(4), 1267–1276. https://doi.org/10.1016/j.tiv.2013.02.016
99. Gao, H., Yang, B. J., Li, N., Feng, L. M., Shi, X. Y., Zhao, W. H., & Liu, S. J. (2015). Bisphenol A and hormone-associated cancers: Current progress and perspectives. *Medicine (Baltimore), 94*(1), e211. PMID: 25569640; PMCID: PMC4602822. https://doi.org/10.1097/MD.0000000000000211
100. Cantonwine, D. E., Hauser, R., & Meeker, J. D. (2013). Bisphenol A and human reproductive health. *Expert Review of Obstetrics & Gynecology, 8*(4). PMID: 24187577; PMCID: PMC3811157. https://doi.org/10.1586/17474108.2013.811939
101. Lü, L., Liu, Y., Yang, Y., He, J., Luo, L., Chen, S., & Xing, H. (2024). Bisphenol A exposure interferes with reproductive hormones and decreases sperm counts: A systematic review and meta-analysis of epidemiological studies. *Toxics, 12*(4), 294. https://doi.org/10.3390/toxics12040294
102. Welch, C., & Mulligan, K. (2022). Does Bisphenol A confer risk of neurodevelopmental disorders? What we have learned from developmental neurotoxicity studies in animal models. *International Journal of Molecular Sciences, 23*(5), 2894. https://doi.org/10.3390/ijms23052894. Published 2022 Mar 7.
103. Costa, H. E., & Cairrao, E. (2024). Effect of bisphenol A on the neurological system: a review update. *Archives of Toxicology, 98*, 1–73. https://doi.org/10.1007/s00204-023-03614-0
104. Gao, X., & Wang, H. S. (2014). Impact of bisphenol a on the cardiovascular system—epidemiological and experimental evidence and molecular mechanisms. *International Journal of Environmental Research and Public Health, 11*(8), 8399–8413. PMID: 25153468; PMCID: PMC4143868. https://doi.org/10.3390/ijerph110808399
105. Khan, N. G., Correia, J., Adiga, D., Rai, P. S., Dsouza, H. S., Chakrabarty, S., & Kabekkodu, S. P. (2021). A comprehensive review on the carcinogenic potential of bisphenol A: Clues and evidence. *Environmental Science and Pollution Research International, 28*(16), 19643–19663. Epub 2021 Mar 5. PMID: 33666848; PMCID: PMC8099816. https://doi.org/10.1007/s11356-021-13071-w
106. Canesi, L., & Fabbri, E. (2015). Environmental effects of BPA: Focus on aquatic species. *Dose-Response, 13*(3), 1559325815598304. https://doi.org/10.1177/1559325815598304
107. Priac, A., Morin-Crini, N., Druart, C., Gavoille, S., Bradu, C., Lagarrigue, C., Torri, G., Winterton, P., & Crini, G. (2017). Alkylphenol and alkylphenol polyethoxylates in water and wastewater: A review of options for their elimination. *Arabian Journal of Chemistry, 10*(Supplement 2), S3749–S3773. ISSN 1878-5352. https://doi.org/10.1016/j.arabjc.2014.05.011

108. Seidel, A. (2004a). *Kirk-Othmer encyclopedia of chemical technology* (Vol. 2, p. 225). Wiley.
109. United States Environmental Protection Agency. (2010). *Nonylphenol (NP) and Nonylphenol Ethoxylates (NPEs) Action Plan [RIN 2070-ZA09]*. Retrieved from https://www.epa.gov/sites/default/files/2015-09/documents/rin2070-za09_np-npes_action_plan_final_2010-08-09.pdf
110. Saravanan, M., Nam, S.-E., Eom, H.-J., Lee, D.-H., & Rhee, J.-S. (2019). Long-term exposure to waterborne nonylphenol alters reproductive physiological parameters in economically important marine fish. *Comparative Biochemistry and Physiology Part C: Toxicology & Pharmacology, 216*, 10–18. ISSN 1532-0456. https://doi.org/10.1016/j.cbpc.2018.11.009
111. Chen, M., Chang, C. C., Shen, Y. J., Hung, J. H., Guo, B. R., Chuang, H. Y., & Mao, I. F. (2008). Quantification of prenatal exposure and maternal-fetal transfer of nonylphenol. *Chemosphere, 73*, 5349–5245.
112. Bechi, N., Ietta, F., Romagnoli, R., Focardi, S., Corsi, I., Buffi, C., et al. (2006). Estrogen-like response to p-nonylphenol in human first trimester placenta and BeWo choriocarcinoma cells. *Toxicological Sciences, 93*(1), 75–81.
113. Bechi, N., Ietta, F., Romagnoli, R., Jantra, S., Cencini, M., Galassi, G., et al. (2010). Environmental levels of para-nonylphenol are able to affect cytokine secretion in human placenta. *Environmental Health Perspectives, 118*(3), 427–431.
114. Ademollo, N., Ferrara, F., Delise, M., Fabietti, F., & Funari, E. (2008). Nonylphenol and octylphenol in human breast milk. *Environment International, 34*, 984–987.
115. EPA. (2009c). *Screening-level hazard characterization. Alkylphenols category*. Retrieved from http://www.epa.gov/chemrtk/hpvis/hazchar/Category_Alkylphenols_Sept2009.pdf
116. Langenbach, B., & Wilson, M. (2021). Per- and polyfluoroalkyl substances (PFAS): Significance and considerations within the regulatory framework of the USA. *International Journal of Environmental Research and Public Health, 18*(21), 11142. PMID: 34769660; PMCID: PMC8583519. https://doi.org/10.3390/ijerph182111142
117. Glüge, J., Scheringer, M., Cousins, I. T., et al. (2020). An overview of the uses of per- and polyfluoroalkyl substances (PFAS). *Environmental Science. Processes & Impacts, 22*, 2345–2373. https://doi.org/10.1039/D0EM00291G
118. Gaines, L. G. T. (2023). Historical and current usage of per- and polyfluoroalkyl substances (PFAS): A literature review. *American Journal of Industrial Medicine, 66*(5), 353–378, Special Issue:Occupational Exposure to PFAS. https://doi.org/10.1002/ajim.23362
119. Simon, J. A., Abrams, S., Bradburne, T., Bryant, D., Burns, M., Cassidy, D., Cherry, J., Chiang, S. Y., Cox, D., Crimi, M., et al. (2019). PFAS Experts Symposium: Statements on regulatory policy, chemistry and analtyics, toxicology, transport/fate, and remediation for per- and polyfluoroalkyl substances (PFAS) contamination issues. *Remediation, 29*, 31–48. https://doi.org/10.1002/rem.21624
120. Wei, Z., Xu, T., & Zhao, D. (2019). Treatment of per- and polyfluoroalkyl substances in landfill leachate: Status, chemistry and prospects. *Environmental Science: Water Research & Technology, 5*, 1814–1835. https://doi.org/10.1039/C9EW00645A
121. Meegoda, J. N., Kewalramani, J. A., Li, B., & Marsh, R. W. (2020). A review of the applications, environmental release, and remediation technologies of per- and polyfluoroalkyl substances. *International Journal of Environmental Research and Public Health, 17*(21), 8117. PMID: 33153160; PMCID: PMC7663283. https://doi.org/10.3390/ijerph17218117
122. Singh, R. K., Fernando, S., Baygi, S. F., Multari, N., Thagard, S. M., & Holsen, T. M. (2019). Breakdown products from perfluorinated alkyl substances (PFAS) degradation in a plasma-based water treatment process. *Environmental Science & Technology, 53*(5), 2731–2738. https://doi.org/10.1021/acs.est.8b07031
123. Li, F., Yang, N., Yang, Z., et al. (2020). Biomimetic degradability of linear perfluorooctanesulfonate (L-PFOS): Degradation products and pathways. *Chemosphere, 259*, 127502. https://doi.org/10.1016/j.chemosphere.2020.127502
124. Wee, S. Y., & Aris, A. Z. (2023). Revisiting the "forever chemicals", PFOA and PFOS exposure in drinking water. *npj Clean Water, 6*, 57. https://doi.org/10.1038/s41545-023-00274-6

125. Habib, Z., Song, M., Ikram, S., & Zahra, Z. (2024). Overview of per- and polyfluoroalkyl substances (PFAS), their applications, sources, and potential impacts on human health. *Pollutants, 4*(1), 136–152. https://doi.org/10.3390/pollutants4010009
126. Dias, D., Bons, J., Kumar, A., Kabir, M. H., & Liang, H. (2024). Forever chemicals, per- and polyfluoroalkyl substances (PFAS), in lubrication. *Lubricants, 12*(4), 114. https://doi.org/10.3390/lubricants12040114
127. Ignat'ev, N. V. (2017). In H. Groult, F. Leroux, & A. Tressaud (Eds.), *Modern synthesis processes and reactivity of fluorinated compounds* (pp. 71–123). Elsevier.
128. Results of the 2006 survey on production and use of PFOS, PFAS, PFOA, PFCA, their related substances and products/mixtures containing these substances. OECD.
129. Siegemund, G., et al. (1999–2016). Fluorine compounds, organic. In *Ullmann's encyclopedia of industrial chemistry* (7th ed.). Wiley.
130. Li, Y., Fletcher, T., Mucs, D., Scott, K., Lindh, C. H., Tallving, P., & Jakobsson, K. (2018). Half-lives of PFOS, PFHxS and PFOA after end of exposure to contaminated drinking water. *Occupational and Environmental Medicine, 75*(1), 46–51. Epub 2017 Nov 13. PMID: 29133598; PMCID: PMC5749314. https://doi.org/10.1136/oemed-2017-104651
131. Nicole, W. (2013). PFOA and cancer in a highly exposed community: new findings from the C8 science panel. *Environmental Health Perspectives, 121*(11-12), A340. PMID: 24284021; PMCID: PMC3855507. https://doi.org/10.1289/ehp.121-A340
132. Fustinoni, S., & Consonni, D. (2023). Historical trend of exposure to perfluoroalkyl surfactants PFOA, ADV, and cC6O4 and its management in two perfluoroalkyl polymers plants, Italy. *Annals of Work Exposures and Health, 67*(4), 518–535. PMID: 36715212; PMCID: PMC10119700. https://doi.org/10.1093/annweh/wxac095
133. White, S. S., Fenton, S. E., & Hines, E. P. (2011). Endocrine disrupting properties of perfluorooctanoic acid. *The Journal of Steroid Biochemistry and Molecular Biology, 127*(1-2), 16–26. Epub 2011 Mar 21. PMID: 21397692; PMCID: PMC3335904. https://doi.org/10.1016/j.jsbmb.2011.03.011
134. Post, G. B., Cohn, P. D., & Cooper, K. R. (2012). Perfluorooctanoic acid (PFOA), an emerging drinking water contaminant: A critical review of recent literature. *Environmental Research, 116*, 93–117. https://doi.org/10.1016/j.envres.2012.03.007
135. National Center for Biotechnology Information. *PubChem compound summary for CID 9554, Perfluorooctanoic acid.* Retrieved June 5,2024, from https://pubchem.ncbi.nlm.nih.gov/compound/Perfluorooctanoic-acid
136. Siddiqi, M. A., Laessig, R. H., & Reed, K. D. (2003). Polybrominated diphenyl ethers (PBDEs): new pollutants-old diseases. *Clinical Medicine & Research, 1*(4), 281–290. PMID: 15931321; PMCID: PMC1069057. https://doi.org/10.3121/cmr.1.4.281
137. Costa, L. G., Giordano, G., Tagliaferri, S., Caglieri, A., & Mutti, A. (2008). Polybrominated diphenyl ether (PBDE) flame retardants: Environmental contamination, human body burden and potential adverse health effects. *Acta Biomed, 79*(3), 172–183.
138. Gascon, M., Fort, M., Martínez, D., Carsin, A. E., Forns, J., Grimalt, J. O., Santa Marina, L., Lertxundi, N., Sunyer, J., & Vrijheid, M. (2012). Polybrominated diphenyl ethers (PBDEs) in breast milk and neuropsychological development in infants. *Environmental Health Perspectives, 120*(12), 1760–1765. Epub 2012 Sep 25. PMID: 23052368; PMCID: PMC3548276. https://doi.org/10.1289/ehp.1205266
139. Costa, L. G., & Giordano, G. (2014). Polybrominated diphenyl ethers. In P. Wexler (Ed.), *Encyclopedia of toxicology* (3rd ed., pp. 1032–1034, ISBN 9780123864550). Academic Press. https://doi.org/10.1016/B978-0-12-386454-3.00422-X
140. Kodavanti, P. R. S., & Loganathan, B. G. (2017). Organohalogen pollutants and human health. In S. R. Quah (Ed.), *International Encyclopedia of Public Health* (2nd ed., pp. 359–366, ISBN 9780128037089). Academic Press. https://doi.org/10.1016/B978-0-12-803678-5.00318-0
141. Beltrán, E. M., González-Doncel, M., García-Mauriño, J. E., Hortigüela, P. G., & Pablos, M. V. (2022). Effects of life cycle exposure to dietary 2,2′, 4,4′-tetrabromodiphenyl ether

(BDE-47) on medaka fish (Oryzias latipes). *Aquatic Toxicology, 245*, 106133. ISSN 0166-445X. https://doi.org/10.1016/j.aquatox.2022.106133

142. Costa, L. G. (2018, ISBN 9780128012383). Polybrominated Diphenyl Ethers (PBDEs). In *Reference module in biomedical sciences*. Elsevier. https://doi.org/10.1016/B978-0-12-801238-3.11151-1
143. Mitan, N. M., Bhaskar, T., Hall, W. J., Muto, A., Williams, P. T., & Sakata, Y. (2008). Effect of decabromodiphenyl ether and antimony trioxide on controlled pyrolysis of high-impact polystyrene mixed with polyolefins. *Chemosphere, 72*(7), 1073–1079. https://doi.org/10.1016/j.chemosphere.2008.04.011
144. National Center for Biotechnology Information. *PubChem compound summary for CID 14410, Decabromodiphenyl oxide*. Retrieved June 6, 2024, from https://pubchem.ncbi.nlm.nih.gov/compound/Decabromodiphenyl-oxide
145. Wenning, R. J., & Martello, L. (2014). Chapter 8. POPs in marine and freshwater environments. In G. O'Sullivan & C. Sandau (Eds.), *Environmental forensics for persistent organic pollutants* (pp. 357–390, ISBN 9780444594242). Elsevier. https://doi.org/10.1016/B978-0-444-59424-2.00008-6
146. Moy, G. G. (2014). Safety of food and beverages: Safety of human milk: chemical aspects. In Y. Motarjemi (Ed.), *Encyclopedia of food safety* (pp. 477–485. ISBN 9780123786135). Academic Press. https://doi.org/10.1016/B978-0-12-378612-8.00444-3
147. García-Fernández, A. J. (2014). Ecotoxicology, Avian. In P. Wexler (Ed.), *Encyclopedia of toxicology* (3rd ed., pp. 289–294. ISBN 9780123864550). Academic Press. https://doi.org/10.1016/B978-0-12-386454-3.00496-6
148. Sparling, D. W. (2016). Chapter 4 - Organochlorine Pesticides. In D. W. Sparling (Ed.), *Ecotoxicology Essentials* (pp. 69–107. ISBN 9780128019474). Academic Press. https://doi.org/10.1016/B978-0-12-801947-4.00004-4
149. Rhee, J. W., & Aks, S. E. (2007). Chapter 81. Organochlorine insecticides. In M. W. Shannon, S. W. Borron, & M. J. Burns (Eds.), *Haddad and Winchester's clinical management of poisoning and drug overdose* (4th ed., pp. 1231–1236, ISBN 9780721606934). W.B. Saunders. https://doi.org/10.1016/B978-0-7216-0693-4.50086-4
150. Araújo, M. F., Castanheira, E. M. S., & Sousa, S. F. (2023). The buzz on insecticides: A review of uses, molecular structures, targets, adverse effects, and alternatives. *Molecules, 28*(8), 3641. PMID: 37110875; PMCID: PMC10144373. https://doi.org/10.3390/molecules28083641
151. National Center for Biotechnology Information. *PubChem compound summary for CID 3036, Clofenotane*. Retrieved June 6, 2024, from https://pubchem.ncbi.nlm.nih.gov/compound/Clofenotane
152. Lewis, R. J., Sr. (2007). *Hawley's condensed chemical dictionary* (15th ed., p. 370). Wiley.
153. Silva, M., & Beauvais, S. (2010). Chapter 16. Risk Assessment for Acute, Subchronic, and Chronic Exposure to Pesticides: Endosulfan. The interpretations expressed are the authors' and do not necessarily reflect policies of the Department of Pesticide Regulation of the California Environmental Protection Agency. In R. Krieger (Ed.), *Hayes' handbook of pesticide toxicology* (3rd ed., pp. 499–522, ISBN 9780123743671). Academic Press. https://doi.org/10.1016/B978-0-12-374367-1.00016-1
154. Bloomquist, J. R. (2003). Chloride channels as tools for developing selective insecticides. *Archives of Insect Biochemistry and Physiology, 54*(4), 145–156. https://doi.org/10.1002/arch.10112
155. O'Neil, M. J. (Ed.). (2006). *The Merck Index—An encyclopedia of chemicals, drugs, and biologicals* (p. 609). Merck.
156. Camacho-Morales, R. L., & Sánchez, J. E. (2016). Chapter 12. Biotechnological use of fungi for the degradation of recalcitrant agro-pesticides. In M. Petre (Ed.), *Mushroom biotechnology* (pp. 203–214, ISBN 9780128027943). Academic Press. https://doi.org/10.1016/B978-0-12-802794-3.00012-6

157. National Center for Biotechnology Information. *PubChem compound summary for CID 3224, Endosulfan*. Retrieved June 3, 2024, from https://pubchem.ncbi.nlm.nih.gov/compound/Endosulfan
158. Menezes, R. G., Qadir, T. F., Moin, A., et al. (2017). Endosulfan poisoning: An overview. *Journal of Forensic and Legal Medicine, 51*, 27–33. https://doi.org/10.1016/j.jflm.2017.07.008
159. Chen, G. (2014). Hexachlorocyclohexanes including lindane. In P. Wexler (Ed.), *Encyclopedia of toxicology* (3rd ed., pp. 874–876. ISBN 9780123864550). Academic Press. https://doi.org/10.1016/B978-0-12-386454-3.00152-4
160. Jackovitz, A. M., & Hebert, R. M. (2015). Chapter 27. Wildlife Toxicity Assessment for Hexachlorocyclohexane (HCH). In M. A. Williams, G. Reddy, M. J. Quinn, & M. S. Johnson (Eds.), *Wildlife toxicity assessments for chemicals of military concern* (pp. 473–497. ISBN 9780128000205). Elsevier. https://doi.org/10.1016/B978-0-12-800020-5.00027-2
161. Muller, F., et al. (1999–2017). Acaricides. In *Ullmann's encyclopedia of industrial chemistry* (7th ed.). Wiley.
162. National Center for Biotechnology Information. *PubChem compound summary for CID 727, Hexachlorocyclohexane*. Retrieved June 2, 2024, from https://pubchem.ncbi.nlm.nih.gov/compound/Hexachlorocyclohexane
163. Wallace, D. R. (2005). Toxaphene. In P. Wexler (Ed.), *Encyclopedia of toxicology* (2nd ed., pp. 207–209. ISBN 9780123694003). Elsevier. https://doi.org/10.1016/B0-12-369400-0/00959-5
164. Wallace, D. R. (2024). Toxaphene. In P. Wexler (Ed.), *Encyclopedia of toxicology* (4th ed., pp. 161–167, . ISBN 9780323854344). Academic Press. https://doi.org/10.1016/B978-0-12-824315-2.00424-3
165. Toxicological profile for Toxaphene. Agency for Toxic Substances and Disease Registry (US); 2014, Health effects. Retrieved from https://www.ncbi.nlm.nih.gov/books/NBK591936/
166. National Center for Biotechnology Information. *PubChem compound summary for CID 102000395, 2,2,5-endo,6-exo,8,9,10-Heptachlorobornane*. Retrieved June 2, 2024, from https://pubchem.ncbi.nlm.nih.gov/compound/102000395
167. Tzanetou, E. N., & Karasali, H. (2022). A comprehensive review of organochlorine pesticide monitoring in agricultural soils: The silent threat of a conventional agricultural past. *Agriculture, 12*(5), 728. https://doi.org/10.3390/agriculture12050728
168. Boucher, O., Simard, M. N., Muckle, G., et al. (2013). Exposure to an organochlorine pesticide (chlordecone) and development of 18-month-old infants. *NeuroToxicology, 35*, 162–168. https://doi.org/10.1016/j.neuro.2013.01.007
169. Sittig, M. (Ed.). (1980). *Pesticide manufacturing and toxic materials control encyclopedia* (p. 171). Noyes Data Corporation.
170. WHO. (1984). *Environmental health criteria 43: Chlordecone* (p. 12).
171. Delannoy, M., Girardet, J.-M., Djelti, F., Yen, F. T., & Cakir-Kiefer, C. (2020). Affinity of chlordecone and chlordecol for human serum lipoproteins. *Environmental Toxicology and Pharmacology, 80*, 103486. ISSN 1382-6689. https://doi.org/10.1016/j.etap.2020.103486
172. Biswas, S., & Ghosh, B. (2014). Chlordecone. In P. Wexler (Ed.), *Encyclopedia of toxicology* (3rd ed., pp. 846–848. ISBN 9780123864550). Academic Press. https://doi.org/10.1016/B978-0-12-386454-3.00275-X
173. National Center for Biotechnology Information. *PubChem compound summary for CID 299, Chlordecone*. Retrieved June 15, 2024, from https://pubchem.ncbi.nlm.nih.gov/compound/Chlordecone
174. Adeyinka, A., Muco, E., Regina, A. C., et al. (2024). Organophosphates. In *StatPearls*. StatPearls. Retrieved November 12, 2023, from https://www.ncbi.nlm.nih.gov/books/NBK499860
175. Gupta, P. K. (2018). Chapter 5. Pesticides (agrochemicals). In P. K. Gupta (Ed.), *Illustrated toxicology* (pp. 165–194, ISBN 9780128132135). Academic Press. https://doi.org/10.1016/B978-0-12-813213-5.00005-5

176. National Center for Biotechnology Information. *PubChem compound summary for CID 3017, Diazinon*. Retrieved June 5, 2024, from https://pubchem.ncbi.nlm.nih.gov/compound/Diazinon
177. National Center for Biotechnology Information. *PubChem compound summary for CID 3289, Ethoprophos*. Retrieved June 5, 2024, from https://pubchem.ncbi.nlm.nih.gov/compound/Ethoprophos
178. National Center for Biotechnology Information. *PubChem compound summary for CID 5125, Tribufos*. Retrieved June 5, 2024, from https://pubchem.ncbi.nlm.nih.gov/compound/Tribufos
179. National Center for Biotechnology Information. *PubChem compound summary for CID 12901, Phosmet*. Retrieved June 5, 2024, from https://pubchem.ncbi.nlm.nih.gov/compound/Phosmet
180. Kannan, K., & Aldous, K. M. (2005). Dioxins. In P. Worsfold, A. Townshend, & C. Poole (Eds.), *Encyclopedia of analytical science* (2nd ed., pp. 274–281. ISBN 9780123693976). Elsevier. https://doi.org/10.1016/B0-12-369397-7/00106-0
181. Young, R. A. (2014). Dioxins. In P. Wexler (Ed.), *Encyclopedia of toxicology* (3rd ed., pp. 190–194. ISBN 9780123864550). Academic Press. https://doi.org/10.1016/B978-0-12-386454-3.00374-2
182. National Center for Biotechnology Information. *PubChem compound summary for CID 15625, 2,3,7,8-Tetrachlorodibenzo-P-dioxin*. Retrieved June 4, 2024, from https://pubchem.ncbi.nlm.nih.gov/compound/2_3_7_8-Tetrachlorodibenzo-P-dioxin
183. Buszewski, B., Bukowska, M., Ligor, M., & Staneczko-Baranowska, I. (2019). A holistic study of neonicotinoids neuroactive insecticides-properties, applications, occurrence, and analysis. *Environmental Science and Pollution Research International, 26*(34), 34723–34740. Epub 2019 Sep 13. PMID: 31520389; PMCID: PMC6900273. https://doi.org/10.1007/s11356-019-06114-w
184. Malhotra, N., Chen, K. H., Huang, J. C., Lai, H. T., Uapipatanakul, B., Roldan, M. J. M., Macabeo, A. P. G., Ger, T. R., & Hsiao, C. D. (2021). Physiological effects of neonicotinoid insecticides on non-target aquatic animals—An updated review. *International Journal of Molecular Sciences, 22*(17), 9591. PMID: 34502500; PMCID: PMC8431157. https://doi.org/10.3390/ijms22179591
185. Ensley, S. M. (2018). Chapter 40. Neonicotinoids. In R. C. Gupta (Ed.), *Veterinary toxicology* (3rd ed., pp. 521–524, ISBN 9780128114100). Academic Press. https://doi.org/10.1016/B978-0-12-811410-0.00040-4
186. National Center for Biotechnology Information. *PubChem compound summary for CID 86287518, Imidacloprid*. Retrieved June 9, 2024, from https://pubchem.ncbi.nlm.nih.gov/compound/Imidacloprid
187. Hodoşan, C., Gîrd, C. E., Ghica, M. V., Dinu-Pîrvu, C. E., Nistor, L., Bărbuică, I. S., Marin, Ş. C., Mihalache, A., & Popa, L. (2023). Pyrethrins and pyrethroids: A comprehensive review of natural occurring compounds and their synthetic derivatives. *Plants (Basel), 12*(23), 4022. PMID: 38068657; PMCID: PMC10707950. https://doi.org/10.3390/plants12234022
188. Riar, N. K. (2014). Bifenthrin. In P. Wexler (Ed.), *Encyclopedia of toxicology* (3rd ed., pp. 449–451, ISBN 9780123864550). Academic Press. https://doi.org/10.1016/B978-0-12-386454-3.01169-6
189. Bradberry, S. M., Cage, S. A., Proudfoot, A. T., & Vale, J. A. (2005). Poisoning due to pyrethroids. *Toxicological Reviews, 24*(2), 93–106. https://doi.org/10.2165/00139709-200524020-00003
190. Glickman, A. H., & Casida, J. E. (1982). Species and structural variations affecting pyrethroid neurotoxicity. *Neurobehavioral Toxicology and Teratology, 4*(6), 793–799.
191. Ramchandra, A. M., Chacko, B., & Victor, P. J. (2019). Pyrethroid poisoning. *Indian Journal of Critical Care Medicine, 23*(Suppl 4), S267–S271. PMID: 32021002; PMCID: PMC6996658. https://doi.org/10.5005/jp-journals-10071-23304

192. National Center for Biotechnology Information. *PubChem compound summary for CID 71310221, Pyrethrins*. Retrieved June 8, 2024, from https://pubchem.ncbi.nlm.nih.gov/compound/Pyrethrins
193. Hejna, M., Kapuścińska, D., & Aksmann, A. (2022). Pharmaceuticals in the aquatic environment: A review on eco-toxicology and the remediation potential of algae. *International Journal of Environmental Research and Public Health, 19*(13), 7717. PMID: 35805373; PMCID: PMC9266021. https://doi.org/10.3390/ijerph19137717
194. Samal, K., Saswat Mahapatra, M. D., & Ali, H. (2022. ISSN 2772-4271). Pharmaceutical wastewater as emerging contaminants (EC): Treatment technologies, impact on environment and human health. *Energy Nexus, 6*, 100076. https://doi.org/10.1016/j.nexus.2022.100076
195. Pradhan, B. L., Lodhi, L., Dey, K. K., & Ghosh, M. (2024). Analyzing atomic scale structural details and nuclear spin dynamics of four macrolide antibiotics: Erythromycin, clarithromycin, azithromycin, and roxithromycin. *RSC Advances, 14*, 17733–17770. https://doi.org/10.1039/D4RA00718B
196. Dinos, G. P. (2017). The macrolide antibiotic renaissance. *British Journal of Pharmacology, 174*(18), 2967–2983. Epub 2017 Aug 10. PMID: 28664582; PMCID: PMC5573421. https://doi.org/10.1111/bph.13936
197. Patel, P. H., & Hashmi, M. F. (2024). Macrolides. In *StatPearls*. StatPearls. Retrieved from May 16, 2023, from https://www.ncbi.nlm.nih.gov/books/NBK551495
198. Diamanti-Kandarakis, E., Bourguignon, J. P., Giudice, L. C., Hauser, R., Prins, G. S., Soto, A. M., Zoeller, R. T., & Gore, A. C. (2009). Endocrine-disrupting chemicals: an Endocrine Society Scientific Statement. *Endocrine Reviews, 30*(4), 293–342. PMID: 19502515; PMCID: PMC2726844. https://doi.org/10.1210/er.2009-0002
199. de Aquino, S. F., Brandt, E. M. F., Bottrel, S. E. C., Gomes, F. B. R., & Silva, S. Q. (2021). Occurrence of pharmaceuticals and endocrine disrupting compounds in Brazilian water and the risks they may represent to human health. *International Journal of Environmental Research and Public Health, 18*(22), 11765. PMID: 34831521; PMCID: PMC8620687. https://doi.org/10.3390/ijerph182211765
200. Yilmaz, B., Terekeci, H., Sandal, S., & Kelestimur, F. (2020). Endocrine disrupting chemicals: exposure, effects on human health, mechanism of action, models for testing and strategies for prevention. *Reviews in Endocrine and Metabolic Disorders, 21*(1), 127–147. https://doi.org/10.1007/s11154-019-09521-z
201. Yazdan, M. M. S., Kumar, R., & Leung, S. W. (2022). The environmental and health impacts of steroids and hormones in wastewater effluent, as well as existing removal technologies: A review. *Ecologies, 3*(2), 206–224. https://doi.org/10.3390/ecologies3020016
202. Wojnarowski, K., Podobiński, P., Cholewińska, P., Smoliński, J., & Dorobisz, K. (2021). Impact of estrogens present in environment on health and welfare of animals. *Animals (Basel), 11*(7), 2152. PMID: 34359280; PMCID: PMC8300725. https://doi.org/10.3390/ani11072152
203. Combalbert, S., & Hernandez-Raquet, G. (2010). Occurrence, fate, and biodegradation of estrogens in sewage and manure. *Applied Microbiology and Biotechnology, 86*(6), 1671–1692. https://doi.org/10.1007/s00253-010-2547-x
204. Naidu, R., Biswas, B., Willett, I. R., Cribb, J., Singh, B. K., Nathanail, C. P., Coulon, F., Semple, K. T., Jones, K. C., Barclay, A., & Aitken, R. J. (2021). Chemical pollution: A growing peril and potential catastrophic risk to humanity. *Environment International, 156*, 106616. ISSN 0160-4120. https://doi.org/10.1016/j.envint.2021.106616
205. Gabby, P. N. (2006). *Lead: In Mineral Commodity Summaries*. U.S. Geological Survey. Retrieved from http://minerals.usgs.gov/minerals/pubs/commodity/lead/lead_mcs05.pdf
206. Flora, S. J. S., Flora, G. J. S., & Saxena, G. (2006). Environmental occurrence, health effects and management of lead poisoning. In S. B. Cascas & J. Sordo (Eds.), *Lead: Chemistry, analytical aspects, environmental impacts and health effects* (pp. 158–228). Elsevier Publication.
207. Centers for Disease Control and Prevention (CDC). (2001). *Managing elevated blood lead levels among young children: Recommendations from the advisory committee on childhood lead poisoning prevention*. CDC.

208. Tchounwou, P. B., Wilson, B., & Ishaque, A. (1999). Important considerations in the development of public health advisories for arsenic and arsenic-containing compounds in drinking water. *Reviews on Environmental Health, 14*(4), 211–229. https://doi.org/10.1515/reveh.1999.14.4.211
209. Agency for Toxic Substances and Disease Registry (ATSDR). (2000). *Toxicological Profile for Arsenic TP-92/09*. Center for Disease Control.
210. Shukla, S. K., Mishra, R. K., Pandey, M., Mishra, V., Pathak, A., Pandey, A., Kumar, R., & Dikshit, A. (2016). Chapter 21. Land reformation using plant growth–Promoting rhizobacteria in the context of heavy metal contamination. In P. Ahmad (Ed.), *Plant metal interaction* (pp. 499–529. ISBN 9780128031582). Elsevier. https://doi.org/10.1016/B978-0-12-803158-2.00021-7
211. Tchounwou, P. B., Centeno, J. A., & Patlolla, A. K. (2004). Arsenic toxicity, mutagenesis and carcinogenesis—A health risk assessment and management approach. *Molecular and Cellular Biochemistry, 255*, 47–55.
212. Bhan, A., & Sarkar, N. N. (2005). Mercury in the environment: Effect on health and reproduction. *Reviews on Environmental Health, 20*(1), 39–56. https://doi.org/10.1515/reveh.2005.20.1.39
213. Guzzi, G., & La Porta, C. A. (2008). Molecular mechanisms triggered by mercury. *Toxicology, 244*(1), 1–12. https://doi.org/10.1016/j.tox.2007.11.002
214. Zahir, F., Rizwi, S. J., Haq, S. K., & Khan, R. H. (2005). Low dose mercury toxicity and human health. *Environmental Toxicology and Pharmacology, 20*(2), 351–360. https://doi.org/10.1016/j.etap.2005.03.007
215. National Center for Biotechnology Information. *PubChem compound summary for CID 23931, Mercury*. Retrieved June 1, 2024, from https://pubchem.ncbi.nlm.nih.gov/compound/Mercury
216. Genchi, G., Carocci, A., Lauria, G., Sinicropi, M. S., & Catalano, A. (2020). Nickel: Human health and environmental toxicology. *International Journal of Environmental Research and Public Health, 17*(3), 679. PMID: 31973020; PMCID: PMC7037090. https://doi.org/10.3390/ijerph17030679
217. Dally, H., & Hartwig, A. (1997). Induction and repair inhibition of oxidative DNA damage by nickel(II) and cadmium(II) in mammalian cells. *Carcinogenesis, 18*(5), 1021–1026. https://doi.org/10.1093/carcin/18.5.1021
218. Fatima, G., Raza, A. M., Hadi, N., Nigam, N., & Mahdi, A. A. (2019). Cadmium in human diseases: It's more than just a mere metal. *Indian Journal of Clinical Biochemistry, 34*(4), 371–378. Epub 2019 Jun 27. PMID: 31686724; PMCID: PMC6801231. https://doi.org/10.1007/s12291-019-00839-8
219. Bagchi, D., Vuchetich, P. J., Bagchi, M., et al. (1997). Induction of oxidative stress by chronic administration of sodium dichromate [chromium VI] and cadmium chloride [cadmium II] to rats. *Free Radical Biology & Medicine, 22*(3), 471–478. https://doi.org/10.1016/s0891-5849(96)00352-8
220. Lin, R. H., Lee, C. H., Chen, W. K., & Lin-Shiau, S. Y. (1994). Studies on cytotoxic and genotoxic effects of cadmium nitrate and lead nitrate in Chinese hamster ovary cells. *Environmental and Molecular Mutagenesis, 23*(2), 143–149. https://doi.org/10.1002/em.2850230212
221. Alexander, J. (2015). Chapter 52. Selenium. In G. F. Nordberg, B. A. Fowler, & M. Nordberg (Eds.), *Handbook on the toxicology of metals* (4th ed., pp. 1175–1208. ISBN 9780444594532). Academic Press. https://doi.org/10.1016/B978-0-444-59453-2.00052-4
222. Agarwal, P., Sharma, S., & Agarwal, U. S. (2016). Selenium toxicity: A rare diagnosis. *Indian Journal of Dermatology, Venereology and Leprology, 82*, 690–693. https://doi.org/10.4103/0378-6323.186476
223. El-Kalliny, A. S., Abdel-Wahed, M. S., El-Zahhar, A. A., et al. (2023). Nanomaterials: A review of emerging contaminants with potential health or environmental impact. *Discover Nano, 18*, 68. https://doi.org/10.1186/s11671-023-03787-8

224. Ferdous, Z., & Nemmar, A. (2020). Health impact of silver nanoparticles: A review of the biodistribution and toxicity following various routes of exposure. *International Journal of Molecular Sciences, 21*(7), 2375. PMID: 32235542; PMCID: PMC7177798. https://doi.org/10.3390/ijms21072375
225. Vyas, T. K., & Vala, A. K. (2022). The impact of magnetic nanoparticles on microbial community structure and function in rhizospheric soils. In S. Thomas & A. Rezazadeh Nochehdehi (Eds.), *Handbook of magnetic hybrid nanoalloys and their nanocomposites*. Springer. https://doi.org/10.1007/978-3-030-90948-2_44
226. Saleem, H., & Zaidi, S. J. (2020). Sustainable use of nanomaterials in textiles and their environmental impact. *Materials (Basel), 13*(22), 5134. PMID: 33203051; PMCID: PMC7696606. https://doi.org/10.3390/ma13225134
227. Khan, Y., Sadia, H., Ali Shah, S. Z., Khan, M. N., Shah, A. A., Ullah, N., Ullah, M. F., Bibi, H., Bafakeeh, O. T., Khedher, N. B., et al. (2022). Classification, synthetic, and characterization approaches to nanoparticles, and their applications in various fields of nanotechnology: A review. *Catalysts, 12*, 1386. https://doi.org/10.3390/catal12111386
228. Ealia, S. A. M., & Saravanakumar, M. P. (2017). A review on the classification, characterisation, synthesis of nanoparticles and their application. In *2017 IOP Conference Series: Materials Science and Engineering* (Vol. 263, p. 032019). https://doi.org/10.1088/1757-899X/263/3/032019
229. Dutta, V., Verma, R., Gopalkrishnan, C., Yuan, M.-H., Batoo, K. M., Jayavel, R., Chauhan, A., Lin, K.-Y. A., Balasubramani, R., & Ghotekar, S. (2022). Bio-inspired synthesis of carbon-based nanomaterials and their potential environmental applications: A state-of-the-art review. *Inorganics, 10*(10), 169. https://doi.org/10.3390/inorganics10100169
230. Qiongchan, G., Jiangtao, L., Xinwei, M., & Xiaoxiao, J. (2022). High aspect ratio metamaterials and their applications. *Sensors and Actuators A: Physical, 335*, 113376. ISSN 0924-4247. https://doi.org/10.1016/j.sna.2022.113376
231. Begines, B., Ortiz, T., Pérez-Aranda, M., Martínez, G., Merinero, M., Argüelles-Arias, F., & Alcudia, A. (2020). Polymeric nanoparticles for drug delivery: Recent developments and future prospects. *Nanomaterials (Basel), 10*(7), 1403. PMID: 32707641; PMCID: PMC7408012. https://doi.org/10.3390/nano10071403
232. Sanmartín-Matalobos, J., Bermejo-Barrera, P., Aboal-Somoza, M., Fondo, M., García-Deibe, A. M., Corredoira-Vázquez, J., & Alves-Iglesias, Y. (2022). Semiconductor quantum dots as target analytes: Properties, surface chemistry and detection. *Nanomaterials, 12*(14), 2501. https://doi.org/10.3390/nano12142501
233. Ehuwa, O., Jaiswal, A. K., & Jaiswal, S. (2021). Salmonella, food safety and food handling practices. *Food, 10*(5), 907. PMID: 33919142; PMCID: PMC8143179. https://doi.org/10.3390/foods10050907
234. Facciolà, A., Riso, R., Avventuroso, E., Visalli, G., Delia, S. A., & Laganà, P. (2017). Campylobacter: From microbiology to prevention. *Journal of Preventive Medicine and Hygiene, 58*(2), E79–E92. PMID: 28900347; PMCID: PMC5584092.
235. Mueller, M., & Tainter, C. R. (2024). *Escherichia coli* infection. In *StatPearls*. StatPearls. Retrieved from July 13, 2023, from https://www.ncbi.nlm.nih.gov/books/NBK564298
236. Mateus, T., Silva, J., Maia, R. L., & Teixeira, P. (2013). Listeriosis during pregnancy: A public health concern. *ISRN Obstetrics and Gynecology, 2013*, 851712. PMID: 24191199; PMCID: PMC3804396. https://doi.org/10.1155/2013/851712
237. Rabbani, G. H., & Greenough, W. B., 3rd. (1999). Food as a vehicle of transmission of cholera. *Journal of Diarrhoeal Diseases Research, 17*(1), 1–9.
238. Deb, S., Mondal, R., Lahiri, D., Shome, G., Roy, A. G., Sarkar, V., Sarkar, S., & Benito-León, J. (2023). Norovirus-associated neurological manifestations: summarizing the evidence. *Journal of Neurovirology, 29*(4), 492–506. Epub 2023 Jul 21. PMID: 37477790; PMCID: PMC10501950. https://doi.org/10.1007/s13365-023-01152-0
239. Sattar, S. A., Jason, T., Bidawid, S., & Farber, J. (2000). Foodborne spread of hepatitis A: Recent studies on virus survival, transfer and inactivation. *Canadian Journal of Infectious Diseases and Medical Microbiology, 11*(3), 159–163. PMID: 18159284; PMCID: PMC2094762. https://doi.org/10.1155/2000/805156

240. Keiser, J., & Utzinger, J. (2009). Food-borne trematodiases. *Clinical Microbiology Reviews, 22*(3), 466–483. PMID: 19597009; PMCID: PMC2708390. https://doi.org/10.1128/CMR.00012-09
241. Cerda, J. R., Buttke, D. E., & Ballweber, L. R. (2018). Echinococcus spp. tapeworms in North America. *Emerging Infectious Diseases, 24*(2), 230–235. PMID: 29350139; PMCID: PMC5782903. https://doi.org/10.3201/eid2402.161126
242. Alemu, G., Nega, M., & Alemu, M. (2020). Parasitic contamination of fruits and vegetables collected from local markets of Bahir Dar City, Northwest Ethiopia. *Research and Reports in Tropical Medicine, 11*, 17–25. PMID: 32280295; PMCID: PMC7127876. https://doi.org/10.2147/RRTM.S244737
243. Awuchi, C. G., Ondari, E. N., Ogbonna, C. U., Upadhyay, A. K., Baran, K., Okpala, C. O. R., Korzeniowska, M., & Guiné, R. P. F. (2021). Mycotoxins affecting animals, foods, humans, and plants: Types, occurrence, toxicities, action mechanisms, prevention, and detoxification strategies—A revisit. *Food, 10*(6), 1279. PMID: 34205122; PMCID: PMC8228748. https://doi.org/10.3390/foods10061279
244. Lyman, E. S. (2011). Nuclear energy and human health. In J. O. Nriagu (Ed.), *Encyclopedia of environmental health* (pp. 185–192. ISBN 9780444522726). Elsevier. https://doi.org/10.1016/B978-0-444-52272-6.00185-9
245. Reddy, S., & Osborne, W. J. (2022). Chapter 1. Effect of pollution on sediments and their impact on the aquatic ecosystem. In J. Samuel, A. Kumar, & J. Singh (Eds.), *Relationship between microbes and the environment for sustainable ecosystem services* (Vol. 2, pp. 1–16. ISBN 9780323899376). Elsevier. https://doi.org/10.1016/B978-0-323-89937-6.00001-2
246. Gaur, N., Sharma, S., & Yadav, N. (2024). Chapter 2. Environmental pollution. In V. K. Garg, A. Yadav, C. Mohan, S. Yadav, & N. Kumari (Eds.), *Advances in green and sustainable chemistry, green chemistry approaches to environmental sustainability* (pp. 23–41, ISBN 9780443189593.). Elsevier. https://doi.org/10.1016/B978-0-443-18959-3.00010-0

Chapter 2
Sources and Causes of Emerging Contaminants

Dikeshwar Kurre and Sandeep Samantaray

1 Introduction

Emerging contaminants (ECs) are new contaminants in the environment that originate from different human activities, such as domestic use, health- care facilities, farming, and industry. The compounds in the environment have gained importance in the last decade because of the availability of advanced instrumental analytical techniques capable of measuring even very low levels of pptv, ppb, and ppb in air, soil, and water, respectively. Additionally, studies have shown that these chemicals have adverse effects on the environment as well as animals. Admixtures include pharmaceuticals, hormones and steroids, disinfectants, personal care products, surfactants, flame retardants, industrial admixtures, household products, pesticides, agrochemicals, microplastics, trace metals, and nanoparticles. These materials are deemed to have the ability to affect the environment because of their effects on natural resource balances, biochemical processes within living organisms, and effects on the human body, including endocrine disruption, mutagenesis, carcinogenicity, and teratogenicity. Rachel Carson and her book Silent Spring, published in 1962, marked a significant leap in the awareness creation process of the existence of pollutants and disasters [1]. Such new EPs require the adoption of new approaches, including genomics, for evaluating the risks in aquatic environments and drinking water resources. The contaminants are described as chemical substances for which there are no standards because they have only recently posed a problem with the help of new methods of analyzing them, new reported effects, or new types of environmental release. Organic chemicals can be divided according to the elements that

D. Kurre · S. Samantaray (✉)
Department of Civil Engineering, National Institute of Technology Srinagar Hazratbal, Srinagar, Jammu and Kashmir, India
e-mail: dikeshwar_2023phaciv015@nitsri.ac.in; sandeep@nitsri.ac.in

M. K. Jindal et al. (eds.), *Emerging Contaminants in Water*, Emerging Contaminants and Associated Treatment Technologies,
https://doi.org/10.1007/978-3-031-89591-3_2

constitute them, for instance, brominates, flame retardants, and, depending on their impacts on organisms, more so endocrine-disrupting substances. The consequences of these chemicals on the aquatic environment are poorly understood, and their release into water sources is unlawful. Over 30,000 compounds are commercially available, but fewer than 3500 have been investigated, meaning that their number needs to rise as well. Therefore, we suggest that improving upon current monitoring systems in terms of the effects of new pollutants on life within water bodies should be accomplished through a whole-genome approach.

The eco-toxicogenomic approach, which has been accelerated by genomic technologies [2], was previously discussed. Until the 1990s, nonpolar hazardous substances, including POPs and heavy metal compounds, were considered relevant pollutants, and before the 1990s, nonpolar dangerous chemicals, including POPs and heavy metals, were considered contaminants. Evaluations of these compound emissions demonstrate that industrialized countries are no longer a concern since the sources of these compounds have been successfully addressed by intervention and eliminated through eradicating key sources of pollution. There are also emerging or "new" pollutants for which legislation has been passed to check them. This section consists of general commodity products, such as pharmaceutical products, cosmetic products, wetting agents, softening agents, and industrial chemicals. These pollutants can produce adverse effects when their concentrations in the environment are high because of their rapid transformation and removal, for which continuous input can neutralize them. Direct inputs from untreated urban wastewater, as well as from the effluents of wastewater treatment plants (WWTPs), are the two sources of developing pollutants. Existing WWTPs are inefficient for processing new chemicals and their metabolites; these compounds can freely pass through sewage effluents to contaminate water systems [3]. Typically, water quality assessments are typically based on nutrients, microbiological contaminants, heavy metals, and priority pollutants. A recent study confirmed that the concentration of organic pollutants decreases water quality. Sources of contaminants include and vary in concentration in nanograms per liter (ng/L). ECs are chemical substances found in water and have recently been found to be of high concern as pollutants. They include chemicals whose presence in the natural environment is not normally tested but are known or believed to cause harm to both humans and other forms of life. Human and wildlife endocrine systems are affected by the PPCPs, insecticides, and hormones found in these compounds. Thus, all of these compounds are termed endocrine-disrupting chemicals (EDCs). Neither natural attenuation nor conventional treatment technologies are effective in eliminating micropollutants in wastewater, surface water, and drinking water. These pollutants have been found to accumulate in macroinvertebrates, other aquatic creatures, and people. Effective water reuse requires both availability and adequate treatment. Despite extensive research on pollutants such as nutrients, hydrocarbons, and heavy metals in stormwater and wastewater, few systematic studies have focused on the presence of ECs and their treatment options. Sustainable reuse of EC-contaminated water remains difficult [4].

2 Background

The ECs do not necessarily have to be newly formed products either. The word pertains to three general classes. The first group includes products that have been in the environment for some time, but their introduction is relatively new, for instance, industrial additives. The second group consists of compounds that might have been present. They have been in the environment for a long time, but their importance has only been realized recently, especially in the manufacture of drugs. The third group includes chemical compounds whose toxic effects on humans and the environment have long been known but have been mentioned only recently, for instance, hormones. ECs are not yet fully understood and documented, and methods for detecting them and related instruments are still being established. Minute concentrations of microcontaminants in the environment were measured through enhanced instruments for analysis and quantitative assessment. Most ECs have uncertain impacts on environmental and human health; monitoring and mitigation options are being designed. We can conclude that only progress in analytical chemistry, instrumentation, and engineering resulted in the discovery of ECs in the environment. Quantitative analytical techniques have improved mainly within the past two decades, from a thousand up to a level of trillion or even quadrillion ppt, where these pollutants can be detected and analyzed in environmental samples. This chapter introduces recently identified emerging pollutants, especially synthetic organic chemicals, in surface and ground waters.

3 Classifications of Sources of Emerging Contaminants

Emergent contaminants enter the environment through industrial processes and emissions, as well as wastewater and effluent, similar to classical pollutants. Contamination can come from either point or nonpoint sources.

3.1 Point Sources of Emerging Contaminants

Point sources of emerging contaminants are identifiable sources that emit pollutants directly into the environment. These sources are usually connected with industrial sites, wastewater treatment plants, and other isolated sources of pollution. Industrial point sources include factories, manufacturing plants, and chemical processing facilities that emit effluents containing a variety of pollutants, including medicines, heavy metals, and industrial chemicals. Another large point source is wastewater treatment plants, which collect and treat domestic and industrial wastewater before it is dumped into bodies of water [5]. While treatment plants are intended to remove conventional pollutants, they may not successfully remove new contaminants, such as medicines and personal care items, resulting in their discharge into receiving waterways. Point sources of emerging contaminants include wastewater treatment plant effluents, artificial recharge, landfills, leakage and septic tanks, industrial, personal care products, hazardous waste sites, and agricultural operations that emit pollutants into the environment via leachate, runoff, or direct discharge (Fig. 2.1). Identifying and managing emerging pollutants at their source is critical for safeguarding water quality and reducing the environmental impact of pollution. Regulatory measures, technical developments, and pollution prevention methods are essential for combating point-source pollution and encouraging long-term environmental stewardship [6]. Emerging contaminants in water and sediments pose a significant environmental risk of anthropogenic and natural origin. Industrial activities, agricultural practices, and urban waste, among others, contribute. Microplastics, per- and polyfluoroalkyl substances (PFASs), pharmaceuticals, and radioactive materials are contaminated by many sources. Microplastics are produced primarily from plastic waste breakdown, cosmetics, and synthetic clothing fibers, which profoundly affect aquatic ecosystems. PFASs, sometimes called "forever chemicals,"

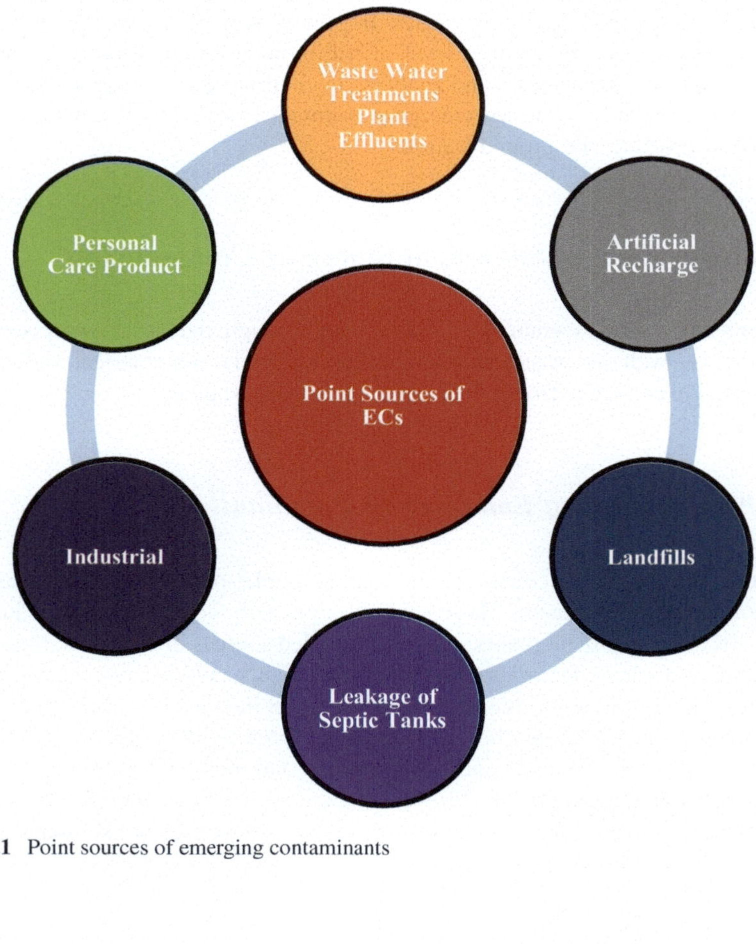

Fig. 2.1 Point sources of emerging contaminants

leach from industrial processes, firefighting foams, and consumer products ranging from nonstick cookware. Pharmaceuticals reach the environment via inappropriate disposal, excretion, and wastewater discharge, contaminating water. Nuclear power plants, medical facilities, and accidental spills release radioactive substances. Industrial and urban sources account for the bulk of this pollution, with agricultural runoff, including pharmaceuticals and PFAS, also playing a major role. The increasing prevalence of these pollutants demands robust mitigation strategies and stricter regulations to minimize their environmental and health impacts [7, 8]. Contamination by emerging substances reported in Colombia, Mexico, and Brazil water compartments from 2013 onward is listed in Table 2.1.

Table 2.1 Contamination by emerging substances was reported in water compartments in Colombia, Mexico, and Brazil from 2013 onward

Sl. no	Country	Water type	Emerging substance	Concentration (ng/L)	Reference
1	Colombia (La Fe)	Drinking water supply reservoir	Ibuprofen	7–39	[9]
2	Colombia (Rio Grande)	Drinking water supply reservoir	Ibuprofen	5–62	[9]
3	Mexico (Mexico City)	Surface water	Ibuprofen	15–45	[10]
4	Mexico (Toluca City)	Hospital wastewater	Ibuprofen	620	[11]
5	Mexico (Morelos)	Surface water	Diclofenac	258–1398	[12]
6	Mexico (Mexico City)	Surface water	Diclofenac	28–32	[10]
7	Mexico (Morelos)	wastewater	Diclofenac	600–2500	[12]
8	Mexico (Cuernavaca)	wastewater	Diclofenac	258–1398	[12]
9	Mexico (Toluca City)	Hospital wastewater	Diclofenac	590	[11]
10	Mexico (Morelos)	Surface water	Naproxen	732–480	[12]
11	Mexico (Tula River)	Surface water	Naproxen	240	[13]
12	Mexico (Mexico City)	Surface water	Naproxen	52–186	[10]
13	Mexico (Mezquital Valley)	Wastewater	Naproxen	11,800	[14]
14	Mexico (Cuernavaca)	Wastewater	Naproxen	732–4880	[72
15	Mexico (Morelos)	Wastewater	Naproxen	800–4200	[12]
16	Mexico (Toluca City)	Hospital wastewater	Naproxen	2660	[11]
17	Brazil (São Paulo)	Surface water	Paracetamol	30,421	[15]

(continued)

Table 2.1 (continued)

Sl. no	Country	Water type	Emerging substance	Concentration (ng/L)	Reference
18	Brazil (São Paulo)	Surface water	17-α ethinylestradiol	777	[16]
19	Brazil (São Paulo)	Surface water	Caffeine	19–127,000	[16]
20	Brazil	Surface water	Caffeine	40–19,000	[17]
21	Brazil (São Paulo)	Surface water	Caffeine	129.585	[16]
22	Brazil	Drinking water	Caffeine	1.8–2000	[17]
23	Brazil (São Roque)	Drinking water	Caffeine	4083	[18]
24	Brazil (São Paulo)	Drinking water	Caffeine	121	[16]
25	Brazil (Rio Grande)	Drinking water supply reservoir	Caffeine	18,828	[17]
26	Brazil (São Paulo)	Surface water	Bisphenol A	2–13,016	[16]
27	Brazil (Rio Grande)	Drinking water supply reservoir	Bisphenol A	11	[17]
28	Brazil (São Paulo)	Drinking water	Bisphenol A	23	[16]
29	Brazil (São Roque)	Drinking water	Bisphenol A	0.5	[18]
30	Brazil (São Paulo)	Surface water	Cocaine	10	[16]
31	Brazil (São Paulo)	Surface water	Benzoylecgonine	133	[16]
32	Colombia (Bogotá)	Wastewater	Benzoylecgonine	4000	[19]
33	Colombia (Bogotá)	Wastewater	Amphetamines	12–68	[19]
34	Brazil (São Paulo)	Drinking water	Triclosan	2–37	[16]
35	Brazil (Rio Grande)	Surface water	Atrazine	5–49	[20]
36	Brazil (São Paulo)	Drinking water	Atrazine	687	[16]
37	Brazil (Rio Grande)	Drinking water	Atrazine	5–37	[20]
38	Brazil	Drinking water	Atrazine	2–6000	[17]
39	Mexico (Yucatan)	Groundwater	Endrin	3.2 mg/L	[21]
40	Mexico (Yucatan)	Groundwater	γ-lindane	10.86 mg/L	[21]
41	Mexico (Yucatan)	Groundwater	Heptachlor	12.54 mg/L	[21]

3.1.1 WWTP Effluents

Wastewater emanating from hospitals and households contains PPCP-EOCs. EOCs are effluents from the soap, pharmaceutical, detergent, tannery, construction, demolition, vehicle, paper, print, furniture, and textile industries. Diseases from hospitals, households, and companies discharge wastewater into WWTPs. Although some of these bacteria are not fully degraded, they spread into the environment, particularly surface water, through effluent discharge from treatment plants. When water on the surface is used in farming systems, the EOCs are transported to the soil, and the plants that feed the consumers and animals feed on the plants. In other words, groundwater pollution begins when water penetrates the subsurface [22].

3.1.2 Artificial Recharge

The amount of water available for drinking is critically low in dry or semi-arid parts of the world because of the overexploitation of ground water and inadequate rainfall. Gradual artificial recharge of groundwater via treated wastewater, rainfall, and surface runoffs has become common in Tunisia to reduce or minimize the level of groundwater [23]. It is concluded that artificial recharge is useful for controlling water quantity but adversely affects aquifer water quality. Most wastewater treatments are known to produce emergent organic compounds (EOCs) in treated wastewater. Rainwater or stormwater can transport EOCs, including pesticides, PAHs, and other pollutants such as roof chemicals or the chemically produced polar compounds capable of polluting groundwater each time they affect the soil. Dry wells are large-diameter, vertical, perforated wells that rely on gravity to recharge aquifers and are often used in areas with less subsoil. In zones of high permeability adjacent to the aquifer, dry wells release replenished water passing through the well screen. Most investigations have shown that artificially recharged water undergoes fewer degrees of attenuation than water naturally recharged through infiltration [7].

3.1.3 Landfills

Improper bottom sealing or liner failure might result in leachates penetrating the soil. Leachates interact with groundwater and affect its quality. Leachate is hazardous and can contaminate groundwater with compounds that exceed permissible amounts. Previously, markers such as BOD, COD, nutrients, and heavy metals were studied only in leachates from landfills. As lifestyles and living standards have improved, landfill garbage has changed significantly. Recent landfill leachate investigations have revealed high levels of phenolic, aromatic, polycyclic, and xenobiotic chemicals that threaten the environment and the wellness of humans. Reports indicate that poorly built landfills have harmed surface and groundwater resources in Nigeria, Morocco, and China [24].

3.1.4 Industrial Contaminants

Industrial emerging contaminants (IECs) are a large group of chemicals and substances that emanate from industries and are relatively new sources of environmental and possibly human health concerns. These contaminants are poorly controlled or even unknown and may not have been previously incorporated into the monitoring of the environment. The sources of industrial contaminants are discharged from factories and processing plants, chemical additives, solvents, degradation products, and leakage from industrial landfills or storage tanks. Examples of IEC Per- and polyfluoroalkyl substances (PFASs) include nonstick coatings; firefighting foams; textiles; pharmaceutical residues; antibiotics, hormones, or analgesics produced during industrial drug manufacturing; microplastics; tiny plastic particles from industrial processes; and the breakdown of larger plastic products. Industrial solvents such as trichloroethylene (TCE) or benzene are used in cleaning or manufacturing.

3.1.5 Personal Care Product

Personal care products, as emerging contaminants, refer to chemicals in products intended for personal hygiene, grooming, and beauty whose presence in the environment as pollution is increasingly acknowledged. They often enter ecosystems via wastewater systems, posing risks to aquatic life and human health.

3.1.6 Leakage of Septic Tanks

Septic tank systems are extremely popular for onsite sewage control globally. Septic tanks are commonly used in places with limited access to municipal sewer systems. Poor placement, unexplained failure, and anaerobic conditions make them less efficient than centralized wastewater treatment plants. Water leakage from onsite treatment tanks is a primary cause of groundwater contamination in China, the United States, and other countries. A survey indicated that 25% of US households utilize septic tanks for waste disposal. Emerging organic contaminants in groundwater are derived primarily from nonylphenol, nonylphenol mono, diethoxy carboxylates, steroid hormones, estrone, caffeine, and fluorescent whitening agents. Septic tank systems can pollute groundwater owing to poor design, installation, maintenance, and concentration. The geology of the location is also significant. Sandy aquifers have high hydraulic conductivity, low specific surface areas, and low organic carbon contents. Sandy aquifers are more vulnerable to septic contamination because pollutants can enter more quickly than other aquifers can. Furthermore, locations with shallow groundwater levels and extremely transmissible aquifers face severe hazards [5, 8].

3.2 Nonpoint Sources of Emerging Contaminants

Identifying diffuse pollution sources of ECs can be challenging because of their spread across broad areas, making it difficult to pinpoint specific sources. This category comprises stormwater runoff from roadways, metropolitan areas, highways, and agricultural land (Fig. 2.2). Compared with point sources, heterogeneous sources release fewer toxins into the surrounding environment. Pesticide use in agriculture is considered the most significant pollution source. Pesticides (DDT metabolites), herbicides (pentazocine and mecoprop), and flame retardants (tris2-chloroethyl)-phosphate and tris (2-Chloroisopropyl-9-phosphate) were found in Lakes Tegel and Wannsee in Berlin, Germany. The water was utilized to replenish the local aquifer artificially [25]. Farming activities produce various ECs, including veterinary medications, hormones, natural toxins, and synthetic chemical compounds. Chemical contaminants may bind to sludge molecules in an activated sludge system. Sludge from wastewater treatment plants can be used as fertilizer in agricultural fields, enabling the introduction of ECs into the environment. When wastewater treatment facilities apply biosolids to land, ECs may contaminate groundwater. Biosolids applied to soils and surface runoff may contain halogenated hydrocarbons (polychlorinated alkanes and perfluorochemicals) that pose a significant risk to groundwater [6]. Nonpoint sources of emerging contaminants, like agriculture runoff, animal farms, surface water-groundwater interactions, atmospheric deposition, recreation, activities, leachate systems, and residential and commercial usage, are given below (Fig. 2.2).

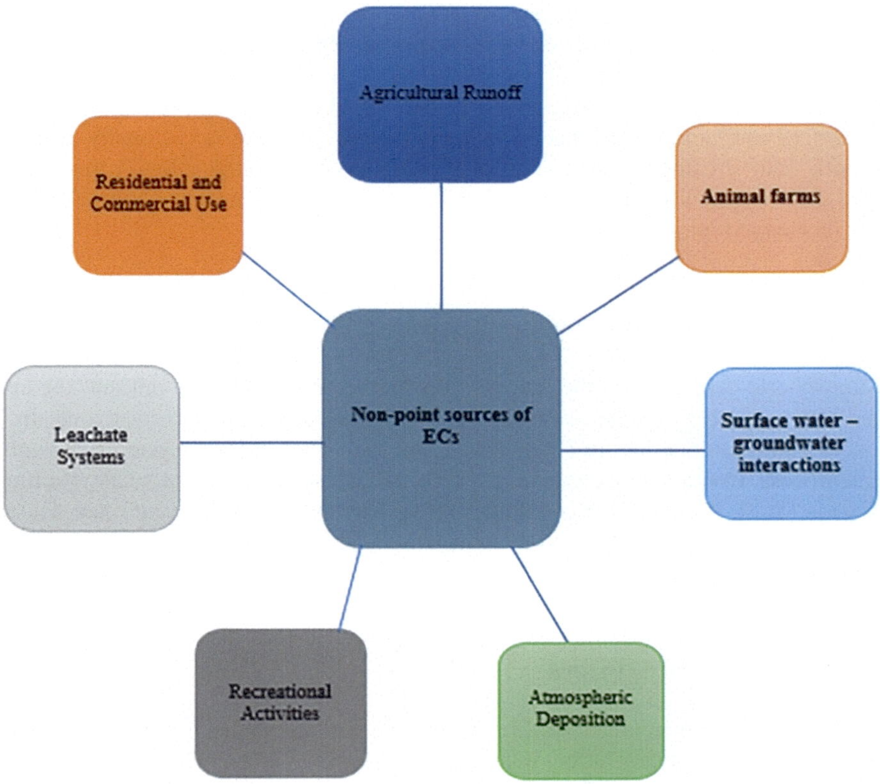

Fig. 2.2 Nonpoint sources of emerging contaminants

3.2.1 Agricultural Sources

Chemical pesticides were widely used in the 1990s to sustain population expansion in emerging countries, contaminating natural streams through nonpoint sources. In 2011, India used up to 5000 tons of pesticides, causing groundwater pollution from agriculture. The inadequate removal of EOCs (e.g., antibiotics, pesticides, perfumes, plasticizers, detergents, and metabolites) from wastewater may leave pollutants in sludge solids, which can then be deposited on the soil. Hydrophobic nonionic organic compounds with high log Kow (n-octanol/water partition coefficient) values bond with sludge and soil organic matter, limiting bioavailability and hindering microbial degradation. Contaminants with a log Kow value of six or above are extremely hydrophobic and bioaccumulative and pose a significant risk [26].

3.2.2 Animal Farms

Animal farms employ many veterinary medications to treat and increase the number of cattle. Animals typically excrete 50–90% of the doses they take. Animal excrement serves as organic fertilizer for agricultural fields. Pharmaceuticals can contaminate groundwater and surface water through agricultural runoff or direct seepage. For surface water samples from the Republic of Korea, seven classes of veterinary antibiotics were used. Veterinary antibiotics are occasionally spilled directly into streams, causing additional contamination. Tetracycline, a regularly used veterinary antibiotic, has been identified in most rivers in China, as 10–90% of it is discharged into water systems [27].

3.2.3 Atmospheric Deposition

Atmospheric deposition is the process by which contaminants are put into the air and settle on land or water surfaces. Pollution can be transported over long distances and introduce harmful substances, not their own, into ecosystems through this process. Contaminants are emitted through industrial processes such as sulfur dioxide (SO_2), nitrogen oxides (NO_x), and heavy metals (e.g., mercury and lead). Volatile organic compounds (VOCs), dioxins, and particulates are released from factories.

3.2.4 Recreational Activities

Emerging contaminants can be directly introduced into the environment through recreational activities, including boating, fishing (synthetic material nets, lines, and lures that remain in the environment for decades), and swimming. These activities give people pleasure and economic benefits but are not without unintended environmental effects, usually through the emission of contaminants to water bodies. Fuels from gasoline or diesel engines in boats can leak and release hydrocarbons and other harmful chemicals. Sources of emerging contaminants such as sunscreen and personal care products: Chemicals from sunscreens, lotions, and insect repellents wash off swimmers' bodies into the water.

3.2.5 Leachate from Landfills and Septic Systems

The poor capture and management of stormwater or inadequate runoff control measures allow stormwater to percolate through landfills or poorly managed septic systems. This carrier can be a host of harmful substances seeping into the surrounding environment. Leachate formation and infiltration are processes with important consequences for soil, water quality, and ecosystems and are often referred to as such. Contaminants release batteries, electronics, and industrial waste from septic systems, detergents, disinfectants, and personal care products.

3.2.6 Residential and Commercial Use

Emerging contaminants are released from routine daily activities such as cleaning, washing, and flushing into the environment. Wastewater treatment plants are designed to remove many pollutants but often fall short of removing all burgeoning contaminants that can be discharged from the system and enter natural water bodies. Examples of residential and commercial contaminants include synthetic fragrances, detergents, cleaning agents (such as acetic acid and ammonia solution), and prescription drugs such as amoxicillin, metformin, and fentanyl .

3.2.7 Surface Water–Groundwater Interactions

A river extension is a watercourse that gains or loses water depending on its water level and nearby groundwater. A developing stream is one where water flows from the aquifer to the stream as base flow through summer; perennial rivers frequently encounter this phenomenon owing to a shortage of primary water sources. The infiltration of river water into a groundwater system is considered a stream loss [28]. Harmful EOCs in rivers can contaminate shallow groundwater, making drinking water unsafe. Excessive groundwater extraction is frequent in densely populated areas to meet human needs. Tracer studies have shown that excessive groundwater pumping can cause contaminants to move vertically into deeper groundwaters. This increases the vulnerability of deep aquifers to various contaminants [29].

4 Major Sources of Emerging Contaminants

Emerging contaminants are found in the environment because they are derived primarily from sources that introduce compounds that have only recently been identified as pollutants. Most of these pollutants have their primary sources in anthropogenic activities and largely escape efficient control by existing regulatory frameworks or conventional wastewater treatment technologies.

4.1 Industrial Activities

Industrial processes, which include various processes from industries such as producing goods, chemical production, electronic devices, and textiles, significantly contribute to the formation of new pollutants. These activities introduce multiple contaminants into the environment, including pharmaceutical residues; plasticizers such as bisphenol A (BPA) and phthalates; industrial solvents such as trichloroethylene (TCE) and perchloroethylene (PCE); heavy metals such as lead, mercury, and cadmium; and pollutants from electronic waste (e-waste) recycling, such as brominated flame retardants and polychlorinated biphenyls (PCBs) [30]. Improper disposal techniques, unintentional discharges during production, and runoff from industrial sites can contaminate land, water, and air. Furthermore, industrial wastewater discharges may contain various chemicals, including persistent organic pollutants (POPs), volatile organic compounds (VOCs), and new contaminants such as microplastics derived from synthetic fibers and nanoparticle additions. Addressing the sources of developing pollutants from industrial operations necessitates strict regulatory measures, the implementation of cleaner manufacturing technology, environmentally friendly waste management procedures, and industry-wide efforts to reduce environmental consequences [31].

4.1.1 ECs in Water: Causes and Effects

ECs are formed of many chemical groups (consisting of many chemical groups that are developing with no sign of ending), and it has been discovered that these compounds constitute components of this classification. During their annual evaluation of ECs, a wide range of substances, including nanomaterials, perfluorinated and other artificial sweeteners, sucralose, and other items that combine to form a group, are included. Disinfectants used in drinking water and swimming pools, along with sun protection products, siloxanes, flame retardants, benzotriazoles, benzothiazoles, and UV filters, are classified as emerging contaminants due to their widespread environmental presence and potential ecological risks. With improvements, the chemical industry's broad range of substances emitted into the environment has expanded. They may cause long-term harm to people and the environment, which can grow dramatically over time. Researchers have demonstrated little consensus on the list of compounds in recent years, and interest in monitoring ECs has increased.

- *Pharmaceuticals*: Medications in drinking water are important for ECs and constitute an important component. This has prompted serious concerns regarding estrogenic and other harmful consequences for humans and animals. There are approximately 3000 distinct chemicals. It is expected to be used in pharmaceutical compounds, such as pain relievers and antibiotics. These include antidiabetics, beta-blockers, contraceptives, and lipid regulators. A few of these ECs have been investigated in the environmental field. Furthermore, their widespread adoption has expanded in urban waters such as surface and groundwater, wastewater, and stormwater runoff zones.
- *Antibiotics, biocides, and pesticides*: The main difficulty with antibiotics, biocides, and pesticides is that they are released into the environment, allowing microorganisms to develop resistance. This disrupts the breakdown of plant materials and the principal food chain of aquatic environments. Pesticides are chemical substances. "Biocide" refers to agricultural chemicals. Bituminous roof sealant is primarily a biocide application. It has multiple applications, including grass management, weed control, membranes, and external facades. Rainfall, biocides, and pesticides all impact surface water and groundwater.
- *Personal care products*: These products are commonly found in urban contexts, including scents, with prospective uses of sunscreens, insect repellents, and antifungal compounds. Because these chemicals do not cause metabolic changes in their chemical structure, they are suitable for external application. They are readily discharged into watery environments. Is their presence in urban runoff? In recent years, groundwater has significantly increased [32, 33].

4.1.2 Chemical Manufacturing

Chemical manufacturing industries that produce chemicals, such as pharmaceuticals, insecticides, plastics, microplastics, heavy metals, flame retardants, endocrine disruptors, agriculture runoff, persistent organic pollutants, and solvents, can discharge various molecules into the environment, which are summarized with its units and levels (Table 2.2). Pharmaceutical residues, plasticizers such as bisphenol A (BPA) and phthalates, and chemical solvents such as trichloroethylene (TCE) and perchloroethylene (PCE) are examples of such compounds. Improper chemical waste disposal and unintentional discharge during production operations can contaminate soil, water, and air [30].

4.1.3 Mining and Extraction Industries

Emerging environmental contamination from mining activities, such as the extraction of metals, minerals, and fossil fuels, is enormous. Traditionally, prominent mining pollutants include heavy metals, such as mercury and cadmium. If they are

Table 2.2 The particular chemical unit and emerging contaminants

S. no	Category	Contaminant example	Typical units	Common levels
1	Pharmace uticals	Ibuprofen, antibiotics	ng/L, μg/L	10–500 ng/L
2	Industrial chemicals	PFAS, BPA	ng/L, μg/L	1–100 ng/L
3	Agricultural runoff	Atrazine, nitrates	mg/L, μg/L	1–50 μg/L
4	Microplastics	Microbeads	particles/m^3	10,000–100,000 particles/m^3
5	Heavy metals	Mercury, cadmium	μg/L	0.1–5 μg/L
6	Flame retardants	PBDEs	ng/g, ng/L	10–100 ng/g
7	Endocrine disruptors	Nonylphenol, synthetic hormones	ng/L, μg/L	5–300 ng/L
8	Persistent organic Pollutants	PCBs, dioxins	pg/L, ng/kg	1–50 pg/L (water)

taken orally, they can contaminate the soil and water in the ecosystem, contaminating entire ecosystems and human health. They can be bioaccumulated through processes such as bioaccumulation and amplified to impact public health and biodiversity.

4.1.4 Textile and Apparel Manufacturing

The textile industry is a major contributor to environmental pollution, and the use of chemicals and synthetic materials in fabric processing is also extensive. Textiles are often chemically colored, polished, and given certain properties, such as water and stain resistance, using dyes, flame retardants, and perfluorinated compounds (PFCs). Wastewater from production that is not treated or is inadequately treated contains these chemicals and is often discharged untreated or inadequately treated into nearby water bodies, introducing toxic substances that can harm aquatic ecosystems and contaminate drinking water sources. Heavy metals such as dyes exist, and flame retardants and PFCs, such as persistent organic pollutants, are rarely degraded but accumulate over time in the environment. More importantly, microplastics are also released during washing processes with synthetic fibers, such as polyester and nylon, which dominate textile production overall, such as during spin rinses. As a result, microscopic plastic fragments pass through sites of conventional wastewater treatment and into aquatic environments, from which marine organisms can ingest and accumulate through the bioaccumulation process and through the food chain. Chemical and microplastic pollution have dual impacts on the environment; thus, the textile industry has become a critical sector where they can change with respect to environmental reform.

4.1.5 Electronic Waste (E-waste) Recycling

The recycling and dismantling of electronic waste, which contains obsolete computers, smartphones, and appliances, among other materials, presents the problem of releasing hazardous substances and rare earth elements (REEs). E-waste contains lead, mercury, brominated flame retardants, and polychlorinated biphenyls (PCBs). Recycling practices worsen the problem in many informal or unregulated recycling sectors, especially when the approaches are inappropriate. For example, once cables are burned to obtain copper or when acid baths are used to obtain valuable metals, toxic fumes and liquid waste are released, contaminating air, soil, and water. They can leach into groundwater or runoff into water bodies, harming aquatic ecosystems and public health. The well-known toxins that e-waste is a source of include neodymium, dysprosium, and lanthanum, which are rare earth elements found in electronic components such as magnets, batteries, and screens. While these elements are necessary for modern technologies, extraction from e-waste can release harmful residues until proper safeguards are taken. Informal methods are often crude, inefficient processes that recover REEs ineffectively with hazardous and ineffective waste generation and no environmental recovery.

4.1.6 Plastic Production and Processing

Plastics are produced and processed via a variety of chemicals, including plasticizers, stabilizers, and additives. Over time, these compounds can leak from plastic objects, releasing new contaminants such as phthalates, bisphenols, and polybrominated diphenyl ethers. Plastic trash contributes to environmental damage, especially microplastic damage, in aquatic ecosystems when it is not managed effectively.

4.1.7 Food Processing and Packaging

Packaging materials like popcorn bags and plastic and cardboard materials, food packaging materials, breast milk, fish liver and muscle, fish liver, vegetables and fish, and fish tissue. Fish homogenates, fish used in food processing, include chemicals such as phthalates, bisphenols, and per- and poly-fluoroalkyl substances (PFASs) with its extraction methods summarized (Table 2.3). These chemicals can contaminate food goods, and if not adequately managed, packaging trash can contribute to environmental pollution [29, 35]. Humans are fed healthy foods that meet all the growth and other developmental process requirements. While the incidence of foodborne diseases increases, many food chain employees worldwide fail to define food safety perfectly. These may include microplastics, tetrabromobisphenol A, agrochemicals/chemical ripening agents, mercury, phthalates, antibiotics, medicinal products, and personal care products that appear in the food chain and cause several harms to human health as well as ecological impacts. Perchlorinated polyfluoroalkyl substances (also known as pers. and polyfluoroalkyl substances, or

Table 2.3 Techniques for extraction and analysis of PFAS in food and food contact materials [34]

Food/food contact material	Compound	Extraction method	LOD/LOQ
Popcorn bag	PFHxA, PFTeDA, PFBA, PFPeA, PFHpA, PFDA, PFNA, PFDoDA, PFOA, PFUnDA, PFTrDA, PFPeDA, PFHxDA, etc.	Focused ultrasonic solid-liquid extraction	0.6–2.2 ng/g
Plastic and cardboard materials	PFHxS, PFBA, PFBS, PFHxA, PFPeA, PFOS, PFOA, PFDA, PFNA, PFHpA, PFOPA, PFOSA, PFHxPA, PFDPA, etc.	Ultrasonic probe-assisted extraction	0.6–2.2 ng/L
Food packaging materials	PFDS, PFPeA, PFTrDA, PFHpA, PFBA, PFNA, PFOA, PFHxA, PFDA, PFTeDA, PFUnDA, PFBS, PFOS, PFDoA, PFHxS, PFODA, PFHxDA	PLE	5–30 ng/g
Breast milk	PFPeA, PFHxA, PFHpA, PFOA, PFNA, PFUnDA, PFTrDA, PFDA, PFHS, PFDoDA, PFTeDA, PFOcDA, PFHxDA, PFBS, PFDS, PFOS	MTBE	0.025–0.01 ng/mL
Fish liver and muscle	PFPeA, PFOPA, PFHpA, PFBA, PFHxA, PFNA, PFDA, PFHxS, PFOSA, PFBS, PFHxPA, PFOS, PFDPA	Focused ultrasonic solid–liquid extraction	1.4–8 ng/g
Fish liver	HFPO-DA	IPE (TBAS + MTBE)	0.05–0.10 ng/g ww
Vegetables and fish	PFBA, PFHxA, PFHxPA PFHpA, PFOPA, PFPeA, PFHxS, PFBS, PFNA	Focused ultrasound solidliquid extraction	0.3–12.4 ng/g
Fish tissue	AFFF	MeOH + NH4OH	0.027–0.54 ng/g ww
Fish	PFOS, PFOA	SPE	5.4–17.2 ng/g
Fish homogenates	FBSA	SLE (ACN + FA)	0.010 ng/g ww
Fish muscle	PFPCPeS, PFECHS	MeOH + KOH	0.22–0.39 ng/g dw

PFNA perfluorononanoic acid, *PFOA* perfluorooctanoic acid, *PFOS* perfluorooctanesulphonate, *PFBS* perfluorobutane sulfonic acid, *PFNS* perfluorononane sulfonic acid, *PFASEs* perfluoroalkane sulfonamidoethanols, *PFHxA* perfluorooctanoic acid

PFASs) have been extensively used in a myriad of products, from nonstick surface repellents to textiles, cosmetics, paper, and firefighting foams, since the 1950s because of their exceptional ability to repel grease, dirt, and water. Because of their chemical properties and uses, PFASs are more likely to bioaccumulate and resist biodegradation, resulting in significant environmental contamination with long-term consequences. A substantial body of information suggests that PFASs have severe repercussions for humans and wildlife. Owing to a lack of access to health, the environment, and laws, it has become a significant source of dental contamination in food and food packaging. They have called upon scientists, researchers,

Table 2.4 List of contaminants emerging from agricultural practices

Particular name ECs	List (examples)	Contribution to emerging contaminants
Agricultural practices	Pesticides (Atrazine)	Pesticides and fertilizers run off into water bodies, causing eutrophication
	Fertilizers (nitrates, phosphates)	Antibiotics promote antimicrobial resistance in aquatic environments
	Antibiotics used in livestock	Persistent residues impact aquatic organisms and contaminate drinking water supplies

public health professionals, and veterinarians to do their work. To play a leading role in assessing the risk of PFAS in food, a worldwide understanding of the implications of consumer behaviors is needed to reveal inadequacies in knowledge [34].

4.2 Agricultural Practices

Agricultural practices are significant sources of developing contaminants, as they cover many activities that release diverse pollutants into the environment. Pesticides and herbicides used to manage pests and weeds, fertilizers containing nutrients such as nitrogen and phosphate, veterinary medications from animal operations, and organic pollutants from manure and agricultural runoff are examples of contaminants [35]. Pesticides, including insecticides, fungicides, and herbicides, can leak into soil and water, contaminating both surface and groundwater sources. Fertilizers can also contribute to nutrient enrichment in water bodies, resulting in eutrophication and toxic algal blooms. Through manure application and runoff, livestock operations release medicines, hormones, and antibiotic residues into the environment, possibly affecting aquatic ecosystems and human health. Furthermore, agricultural runoff can transport debris, pathogens, and organic matter, all of which contribute to poor water quality. Environmentally friendly agricultural practices, soil conservation measures, precision farming techniques, and nutrient management strategies are critical for lowering the input of new pollutants from agriculture while fostering the preservation of the environment [34]. Agricultural practices associated with emerging contaminants are listed in Table 2.4, and their contributions to emerging contaminants, with examples such as pesticides (atrazine), fertilizers (nitrates, phosphates), and antibiotics used in livestock, are as follows:

4.3 Urban Runoff

Urban runoff is a major source of emerging contaminants, which accumulate various pollutants when they travel over impermeable surfaces in urban areas. These contaminants originate from diverse sources, including sediments carrying soil, silt,

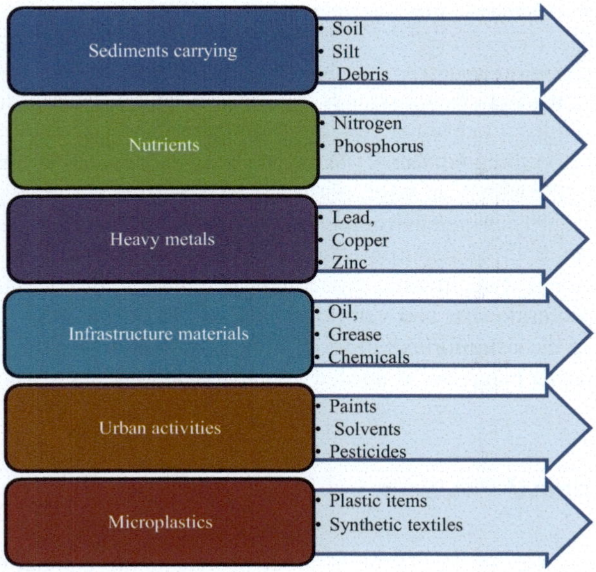

Fig. 2.3 Urban runoff contaminants

and debris from construction sites and erosion-prone regions; nutrients such as nitrogen and phosphorus from fertilizers used in lawns and gardens; heavy metals such as lead, copper, and zinc from vehicle emissions and infrastructure materials; oil, grease, and chemicals from roadways and urban activities; and microplastics originating from plastic litter, synthetic textiles, and personnel. Furthermore, urban runoff can transfer pathogens, pesticides, and pharmaceutical residues from residential, commercial, and industrial zones, endangering water quality and ecosystem health in receiving water bodies [36]. Climate change exacerbates these difficulties by modifying precipitation patterns and increasing storm events, resulting in greater runoff volumes and possible contamination. Green infrastructure projects, stormwater treatment methods, pollution prevention measures, and public awareness campaigns are all effective urban runoff management strategies for reducing new pollutants and mitigating their environmental impacts [37]. Figure 2.3 shows that urban runoff includes contaminants such as sediments that carry runoff (soil, silt, and debris), nutrient runoff (nitrogen, phosphorus), heavy metal runoff (lead, copper, zinc), infrastructure development materials (oil, grease, and chemicals), and other urban activities (paints, solvents, and pesticides), such as microplastics (plastic items, synthetic textiles).

4.3.1 Contaminants in Urban Runoff Sediments

Runoff collects soil, silt, and debris from paved surfaces, construction sites, and erosion-prone locations. These sediments can obscure streams, alter aquatic habitats, and degrade water quality by limiting light penetration and oxygen levels.

Nutrients: Runoff can carry fertilizers from lawns, gardens, and agricultural regions, resulting in nutrient enrichment in bodies of water. Excessive nutrients, such as nitrogen and phosphorus, can lead to algal blooms, oxygen depletion, and eutrophication, all of which have deleterious influences on aquatic ecosystems. Heavy metals: Vehicle emissions, brake pads, roofing materials, and industrial operations can all contribute to urban runoff, which can contain heavy metals, including lead, copper, zinc, and cadmium. These metals harm aquatic life and can accumulate in sediments, causing long-term threats to ecosystem health. Oil, grease, and chemicals: Runoff from highways transport contaminants such as oil, grease, petrol, and motor oil from automobiles. Chemicals from urban activities, such as paints, solvents, pesticides, and cleaning agents, can also reach water bodies through stormwater runoff, leading to pollution and toxicity. Microplastics: Urban runoff may contain microplastics, which are microscopic plastic particles less than 5 mm in size and are generated from scattered plastics, deteriorated plastic items, and synthetic textiles. Microplastics can remain in the environment, accumulate in aquatic organisms, and damage human and environmental health [37, 38].

4.4 Wastewater Treatment Plants

Wastewater treatment plants (WWTPs) are crucial parts of municipal infrastructure; they can also serve as sources of emerging contaminants in the environment. These contaminants come from various sources in the wastewater streams that WWTPs process. Notable contributions are pharmaceutical and personal care products (PPCPs), which find their way into WWTPs through domestic wastewater from homes, clinics, and hospitals. Pharmaceuticals and personal care products (PPCPs), which enter WWTPs through domestic wastewater from residences, clinics, and hospitals, are noteworthy contributors [36]. Furthermore, the discharge of industrial wastewater into WWTPs introduces heavy metals, organic contaminants, and industrial chemicals into the treatment process. Manufacturing, chemical manufacturing, textile processing, and other industrial operations might be the sources of these toxins. With wastewater and WWTP effluents, microplastic particles smaller than 5 mm are becoming more widely acknowledged as emerging contaminants. Plastic litter, synthetic textiles, and microbeads in personal care items are microplastic sources. While traditional WWTP methods can remove some emerging contaminants, some may linger or disintegrate, releasing microplastics into receiving water bodies [39].

4.5 Landfills and Waste Sites

Through leachate and gas emissions, landfills and waste sites are important sources of emerging contaminants in the environment. Numerous waste streams, such as municipal solid trash, industrial waste, building and demolition debris, electronic waste (e-waste), and hazardous materials, are sources of these emerging contaminants. Water seeping through garbage produces a liquid known as leachate, which is loaded with heavy metals, organic and inorganic compounds, persistent organic pollutants (POPs), etc. These substances can contaminate aquifers and surface water bodies by seeping into the soil and groundwater. Leachate from decomposing organic waste and industrial leftovers may contain organic contaminants such as volatile organic compounds (VOCs), polycyclic aromatic hydrocarbons (PAHs), and chlorinated solvents.

Water contamination occurs when water leaches into water sources. Heavy metals, such as lead, cadmium, mercury, and arsenic, present very high risks to human health and ecosystems. Most of these metals originate from improperly discarded batteries, electronics, paints, and other metal rubbish products. They can contaminate soil and groundwater and often reach drinking water supplies or aquatic habitats once in the environment. This contamination can make humans very sick from neurological damage to kidney disease and cancer and ecosystems in the environment through bioaccumulation and toxicity.

Air contamination emissions at landfills include harmful gases, including methane (CH_4), carbon dioxide (CO_2), volatile organic compounds (VOCs), semivolatile organic compounds (SVOCs), methane, and carbon dioxide, which all contribute to climate change as greenhouse gases. VOCs, SVOCs, and HAPs, such as respiratory problems, hormonal disruption, and cancer, directly impact human health. Furthermore, they can be deposited in the soil and water, affect these resources indirectly, and extend their environmental reach even further hazardous air pollutants (HAPs). These gases are released during the decomposition of organic waste under anaerobic conditions [40].

4.6 Atmospheric Deposition

One of the important sources of emerging contaminants that can impact terrestrial and aquatic ecosystems is atmospheric deposition. This process entails the wet (rain and snow) and dry (particulate settling) deposition of pollutants and particulate matter from the atmosphere onto land and aquatic surfaces [41]. Contaminants are released into the atmosphere by various human activities, industrial processes, transportation, and natural events, and these emissions eventually lead to their deposition. Heavy metals from industrial emissions, such as mercury and cadmium, are frequently emerging contaminants linked to atmospheric deposition, fossil fuel combustion, and mining processes. Through bioaccumulation and biomagnification in food chains, these metals can accumulate in soils, sediments, and water bodies, compromising the health of people, creatures, and plants [42]. Persistent organic pollutants (POPs), dioxins, furans, and polycyclic aromatic hydrocarbons, or PAHs, are airborne pollutants that can travel large distances and settle on aquatic and terrestrial surfaces. Combustion activities, industrial emissions, agricultural burning, and trash incineration are common sources of these contaminants. After being deposited, they can eliminate in their surroundings, bioaccumulate in living organisms, and contaminate the environment [43].

4.7 Land Application of Biosolids

Biosolids are organic materials rich in nutrients that are obtained from treated sewage sludge. They are applied to land and maybe a source of emerging contaminants that need to be monitored. Biosolids are frequently added to soil in landscaping, agriculture, and land reclamation because of their high nutritional content and possible advantages for improving soil fertility and structure. Additionally, biosolids may contain various contaminants that, if improperly managed, could endanger human health and the environment. Personal care products and medications are among the types of emerging contaminants found in biosolids (PPCPs) [44]. Certain substances, including hormones, antibiotics, and chemicals found in personal care products, can enter wastewater treatment plants (WWTPs) through household sewage and linger in biosolids following treatment. PPCPs can move to groundwater after being applied to land as biosolids, raising concerns about the development of antibiotic resistance, alteration of hormones, and effects on the environment and ecosystems [45].

4.8 Household Activities

Household activities, including a wide range of routines and goods contributing to environmental contamination, are important sources of emerging contaminants. Personal care and pharmaceutical products are a significant group of contaminants that are emerging from homes (PPCPs), which consist of cleaning supplies, cosmetics, overtime counter (OTC) drugs, and prescription pharmaceuticals [44]. When prescription drugs are flushed down the drain or are not properly disposed of, they can end up in wastewater and eventually make their way to water treatment plants. PPCPs can withstand treatment procedures and find their way into water bodies, endangering aquatic ecosystems and potentially harming human health if ingested through seafood or drinking water. Furthermore, several compounds included in household cleaners, detergents, and disinfectants, such as perfumes, surfactants, and antibacterial agents, can contaminate water when they are washed down drains. These substances might not be adequately eliminated throughout the wastewater treatment process, which could result in their release into the environment and possible ecological effects. Indoor air pollution, which can be caused by household goods containing formaldehyde, particulate matter, and volatile organic compounds (VOCs), is another source of new toxins from domestic activities. The contaminants released into indoor air by household cleaners, paints, air fresheners, and tobacco smoke can affect the indoor air quality. They can cause respiratory disorders and other health concerns, especially in susceptible groups [46].

4.9 Natural Sources

Through various processes and occurrences, natural sources contribute to the presence of significant sources of emerging contaminants in the environment as industrial activities, agriculture practices, urban runoff, wastewater treatment plants, landfill and waste sites, atmospheric deposition, land application of biosolids, household activities, and natural resources with its examples summarized (Fig. 2.4). One significant source is naturally occurring chemicals and compounds, such as certain heavy metals (e.g., mercury, cadmium selenium, and uranium) and other

Fig. 2.4 Major sources of emerging contaminants with examples

radionuclides, which can be released into soil, water, and air through geological processes, rock and mineral weathering, and volcanic activity. While these compounds naturally occur on Earth, anthropogenic activities can increase their release and mobility, resulting in environmental contamination and possibly adverse health effects. Another natural source of emerging contaminants is the release of organic molecules from vegetation, soil organic matter, and decomposition. These components, including humic substances, tannins, and natural phenols, can impact water quality, contribute to the creation of disinfection byproducts during drinking water treatment, and influence ecosystem dynamics [47].

5 Managing ECs and Their Challenges

5.1 Identification

The first challenge is to identify substances that can cause harm to receptors. The main goal is to develop analytical methods for quantifying ECs in different matrices at the subparts per billion levels. After completing this stage, it is possible to characterize the source and pathways, identify new avenues for human exposure, and determine the transit and destiny of the substance in the environment [48].

5.2 Risk-Based Analysis

Progress in instrumental analytical chemistry has made it possible to assess the toxicity of EPs since they can barely be detected at low concentrations [49], and a decision analysis model for assessing candidates for remediation solutions in terms of their capacity to minimize health risk and material cost has been designed. This technique makes it possible to make justified remediation concerning hazards (population, individual, and residual) and costs (reducing risks and improving efficiency) possible [48]. Although the present-day emphasis is on risk prewarning for emerging pollutants, the concern of building methods for pollution control has recently become the focus of ecological and environmental protection. Newly generated organic pollutants are biotoxic, relatively resistant to biodegradation, and potential accumulators. Additionally, they are dispersing and can be hard to control. With time and improvements in our understanding of chemical risks and advancements in environmental assessment techniques, the range and number of emerging pollutants will continue to evolve. This dynamic is illustrated in the flowchart of risk-based decision analysis for emerging contaminants (Fig. 2.5), which outlines key steps including environmental concentration assessment—comprising the pathway and receptor models—followed by risk assessment, multicriteria decision analysis, sensitivity analysis, and, ultimately, the determination of environmental limits. Risk communication involves transmitting information regarding health and

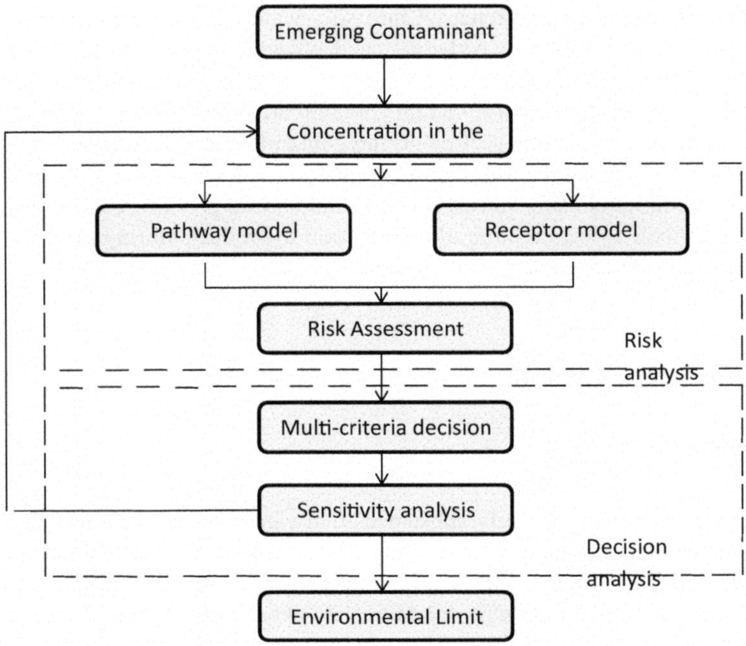

Fig. 2.5 Risk-based decision analysis processes

environmental risks to risk assessors, management, stakeholders, the public, and the media. By considering every angle, more effective, clear, and improved methodologies have been developed through this process. Information transfer may assist in determining site-specific short-term and long-term management goals and objectives and early decisions on applicable remediation options. Risk communication is ideal when managing ECs since their fate and transportation remain unpredictable, making decision-making thorough. Therefore, environmental exposure is a result of ECs and other chemical contaminants. MTS mixes 3 and 4 are expected to have greater effects [50].

5.3 Reducing Releases

To reduce the environmental impact of ECs, better chemical handling should be considered, "green" alternatives should be used, and their use in manufacturing should be reduced. This can reduce waste, ensure proper disposal, minimize chemical discharge, and improve water treatment systems. Improving existing processes and developing cost-effective technologies to eliminate a wide range of pollutants are significant problems [48].

5.4 *Improved Monitoring*

The development of new analytical methods to detect emerging pollutants is crucial. Assays that assess the toxicological activity of a contaminant combination rather than individual chemicals should be used. Biosensors offer advantages in detecting pharmaceuticals and endocrine disruptors because of their high sensitivity and selectivity for specific substances. They are also easily automated and cost-effective. Multiresidue approaches, such as targeted ultra-performance liquid chromatography (UPLC), may detect hundreds of ECs simultaneously in a variety of environmental matrices with a feasible analysis time of approximately 10 min [51, 52].

5.5 *Development of Novel Remediation Technologies*

Some ECs, such as 1,4-dioxane and PFOS/PFOA, are ineffectual to conventional remediation strategies, such as air stripping, carbon adsorption, and in situ chemical oxidation. Therefore, unique or combination remediation solutions must be developed and demonstrated. Undoubtedly, nanotechnology has achieved considerable advancements. This has made the production of more efficient and less expensive remediation technologies possible. Nanoscale zero-valent iron (NZVI) particles effectively destroy organic pollutants, particularly chlorinated hydrocarbons. However, this approach may not effectively treat certain ECs, such as perfluorinated chemicals, indicating the need for future development. CRC CARE developed a mobile system that uses modified clay to repair film-forming foam (AFFF)-impacted wastewater and soil. CRC CARE technology has been successfully implemented at Australian Department of Defense sites and is now being commercialized under the trade name matCARE™. This method can eliminate more than 99% of PFOS and PFOA from soil and water samples. Although it has not been thoroughly tested for the treatment of other perfluorinated compounds, it is expected that many more compounds may be able to be used to treat and modify clay to further expand that therapy list [53, 54].

6 Major Emerging Contaminant Groups and Their Examples

Emerging contaminants have recently received attention for their potential environmental and health effects; major contaminant groups (pharmaceuticals, human antibiotics, veterinary, analgesics, anti-inflammatory drugs, psychiatric drugs, hormones, and steroids, perfluorinated compounds, etc.), examples, and concentrations of surface water, wastewater effluents, and groundwater with its concentration unit of μg/L summarized in (Table 2.5). They frequently comprise substances that

Table 2.5 Major contaminant groups and their examples and their concentrations

Sl. no	Emerging contaminant groups	Examples	Concentration in surface water (μg/L)	Wastewater effluent (μg/L)	Groundwater (μg/L)
1	Pharmaceuticals	Ibuprofen, Diclofenac, Paracetamol	0.01–2.0	0.1–50.0	Rarely detected (<0.1)
		Ciprofloxacin, Amoxicillin, Sulfamethoxazole	0.01–10.0	0.05–20.0	Typically, <0.1
		Fluoxetine (Prozac), Sertraline (Zoloft)	0.01–1.0	0.1–2.0	Rarely detected (<0.1)
		17β-Estradiol, Ethinylestradiol	<0.001–0.1	0.001–0.05	Rarely detected (<0.01)
		Carbamazepine	0.1–1.0	0.05–2.0	0.01–0.1
		Atenolol, Propranolol	0.01–0.5	0.05–1.0	Rarely detected (<0.1)
2	Human antibiotics and veterinary	Amoxicillin, Ciprofloxacin, Azithromycin	0.01–10.0	0.05–20.0	<0.1
3	Analgesics, anti-inflammatory drugs	Paracetamol	0.01–1.0	0.1–10.0	Rarely detected (<0.1)
4	Psychiatric drugs	Diazepam	<0.001–0.3	0.01–0.5	Rarely detected (<0.01)
5	β-blockers	Metoprolol	0.01–0.8	0.1–1.5	Rarely detected (<0.1)
		Propranolol	0.01–0.3	0.05–0.5	Rarely detected (<0.05)
6	Lipid regulators	Clofibric Acid	0.01–2.0	0.1–8.0	0.01–0.1
		Bezafibrate	0.01–1.5	0.1–3.0	Rarely detected (<0.05)
7	X-ray contrast agents	Iopamidol	0.01–8.0	0.05–30.0	Rarely detected (<0.05)
8	Personal care products (PCPs)	Parabens	0.01–1.0	0.1–5.0	Rarely detected (<0.1)
		Triclosan	0.01–2.0	0.1–5.0	0.01–0.1
9	Fragrances	Musk	0.01–0.5	0.1–5.0	Rarely detected (<0.1)
10	Sunscreen agents	Oxybenzone,Octinoxate,	0.01–1.0	0.05–2.0	Rarely detected (<0.1)
		DEET (N,N-Diethyl-meta-toluamide), Picaridin,	0.01–1.0	0.1–10.0	Rarely detected (<0.1)
11	Endocrine-d isrupting chemicals (EDCs)	Nonylphenol	0.01–0.5	0.1–10.0	Rarely detected (<0.1)

(continued)

Table 2.5 (continued)

Sl. no	Emerging contaminant groups	Examples	Concentration in surface water (μg/L)	Wastewater effluent (μg/L)	Groundwater (μg/L)
12	Hormones and steroids	17β-Estradiol, Estrone, Ethinylestradiol	0.01–0.5	0.1–2.0	Rarely detected (<0.1)
13	Perfluorinated compounds (PFCs)	Perfluorooctanoic acid (PFOA), Perfluorooctanesulfonic acid (PFOS)	0.01–1.0	0.1–10.0	0.01–0.5
14	Surfactants and surfactant metabolites	Alkylphenol ethoxylates (APEs)	0.01–0.5	0.1–5.0	Rarely detected (<0.1)
15	Flame retardants	HBCD (α, β, γ isomers)	0.01–0.2	0.1–3.0	Rarely detected (<0.05)
16	Industrial additives and agents	Di (2 ethylhexyl) phthalate (DEHP), Dibutyl phthalate (DBP)	0.01–1.0	0.05–5.0	0.01–0.1
17	Gasoline additives	Methyl tert-butyl ether (MTBE), Ethyl tert-butyl ether (ETBE)	0.1–50.0	0.05–5.0	0.1–10.0
18	Antiseptics	Chlorhexidine gluconate	0.01–0.1	0.1–5.0	Rarely detected (<0.1)
19	Insect repellents	Picaridin (also known as icaridin)	0.01–0.3	0.05–0.5	Rarely detected (<0.1)

have yet to be previously regulated or thoroughly examined. Pharmaceuticals and personal care products (PPCPs), including antibiotics, hormones, and perfumes, are key sources of new pollutants. These substances reach the environment through wastewater, agricultural runoff, and poor drug disposal. Another prominent group is per- and polyfluoroalkyl compounds (PFASs), which are employed in numerous industrial and consumer products because of their water and oil-repellent qualities [34]. PFAS chemicals can remain in the environment for long durations and are related to harmful health effects. Furthermore, microplastics, minute plastic particles that result from the breakdown of larger plastic items, are receiving attention as emerging contaminants because of their ubiquitous presence in aquatic ecosystems and possible harm to marine life. Monitoring and understanding new contaminant groups is critical for creating effective mitigation strategies and protecting the environment and human health [51].

7 Examples of Recent Advances in EC Analytical Instruments for Water Samples

Recent advances in analytical tools for detecting emerging contaminants (ECs) in water samples have considerably increased our ability to detect and quantify these compounds with improved precision and sensitivity, as shown in the summarized table below types of emerging contaminants and recent technological advances in ECs analytical instruments for water sample given significant details for sample pretreatment, equipment, and analysis conditions (Table 2.6). One significant innovation is the widespread use of high-performance liquid chromatography coupled with tandem mass spectrometry (HPLC-MS/MS). This technique separates and identifies complex combinations of ECs in water samples, allowing researchers and environmental experts to detect a wide spectrum of contaminants at low levels. Additionally, gas chromatography–mass spectrometry (GC–MS) has advanced, particularly in its ability to detect volatile organic compounds (VOCs) and semi-volatile compounds (SVOCs) in aqueous matrices. The recent development of novel sample preparation procedures, such as solid phase microextraction (SPME) and passive sampling techniques, has made it easier to extract and analyze ECs from water samples, increasing their efficiency and reliability [54].

Table 2.6 Examples of recent advances in EC analytical instruments for water samples

S. no	ECs number/ type	Sample pretreatment	Equipment and analysis conditions	Notes	References
1	27 ECs including pharmaceuticals, sunscreens compounds, fragrances, antiseptics, fire retardants	Methylation of carboxyl groups with trimethylsulfonium hydroxide at 270 °C	Gas chromatography–mass spectrometry (GC–MS) with electron impact mode (70 eV ionization energy) using a TRB5-MS column (5% diphenyl-95% dimethylpolisoxane)	Sample type: wastewater. LOD and LOQ in the range 1–40 and 3–80 ng/L, respectively. Recoveries and repeatability were greater than 80%	[3]
2	105 pharmaceuticals and lifestyle products; 21 drugs of abuse, metabolites	Solid-phase extraction (SPE) using polymeric cartridges	Liquid chromatography-electrospray time-of-flight mass spectrometry in positive and negative modes	Sample type: surface and wastewater LOQ raged 0.2–777.9 mg/L	[4]
3	8 artificial sweeteners	Reversed-phase SPE extraction for sample preconcentration and cleanup	LC–MS and LC–MS/MS with electrospray interface (ESI)	Sample type: surface and drinking water. LOD in the range of 0.82 and 2.8 mg/L	[29]
4	44 pharmaceuticals and 13 EDCs	Pressurized liquid extraction, SPE purification	Ultraperformance liquid chromatograph coupled to a mass spectrometer (UPLC-MS/MS)	Sample type: river water and biofilms. LOD in the range 0.2–2.4 ng/g for EDCs and 0.07–6.7 ng/g for pharmaceuticals	[30]

(continued)

Table 2.6 (continued)

S. no	ECs number/ type	Sample pretreatment	Equipment and analysis conditions	Notes	References
5	90 ECs including pharmaceuticals andestrogens	Microwave assisted extraction (MAE) protocol followed by SPE	UPLC–MS/MS with ESI and using an Acquity BEH C18 column for both positive and negative modes	Recoveries in the range 40–152%. LOQ from 0.1 to 24.1 ng/L	[31]
6	8 ECs including perfluorinated compounds, pharmaceuticals	Polar organic chemical integrative samplers (POCIS) forin situ pre concentration	Fast LC–MS/MS with ESI and using a Zorbax XDB-C18 column	Sample type: drinking water Concentration in the range 4.2–15.9 ng/L	[35]

8 Detection and Quantification of Emerging Contaminants

Chemically, most ECs can break into water and harm aquatic life, the environment, and humans. Most recently, the number of environmental contaminants (ECs) in drinking water, groundwater, wastewater, and surface water has gradually increased. The recognition and assessment of ECs in water, which require sensitive and systematic procedures at ng/L scales, has become an important scientific endeavor. Effectively performing rapid and responsive analytical processes is essential for monitoring and determining various ECs. New analytical techniques have been developed to detect and identify new pollutants in water. Major methodologies include liquid or gas chromatography and mass spectrometry. Mass spectroscopy can be used to test EC samples from wastewater accurately because of its high sensitivity and specificity. To estimate ECs at the ng L^{-1} level, advanced mass spectroscopic methods such as ion traps and triple quadrupoles are used. More advanced techniques, such as linear ion traps, quadrupole-time of flight, triple quadrupole, and quadrupole-linear ion traps, are employed for morphological analysis of changed products. Additional means of measuring EC include immunoanalytical methods and capillary electrophoresis. Capillary electrophoresis is less sensitive than liquid chromatography is; however, immunoanalytical methods require specific antibodies, and it can be challenging to perform multiple analyses simultaneously. Microbiological assays are also highly dependent on the sample type [55]. The concentration and physicochemical properties of ECs are Log $K_{ow,}$ p$K_{a,}$ water solubility (mg/L), country, and mean biosolid concentration (mg kg^{-1}dry weight) presented in (Table 2.7).

However, in addition to enhancing the detection and evaluation of unknown substances in a more comprehensive way, modern analytical techniques have progressed. This is being done through research and development. Therefore, establishing modern means of identifying unknown ECs as analyses is important. Precise techniques can be faster and less expensive than costly and elaborate procedures for identifying genuineness. Moreover, problem samples often require

Table 2.7 Concentration and physicochemical properties of ECs

Emerging contaminants	Log K_{ow}	pK_a	Water solubility (mg/L)	Country	Mean biosolid conc. (mg/kg dry weight)	Reference
Estrone	3.13	10.3	30.0	Germany	<MQL-0.02	[56]
				Canada	0.06	[57]
				Australia	<MQL-0.28	[58]
Ibuprofen	3.97	4.91	21.0	Spain	0.30	[59]
				Canada	0.15[a]	[60]
				Spain	0.02e0.44	[61]
				Canada	0.17	[62]
				Canada	0.06	[63]
				Spain	0.09	[60]
Naproxen	3.18	4.15	15.9	Canada	0.02[a]	[61]
				Spain	<MQL-0.02	[57]
				Canada	0.01	[58]
Diclofenac	4.51	4.15	2.37	Spain	0.19	[59]
				Spain	<MQL-0.63	[61]
				Spain	0.06	[63]
				Canada	<MQL-0.02	[64]
Acetaminophen	0.460	9.38	1.40×10^4	Spain	0.03	[59]
				USA	<MQL-0.37	[65]
				Canada	0.02[a]	[60]
				Spain	0.02–0.29	[61]
Gemfibrozil	3.40	4.75	–	Spain	0.12	[59]
				Spain	<MQL-0.07	[61]
				Canada	0.008	[57]
				Canada	<MQL-0.003	[64]
Carbamazepine	2.45	13.9	17.7	Spain	0.08	[59]
				USA	<MQL-0.02	[65]
				Canada	0.26	[66]
				Canada	0.18	[63]
				Canada	0.09	[57]
				Spain	0.03	[63]
				Canada	0.01	[64]
Fluoxetine	4.05	3.95	5.00×10^4	Spain	0.12	[59]
				Canada	0.09	[57]
				Canada	0.11	[62]
Propranolol	3.48	9.42	61.7	Spain	0.03	[59]
				Canada	0.04	[57]
				Canada	0.12	[62]
				Canada	0.36[a]	[60]
				Canada	0.23	[62]

(continued)

Table 2.7 (continued)

Emerging contaminants	Log K_{ow}	pK_a	Water solubility (mg/L)	Country	Mean biosolid conc. (mg/kg dry weight)	Reference
Ciprofloxacin	0.280	6.09	3.00×10^4	Switzerland	2.27–2.42	[67]
				Sweden	0.50–4.80	[68]
				Canada	6.50[a]	[60]
				Canada	5.87	[57]
				Canada	5.87	[62]
				Canada	<MQL-0.05	[64]
Norfloxacin	−1.03	2.75	1.78×10^5	Switzerland	2.13–2.37	[67]
				Sweden	0.10–4.20	[68]
				Canada	1.75	[57]
				Canada	1.01	[62]
Ofloxacin	−0.390	0.390	2.83×10^4	Sweden	<MQL-2.00	[68]
				Spain	0.07	[59]
				Canada	0.69[a]	[60]
				Canada	1.07	[57]
				Canada	1.40	[62]
Tetracycline	−1.30	3.30	5.00×10^4	USA	<MQL-0.56	[65]
				Canada	0.24[a]	[60]
				Canada	0.34	[57]
				Canada	0.51	[62]
				Canada	<MQL-0.01	[64]
Triclosan	4.80	7.90	10.0	Greece	3.21	[69]
				USA	0.53–15.6	[70]
				Spain	1.51	[71]
				Canada	0.68–11.55	[72]
				Australia	5.58	[73]
				USA	4.89–9.28	[72]
				Canada	6.80[a]	[60]
				Canada	4.68	[57]
				Canada	10.9	[62]
				Australia	0.22–9.89	[58]
Triclocarban	4.20	12.7	0.110	Canada	3.05–5.97	[72]
				USA	0.09–7.06	[32]
				USA	51.0	[33]
				Canada	2.90[a]	[60]
				Canada	6.03	[57]
				Canada	4.94	[62]
Bisphenol A				Greece	0.53	[69]
				China	0.01	[56]
				Australia	0.06–1.37	[58]

Note: *Log* K_{ow} octanole–water coefficient, pK_a acid dissociation coefficient, *MQL* method quantitation limit

[a]Reported as the median

significant pretreatment before the ECs can be measured with reasonable accuracy. Several authors have focused only on liquid–liquid extraction in the past. Solid-phase extraction has been embraced because of the benefits associated with it over liquid–liquid extraction, which include accuracy, applicability, and easy accessibility. To achieve these levels, the concentration of some water samples before analysis should be determined via a suggested method to determine the lower limits of detection (LODs) and quantification (LOQs). Several preconcentration techniques have been described, such as magnetic online SPE, automated SPE, and grafted molecular polymers [74]. These strategies have enhanced sample processing since they reduce the process's cross-contamination, effort, and costs. This reduces labor, lowers LODs and selectivity, and is observed in sample pretreatment. Nanostructured materials could be one of the possibilities for future studies in this area. Magnetic nanomaterials, such as iron nanoparticles, are used in solid-phase applications [74, 75].

9 Causes and Effects of Emerging Contaminants

Emerging contaminants (ECs) have various origins, including human activities and industrial operations. One major factor is the widespread use and disposal of pharmaceuticals and personal care products (PPCPs). Prescription pharmaceuticals, over-the-counter medications, and cosmetics enter the environment via wastewater effluents, sewage leaks, and inappropriate disposal procedures; adding to ECs in water sources is an industrial process, and industrial activities contribute significantly to the emergence of pollutants. Chemical manufacturing activities, such as creating plastics, insecticides, and flame retardants, emit various synthetic substances into the environment. These pollutants can seep into soil and water, contaminating natural resources. Agricultural activities, particularly the use of fertilizers, pesticides, and animal manure, all contribute to the occurrence of ECs in water bodies. Runoff from agricultural areas can transport contaminants such as nitrates, phosphates, pesticides, and antibiotic residues into rivers, lakes, and groundwater, affecting aquatic ecosystems and human health. Pharmaceuticals and personal care products (PPCPs): The widespread use and inappropriate disposal of pharmaceuticals, cosmetics, and hygiene products add PPCPs to water sources. Electronic waste (E-waste) disposal: The disposal of electronic trash (e-waste) and the breakdown of consumer products that include hazardous substances such as heavy metals, brominated flame retardants, and per- and polyfluoroalkyl substances (PFASs) all contribute to the development of pollutants in the environment. Improper disposal methods, such as landfill leaching and incineration, can release toxic compounds into the atmosphere, water, and soil, endangering ecosystems and humans. Climate change effects: Climate change and urbanization exacerbate this problem by disrupting natural processes and increasing environmental pollutant emissions. For example, extreme weather events such as floods and droughts can transport toxins from soil and sediments into bodies of water. In contrast, urban expansion can increase pollutant runoff from paved surfaces into neighboring rivers. Urbanization impact: As cities grow, the number of paved surfaces increases, allowing more pollutants to run off into surrounding rivers. This increase in runoff increases aquatic habitat pollution levels, creating environmental and public health risks [41].

9.1 Emerging Contaminants in Developed and Developing Countries

The reasons for the emergence of contaminants in developed and developing countries are significantly different, as there are differences in industrial practices, daily lifestyles, and infrastructure. The principal drivers in developed countries are high pharmaceutical and personal care product (PPCP) consumption, high-tech manufacturing, and aging wastewater treatment infrastructure, which results in the removal of complex contaminants such as microplastics and hormones. These nations, however, with robust environmental regulations, have difficulty overcoming legacy pollutants and continuous innovation, which are increasing the number of new chemicals in ecosystems. However, developing countries face more adverse effects from poor waste management systems, uncontrolled industrial growth, and the massive use of agrochemicals (pesticides and fertilizers). This difficulty is exacerbated by poor governance, limited awareness, and virtually no advanced treatment facilities that might prevent contaminants from entering water bodies or soils. Analysis of these disparities highlights the need for context-specific strategies to properly combat the sources of emerging contaminants in both contexts [50].

9.2 Effects of Emerging Contaminants

The presence of many new compounds in surface water, ground water, and drinking water is troublesome even if guidance and standard concentrations of regulation continue to focus on individual risks, and whether the identified concentrations portray any health risks is an issue of great controversy. The effects of emerging contaminants on the marine environment urgently need to be addressed because of the increasing number of scientific studies linking exposure to new chemicals to their reproductive and health consequences on users and other life forms. This is because there are still other types of pollutants that these aquatic ecosystems will influence. It has great importance for the health of people and the conditions on Earth [76].

9.2.1 Human Health

Emerging organic contaminants, such as modern brominated flame retardants, phthalate substitutes, organophosphate flame retardants, triclosan, bisphenol-A, synthetic muscles, perchlorate, and polycyclic siloxanes, can harm human health. Bisphenol A is among the most extensively used plasticizers worldwide. It causes endocrine abnormalities in multiple systems, including the salivary glands, thyroid, and men's genetic system. Bisphenol A increases the incidence of breast cancer and works as an anti-androgen, leading to feminizing effects in males. Pesticide poisoning directly impacts the endocrine cycle. Occupational exposure can lead to neurotoxicity, genotoxicity, cytogenetic disruption, cancer, and microbial induction. Brominated flame retardants have negative effects, including pulmonary, impermeable, neuroprotective, endocrine, and cancer-causing effects. Tetrabromodiphenyl ether can affect the endocrine system, lead to hormonal imbalances, and damage DNA and mitochondria; pentabromodiphenyl ether and hexabromodiphenyl ether can harm neurodevelopment and mitochondria; and decabromodiphenyl ether is carcinogenic and can impact thyroid function, brain development, neural function, fertility, and growth. Brominated flame retardants may cause genetic abnormalities in female anatomy, the breast, and the liver, leading to increased thyroid hormone, androgen, progestin, and estrogen levels due to immune cell disruption [77]. Excessive phthalate exposure has been linked to pregnancy complications and termination. Chemical pesticides can have hormonal effects, including eggshell shrinking, reduced male productivity, thyroid issues, ovarian tumors, and behavioral abnormalities. In a rat bacterial genome, musk xylol, commonly used for fragrance, has been linked to cancer and found in human tissue. Additionally, it can cause brain damage [77, 78]. Depicts the flowchart of principal health impacts of emerging pollutants and additional effects like hormonal imbalance, cancer, fertility, endocrine disruption, thyroid problems, ovarian tumors, gastrointestinal toxicity, DNA damage, damage in neural development, damage to male reproductivity, etc. (Fig. 2.6).

9.2.2 Environment

Establishing direct cause-and-effect relationships between contaminant exposure and human health is difficult. Analyzing field observations at the environmental level is even more difficult because of the presence of multiple organisms and external conditions. Surfactants have negative effects on organisms. High levels of these chemicals can harm the ecosystem. Biphenyl A is neurotoxic and negatively impacts cell division and plant growth [77]. Alkylphenols have been linked to negative effects on spermatogenesis in mice. Alkylphenols can have estrogenic effects on marine creatures, including reduced male sexual competence. Brominated dioxins, polychlorinated naphthalenes, and furans exhibit toxicity-like behavior, causing newborn animal deformities and potentially harming reproductive growth, immunity, and hormonal systems. At relatively low concentrations, perchlorate has a negative influence on brain and thyroid development. Rodent pups born to

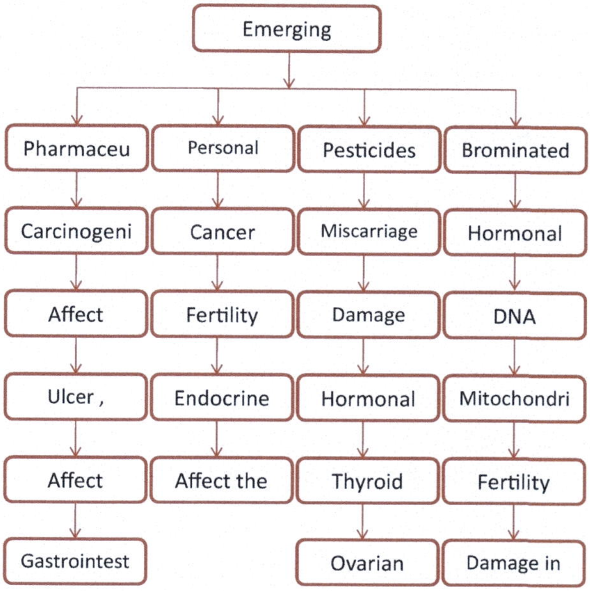

Fig. 2.6 Effects of emerging contaminants on humans

contaminated mothers have a greater risk of damage. Perfluorochemicals can disrupt the growth of oxidative metabolism and outer membrane proteins in aquatic organisms. Exposing fetuses to phthalates has resulted in severe cognitive toxic consequences in research animals, particularly with respect to the fertility of their organs. Phthalates can cause mammary gland, thyroid, kidney, and liver failure in elderly animals. Polybrominated diphenyl ethers have been linked to negative health consequences such as thyroxine disruption, memory issues, behavior changes, late onset of adolescence, lower sperm quality, and fetal malformations in animal studies. Neuromuscular and polycyclic Musk compounds block meiosis cell membrane resistance detection systems caused by specific carrier proteins in marine invertebrates for extended periods. Triclosan is known to be extremely toxic to certain marine species. Triclosan and its breakdown metabolites, including methyl triclosan, can pollute biodiversity, particularly in fish with high fat contents [79].

References

1. Ramírez-Malule, H., Quinones-Murillo, D. H., & Manotas-Duque, D. (2020). Emerging contaminants as global environmental hazards. A bibliometric analysis. *Emerging Contaminants, 6*, 179–193.
2. Poynton, H. C., & Vulpe, C. D. (2009). Ecotoxicogenomics: emerging technologies for emerging contaminants. *Journal of the American Water Resources Association (JAWRA), 45*(1), 83–96.

3. Petrovic, M., Gonzalez, S., & Barcelo, D. (2003). Analysis and removal of emerging contaminants in wastewater and drinking water. *Trends in Analytical Chemistry, 22*(10), 685.
4. Rodriguez-Narvaez, O. M., Peralta-Hernandez, J. M., Goonetilleke, A., & Bandala, E. R. (2017). Treatment technologies for emerging contaminants in water: A review. *Chemical Engineering Journal, 323*, 361–380.
5. Parida, V. K., Saidulu, D., Majumder, A., Srivastava, A., Gupta, B., & Gupta, A. K. (2021). Emerging contaminants in wastewater: A critical review on occurrence, existing legislations, risk assessment, and sustainable treatment alternatives. *Journal of Environmental Chemical Engineering, 9*, 105966.
6. Jones-Lepp, T. L., Sanchez, C., Alvarez, D. A., Wilson, D. C., & Taniguchi-Fu, R.-L. (2012). Point sources of emerging contaminants along the Colorado River Basin: Source water for the arid Southwestern United States. *Science of the Total Environment, 430*, 237–245.
7. Edwards, E. C., Nelson, C., Harter, T., Bowles, C., Li, X., Lock, B., Fogg, G. E., & Washburn, B. S. (2022). Potential effects on groundwater quality associated with infiltrating stormwater through dry wells for aquifer recharge. *Journal of Contaminant Hydrology, 246*, 103964.
8. Indira, D. P., Judy, T., & Raju, R. K. (2017). Pesticide consumption in India: A spatiotemporal analysis. *Agricultural Economics Research Review, 30*(1), 163–172.
9. Aristizabal-Ciro, C., Botero-Coy, A. M., López, F. J., & Peñuela, G. A. (2017). Monitoring pharmaceuticals and personal care products in reservoir water used for drinking water supply. *Environmental Science and Pollution Research, 24*, 7335–7347.
10. Félix-Cañedo, T. E., Durán-Álvarez, J. C., & Jiménez-Cisneros, B. (2013). The occurrence and distribution of a group of organic micropollutants in Mexico City's water sources. *Science of the Total Environment, 454*, 109–118.
11. Pérez-Alvarez, I., Islas-Flores, H., Gómez-Oliván, L. M., Barceló, D., De Alda, M. L., Solsona, S. P., et al. (2018). Determination of metals and pharmaceutical compounds released in hospital wastewater from Toluca, Mexico, and evaluation of their toxic impact. *Environmental Pollution, 240*, 330–341.
12. Rivera-Jaimes, J. A., Postigo, C., Melgoza-Alemán, R. M., Aceña, J., Barceló, D., & de Alda, M. L. (2018). Study of pharmaceuticals in surface and wastewater from Cuernavaca, Morelos, Mexico: Occurrence and environmental risk assessment. *Science of the Total Environment, 613*, 1263–1274.
13. Díaz, A., & Peña-Alvarez, A. (2017). A simple method for the simultaneous determination of pharmaceuticals and personal care products in river sediment by ultrasound-assisted extraction followed by solid-phase microextraction coupled with gas chromatography—Mass spectrometry. *Journal of Chromatographic Science, 55*(9), 946–953.
14. Lesser, L. E., Mora, A., Moreau, C., Mahlknecht, J., Hernández-Antonio, A., Ramírez, A. I., & Barrios-Piña, H. (2018). Survey of 218 organic contaminants in groundwater derived from the world's largest untreated wastewater irrigation system: Mezquital Valley, Mexico. *Chemosphere, 198*, 510–521.
15. Campanha, M. B., Awan, A. T., de Sousa, D. N., Grosseli, G. M., Mozeto, A. A., & Fadini, P. S. (2015). A 3-year study on occurrence of emerging contaminants in an urban stream of São Paulo State of Southeast Brazil. *Environmental Science and Pollution Research, 22*, 7936–7947.
16. Montagner, C. C., Sodré, F. F., Acayaba, R. D., Vidal, C., Campestrini, I., Locatelli, M. A., et al. (2019). Ten years-snapshot of the occurrence of emerging contaminants in drinking, surface and ground waters and wastewaters from São Paulo State, Brazil. *Journal of the Brazilian Chemical Society, 30*(3), 614–632.
17. Machado, K. C., Grassi, M. T., Vidal, C., Pescara, I. C., Jardim, W. F., Fernandes, A. N., et al. (2016). A preliminary nationwide survey of the presence of emerging contaminants in drinking and source waters in Brazil. *Science of the Total Environment, 572*, 138–146.
18. Peña-Guzmán, C., Ulloa-Sánchez, S., Mora, K., Helena-Bustos, R., Lopez-Barrera, E., Alvarez, J., & Rodriguez-Pinzón, M. (2019). Emerging pollutants in the urban water cycle in Latin America: A review of the current literature. *Journal of Environmental Management, 237*, 408–423.

19. Bijlsma, L., Botero-Coy, A. M., Rincón, R. J., Peñuela, G. A., & Hernández, F. (2016). Estimation of illicit drug use in the main cities of Colombia by means of urban wastewater analysis. *Science of the Total Environment, 565*, 984–993.
20. Caldas, S. S., Arias, J. L. O., Rombaldi, C., Mello, L. L., Cerqueira, M. B., Martins, A. F., & Primel, E. G. (2019). Occurrence of pesticides and PPCPs in surface and drinking water in southern Brazil: Data on 4-year monitoring. *Journal of the Brazilian Chemical Society, 30*(1), 71–80.
21. Rodríguez, A. G. P., López, M. I. R., Casillas, Á. D., León, J. A. A., & Banik, S. D. (2018). Impact of pesticides in karst groundwater. Review of recent trends in Yucatan, Mexico. *Groundwater for Sustainable Development, 7*, 20–29.
22. Mukhopadhyay, A., Duttagupta, S., & Mukherjee, A. (2022). Emerging organic contaminants in global community drinking water sources and supply: A review of occurrence, processes and remediation. *Journal of Environmental Chemical Engineering, 10*, 107560.
23. Mahmoudi, M., Aydi, A., & Ibrahim, H. (2021). Site selection for artificial recharge with treated wastewater with the integration of multicriteria evaluation and ELECTRE III. *Environmental Science and Pollution Research, 28*, 46748–46763.
24. Han, Z., HainingMa, G. S., He, L., & LuoyuWei, Q. S. (2016). A review of groundwater contamination near municipal solid waste landfill sites in China. *Science of the Total Environment, 10*, 569–570.
25. Yuan, L., Sinshaw, T., & Forshay, K. J. (2020). Review of watershed-scale water quality and nonpoint source pollution models. *Geosciences, 10*, 25.
26. Alygizakis, N. A., Urík, J., Beretsou, V. G., Kampouris, I., Galani, A., Oswaldova, M., Berendonk, T., Oswald, P., Thomaidis, N. S., Slobodnik, J., Vrana, B., & Fatta-Kassinos, D. (2020). Evaluation of chemical and biological contaminants of emerging concern in treated wastewater intended for agricultural reuse. *Environment International, 138*, 105597.
27. Tang, J., SaiWang, J. F., Long, S., LinWang, C. T., Tam, N. F., & Yang, Y. (2019). Predicting distribution coefficients for antibiotics in a river water-sediment using quantitative models based on their spatiotemporal variations. *Science of the Total Environment, 655*, 1301–1310.
28. Ntona, M. M., Busico, G., Mastrocicco, M., & Kazakis, N. (2022). Modeling groundwater and surface water interaction: An overview of current status and future challenges. *Science of the Total Environment, 846*, 157355.
29. Li, B., Yang, L., Song, X., & Diamantopoulos, E. (2023). Identifying surface water and groundwater interactions using multiple experimental methods in the riparian zone of the polluted and disturbed Shaying River, China. *Science of the Total Environment, 875*, 162616.
30. Khalidi-Idrissi, A., Madinzi, A., Anouzla, A., Pala, A., Mouhir, L., Kadmi, Y., & Souabi, S. (2023). Recent advances in the biological treatment of wastewater rich in emerging pollutants produced by pharmaceutical industrial discharges. *International Journal of Environmental Science and Technology, 20*, 11719–11740.
31. Saini, N., Dhull, P., Pal, M., Manzoor, I., Rao, R., Mushtaq, B., & Aamir, M. (2024). Algal membrane photo-bioreactors for efficient removal of emerging contaminants and resource recovery: Current advances and future outlook. *Journal of Environmental Chemical Engineering, 12*, 112669.
32. Cha, J., & Cupples, A. M. (2009). Detection of the antimicrobials triclocarban and triclosan in agricultural soils following land application of municipal biosolids. *Water Research, 43*(9), 2522–2530.
33. Heidler, J., Sapkota, A., & Halden, R. U. (2006). Partitioning, persistence, and accumulation in digested sludge of the topical antiseptic Triclocarban during wastewater treatment. *Environmental Science & Technology, 40*(11), 3634–3639.
34. Eze, C. G., Okeke, E. S., Nwankwo, C. E., Nyaruaba, R., Anand, U., Okoro, O. J., & Bontempi, E. (2024). Emerging contaminants in food matrices: An overview of the occurrence, pathways, impacts and detection techniques of per- and polyfluoroalkyl substances. *Toxicology Reports, 12*, 436–447.

35. Zhou, H., Yue, X., Chen, Y., & Liu, Y. (2024). Source-specific probabilistic contamination risk and health risk assessment of soil heavy metals in a typical ancient mining area. *Science of the Total Environment, 906*, 167772.
36. Pamuru, S. T., Forgione, E., Croft, K., Kjellerup, B. V., & Davis, A. P. (2023). Chemical characterization of urban stormwater: Traditional and emerging contaminants. *Science of the Total Environment, 813*, 151887.
37. Xiao, R., Tian, O., Ding, S., Fang, C., Xu, Z., & Chu, W. (2022). Disinfection byproducts as environmental contaminants of emerging concern: A review on their occurrence, fate and removal in the urban water cycle. *Critical Reviews in Environmental Science and Technology, 53*, 19–46.
38. Nwokediegwu, Z. Q. S., Daraojimba, O. H., Oliha, J. S., Obaigbena, A., Dada, M. A., & Majemite, M. T. (2024). Review of emerging contaminants in water: USA and African perspectives. *International Journal of Science and Research Archive, 11*(01), 350–360.
39. Tran, N. H., Reinhard, M., & Gin, K. Y.-H. (2018). Occurrence and fate of emerging contaminants in municipal wastewater treatment plants from different geographical regions-a review. *Water Research, 133*, 182–207.
40. Nika, M.-C., Alygizakis, N., Arvaniti, O. S., & Thomaidis, N. S. (2023). Nontarget screening of emerging contaminants in landfills: A review. *Current Opinion in Environmental Science & Health, 32*, 100430.
41. Tong, X., Mohapatra, S., Zhang, J., Tran, N. H., You, L., He, Y., & Gin, K. Y.-H. (2023). Source, fate, transport, and modeling of selected emerging contaminants in the aquatic environment: Current status and future perspectives. *Water Research, 217*, 118418.
42. Parashar, N., & Hait, S. (2023). Plastic rain—Atmospheric microplastics deposition in urban and peri-urban areas of Patna City, Bihar, India: Distribution, characteristics, transport, and source analysis. *Journal of Hazardous Materials, 458*, 131883.
43. Adhikari, K., Pearce, C. I., Sanguinet, K. A., Bary, A. I., Chowdhury, I., Eggleston, I., Xing, B., & Flury, M. (2024). Accumulation of microplastics in soil after long-term application of biosolids and atmospheric deposition. *Science of the Total Environment, 912*, 168883.
44. Marchuk, S., Tait, S., Sinha, P., Harris, P., Antille, D. L., & McCabe, B. K. (2023). Biosolids-derived fertilizersfertilisers: A review of challenges and opportunities. *Science of the Total Environment, 875*, 162555.
45. Dennis, N. M., Braun, A. J., & Gan, J. (2024). A high-throughput analytical method for complex contaminant mixtures in biosolids. *Environmental Pollution, 345*, 123517.
46. Choi, Y., Lee, J.-H., Kim, K., Mun, H., Park, N., & Jeon, J. (2021). Identification, quantification, and prioritization of new emerging pollutants in domestic and industrial effluents, Korea: Application of LC-HRMS based suspect and non-target screening. *Journal of Hazardous Materials, 402*, 123706.
47. Samal, K., Saswat Mahapatra, M., & Ali, H. (2022). Pharmaceutical wastewater as Emerging Contaminants (EC): Treatment technologies, impact on environment and human health. *Energy Nexus, 6*, 100076.
48. Stuart, M., Lapworth, D., Crane, E., & Hart, A. (2012). Review of risk from potential emerging contaminants in UK groundwater. *Science of the Total Environment, 416*, 1–21.
49. Camargo-Perea AL., Rubio-Clemente A., Peñuela GA. (2020). Use of ultrasound as an advanced oxidation process for the degradation of emerging pollutants in water Water, 12 (4), p. 1068
50. Khadam, I. M., & Kaluarachchi, J. J. (2003). Multi-criteria decision analysis with probabilistic risk assessment for the management of contaminated ground water. *Environmental Impact Assessment Review, 23*, 683–721.
51. López-Serna, R., Petrovic, M., & Barcelo, D. (2011). Development of a fast instrumental method for the analysis of pharmaceuticals in environmental and wastewaters based on ultra high performance liquid chromatography (UHPLC)–tandem mass spectrometry (MS/MS). *Chemosphere, 85*, 1390–1399.
52. Gros, M., Rodríguez-Mozaz, S., & Barcelo, D. (2012). Fast and comprehensive multi-residue analysis of a broad range of human and veterinary pharmaceuticals and some of their metabo-

lites in surface and treated waters by ultra-high-performance liquid chromatography coupled to quadrupole-linear ion trap tandem mass spectrometry. *Journal of Chromatography A, 1248*, 104–121.

53. Das, P., Arias, V. A., Kambala, V., Mallavarapu, M., & Naidu, R. (2024). Remediation of perfluorooctane sulfonate in contaminated soils by modified clay adsorbent—A risk-based approach. *Water, Air, and Soil Pollution, 224*, 1714.
54. Morin-Crini, N., Lichtfouse, E., Liu, G., Balaram, V., Ribeiro, A. R. L., Lu, Z., & Crini, G. (2022). Worldwide cases of water pollution by emerging contaminants: A review. *Environmental Chemistry Letters, 20*(4), 2311–2338.
55. Shamshad Khan, M., Naushad, M. G., Iqbal, J., & Alfadul, S. M. (2022). Emerging contaminants of high concern for the environment: Current trends and future research. *Environmental Research, 207*, 112609.
56. Ternes, T. A., Andersen, H., Gilberg, D., & Bonerz, M. (2002). Determination of estrogens in sludge and sediments by liquid extraction and GC/MS/MS. *Analytical Chemistry, 74*, 14.
57. Sabourin, L., Duenk, P., Bonte-Gelok, S., Payne, M., Lapen, D. R., & Topp, E. (2012). Uptake of pharmaceuticals, hormones and parabens into vegetables grown in soil fertilized with municipal biosolids. *Science of the Total Environment, 431*, 233–236.
58. Langdon, K. A., Warne, M. J., Smernik, R. J., Shareef, A., & Kookana, R. S. (2011). Selected personal care products and endocrine disruptors in biosolids: An Australia-wide survey. *Science of the Total Environment, 409*(6), 1075–1081.
59. Radjenović, J., Jelić, A., Petrović, M., & Barceló, D. (2009). Determination of pharmaceuticals in sewage sludge by pressurized liquid extraction (PLE) coupled to liquid chromatography-tandem mass spectrometry (LC-MS/MS). *Analytical and Bioanalytical Chemistry, 393*, 1685–1695.
60. Guerra, P., Kim, M., Shah, A., Alaee, M., & Smyth, S. A. (2014). Occurrence and fate of antibiotic, analgesic/anti-inflammatory, and antifungal compounds in five wastewater treatment processes. *Science of the Total Environment, 473–474*, 235–243.
61. Albero, B., Sánchez-Brunete, C., Miguel, E., Aznar, R., & Tadeo, J. L. (2014). Determination of selected pharmaceutical compounds in biosolids by supported liquid extraction and gas chromatography—Tandem mass spectrometry. *Journal of Chromatography A, 1336*, 52–58.
62. Gottschall, N., Topp, E., Metcalfe, C., Edwards, M., Payne, M., Kleywegt, S., Russell, P., & Lapen, D. R. (2012). Pharmaceutical and personal care products in groundwater, subsurface drainage, soil, and wheat grain, following a high single application of municipal biosolids to a field. *Chemosphere, 87*(2), 194–203.
63. Morais, S. A., Delerue-Matos, C., Gabarrell, X., & Blánquez, P. (2013). Multimedia fate modeling and comparative impact on freshwater ecosystems of pharmaceuticals from biosolids-amended soils. *Chemosphere, 93*(2), 252–262.
64. Sponberg, A. L., & Witter, J. D. (2008). Pharmaceutical compounds in the wastewater process stream in Northwest Ohio. *Science of the Total Environment, 397*(1–3), 148–157.
65. Ding, Y., Zhang, W., Cheng, G., Xagoraraki, I., & Li, H. (2011). Determination of pharmaceuticals in biosolids using accelerated solvent extraction and liquid chromatography/tandem mass spectrometry. *Journal of Chromatography A, 1218*(1), 10–16.
66. Miao, X.-S., Yang, J.-J., & Metcalfe, C. D. (2005). Carbamazepine and its metabolites in wastewater and in biosolids in a municipal wastewater treatment plant. *Environmental Science & Technology, 39*(19), 7469–7475.
67. Golet, E. M., Strehler, A., Alder, A. C., & Giger, W. (2002). Determination of fluoroquinolone antibacterial agents in sewage sludge and sludge-treated soil using accelerated solvent extraction followed by solid-phase extraction. *Analytical Chemistry, 74*(21), 5455–5462.
68. Lindberg, R. H., Wennberg, P., Johansson, M. I., Tysklind, M., & Andersson, B. A. (2005). Screening of human antibiotic substances and determination of weekly mass flows in five sewage treatment plants in Sweden. *Environmental Science & Technology, 39*(10), 3421–3429.
69. Stasinakis, A. S., Gatidou, G., Mamais, D., Thomaidis, N. S., & Lekkas, T. D. (2008). Occurrence and fate of endocrine disrupters in Greek sewage treatment plants. *Water Research, 42*(6–7), 1796–1804.

70. McAvoy, D. C., Schatowitz, B., Jacob, M., Hauk, A., & Eckhoff, W. S. (2009). Measurement of triclosan in wastewater treatment systems. *Environmental Toxicology and Chemistry, 21*(7), 1323–1329.
71. Morales, S., Canosa, P., Rodríguez, I., Rubí, E., & Cela, R. (2005). Microwave assisted extraction followed by gas chromatography with tandem mass spectrometry for the determination of triclosan and two related chlorophenols in sludge and sediments. *Journal of Chromatography A, 1082*(2), 128–135.
72. Chu, S., & Metcalfe, C. D. (2007). Simultaneous determination of triclocarban and triclosan in municipal biosolids by liquid chromatography tandem mass spectrometry. *Journal of Chromatography A, 1164*(1–2), 212–218.
73. Ying, G.-G., & Kookana, R. S. (2007). Triclosan in wastewaters and biosolids from Australian wastewater treatment plants. *Environment International, 33*(2), 199–205.
74. Moliner-Martínez, Y., Ribera, A., Eugenio Coronado, P., & Campíns-Falcó. (2011). Preconcentration of emerging contaminants in environmental water samples by using silica-supported Fe3O4 magnetic nanoparticles for improving mass detection in capillary liquid chromatography. *Journal of Chromatography A, 1218*, 2276–2283.
75. Thomaidis, N. S., Asimakopoulos, A. G., & Bletsou, A. A. (2012). Emerging contaminants: A tutorial mini-review. *Global NEST Journal, 14*(1), 72–79.
76. Senthil Rathi, B., Senthil Kumar, P., & Show, P.-L. (2021). A review on effective removal of emerging contaminants from aquatic systems: Current trends and scope for further research. *Journal of Hazardous Materials, 409*, 124413.
77. Pereira, L. C., Oliveira, A., de Souza, M., Bernardes, F. F., Pazin, M., Tasso, M. J., Pereira, P. H., & Dorta, D. J. (2015). A perspective on the potential risks of emerging contaminants to human and environmental health. *Environmental Science and Pollution Research, 22*, 13800–13823.
78. Bolong, N., Ismail, A. F., Salim, M. R., & Matsuura, T. (2009). A review of the effects of emerging contaminants in wastewater and options for their removal. *Desalination, 239*, 229–246.
79. Smital, T. (2008). Acute and chronic effects of emerging contaminants. In *The handbook of environmental chemistry* (Vol. 5, pp. 105–142).

Part II
Detection Techniques

Chapter 3
Detection Techniques for Comprehensive Identification of Emerging Contaminants

Nitish Karn, Vivek Kumar, Shivali Bhandari, Atin Kumar, and Sharad Sachan

1 Introduction

Emerging contaminants (ECs) can be described as chemicals, compounds, or materials that have entered the environment in the past decade but for which there are no environmental standards or limited scientific data to compare their potential impact on the environment and the health of living organisms. It includes various enforcing agents and compounds that are spurred off from industrial, horticultural, and domestic tasks; these compounds are being increasingly recognized because of their enhanced ability to detect these contaminants [1]. The category of ECs can include a wide range of substances, such as pharmaceuticals and personal products, PPCPs, endocrine-disrupting chemicals, EDCs, microplastics, pesticides, industrial chemicals, and even some that occur naturally but are brought into the environment by humans, as shown in Fig. 3.1 [2].

Among all the ECs identified, the most worrisome are those that are used in pharmaceuticals and personal care. These include drugs and pharmaceuticals, UV protectants, fragrances, body care products and others, which find their way into the environment through sewer effluents, leachate wastewater from agriculture, and direct disposal. Pharmaceuticals are of greatest concern in water systems because they are bioactive and can have effects on organisms in aquatic ecosystems and even in the food chain. For example, the way in which antibiotics are made available through water increases the risk of antibiotic-resistant bacteria being dangerous to

N. Karn · V. Kumar · S. Bhandari · A. Kumar (✉)
School of Agriculture, Uttaranchal University, Dehradun, Uttarakhand, India

S. Sachan
School of Agriculture, Lovely Professional University, Phagwara, Punjab, India
e-mail: sharad.19461@lpu.co.in

M. K. Jindal et al. (eds.), *Emerging Contaminants in Water*, Emerging Contaminants and Associated Treatment Technologies,
https://doi.org/10.1007/978-3-031-89591-3_3

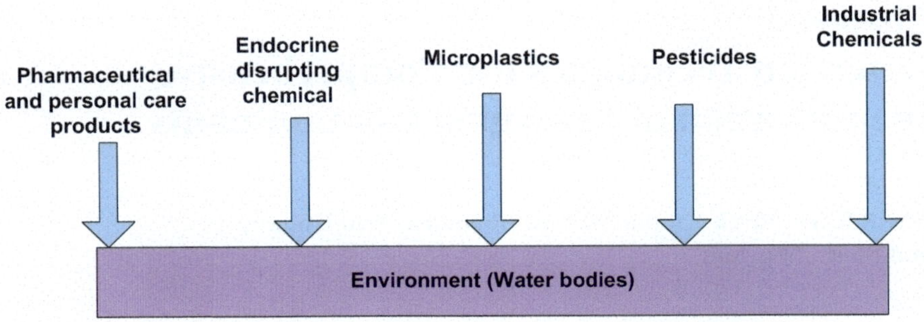

Fig. 3.1 Movement of different emerging contaminants in water bodies

human life [3]. Another essential type of EC is EDCs or endocrine-disruptive chemicals, which are agents that interfere with the body's hormonal system. These substances hinder the usual operations of the endocrine system, the body program that controls the use of hormones in people and animals. EDCs are associated with different diseases, such as reproductive disorders, developmental abnormalities, and cancers. Some EDCs include BPA (bisphenol A), phthalates, and several pesticides, nearly all of which infiltrate water and soil through industrial, agricultural, and household use. This type of exposure has been increasing due to evidence suggesting that they are injurious to wildlife and even human health [4]. They also include microplastics, which are now a cause for increased concern because of their effects on the environment. Microplastics refers to plastic pieces that are less than 5 mm in size, which are formed by the breakdown of larger plastics or are intentionally manufactured as small as they are used in items such as scrubbing powders and detergents. Current evidence shows that plastically bound particles are chemically stable, resistant to dilution, can be ingested by organisms in the water chain, can accumulate in body tissues, and may be toxic. Microplastic pollution of water, soil, and air has become a great concern to society as research aims to determine the extent of pollution and the possibility of accumulation [5]. Pesticides have long been recognized as ECs for the same purpose, and many of the newer, more efficient forms are either bioaccumulative or more toxic than earlier formulations. Pesticides are useful for present-day farming, but their byproducts linger in the soil and water, causing harm to wildlife and humans. For example, neonicotinoid insecticides that are commonly used in crop production, especially honeybees, have been blamed for pollinator loss. Because these chemical toxins accumulate in the environment, constant identification and evaluation of pesticide residues to measure their effects are crucial [6].

These ECs are now present in very high concentrations in environmental matrices, including soil, water, and air, and pose serious problems for both the environment and human health. Concentration levels in the environment are normally low; therefore, their detection is not easy, and many ECs are present in ecosystems, resulting in long-term exposure. This persistence, together with the ability of many ECs to bioaccumulate in biological systems, has consequently given rise to increasing concerns about the synergistic effects of such compounds on both environmental and biological health. High-resolution mass spectrometry (HRMS), liquid chromatography–mass spectrometry (LC–MS) and gas chromatography–mass spectrometry (GC–MS) are mandatory for the identification and quantification of ECs in environmental samples [7]. However, sample preparation via these methods might be lengthy and sometimes labor intensive. Moreover, many ECs, particularly when they react in the environment, are not sufficiently studied, and new data are required to create more efficient and unified detection protocols. The use of unselective strategies, which may include the recognition of new peaks and their transformation products, is becoming more significant, as such approaches allow the consideration of a wider range of contaminants and potential hazards [3, 8].

With respect to the limitations of the present study, it is crucial that future studies aim at refining detection strategies, establishing the long-term effects of ECs on the environment and determining safe levels of contaminants or formulating appropriate policies that may assess the spread of pollutants across the world. Such attempts to define and assess the extent of ECs will also be important for enabling risk awareness and mitigation to populations as well as helping policy makers develop environmental laws and standards to preserve ecosystems.

2 Analytical Techniques for the Detection of Emerging Contaminants

The identification and quantification of ECs in environmental samples require highly selective and sensitive methods to identify and quantify these substances, which are present at very low concentrations. Very high selectivity and sensitivity are required to detect and quantify ECs among the large number of other substances. This process starts with sample preparation, where complicated and bulky elements of the sample are removed through methods such as SPE (solid-phase extraction), LLE (liquid–liquid extraction), UAE (ultrasound-assisted extraction), microextraction techniques such as MSPE (microsolid-phase extraction), or SPME (solid-phase microextraction) to increase the concentration of the ECs and minimize extra interferences. Analytical techniques play a major role, with chromatography now in its gas and liquid forms and mass spectrometry (MS), including tandem MS/MS and high-temperature-resolution MS in particular, at the very heart of the separation and quantitation processes. Instrumental methods such as capillary electrophoresis and nuclear magnetic resonance spectroscopy provide added convenience for the examination of ECs; samples require ionic differentiation, structural elucidation, and so forth. The approach is comprehensive, including screening of nontarget and suspected compounds for any unknown or suspected contaminant, whereas enhancement of quality control and method validation makes the results as accurate as possible and conforms to regulating agency standards. Finally, advanced analytical methods are applied for data handling and analysis to preserve and increase the quality of the data, as well as to extract useful information precisely. This integrated methodology is highly compliant with regulatory standards and advances what can be measured and learned about the ECs in the environment [9].

2.1 *Gas Chromatography–Mass Spectrometry (GC–MS)*

The primary advantage of GC–MS is that it is highly suitable for determining volatile and semivolatile organic compounds. One is the merging of the separation techniques of gas chromatography with the excellent detection capability of mass spectrometry [10]. New and improved studies have also provided further support for the effectiveness of GC–MS in numerous environmental and analytical chemistry applications and have created a new HS-SPME-GC–MS method for the detection of volatile ECs in fruit, with LODs (limits of detection) and LOQs (limits of quantitation) of 0.011–0.033 ng/g and 0.037–0.098 ng/g, respectively. Some researchers have established a multiresidue method for the analysis of 17 ECs in vegetables via the use of an ultrasound-assisted matrix solid-phase dispersion, with the concentrations ranging from 0.1 to 0.4 ng/g in the various vegetables studied. Some researchers have applied the portability of GC–MS in fire investigations and the identification of hazardous organic compounds via this procedure, emphasizing its ability to establish concentrations of volatile and semivolatile alternatives. In another study, micro-SPE/GC–MS was employed for the determination of geosmin and 2-methyl isoborneol in drinking water, with distinguishabilities of 2.0 ng/g and 4.3 ng/g, respectively [9]. However, GC–MS, which is highly sensitive and specific for detecting and analyzing identified ECs, has several limitations. The major disadvantage of GC–MS is the applicability of the method only for the determination of volatile and semivolatile compounds. Polar or high-molecular-weight emerging contaminants may be less efficient with GC–MS analysis if they are not first chemically derivatized; the alteration process can be complex and may lead to undesirable errors. However, the technique entails proper calibration and main tenance of its associated equipment since extreme accuracy is paramount; however, it is very time consuming and costly. The technical sophistication of the GC–MS systems also involve a high skill level among the operating personnel and the data analyst; this can be a challenge in low-income settings where it is difficult for qualified personnel to come by. Moreover, matrix effects can influence the analysis in cases where other substances in the sample matrix interfere with the identification of target contaminants, which requires complex sample preparation. However, the current operational and technical limitations of GC–MS, as demonstrated in this work, limit its applicability in the broad monitoring of ECs.

2.1.1 Sample Preparation Process for Gas Chromatography–Mass Spectrometry (GC–MS)

Most emerging contaminants are frequently analyzed via GC–MS; therefore, sample preparation methods are critical starting points at which contaminants are prepared appropriately measurable. New contaminants include a wide range of compounds with relatively unknown consequences for human health and the environment, including pharmaceutical products, personal care products, and industrial products, among others. The following is a detailed explanation of the sample preparation methods specifically tailored for GC–MS analysis [11].

2.1.1.1 Liquid Extraction Methods

This method is intended to extract contaminants from solids and semisolid mixtures with liquid solvents that are suitable for GC–MS analysis. The different liquid extraction methods are shown in Fig. 3.2.

Sources of extraction techniques include LLE, Soxhlet extraction, SLE, MAE, USE, ASE, and SFE, which are used to isolate contaminants prior to GC–MS analysis. LLE is a technique in which samples are extracted with water-immiscible solvents in which contaminants partition to the organic phase for analysis. Soxhlet extraction involves the cyclic use of a solvent in solid food samples with unique effectiveness in the extraction of volatile and semivolatile interferents. Likewise, in SLE, solvents are employed to desorb pollutants from contaminated solid matrices, followed by gas chromatography–mass spectrometry (GC–MS) injection of the extracted solution, which could need concentration or purification. MAE, which is usually performed with microwave energy, heats the solvent and sample, increasing the extraction efficiency of the vaporized substances for GC analysis. With the help of ultrasonic waves, USE improves the mobility of solvents and facilitates the extraction of contaminants from the food matrix. ASE uses a combination of high pressure and high temperature to effectively eliminate contaminants, especially TVOCs and SVOCs, from the sample surface. Finally, SFE employs supercritical fluids, most commonly supercritical carbon dioxide, to extract compounds that are nonpolar in nature with high discriminating power for the solvent used as well as less solvent usage than other methods do. In combination, these techniques provide adequate isolation of analytes to facilitate proper identification of contaminants [11, 12].

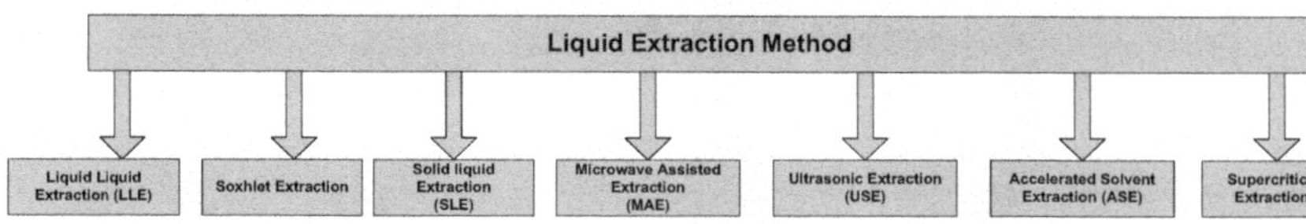

Fig. 3.2 Different liquid extraction methods for the detection of emerging contaminants

2.1.1.2 Solid Extraction Methods

These methods involve the extraction of contaminants from samples of solid-phase materials that can be analyzed via GC–MS [13]. Figure 3.3 shows different solid extraction methods for the identification and detection of solid contaminants.

SPE and its modifications are applied to concentrate contaminants in different matrices before their analysis via GC–MS. SPE involves the flow of a liquid sample through a solid matrix, where target analytes are sequestered, while interferences are washed out, and analytes are desorbed with a solvent suitable for GCMS. Micro-SPE is essentially a small-scale version of SPE, as it operates with lower volumes of sample and adsorbent phases and is particularly useful for the extraction of relatively active compounds. Dispersive-MSPE (DMSPE) improves extraction efficiency because of the direct dispersion of the sorbent in the sample matrix, which speeds up the isolation of analytes, followed by GC–MS analysis. Finally, matrix solid–phase dispersion (MSPD) combines a sample with a dispersive matrix and SPE, and this method helps in the extraction of analytes from complex food matrices, as observed in multiple complicated samples. Stir bar sorptive extraction (SBSE) is also a solventless approach; the stir bar is coated with a polymer that extracts nonpolar compounds from the liquid matrix. SBSE is a powerful and eco-friendly technique because these analytes are thermally desorbed and analyzed via GC–MS.

This is why the choice of extraction technique mainly depends upon the nature of the sample matrix and the properties of the target analytes as well as the desired limit of detection. Moreover, owing to their efficiency and capacity for handling multiple residue analysis, QuEChERS (Quick, Easy, Cheap, Effective, Rugged and Safe) is widely implemented in current approaches. These techniques are intended to offer concentration, preconcentration, and purification of specific contaminants, which enables instrumental identification as well as quantification via GC–MS technology [14].

2.2 *Liquid Chromatography–Mass Spectrometry (LC–MS)*

Techniques such as LC, LC–MS, and LC–MS/MS are very efficient for the identification of low levels of contaminants. Today's developments have shown the enhanced suitability of LC, LC–MS, and LC–MS/MS in MRAs of environmental

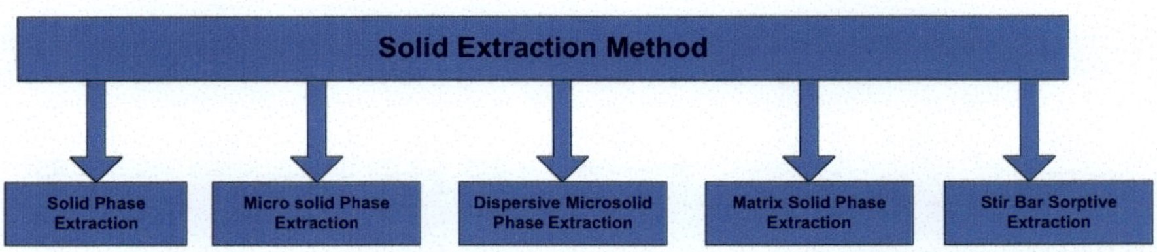

Fig. 3.3 Different solid extraction methods for the detection of solid emerging contaminants

samples for qualifying ECs, including bisphenols, phthalates, and pharmaceuticals. A new reversed-phase high-performance liquid chromatography (RP-HPLC) methodology was developed for the individual identification of products in connection with the evaluation of microalgae bioremediation efficiency, with LODs and LOQs equal to 0.03 and 0.10 ng/g for paracetamol, 0.03 and 0.09 ng/g for ibuprofen, and 0.04 and 0.13 ng/g. High-throughput extraction via freeze-centrifugation was customized via HPLC–MS/MS to quantify 11 ECs in three model crops, corn, tomato, and wheat, where the LOD range was from 0.01 to 2.0 ng/g. Compared with other methods, LC–MS/MS has been widely used because of its ability to distinguish analytes against complex backdrops. LC–MS/MS has been widely used since it can sensitively detect analytes in complex matrices. LC–MS/MS detection and analysis of ECs are characterized by high sensitivity and selectivity; however, this method has several drawbacks. The first issue is the matrix effects that can affect LC–MS–MS/MS analysis and generate highly variable and inaccurate data. These effects arise when other species in the sample matrix cause suppression or enhancement of the ionization of the analytes, in turn distorting quantitation. Moreover, the performance of LC–MS/MS is very accurate for most compounds, but the applicability of LC–MS/MS is quite sensitive to the chemical nature of some ECs, such as very large-molecular-weight molecules or highly polar molecules, which may require special treatment or derivatizations [14].

2.2.1 Sample Preparation Process for Liquid Chromatography–Mass Spectrometry (LC–MS)

Sample preparation techniques for LC–MS/MS analysis of emerging contaminants remain essential for precision of the results obtained. Some commonly used sample preparation techniques are shown in Fig. 3.4.

Different extraction and sample preparation techniques are applied to obtain high and selective enrichment of target compounds for efficient determination of contaminants via LC–MS/MS. SPE is an extraction technique where sample components with affinity to the solid phase and eluting solvents are selectively adsorbed on the stationary phase while interferences are washed out with appropriate solvents. The analytes are then eluted in a suitable solvent, making SPE useful for concentration and purification, especially from environmental samples. Liquid–liquid extraction (LLE) depends on the distribution of analytes between two immiscible liquids in which solutes are selectively dissolved in the selected solvent.

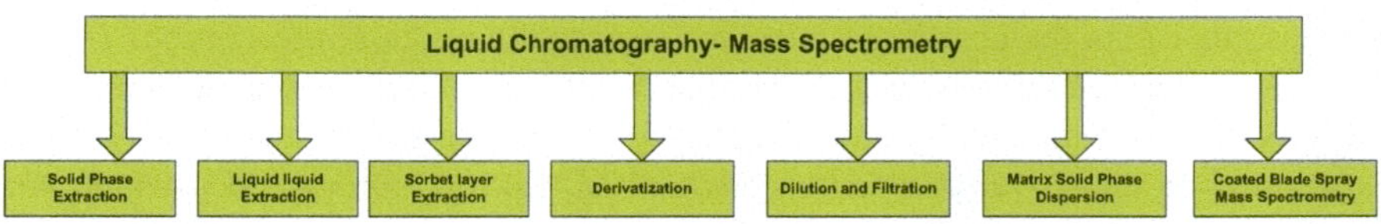

Fig. 3.4 Sample preparation process for liquid chromatography and mass spectrometry

However, the use of the LLE method may be time consuming and may require several cycles before the pure product is obtained. SLE is similar to LLE, but instead of an organic layer holding the analyte, a sorbent layer such as diatomaceous earth holds it. The analytes are then eluted with an organic solvent, and the method is better in terms of throughput and extraction purity than LLE. Derivatization makes the analytes more amenable to separation and increases the ionization of compounds, which are polar or of low volatility to MS. In general, dilution and filtration are easy and efficient processes for handling samples to minimize sediment content prior to LC–MS/MS analysis, for instance, the use of a mobile phase or water for dilution of urine, serum, and other similar samples and subsequent sample-centrifugation or filtration. MSPD integrates sample preparation in the form of SPE, in which sorbents are employed for the extraction of analytes of interest without containing solid matrix interferences. Another relatively recent method, called coated blade spray mass spectrometry, where direct sampling and introduction of complex samples are performed with the minimum and often no pretreatment, is used, as the coated blade is used for the extraction of analytes. The method used in this process depends on the characteristics of the compounds and their number in the matrix. In general, sample preparation can influence the results of LC–MS–MS/MS analytical processes, particularly for complex biological matrices; hence, efficient sample preparation is important for reducing matrix interference [15].

2.3 *Enzyme-Linked Immunosorbent Assay (ELISA)*

ELISA is a rather popular analytical method used to determine specific impurities in different samples. This technique employs antibody–antigen reactions to measure and identify analytes of interest, such as selected pesticides, hormones, and certain metals. Some of the contemporary uses of the ELISA technique include testing water quality to determine if it contains and/or represents substances such as pesticides, herbicides, and endocrine disruptor components. Silva et al. used ELISA to trace the concentrations of 17α-ethinylestradiol and 17β-estradiol, carbamazepine, cetirizine, and caffeine in water samples. The preparation of ELISA reagents available for purchase as kits serves as an efficient method for easily and effectively assessing various factors, including hormone-disrupting factors, pesticides, and other regulatory compounds. Although ELISA has received significant interest in environmental analysis because of its high specificity, simplicity, and economy, the method has the following limitations. The first is the possibility of cross-reactivity, where the immunoreactivity of the antibodies used in the assay reacts with substances other than the analyte of interest, thus producing faux information. This is especially true for multiple related compounds often found in complicated matrices, such as the environment. Additionally, ELISA is comparatively less sensitive than more modern analytical procedures, such as LC–MS/MS or GC–MS, which are capable of identifying target substances at significantly lower concentrations. Additionally, ELISA commonly demands the presence of specific antibodies for each target contaminant and therefore can be easily applied only for the detection of further novel or unidentified ECs if specific antibodies already exist or if they can be produced. In addition, the results achieved by ELISA based on quantitative measurements may be affected by the sample preparation technique and matrix interference, which may affect the data quality and reliability and the interassay reproducibility [16].

2.3.1 Sample Preparation Methods for Enzyme-Linked Immunosorbent Assays (ELISAs) Related to Emerging Contaminants

The major techniques used for sample preparation for enzyme-linked immunosorbent assays (ELISAs) of emerging contaminants include the following steps. A detailed explanation of the sample preparation methods typically used is shown in Fig. 3.5.

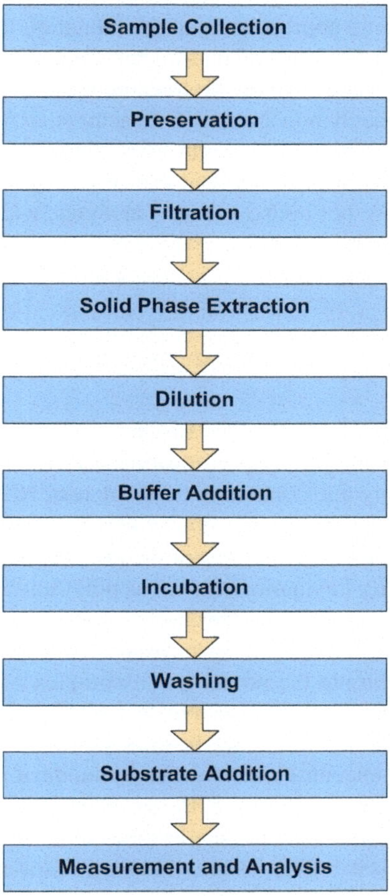

Fig. 3.5 Procedure for the detection of emerging contaminants via ELISA

Sample preparation for the detection of emerging contaminants in water samples via ELISA is highly sensitive for providing accurate and reproducible data. The procedure starts from sample collection, ensuring that the containers are free of contaminant impurities. To reduce chemical leaching, high-quality glass or plastics are used to prepare food. Courtesy of conservation techniques such as freezing or the use of a preservative chemically, sample degradation prior to analysis is averted. Filtration is used to eliminate compacted suspended materials via membrane filters, whereby particle sizes, such as 0.455 or 0.2 μm, are expected. This is followed by the removal of sediment and organic matter, which, if present, interferes with the assay. In the case of large samples, the extraction of contaminants through concentration is performed via a method known as SPE. Water is passed through a cartridge that consists of material that selectively removes contaminants of interest. Additionally, samples may need to be concentrated by removing a specific solvent from the sample through a rotary evaporator or nitrogen stream. When concentrations are too high, dilution of the samples with an appropriate buffer (e.g., PBS) brings the contaminant levels to a range that can be detected with an ELISA kit. It also helps buffer addition in preparing the sample to ensure that the contaminants bind appropriately to the kit antibodies, whereby mixing must be consistent.

During incubation, the prepared sample is again tested using antibodies furnished with the kit to capture the target contaminants. The temperature and time of incubation are sensitive factors that must be considered in accordance with the manufacturer's instructions so that the optimum assay can be obtained. Another step involves subsequent washing with a buffer to minimize contact with antibodies that are not bound to target analytes, which contributes to low background noise in the assay [17, 18].

2.4 *Advanced Spectroscopic Techniques*

Nuclear magnetic resonance (NMR) and Fourier transform-infrared (FTIR) spectroscopy provide information on the structure and concentration of the identified organic contaminants of interest. NMR spectroscopy is useful for the identification and characterization of closely related contaminants and precise determination of the concentrations of the target components in complex matrices. It is used to identify contaminants such as polyfluoroalkyl substances, olmesartan acid, and heparin. Likewise, for the identification and determination of contaminants such as nanoplastics, FTIR spectroscopy is rather useful. FTIR and NMR are crucial in determining the properties of complex organic material contaminants that are found in various materials. NMR, which is used for the precise determination of the molecular structure, has the disadvantage that relatively large amounts of sample and high concentrations of the compounds of interest are necessary to obtain sufficient signal intensities. Often, the problem consists of determining trace contaminants in environmental samples. Furthermore, NMR has lower sensitivity than mass spectrometry-based methods and its use is time- and cost-consuming; therefore, it is not widely used in the routine monitoring of the environment. In contrast, FTIR is normally preferable because of the noninvasive nature of the analysis as well as the short turnaround time. FTIR may have problems with sample complexity, and in many cases, water and other compounds that are always present in the environment mask the spectra of contaminants. Dissection, polarized, fluorescence, scanning electron, and atomic force microscopy are the most widely used methods to quantify micro- and nanoplastics. However, these methods have several disadvantages related to the analysis of small particles, and these methods are ineffective at obtaining the complete range of results. The larger sample sizes required to provide meaningful results proportional to the number of cells present in a given tissue have become a significant issue when attempting to use microscopy in combination with other chemical analysis techniques such as spectroscopy. Additionally, new methods of analytics that have emerged late present some other viable methodologies. For example, using holography in a microscope form, we can directly image microplastics in raw water samples and distinguish between microplastics and diatoms and between the different sizes of microplastics, shapes, and types of plastics. A further evolution of analytical techniques or their combinations in a single analytical instrument or hybrid with conventional and advanced microscopy can present prospects for eliminating the existing drawbacks in the detection of micro- and nanoplastics. In this context, researchers can enhance the detection and characterization of micro- and nanoplastics through these innovations [19].

2.4.1 Sample Preparation for Advanced Spectroscopic Techniques

Figure 3.6 shows that different sample preparation procedures are used for the detection of emerging contaminants via advanced spectroscopic techniques.

Sample preparation is important for determining the concentration of possible contaminants in samples that may be in water, soil, air, etc. The procedure starts with sample collection, and the containers used to store the samples are kept free from contamination. The next step is prefiltering to remove large particles and Werne Whatman filters for clarity to permit further analysis. To increase the ability to find trace contaminants, concentration methods such as solid-phase extraction or liquid–liquid extraction are used during the sample preparation stage. Conformity adjustments include making changes to the sample to best prepare it for the desired technique, for example, by preventing metals from precipitating at low pH values. In addition, the alteration of contaminants to forms that would enhance their identification via methods such as GC–MS is needed. The sample is then purified by filtration or centrifugation to selectively reject even finer particles so that the sample can be subjected to instrumental analysis.

Dilution or matrix matching modifies the concentration of a sample so that the analysis is not skewed by any given factor. For example, the introduction of salts can secure the sample matrix for the analytical technique of interest, which was practiced in the research employing ICP–MS. Both storage and transportation are critical, as containers (amber glass for the compounds sensitive to light) and

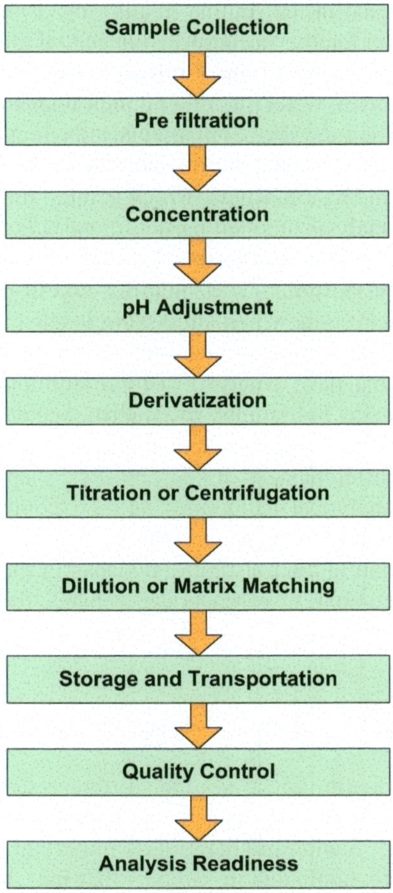

Fig. 3.6 Procedure for the detection of emerging contaminants via advanced spectroscopy

conditions (refrigeration for volative samples) also affect sample stability. Quality control accredits the capacity of analytical precision through blanks, standards, and duplicates that try to offer ratios of possible contamination or mistakes. Finally, analysis is performed by using prepared samples in spectroscopic or chromatographic methods, such as UV–Vis, FTIR, or mass spectrometry, to identify contaminants and assess their concentration levels.

The use of this systematic method also ensures that accurate analysis of the contaminant is performed across any of the analytical methods.

2.5 *Biosensors*

Innovative biosensor solutions are currently being developed to allow easy onsite screening of contaminating agents. New generation biosensors play a role in biological reactions for accurate measurement and hence can be considered a trans-formative process for several environmental and health contaminants, including microcystins, mycotoxins, proteins, and RNAs. Graphene-based biosensors have shown potential in the sensing of chemical contaminants and profiling for portable use. Furthermore, improvements in electrochemical biosensing have improved the accuracy of screening for contaminants present in foods and water sources. For contamination monitoring, the high selectivity of biological recognition is used in these emerging platforms. Its development enables testing at various times on different commodities, such as crop produce, food items, and water, to be used in other production processes. Biosensors are contemporary analytical tools that exploit a biological recognition element associated with a physicochemical transducer to detect and measure ECs in different ecosystems. First, despite the great advantages of biosensors, such as their short response time, mobility, and expanded applications of real-time monitoring, multiple challenges arise in the biosensing of ECs. A major advantage is the ability of the environment to alter characteristics such as temperature and pH, which causes instability and reduces the functionality of the biological component to produce reliable results. Furthermore, biosensors have low selectivity; they need to be fine-tuned for the kind of pollutants being detected, which is time-consuming and costly. Interference with nontarget substances within a sample also causes variability and contributes to false positive or negative results that are only true of the selected target analyte in a rich matrix [20].

2.5.1 Sample Preparation Techniques for Biosensors

Figure 3.7 shows various sample preparation techniques that use biosensors for the estimation of emerging contaminants.

Aptamer-based biosensors are advanced, miniaturized analytical devices capable of the selective identification of emerging pollutants due to target complementarity and new approaches to signal conversion. The process starts with the choice of

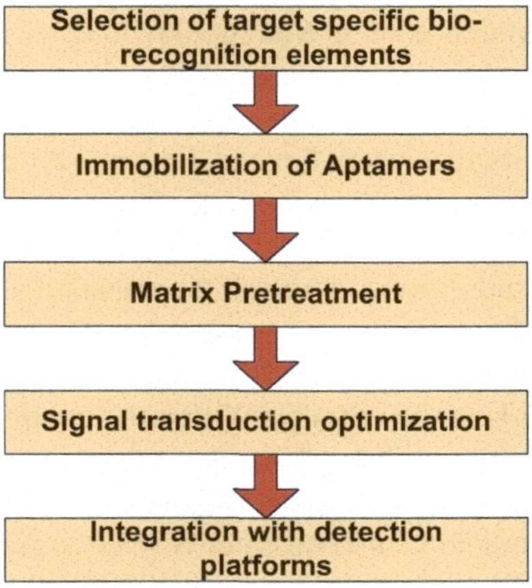

Fig. 3.7 Procedure for estimation of emerging contaminants via biosensors

target-specific biorecognition elements, such as aptamers specific to pharmaceuticals, endocrine agents, pesticides, and heavy metals. These aptamers are generated via Systematic Evolution of Ligands by Exponential Enrichment (SELEX) to guarantee high selectivity and a binding constant for the target contaminating species. The aptamers are then immobilized on the substrate base, such as electrodes, nanoparticles or nanomaterials. It can be performed chemically or through functionalization (e.g., through thiol group conjugation) or the use of linkers, such as streptavidin-biotin, for the attachment of aptamers onto the detection platform. The next step is matrix pretreatment, which involves the preparation of environmental samples, whether water, soil or food, by filtration, changing the pH or adding agents for the extraction of pollutants. For the sake of measurable detection, signal transduction must be optimized. These aptamers contain fluorescent dyes, metallic nanoparticles, or quenchers to generate signals. Methods such as fluorescence, colorimetric assays, or electrochemical procedures reflect binding processes with high reliability in terms of measurable signals. Finally, biosensor system integration follows an application with detection platforms that reveal the conversion of interactions between the aptamer and target pollutants. This approach results in high sensitivity, selectivity, and low interference, thus allowing rapid identification and determination of the concentration of pollutants in any matrix.

This method presents one of the most promising approaches for the detection of new contaminants in water and food matrices while using new concepts in molecular recognition and signal enhancement [21].

3 Case Studies

3.1 *Detection of Pharmaceuticals in Urban Wastewater Treatment Plants*

The monitoring study involved four seasons for a one-year monitoring period from March 2006 to March 2007 to determine the residues of 11 PPCPs (pharmaceutical and personal care products) from different therapeutic groups. The study was carried out in the municipal and hospital WWTPs (water treatment plants) of Ioannina city in western Greece. Like most pharmaceuticals, the extraction method employed prior to analysis is solid-phase extraction (SPE). The samples were subjected to gas chromatography–mass spectrometry (GC–MS). The analysis of the monitoring study revealed that all the intended compounds were present in the wastewater samples. The concentrations identified in the studied municipal WWTP varied from 0.3 (±0.1) to 164.4 (±26.7) g/L in the influent as well as from 0.5 (± 0.2) to 13.9 (± 2.8) g/L in the effluent. Within the hospital, the concentrates of WWTP varied between 0.6 and 70.1 g/L in the influent and 0.5 and 14.6 g/L in the effluent. The mean removal efficiencies of the three parameters varied from 13–97% to 9–87% in the municipal and hospital WWTPs, respectively. The removal efficiencies were generally higher in municipal WWTPs than in hospital WWTPs [21].

3.2 *Transboundary River Water Pesticide Pollution in Cote d'Ivoire, West Africa*

For this purpose, the concentrations of pesticides in water samples collected from the Bia, Tanoé, and Comoe Rivers in the western African state of Côte d'Ivoire were investigated. Pesticides were extracted via solid-phase microextraction (SPME) via the US EPA Method 3510C, and 500 mL of water obtained after being filtered through 0.45-μm pore filters was successively shaken three times with an equal volume of dichloromethane in a 2-L separating funnel. The extracts were spiked with surrogate standards for recovery prior to concentration by a rotary evaporator and purification through SPE cartridges with methanol and water conditioning. The eluates were evaporated under nitrogen at 40 °C, reconstituted with hexane and subjected to high-performance liquid chromatography (HPLC) analysis via a UV/visible detector. The system utilized a VP-ODS column and an acetonitrile-water mobile phase (20:80) at 40 °C. A total of 30 pesticides from five chemical families, including organophosphates, urea, carbamates, triazines, and chloroacetanilides, were identified for determination.

Among the 30 pesticides, 26 were found in the three river estuaries with concentration levels above the recommended international benchmark. Among all the target chemicals, triazinone herbicides were identified most often, while pyrimidine rodenticides were second most common. The herbicide simazine, a triazine, was detected at the highest concentration of 120.7 μg/L. The Tanoé River, which is shared by Côte d'Ivoire and Ghana, presented the highest levels of pesticide detection per water sample, whereas the Bia River, which is proximal to large cocoa and rubber plantations, presented the highest concentrations of those pesticides, particularly during the rainy season. There was otherwise no distinguishable seasonality within the water quality data of the Comoe and the Tanoé Rivers. The calculated pesticide risks for the rivers through risk quotient analysis revealed that the pesticide level has put aquatic organisms in these rivers at high chronic risk and that the Bia River may pose chronic health risks to children in the age bracket of 0–11 years [22].

3.3 *Microplastics in Soil: A Study of Bhopal, Madhya Pradesh*

Duplicate 1 kg samples of surface soil were collected from each site via the quadrat method (quadrat size of 1 m × 1 m) in May 2022, as there is currently no standardized method for sampling microplastics in soil. The soil samples were collected from the upper 5 cm of depth via a metallic spade and SSC Scoop, placed in cotton bags, and then transported to the laboratory. Since sampling was performed during the dry period at a mean temperature of 41 °C, the samples were subjected to further drying at 40 °C for 48–72 h to minimize any residual humidity. First, large soil particles were broken into smaller particles and mixed well; then, the samples were powdered and passed through stainless steel meshes of 5 mm, 1 mm, and 500 μm sizes. Any particles greater than 5 mm in size were removed, and a balance divided according to the following size fractions was used for testing. Peroxide oxidation of the wet sample was ruled out because no chunky particles were observed, and density separation was performed with a NaCl solution with a density of 1.2 g/mL. Samples of microplastic particles were identified via a stereomicroscope made by Zeiss Company, Stemi 305, with an integrated camera (magnification 8×–40×), and identification of the size, shape, color, and chemical composition of the collected samples was conducted via FTIR spectroscopy performed by Shimadzu Company, model IR Affinity-1S. Measures such as the wearing of cotton lab coats, nitrile gloves, and aluminum foil covers helped reduce contamination. The airborne contamination was not considered high because any particle measuring less than 500 μm was not sampled. From a total of 576 microplastic particles detected via FTIR analysis, the majority were made up of polymer classes, including polyethylene (PE), polypropylene (PP), polyvinyl chloride (PVC), and polyethylene terephthalate (PET), while the rest were polyvinyl acetate (PVAc). The fundamental absorption bands for PE in the regions of alkanes were found to be 2914/cm and 2847/cm; for PP, they were identified at 2956/cm, 2921/cm, 2875/cm, and 2840/cm. The other bands for PE were at 1470/cm and 718 cm. The weak bands at 3423/cm were due to hydroxyl (O–H) groups, which may originate from possible degradation of the material within the soil environment; the strong broad band at 1032/cm may be attributed to degradation processes or the inclusion of silane-based compounds often used in polymerization processes [23].

4 Challenges

The detection of emerging contaminants poses several challenges:

New-generation chemicals, which include pharmaceuticals and personal care products, pesticides, EDCs, microplastics, and industrial chemicals, are quite challenging to detect because they are present at low concentrations, have complex interactions with the environment, and require sensitive and efficient methods of analysis [14].

4.1 Diverse Nature and Low Concentrations of ECs

The general category of ECs can include small organic molecules such as pharmaceuticals, large polymers such as microplastics, or complex mixtures such as pesticides. The members of each class possess unique molecular compositions; characteristic physical attributes such as volatility, polarity and solubility; and characteristics of how they interact with their environment, such as persistence and degradation. Most ECs are detected in environmental samples at concentrations ranging from ng/L or g to μg/L or g, making the required methods highly sensitive and accurate. Owing to the low concentrations of ECs, selective methods for the identification of target ECs coupled with complex and diverse matrices are needed. Routine conventional wet chemical techniques could be inadequate for identifying contaminants at such small concentrations at any time without some alteration. For high-sensitivity methods, there are certain instruments, including liquid chromatography–mass spectrometry (LC–MS/MS), gas chromatography–mass spectrometry (GC–MS), and high-resolution mass spectrometry (HRMS), but all of them are very costly, time-consuming, and usually performed and operated by professionals [24]. The pharmaceuticals examined are typically found at a concentration of even ng/L (nanograms per liter), and LC–MS/MS is utilized as the technique of elucidation and quantification, although it is very expensive and lacks usability.

4.2 Complex Environmental Matrices

Solids such as soils, water and air, and biota are also rich samples with compounding and fluctuating matrices that include organic matter such as ions, colloids, and other intimate interferences that may hide ECs. These interferences can lead to the suppression of signals from target contaminants or modified signals. Sample preparation is important for separating ECs from other materials and can therefore involve several procedural steps (filtration, extraction, etc.). However, these processes are very slow, may cause loss of samples in the process, and may also be inconsistent between samples. Other compounds in the matrix may interfere; for example, ion suppression or enhancement presents a challenge to the quantification of ECs. The qualitative determination of pesticide residues in soil involves two techniques: solid-phase extraction and liquid–liquid extraction. This can be difficult or even impossible when other organic or inorganic compounds that may react with the pesticide interfere with the activity of the functional groups [7].

4.3 Lack of Standardized Methods

ECs are a relatively young class of pollutants; therefore, generic approaches for sample preparation and extraction from environmental samples are still being developed. ECs are not classical pollutants, such as heavy metals, and the literature lacks standard methods regarding the sampling, extraction and concentration of such compounds. A lack of standardization regarding the approach causes an enormous degree of variation in the sampling strategy, analysis procedures, and interpretation of the results in various investigations or laboratories. This poses problems when attempting to compare findings within and across the identified investigations and to develop sound reference levels for the prevalence and geographic distribution of ECs. In the absence of universal procedures, regulatory authorities are unable to set tolerability thresholds for ECs, which makes risk analyses and initiatives difficult. There is yet no well-defined method for sampling microplastics in the soil. This is because various methods, such as wet sieving, density separation, and different methods of chemical digestion, are used to extract macrofossils, and different researchers interpret the results differently.

4.4 High Analytical Costs and Resource Intensity

The methods needed for the identification of ECs, such as LC–MS/MS, GC–MS, and FTIR, are rather complex and expensive. Moreover, these techniques involve expensive equipment, such as laboratories and skilled personnel. The relatively high cost of instruments and consumables and the requirements for skilled technicians restrict their use, particularly in developing laboratories. Consequently, few or no surveillance studies of ECs have been reported in many regions of the world. Sophisticated methods for detection are time-consuming and require many resources, which leads to longer times to release information and elevated operating costs to provide regular environment check-ups. Owing to the scarcity of environmental monitoring setups in developing states, some of the most advanced tools, such as liquid chromatography–mass spectrometry (LC–MS)/mass spectrometry (MS) and gas chromatography–mass spectrometry (GC–MS), are not frequently used when analyzing pharmaceuticals in water, leading to information deficiencies.

4.5 Transformation Products and Nontargeted Analysis

Chemicals found in ECs, for example, pesticides and pharmaceutical chemicals, transit through change processes in the environment and form metabolites and byproducts. Some of these transformation products may be far more toxic or persistent than the parent contaminants. For these reasons, target analysis has limitations, which means that nontargeted analyses, such as high-resolution mass spectrometry (HRMS), are necessary. However, these methods require large amounts of data, which normally require high-level bioinformatics tools for analysis and comprehension. Identification of Unknowns: Transformation products may not be well defined, and identification may be limited by a poor ability to match them with known standards. New methods and algorithms are needed for the detection of all unknown contaminants. A common herbicide is atrazine; its metabolites are desethylatrazine and desisopropylatrazine, which have different environmental characteristics and toxicities from the parent compound and thus should be separately calibrated and detected [25].

4.6 Matrix Contaminants

Matrix interference may occur in the determination of ECs in environmental samples since the presence of other compounds may suppress or enhance the signal. These matrix effects lead to inaccurate estimations, which are particularly troublesome in complex samples. ECs are frequently found with other contaminants (such as heavy metals or other pesticides), and they can mask or suppress each other when analyzed. For this reason, accurate quantification of these elements is a challenge because interference from other species and matrix interferences such as isopycnic, temperature, and pH interferences can be alleviated only by techniques such as isotope dilution or matrix matching calibration. Interferences manifest themselves in LC–MS/MS pesticide analysis through the matrix, causing false results; to overcome these matrix interferences, matrix-matched calibration standards must be used.

4.7 Airborne ECs

Many airborne ECs include microplastics and oils or VOCs in a particulate or gaseous state in the environment. These contaminants are very difficult to sample via conventional sampling techniques. Sampling airborne ECs therefore involves the use of different approaches: thermal desorption GC–MS for the detection of VOCs and airborne microplastic sampling equipment. These techniques require refinement to make an effective search for ECs possible. Microplastic aerosols are difficult to estimate when usual water and soil sampling methodologies are employed; detecting them is challenging because of the scarcity of sampling equipment and the absence of effective identification methods.

4.8 Dynamic Nature of the Issue

The incorporation of new chemicals into manufactured articles and industries makes ECs fairly dynamic. It might take some time to develop a method for detecting a new chemical or a product. New forms of pesticides and other products, such as personal care products and industrial chemical formulations, have no analytical standards, so the necessary methods for detection have not been developed in time. Vigorous advancements in diagnostic technologies are needed to address such alterations. Silver nanoparticles used in consumer products are often released, but their identification poses some challenges, primarily because there are no standard procedures and because the chemical characteristics may vary.

4.9 Lack of Reference Standards

A considerable number of ECs, particularly new emerging substances, have difficulty obtaining reference standards on the market for calibration and quantification purposes. There are usually no reference standards, which complicates the process of calibrating the instruments used in providing results. This is particularly the case in the case of new ECs for which there is no earlier research to benchmark or to set the standard. PVC microplastics are newly emerged contaminants; they have no reference material; thus, identification through FTIR spectroscopy is difficult [25].

5 Future Research

Given that ECs have become a focal point of concern in the ever-evolving world, there is a dire need to develop efficient techniques for detecting these pollutants in the environment. Future research on the detection of ECs may focus on the following areas: increasing the sensitivity of the technology, working on sample preparation procedures, developing targeted and nontargeted approaches and confronting current technologies with the complexities of environmental samples.

One important area that may be further developed is the enhancement of the sensitivity and specificity of detection methods. A number of ECs are present in the environment at the pg level, which cannot be detected by most conventional techniques. As new technologies emerge from sample preparation techniques such as high-resolution mass spectrometry (HRMS), liquid chromatography–mass spectrometry (LC–MS–MS), and method development, detection capabilities should increase. Investigators are now beginning to work on improving the selectivity of the sensors and to be able to detect the ECs at lower concentrations and to investigate the effects of these chemicals on the environment and possible effects on the health of those living within the environment. For example, the integration of aptamers as biorecognition elements in biosensors increases the sensitivity and specificity of EC identification in water and soil matrices [4].

Other research focuses include the development of screening methods that have the ability to identify a large number of ECs and their metabolites. Targeted and nontargeted methods, including high-resolution and accurate mass spectrometry (HR/AMS), can detect both targeted and claimed but unknown ECs, and their degradation products are also covered. This is particularly important because many ECs transform in the environment, forming new compounds that could be more toxic than indigenous chemicals. For example, measuring the degradation of pesticides such as atrazine and chlorpyrifos is important not only to determine the fate of the parent molecules but also that of their metabolites, which often have different toxicities [3].

Enhancing sample preparation methods is another significant domain of the future development of these approaches. Soil and water samples as well as some types of biota are usually complex and contain organic matter and ions as well as other pollutants, which complicates the determination of ECs. Recent developments in extraction techniques such as solid-phase microextraction (SPME) and supercritical fluid extraction (SFE) are likely to improve sample preparation techniques since matrix effects can be controlled. These methods can increase the effectiveness of low-concentration contaminant analysis and play important roles in environmental monitoring [8]. In addition, studies on point-of-use assessment and field-based sensors for the detection of ECs in real time are expected to increase. Educational toys that warn when they are exposed to high levels of microplastics or toys that can detect pesticides and pharmaceuticals in food and water can be useful because they help people make decisions regarding the health risks of polluted foods and water without necessarily having to visit laboratories. These organs could also be connected to the IoT to develop several EC monitoring devices that deliver current details regarding fluctuations in the levels of EC in different areas [4]. Finally, standardization of detection protocols and the creation of EC reference materials are needed to harmonize the measurement data acquired in different studies and lab settings. Standardizing the techniques used in sampling, extraction, and analysis will make it easier to develop a baseline for ECs across the world. The same advancement in reference standards will help enhance the reliability of the analytical results. Scientists are working primarily on developing compendial studies for materials such as plastics at the microscale and endocrine-active substances [26].

Finally, the development of future research in detecting emerging contaminants will have to focus on improving the method's selectivity, sensitivity, and directness together with other advancements that can be put forward as limitations regarding the environmental matrix and the future nature of ECs. By doing so, it will be easier to establish better ways of tracking ECs; hence, the environment and health effects related to these contaminants will be well addressed.

6 Conclusion

Many new substances, collectively referred to as emerging contaminants (ECs), have recently been reported in the environment, and the impacts of these substances on human and ecological health are largely unknown and remain an enormous concern to society and science worldwide, requiring prompt scientific, technology, and policy actions. Pharmaceuticals, pesticides, endocrine disruptors, microplastics, and industrial emissions are diverse, persistent, and bioaccumulable, endangering ecosystems and human health. These compounds enter environmental vehicles, such as soil, water, and air, through various routes; hence, total identification and control measures are needed. This chapter focuses on the fact that ECs are numerous but present at relatively low concentrations and are therefore best analyzed using what is truthfully considered recognized sophisticated techniques. Purposes such as gas chromatography–mass spectrometry (GC–MS), liquid chromatography–mass spectrometry (LC–MS/MS), and high-resolution mass spectrometry

(HRMS) offer sensitivity and specificity for the identification of ECs. Other preconcentration techniques, such as SPE, LLE, SPME, and SFE, increase the ability of the sample matrix to prevent interference. However, there are inevitably matrix effects and high costs, and often, there are no reference materials available for the more recent ECs, making the process of analysis very challenging; thus, more research and development are still needed. Other methods, including biosensors, NMR and FTIR spectroscopy, and other techniques, hold promise for monitoring EC. The successful integration of IoT field sensors and aptamer-based biosensors is another promising area, as communities are then able to independently track their environments and react to contaminant episodes. Such innovations improve awareness within a particular community and allow for quick responses to possible harm to health. The chapter also establishes the dynamism of ECs due to the existence of new chemicals and products, hence the need to develop standards. To achieve better comparability, the creation of standard detection procedures and reference materials is crucial. These efforts will help researchers and policymakers develop a systematic knowledge stock of ECs and their transformation products and their effects on health and the environment. Even so, obstacles exist, for example, in the absence of standardized operations, time-consuming methods of detection, and the high cost of analysis, which is unaffordable for many low-income countries. These challenges can be met only by contributions from scientists, policymakers, and the industry in terms of accessibility, affordability, and sustainability. It is therefore important to increase awareness, employ regulation and tendencies, and develop environmentally mindful legislation, green chemistry, and advanced policies that support technological progress.

Emerging contaminants help mitigate the harm they present to the environment and the world populace; thus, solving these problems should be a top priority. The incorporation of advanced detection technologies, the promotion of interdisciplinary research cooperation, and the establishment of effective legislation and regulations can effectively reduce the influence of budding ECs and provide a backbone for research on environmentally sustainable management. By realizing collective work and preventive strategies, we can create a future in which threats originating from ECs can be controlled and make ecosystems and communities healthier.

References

1. Kosma, C. I., Lambropoulou, D. A., & Albanis, T. A. (2010). Occurrence and removal of PPCPs in municipal and hospital wastewaters in Greece. *Journal of Hazardous Materials, 179*(1–3), 804–817. https://doi.org/10.1016/j.jhazmat.2010.03.075
2. Zambrano Flores, J. V. (2021). *Biological treatment of swine manure with microalgae-bacteria consortium. Removal and fate of emerging pollutants present in pig manure wastewater.* Thesis. Retrieved from https://uvadoc.uva.es/bitstream/handle/10324/62629/Tesis%20publicable.pdf?isAllowed=y&sequence=1
3. Luo, Y., Guo, W., Ngo, H. H., Nghiem, L. D., Hai, F. I., Zhang, J., Liang, S., & Wang, X. C. (2014). A review on the occurrence of micropollutants in the aquatic environment and

their fate and removal during wastewater treatment. *Science of the Total Environment, 473*, 619–641. https://doi.org/10.1016/j.scitotenv.2013.12.017
4. Sajjad, M., Anwar, S., & Qadir, A. (2022). Advances in the monitoring of emerging contaminants in soil and water. *Environmental Research, 204*, 111908. https://doi.org/10.1016/j.envres.2021.111908
5. Lusher, A. L., Hollman, P. C., & Mendoza-Hill, J. J. (2017). Microplastics in fisheries and aquaculture: Status of knowledge on their occurrence and implications for aquatic organisms. In *FAO Fisheries and Aquaculture Technical Paper* (Vol. 615, pp. 33–50).
6. Dasgupta, S., Khillare, P. S., & Sharma, S. (2011). Environmental monitoring of pesticides: Method development and challenges. *Environmental Pollution, 159*(1), 12–18. https://doi.org/10.1016/j.envpol.2010.10.031
7. *High-performance liquid chromatography in pesticide residue analysis*. 2015. doi:https://doi.org/10.1201/b18481
8. Pereira, A. L., Franco, D. L., & Ramos, S. J. (2018). Matrix effects in pesticide residue analysis using LC–MS/MS: Mitigation strategies. *Analytical Chemistry, 90*(12), 7012–7020. https://doi.org/10.1021/acs.analchem.8b01386
9. Li, X., Shen, X., Jiang, W., Xi, Y., & Li, S. (2024). Comprehensive review of emerging contaminants: Detection technologies, environmental impact, and management strategies. *Ecotoxicology and Environmental Safety, 278*, 116420. https://doi.org/10.1016/j.ecoenv.2024.116420
10. Applications of Gas Chromatography [appl, Chem] [PDF] [6ormctgsehs0]. Retrieved from https://vdoc.pub/documents/applications-of-gas-chromatography-appl-chem-6ormctgsehs0
11. Xu, M.-L., Gao, Y., Wang, X., Han, X. X., & Zhao, B. (2021). Comprehensive strategy for sample preparation for the analysis of food contaminants and residues by GC–MS/MS: A review of recent research trends. *Food, 10*(10), 2473. https://doi.org/10.3390/foods10102473
12. Crompton, T. R. (2012). *Organic compounds in soils, sediments & sludges*. https://doi.org/10.1201/b13741
13. Nollet, L. M. L. (2013). *Handbook of water analysis*. https://doi.org/10.1201/b15314
14. Díaz-Cruz, M. S., & Barceló, D. (2015). *Personal care products in the aquatic environment*. https://doi.org/10.1007/978-3-319-18809-6
15. Jung, K. (2018). *Analysis and sensory evaluation of volatile constituents of blackcurrant (Ribes nigrum L.) and redcurrant (Ribes rubrum L.) fruits*. [Online]. Retrieved from https://mediatum.ub.tum.de/doc/1427712/1427712.pdf
16. Ong, J. Y., Pike, A., & Tan, L. L. (2021). Recent advances in conventional methods and electrochemical aPTasensors for mycotoxin detection. *Food, 10*(7), 1437. https://doi.org/10.3390/foods10071437
17. Hyotylainen, T. (2009). Sample preparation for chromatographic analysis of environmental samples. *Advances in Chromatography, 48*, 329–371. https://doi.org/10.1201/9781420084542-c8
18. Extraction Technologies for medicinal and Aromatic plants [PDF|TXT]. Retrieved from https://pdfcookie.com/documents/extraction-technologies-for-medicinal-and-aromatic-plants-9mlx1rj9epv7
19. Kassim, T. A., Simoneit, B. R. T., & Williamson, K. J. (2005). Forensic investigation of leachates from recycled solid wastes: An environmental analysis approach. In *Springer eBooks* (pp. 321–400). https://doi.org/10.1007/b98265
20. Awasthi, L. P., Das, S., Lee, R. F., & Pattanayak, S. (2024). *Plant pathology*. CRC. https://doi.org/10.1201/9781003326915
21. Flores-Contreras, E. A., González-González, R. B., González-González, E., Melchor-Martínez, E. M., Parra-Saldívar, R., & Iqbal, H. M. N. (2022). Detection of emerging pollutants using AptAMer-based biosensors: Recent advances, challenges, and outlook. *Biosensors, 12*(12), 1078. https://doi.org/10.3390/bios12121078
22. Siriki, K., Yao, K. M., Gnonsoro, U. P., & Trokourey, A. (2021). Transboundary river water pesticide pollution in historical agriculture areas in West Africa: A case study in the Comoe,

Bia, and Tanoe rivers (Cote d'Ivoire). *Arabian Journal of Geosciences, 14*(20), 1–15. https://doi.org/10.1007/s12517-021-08294-7

23. Singh, S., Chakma, S., Alawa, B., Kalyanasundaram, M., & Diwan, V. (2022). Identification, characterization, and implications of microplastics in soil—A case study of Bhopal, Central India. *Journal of Hazardous Materials Advances, 9*, 100225. https://doi.org/10.1016/j.hazadv.2022.100225
24. Méndez, J. M., et al. (2023). Pesticide residues in bananas from the Canary Islands. *Food, 12*(3), 437. https://doi.org/10.3390/foods12030437
25. Klingberg, J., Cawley, A., Shimmon, R., & Fu, S. (2019). Collision-induced dissociation studies of synthetic opioids for non-targeted analysis. *Frontiers in Chemistry, 7*, 331. https://doi.org/10.3389/fchem.2019.00331
26. Xie, J., et al. (2020). Simultaneous removal of arsenic and antimony from mining wastewater. *Journal of Environmental Sciences, 93*, 117–119. https://doi.org/10.1016/j.jes.2020.04.003

Chapter 4
Emerging Contaminants of High Concern in the Water Ecosystem: Economic Aspects and a Perspective for Sustainable Detection Approach

Preeti A., Kanakalakshmi A., Renugadevi K., Archana C. M., Anbarasi G., and Kaarunya E.

1 Introduction

Microplastic (MP) pollution has become a global issue because of its increasing accumulation in aquatic environments. A study revealed that microplastics also serve as carriers of various hazardous pollutants from biomedical waste, cosmetic products, and some organic contaminants, such as polychlorinated biphenyls (PCBs), dichlorodiphenyltrichloroethane (DDT), and polycyclic aromatic hydrocarbons (PAHs) [1]. Microplastics tend to accumulate other pollutants from the ecosystem, primarily because of their polarities and functional groups, where 18 different perfluoroalkyl (PFAS) compounds are found on three different microplastic *materials: polystyrene, polyethylene*, and carboxylate polystyrene [2]. Microplastics can enter the environment through various routes, and based on their origin and usage, they are classified as primary or secondary MPs. Primary MPs are materials deliberately fabricated for a particular application, whereas secondary MPs are formed as a byproduct of the fragmentation and breakdown of larger microplastics by hydrolysis, UV rays, mechanical friction, etc. [3]. Primary microplastics are derived from several sources, such as plastic pellets and flakes, vectors for drugs, and cosmetic products [4]. The potential for accidental release of these MPs into the environment exists, particularly during transportation or processing, if

Preeti A. · Kanakalakshmi A. · Archana C. M. · Kaarunya E. (✉)
Department of Microbiology, Thiagarajar College, Madurai, Tamil Nadu, India

Renugadevi K.
Department of Biotechnology, Thiagarajar College, Madurai, Tamil Nadu, India

Anbarasi G.
Department of Biotechnology, PSGR Krishnammal College for Women, Coimbatore, Tamil Nadu, India

M. K. Jindal et al. (eds.), *Emerging Contaminants in Water*, Emerging Contaminants and Associated Treatment Technologies,
https://doi.org/10.1007/978-3-031-89591-3_4

proper handling practices are not followed, which could lead to contamination of ecosystems [5]. Personal care products such as hand sanitizers, facial cleansers, sunscreen, and toothpaste on the market have gradually replaced natural materials with microbeads as exofoliants [6, 7]. The extensive use of microplastics in medical applications, cosmetics, and tooth polishing for dentistry, and as pharmaceutical carriers, is released into the natural environment, leading to the worsening of MP pollution [8]. As a result of fragmentation or degradation of plastic waste due to natural weathering processes such as UV irradiation, pH changes, biological activities, and exposure to chemicals, it leads to accumulation of secondary MPs [9]. By 2030, plastic manufacturing is expected to be the primary source of environmental pollution in terms of toxic chemicals and the carbon footprint. Compared with macroplastic waste management, microplastic waste management is much more challenging because of their tiny size and less noticeable.

PFAs include a group of related compounds, such as perfluorooctanoic acid (PFOA), perfluorobutane sulfonic acid (PFBS), perfluorononanoic acid (PFNA), perfluorooctane sulfonic acid (PFOS), perfluorobutanoic acid (PFBA), and perfluorohexane sulfonic acid (PFHxS), which are emerging contaminants often found in the environment [10]. The major concern for the transmission of PFAS into aquatic environments is the large emission volume and high solubility of ionic PFAs that are discharged into rivers or seas from land. It is crucial to gain knowledge on low-cost detection methods and identify the adverse effects of PFASs in aquatic ecosystems, as this is currently a pressing necessity. This chapter features the prevalence of MPs and PFAs with case studies and their co-occurrence as secondary pollutants. New detection techniques and current challenges for the sustainable removal of microplastics from aquatic systems to ensure water safety are emphasized.

2 Microplastics in Fresh Water

Microplastics enter the environment through various anthropogenic resources. Activities such as burning, landfills, traffic pollution, and several other industrial and urban practices deposit large numbers of MPs in freshwater environments [11, 12]. The primary discharge of MPs into freshwater ecosystems occurs through domestic sewage release [13]. Wastewater treatment plants discharge MPs in the form of fibers, which obtain their access mostly from textiles [14]. The fibrous form of plastics usually comes from washing clothes, which are mainly composed of polyamide, acrylic, and polyesters with a level >100 fibers/L [15]. Compared with new clothes, which eventually settle and sink in nearby aquatic bodies, air has been found to be a very good source for discharging MPs where old clothes produced greater amounts of fiber loss during abrasion treatment [16]. Airborne MPs are likely to be rich in heavy metal contaminants, which pose a serious threat to aquatic organisms [17]. Fishing activities in lakes and rivers result in the deposition of nylon particles from fishing equipment [18]. Analyzing freshwater fish species provides a better understanding of the interactions and risks associated with MP contamination. In an indicial fish species, 142 microplastic particles were identified [19]. Similarly, another study reported the percentage of polymeric compounds and reported that 58% polyethylene, 38% polyamide, and 8% polypropylene were present in the edible fish *Oreochromis mossambicus* [20]. MP pollution in urban and rural areas was compared, and it was found that urban rivers have 11.6 trillion particles (i.e., 0.4 MP particles/L, whereas rural rivers are likely to have only 0.2 particles/L [21]. Another way for MPs to enter fresh water is through agricultural practices. Researchers have reported that per kg of soil in agricultural land, 0.34 ± 0.36 MPs are present [22]. Moreover, density, shape and size play significant roles in the translocation of MPs from the soil to freshwater ecosystems [23]. Finally, surface runoff contributes substantially to the transfer of MPs from terrestrial environments to freshwater environments. Every year, 7500–13,700 tons are deposited [24]. Annually, 28,000 kg of floating debris enters the river annually [25]. Moreover, storms and floods also subsidize plastic debris into the water. Finally, evaluating MP contamination has some substantial erraticism due to their spatial constraints.

2.1 Occurrence and Distribution of MPs and PFAS

Plastic has appeared since the Second World War [26]. The industrial and domestic use of plastics is growing exponentially because their durability, light weight, and cost-effectiveness led to the production of 391 million tons in 2021 [27]. Plastics and their components may occur in various forms based on their size ranges. Plastics above 25 mm are considered macroplastics; similarly, those in the 5–25 mm range are mesoplatics, whereas those in the 1–5 mm range are considered microplastics, which poses a serious threat worldwide, especially in terrestrial and aquatic environments [28]. The occurrence of microplastics in the environment is due to irregular anthropogenic activities, and with the impact of several ecological factors, their concentrations vary greatly. Places such as surface water and residues are found in coastal environments. Substantial amounts of microplastics are also present in deep environments, and microplastics are also present in rain and snow. Chemical and physical properties help their transportation and increase their ability to accumulate various other components and emerging contaminants, such as polyfluoroalkyl substances, polychlorinated biphenols, polyaromatic hydrocarbons, and other emerging contaminants, which results in secondary pollution [29, 30]. In the year 2004, microplastics were used for particles with a size of 20 μm, whereas in 2008, the US National Oceanographic and Atmospheric Agency (NOSAA) organized the first workshop on microplastics and in the event that proper definitions were given to microplastics, such as particles that are likely to be smaller than 5 mm in size [26]. The sorption behavior of MPs attracts emerging contaminants such as PFAs, which pose serious threats to the environment. PFAs are complex compounds composed of alkyl-based substances that contain $C_{11}F_{2n+1}$, and some contain fluorine atoms [31, 32].

2.2 PFAS in Groundwater

Per- and polyfluoroalkyl substances are rarely known compounds, but they pose a serious threat to the environment and to human health through effective interventions, such as endocrine disruption and immunotoxicity, and affect metabolic processes [33–35]. PFAs are synthetic chemicals that are universally utilized because of their unique characteristics, such as water resistance, strain resistance, and chemical stability. Owing to their unique characteristics, these materials can be used in various products, such as water repellent coatings for nonstick utensils and countless industrial applications. Polyfluoroalkyl substances are not completely fluorinated and have a single C atom in the carbon chain [36]. The production of PFOAs was stopped in 2015 by the USEPA, but manufacturers have adopted other alternatives with short-chain analogs, such as the GenX commercial name given to the ammonium salt of the hexafluoropropylene oxide dimer [37].

Groundwater is likely contaminated by a variety of chemical and biological contaminants. Among the contaminants, nonserious PFAS enters groundwater through wastewater treatment plants, industrial runoff, and contaminated soils. Groundwater is connected to lakes and rivers, where PFAS seeps into domestic water sources, posing threats [38–40]. Currently, there are no drinking water standards for PFAS, but in 2015, the Swedish National Food Agency set a limit for derivatives of PFAS of 90 ng/L [36]. In 2016, the US Environmental Protection Agency set the limits for PFOS (perfluorooctane sulfonate) and PFOA (perfluorooctanoate) as 70 ng/L, and these PFAS are commonly found PFAS in groundwater [41]. In general, these contaminants in groundwater exist in anionic form with low vapor pressure and are water soluble. The presence of C–F bonds make them resistant to chemical and biological degradation [42]. In one study, the amount of PFAS present in ground water near the municipal landfill site was found to be 26–5200 ng/L and was dominated by PFOA [43]. Additionally, these domestic wells and the public supply were found to contain PFAS with other contaminants, such as tritium, chloride, sulfate, and pharmaceutical waste. It was determined that activated bituminous coal-based carbons were used to adsorb PFAS and that pores on the surface of activated carbon exhibited a very good adsorption capacity for both hydrophobic and hydrophilic PFAS [44].

2.3 Microplastics in Coastal Waters

It has been reported that annually, 4.8–12.7 million tons of waste plastic material are dumped into marine areas [45]. Microplastics have attracted increasing interest from researchers because the serious effects of microplastics are completely unspoken and are still a hot topic of ongoing research. Researchers have focused on developing observational and numerical methodologies to identify where these microplastics accumulate in the ocean and how to eliminate them [46–48]. In coastal environments, salinity and the humic acid content are crucial for the migration and maintenance of the stability of microplastics [49]. The reason behind this is that microplastics usually exhibit weak electrostatic repulsion, which helps in vertical migration, whereas the humic acid content, which results in electrostatic and steric repulsion, helps increase the stability of microplastics, and prevents vertical migration. Polymeric compounds easily enter a wide range of marine organisms, from small zooplanktons to higher mammals. Furthermore, when it enters the food chain, it poses serious health-related issues to humans. The impact of microplastics was greater in 38% of fish, and it was found to be 1% in turtles [50]. The detrimental effects of microplastic exposure on various aquatic species have been investigated by many researchers. In the South China Sea, 24 fish species were taken for the experiment, and microplastics were found in their gastrointestinal tract [51]. Similarly, in South Africa, 91 fish samples were found to have ingested microplastics due to confusion as prey [52]. Additionally, 0.28 g/dw microplastic debris has been reported in polychaetes [53]. MPs are deposited or present in these invertebrates, as they are mostly found adjacent to stormwater. Usually, stormwater is likely to carry a greater amount of plastic debris. Moreover, microplastics are exposed to zooplankton in Delaware coastal waters [54]. Researchers have reported that three mechanisms are responsible for increasing microplastic concentrations: the transport of zooplankton into microplastic tidelines, the movement of microplastics into the floating efflux current, and their accumulation off the coast. Another study, which was conducted along the shoreline of Cape town, analyzed microplastic concentrations in summer and cold seasons and focused particularly on stormwater outlet areas [55], where the amount of microplastics was greater in water and sediments during the winter, whereas the amount of MPs was greater in marine biota such as mussels, whelks, and sea urchins in the summer. They also reported that mussels are very good bioindicators for determining microplastics because, compared with other test organisms, mussels had higher concentrations of microplastics during analysis [56]. Other invertebrates, such as urchins and whelks, are likely to be more susceptible to microplastics.

2.4 PFAS in Marine Environments

Discharge of industrial leftovers into the marine ecosystem results in the entry of PFAS; thus, discharge automatically affects marine PFAS life. PFAS are released into marine ecosystems through various sources, and it is difficult to predict the amount of discharge worldwide. According to the literature, the concentration of PFAS is diluted, which is referred to as the dilution effect [57–61]. Usually, their level decreases when the depth increases. Marine animals with longer life spans tend to accumulate more PFAS during their lifetime [62]. The long-lasting effects of PFAS in marine environments occur through bioaccumulation and biomagnification effects [63]. Biomagnification refers to greater concentrations of PFAS at higher trophic levels, whereas bioaccumulation occurs only at lower levels because of the uptake of contaminated seawater and sediments. The presece of PFAS are highly persistent in the environment, and can remain in the ocean for many decades, potentially causing long-term damage to the ecosystem. Recently, unhatched eggs of the vulnerable species *Caretta caretta* from the Mediterranean Sea were reported to contain PFAS, and 3.34 ng/g PFAS were detected in fresh eggs via liquid chromatography coupled with hybrid high-resolution mass spectrometry [64]. Similarly, bioaccumulation and tropic transfer PFAS contamination in seven fish species have also been reported [65]. Higher PFAS concentrations are found in the liver and muscle tissues, and the concentration ratio is linked with the feeding ecology of fish because they usually feed on mollusks, fish larvae, and plankton. Moreover, the median absorption of PFAS in Korean bird liver tissues was found to be 294 ng/g of weight [66]. Furthermore, the detrimental effects of PFAS lead to immunological dysfunction; research on bottlenose dolphins (*Tursiops truncatus*) revealed that they produced more proinflammatory cytokines and had greater proliferation of CD^{4+} and CD^{8+} T cells [67]. The geographic characteristics that contribute to PFAS accumulation are a greater focus [62]. Areas such as tidal flats, mangroves, coral reefs, sea grasses, and even brackish marshes tend to bioaccumulate more PFAS because of their longer exposure. According to the data of PFA in marine environments, areas such as the North Atlantic Ocean (4.0 ng/L), North Pacific Ocean (12.8 ng/L), and Indian Ocean (41.1 ng/L) have varying levels of PFAS. The fate of PFAS in the marine cryospheric environment under laboratory conditions was studied [68], and NaCl solutions were used as artificial seawater for ice formation and incorporated PFAS during bulk ice formation. As a result, they reported that the concentration of PFAS in sea ice was linearly related to the salinity of the water ($r^2 = 0.30$ to 0.88, $n = 18$, $p < 0.05$). However, a substantial increase in the PFAS concentration is certain for the future. After the COVID-19 pandemic, the PFAS level increased in the environment due to the use of single PFAS-use containing products. Therefore, additional studies are needed to understand the hidden aspects of "forever chemical." Hence, a coastal management program is needed to monitor the marine litter that causes damage to marine biodiversity.

3 Methods for the Detection of Microplastics and PFAS

3.1 Common Detection Methods for MP Detection

Optical microscopy, which includes conventional light microscopy, fluorescence microscopy (FM), scanning electron microscopy (SEM), transmission electron microscopy (TEM), and atomic force microscopy (AFM), is the most widely used basic microscopic tool for MP analysis. The ability to visually evaluate the size, shape, and morphology of particles makes microscopic approaches indispensable for the investigation of MPs [69]. Moreover, these methods require minimal amounts of reagents, solvents, and energy [70].

Optical Microscopic Methods Optical microscopic methods are inexpensive, practical, easy-to-use operating strategies with minimal chemical requirements. The wavelength of the light being used determines the limitations of optical microscopy. MP detection under a light microscope has an ambiguity of approximately 20%, and optical microscopy has a relatively low detection accuracy.

Fluorescence Microscopy Analyses of MPs in coastal marine habitats [71], zooplankton [72, 73], marine mammals [74], and biological samples [75, 76] have been conducted via this technique. FM made it possible to visualize and characterize MPs via the fluorescence qualities of the dyes that were used to mark them.

SEM The most popular method for examining the physical properties of MP particles is SEM. For the detailed characterization of MPs [77–80], Fourier transform infrared spectroscopy (μ-FTIR), Raman microscopy (-RM), SEM coupled with Py-GC/MS, and energy dispersive X-ray spectroscopy (EDS) were utilized. To distinguish between inorganic and organic MPs and determine their chemical composition, SEM and EDS are used in tandem [81].

TEM Because the electron densities of various plastics are similar, distinguishing between different types of plastics can be difficult. A more thorough characterization of MPs is sometimes achieved by combining TEM with additional analytical methods such as FTIR and Raman spectroscopy [82, 83]. For morphological characterization and MP detection in soil amended with municipal solid waste, TEM in conjunction with EDX was employed [83].

AFM AFM is generally more effective when used in conjunction with other techniques, such as tip-enhanced Raman spectroscopy (TERS), AFM-IR (also known as photothermal-induced resonance), and nano-FTIR. However, for extensive environmental monitoring, AFM might not be the most feasible choice. AFM-IR provides both chemical and topographical information about the polymer nanostructures at the same time, which increases the thoroughness and precision of investigations.

Vibrational Spectroscopy For particle sizes up to 500 μm, vibrational spectroscopy (VS), a particle-based, nondestructive imaging technique, is extensively utilized in MP analysis [84, 85].

NIR Spectroscopy Quantitative analysis can be performed via NIR spectroscopy. The focus of NIR spectroscopy is just beyond the visible range of the electromagnetic spectrum. The distinctive absorption characteristics of certain MP materials in the near-infrared (NIR) band (usually between 780 and 2500 nm) is due to functional groups like C–H, N–H, O–H, and C–O which absorbs NIR light at specific wavelengths, creating unique spectral fingerprints. NIR spectroscopy may be used with other methods, such as microscopy or Raman spectroscopy, to provide a thorough investigation of MPs.

To identify MPs, PE, PS, PP, and PET near-infrared hyperspectral imaging (NIR-HSI) at wavelengths between 1000 and 2500 nm were used [86]. Furthermore, NIR spectroscopy has been used to quantify MPs in soil [87]. For environmental monitoring, NIR spectroscopy can be a quick and nondestructive technique.

Raman Spectroscopy Through the interaction of light with molecular rotations and vibrations, Raman spectroscopy generates a spectrum that is distinctively fingerprint-like. The inelastic scattering of laser light by molecules serves as the foundation for Raman spectroscopy. When a laser is pointed at a sample, the majority of the dispersed light has the same energy (frequency) as the incident light. To determine the types of polymers contained in the MPs, three methods are utilized: reflectance, transmission, and attenuated total reflection (ATR). A small fraction of these interacts with each other to cause energy shifts [88, 89]. A novel approach to chemically characterize NMPs via thermal desorption−proton transfer reaction−mass spectrometry (TD-PTR/MS) in which analytes are ionized via hydronium ions produced from water vapor [90].

GC–MS The combination of thermogravimetric analysis with Fourier transforms infrared spectroscopy and gas chromatography–mass spectrometry (GC–MS) (TGA–FTIR–GC–MS; DB–5MS, SQ) has been used only once to measure MPs in mussels precisely. The average plastic content of the tissue was found to be 0.58 mg/kg, with a range of 0.16–1.71 mg/kg [91].

LC–MS MPs in water can be detected with a lower detection limit of up to 10 ng/L than those used for soil or sediment analyses [92].

MALDI-ToF/MS Soft ionization and extensive mass detection can be achieved via a matrix-aided laser desorption/ionization time-of-flight mass spectrometer (MALDI-ToF/MS), which was previously recognized as a powerful technique for the characterization of biopolymers [93]. Recently, the prospect of using MALDI-ToF/MS for the analysis of MPs was considered [94]. An alternative method of analyzing MPs focuses on identifying and measuring each of their constituent chemical elements, such as trace elements.

ICP Inductively coupled plasma–mass spectrometry (ICP-MS) and ICP optical emission spectrometry (ICP–OES) were used for quantifying microplastics in water, the environment, and food chemical components [95–100].

3.2 Recently Developed New Technologies to Detect MPs

The numerous anthropological activities caused by an expanding population can contribute to environmental damage. Because of their superior performance, plastics are widely used in modern human existence. Every region of the planet and every type of life on it has been found to contain MPs. There are serious repercussions when microplastics contaminate matrices such as rivers, lakes, and the ocean.

One study reported that microplastics, such as those of fish and zooplankton [101–103], have been discovered on the ocean floor and inside Earth. The widespread existence of these microplastics (MPs) in our surroundings necessitates immediate attention for their identification in environmental samples and a coordinated attempt to lessen their effects. Despite advancements in detection techniques, false-positive rates and prolonged detection durations are challenging. To detect MPs in soil and water in the environment, appropriate sample locations must be chosen; potential MP sources must be identified; biotic and abiotic factors such as human action, water current, and wind patterns need to be considered; and clean sampling tools and containers must be used.

3.2.1 Pyrolysis—Gas Chromatography with Differential Mobility Spectroscopy

Pyrolysis gas chromatography, when combined with mass spectrometry, is a powerful analytical method for polymer classification that relies on the release of oxygen by thermal decomposition (pyrolysis; Pyr) at a particular temperature and breaks down the complex polymers of micro- and nanoplastics into simpler organic molecules. Additionally, it helps characterize contaminants, additives, monomers, and comonomers. Large molecules undergo pyrolysis, which is typically carried out at a temperature of approximately 700 °C and is void of oxygen. The smaller molecules are subsequently separated by GC and identified by MS. The volatile pyrolysis products of MPs are exhibited in the form of polymer fingerprints or Pyrograms in GC, which indicates the corresponding microplastics. The first report of plastic polymers in soil and sediments was made via Py-GC–MS [104]. Twelve common plastic polymers, including acrylonitrile butadiene styrene copolymer (SBR), PMMA, poly urethane (PUR), nylon 6, nylon 66, PE, PS, PVC, PC, and PET, were recently investigated via qualitative and quantitative approaches [105]. Various operational modes for Py-GC–MS have been suggested [106]. The modes that fall under this category evolved gas analysis (EGA), heart-cut analysis, “single-shot,” “double shot or multi shot,” reactive pyrolysis, and thermochemolysis.

3.2.2 Differential Mobility Spectroscopy (DMS)

A continuous electric field and a radiofrequency are delivered between two parallel plates in differential mobility spectroscopy (DMS). Based on their mobility, ions become spatially separated as a result of this superposition of fields. Ions that move through the air without touching the detector plates are picked up by an electrometer, whereas ions that come into contact with the plates are neutralized and therefore missed. Some ions flow through the plates, whereas others are absorbed, making the sensor a tunable ion at specific compensation voltages (VC). This technique creates two-dimensional graphs that are represented as heatmaps by sweeping the compensation voltages for DMS data. Plotting the electrometer’s ion intensity (V) against retention time at a specific compensating voltage produces numbers that are similar to those found in conventional GC chromatograms [107]. To explore Py-GC/DMS for MP analysis, a microanalyzer that integrates three integrated technologies—a gas chromatography (GC) column, a DMS sensor, and a preconcentrator—can be employed. Microanalyzer technology has been applied in various fields, such as infectious disease diagnostics [107] and air quality monitoring at the International Space Station [108, 109].

3.2.3 Electrochemical Sensors for MP Determination

Everywhere on Earth, MPs are a part of everyday existence. The use of electrochemical sensing techniques simplifies the process of quickly determining MPs to manage the risks posed by the presence of these pollutants in the environment. Therefore, electrochemical techniques are practical tools that can transform our understanding of microplastics (MPs) in the environment and expedite their detection and removal. High sensitivity, low operating costs, and simplicity of use are advantages of electrochemical technologies. Many self-powered devices and single-use, disposable electrodes are available that are suitable platforms for effective, economical, and onsite contamination detection in decentralized labs. The selectivity, stability, shelf-life, and detection capabilities of sensors have improved with the advent of new materials, microfabrication, and additive manufacturing techniques such as 3D printing and inkjet printing. Despite these developments, a few more improvements are needed to take this technology from the laboratory to the real world. Simple operations and the development of small circuits as portable devices for in situ contaminant sample monitoring are made possible by these electrochemical approaches. Furthermore, these techniques are faster to analyze than conventional spectroscopic techniques are, enabling real-time sample monitoring at precise locations. The sensitivity and limits of detection (LODs) of these methods are lower than those of other spectroscopic and optical methods. By altering the electrode material via a variety of electrochemical techniques, different biosensors can be used to increase sensitivity and restrict detection capabilities. Low-cost sensors and electrochemical approaches offer promising analytical tools for detecting MPs [110, 111]. Chemical sensors are often built by combining an electrochemical transducer with embedded identifying elements and a sensing material [112, 113]. The interaction of a target molecule with the sensor surface is due to the covalent or noncovalent binding of the analyte, resulting in a chemical or physical change. The binding is converted into a readable signal that can be electrochemically assessed, allowing for both qualitative and quantitative identification of the target. As a result, while portable electrochemical sensors are being built, care must be taken to select and understand the materials that will be used to build the sensor, as well as the chemical or biological processes that occur at the contact point [114, 115]. On the basis of their compatibility and affinity with the target analyte, recognition elements are produced and chosen on the electrode surface.

During the identification process, physical parameters such as light absorbance, electrical conductivity, permittivity, and work tasks can be modified. These deviations are interpreted as measurable signals, such as changes in color, voltage, current, resistance, or impedance [116]. The work function shows how much thermodynamic effort is required to remove one electron from the material's Fermi level, whereas permittivity indicates how easily electromagnetic radiation can be transmitted. In the recent era, electrochemical sensors have been widely utilized to detect environmental toxins, including leachates from MPs [117]; nevertheless, their applications for MP detection are very rare and have just recently emerged. The development of electrochemical sensors for MPs is particularly attractive because of their reduced cost, rapid response time, simple use, and flexibility. This technique does not require any preliminary isolation or extraction of microplastics and can be applied to various sample types. To date, monitoring of MPs has been performed via amperometry, voltammetry, and label-free electrochemical impedance spectroscopy (EIS). Hence, electrochemical sensor approaches are vital in the detection of MPs in the environment.

3.2.4 Novel Hydrogels for the Detection and Removal of Microplastics

To detect and remove microplastics from aquatic environments, 3D hydrogel polymers made of chitosan (CS), polyvinyl alcohol (PVA), and polyaniline (PANI) and copper-infused polyoxometalate (Cu-POM) nanoclusters have been developed [118]. For the detection of microplastic compound pollutants (MCPs), covalently bonded polyethyleneimine and polydopamine copolymers, graphene oxide nanosheets, and poly(N-isopropylacrylamide) hydrogels, which are self-powered soft robots, identify and adsorb MCPs, with 97.09% adsorbing selectivity and multiple, unrestricted photothermal actuations [119]. Nanocellulose hydrogels derived from plants efficiently absorb nano- and microplastics because of their exceptional adsorption capabilities and environmentally beneficial properties [120]. The polyvinyl chloride (PVC) and polypropylene (PP) microplastics at pH -6.5 were effectively removed via pGel@IPN and could be reprocessed for up to five cycles, whereas the Cu–POM nanoclusters acted as catalysts and, with the aid of UV, disintegrated the MPs. The combination of polymers and nanoclusters results in a strong hydrogel with the ability to adsorb and degrade large amounts of microplastics in different water bodies, offering a sustainable strategy to remove MPs from water bodies through adsorption [118].

4 Impact of the COVID-19 Pandemic on Microplastic Abundance in Aquatic Environments

The aquatic environment's microplastic loads decreased because of fewer anthropogenic activities at the start of the pandemic. The variations in microplastics across the five sampling sites along the river were evaluated [121]. The microplastic levels were lower in 2021 than in 2019, with the highest microplastic abundance observed during the second lockdown. In personal care products, polyvinyl chloride was the most commonly identified microplastic, followed by polystyrene and plastics. COVID-19-induced lockdown in Cyprus resulted in a drop in micro-, meso-, and macroplastics on the beach. The microplastic concentration decreased from 4.7% to 1.7% between 2019 and 2020 [122]. A long-term assessment of microparticles in Elliott Bay, Washington, USA, revealed an 81% decrease in microparticle concentration with the onset of COVID-19. The reduction in tourism and waterfront activities was responsible for this trend [123]. A comparative study reported the occurrence of MPs before and after the pandemic in the sediments and surface water of the Goan rivers Mandovi and Zuari [124]. There was a steep decline in microplastic concentrations immediately following the pandemic. Prior to the pandemic, polyacetylene, polyacrylamide, and polyvinyl pyrrolidone were prevalent, whereas after the occurrence of polyamide, polyvinyl alcohol, and polyvinyl chloride were dominant. Similarly, a study of spatiotemporal microplastic variations in the Netravathi River of Karnataka reported a significant reduction in microplastic abundance [125]. The authors reasoned that the COVID-19 lockdown was an attribute of this decline. Similarly, the studies listed above indicate that postpandemic microplastic concentrations have decreased considerably. Notably, these studies investigated the immediate postpandemic period, but long-term pandemic evaluations remain elusive. However, as humanity rose from the pandemic, plastic utilization increased synchronously. The upsurge is largely contributed by the polymeric materials used in personal protective equipment (PPE), including face masks, gloves, shoe covers, protective suits, hand sanitizer bottles, disinfectant wipes, and incineration of a COVID-19 PCR diagnostic kit [126]. Additionally, with the onset of COVID-19, the consumption of plastic has increased worldwide. It is speculated that inefficient waste management strategies will pave the way for microplastics in the listed objects to make their way into coastal environments [127]. Using data on mismanaged plastic waste production from various countries, a hydrological model was used to predict rivers with the highest pandemic-associated plastic waste discharge. The Asian rivers Shatt al Arab ranked first, followed by Indus, the Yangtze River, Ganges, and Brahmaputra. The model further projected that within 3 years, the spatial distribution of this discharge would rapidly change in international waters. A large fraction of the plastic waste will first settle on the beach and later the seabed, and a plastic accumulation zone will develop around the Arctic [128].

5 Cooccurrence of MPs and PFASs in Aquatic Environments

Appreciable quantities of MPs and PFAS have been reported in various water bodies. For example, studies from Taihu Lake, the Pearl River, Bohai Sea, and the Atlantic Ocean reported concentrations of 3–26 MPs/L and PFAS (5–78 ng/L), 0.57 ± 0.7 MPs/L and PFAS (6–14 ng/L), 0.33 ± 0.34 MPs/L and PFAS (4–118 ng/L), and 2.46 ± 2.43 MPs/L and PFAS (0.011–0.069 pg/L, respectively [129]. The coexistence of MPs and PFASs in aquatic systems facilitates their mutual interaction, which, in turn, has manifold outcomes. The physical and chemical features of MPs and PFAS determine the extent of PFAS adsorption on MPs. The adsorption efficiency is greater for smaller MPs than for larger ones [130]. However, long-chain PFASs adsorb more readily to MP surfaces than do short-chain PFASs. Certain MP shapes (foam, fibers, films, fragments, and beads) are more vulnerable to the creation of pores and cracks upon weathering/biodegradation. These imperfections are the sites for PFAS accumulation. Furthermore, rough and porous MP surfaces also enhance PFAS adsorption. More importantly, the greater hydrophobicity of both MPs and PFASs has a positive effect on their interactions. The forces governing the adsorption of PFASs on MPs involve hydrophobic and electrostatic interactions [131]. Environmental parameters also play crucial roles in influencing PFAS and MP interactions. The concentration of PFAS adsorbed on MPs can vary greatly because of exposure duration and the concentration of PFAS in the surrounding environment. The presence of inorganic and/or organic matter in water bodies enhances PFAS adsorption. Alterations in the surrounding pH change the surface charge on MPs and PFASs, thereby modulating their extent of interactions [132]. Undoubtedly, individual MPs and PFASs are harmful to aquatic life forms. However, when these compounds are present in combination, they have different toxic effects. Furthermore, MPs may become more resistant to degradation as a result of PFAS binding, increasing their bioavailability. A three-week study involving zebrafish that investigated the effects of feed supplemented with MPs alone, perfluorinated compounds alone, and MPs sorbed with PFAS revealed that the adsorbed feed disrupted the homeostasis of the fish liver to a greater degree than either contaminant alone [133].

6 Impact of Microplastics and PFAS in Aquatic Organisms and Their Potential Toxicity to Humans

The occurrence of MPs in aquatic environments results in their bioaccumulation in the aquatic food chain. MP occurrence has been reported in plankton, mollusks, crustaceans, and fishes. Approximately 49% of the fish examined from the Northeast Atlantic Ocean had various concentrations of MPs in the gastrointestinal tract, gills, and dorsal muscles [134]. In situ toxicity studies on marine organisms that are fed MPs have revealed multiple toxic effects on these organisms. At the molecular level, DNA damage, alterations in gene expression, and protein transcription were observed. Cellular damage, ranging from coagulation, apoptosis, and necrosis to alterations in cell division, the stress response, the loss of cell viability, the production of reactive oxygen species, changes in fatty acid metabolism, and increased calcium ions, has been reported. Furthermore, tissue damage (fibrosis, inflammation, and osteolysis) and organ damage (lesions and carcinogenesis) have been observed [135]. The holistic outcomes of this damage to aquatic species include metabolic disturbances, slower growth, decreased fertility, a reduced brood survival rate, and a smaller population size. The microplastics present in aquatic ecosystems not only affect aquatic organisms but also make their way into humans. The consumption of MP-contaminated seafood and drinking or dermal contact with MP-contaminated water are the major routes of human accumulation. Data on the effects of MPs on human health are limited; however, experiments on the effects of MP exposure in cell, organoid and animal models will help us understand the potential toxicity of MPs in humans. Cell-based studies have shown that the toxic effects of MPs vary depending on the concentration of MPs, exposure duration, chemical nature, and size [136]. Additionally, high variability in MP-induced damage to DNA and the antioxidant system has been observed between different hematopoietic cell lines [137]. Following microplastic exposure, human organoids, namely, airway, forebrain, intestinal, liver, and functional disorders have been observed [138]. Toxicity studies in animal models (mice and fish) revealed that microplastic exposure leads to the activation of the immune response, metabolic dysfunction, and reproductive and developmental disorders [139]. Like MPs, PFAS accumulates in water bodies and biomagnify in food chains over time. In aquatic ecosystems, PFAS-related toxicity effects have been reported in aquatic plankton (*Daphnia magna*, *Brachionus calyciflorus*, and phytoplankton), aquatic invertebrates (Green mussel, Pearl mussel, *Dugesia japonica*, *Perinereis nuntia*, crustaceans, *Echinidae*, and *Tubifex*), and microorganisms [140].

Perfluoroalkyl substances are suggested to have immunosuppressive effects. In humans, exposure to PFASs has been shown to modulate the immune response elicited by vaccination. In a study involving 411 adults, high perfluorooctanoate serum concentrations were correlated with a reduction in the A/H3N2 influenza virus antibody titer [141]. Alarmingly, prenatal exposure to PFASs has been shown to negatively affect the response to pediatric vaccines in children up to 3 years of age [142]. In the same scenario, the accumulation of PFAS by children is greater than that by parents because of their small body mass and transfer via the placenta and lactation. In humans, PFASs have extended half-lives due to greater renal tubular reabsorption. This leads to adverse outcomes such as histologic and cellular derangements in the proximal tubules, diminished kidney function, and impaired regulation of metabolic pathways associated with kidney diseases [143]. Reproductive dysregulation induced by PFAS in men includes impaired sperm motility and compromised penetration into viscous media [144, 145]. In young women, PFAS changes endometrial regulation [146]. In adults and children, associations among PFASs and elevated serum and low-density lipoprotein cholesterol have been reported. Furthermore, studies indicate that PFASs play a role in modulating thyroid and hepatic function [147].

7 The Economic Impact of Microplastic Pollution, Including PFASs

According to the UNEP, plastic pollution could triple by the year 2040 [148]. A necessary action is to eliminate plastic waste to meet the 26th UN Climate Change Conference of the Parties (COP26) agreement on net zero by 2060 [149]. As a response to microplastic threats, the UNEA approved the motion of End Plastic Pollution on March 2, 2022, aiming at promoting a global strategy against plastic pollution and posing new challenges to the study of environmental and human health risks. The approval of the UNEP's End Plastic Pollution would accelerate plastic recycling and thereby help achieve economic growth and ecosystem health [150]. Since microplastics are smaller and less visible than macroplastics are, managing their waste is far more difficult. This is the definite moment for all governments to take notice and implement this resolution. International organizations and governments are endorsing concepts such as the circular economy and the six "Rs"—Reduce (raw material usage), Redesign (designing reusable and recyclable products), Remove (avoiding the usage of single-use plastics), Reuse (refurbishment of old products), Recycle (repeated usage of products), and Recover (regeneration and resynthesis) for sustainable growth and development [151]. Microplastic occurrence and release into water bodies must be closely monitored, and effective measures to minimize their unfavorable impacts on ecosystems have become an inevitable burden on the global economy. Monetary tools such as penalties, taxes, dues, subsidizations, deposit-refund schemes, and sponsors have also been demonstrated to be helpful in encouraging the reduction of MP pollutants in aquatic ecosystems. The removal of MPs in marine ecosystems is a focal point of goal no. 14—Life Under Water. The emphasis on the health of marine ecosystems was proposed by the Sustainable Development Goals of the UN to be achieved by 2025. According to the UNEP, in 2018, approximately 5.3 million tons of plastic waste from neglected waste and litter became secondary microplastics over the course of time. Globally, 80% of the plastic that ends up in the ocean is produced by the discharge of more than 1000 rivers, with small rivers that flow through densely populated urban areas, especially in Asia, being most polluted by plastic waste [152]. To eradicate the harmful effects of plastic pollution, a thorough account of the ecological, social, and economic factors that contribute to redefining the global shift in how we produce, use, and recycle plastics through the circular economy is essential. This aspect is essential for laying the groundwork for successful and efficient international negotiations about plastic management, disposal, treatment, and sustainable use.

7.1 Global Economic Burden of Microplastic Recovery

The increasing presence of emerging contaminants in the environment is increasingly concerning because of their bioaccumulation, toxicity, and persistence (Sarkar et al., 2021). Additionally, sewage and wastewater treatment plants are not entirely

Table 4.1 Various recently developed techniques to recover microplastics from water matrices

Electrocoagulation	>90% removal of polyethylene microbeads and microplastics	[155, 156]
Granular filtration	80%	[157, 158]
Membrane processes	>90%	[159]
Froth floatation	95%	[160, 161]
Adsorption by MOF	90%	[162]
Biochar sand filter	95%	[163, 164]
Magnetic separation	95%	[165]

Table 4.2 The factors contributing to the high costs of reducing microplastic pollution

Detection and monitoring	• Technology investment, skilled personnel for monitoring and data analysis and infrastructure
Removal and treatment	• Upgraded wastewater treatment plants with technology to filter requires significant investment • Many removal processes are energy-intensive, increasing costs over time
Environmental rehabilitation	• Cleaning up affected ecosystems and restoring biodiversity demands long-term funding
Public health impacts	• Exposure to microplastics through food, water, and air may lead to health issues, indirectly increasing public healthcare expenditures • Investigating long-term health implications of microplastics requires sustained funding
Policy and governance	• Enacting and enforcing bans on microplastic-producing products (e.g., microbeads in cosmetics) involves administrative costs
Regulation enforcement	• Incentives for alternatives: Governments often subsidize alternatives to microplastics, adding to public spending
Economic losses	• Contaminated marine life affects market value, leading to reduced income • Polluted beaches and water bodies deter tourists, impacting local economies • Microplastic-contaminated soils can affect crop yield and quality, leading to economic losses for farmers
Mitigation strategies to reduce costs	• Promoting sustainable packaging and biodegradable materials to minimize microplastic generation • Educating the public about reducing plastic use and improving waste management can lessen the need for expensive cleanup operations • Developing systems for recycling and reusing materials efficiently can reduce overall waste and its environmental impact

effective at removing all microplastics, leading to significant amounts being released in their effluent [153], which can subsequently contaminate freshwater sources and enter drinking water supplies. Microplastics are persistent in environmental and drinking water samples, and treatment processes such as coagulation, sand filtration, and biological treatment are needed to manage their presence in the environment [154]. Advanced water treatment methods are crucial for enhancing effluent quality and effectively removing microplastics from wastewater, and different physical removal methods and the factors influencing their efficiency are discussed (Table 4.1).

The universal challenge of confronting microplastic PFAS pollution depends on each county's economic conditions and priorities. The United States and Europe have initiated significant funding efforts targeting PFAS detection and removal. The

US Environmental Protection Agency (EPA) assigned generous resources in 2023 under the PFAS Strategic Roadmap, focusing on water filtration technology and research on safe disposal. Europe has implemented the Green Deal framework, with funding directed toward innovation in recycling and reducing microplastics in industrial processes, in line with the REACH regulation. Despite facing a financial crisis, India is aware of emerging contaminants and has adopted programs such as the Namami Ganges Program for river cleanup. India's allocation is limited in comparison with that of Western countries because of economic constraints and conflicting priorities. The economic burden for a country stems from the complexities of detection, removal, and mitigation of microplastic contaminants as well as broader societal impacts. Table 4.2 hows the factors contributing to the high costs of reducing microplastic pollution.

7.2 *Economic Challenges for the Sustainable Removal of PFAS*

As the current situation demands sustainable treatment technologies to eliminate persistent PFAS from water sources, comprehensive sustainability studies are needed. The initial step in the sustainable removal of PFAS involves identifying the type of water (e.g., groundwater and drinking water) that requires treatment. On the basis of the water type, appropriate treatment methods (ion exchange resin, granular activated carbon (GAC), reverse osmosis, and electrochemical oxidation) and parameters are considered. Sustainability is assessed across three primary indicators: environmental sustainability, financial viability, and social impacts. The prioritization of indicators aids in developing frameworks and policies for the removal of such emerging contaminants from water bodies [166]. The Biden–Harris administration in 2024 implemented several important aspects of its PFAS plan, including drinking water standards to detect six PFAS, a new framework for evaluating PFAS usage, and $10 billion grants and loans to address PFAS and emerging contaminants. The European Union has strengthened regulatory frameworks to limit PFAS usage and fund research into their removal. Key efforts include wastewater management and restricting the use of PFAS in consumer goods. **These programs support both the EU Green Deal and more general sustainability objectives.** India is in the early stages of addressing microplastic and PFAS contamination. While specific funding initiatives are limited, the government collaborates with international bodies to research removal technologies. Efforts to enhance waste management and reduce single-use plastics aim to address this problem indirectly. With varying levels of investment in reducing PFAS and microplastic pollution, each country has unique economic and environmental difficulties. For sustainable solutions, support and originality in technological and regulatory approaches are essential.

8 Conclusions

Each year, up to $2.5 trillion is lost due to marine plastic pollution. Recently, research has focused on measuring economic losses due to microplastics and secondary pollutant PFASs. Owing to annual increases in plastic production and the tenacious presence of microplastics, they are predicted to have been added to the marine environment from land-based sources. The negative ecological, social, and economic impacts of microplastic pollution are likely to continue to increase. In-depth knowledge of the environmental fate and potential adverse effects of microplastics and PFASs in aquatic environments is needed. The public community and front lines should advocate the implementation of policies to mitigate these emerging contaminants and restore balance in aquatic ecosystems. Although researchers are exploring sustainable ways to detect and remove microplastics and PFASs in most developing countries, convincing data on harmful ecotoxicological effects are still lacking, which may help policy makers curb plastic pollution in the future.

References

1. Anderson, J. C., Park, B. J., & Palace, V. P. (2016). Microplastics in aquatic environments: Implications for Canadian ecosystems. *Environmental Pollution, 218*, 269–280.
2. Llorca, M., Schirinzi, G., Martínez, M., Barceló, D., & Farré, M. (2018). Adsorption of perfluoroalkyl substances on microplastics under environmental conditions. *Environmental Pollution, 235*, 680–691.
3. Stolte, A., Forster, S., Gerdts, G., & Schubert, H. (2015). Microplastic concentrations in beach sediments along the German Baltic coast. *Marine Pollution Bulletin, 99*(1–2), 216–229.
4. An, L., Liu, Q., Deng, Y., Wu, W., Gao, Y., & Ling, W. (2020). Sources of microplastic in the environment. In *Microplastics in terrestrial environments: Emerging contaminants and major challenges* (pp. 143–159).
5. Karlsson, T. M., Arneborg, L., Broström, G., Almroth, B. C., Gipperth, L., & Hassellöv, M. (2018). The unaccountability case of plastic pellet pollution. *Marine Pollution Bulletin, 129*(1), 52–60.
6. Napper, I. E., Bakir, A., Rowland, S. J., & Thompson, R. C. (2015). Characterization, quantity and sorptive properties of microplastics extracted from cosmetics. *Marine Pollution Bulletin, 99*(1–2), 178–185.
7. Atugoda, T., Vithanage, M., Wijesekara, H., Bolan, N., Sarmah, A. K., Bank, M. S., You, S., & Ok, Y. S. (2021). Interactions between microplastics, pharmaceuticals and personal care products: Implications for vector transport. *Environment International, 149*, 106367.
8. Gouin, T., Avalos, J., Brunning, I., Brzuska, K., De Graaf, J., Kaumanns, J., Koning, T., Meyberg, M., Rettinger, K., Schlatter, H., & Thomas, J. (2015). Use of microplastic beads in cosmetic products in Europe and their estimated emissions to the North Sea environment. *SOFW Journal, 141*(4), 40–46.
9. Barnes, D. K., Galgani, F., Thompson, R. C., & Barlaz, M. (2009). Accumulation and fragmentation of plastic debris in global environments. *Philosophical Transactions of the Royal Society B: Biological Sciences, 364*(1526), 1985–1998.
10. Xu, B., Liu, S., Zhou, J. L., Zheng, C., Weifeng, J., Chen, B., Zhang, T., & Qiu, W. (2021). PFAS and their substitutes in groundwater: Occurrence, transformation and remediation. *Journal of Hazardous Materials, 412*, 125159.

11. Ding, R., Tong, L., & Zhang, W. (2021). Microplastics in freshwater environments: Sources, fates and toxicity. *Water, Air, and Soil Pollution, 232*, 181.
12. Abbasi, S., Keshavarzi, B., Moore, F., Delshab, H., Soltani, N., & Sorooshian, A. (2017). Investigation of microrubbers, microplastics and heavy metals in street dust: A study in Bushehr City, Iran. *Environmental Earth Sciences, 76*(23), 798.
13. Fendall, L. S., & Sewell, M. A. (2009). Contributing to marine pollution by washing your face: Microplastics in facial cleansers. *Marine Pollution Bulletin, 58*(8), 1225–1228.
14. Lambert, S., & Wagner, M. (2018). *Microplastics are contaminants of emerging concern in freshwater environments: An overview* (pp. 1–23). Springer International.
15. Browne, M. A., Crump, P., Niven, S. J., Teuten, E., Tonkin, A., Galloway, T., & Thompson, R. (2011). Accumulation of microplastic onshorelines worldwide: Sources and sinks. *Environmental Science & Technology, 45*(21), 9175–9179.
16. Astrom, L. (2016). Shedding of synthetic microfibers from textiles. *Göteborgs Universitet, 427*, 428.
17. Dehghani, S., Moore, F., & Akhbarizadeh, R. (2017). Microplastic pollution in deposited urban dust, Tehran metropolis, Iran. *Environmental Science and Pollution Research, 24*, 20360–20371.
18. Yuan, W., Liu, X., Wang, W., Di, M., & Wang, J. (2019). Microplastic abundance, distribution and composition in water, sediments, and wild fish from Poyang Lake, China. *Ecotoxicology and Environmental Safety, 170*, 180–187.
19. Akter, M. S., Chakraborty, T. K., Ghosh, G. C., Nice, M. S., Zaman, S., & Khan, A. S. (2024). Microplastics and heavy metals in freshwater fish species in the southwestern region of Bangladesh: An emerging concern for public health. *Emerging Contaminants, 10*(3), 100325.
20. Anandhan, K., Thangal, S. H., Yogeshwaran, A., Kaaran, S., Ajith Kumar, T. T., & Muralisankar, T. (2023). Microplastics contamination in the edible fish Mozambique tilapia (Oreochromis mossambicus) from the Selvampathy wetland of Coimbatore, Tamil Nadu, India. *Bulletin of Environmental Contamination and Toxicology, 112*(1), 7.
21. Lechthaler, S., Waldschläger, K., Sandhani, C. G., Sannasiraj, S. A., Sundar, V., Schwarzbauer, J., & Schüttrumpf, H. (2021). Baseline study on microplastics in Indian Rivers under different anthropogenic influences. *Water, 13*(12), 1648.
22. Lwanga, E. H., Beriot, N., Corradini, F., Silva, V., Yang, X., Baartman, J., Rezaei, M., van Schaik, L., Riksen, M., & Geissen, V. (2022). Review of microplastic sources, transport pathways and correlations with other soil stressors: A journey from agricultural sites into the environment. *Chemical and Biological Technologies in Agriculture, 9*(1), 20.
23. Nizzetto, L., Bussi, G., Futter, M. N., Butterfield, D., & Whitehead, P. G. (2016). A theoretical assessment of microplastic transport in river catchments and their retention by soils and river sediments. *Environmental Science. Processes & Impacts, 18*(8), 1050–1059.
24. Van Emmerik, T., Kieu-Le, T. C., Loozen, M., Van Oeveren, K., Strady, E., Bui, X. T., et al. (2018). A methodology to characterize riverine macroplastic emission into the ocean. *Frontiers in Marine Science, 5*, 372.
25. Gasperi, J., Dris, R., Bonin, T., Rocher, V., & Tassin, B. (2014). Assessment of floating plastic debris in surface water along the Seine River. *Environmental Pollution, 195*, 163–166.
26. Ziani, K., Ioniță-Mîndrican, C. B., Mititelu, M., Neacşu, S. M., Negrei, C., Moroşan, E., Drăgănescu, D., & Preda, O. T. (2023). Microplastics: A real global threat for environment and food safety: A state of the art review. *Nutrients, 15*(3), 617.
27. Salawu, O. A., Olivares, C. I., & Adeleye, A. S. (2024). Adsorption of PFAS onto secondary microplastics: A mechanistic study. *Journal of Hazardous Materials, 470*, 134185.
28. da Costa, J. P., Santos, P. S. M., Duarte, A. C., & Rocha-Santos, T. (2016). (Nano) plastics in the environment—Sources, fates and effects. *The Science of the Total Environment, 566–567*, 15–26.
29. Ateia, M., Zheng, T., Calace, S., Tharayil, N., Pilla, S., & Karanfil, T. (2020). Sorption behavior of real microplastics (MPs): Insights for organic micropollutants adsorption on a large set of well-characterized MPs. *The Science of the Total Environment, 720*, 137634.

30. Parashar, N., Mahanty, B., & Hait, S. (2023). Microplastics as carriers of per- and polyfluoroalkyl substances (PFAS) in aquatic environment: Interactions and ecotoxicological effects. *Water Emerging Contaminants & Nanoplastics, 2*(7), 15.
31. Lu, D., Sha, S., Luo, J., Huang, Z., & Zhang Jackie, X. (2020). Treatment train approaches for the remediation of per- and polyfluoroalkyl substances (PFAS): A critical review. *Journal of Hazardous Materials, 386*, 121963.
32. Daly, E. R., Chan, B. P., Talbot, E. A., Nassif, J., Bean, C., Cavallo, S. J., Metcalf, E., Simone, K., & Woolf, A. D. (2018). Per- and polyfluoroalkyl substance (PFAS) exposure assessment in a community exposed to contaminated drinking water, New Hampshire, 2015. *International Journal of Hygiene and Environmental Health, 221*(3), 569–577.
33. Nelson, J. W., Hatch, E. E., & Webster, T. F. (2010). Exposure to polyfluoroalkyl chemicals and cholesterol, body weight, and insulin resistance in the general U.S. population. *Environmental Health Perspectives, 118*(2), 197–202.
34. Grandjean, P., Andersen, E. W., Budtz-Jørgensen, E., Nielsen, F., Mølbak, K., Weihe, P., & Heilmann, C. (2012). Serum vaccine antibody concentrations in children exposed to perfluorinated compounds. *JAMA, 307*(4), 391–397.
35. Sun, Q., Zong, G., Valvi, D., Nielsen, F., Coull, B., & Grandjean, P. (2018). Plasma concentrations of Perfluoroalkyl substances and risk of type 2 diabetes: A prospective investigation among U.S. women. *Environmental Health Perspectives, 126*(3), 037001.
36. Ahrens, L., Hedlund, J., Tröger, R., Wiberg, K., & Dürig, W. (2016). Screening of PFASs in groundwater and surface water. *Rapport/Sveriges lantbruksuniversitet, Institutionen för vatten och miljö, 2016*, 2.
37. Gomis, M. I., Vestergren, R., Borg, D., & Cousins, I. T. (2018). Comparing the toxic potency in vivo of long-chain perfluoroalkyl acids and fluorinated alternatives. *Environment International, 113*, 1–9.
38. Pétré, M. A., Genereux, D. P., Koropeckyj-Cox, L., Knappe, D. R. U., Duboscq, S., Gilmore, T. E., & Hopkins, Z. R. (2021). Per- and Polyfluoroalkyl substance (PFAS) transport from groundwater to streams near a PFAS manufacturing Facility in North Carolina, USA. *Environmental Science & Technology, 55*(9), 5848–5856.
39. Briggs, M. A., Tokranov, A. K., Hull, R. B., LeBlanc, D. R., Haynes, A., & Lane, J. W., Jr. (2020). Hillslope groundwater discharges provide localized ecosystem buffers from regional PFAS contamination in a gaining coastal stream. *Hydrological Processes, 34*(10), 2281–2291.
40. Sharma, B. M., Bharat, G. K., Tayal, S., Larssen, T., Bečanová, J., Karásková, P., Whitehead, P. G., Futter, M. N., Butterfield, D., & Nizzetto, L. (2016). Perfluoroalkyl substances (PFAS) in river and ground/drinking water of the Ganges River basin: Emissions and implications for human exposure. *Environmental Pollution, 208*(Pt B), 704–713.
41. McMahon, P. B., Tokranov, A. K., Bexfield, L. M., Lindsey, B. D., Johnson, T. D., Lombard, M. A., & Watson, E. (2022). Perfluoroalkyl and polyfluoroalkyl substances in groundwater used as a source of drinking water in the eastern United States. *Environmental Science & Technology, 56*(4), 2279–2288.
42. Nzeribe, B. N., Crimi, M., Thagard, S. M., & Holsen, T. M. (2019). Physico-chemical processes for the treatment of per- and polyfluoroalkyl substances (PFAS): A review. *Critical Reviews in Environmental Science and Technology, 49*(10), 866–915.
43. Hepburn, E., Madden, C., Szabo, D., Coggan, T. L., Clarke, B., & Currell, M. (2019). Contamination of groundwater with per- and polyfluoroalkyl substances (PFAS) from legacy landfills in an urban redevelopment precinct. *Environmental Pollution, 248*, 101–113.
44. Park, M., Wu, S., Lopez, I. J., Chang, J. Y., Karanfil, T., & Snyder, S. A. (2020). Adsorption of perfluoroalkyl substances (PFAS) in groundwater by granular activated carbons: Roles of hydrophobicity of PFAS and carbon characteristics. *Water Research, 170*, 115364.
45. Jambeck, J. R., Geyer, R., Wilcox, C., et al. (2015). Plastic waste inputs from land into the ocean. *Science, 347*(6223), 768–771.

46. Kulka, T., & Brunner, K. (2015). Passive buoyant tracers in the ocean surface boundary layer: 1. Influence of equilibrium wind-waves on vertical distributions. *Journal of Geophysical Research: Oceans, 120*(5), 3837–3858.
47. Hale, R. C., Seeley, M. E., La Guardia, M. J., Mai, L., & Zeng, E. Y. (2020). A global perspective on microplastics. *Journal of Geophysical Research: Oceans, 125*(1), e2018JC014719.
48. Mason, R., Kukulka, T., & Cohen, T. (2022). Effects of particle buoyancy, release location, and diel vertical migration on exposure of marine organisms to microplastics in Delaware Bay. *Estuarine, Coastal and Shelf Science, 275*, 107990.
49. Yang, X., Huang, G., Chen, Z., Feng, Q., An, C., Lyu, L., Bi, H., & Zhou, S. (2024). Spotlight on the vertical migration of aged microplastics in coastal waters. *Journal of Hazardous Materials, 469*, 134040.
50. Ajith, N., Arumugam, S., Parthasarathy, S., Manupoori, S., & Janakiraman, S. (2020). Global distribution of microplastics and its impact on marine environment-a review. *Environmental Science and Pollution Research International, 27*(21), 25970–25986.
51. Koongolla, J. B., Lin, L., Pan, Y. F., Yang, C. P., Sun, D. R., Liu, S., Xu, X. R., Maharana, D., Huang, J. S., & Li, H. X. (2020). Occurrence of microplastics in gastrointestinal tracts and gills of fish from Beibu gulf, South China Sea. *Environmental Pollution, 258*, 113734.
52. Naidoo, T., Sershen, Thompson, R. C., & Rajkaran, A. (2020). Quantification and characterization of microplastics ingested by selected juvenile fish species associated with mangroves in KwaZulu-Natal, South Africa. *Environmental Pollution, 257*, 113635.
53. Nel, H. A., & Froneman, P. W. (2018). Presence of microplastics in the tube structure of the reef-building polychaete *Gunnarea gaimardi* (Quatrefages 1848). *African Journal of Marine Science, 40*(1), 87–89.
54. Thoman, T. X., Kukulka, T., Cohen, J. H., & Boettcher, H. (2023). Zooplankton-microplastic exposure in Delaware coastal waters: Atlantic blue crab (Callinectes sapidus) larvae case study. *Marine Pollution Bulletin, 196*, 115541.
55. Ariefdien, R., Pfaff, M., Awe, A., & Sparks, C. (2024). Stormwater outlets: A source of microplastics in coastal zones of Cape Town, South Africa. *Marine Pollution Bulletin, 198*, 115800.
56. Sparks, C., & Immelman, S. (2020). Microplastics in offshore fish from the Agulhas Bank, South Africa. *Marine Pollution Bulletin, 156*, 111216.
57. Yamashita, N., Taniyasu, S., Petrick, G., Wei, S., Gamo, T., Lam, P. K., & Kannan, K. (2008). Perfluorinated acids as novel chemical tracers of global circulation of ocean waters. *Chemosphere, 70*(7), 1247–1255.
58. Benskin, J. P., Muir, D. C., Scott, B. F., Spencer, C., De Silva, A. O., Kylin, H., Martin, J. W., Morris, A., Lohmann, R., Tomy, G., Rosenberg, B., Taniyasu, S., & Yamashita, N. (2012). Perfluoroalkyl acids in the Atlantic and Canadian Arctic oceans. *Environmental Science & Technology, 46*(11), 5815–5823.
59. González-Gaya, B., Dachs, J., Roscales, J. L., Caballero, G., & Jiménez, B. (2014). Perfluoroalkylated substances in the global tropical and subtropical surface oceans. *Environmental Science & Technology, 48*(22), 13076–13084.
60. Yeung, L. W. Y., Dassuncao, C., Mabury, S., Sunderland, E. M., Zhang, X., & Lohmann, R. (2017). Vertical profiles, sources, and transport of PFASs in the Arctic Ocean. *Environmental Science & Technology, 51*(12), 6735–6744.
61. Shan, G., Qian, X., Chen, X., Feng, X., Cai, M., Yang, L., Chen, M., Zhu, L., & Zhang, S. (2021). Legacy and emerging per- and poly-fluoroalkyl substances in surface seawater from northwestern Pacific to Southern Ocean: Evidence of current and historical release. *Journal of Hazardous Materials, 411*, 125049.
62. Tansel, B. (2024). Geographical characteristics that promote persistence and accumulation of PFAS in coastal waters and open seas: Current and emerging hot spots. *Environmental Challenges, 14*, 100861.
63. Bilela, L. L., Matijošytė, I., Krutkevičius, J., Alexandrino, D. A., Safarik, I., Burlakovs, J., Gaudêncio, S. P., & Carvalho, M. F. (2023). Impact of per-and polyfluorinated alkyl substances (PFAS) on the marine environment: Raising awareness, challenges, legislation, and mitigation approaches under the one health concept. *Marine Pollution Bulletin, 194*, 115309.

64. Moretti, S., Brambilla, G., Maffucci, F., Barola, C., Bucaletti, E., Hochscheid, S., Canzanella, S., Galarini, R., & Esposito, M. (2024). Occurrence and pattern of legacy and emerging per- and poly-FluoroAlkyl substances (PFAS) in eggs of loggerhead turtle Caretta caretta from western Mediterranean. *Environmental Pollution, 343*, 123257.
65. Cara, B., Lies, T., Thimo, G., Robin, L., & Lieven, B. (2022). Bioaccumulation and trophic transfer of perfluorinated alkyl substances (PFAS) in marine biota from the Belgian North Sea: Distribution and human health risk implications. *Environmental Pollution, 311*, 119907.
66. Barghi, M., Jin, X., Lee, S., Jeong, Y., Yu, J.-P., Paek, W.-K., & Moon, H.-B. (2018). Accumulation and exposure assessment of persistent chlorinated and fluorinated contaminants in Korean birds. *Science of the Total Environment, 645*, 220–228.
67. Soloff, A. C., Wolf, B. J., White, N. D., Muir, D., Courtney, S., Hardiman, G., Bossart, G. D., & Fair, P. A. (2017). Environmental perfluorooctane sulfonate exposure drives T-cell activation in bottlenose dolphins. *Journal of Applied Toxicology, 37*(9), 1108–1116.
68. Garnett, J., Halsall, C., Thomas, M., Crabeck, O., France, J., Joerss, H., Ebinghaus, R., Kaiser, J., Leeson, A., & Wynn, P. M. (2021). Investigating the uptake and fate of poly- and perfluoroalkylated substances (PFAS) in sea ice using an Experimental Sea ice chamber. *Environmental Science & Technology, 55*(14), 9601–9608.
69. Rocha-Santos, T., & Duarte, A. C. (2015). A critical overview of the analytical approaches to the occurrence, the fate and the behavior of microplastics in the environment. *TrAC Trends in Analytical Chemistry, 65*, 47–53.
70. Wang, W., & Wang, J. (2018). Investigation of microplastics in aquatic environments: An overview of the methods used, from field sampling to laboratory analysis. *TrAC Trends in Analytical Chemistry, 108*, 195–202.
71. Okubo, N., Takahashi, S., & Nakano, Y. (2018). Microplastics disturb the anthozoan-algae symbiotic relationship. *Marine Pollution Bulletin, 135*, 83–89.
72. Cole, M., Lindeque, P., Fileman, E., Halsband, C., Goodhead, R., Moger, J., & Galloway, T. S. (2013). Microplastic ingestion by zooplankton. *Environmental Science & Technology, 47*(12), 6646–6655.
73. Lee, K. W., Shim, W. J., Kwon, O. Y., & Kang, J. H. (2013). Size-dependent effects of micro polystyrene particles in the marine copepod Tigriopus japonicus. *Environmental Science & Technology, 47*(19), 11278–11283.
74. Fu, W., Min, J., Jiang, W., Li, Y., & Zhang, W. (2020). Separation, characterization and identification of microplastics and nanoplastics in the environment. *Science of the Total Environment, 721*, 137561.
75. Löder, M. G., Imhof, H. K., Ladehoff, M., Löschel, L. A., Lorenz, C., Mintenig, S., et al. (2017). Enzymatic purification of microplastics in environmental samples. *Environmental Science & Technology, 51*(24), 14283–14292.
76. Prudent, E., & Raoult, D. (2019). Fluorescence in situ hybridization, a complementary molecular tool for the clinical diagnosis of infectious diseases by intracellular and fastidious bacteria. *FEMS Microbiology Reviews, 43*(1), 88–107.
77. Ter Halle, A., Ladirat, L., Gendre, X., Goudouneche, D., Pusineri, C., Routaboul, C., Tenailleau, C., Duployer, B., & Perez, E. (2016). Understanding the fragmentation pattern of marine plastic debris. *Environmental Science & Technology, 50*(11), 5668–5675.
78. Wagner, J., Wang, Z. M., Ghosal, S., Rochman, C., Gassel, M., & Wall, S. (2017). Novel method for the extraction and identification of microplastics in ocean trawl and fish gut matrices. *Analytical Methods, 9*(9), 1479–1490.
79. Karami, A., Golieskardi, A., Choo, C. K., Larat, V., Karbalaei, S., & Salamatinia, B. (2018). Microplastic and mesoplastic contamination in canned sardines and sprats. *Science of the Total Environment, 612*, 1380–1386.
80. Dong, M., Zhang, Q., Xing, X., Chen, W., She, Z., & Luo, Z. (2020). Raman spectra and surface changes of microplastics weathered under natural environments. *Science of the Total Environment, 739*, 139990.

81. Shi, B., Patel, M., Yu, D., Yan, J., Li, Z., Petriw, D., Pruyn, T., Smyth, K., Passeport, E., Miller, R. D., & Howe, J. Y. (2022). Automatic quantification and classification of microplastics in scanning electron micrographs via deep learning. *Science of the Total Environment, 825*, 153903.
82. Felts, J. R., Kjoller, K., Lo, M., Prater, C. B., & King, W. P. (2012). Nanometer-scale infrared spectroscopy of heterogeneous polymer nanostructures fabricated by tip-based nanofabrication. *ACS Nano, 6*(9), 8015–8021.
83. Pico, Y., Alfarhan, A., & Barcelo, D. (2019). Nanoand microplastic analysis: Focus on their occurrence in freshwater ecosystems and remediation technologies. *TrAC Trends in Analytical Chemistry, 113*, 409–425.
84. Simon, M., Vianello, A., Shashoua, Y., & Vollertsen, J. (2021). Accelerated weathering affects the chemical and physical properties of marine antifouling paint microplastics and their identification by ATR-FTIR spectroscopy. *Chemosphere, 274*, 129749.
85. Enyoh, C. E., & Wang, Q. (2024). Automated classification of Undegraded and aged polyethylene terephthalate microplastics from ATR-FTIR spectroscopy using machine learning algorithms. *Journal of Polymers and the Environment, 32*, 1–16.
86. Faltynkova, A., & Wagner, M. (2023). Developing and testing a workflow to identify microplastics using near infrared hyperspectral imaging. *Chemosphere, 336*, 139186.
87. Wander, L., Lommel, L., Meyer, K., Braun, U., & Paul, A. (2022). Development of a low-cost method for quantifying microplastics in soils and compost using near-infrared spectroscopy. *Measurement Science and Technology, 33*(7), 075801.
88. Thompson, R. C., Olsen, Y., Mitchell, R. P., Davis, A., Rowland, S. J., John, A. W., McGonigle, D., & Russell, A. E. (2004). Lost at sea: Where is all the plastic? *Science, 304*(5672), 838–838.
89. Wu, M., Yang, C., Du, C., & Liu, H. (2020). Microplastics in waters and soils: Occurrence, analytical methods and ecotoxicological effects. *Ecotoxicology and Environmental Safety, 202*, 110910.
90. Materic, D., Kasper-Giebl, A., Kau, D., Anten, M., Greilinger, M., Ludewig, E., van Sebille, E., Roeckmann, T., & Holzinger, R. (2020). Micro and Nanoplastics in alpine snow: A new method for chemical identification and (semi)quantification in the Nanogram range. *Environmental Science & Technology, 54*, 2353–2359.
91. Liu, Y., Li, R., Yu, J., Ni, F., Sheng, Y., Scircle, A., Cizdziel, J. V., & Zhou, Y. (2021). Separation and identification of microplastics in marine organisms by TGA-FTIR-GC/MS: A case study of mussels from coastal China. *Environmental Pollution, 272*, 115946.
92. John, J., Coulon, F., & Chellam, P. V. (2022). Detection and treatment strategies of per- and polyfluoroalkyl substances (PFAS): Fate of PFAS through DPSIR framework analysis. *Journal of Water Process Engineering, 45*, 102463.
93. Ren, H., Hu, H., Yu, B., Du, Y., & Wu, T. (2018). Identification of polymer building blocks by Py–GC/MS and MALDI-TOF MS. *International Journal of Polymer Analysis and Characterization, 23*(1), 9–17.
94. Lin, Y., Huang, X., Liu, Q., Lin, Z., & Jiang, G. (2020). Thermal fragmentation enhanced identification and quantification of polystyrene micro/nanoplastics in complex media. *Talanta, 208*, 120478.
95. Imhof, H. K., Laforsch, C., Wiesheu, A. C., Schmid, J., Anger, P. M., Niessner, R., & Ivleva, N. P. (2016). Pigments and plastic in limnetic ecosystems: A qualitative and quantitative study on microparticles of different size classes. *Water Research, 98*, 64–74.
96. Wang, J., Peng, J., Tan, Z., Gao, Y., Zhan, Z., Chen, Q., & Cai, L. (2017). Microplastics in the surface sediments from the Beijiang River littoral zone: Composition, abundance, surface textures and interaction with heavy metals. *Chemosphere, 171*, 248–258.
97. Kühn, S., Van Oyen, A., Booth, A. M., Meijboom, A., & Van Franeker, J. A. (2018). Marine microplastic: Preparation of relevant test materials for laboratory assessment of ecosystem impacts. *Chemosphere, 213*, 103–113.

98. Acosta-Coley, I., Mendez-Cuadro, D., Rodriguez-Cavallo, E., de la Rosa, J., & Olivero-Verbel, J. (2019). Trace elements in microplastics in Cartagena: A hotspot for plastic pollution at the Caribbean. *Marine Pollution Bulletin, 139*, 402–411.
99. Klöckner, P., Reemtsma, T., Eisentraut, P., Braun, U., Ruhl, A. S., & Wagner, S. (2019). Tire and road wear particles in road environment–quantification and assessment of particle dynamics by Zn determination after density separation. *Chemosphere, 222*, 714–721.
100. Klöckner, P., Reemtsma, T., & Wagner, S. (2021). The diverse metal composition of plastic items and its implications. *Science of the Total Environment, 764*, 142870.
101. Auta, H. S., Emenike, C. U., & Fauziah, S. H. (2017). Distribution and importance of microplastics in the marine environment: A review of the sources, fate, effects, and potential solutions. *Environment International, 102*, 165–176.
102. Botterell, Z. L., Beaumont, N., Dorrington, T., Steinke, M., Thompson, R. C., & Lindeque, P. K. (2019). Bioavailability and effects of microplastics on marine zooplankton: A review. *Environmental Pollution, 245*, 98–110.
103. Pereira, R., Rodrigues, S. M., Silva, D., Freitas, V., Almeida, C. M. R., & Ramos, S. (2023). Microplastic contamination in large migratory fishes collected in the open Atlantic Ocean. *Marine Pollution Bulletin, 186*, 114454.
104. De Leeuw, J. W., De Leer, E. W. B., Damste, J. S., & Schuyl, P. J. W. (1986). Screening of anthropogenic compounds in polluted sediments and soils by flash evaporation/pyrolysis gas chromatography–mass spectrometry. *Analytical Chemistry, 58*(8), 1852–1857.
105. Santos, L. H., Insa, S., Arxé, M., Buttiglieri, G., Rodríguez-Mozaz, S., & Barceló, D. (2023). Analysis of microplastics in the environment: Identification and quantification of trace levels of common types of plastic polymers using pyrolysis-GC/MS. *MethodsX, 10*, 102143.
106. Picó, Y., & Barceló, D. (2020). Pyrolysis gas chromatography–mass spectrometry in environmental analysis: Focus on organic matter and microplastics. *TrAC Trends in Analytical Chemistry, 130*, 115964.
107. Nazarov, E. G., Postlethwaite, T., Markoski, K., Koo, S., & Borenstein, J. T. (2017). Planar differential mobility spectrometry for clinical breath diagnostics. In *Diagnostic devices with microfluidics*. CRC.
108. Limero, T., Reese, E., Cheng, P., & Trowbridge, J. (2011). Preparation of a gas chromatograph-differential mobility spectrometer to measure target volatile organic compounds on the international space station. *International Journal for Ion Mobility Spectrometry, 14*, 81–91.
109. Limero, T., Reese, E., Wallace, W. T., Cheng, P., & Trowbridge, J. (2012). The results from the air quality monitor (gas chromatograph-differential mobility spectrometer) experiment on board the international space station. *International Journal for Ion Mobility Spectrometry, 15*(3), 189–198.
110. Aralekallu, S., Palanna, M., Hadimani, S., Prabhu, C. P. K., Sajjan, V. A., Thotiyl, M. O., & Sannegowda, L. K. (2020). Biologically inspired catalyst for electrochemical reduction of hazardous hexavalent chromium. *Dalton Transactions, 49*(42), 15061–15071.
111. Palanna, M., Aralekallu, S., Prabhu, C. K., Sajjan, V. A., & Sannegowda, L. K. (2021). Nanomolar detection of mercury (II) using electropolymerized phthalocyanine film. *Electrochimica Acta, 367*, 137519.
112. Aralekallu, S., Mohammed, I., Manjunatha, N., Palanna, M., & Sannegowda, L. K. (2019). Synthesis of novel azo group substituted polymeric phthalocyanine for amperometric sensing of nitrite. *Sensors and Actuators B: Chemical, 282*, 417–425.
113. Khaire, S., Gaikwad, P., Aralekallu, S., Bhat, Z. M., Kottaichamy, A. R., Devendrachari, M. C., Thimmappa, R., Shafi, S. P., Gautam, M., & Thotiyl, M. O. (2018). An Interface-controlled redox switch for wastewater remediation. *ChemElectroChem, 5*(2), 362–366.
114. Meng, Z., Stolz, R. M., Mendecki, L., & Mirica, K. A. (2019). Electrically transduced chemical sensors based on two-dimensional nanomaterials. *Chemical Reviews, 119*(1), 478–598.
115. Mamun, M. A. A., & Yuce, M. R. (2020). Recent progress in nanomaterial enabled chemical sensors for wearable environmental monitoring applications. *Advanced Functional Materials, 30*(51), 2005703.

116. Chandran, G. T., Li, X., Ogata, A., & Penner, R. M. (2017). Electrically transduced sensors based on nanomaterials (2012–2016). *Analytical Chemistry, 89*(1), 249–275.
117. Oliveira, T. M., Ribeiro, F. W., Morais, S., de Lima-Neto, P., & Correia, A. N. (2022). Removal and sensing of emerging pollutants released from (micro) plastic degradation: Strategies based on boron-doped diamond electrodes. *Current Opinion in Electrochemistry, 31*, 100866.
118. Dutta, S., Misra, A., & Bose, S. (2024). Polyoxometalate nanocluster-infused triple IPN hydrogel for excellent microplastics alleviation from contaminated water: Detection, photodegradation, and upcycling. *Nanoscale, 16*, 5188.
119. Guo, Q., Liu, Y., Liu, J., Wang, Y., Cui, Q., Song, P., Zhang, X., & Zhang, C. (2022). Hierarchically structured hydrogel actuator for microplastic pollutant detection and removal. *Chemistry of Materials, 34*(11), 5165–5175.
120. Leppänen, I., Lappalainen, T., Lohtander, T., Jonkergouw, C., Arola, S., & Tammelin, T. (2022). Capturing colloidal nanoand microplastics with plant-based nanocellulose networks. *Nature Communications, 13*(1), 1814.
121. Devereux, R., Ayati, B., Westhead, E. K., Jayaratne, R., & Newport, D. (2023). Impact of the Covid-19 pandemic on microplastic abundance along the river Thames. *Marine Pollution Bulletin, 1*(189), 114763.
122. Loizia, P., Voukkali, I., Chatziparaskeva, G., Navarro-Pedreno, J., & Zorpas, A. A. (2021). Measuring the level of environmental performance on coastal environment before and during the COVID-19 pandemic: A case study from Cyprus. *Sustainability, 13*(5), 2485.
123. Harris, L. S., La Beur, L., Olsen, A. Y., Smith, A., Eggers, L., Pedersen, E., Van Brocklin, J., Brander, S. M., & Larson, S. (2022). Temporal variability of microparticles under the Seattle aquarium, Washington state: Documenting the global Covid-19 pandemic. *Environmental Toxicology and Chemistry, 41*(4), 917–930.
124. Gupta, P., Saha, M., Suneel, V., Rathore, C., Ray, D., & Naik, A. (2024). The consequences of reduced anthropogenic activities during the COVID-19 pandemic on microplastic abundance in a tropical estuarine region: Goa, India. *Science of the Total Environment, 912*, 169041.
125. Amrutha, K., Warrier, A. K., & Rangel-Buitrago, N. (2023). Did the COVID-19 pandemic play a role in the spatial and temporal variations of microplastics? Evidence from a tropical river in southern India. *Marine Pollution Bulletin, 192*, 115088.
126. Celis, J. E., Espejo, W., Paredes-Osses, E., Contreras, S. A., Chiang, G., & Bahamonde, P. (2021). Plastic residues produced with confirmatory testing for COVID-19: Classification, quantification, fate, and impacts on human health. *Science of the Total Environment, 760*, 144167.
127. Kassam, A. (2020). More masks than jellyfish: Coronavirus waste ends up in ocean. *The Guardian, 8*.
128. Peng, Y., Wu, P., Schartup, A. T., & Zhang, Y. (2021). Plastic waste release caused by COVID-19 and its fate in the global ocean. *Proceedings of the National Academy of Sciences, 118*(47), e2111530118.
129. Parashar, N., & Hait, S. (2023). Recent advances on microplastics pollution and removal from wastewater systems: A critical review. *Journal of Environmental Management, 340*, 118014.
130. Cormier, B., Borchet, F., Kärrman, A., Szot, M., Yeung, L. W., & Keiter, S. H. (2022). Sorption and desorption kinetics of PFOS to pristine microplastic. *Environmental Science and Pollution Research, 29*, 4497–4507.
131. Dai, Y., Zhao, J., Sun, C., Li, D., Liu, X., Wang, Z., Yue, T., & Xing, B. (2022). Interaction and combined toxicity of microplastics and per-and polyfluoroalkyl substances in aquatic environment. *Frontiers of Environmental Science & Engineering, 16*(10), 136.
132. Mejías, C., Martín, J., Santos, J. L., Aparicio, I., & Alonso, E. (2023). Adsorption of perfluoroalkyl substances on polyamide microplastics: Effect of sorbent and influence of environmental factors. *Environmental Research, 216*, 114834.
133. Rainieri, S., Conlledo, N., Larsen, B. K., Granby, K., & Barranco, A. (2018). Combined effects of microplastics and chemical contaminants on the organ toxicity of zebrafish (*Danio rerio*). *Environmental Research, 162*, 135–143.

134. Zolotova, N., Kosyreva, A., Dzhalilova, D., Fokichev, N., & Makarova, O. (2022). Harmful effects of the microplastic pollution on animal health: A literature review. *PeerJ, 10*, e13503.
135. Yuan, Z., Nag, R., & Cummins, E. (2022). Human health concerns regarding microplastics in the aquatic environment-from marine to food systems. *Science of the Total Environment, 823*, 153730.
136. Banerjee, A., & Shelver, W. L. (2021). Microand nanoplastic induced cellular toxicity in mammals: A review. *Science of the Total Environment, 755*, 142518.
137. Rubio, L., Barguilla, I., Domenech, J., Marcos, R., & Hernández, A. (2020). Biological effects, including oxidative stress and genotoxic damage, of polystyrene nanoparticles in different human hematopoietic cell lines. *Journal of Hazardous Materials, 398*, 122900.
138. Cheng, W., Zhou, Y., Xie, Y., Li, Y., Zhou, R., Wang, H., Feng, Y. & Wang, Y. (2023). Combined effect of polystyrene microplastics and bisphenol A on the human embryonic stem cells-derived liver organoids: The hepatotoxicity and lipid accumulation. Science of The Total Environment, 854, 158585.
139. Li, Y., Tao, L., Wang, Q., Wang, F., Li, G., & Song, M. (2023). Potential health impact of microplastics: A review of environmental distribution, human exposure, and toxic effects. *Environment & Health, 1*(4), 249–257.
140. Ma, T., Ye, C., Wang, T., Li, X., & Luo, Y. (2022). Toxicity of per-and polyfluoroalkyl substances to aquatic invertebrates, planktons, and microorganisms. *International Journal of Environmental Research and Public Health, 19*(24), 16729.
141. Looker, C., Luster, M. I., Calafat, A. M., Johnson, V. J., Burleson, G. R., Burleson, F. G., & Fletcher, T. (2014). Influenza vaccine response in adults exposed to perfluorooctanoate and perfluorooctanesulfonate. *Toxicological Sciences, 138*(1), 76–88.
142. Granum, B., Haug, L. S., Namork, E., Stølevik, S. B., Thomsen, C., Aaberge, I. S., van Loveren, H., Løvik, M., & Nygaard, U. C. (2013). Prenatal exposure to perfluoroalkyl substances may be associated with altered vaccine antibody levels and immune-related health outcomes in early childhood. *Journal of Immunotoxicology, 10*(4), 373–379.
143. Stanifer, J. W., Stapleton, H. M., Souma, T., Wittmer, A., Zhao, X., & Boulware, L. E. (2018). Perfluorinated chemicals as emerging environmental threats to kidney health: A scoping review. *Clinical Journal of the American Society of Nephrology, 13*(10), 1479–1492.
144. Šabović, I., Cosci, I., De Toni, L., Ferramosca, A., Stornaiuolo, M., Di Nisio, A., Dall'Acqua, S., Garolla, A., & Foresta, C. (2020). Perfluoro-octanoic acid impairs sperm motility through the alteration of plasma membrane. *Journal of Endocrinological Investigation, 43*, 641–652.
145. Yuan, Y., Ding, X., Cheng, Y., Kang, H., Luo, T., Zhang, X., Kuang, H., Chen, Y., Zeng, X., & Zhang, D. (2020). PFOA evokes extracellular Ca2+ influx and compromises progesterone-induced response in human sperm. *Chemosphere, 241*, 125074.
146. Di Nisio, A., Rocca, M. S., Sabovic, I., Ponce, M. D. R., Corsini, C., Guidolin, D., Zanon, C., Acquasaliente, L., Carosso, A. R., De Toni, L., & Foresta, C. (2020). Perfluorooctanoic acid alters progesterone activity in human endometrial cells and induces reproductive alterations in young women. *Chemosphere, 242*, 125208.
147. Fenton, S. E., Ducatman, A., Boobis, A., DeWitt, J. C., Lau, C., Ng, C., Smith, J. S., & Roberts, S. M. (2021). Per-and polyfluoroalkyl substance toxicity and human health review: Current state of knowledge and strategies for informing future research. *Environmental Toxicology and Chemistry, 40*(3), 606–630.
148. United Nations Environment Programme (UNEP). (2021). From pollution to solution: *A global assessment of marine litter and plastic pollution*. ISBN: 978-92-807-3881-0 Retrieved from https://www.unep.org/resources/pollution-solution-global-assessment-marine-litter-and-plastic-pollution
149. Ritchie, H., & Roser, M. (2024). How much of global greenhouse gas emissions come from food? *Our World in Data*.
150. Dang, F., Wang, Q., Huang, Y., Wang, Y., & Xing, B. (2022). Key knowledge gaps for one health approach to mitigate nanoplastic risks. *Eco-Environment & Health, 1*(1), 11–22.

151. Kumar, A., Mishra, S., Pandey, R., Yu, Z. G., Kumar, M., Khoo, K. S., Thakur, T. K., & Show, P. L. (2023). Microplastics in terrestrial ecosystems: Unignorable impacts on soil characterizes, nutrient storage and its cycling. *TrAC Trends in Analytical Chemistry, 158*, 116869.
152. Meijer, L. J., Van Emmerik, T., Van Der Ent, R., Schmidt, C., & Lebreton, L. (2021). More than 1000 rivers account for 80% of global riverine plastic emissions into the ocean. *Science Advances, 7*(18), eaaz5803.
153. Amrutha, K., & Warrier, A. K. (2020). The first report on the source-to-sink characterization of microplastic pollution from a riverine environment in tropical India. *Science of the Total Environment, 739*, 140377.
154. Shen, M., Song, B., Zhu, Y., Zeng, G., Zhang, Y., Yang, Y., Wen, X., Chen, M., & Yi, H. (2020). Removal of microplastics by drinking water treatment: Current knowledge and future directions. *Chemosphere, 251*, 126612.
155. Kim, K. T., & Park, S. (2021). Enhancing microplastics removal from wastewater using electrocoagulation and granule-activated carbon with thermal regeneration. *PRO, 9*(4), 617.
156. Shen, M., Zhang, Y., Almatrafi, E., Hu, T., Zhou, C., Song, B., Zeng, Z., & Zeng, G. (2022). Efficient removal of microplastics from wastewater by an electrocoagulation process. *Chemical Engineering Journal, 428*, 131161.
157. Pivokonský, M., Pivokonská, L., Novotná, K., Čermáková, L., & Klimtová, M. (2020). Occurrence and fate of microplastics at two different drinking water treatment plants within a river catchment. *Science of the Total Environment, 741*, 140236.
158. Sembiring, E., Fajar, M., & Handajani, M. (2021). Performance of rapid sand filter–single media to remove microplastics. *Water Supply, 21*(5), 2273–2284.
159. Di Bella, G., Corsino, S. F., De Marines, F., Lopresti, F., La Carrubba, V., Torregrossa, M., & Viviani, G. (2022). Occurrence of microplastics in waste sludge of wastewater treatment plants: Comparison between membrane bioreactor (MBR) and conventional activated sludge (CAS) technologies. *Membranes, 12*(4), 371.
160. Bian, K., Hu, B., Jiang, H., Zhang, Y., Wang, H., & Wang, C. (2022). Is the presence of cu (II) and p-benzoquinone a challenge for the removal of microplastics from landfill leachate? *Science of the Total Environment, 851*, 158395.
161. Jiang, H., Zhang, Y., Bian, K., Wang, H., & Wang, C. (2022). Insight into the effect of aqueous species on microplastics removal by froth flotation: Kinetics and mechanism. *Journal of Environmental Chemical Engineering, 10*(3), 107834.
162. Wan, H., Wang, J., Sheng, X., Yan, J., Zhang, W., & Xu, Y. (2022). Removal of polystyrene microplastics from aqueous solution using the metal–organic framework material of ZIF-67. *Toxics, 10*(2), 70.
163. Wang, Z., Lin, T., & Chen, W. (2020). Occurrence and removal of microplastics in an advanced drinking water treatment plant (ADWTP). *Science of the Total Environment, 700*, 134520.
164. Hsieh, L., He, L., Zhang, M., Lv, W., Yang, K., & Tong, M. (2022). Addition of biochar as thin preamble layer into sand filtration columns could improve the microplastics removal from water. *Water Research, 221*, 118783.
165. Martin, L. M., Sheng, J., Zimba, P. V., Zhu, L., Fadare, O. O., Haley, C., Wang, M., Phillips, T. D., Conkle, J., & Xu, W. (2022). Testing an iron oxide nanoparticle-based method for magnetic separation of nanoplastics and microplastics from water. *Nanomaterials, 12*(14), 2348.
166. Tushar, M. M. R., Pushan, Z. A., Aich, N., & Rowles, L. S. (2024). Balancing sustainability goals and treatment efficacy for PFAS removal from water. *npj Clean Water, 7*(1), 130.

Part III
Current Status of Contaminants

Chapter 5
Global Occurrence and Distribution of PFAS in Groundwater with Emphasis on PFOA and PFOS

K. Brindha and Feifei Cao

1 Introduction

Legacy and emerging organic contaminants are increasingly a source of concern in global aquatic systems. These contaminants are newly identified, naturally occurring or synthetic chemicals or biological substances that are harmful to the environment and human health [1, 2]. Per-and polyfluoroalkyl substances (PFAS) are a class of these emerging contaminants encompassing over 15,000 chemicals that have extensive applications in industry and consumer products [3]. Due to their oil repellent, waterproof and heat-tolerant properties, they are used in cookware, textiles, packaging, surface protection, firefighting foam, etc. Perfluorooctanoic acid (PFOA) and perfluorooctane sulfonate (PFOS) are long-chain PFAS that are widely used, commonly detected in the environment, and subject to international regulations, making them focal points in aquatic studies. These substances with the strongest carbon–fluoride bonds are extremely resistant to chemical, biological, and thermal degradation and are therefore persistent. This also makes the removal of these substances from water or wastewater during treatment processes difficult. The presence of these substances in environmental media has been widely reported in air [4, 5], surface water [6, 7], groundwater [8, 9], pore water [10], rainwater [11, 12], tap water [13, 14], soil [15, 16], sediments, landfill leachate [17, 18], wastewater [19, 20], fish and other aquatic organisms [21, 22], seawater [23], vegetables [24],

K. Brindha (✉)
Department of Water Resources and Ecosystems, IHE Delft Institute for Water Education, Delft, The Netherlands
e-mail: b.karthikeyan@un-ihe.org

F. Cao
IGRAC—International Groundwater Resources Assessment Centre, Delft, The Netherlands
e-mail: feifei.cao@un-igrac.org

M. K. Jindal et al. (eds.), *Emerging Contaminants in Water*, Emerging Contaminants and Associated Treatment Technologies,
https://doi.org/10.1007/978-3-031-89591-3_5

eggs [25], and human blood and serum [26]. This has brought more attention to PFAS to the general public.

Compared with the number of studies available on other environmental media, those on water, especially groundwater, are relatively small. This is evident from the 349 scientific articles published on surface waters since 2001 compared with the 65 papers published on groundwater globally [27]. One reason is that groundwater is commonly less monitored than surface water [28, 29]. Surface water is often the focus, as PFAS contamination is most certain and quantifiable at relatively high concentrations around known industrial areas that discharge treated wastewater to surface water bodies. Groundwater contamination is expected to be present in trace amounts or below the detection limits and is not often included in these surface water studies. This is partially true when confined aquifers are the major source of water in the vicinity. Shallow unconfined aquifers are vulnerable to these contaminants, although only low levels have been reported at nonindustrial, and nonaqueous fire-forming foam (non-AFFF) use sites. This is one of the causes for the poor understanding of the fate and transport of PFASs in groundwater.

Groundwater quality is crucial, as it serves as the primary source of domestic and drinking water in many countries. In the United States, 38% of drinking water is supplied from groundwater [30], and in the European Union, groundwater accounts for 65% of the drinking water source [31]. Few countries in Africa, such as Chad, Malawi, South Sudan, Uganda etc. and Bangladesh in Asia depend more than 70% on groundwater as the main drinking water source [32]. These are also some of the countries where PFAS contamination in groundwater has not been studied. Therefore, there is a large knowledge gap in the studies reported from various regions. For example, the United States, several European countries and Australia have realized the importance and extent of contamination; therefore, more studies are reported. Recently, there has been a significant increase in groundwater studies originating from China. However, this does not relate proportionally to the efforts taken to contain the existing contamination. Countries, including Australia, the United States, Canada, Japan, and members of the European Union, have restricted or banned the use of certain PFAS substances and have taken measures to remediate contamination, while many other countries have yet to implement measures to address this issue.

The ubiquitous presence of PFAS, including those in tap water and their documented adverse health risks, has led to regulations for PFASs or, in some countries, for selected PFASs such as PFOA or PFOS in drinking water. The European Union established guideline values in its Drinking Water Directive (DWD) for PFAS, setting a maximum of 500 ng L^{-1} for total PFAS and a limit of 100 ng L^{-1} for a total of 20 selected PFAS [33]. Australian drinking water guidelines [34] set a national guideline value of 560 ng L^{-1} for PFOA and 70 ng L^{-1} for the sum of PFOS and another PFAS (perfluorohexane sulfonate (PFHxS)). The United States Environmental Protection Agency (USEPA) has proposed stringent guidelines of 4 ng L^{-1} for PFOA and PFOS individually in treated drinking water [35]. This is by far the most restrictive guideline for drinking water. The sum of all PFASs in drinking water should not exceed 30 ng L^{-1} in Canada [36]. The provisional target value for PFOA and PFOS in drinking water in Japan is <50 ng L^{-1} [37].

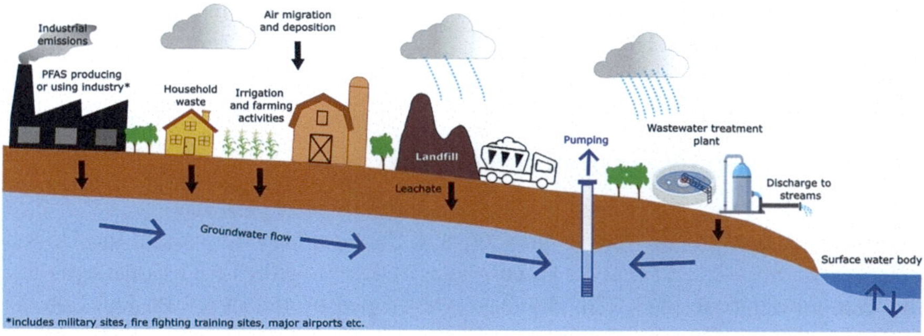

Fig. 5.1 PFAS pathway from various sources to groundwater

Sources of PFASs in environmental media and their pathway to groundwater can be broadly categorized as (i) primary sources: aqueous fire-forming foam (AFFF) discharge sites such as firefighting training or fire suppression sites and PFAS producing or using industries and (ii) secondary sources: landfills, wastewater treatment plants (WWTPs), and the use of treated municipal wastewater or digested sewage sludge on the land surface for agriculture (Fig. 5.1). Other sources include atmospheric deposition, surface runoff, leakage and accidents from industries. Especially in groundwater, multiple pathways exist for PFAS contamination, and the fate and transport of these compounds, along with the transformation of their precursors, pose a complex challenge for effective monitoring and remediation efforts.

Despite the increasing awareness of PFAS concentrations in the environment, groundwater contamination remains understated. The lack of information on PFAS contents in groundwaters worldwide makes addressing existing knowledge gaps difficult. The suite of PFAS substances is numerous; hence, we review the total PFAS and focus on the two most commonly investigated and identified PFASs, that is, PFOA and PFOS, depending on data availability. The objective of this chapter is to review in detail the global distributions of PFOA and PFOS in groundwater and its sources. Existing reviews [27, 38–40] emphasize not only groundwater but also other media, such as surface water, wastewater, and soil; hence, these studies often do not cover all reported studies on groundwater. Past reviews also rely only on peer-reviewed scientific articles, even when a nationwide database is available. In this study, we compiled data from two extensive existing databases to provide a comprehensive and quantitative assessment of PFOA and PFOS in global groundwater.

2 Review Methods

The relevant literature on PFOA and PFOS in global groundwater was collected via various methods. Data for the United States and several countries in Europe are collected and compiled from existing databases. Reviews of other regions lacking an existing database are primarily from literature. The two approaches used in the present study are briefly described below.

2.1 Data Analysis for the United States

The occurrence of PFASs in groundwater and their primary sources in the United States were compiled and analyzed on the basis of data from the PFAS Contamination Site Tracker database [41]. This database, developed by the PFAS Project Lab, as part of the Social Science Environmental Health Research Institute (SSEHRI) at Northeastern University, includes detailed information on PFAS levels, sources, timelines of discovery, health impacts, and responses from both communities and governments at each contamination site. The data and information in the database mainly come from government websites, news articles, or publicly available documents, with all the data sources well referenced in the database. The database encompasses over 2000 known PFAS contamination sites across all 50 states, with PFAS concentrations measured in various environmental matrices, including surface water, groundwater, wastewater, and solids (such as soil, sludge, and compost). This extensive dataset provides a comprehensive and representative overview of PFAS contamination across the United States and was therefore used in this study to review the status of PFAS contamination in groundwater in the United States.

2.2 Data Analysis for Europe

In 2023, the Forever Pollution Project [42], a unique collaborative cross-border and cross-field expert-reviewed journalism investigation, identified approximately 23,000 sites across Europe contaminated by PFAS. Additionally, approximately 21,500 suspected contaminated sites have been reported, linked to current or past industrial activities, underscoring the widespread presence of PFAS throughout Europe. The project developed a "Map of Forever Pollution in Europe" [43], providing the first extensive visualization of environmental contamination by PFAS compounds across the continent. The open and reusable dataset [44] of PFAS contamination from the Forever Pollution Project, compiled from multiple sources, including national authorities and scientific publications, offers a comprehensive and representative picture of PFAS contamination across Europe. This database was used in this study to review the status of PFAS contamination in groundwater in Europe. This data includes information from various environmental matrices, such as drinking water, surface water, groundwater, wastewater, soil, sediment, and biota. The methodology for identifying suspected contamination sites was adapted from The PFAS Project Lab [41] approach used in the United States, which is supported by expert input. Data on PFAS contamination in groundwater were extracted from the PFAS Data Hub [44] database and analyzed at the regional scale for each country, revealing more than 20,000 groundwater-contaminated sites across Europe. These data were provided by national and/or local authorities for ten countries, including Belgium, Denmark, France, Germany, Italy, Latvia, Luxembourg, the Netherlands, Switzerland, and the United Kingdom. Data for the other ten countries (Finland, Sweden, Austria, Bulgaria, Croatia, Greece, Hungary, Romania, Serbia, and Slovakia) were compiled mainly from scientific publications, documents, and technique reports. All the data sources were well referenced in the database.

2.3 Data Analysis for Other Regions

For other regions, a national or international database was not available. We meticulously collected data from peer-reviewed articles and reports via the Scopus and Google Scholar search engines. The searches included a combination of terms, including "PFAS" "perfluoroalkyl," "polyfluoroalkyl," "PFOA," "PFOS," "groundwater," "ground water," "occurrence," and "country name," for the period from 1990 to present. For Asia, Africa, and Oceania, we primarily used these specified search options. Scientific articles and their supplementary materials were screened based on relevance and the availability of all required information for analysis. For example, articles reporting only PFAO or PFOS were included, but those reporting only total PFAS without specifying individual concentrations were excluded. The statistical data table for Asia and Oceania includes only articles that have reported data precisely, that is, we do not infer concentrations from graphs but report only if the authors have explicitly stated them.

3 PFOA and PFOS in Global Groundwater

A recent study compiling PFAS sources in groundwater reported maximum PFOA and PFOS concentrations in groundwater of 6.57 mg L^{-1} and 9.7 mg L^{-1}, respectively [38]. Most of this was attributed to sites using AFFFs. The present review reports that the maximum PFOA concentration determined in groundwater is 6.57 mg L^{-1} from the United States, and the maximum PFOS concentration is 68.9 mg L^{-} from Belgium.

3.1 Occurrence in North America

Groundwater is the primary matrix sampled and tested at contaminated sites [41], as PFAS from firefighting foams and other sources can leach through soil and into groundwater. Among the identified sites, PFAS have been detected in groundwater at 1585 locations [41]. They were detected in all 50 states across the country, underscoring the widespread nature of this environmental issue. Table 5.1 provides a summary of PFAS occurrences in groundwater across various states in the United States on the basis of the PFAS contamination site tracker database. This table highlights the number of contaminated sites, types of PFAS contamination sources, and PFAS concentration statistics (minimum, maximum, and median values) for each state. The sources of contamination are categorized into three main types according to the literature [45]: (1) AFFF discharge sites, including military sites, major airports, and other firefighting training sites; (2) industrial facilities that produce and/or use PFAS; and (3) PFAS-containing waste sites, such as landfills and WWTPs. Among the 1585 PFAS-contaminated groundwater sites, 408 sites (26%) are related to AFFF discharge sites, 385 sites (24%) are related to industrial facilities, 661 sites (42%) are related to PFAS-containing waste, and 131 sites (8%) are related to other or unknown sources.

The states of New Hampshire, Michigan, and California reported the highest number of PFAS-contaminated sites among all the states, with 466, 327 and 244 sites, respectively. With the highest number of detected contaminated sites, the state of New Hampshire shows substantial PFAS presence in groundwater originating from industrial facilities (29%). Among the 45% of the sites related to PFAS-containing waste, 3% were contaminated by AFFF-related discharge. The median PFAS, PFOA, and PFOS concentrations were 155 ng L^{-1}, 28 ng $L^{-1}$and 30 ng L^{-1}, respectively (Table 5.1). Similarly, the sources of PFAS contamination in the state of Michigan also come mainly from industrial facilities and waste sites, with a minority of sites (11%) contaminated from AFFF discharges. The relevant median PFAS, PFOA, and PFOS concentrations are much higher than those in New Hampshire, with values of 1062 ng L^{-1}, 53 ng L^{-1}, and 130 ng L^{-1}, respectively. In California, PFAS contamination in groundwater is mostly related to waste sites (74%) as well as AFFF discharge sites (21%, particularly airports and military sites), whereas only 5% of contaminated sites are related to industrial facilities. The median PFAS, PFOA, and PFOS concentrations were 189 ng L^{-1}, 39 ng L^{-1}, and 18 ng L^{-1}, respectively (Table 5.1). Minnesota (98 sites) and Florida (84 sites) also presented substantial PFAS presence in groundwater, with median PFAS concentrations of 3795 ng L^{-1} and 322 ng L^{-1}, respectively. The highest PFAS concentration in groundwater in the United States was measured at the England Air Force Base near Alexandria, Louisiana, which was closed in 1992. This site recorded an astonishing PFAS level of 33,502,500 ng L^{-1} in groundwater. This was followed by a PFAS concentration of 8,770,829 ng L^{-1} measured in groundwater at the Santa Barbara Airport in California and a concentration of 8,770,000 ng L^{-1} measured at the Naval Air Station Fallon in Nevada.

In addition to states with a high number of PFAS-contaminated groundwater sites, 35 states reported fewer than 10 known contaminated sites, and 20 states reported fewer than 5. Despite the lower number of sites in these states, some still report extremely high PFAS concentrations. For example, Louisiana has only two known contaminated sites, yet one of them is among the most contaminated former or current military sites in the country. Notably, the number of identified sites in each state reflects not only the extent of contamination but also the level of testing

Table 5.1 Concentrations of PFAS, PFOA, and PFOS in groundwater from various states of the United States (compiled from The PFAS Project Lab [41])

State	Number of sites	AFFF discharge sites	Industrial facilities that produce and/or use PFAS	PFAS-containing waste sites	Other	Unknown	PFAS ($ng\ L^{-1}$)			PFOA ($ng\ L^{-1}$)			PFOS ($ng\ L^{-1}$)		
							Min	Median	Max	Min	Median	Max	Min	Median	Max
New Hampshire	466	12	137	211	0	106	2	155	651,290	1	28	130,000	1	30	490,000
Michigan	327	37	156	123	6	5	15	1062	1,181,780	2	53	170,000	3	130	1,280,000
California	244	52	11	180	0	1	<LOD[a]	189	8,770,829	<LOD	39	1,071,530	<LOD	18	13,407,957
Minnesota	98	3	4	91	0	0	40	3795	582,550	3	555	226,000	5	540	33,300
Florida	84	44	34	6	0	0	2	322	4,856,453	<LOD	109	280,000	1	190	4,300,000
Vermont	38	4	7	27	0	0	20	85	46,332	2	44	7200	5	32.5	34,000
Alaska	30	27	3	0	0	0	186	30,676	740,000	18	2800	1,300,000	86	11,000	1,400,000
New York	28	17	8	3	0	0	14	7477	1,341,000	6	1600	152,000	5	2100	1,200,000
Maine	25	8	5	9	0	3	10	205	41,861	2	34.5	150,000	1	57.5	24,000
Wisconsin	24	10	8	2	3	1	15	20,905	169,467	1	580	43,000	3	442.5	120,000
Colorado	18	17	1	0	0	0	<LOD	4	299,000	<LOD	100	190,000	<LOD	430.5	90,000
Texas	15	14	1	0	0	0	5633	110,215	3,502,600	2260	62,000	384,000	1470	64,200	1,200,000
Washington	13	8	1	0	4	0	15	2354	176,200	1	350	58,500	3	8600	150,000
Massachusetts	11	10	0	0	0	1	952	8718.5	313,000	33	2080	270,000	120	2500	140,000
Virginia	10	9	1	0	0	0	6732	515,624	2,275,000	69	12,550	506,000	332	15,221	2,200,000
Ohio	9	9	0	0	0	0	1831	7861	135,533	34	2200	45,800	5	2100	54,000
North Carolina	9	8	1	0	0	0	4000	10,665	47,825	133	3780	100,000	221	23,550	300,000
New Jersey	8	6	2	0	0	0	312	4220	13,000	29	250	2200	48	2272.5	34,000
Maryland	8	8	0	0	0	0	1158	11,186	40,100	135	8280	28,000	38	23,350	234,000

(continued)

Table 5.1 (continued)

State	Number of sites	AFFF discharge sites	Industrial facilities that produce and/or use PFAS	PFAS-containing waste sites	Other	Unknown	PFAS ($ng\ L^{-1}$)			PFOA ($ng\ L^{-1}$)			PFOS ($ng\ L^{-1}$)		
							Min	Median	Max	Min	Median	Max	Min	Median	Max
Delaware	8	2	2	3	0	1	6156	70,848	335,000	12	545	20,000	130	2820	270,000
Pennsylvania	7	6	1	0	0	0	6591	74,483	388,768	607	889	24,400	390	4174	320,000
Mississippi	7	7	0	0	0	0	798	90,041	1,446,118	88	6600	70,700	69	67,000	740,000
Arizona	6	6	0	0	0	0	205	1221	13,251	23	105	4700	55	485	3400
Tennessee	6	6	0	0	0	0	183	7355	390,064	13	630	16,000	66	4300	159,000
Alabama	6	3	0	3	0	0	9564	86,060	540,637	230	34,200	84,000	260	25,000	45,000
Oklahoma	6	6	0	0	0	0	12,694	208,000	1,180,000	410	22,000	250,000	64	167,000	900,000
Oregon	5	5	0	0	0	0	1600	42,277	484,740	21,000	24,000	30,900	25,000	203,000	380,000
Illinois	5	5	0	0	0	0	18,707	105,395	2,920,500	1620	13,600	151,000	1520	81,000	1,960,000
Georgia	5	5	0	0	0	0	121,000	137,990	665,000	3900	52,000	80,000	41,000	250,000	320,000
South Carolina	5	5	0	0	0	0	13,388	697,000	3,479,800	550	13,000	790,000	9500	195,000	2,490,000
Connecticut	4	1	0	3	0	0	518	518	518	19	140	347	60	355	2500
Indiana	4	4	0	0	0	0	270	14,620	28,970	40	44	5600	110	428	15,000
New Mexico	4	4	0	0	0	0	993	27,040	1,094,300	9	1112.5	254,000	25	12,490	1,040,000
Missouri	4	4	0	0	0	0	1509	88,483	571,288	340	24,200	60,000	570	31,500	329,000
North Dakota	4	4	0	0	0	0	83,500	127,965	172,429	1400	18,000	40,000	4000	46,500	440,000
Nevada	4	4	0	0	0	0	165,400	4,467,700	8,770,000	200	7550	6,570,000	21,000	76,000	380,000
Kentucky	3	2	1	0	0	0	1071	1449	230,021	72	115	5230	105	240	128,000
Kansas	3	3	0	0	0	0	48,967	48,967	48,967	8100	17,050	26,000	2930	30,000	430,000
West Virginia	3	2	1	0	0	0	20,898	90,530	160,162	1170	2635	4100	6930	26,465	46,000

State	Number of sites	AFFF discharge sites	Industrial facilities that produce and/or use PFAS	PFAS-containing waste sites	Other	Unknown	PFAS (ng L^{-1})			PFOA (ng L^{-1})			PFOS (ng L^{-1})		
							Min	Median	Max	Min	Median	Max	Min	Median	Max
Wyoming	3	3	0	0	0	0	122,440	134,720	147,000	8500	8500	72,000	64,000	89,000	89,000
Arkansas	3	3	0	0	0	0	468,340	784,997	1,987,820	116,000	285,000	360,000	52,000	105,000	164,000
Rhode Island	2	2	0	0	0	0	3961	3961	3961	1110	1110	1110	2680	2680	2680
Montana	2	2	0	0	0	0	6730	6730	6730	5500	22,250	39,000	1100	1500	1900
Iowa	2	2	0	0	0	0	17,811	25,406	33,000	850	1375	1900	7800	10,400	13,000
Nebraska	2	2	0	0	0	0	32,651	32,651	32,651	1700	205,850	410,000	13,000	27,500	42,000
Utah	2	2	0	0	0	0	64,900	303,300	541,700	11,000	20,000	29,000	54,000	82,000	110,000
Louisiana	2	2	0	0	0	0	33,502,500	33,502,500	33,502,500	200,000	2,010,000	3,820,000	1,100,000	4,125,000	7,150,000
Idaho	1	1	0	0	0	0	1240	1240	1240	160	160	160	118	118	118
South Dakota	1	1	0	0	0	0	633,400	633,400	633,400	14,000	14,000	14,000	370,000	370,000	370,000
District of Columbia	1	1	0	0	0	0	n.a.[b]	n.a.	n.a.	69	69	69	58	58	58

Note: Detailed data sources and references are not included in this table. For comprehensive information, please refer to the original dataset of The PFAS Project Lab [41]

[a] LOD = Limit of detection

[b] n.a. = not available

conducted and the availability of publicly accessible information. Therefore, the actual extent of PFAS contamination in groundwater is likely worse than what is indicated in the current database. There is a strong likelihood that many other contaminated sites remain unknown or undetected. The PFAS Project Lab database has also identified 57,412 sites suspected of PFAS contamination, including industrial facilities, airports, current and former military sites, and WWTPs. These presumptive contamination sites were identified by applying a conceptual model that includes three categories of sites (AFFF sites, certain industrial sites, and PFAS-containing waste sites) and operationalizing it with high-quality nationwide data [45]. However, many types of presumptive PFAS contamination sites lack high-quality nationwide data e.g., industrial sludge application sites and firefighting training sites, meaning that the estimate of 57,412 sites likely underrepresents the true number of presumptive sites.

Data on PFAS in Canadian groundwater are relatively limited (Table 5.2). In 2023, the Environment and Climate Change Canada (ECCC) and Health Canada (HC) released a draft report on the state of PFAS to inform decision-making regarding PFAS in Canada [46]. This report identified over 100 federal contaminated sites with confirmed (91 sites) or suspected (22 sites) PFAS contamination. These sites are listed on the Federal Contaminates Sites inventory [47] and can be accessed by performing a search for PFAS under the keyword function. The sites span all provinces and territories and are predominantly linked to the past and/or current use of AFFFs, such as in fire training activities and maintenance of firefighting equipment at airports and military facilities. Several PFASs have been detected in groundwater in former firefighting training areas in British Columbia, Alberta, Nova Scotia, and Ontario [48, 49]. Other sources of PFAS at contaminated federal sites include landfills and WWTPs. For example, leachate-impacted groundwater was sampled from 20 historic landfills in Ontario. Some landfills closed in the 1960s had total PFAS concentrations similar to those of modern landfills, with a maximum level of 12,700 ng L^{-1} in groundwater [46].

PFAS occurrence in Canadian drinking water was also documented in the draft report. Source and treated water from groundwater and surface water sources were monitored across Canada in 2009 and 2010. The investigation revealed that PFOA was the most frequently detected, with an average level of approximately 0.06 ng L^{-1}, whereas PFOS was rarely detected. Two more recent studies also investigated PFAS in drinking water systems. To assess PFAS occurrence and potential risk in Ontario, the Ontario Ministry of the Environment measured PFAS levels in source and treated drinking water [50]. Twenty-five sites were sampled, including river, lake, and groundwater sources. The levels in groundwater (median 3.8 ng L^{-1}) were lower than those in both lakes (median 5.5 ng L^{-1}) and rivers (median 5.9 ng L^{-1}). Nevertheless, the presence of PFASs in groundwater indicates that these compounds can be transferred throughout the environment to impact groundwater sources [50]. In Quebec, PFAS were analyzed in tap water samples between 2018 and 2020 [51]. This study covered a large spatial scale, with 172 samples from groundwater and 165 from surface water. Overall, 99.3% of the tap water samples contained at least one PFAS. The groundwater-sourced samples had a median total PFAS level of

Table 5.2 Concentrations of PFAS, PFOA, and PFOS in groundwater from Canada and countries in South America

Country	Region	Year	Site info	PFAS (ng L^{-1})	PFOA (ng L^{-1})	PFOS (ng L^{-1})	Reference
Canada	Newfoundland	2016	Commercial airport-fire-training area (decommissioned)	n.a.[a]	620–4100	n.a.	Martin et al. [52]
Canada	Ontario	2016	Commercial airport-fire equipment testing (in activity)	n.a.	36,000–85,000	840 to 1500	Martin et al. [52]
Canada	Ontario	2012–2016	Drinking water source	median 3.8	n.a.	n.a.	Kleywegt et al. [50]
Canada	Central and Eastern	2016–2017	Airport	29.2–10,800,000	n.a.	n.a.	Liu et al. [53]
Canada	Quebec	2018–2020	Tap water based on groundwater source	median 0.41	median 0.043	median 0.022	Munoz et al. [51]
Canada	Ontario	n.a.	Landfill	max 12,700	n.a.	n.a.	ECCC & HC [46]
Guadeloupe	–	2012	Likely from former and/or current AFFF operations	<LOD[b] to 638	n.a.	n.a.	Munoz et al. [54]
French Guiana	–			<LOD to 400	n.a.	n.a.	
Martinique	–			<LOD to approx. 300	n.a.	n.a.	
Brazil	Bahia State	2016	Likely from sulfluramid use	median 5.73	n.a.	n.a.	Nascimento et al. [56]
Guatemala	–	2021	Tap water	2.1–91.5	n.a.	n.a.	Hoponick Redmon et al. [58]
Brazil	Porto Alegre	n.a.	Wells in urban areas	22–718	<LOD to 249	<LOD to 15.8	Stefano et al. [57]

[a] n.a. = not available
[b] LOD = Limit of detection

0.41 ng L^{-1}, whereas the median PFAS level in the surface water-sourced samples was 5 ng L-1. PFOA and PFOS also had high detection rates, with median values of 0.043 ng L^{-1} and 0.022 ng L^{-1}, respectively, in groundwater, which were lower than those in surface water sources. Despite higher median PFAS levels in surface water, six of the top ten contaminated locations were groundwater- [51].

Several recent studies have investigated groundwater contamination by PFAS at airports and fire training sites, which are commonly related to AFFF discharges. In a 2019 study, PFAS were analyzed in contaminated groundwater samples from two commercial airports in Canada [52]. A decommissioned fire training area within an airport in Newfoundland had PFOA concentrations ranging from 620 to 4100 ng L^{-1}, whereas an active site in Ontario had concentrations ranging from 36,000 to 85,000 ng L^{-1}. The Ontario site also presented PFOS concentrations ranging from 840 to 1500 ng L^{-1}. A study conducted by Liu et al. [53] investigated PFAS occurrence in four fire training areas within airports in central and eastern Canada. Among the 70 groundwater samples analyzed, 66 presented detectable levels of at least one PFAS. The highest total PFAS level was detected in the source zone of an active firefighting training site, with a value of 10,800,000 ng L^{-1}. This study highlighted the risk of PFAS transport from source zones to downgradient soil and groundwater, even in low-permeability areas [53].

3.2 *Occurrence in South America and the Caribbean*

In South America and the Caribbean, there have been limited efforts to monitor PFAS levels in the environment. Despite this, there is growing concern regarding the occurrence of PFAS in the environment. Table 5.2 provides a summary of PFAS occurrences in groundwater in South America and the Caribbean. The PFAS contamination site tracker database from the United States also identified several PFAS contamination sites in the Caribbean region, including the Marshall Islands and Puerto Rico. In the United States Kwajalein Atoll military base in the Marshall Islands, 380 gallons of AFFFs were discovered to have leaked into the ground by March 2022. PFASs were not detected in drinking water, and the available data revealed a PFAS level of 1.7 ng L^{-1} in groundwater measured in April 2022. Details about the accident remain unknown or unavailable publicly, with no updated data on PFAS levels in the environment matrix of the site. Another PFAS-contaminated site was identified at a military site in Puerto Rico, with groundwater levels of 351,375 ng L^{-1} for PFOA and 491,401 ng L^{-1} for PFOS [41].

In French overseas territories such as French Guiana, Guadeloupe, and Martinique, PFAS levels were investigated in groundwater, surface water, and sediments [54]. Generally, PFAS concentrations in groundwater were low. For example, limestone aquifers of the Grande-Terre and Marie-Galante islands (Guadeloupe) presented total PFAS levels below 2 ng L^{-1}, and some remote regions of French Guiana presented PFAS concentrations ranging from 0.1 to 0.4 ng L^{-1}. However, some PFAS hotspots were identified: a well near an electricity power plant in Guadeloupe had total PFAS concentrations between 614 and 638 ng L^{-1}; areas near an oil refinery and composting facility in French Guiana had total PFAS levels of up to nearly 400 ng L^{-1}. Martinique's capital city (Fort-de-France) also reported approximately 300 ng L^{-1} PFAS in groundwater. The PFAS contamination sources at these sites seem highly congruent with former and/or current AFFF operations. The study also highlighted that shorter-chain compounds of PFAS prevailed in groundwater because of their higher transfer potential through infiltration processes and the possible influence of precursors.

In Guatemala City, PFAS were detected in tap water, marking the first such analysis in Central America [55]. Among the 113 households studied, 62% were supplied with groundwater. PFASs were detected in 19% of the water samples, with concentrations ranging from 2.1 to 91.5 ng L^{-1}. This study revealed a significant association between PFAS occurrence and the use of plastic water storage tanks, highlighting the need for further investigation and public engagement to understand contamination sources and exposure risks.

In Brazil, PFAS occurrence has been investigated in several regions. Nascimento et al. [56] studied PFAS in soil, sediments, plants, surface water, and groundwater in an agricultural area of Bahia State and detected total PFAS concentrations of 5.7 ng L^{-1} in groundwater and river water concentrations of up to 8.9 ng L^{-1}. The use of sulfluramid, a pesticide largely used in agriculture with EtFOSA as the active ingredient, is suspected to contribute significantly to PFAS contamination in this region. A later study examined PFAS in surface water and groundwater in Porto Alegre, the capital of Rio Grande do Sul [57]. The groundwater PFAS concentrations ranged from 22 to 718 ng L^{-1}, with PFOA being the dominant PFAS compound in the groundwater (below the detection limit to 249 ng L^{-1}). Industrial activities and untreated domestic sewage were suspected as major contamination sources in this region.

3.3 Occurrence in Europe

Table 5.3 provides a summary of PFAS occurrences in groundwater across various countries in Europe on the basis of the Forever Pollution Project [42] database, highlighting the number of contaminated sites and PFAS concentration statistics (% of detection, minimum and maximum in regions of each country).

In Belgium, PFAS were detected in 100%, 94%, and 40% of the groundwater samples collected in the regions of Flanders (n = 90), Brussels (n = 176), and Wallonia (n = 202), respectively. The highest PFAS level in groundwater (68,900,000 ng L^{-1}) was measured in the proximity of 3M's PFAS manufacturing site in Zwijndrecht, Flanders, which is thought to be one of the most PFAS-contaminated sites in the world. The region of Brussels also showed a substantial presence of PFAS in groundwater, with 22% of the sampled sites (n = 38) showing PFAS levels higher than 500 ng L^{-1}. The highest concentration of PFAS in

Table 5.3 Concentrations of PFAS, PFOA, and PFOS in European groundwater (compiled from the Forever Pollution Project [42])

Country	Region	Number of measurements	Number of detections	% detection	PFAS ($ng\ L^{-1}$)		PFOA ($ng\ L^{-1}$)		PFOS ($ng\ L^{-1}$)	
					Min	Max	Min	Max	Min	Max
Belgium	Brussels	176	165	94%	5.5	122,836.07	<LOD[a]	11,000	<LOD	42,000
	Flanders	90	90	100%	1	68,900,000	1	110	1	68,900,000
	Wallonia	202	81	40%	0.2	1091	<LOD	67	<LOD	95
Denmark	Midtjyllands	2680	1659	62%	0.54	60,000	<LOD	8700	<LOD	39,000
	Nordjylland	1786	714	40%	0.68	62,703	<LOD	3300	<LOD	4940
	Sjælland	847	317	37%	0.34	3218.8	<LOD	380	<LOD	1300
	Syddanmark	2695	1039	39%	0.3	247,586	<LOD	1000	<LOD	16,000
Finland	North Karelia	52	48	92%	0.11	3997.27	0.11	191	0.11	3500
France	Auvergne-Rhône-Alpes	24,021	1405	6%	1.3	7340	<LOD	2560	<LOD	1110
	Grand Est	36,489	2845	8%	0.001	12,108	<LOD	1050	<LOD	503
	Bourgogne-Franche-Comté	16,200	564	3%	0.2	18,700	<LOD	200	<LOD	80
	Île-de-France	5223	523	10%	0.2	31,151	<LOD	10,200	<LOD	6200
	Provence-Alpes-Côte d'Azur	6703	270	4%	1	5527.6	<LOD	138	<LOD	1540
	Bretagne	1657	192	12%	0.2	1188.4	<LOD	53	<LOD	351
	Centre-Val de Loire	3538	235	7%	0.2	382.7	<LOD	31	<LOD	50
	Corse	353	24	7%	2	156.6	<LOD	6.7	<LOD	65.4
	Hauts-de-France	6108	1029	17%	0.2	6700	<LOD	38	<LOD	115.5
	Normandie	9620	477	5%	0.2	5830	<LOD	341	<LOD	539
	Nouvelle-Aquitaine	5776	615	11%	0.2	2482	<LOD	88	<LOD	1760
	Occitanie	11,218	467	4%	0.3	810	<LOD	24.2	<LOD	586
	Pays de la Loire	6226	235	4%	1	319.2	<LOD	21.6	<LOD	13

Germany	Baden-Württemberg	388	388	100%	1	5708	<LOD	4210	<LOD	600
	Bavaria	21	21	100%	20	1,361,000	<LOD	114,000	10	1,100,000
	Hamburg	33	33	100%	10.1	681	1.23	170	0.476	511
	Hesse	434	434	100%	600	733,800	<LOD	248,400	<LOD	206,200
	Rhineland-Palatinate	46	46	100%	0.01	8.55	<LOD	0.29	<LOD	0.16
	Saarland	1	1	100%	34,540	34,540	230	230	34,000	34,000
	Saxony	12	12	100%	10	41	2	31	1	26
	Saxony-Anhalt	10	10	100%	10.5	1700	4.2	1700	0.6	23
	Schleswig-Holstein	28	28	100%	1	86	<LOD	16	<LOD	13
Italy	Abruzzo	5	0	0%	NA[b]	NA	NA	NA	NA	NA
	Basilicata	6	2	33%	0.35	0.57	<LOD	<LOD	0.35	0.57
	Calabria	2	1	50%	0.4	0.4	<LOD	<LOD	0.4	0.4
	Campania	59	18	31%	0.2	12.3	<LOD	<LOD	0.2	3.6
	Emilia Romagna	6	3	50%	0.4	51	<LOD	10.7	<LOD	0.8
	Friuli VG	6	5	83%	0.5	818.8	<LOD	560	0.5	119.1
	Lazio	13	4	31%	1.18	30	0.2	0.6	0.3	0.8
	Liguria	5	5	100%	0.3	28	<LOD	11	0.3	17
	Lombardia	128	79	62%	0.21	190.4	<LOD	68	<LOD	44
	Marche	5	1	20%	334.1	334.1	316	316	1.1	1.1
	Piemonte	92	64	70%	5	410	<LOD	410	<LOD	60
	Puglia	2	2	100%	0.3	32.1	<LOD	10	0.3	2.1
	Sardegna	1	1	100%	0.4	0.4	<LOD	<LOD	0.4	0.4
	Sicilia	114	49	43%	0.2	451	<LOD	320	<LOD	102.2
	Toscana	23	16	70%	1.11	58.471	<LOD	56.03	<LOD	4.505
	Trentino-Alto Adige	8	6	75%	0.7	6.2	<LOD	<LOD	0.7	6.2
	Umbria	285	106	37%	5	121	<LOD	14	<LOD	27
	Valle d'Aosta	9	8	89%	0.2	260	<LOD	36	0.2	36
	Veneto	18,977	10,870	57%	0.2	149,190	<LOD	82,400	<LOD	31,150

(continued)

Table 5.3 (continued)

Country	Region	Number of measurements	Number of detections	% detection	PFAS ($ng\ L^{-1}$)		PFOA ($ng\ L^{-1}$)		PFOS ($ng\ L^{-1}$)	
					Min	Max	Min	Max	Min	Max
Latvia	Kurzeme	14	13	93%	0.166	20.195	<LOD	12.5	<LOD	<LOD
	Riga	19	16	84%	0.128	62.4	<LOD	15.6	<LOD	2.39
	Vidzeme	16	15	94%	0.052	40.866	<LOD	2.067	<LOD	1.45
	Zemgale	13	13	100%	0.121	34.2	<LOD	2.54	<LOD	3.67
Luxembourg		5	2	40%	31	153	<LOD	<LOD	<LOD	11
Netherlands	South	1	1	100%	3030	3030	3030	3030	n.a.[c]	n.a.
	North	12	12	100%	900	4857.14	1350	2761.9	900	2095.24
Sweden	Blekinge	5	4	80%	4.9	14.1	<LOD	2.2	<LOD	13
	Dalarna	2	2	100%	0.2	0.2	<LOD	<LOD	<LOD	<LOD
	Gävleborg	19	13	68%	0.2	33.4	<LOD	28	<LOD	0.4
	Gotland	25	16	64%	0.2	11	<LOD	1.1	<LOD	<LOD
	Halland	22	22	100%	0.16	269.29	<LOD	52	<LOD	24
	Jämtland	10	5	50%	0.1	3.1	<LOD	<LOD	<LOD	1.1
	Jönköping	2	0	0%	NA	NA	NA	NA	NA	NA
	Kalmar	6	4	67%	0.3	48.9	<LOD	6.9	<LOD	8
	Kronoberg	2	2	100%	0.1	0.4	<LOD	<LOD	<LOD	<LOD
	Norrbotten	9	5	56%	0.2	101.5	<LOD	41	<LOD	2.6
	Örebro	2	2	100%	0.4	2.7	<LOD	<LOD	<LOD	1.9
	Östergötland	4	4	100%	0.27	10.6	<LOD	1.2	<LOD	<LOD
	Skåne	7	7	100%	0.2	159.4	<LOD	26	<LOD	29
	Södermanland	4	3	75%	0.1	20.7	<LOD	0.6	2.8	2.8
	Stockholm	10	9	90%	4.5	45	0.5	1.4	1.3	45
	Uppsala	6	5	83%	0.1	5.6	<LOD	<LOD	<LOD	1
	Värmland	3	2	67%	0.07	0.18	<LOD	<LOD	<LOD	<LOD
	Västerbotten	1	0	0%	NA	NA	NA	NA	NA	NA
	Västernorrland	4	4	100%	0.05	6.5	<LOD	<LOD	0.3	0.3
	Västmanland	6	5	83%	1	32.5	<LOD	2.6	<LOD	2.3
	Västra Götaland	27	16	59%	0.1	6423.4	<LOD	436	<LOD	8.6

Switzerland	Central Switzerland	2	0	0%	NA	NA	NA	NA	NA	NA
	Eastern Switzerland	10	3	30%	0.6	21.8	<LOD	2.5	<LOD	12.1
	Espace Mittelland	11	3	27%	2.4	17.6	<LOD	8.8	1	5.7
	Lemanic Region	9	3	33%	1	242.8	<LOD	15	1	123
	Northwestern Switzerland	8	6	75%	4.5	64.2	<LOD	15.7	<LOD	22.6
	Ticinp	4	2	50%	5.7	17.9	<LOD	1.8	2	2
	Zurich	5	4	80%	4.4	70.9	3.3	15	1.1	39
United Kingdom	East of England	106	98	92%	0.1	31,000	<LOD	50	<LOD	1000
	London	13	13	100%	1	13,035.4	0.2	9.6	2.7	6.1
	Midlands	105	100	95%	0.1	31,000	<LOD	1000	<LOD	1000
	North East and Yorkshire	132	132	100%	0.1	16,000	<LOD	23	<LOD	1000
	North West	98	98	100%	0.2	4676.9	<LOD	257.2	<LOD	710
	South East	228	228	100%	0.1	30,000	<LOD	1000	<LOD	1000
	South West	59	59	100%	0.1	20,000	<LOD	1000	<LOD	1000
Austria	Danube Basin	1	NA	NA	2.8		n.a.[c]		n.a.	
Bulgaria	Danube Basin	1	NA	NA	3.2		1.1		n.a.	
Croatia	Danube Basin	1	NA	NA	28.9		4.8		18	
Greece	Municipality of Oropos	1	NA	NA	229.6		73.4		30.4	
Hungary	Danube Basin	1	NA	NA	13.1		2.8		2.7	
Romania	Danube Basin	1	NA	NA	201.7		n.a.		n.a.	
Serbia	Danube Basin	1	NA	NA	7.2		4.3		n.a.	
Slovakia	Danube Basin	1	NA	NA	8.5		1.4		1.6	

Note: The statistics (min and max) for PFAS, PFOA, and PFOS were analyzed based on samples where PFAS was detected, with detection percentages provided for all regions. For regions with no PFAS detection, statistics are marked as NA (not applicable). Detailed data sources and references are not included in this table. For comprehensive information, please refer to the original dataset of the Forever Pollution Project [42].

[a] LOD = Limit of detection

[b] NA = not applicable

[c] n.a. = not available

groundwater reached 122,836 ng L^{-1} in this region, with maximum levels of PFOA and PFOS reaching 11,000 ng L^{-1} and 42,000 ng L^{-1}, respectively. The region of Wallonia was the least contaminated among the three regions in Belgium. Nevertheless, PFAS was detected in more than one-third of the groundwater samples in this region, with levels of up to 1091 ng L^{-1}.

PFAS was also detected in all four regions (Midtjyllands, Nordjylland, Sjælland, and Syddanmark) in Denmark, with detection rates ranging from 37% to 62%. The highest PFAS concentration in groundwater (247,586 ng L^{-1}) was measured at Sønderborg, Syddanmark. The region of Sjælland seems to be the least impacted region, with 37% of the groundwater samples detected with PFAS, with a maximum concentration of 3219 ng L^{-1}.

PFAS pollution was widespread in French groundwater. All 13 metropolitan regions of France presented PFASs in groundwater, with detection rates ranging from 3% to 17%. The Haut-de-France region presented the highest detection rate, with a maximum PFAS level of 6700 ng L^{-1}. This was followed by the region of Bretagne, with a detection rate of 12% and a maximum concentration of 1188 ng L^{-1}. The Nouvelle-Aquitaine and Ile-de-France regions also presented a substantial presence of PFAS in groundwater, with detection rates of 11% and 10%, respectively. The highest PFAS concentration in groundwater (31,151 ng L^{-1}) was measured in Villeneuve-la-Garenne (Ile-de-France). Although the dataset does not associate the detected PFAS in groundwater with relevant contaminated sources, the Forever Pollution Project highlighted that Europe's best-known hotspots of massive pollution all have PFAS production facilities at their epicenter. A total of 20 large PFAS manufacturing facilities were identified within the project, five of which are located in France, including the Arkema and Daikin plants and the Solvay facilities in eastern France, Chemours in northern France, and Salindres in southern France.

In Germany, PFAS compounds were detected in all groundwater samples ($n = 973$) throughout all nine regions of the country. The highest total PFAS concentration in groundwater (1,361,000 ng L^{-1}) was measured at the Nürnberg airport in the region of Bavaria. The contamination source was related to AFFF discharges from firefighting trains/incidents at the site. The Hesse region also presented substantially high concentrations of PFAs in groundwater, ranging from 600 to 733,800 ng L^{-1}. According to investigations by the Forever Pollution Project investigation, there are no fewer than six PFAS manufacturing plants in Germany, constituting potential risks of PFAS contamination to the environment.

Among all the regions in Italy, the Veneto region presented significantly higher concentrations of PFAS in groundwater than all the other regions. The maximum level of PFAS in groundwater (149,190 ng L^{-1}) was measured in Trissino, Veneto, where the Miteni company synthesized and emitted a range of PFAS compounds for half a century. As discovered in 2013, the contamination of water and soil extends over 200 square kilometers and is believed to affect up to 350,000 people in the Veneto region. Other high levels of PFAS in groundwater were also detected in the regions of Friuli VG, Sicilia, and Piemonte, with PFAS concentrations of up to 819 ng L^{-1}, 451 ng L^{-1}, and 410 ng L^{-1}, respectively.

In the Netherlands, high concentrations of PFAS were detected in groundwater in both the South and North Holland regions. In Dordrecht, South Holland, the Chemours Company has caused severe pollution around the PFAS plant. The PFAS concentration reached 3030 ng L^{-1} in groundwater at the site. Investigations have shown that not only water but also soil and air are contaminated by PFASs, which are dominated by PFOA and GenY, which are short-chain substitute long-chain PFASs with high water solubility and stability with high transfer potential in the environment. High concentrations of PFAS were also detected in Hoofddorp in North Holland, with concentrations ranging from 900 to 4857 ng L^{-1} in groundwater.

In England, the highest level of PFAS in groundwater was 31,000 ng L^{-1}, which was measured in Becton Bunny in East England as well as in Derwent Upper Derbyshire in the Midlands. The other regions also showed substantial occurrence of PFAS in groundwater, with levels reaching up to 30,000 (South East), 20,000 (South West), 16,000 (North East and Yorkshire), 13,035 (London), and 4677 ng L^{-1} (North West) in different regions.

PFASs were detected in groundwater in Latvia, with the highest level (62.4 ng L^{-1}) measured in Riga, followed by Vidzeme, with a concentration of 40.9 ng L^{-1}. In Luxembourg, PFAS levels of up to 153 ng L^{-1} were measured in groundwater. In Switzerland, PFAS were detected in groundwater in all regions except for the region of Central Switzerland. The highest concentration (242.8 ng L^{-1}) was measured in the Lemanic region, followed by 70.9 ng L^{-1} in Zurich and 64.2 ng L^{-1} in northwestern Switzerland.

A study in Finland investigated the occurrence and risks of PFAS at four AFFF-related sites (three firefighting sites and one industrial site) [59]. PFASs were detected at these sites at concentrations of up to nearly 4000 ng L^{-1} (median 20.1 ng L^{-1}). PFOS was the dominant compound, with concentrations of up to 3500 ng L^{-1}, whereas PFOA had a maximum concentration of 191 ng L^{-1}. The study highlighted that even a single fire suppression event can lead to long-term PFAS discharges. Gobelius et al. [60] investigated the occurrence of PFAS in groundwater and surface water across Sweden. The average groundwater ($n = 161$) concentration of PFAS was 49 ng L^{-1} (median 0.04 ng L^{-1}), with levels of up to 6400 ng L^{-1}, which was most likely related to an adjacent fire training site. The low median values of PFASs revealed that the majority of the groundwater samples were unaffected by PFAS contamination. However, a few elevated concentrations can be found in the southern part of Sweden, which are correlated with increasing anthropogenic activities in that region. Ng et al. [22] conducted a study to investigate PFAS occurrence in river water, wastewater, groundwater, and biota samples in the Danube River basin. Seven groundwater samples were collected from Austria, Bulgaria, Croatia, Hungary, Romania, Serbia, and Slovakia within the Danube River basin. PFASs were detected in all seven groundwater samples, with concentrations ranging from 2.8 ng L^{-1} (Austria) to 201.7 ng L^{-1} (Romania). PFASs were also detected in all the other studied environmental matrices, indicating potentially large-scale migration of PFASs in Europe, which might require further systematic regulatory monitoring.

By adapting the conceptual approach for identifying suspected contamination sites from The PFAS Project Lab [41] approach used in the United States [45], Forever Pollution Project [42] also identified over 21,500 presumptive contamination sites in Europe with current or past industrial activity documented as both using and emitting PFAS. Further monitoring is needed to confirm the situation of contamination at these sites.

3.4 *Occurrence in Africa*

This is one of the regions where PFAS monitoring is not included in the national groundwater monitoring plan, and there are no existing records with national departments or international agencies on the measured PFAS concentrations in groundwater. Africa is also one of the regions with very few studies on PFAS, and notably, there are no reported studies specifically on groundwater contamination. The existence of large data gaps in Africa has been highlighted by many researchers [27], yet this has not yet been addressed by the scientific community. One of the reasons for this lack of data is the limited capacity in terms of trained manpower and laboratory facilities [61]. The high concentrations of PFOA and PFOS found in other environmental media, including surface water (particularly freshwater) and soil, indicate the potential vulnerability of groundwater to these contaminants.

A recent study from Kabwe, Zambia, tested urban groundwater for a wide range of emerging contaminants potentially contributed from sources, including agrochemicals, flame retardants, plasticizers, food additives, etc. [62]. This suggests the need for more studies on PFAS in the groundwater of this region. Other studies from Kenya, Ethiopia, Nigeria, South Africa, and Uganda have identified PFOA and PFOS in surface waters [63–71]. Soil [70] and sediment [72, 73] contaminated with PFAS have also been reported. These studies do not report the PFAS concentrations in groundwater, but they suggest that contamination from surface water, soil, or sediments can affect groundwater. Studies that are close to groundwater are those that report concentrations from tap water or drinking water. Although the source of contamination is not clearly defined, the potential of using groundwater for water supply cannot be underestimated. PFOA (1.78–321 ng L^{-1}) and PFOS (95.4–277 ng L^{-1}) were detected in tap waters in Ghana, but the source of the tap water was not clarified by Essumang et al. [74]. Tap water from Uganda contained 1.05–5.77 ng L^{-1} PFOA and 0.02–0.62 ng L^{-1} PFOS [75]. PFAS research in other environmental media has been conducted in this region; however, the severity of the impact and link of these media on groundwater is not fully understood.

3.5 Occurrence in Asia

The present review reports the largest number of published studies in Asia, including global reviews and reviews on PFASs in Asian groundwater. Overall, the concentrations of PFOA and PFOS reported in groundwater in Asian countries range from <0.05 to 660,000 ng L^{-1} and from 0.008 to 101,110 ng L^{-1}, respectively (Table 5.4).

China had the greatest number of reported cases in groundwater (n = 23) from both rural and urban settings. Firefighting training sites and military sites with AFFFs were often not the primary sources of PFOA and PFOS in groundwater in China. Instead, secondary sources such as municipal wastewater, industrial wastewater, and contaminated surface water were the main contributors. PFOAs were much more common in groundwater from industrial areas, whereas PFOS was more common in nonindustrial areas in 13 cities of Jiangsu Province, China. The total PFAS concentrations in the groundwater samples ranged from 2.69 to 556 ng L^{-1} [76]. Fluorochemical industries have been frequently reported as sources of groundwater contamination with PFOA and PFOS, with the extent of pollution increasing over the years. For example, studies around the Fuxin fluorochemical park revealed a fivefold increase in PFOA, that is, 524 ng L^{-1} in 2009 to 2510 ng L^{-1} in 2017 [77, 78]. This industry was the major source, along with household wastes, farming drains, and atmospheric precipitation, in the rural areas studied by Chen et al. [79]. Very high concentrations of PFOA have been detected in groundwater from fluorochemical industries in other parts of China [80]. A maximum PFOA concentration of 239,644 ng L^{-1} from groundwater near a fluorochemical industrial park was reported by Liu et al. [81]. The highest PFOA concentration reported in Asian groundwaters to date from the same industrial source is 660,000 ng L^{-1} [82].

Compared with surface water, groundwater monitored around the Yuqiao Reservoir, Tianjin city, China, has low detection rates [83], and the total PFAS concentration ranges from 1.4 to 17.1 ng L^{-1}, which is still higher than that of soil and sediment samples. In most cases, in Chinese groundwater, PFOA is dominant, and PFOS is often the least common PFAS compound identified [84–87]. Multiple primary and secondary sources responsible for PFOA and PFOS in groundwater are often reiterated in many studies [79, 88, 89].

A study on the bioaccumulation of PFAS in vegetables also determined the PFOA and PFOS in groundwater that is used for irrigation [90]. The groundwater used for irrigation in this study was located near the fluorochemical plant, and this contaminated groundwater, when used for irrigation, percolated through the soil and reached the shallow aquifer again. In the case of PFOA, the concentration after infiltration through the soil was 2470 ng L^{-1}, and the concentration of PFOS reached a maximum of 503 ng L^{-1} [90]. The potential human health risk caused by these contaminants through drinking groundwater is lower in rural areas than in urban areas, except for rural areas with fluorochemical industrial plants [91, 92]. Few other studies have also reported PFOA and PFOS in groundwater, but the concentrations could not be retrieved from the figs [93]. and therefore are not included in Table 5.4.

Table 5.4 Concentrations of PFOA and PFOS in groundwater from Asia and Oceania

Continent	Country	Sampling year	PFOA (ng L^{-1})		PFOS (ng L^{-1})		Number of PFAs reported	Source	Reference
			Min	Max	Min	Max			
Asia	China	2002	n.a.[a]	<0.1	n.a.	0.3	2	Not specified	Jin et al. [107]
Asia	China	2009	4.85	524	<0.1	0.73	5	Fluorochemical industry	Bao et al. [78]
Asia	China	2011	0.24	6.2	0.06	5.6	12	Rainfall, wastewater	Zhu et al. [108]
Asia	China	2013	0.07	7.1	n.a.	0.21	7	Not specified	Li et al. [86]
Asia	China	2013	0.1	3.9	0.1	0.2	13	Industrial and nonindustrial sources	Qi et al. [85]
Asia	China	2014	64.5	4150	0.15	1.05	10	Fluorochemical industry	Wang et al. [80]
Asia	China	2014	<0.13	4.27	<0.03	0.1	10	Industrial sources	Liu et al. [84]
Asia	China	2014	0.52	239,644	0.02	37.8	12	Fluorochemical industry	Liu et al. [81]
Asia	China	2014–2015	<0.1	475	<0.5	94.9	17	Industrial wastewater, household and farming sewage, atmospheric deposition	Chen et al. [79]
Asia	China	2014–2016	<LOD[b]	107.67	n.a.	0.83 (mean)	10	Industrial, mining, atmospheric deposition	Qiao et al. [92]
Asia	China	2014–2017	3.9	22.1	0.4	48.3	11	Contaminated river	Liu et al. [109]
Asia	China	2015	0.73	87.7	<0.04	776	10	Industrial wastes	Gao et al. [14]
Asia	China	2015	21.89	104.66	2.89	13.07	10	Industrial sources	Sun et al. [110]
Asia	China	2016	<0.2	453	<0.2	79	14	Industrial and nonindustrial sources	Wei et al. [76]
Asia	China	2016	<0.03	54.5	<0.01	33.75	18	Industrial, consumer products, local point sources and others	Wang et al. [89]
Asia	China	2016	2	660,000	n.a.	n.a.	1	Fluorochemical industry	Lu et al. [82]
Asia	China	2017	<0.05	5.11	<0.05	0.5	16	Sewage and atmospheric precipitation	Cao et al. [83]
Asia	China	2017	105	2510	<0.1	403	10	Fluorochemical industry	Bao et al. [77]
Asia	China	2017	0.25	566	0.09	24.52	17	Fluorochemical industry	Liu et al. [111]

Asia	China	2018	48.3	176.9	47.8	49.5	11	Domestic wastewater and other	Li et al. [112]
Asia	China	2018	<0.005	48	<LOD	2.9	35	AFFF, industrial and municipal	Chen et al. [113]
Asia	China	2018	2.78	53.72	0.16	2.44	15	Landfill and wastewater treatment plant	Xu et al. [91]
Asia	China	2019	n.a.	2470	500	503	8	Fluorochemical industry	Bao et al. [90]
Asia	China	2019	0.41	99.9	n.a.	0.07 (mean)	25	Industrial sources	Zhou et al. [87]
Asia	China	2019	0.03	3540.9	<LOD	9.31	36	Industrial and domestic wastewater, food packaging sources, AFFF, metal plating, surface runoff	Li et al. [88]
Asia	India	2014	<0.03	0.76	<0.01	1.13	14	Surface water receiving wastewater from industrial and domestic sources	Sharma et al. [94]
Asia	India	2019	n.a.	0.0003	0.0001	0.0008	5	Not clearly specified	Richards et al. [95]
Asia	India	2022–2023	17.6	77.6	1.48	2.72	8	Landfills and domestic	Koulini and Nambi [96]
Asia	Japan	2006	0.47	60	0.28	133	9	Urban wastewater, street runoff	Murakami et al. [97]
Asia	Japan	2006–2007	20	58	2.9	6.1	2	Not known	Takagi et al. [13]
Asia	Japan	2015–2016	45.2	7440	n.a.	n.a.	6	Fluoropolymer industry	Shiwaku et al. [98]
Asia	Japan	2007–2009	<0.25	1800	<0.25	990	13	Sewage and point sources	Kuroda et al. [99]
Asia	Vietnam	2013	<0.05	4.5	<0.05	8.2	7	Untreated municipal wastewater and storm water	Duong et al. [100]
Asia	Vietnam	2013–2015	<0.37	5.48	<0.03	1.42	13	Wastewater treatment plant	Lam et al. [101]
Asia	Thailand	n.a.	0.65	34.96	<0.37	25.88	10	Municipal and industrial waste	Hongkachok et al. [102, 103]

(continued)

Table 5.4 (continued)

Continent	Country	Sampling year	PFOA ($ng\ L^{-1}$)		PFOS ($ng\ L^{-1}$)		Number of PFAs reported	Source	Reference
			Min	Max	Min	Max			
Asia	South Korea	2019	1.19	6.72	<LOD	2.35	8	Leakage from industrial plant	Yong et al. [104]
Asia	Taiwan	2013	0.1	40.3	0.6	76.8	9	Mix of various secondary sources	Lin et al. [106]
Oceania	Australia	2015–2016	<50	580	<170	13,000	10	AFFF	Bräunig et al. [9]
Oceania	Australia	2019	168	5433	550	49,144	17	AFFF	Kaserzon et al. [114]
Oceania	Australia	n.a.	n.a.	6920	n.a.	101,110	11	AFFF	Turner et al. [115]
Oceania	Australia	2017–2018	<0.09	6.9	<0.03	34	20	Irrigation with recycled water from WWTP	Szabo et al. [116]
Oceania	Australia	2017	1.7	74	1.3	4800	14	AFFF	Hepburn et al. [117]
Oceania	Australia	2020	7	66	15	170	15	AFFF	Currell et al. [118]
Oceania	New Zealand	2022	<1	2.8	<1	17	10	Human and industrial activities	Close and Banasiak [119]
Oceania	New Zealand	2018	1.2	340	0.2	110	8	Multiple sources, not yet well-defined	Moreau et al. [8]

[a] n.a. = not available
[b] LOD = Limit of detection

The Ganges River, India, is contaminated with wastewater inputs from various sources, with 15 PFAS compounds detected, which, through infiltration, also contaminate groundwater with 14 PFAS [94]. Sharma et al. [94] reported PFOA concentrations ranging from <0.03 to 0.76 ng L^{-1} in groundwater and PFOS concentrations ranging from <0.01 to 1.13 ng L^{-1}. PFOA and PFOS were detected at very low frequencies along with 3 other PFASs by Richards et al. [95] from the middle Gangetic Plain, India. However, these compounds were found at trace levels, that is, PFOA at 3×10^{-4} ng/L and a maximum PFAS of 8×10^{-4} ng/L in groundwater [95]. Eight PFASs, including PFOA (17.6–77.6 ng L^{-1}) and PFOS (1.48–2.72 ng L^{-1}), were detected in groundwater from three locations in Chennai, India [96]. In studies by Koulini and Nambi [96] and Sharma et al. [94], the presence of PFOA and PFOS in groundwater was attributed not only to direct seepage from contaminated rivers but also to the use of contaminated river water for irrigation and other purposes, leaching from unpaved surfaces, leachate from landfills, and inputs from municipal and industrial wastewater.

In one of the first studies on PFAS in groundwater, Murakami et al. [97] identified sources of contamination from urban wastewater and street runoff in Tokyo, Japan. Although the concentrations were not as high as the values reported from industrial areas, the PFOA concentration was a maximum of 60 ng L^{-1}, and the PFOS concentration was 133 ng L^{-1} [97]. Osaka, Japan, reported similar maximum concentrations of PFOA (58 ng L^{-1}), but the concentration of PFOS was much lower (6.1 ng L^{-1}) [13]. Ahrens et al. [10] measured PFAS in pore water extracted from sediment cores from Tokyo Bay, Japan, to a depth of 79 cm. These findings indicate that the sediments act as sinks and retain PFOA and PFOS, which are eventually transported to groundwater. PFOA in pore water ranges between 0.44 and 11 ng L^{-1}, and the measured PFOS concentration ranges between <0.1 and 3.49 ng L^{-1} [10]. The groundwater near a fluoropolymer industrial plant had a maximum PFOA concentration of 7440 ng L^{-1} [98], which is lower than the levels reported at similar industrial sites in China. Both shallow young groundwater and deep old groundwater recharged before the 1950s were contaminated in Tokyo, and unconfined aquifers were contaminated chiefly by sewage and point sources [99].

Sixteen PFASs were tested in the groundwaters of Hanoi and Ho Chi Minh City, Vietnam. Ten PFAS were the major species reported, with PFOA and PFOS being the most common, that is, in 45% and 36% of the samples, respectively [100]. The maximum PFOA concentration determined was 4.5 ng L^{-1}, and the maximum PFOS concentration was 8.2 ng L^{-1}. Lam et al. [101] reported a slightly higher PFOA concentration of 5.48 ng L^{-1} and a much lower PFOS concentration of 1.42 ng L^{-1} from central Vietnam. This study also revealed PFOA (<0.37–0.48 ng L^{-1}) and PFOS in tap water (<0.03 ng L^{-1}), which is a mixture of groundwater and surface water. Studies from India and Vietnam noted that the degree of contamination by PFOA and PFOS is much lower in these countries than in developed nations such as North America or Europe [94, 100].

Other Asian countries have rarely published studies on PFAS in peer-reviewed journals or as gray literature. In Thailand, there are two published scientific articles, but the data appear to be identical. Hongkachok et al. [102] and Hongkachok et al.

[103] reported that municipal and industrial waste from two cities in Thailand had contaminated groundwater with a maximum PFOA concentration of 34.96 ng L^{-1} and a maximum PFOS concentration of 25.88 ng L^{-1}. The effects of a PFAS leakage incident from an industrial plant in the Nakdong River basin, South Korea, revealed evidence of seepage into groundwater [104]. Differences in the occurrence of various PFASs from different land uses, that is, industrial, urban, and agricultural, and from groundwater samples at different reaches of the river were also noted. The maximum recorded PFOA concentration was 6.72 ng L^{-1}, and the PFOS concentration was 2.35 ng L^{-1} [104]. A recent study from South Korea reported a total maximum PFAS of 45.2 ng L^{-1}, and PFOA was the dominant PFAS, but the results of each of the PFAS compounds were not elaborated [105].

A suite of pharmaceuticals, hormones, and perfluorinated compounds were studied by Lin et al. [106] in groundwater from two cities in Taiwan. Relatively high concentrations were reported: PFOA from 0.1 to 40.3 ng L^{-1} and PFOS from 0.6 to 76.8 ng L^{-1}. Although the sources were not clearly demarcated, they were indicated to be from a mixture of several secondary sources, such as municipal or industrial wastewater, and due to the interaction between contaminated surface water and groundwater.

3.6 *Occurrence in Oceania*

Studies in Oceania are available only for Australia, with six peer-reviewed studies reporting PFAS in groundwater (Table 5.4). Most of the reported studies have been conducted around legacy landfill sites, which are known secondary sources of PFAS. Shallow wells around the Fishermans Bend legacy landfill in Melbourne, Australia, detected a maximum of 14 PFAS, where PFOS ranged between 1.3 and 4800 ng L^{-1}, and PFOA ranged from 1.7 to 74 ng L^{-1} [117]. Additionally, PFOA was the dominant PFAS detected, with one sample recording the highest PFOS concentration of 4800 ng L^{-1}, which was attributed to point-source contamination from an industrial site. A follow-up assessment conducted in the same area revealed a localized increase in PFAS contents near the landfill [118]. PFOS became the dominant contaminant in this recent study, in contrast to the results of Hepburn et al. [117]. Currell et al. [118] reported PFOA concentrations ranging from 7 to 66 ng L^{-1} and PFOS concentrations ranging from 15 to 170 ng L^{-1} in groundwater around a landfill site. Resampling from a few boreholes from Hepburn et al. [117] indicated an increase in contamination, especially with PFOS and total PFAS.

Contamination from one of the primary sources of PFAS, that is, fire-fighting training areas, is well-known to contaminate the aquifer underground. In anticipation of the effect of the AFFF on groundwater PFAS in Queensland, Australia, Bräunig et al. [9] analyzed groundwater samples from a military airfield. PFOS was the most dominant of the analyzed PFAS (maximum of 13,000 ng L^{-1}), and the maximum PFOA reported was 580 ng L^{-1}. The subsequent use of contaminated groundwater for livestock and irrigation and the consequent detection of PFAS in other environmental and biological samples were also reported in this study. The range of detected PFOA was 168–5433 ng L^{-1}, and that of PFAS was 550–49,144 ng L^{-1}, which is near yet another firefighting training ground. The concentrations determined by Kaserzon et al. [114] were from shallow wells with depths <7.5 m, and deep bore wells with depths >12 m were not detected for the targeted contaminants. The reported concentrations of the targeted contaminants in the groundwater of these studies in Australia were much lower than the maximum PFOS and PFOA values of 101,110 ng L^{-1} and 6920 ng L^{-1}, respectively, reported by Turner et al. [115] from groundwater affected by an AFFF site in Williamtown Airforce Base, Australia.

The recycled water from the WWTPs used for irrigation eventually percolated to the shallow aquifers in Werribee South, Australia. Twenty PFAs were determined to range from <0.03 to 74 ng L^{-1} [116]. PFOS (<0.03–34 ng L^{-1}) and PFOA (<0.09–6.9 ng L^{-1}) were detected in 96% and 92% of the groundwater samples, respectively, which, according to Szabo et al. [116], is higher than the background levels detected in groundwater from areas irrigated with treated wastewater.

New Zealand has conducted systematic surveys at the national level on PFASs in groundwater. In 2022, groundwater contains PFOA concentrations ranging from <1 to 2.8 ng L^{-1} [119], which is much lower than the range of 1.2 to 340 ng L^{-1} measured in 2018 [8]. A similar pattern was also observed for PFOS (Table 5.4). These are baseline studies in the country, and the sources are not yet clearly defined. Multiple sources from industrial wastewater, municipal wastewater, and non-wastewater need to be differentiated in future studies [8].

4 Source Distribution of PFASs

The detailed assessment of the extent and importance of PFASs in groundwater in this study indicates the widespread occurrence of PFASs across regions on the five continents, originating from various contamination sources. The release of PFAS into the groundwater system occurs through either point sources or nonpoint (diffuse) sources. Point sources are usually stationary and discrete, whereas nonpoint sources are diffuse and often involve releases from multiple origins. These sites can be mainly divided into AFFF discharge sites, industrial facilities that produce and/or use PFAS, and sites related to PFAS-containing waste disposal (Fig. 5.1). Point sources are usually associated with hot spots of contamination where high concentrations of PFAS can be detected at a level that poses a risk to human health.

More specifically, AFFF discharge sites are places where fluorinated AFFFs have been largely used for extinguishing fires or fire training. This includes (1) military sites where AFFFs can be routinely discharged from training, testing, and firefighting operations; (2) major airports with regular testing and discharge of AFFFs; (3) other firefighting training sites, such as fire departments; and (4) fire suppression sites using AFFFs, such as railroad crash sites, oil and gas extraction sites, and petroleum refineries [45]. In the United States, 26% of the 1585 identified

PFAS-contaminated groundwater sites are related to AFFF discharge sites (Table 5.1), particularly military sites and airports [41]. In Canada, over 100 identified federal contaminated sites with PFAS contamination are also predominantly related to the past and/or current use of AFFF [46]. In addition, several groundwater-contaminated sites in South America and the Caribbean region are associated with AFFF discharges (Table 5.2), including the Marshall Islands and Puerto Rico [41], as well as several French overseas territories (French Guiana, Guadeloupe, and Martinique) [54]. Significant AFFF-related groundwater contamination has also been reported in the European Union (e.g., Germany, Finland, Sweden; Table 5.3), China, and Australia (Table 5.4). The contaminated soil and vadose zones at these sites can become large sinks of PFAS contaminants, impacting groundwater for decades through continuous leaching [120].

Industrial facilities that manufacture and/or use PFAS are another substantial source of PFAS contamination in groundwater. In the United States, approximately 24% of the identified PFAS-contaminated groundwater sites are industrial facilities [41]. Within the framework of the Forever Pollution Project [42] in Europe, 20 large PFAS manufacturing plants have been identified that span the continent, including six plants in Germany, five in France, three in the United Kingdom, two in Italy, and one each in Poland, Spain, the Netherlands, and Belgium. The investigation revealed that the best-known hotspots of pollution in the European Union all have PFAS manufacturing plants at their epicenter. In addition to these PFAS producers, another 232 industrial sites across Europe, which use PFASs to synthesize plastics, paints, pesticides, waterproof textiles, and other chemicals, have also been identified as suspected sources of PFAS contamination [42]. In China, industrial activities are the predominant documented sources of PFAS contamination in groundwater, with over two-thirds of the reviewed studies (18 out of 25) related to pollution from industrial sites (Table 5.4). Other countries with identified groundwater contamination sites related to industrial facilities include India, Thailand, Japan, South Korea, and New Zealand (Table 5.4).

The aforementioned AFFF-related discharge sites and industrial facilities are typically categorized as primary sources of PFAS contamination. Another significant source of PFASs is related to PFAS-containing waste disposal sites, which are categorized as secondary sources. These include WWTPs with PFAS-contaminated effluent and sludge, landfills, land application of biosolids, etc. In the United States, approximately 42% of the identified PFAS-contaminated groundwater sites are related to PFAS-containing waste, making it a predominant source of PFAS contamination [41]. In Canada, groundwater contamination related to landfills has been investigated, highlighting that some landfills closed decades ago still present PFAS concentrations similar to those of modern landfills, indicating the long-lasting impact of these sites on groundwater quality [46]. In Australia, studies around legacy landfill sites have detected high concentrations of PFAS in surrounding groundwater systems, highlighting waste disposal sites as significant sources of PFAS contamination [117, 118]. Similar issues have been reported in China, India, Japan, Vietnam, and Thailand (Table 5.4).

Notably, if research was undertaken in more locations worldwide at the aforementioned sites with suspected point sources of PFAS, comparable levels of contamination would likely be found. In the absence of widespread testing data, a presumptive contamination model approach, as demonstrated by Salvatore et al. [45] and already applied by The PFAS Project Lab [41] in the United States and the Forever Pollution Project [42] in Europe, could help governments, industries, and communities identify potential sources and take data-informed steps to investigate and address PFAS contamination. This approach can facilitate the identification and prioritization of locations for monitoring, regulation, and remediation. Additionally, site-by-site verification can be conducted at smaller scales to pinpoint specific contamination sites.

Compared with point sources of PFASs, which are specific and identifiable locations, diffuse sources of PFAS contamination, which often involve multiple sources, are more dispersed, and identifying the origin(s) or distinguishing the location(s) of contamination is more challenging. Diffuse sources of PFAS into groundwater include atmospheric deposition, surface runoff, agricultural activities, interactions with contaminated surface water, etc. Studies conducted by Zhu et al. [108] and Qiao et al. [92] in China (Table 5.4) have illustrated the contributions of atmospheric or rainfall deposition of PFAS into groundwater systems. These studies highlight how PFASs can be transported through the atmosphere and deposited over large areas, ultimately infiltrating groundwater. Surface runoff can be a significant contributor to PFAS contamination, especially in regions where intense rainfall events or frequent storms generate large volumes of runoff. This phenomenon has been reported in several Asian countries, including China, Japan, and Vietnam (Table 5.4). Agricultural practices are another common source of PFAS contamination. This includes the application of PFAS-containing pesticides or herbicides, as reported in an agricultural area in Brazil [56]. Additionally, irrigation with contaminated water, especially water recycled from WWTPs, can lead to PFAS leaching into groundwater (e.g., Szabo et al. [116]). Surface water bodies that are contaminated with PFAS can also infiltrate adjacent groundwater, especially during high-flow conditions or flooding events, for example, Liu et al. [109] and Sharma et al. [94]. This process can extend the reach of PFAS contamination from surface water to groundwater systems.

To address diffuse sources of PFAS contamination effectively, regional and national-scale screening and monitoring programs are essential. Comprehensive sampling of raw groundwater and tap water can help identify potential hotspots and plumes, providing a general understanding of the extent of PFAS contamination in the environment. To better identify the sources of contamination in groundwater, measurements in other environmental matrices, such as surface water, soil, sludge, and precipitation, are also crucial. This can facilitate an understanding of the pathways and extent of PFAS contamination, allowing for more targeted and effective remediation efforts.

5 Conclusion and Future Directions

A detailed, quantitative assessment of PFOA and PFOS concentrations in groundwater was conducted. Compared with other environmental media, groundwater has received relatively little attention. Identifying sources from primary and secondary sources was not as easy as identifying sources from surface water, and many studies have investigated unidentified sources for groundwater. Large geographical data gaps exist with many countries from Africa, the Middle East, South America, Southeast Asia, and Oceania. National-level monitoring has been initiated in several countries but few in developing and underdeveloped countries. A full-scale inventory of PFASs, including PFOA and PFOS, is needed to develop a regulatory framework and guidelines for industrial wastewater, WWTPs, and drinking water. This, in turn, protects vulnerable groundwater resources and prevents costs and extensive manpower involved in remediation. In geographical regions lacking national monitoring surveys, developing a database from published literature can be beneficial as a first step toward water resources management. Primary sources from firefighting training sites, military sites, major airports and industrial wastewater are the major contributors to groundwater PFOA and PFOS in developed countries such as North America and Europe. Secondary sources from consumer products ending in landfills, leachates, WWTPs, biosolids used in land application, and recycled water used for irrigation are the leading causes of contamination in other regions of the world.

Despite significant efforts to compile data on PFAS contamination sites, particularly in the United States and Europe, there are still limitations to the current databases. Importantly, the databases used for analysis in this study may not capture all known contamination sites since there are certain sites for which testing data either do not exist or are not publicly available. Furthermore, the extent of known contamination is related to the scope and focus of existing testing efforts. Many sites likely remain undetected or untested, resulting in an underestimation of the true scale and severity of PFAS contamination. Conceptual approaches based on likely sources to pinpoint presumptive contamination could be effective in the absence of groundwater monitoring data, allowing for the rapid and systematic identification of potential exposure sources, as illustrated by studies conducted in the United States and the European Union. The PFOA and PFOS concentrations in groundwater from many of the reported studies exceed the guidelines available from the European Union, the United States, Australia, and Japan. This study highlights the widespread presence of PFOA and PFOS in global groundwater and the pressing need to protect these resources, which serve as sources of potable water. The need for systematic monitoring and inclusion of these parameters in drinking water guidelines is emphasized, as the human health impacts and environmental burdens from these substances are evident.

Acknowledgments The authors want to thank Dr. Alissa Cordner, Associate Professor of Sociology and Paul Garrett Fellow at Whitman College (USA), for her valuable review and feedback on the data analysis for the United States. The authors also acknowledge the support of IHE Delft for providing open-access funding. This work was carried out within the framework of the International Association of Hydrogeologists' Commission on Groundwater Quality.

References

1. Wang, F., Xiang, L., Sze-Yin Leung, K., Elsner, M., Zhang, Y., Guo, Y., Pan, B., Sun, H., An, T., Ying, G., Brooks, B. W., Hou, D., Helbling, D. E., Sun, J., Qiu, H., Vogel, T. M., Zhang, W., Gao, Y., et al. (2024). Emerging contaminants: A one health perspective. *The Innovation, 5*(4), 100612.
2. Ruan, T., Field, J., Cousins, I., Lohmann, R., & Jiang, G. (2022). Emerging Contaminants: Fluorinated Alternatives to Existing PFAS. *Environmental Science & Technology, 56*(10), 6001–6003.
3. CompTox. (2023) *The CompTox Chemistry Dashboard: a community data resource for environmental chemistry.* Retrieved Auguest 1, 2024 from https://comptox.epa.gov/dashboard/
4. Morales-McDevitt, M. E., Becanova, J., Blum, A., Bruton, T. A., Vojta, S., Woodward, M., & Lohmann, R. (2021). The Air That We Breathe: Neutral and Volatile PFAS in Indoor Air. *Environmental Science & Technology Letters, 8*(10), 897–902.
5. Li, W.-L., & Kannan, K. (2024). *Determination of Legacy and Emerging Per-and polyfluoroalkyl substances (PFAS) in indoor and outdoor air* (Vol. 1, p. 1147). ACS ES&T Air.
6. Podder, A., Sadmani, A. H. M. A., Reinhart, D., Chang, N.-B., & Goel, R. (2021). Per and poly-fluoroalkyl substances (PFAS) as a contaminant of emerging concern in surface water: A transboundary review of their occurrences and toxicity effects. *Journal of Hazardous Materials, 419*, 126361.
7. Möller, A., Ahrens, L., Surm, R., Westerveld, J., van der Wielen, F., Ebinghaus, R., & de Voogt, P. (2010). Distribution and sources of polyfluoroalkyl substances (PFAS) in the River Rhine watershed. *Environmental Pollution, 158*(10), 3243–3250.
8. Moreau, M., Hadfield, J., Hughey, J., Sanders, F., Lapworth, D. J., White, D., & Civil, W. (2019). A baseline assessment of emerging organic contaminants in New Zealand groundwater. *Science of the Total Environment, 686*, 425–439.
9. Bräunig, J., Baduel, C., Heffernan, A., Rotander, A., Donaldson, E., & Mueller, J. F. (2017). Fate and redistribution of perfluoroalkyl acids through AFFF-impacted groundwater. *Science of the Total Environment, 596-597*, 360–368.
10. Ahrens, L., Yamashita, N., Yeung, L. W. Y., Taniyasu, S., Horii, Y., Lam, P. K. S., & Ebinghaus, R. (2009). Partitioning Behavior of Per-and Polyfluoroalkyl Compounds between Pore Water and Sediment in Two Sediment Cores from Tokyo Bay, Japan. *Environmental Science & Technology, 43*(18), 6969–6975.
11. Jansson, F., *Occurrence of per-and polyfluorinated alkyl substances (PFAS), including ultra-short-chain compounds. Seasonal variation in rainwater from the Swedish west coast* Örebro University 2019. p. 29.
12. Sammut, G., Sinagra, E., Helmus, R., & de Voogt, P. (2017). Perfluoroalkyl substances in the Maltese environment—(I) surface water and rain water. *Science of the Total Environment, 589*, 182–190.
13. Takagi, S., Adachi, F., Miyano, K., Koizumi, Y., Tanaka, H., Mimura, M., Watanabe, I., Tanabe, S., & Kannan, K. (2008). Perfluorooctanesulfonate and perfluorooctanoate in raw and treated tap water from Osaka, Japan. *Chemosphere, 72*(10), 1409–1412.
14. Gao, Y., Liang, Y., Gao, K., Wang, Y., Wang, C., Fu, J., Wang, Y., Jiang, G., & Jiang, Y. (2019). Levels, spatial distribution and isomer profiles of perfluoroalkyl acids in soil, groundwater and tap water around a manufactory in China. *Chemosphere, 227*, 305–314.
15. Bolan, N., Sarkar, B., Vithanage, M., Singh, G., Tsang, D. C. W., Mukhopadhyay, R., Ramadass, K., Vinu, A., Sun, Y., Ramanayaka, S., Hoang, S. A., Yan, Y., Li, Y., Rinklebe, J., Li, H., & Kirkham, M. B. (2021). Distribution, behaviour, bioavailability and remediation of poly-and per-fluoroalkyl substances (PFAS) in solid biowastes and biowaste-treated soil. *Environment International, 155*, 106600.
16. Brusseau, M. L., Anderson, R. H., & Guo, B. (2020). PFAS concentrations in soils: Background levels versus contaminated sites. *Science of the Total Environment, 740*, 140017.

17. Yan, H., Cousins, I. T., Zhang, C., & Zhou, Q. (2015). Perfluoroalkyl acids in municipal landfill leachates from China: Occurrence, fate during leachate treatment and potential impact on groundwater. *Science of the Total Environment, 524-525*, 23–31.
18. Gallen, C., Drage, D., Eaglesham, G., Grant, S., Bowman, M., & Mueller, J. F. (2017). Australia-wide assessment of perfluoroalkyl substances (PFASs) in landfill leachates. *Journal of Hazardous Materials, 331*, 132–141.
19. Kim, J., Xin, X., Mamo, B. T., Hawkins, G. L., Li, K., Chen, Y., Huang, Q., & Huang, C.-H. (2022). Occurrence and fate of ultrashort-chain and other Per-and polyfluoroalkyl substances (PFAS) in wastewater treatment plants. *ACS ES&T Water, 2*(8), 1380–1390.
20. Gallen, C., Eaglesham, G., Drage, D., Nguyen, T. H., & Mueller, J. F. (2018). A mass estimate of perfluoroalkyl substance (PFAS) release from Australian wastewater treatment plants. *Chemosphere, 208*, 975–983.
21. Goodrow, S. M., Ruppel, B., Lippincott, R. L., Post, G. B., & Procopio, N. A. (2020). Investigation of levels of perfluoroalkyl substances in surface water, sediment and fish tissue in New Jersey, USA. *Science of the Total Environment, 729*, 138839.
22. Ng, K., Alygizakis, N., Androulakakis, A., Galani, A., Aalizadeh, R., Thomaidis, N. S., & Slobodnik, J. (2022). Target and suspect screening of 4777 per-and polyfluoroalkyl substances (PFAS) in river water, wastewater, groundwater and biota samples in the Danube River Basin. *Journal of Hazardous Materials, 436*, 129276.
23. Yeung, L. W. Y., Dassuncao, C., Mabury, S., Sunderland, E. M., Zhang, X., & Lohmann, R. (2017). Vertical profiles, sources, and transport of PFASs in the Arctic Ocean. *Environmental Science & Technology, 51*(12), 6735–6744.
24. Sznajder-Katarzyńska, K., Surma, M., Cieślik, E., & Wiczkowski, W. (2018). The perfluoroalkyl substances (PFASs) contamination of fruits and vegetables. *Food Additives & Contaminants: Part A, 35*(9), 1776–1786.
25. Wilson, T. B., Stevenson, G., Crough, R., de Araujo, J., Fernando, N., Anwar, A., Scott, T., Quinteros, J. A., Scott, P. C., & Archer, M. J. G. (2021). Evaluation of Residues in Hen Eggs After Exposure of Laying Hens to Water Containing Per-and Polyfluoroalkyl Substances. *Environmental Toxicology and Chemistry, 40*(3), 735–743.
26. Poothong, S., Thomsen, C., Padilla-Sanchez, J. A., Papadopoulou, E., & Haug, L. S. (2017). Distribution of Novel and Well-Known Poly-and Perfluoroalkyl Substances (PFASs) in Human Serum, Plasma, and Whole Blood. *Environmental Science & Technology, 51*(22), 13388–13396.
27. Sims, J. L., Stroski, K. M., Kim, S., Killeen, G., Ehalt, R., Simcik, M. F., & Brooks, B. W. (2022). Global occurrence and probabilistic environmental health hazard assessment of per-and polyfluoroalkyl substances (PFASs) in groundwater and surface waters. *Science of the Total Environment, 816*, 151535.
28. Misstear, B., Vargas, C. R., Lapworth, D., Ouedraogo, I., & Podgorski, J. (2023). A global perspective on assessing groundwater quality. *Hydrogeology Journal, 31*(1), 11–14.
29. UNEP. (2011). *Progress on ambient water quality. Tracking SDG 6 series: global indicator 6.3.2 updates and acceleration needs. Nairobi*. UNEP.
30. NNGWA. (2024). *National Ground Water Association, United States*. Retrieved Auguest 1, 2024 from https://www.ngwa.org/what-is-groundwater/About-groundwater/groundwater-facts
31. European Commission. (2024). *Groundwater: Ensuring groundwater quantity and quality across the EU*. Retrieved Auguest 1, 2024 from https://environment.ec.europa.eu/topics/water/groundwater_en
32. WHO/UNICEF JMP. (2019). *Global data on Water Supply, Sanitation and Hygiene (WASH), WHO/UNICEF Joint Monitoring Programme (JMP)*. Retrieved Auguest 1, 2024 from https://washdata.org/data
33. EU. (2020). *Directive (EU) 2020/2184 of the European Parliament and of the Council of 16 December 2020 on the quality of water intended for human consumption (recast) (OJ L 435, 23.12.2020)* (pp. 1–62). EU.

34. NHMRC-NRMMC. (2011). Australian Drinking Water Guidelines Paper 6 National Water Quality Management Strategy. In *National Health and Medical Research Council, National Resource Management Ministerial Council, Commonwealth of Australia, Canberra*. NHMRC-NRMMC.
35. USEPA, (2024). *Per-and Polyfluoroalkyl Substances (PFAS). Final PFAS National Primary Drinking Water Regulation.* Retrieved Auguest 1, 2024 from https://www.epa.gov/sdwa/and-polyfluoroalkyl-substances-pfas
36. Health Canada. (2023). *Per-and polyfluoroalkyl substances (PFAS) in drinking water.* Retrieved Auguest 1, 2024 from https://www.canada.ca/en/health-canada/services/environmental-workplace-health/reports-publications/water-quality/water-talk-per-polyfluoroalkyl-substances-drinking-water.html
37. PWD Environmental. (2022). *Consumer confidence report drinking water systems.* Retrieved Auguest 1, 2024, from https://cnrj.cnic.navy.mil/Portals/80/CNRJ/Documents/environmental/2022-CFAY-Drinking-Water-CCR-(EN).pdf?ver=srvGcOEJfztcQzzpeBT68Q%3D%3D
38. Johnson, G. R., Brusseau, M. L., Carroll, K. C., Tick, G. R., & Duncan, C. M. (2022). Global distributions, source-type dependencies, and concentration ranges of per-and polyfluoroalkyl substances in groundwater. *Science of the Total Environment, 841*, 156602.
39. Kurwadkar, S., Dane, J., Kanel, S. R., Nadagouda, M. N., Cawdrey, R. W., Ambade, B., Struckhoff, G. C., & Wilkin, R. (2022). Per-and polyfluoroalkyl substances in water and wastewater: A critical review of their global occurrence and distribution. *Science of the Total Environment, 809*, 151003.
40. Ackerman Grunfeld, D., Gilbert, D., Hou, J., Jones, A. M., Lee, M. J., Kibbey, T. C. G., & O'Carroll, D. M. (2024). Underestimated burden of per-and polyfluoroalkyl substances in global surface waters and groundwaters. *Nature Geoscience, 17*(4), 340–346.
41. The PFAS Project Lab. (2024). *PFAS Contamination site tracker.* Retrieved July 31, 2024, from https://pfasproject.com/pfas-sites-and-community-resources/
42. Forever Pollution Project. (2024) Retrieved July 31, 2024, from https://foreverpollution.eu/
43. Le Monde. (2024). Retrieved July 31, 2024, from https://www.lemonde.fr/en/les-decodeurs/article/2023/02/23/forever-pollution-explore-the-map-of-europe-s-pfas-contamination_6016905_8.html
44. PFAS Data Hub. (2024). Retrieved July 31, 2024, from https://pdh.cnrs.fr/en/download/
45. Salvatore, D., Mok, K., Garrett, K. K., Poudrier, G., Brown, P., Birnbaum, L. S., Goldenman, G., Miller, M. F., Patton, S., Poehlein, M., Varshavsky, J., & Cordner, A. (2022). Presumptive contamination: A new approach to PFAS contamination based on likely sources. *Environmental Science & Technology Letters, 9*(11), 983–990.
46. ECCC & HC. (*May* 2023). *Draft State of Per-and Polyfluoroalkyl Substances (PFAS) Report. Environment and Climate Change Canada, Health Canada.* Retrieved Auguest 8, 2024, from https://www.canada.ca/en/environment-climate-change/services/evaluating-existing-substances/draft-state-per-polyfluoroalkyl-substances-report.html
47. FCSI, Federal Contaminates Sites inventory. (2021) Retrieved July 31, 2024, from https://www.tbs-sct.gc.ca/fcsi-rscf/home-accueil-eng.aspx.
48. Battye, N., Loock, D., Reimer, K., Koch, I., & Weber, K. (2015). *Investigation of Environmental PFAS Contamination: Sampling and analysis*. Environmental Sciences Group, Royal Military College of Canada. Report number: RMC-CCE-ES-15-05.
49. Paterson, L., Kennedy, T. S., & Sweeney, D. (2008). Remediation of perfluorinated alkyl compounds at a former fire fighting training area. In *Proceedings of the Remediation Technologies Symposium, Banff, AB*.
50. Kleywegt, S., Raby, M., McGill, S., & Helm, P. (2020). The impact of risk management measures on the concentrations of per-and polyfluoroalkyl substances in source and treated drinking waters in Ontario, Canada. *Science of The Total Environment, 748*, 141195.
51. Munoz, G., Liu, M., Vo Duy, S., Liu, J., & Sauvé, S. (2023). Target and nontarget screening of PFAS in drinking water for a large-scale survey of urban and rural communities in Québec, Canada. *Water Research, 233*, 119750.

52. Martin, D., Munoz, G., Mejia-Avendaño, S., Duy, S. V., Yao, Y., Volchek, K., Brown, C. E., Liu, J., & Sauvé, S. (2019). Zwitterionic, cationic, and anionic perfluoroalkyl and polyfluoroalkyl substances integrated into total oxidizable precursor assay of contaminated groundwater. *Talanta, 195*, 533–542.
53. Liu, M., Munoz, G., Vo Duy, S., Sauvé, S., & Liu, J. (2022). Per-and polyfluoroalkyl substances in contaminated soil and groundwater at Airports: A Canadian case study. *Environmental Science & Technology, 56*(2), 885–895.
54. Munoz, G., Labadie, P., Botta, F., Lestremau, F., Lopez, B., Geneste, E., Pardon, P., Dévier, M.-H., & Budzinski, H. (2017). Occurrence survey and spatial distribution of perfluoroalkyl and polyfluoroalkyl surfactants in groundwater, surface water, and sediments from tropical environments. *Science of the Total Environment, 607-608*, 243–252.
55. Redmon, J. H., Mulhern, R. E., Castellanos, E., Wood, E., McWilliams, A., Herrera, I., Liyanapatirana, C., Weber, F., Levine, K., Thorp, E., Bynum, N., Amato, K., Najera Acevedo, M. A., Baker, J., Van Houtven, G., Henry, C., Wade, C., & Kondash, A. J. (2022). A Participatory Science Approach to Evaluating Factors Associated with the Occurrence of Metals and PFAS in Guatemala City Tap Water. *International Journal of Environmental Research and Public Health, 19*(10), 6004.
56. Nascimento, R. A., Nunoo, D. B. O., Bizkarguenaga, E., Schultes, L., Zabaleta, I., Benskin, J. P., Spanó, S., & Leonel, J. (2018). Sulfluramid use in Brazilian agriculture: A source of per-and polyfluoroalkyl substances (PFASs) to the environment. *Environmental Pollution, 242*(Pt B), 1436–1443.
57. Stefano, P. H. P., Roisenberg, A., D'Anna Acayaba, R., Roque, A. P., Bandoria, D. R., Soares, A., & Montagner, C. C. (2023). Occurrence and distribution of per-and polyfluoroalkyl substances (PFAS) in surface and groundwaters in an urbanized and agricultural area, Southern Brazil. *Environmental Science and Pollution Research International, 30*(3), 6159–6169.
58. Hoponick Redmon, J., Mulhern, R. E., Castellanos, E., Wood, E., McWilliams, A., Herrera, I., Liyanapatirana, C., Weber, F., Levine, K., Thorp, E., Bynum, N., Amato, K., Najera Acevedo, M. A., Baker, J., Van Houtven, G., Henry, C., Wade, C., & Kondash, A. (2022). A Participatory Science Approach to Evaluating Factors Associated with the Occurrence of Metals and PFAS in Guatemala City Tap Water. *International Journal of Environmental Research and Public Health, 19*(10), 6004.
59. Reinikainen, J., Perkola, N., Äystö, L., & Sorvari, J. (2022). The occurrence, distribution, and risks of PFAS at AFFF-impacted sites in Finland. *Science of the Total Environment, 829*, 154237.
60. Gobelius, L., Hedlund, J., Dürig, W., Tröger, R., Lilja, K., Wiberg, K., & Ahrens, L. (2018). Per-and Polyfluoroalkyl Substances in Swedish Groundwater and Surface Water: Implications for Environmental Quality Standards and Drinking Water Guidelines. *Environmental Science & Technology, 52*(7), 4340–4349.
61. Ssebugere, P., Sillanpää, M., Matovu, H., Wang, Z., Schramm, K. W., Omwoma, S., Wanasolo, W., Ngeno, E. C., & Odongo, S. (2020). Environmental levels and human body burdens of per-and poly-fluoroalkyl substances in Africa: A critical review. *Science of the Total Environment, 739*, 139913.
62. Sorensen, J. P. R., Lapworth, D. J., Nkhuwa, D. C. W., Stuart, M. E., Gooddy, D. C., Bell, R. A., Chirwa, M., Kabika, J., Liemisa, M., Chibesa, M., & Pedley, S. (2015). Emerging contaminants in urban groundwater sources in Africa. *Water Research, 72*, 51–63.
63. Orata, F., Quinete, N., Werres, F., & Wilken, R. D. (2009). Determination of perfluorooctanoic acid and perfluorooctane sulfonate in Lake Victoria Gulf water. *Bulletin of Environmental Contamination and Toxicology, 82*(2), 218–222.
64. Shafique, U., Schulze, S., Slawik, C., Böhme, A., Paschke, A., & Schüürmann, G. (2017). Perfluoroalkyl acids in aqueous samples from Germany and Kenya. *Environmental Science and Pollution Research, 24*(12), 11031–11043.
65. Ahrens, L., Gashaw, H., Sjöholm, M., Gebrehiwot, S. G., Getahun, A., Derbe, E., Bishop, K., & Åkerblom, S. (2016). Poly-and perfluoroalkylated substances (PFASs) in water, sediment and fish muscle tissue from Lake Tana, Ethiopia and implications for human exposure. *Chemosphere, 165*, 352–357.

66. Ololade, I. A., Oladoja, N. A., Ololade, O. O., Oloye, F. F., Adeola, A. O., Alabi, A. B., Ademila, O., Adanigbo, P., & Owolabi, M. B. (2018). Geographical distribution of perfluorooctanesulfonate and perfluorooctanoate in selected rivers from Nigeria. *Journal of Environmental Chemical Engineering, 6*(4), 4061–4069.
67. Groffen, T., Wepener, V., Malherbe, W., & Bervoets, L. (2018). Distribution of perfluorinated compounds (PFASs) in the aquatic environment of the industrially polluted Vaal River, South Africa. *Science of the Total Environment, 627*, 1334–1344.
68. Fagbayigbo, B. O., Opeolu, B. O., Fatoki, O. S., & Olatunji, O. S. (2018). Validation and determination of nine PFCS in surface water and sediment samples using UPLC-QTOF-MS. *Environmental Monitoring and Assessment, 190*(6), 346.
69. Mudumbi, J. B. N., Ntwampe, S. K. O., Muganza, F. M., & Okonkwo, J. O. (2013). Perfluorooctanoate and perfluorooctane sulfonate in South African river water. *Water Science and Technology, 69*(1), 185–194.
70. Dalahmeh, S., Tirgani, S., Komakech, A. J., Niwagaba, C. B., & Ahrens, L. (2018). Per-and polyfluoroalkyl substances (PFASs) in water, soil and plants in wetlands and agricultural areas in Kampala, Uganda. *Science of The Total Environment, 631-632*, 660–667.
71. Fauconier, G., Groffen, T., Wepener, V., & Bervoets, L. (2020). Perfluorinated compounds in the aquatic food chains of two subtropical estuaries. *Science of the Total Environment, 719*, 135047.
72. Fagbayigbo, B. O., Opeolu, B. O., Fatoki, O. S., Olatunji, O. S., Akharame, M. O., & Human, I. S. (2022). Sorption and partitioning of perfluorooctanoic acid (PFOA) and perfluorooctane sulfonate (PFOS) onto sediments of Diep and Plankenburg river systems Western Cape, South Africa. *Environmental Technology & Innovation, 25*, 102110.
73. Batayi, B., Okonkwo, J. O., Daso, P. A., & Rimayi, C. C. (2020). Poly-and perfluoroalkyl substances (PFASs) in sediment samples from Roodeplaat and Hartbeespoort Dams, *South Africa. Emerging Contaminants, 6*, 367–375.
74. Essumang, D. K., Eshun, A., Hogarh, J. N., Bentum, J. K., Adjei, J. K., Negishi, J., Nakamichi, S., Habibullah-Al-Mamun, M., & Masunaga, S. (2017). Perfluoroalkyl acids (PFAAs) in the Pra and Kakum River basins and associated tap water in Ghana. *Science of the Total Environment, 579*, 729–735.
75. Arinaitwe, K., Keltsch, N., Taabu-Munyaho, A., Reemtsma, T., & Berger, U. (2021). Perfluoroalkyl substances (PFASs) in the Ugandan waters of Lake Victoria: Spatial distribution, catchment release and public exposure risk via municipal water consumption. *Science of the Total Environment, 783*, 146970.
76. Wei, C., Wang, Q., Song, X., Chen, X., Fan, R., Ding, D., & Liu, Y. (2018). Distribution, source identification and health risk assessment of PFASs and two PFOS alternatives in groundwater from non-industrial areas. *Ecotoxicology and Environmental Safety, 152*, 141–150.
77. Bao, J., Yu, W.-J., Liu, Y., Wang, X., Jin, Y.-H., & Dong, G.-H. (2019). Perfluoroalkyl substances in groundwater and home-produced vegetables and eggs around a fluorochemical industrial park in China. *Ecotoxicology and Environmental Safety, 171*, 199–205.
78. Bao, J., Liu, W., Liu, L., Jin, Y., Dai, J., Ran, X., Zhang, Z., & Tsuda, S. (2011). Perfluorinated Compounds in the Environment and the Blood of Residents Living near Fluorochemical Plants in Fuxin, *China. Environmental Science & Technology, 45*(19), 8075–8080.
79. Chen, S., Jiao, X.-C., Gai, N., Li, X.-J., Wang, X.-C., Lu, G.-H., Piao, H.-T., Rao, Z., & Yang, Y.-L. (2016). Perfluorinated compounds in soil, surface water, and groundwater from rural areas in eastern China. *Environmental Pollution, 211*, 124–131.
80. Wang, P., Lu, Y., Wang, T., Zhu, Z., Li, Q., Meng, J., Su, H., Johnson, A. C., & Sweetman, A. J. (2016). Coupled production and emission of short chain perfluoroalkyl acids from a fast developing fluorochemical industry: Evidence from yearly and seasonal monitoring in Daling River Basin, China. *Environmental Pollution, 218*, 1234–1244.
81. Liu, Z., Lu, Y., Wang, T., Wang, P., Li, Q., Johnson, A. C., Sarvajayakesavalu, S., & Sweetman, A. J. (2016). Risk assessment and source identification of perfluoroalkyl acids in surface and ground water: Spatial distribution around a mega-fluorochemical industrial park, China. *Environment International, 91*, 69–77.

82. Lu, H., Ma, Z., & Zhang, X. (2018). Distribution characteristics of perfluorooctanoic acid (PFOA) in shallow groundwater of a Fluorine Industrial Park in Shandong Province. *IOP Conference Series: Earth and Environmental Science, 189*(3), 032064.
83. Cao, X., Wang, C., Lu, Y., Zhang, M., Khan, K., Song, S., Wang, P., & Wang, C. (2019). Occurrence, sources and health risk of polyfluoroalkyl substances (PFASs) in soil, water and sediment from a drinking water source area. *Ecotoxicology and Environmental Safety, 174*, 208–217.
84. Liu, Y., Li, X., Wang, X., Qiao, X., Hao, S., Lu, J., Duan, X., Dionysiou, D. D., & Zheng, B. (2019). Contamination Profiles of Perfluoroalkyl Substances (PFAS) in Groundwater in the Alluvial–Pluvial Plain of Hutuo River, China. *Water, 11*(11), 2316.
85. Qi, Y., Huo, S., Hu, S., Xi, B., Su, J., & Tang, Z. (2016). Identification, characterization, and human health risk assessment of perfluorinated compounds in groundwater from a suburb of Tianjin, China. *Environmental Earth Sciences, 75*(5), 432.
86. Li, X., Shang, X., Luo, T., Du, X., Wang, Y., Xie, Q., Matsuura, N., Chen, J., & Kadokami, K. (2016). Screening and health risk of organic micropollutants in rural groundwater of Liaodong Peninsula, *China. Environmental Pollution, 218*, 739–748.
87. Zhou, J., Li, S., Liang, X., Feng, X., Wang, T., Li, Z., & Zhu, L. (2021). First report on the sources, vertical distribution and human health risks of legacy and novel per-and polyfluoroalkyl substances in groundwater from the Loess Plateau, China. *Journal of Hazardous Materials, 404*, 124134.
88. Li, J., Peng, G., Xu, X., Liang, E., Sun, W., Chen, Q., & Yao, L. (2022). Per-and polyfluoroalkyl substances (PFASs) in groundwater from a contaminated site in the North China Plain: Occurrence, source apportionment, and health risk assessment. *Chemosphere, 302*, 134873.
89. Wang, Q., Song, X., Wei, C., Ding, D., Tang, Z., Tu, X., Chen, X., & Wang, S. (2022). Distribution, source identification and health risk assessment of PFASs in groundwater from Jiangxi Province, China. *Chemosphere, 291*, 132946.
90. Bao, J., Li, C.-L., Liu, Y., Wang, X., Yu, W.-J., Liu, Z.-Q., Shao, L.-X., & Jin, Y.-H. (2020). Bioaccumulation of perfluoroalkyl substances in greenhouse vegetables with long-term groundwater irrigation near fluorochemical plants in Fuxin, China. *Environmental Research, 188*, 109751.
91. Xu, C., Liu, Z., Song, X., Ding, X., & Ding, D. (2021). Legacy and emerging per-and polyfluoroalkyl substances (PFASs) in multi-media around a landfill in China: Implications for the usage of PFASs alternatives. *Science of the Total Environment, 751*, 141767.
92. Qiao, X., Jiao, L., Zhang, X., Li, X., Hao, S., Kong, M., & Liu, Y. (2020). Contamination profiles and risk assessment of per-and polyfluoroalkyl substances in groundwater in China. *Environmental Monitoring and Assessment, 192*(2), 76.
93. Li, Y., Niu, Z., & Zhang, Y. (2022). Occurrence of legacy and emerging poly-and perfluoroalkyl substances in water: A case study in Tianjin (China). *Chemosphere, 287*, 132409.
94. Sharma, B. M., Bharat, G. K., Tayal, S., Larssen, T., Bečanová, J., Karásková, P., Whitehead, P. G., Futter, M. N., Butterfield, D., & Nizzetto, L. (2016). Perfluoroalkyl substances (PFAS) in river and ground/drinking water of the Ganges River basin: Emissions and implications for human exposure. *Environmental Pollution, 208*, 704–713.
95. Richards, L. A., Kumari, R., White, D., Parashar, N., Kumar, A., Ghosh, A., Kumar, S., Chakravorty, B., Lu, C., Civil, W., Lapworth, D. J., Krause, S., Polya, D. A., & Gooddy, D. C. (2021). Emerging organic contaminants in groundwater under a rapidly developing city (Patna) in northern India dominated by high concentrations of lifestyle chemicals. *Environmental Pollution, 268*, 115765.
96. Koulini, G. V., & Nambi, I. M. (2024). Occurrence of forever chemicals in Chennai waters, India. *Environmental Sciences Europe, 36*(1), 60.
97. Murakami, M., Kuroda, K., Sato, N., Fukushi, T., Takizawa, S., & Takada, H. (2009). Groundwater Pollution by Perfluorinated Surfactants in Tokyo. *Environmental Science & Technology, 43*(10), 3480–3486.

98. Shiwaku, Y., Lee, P., Thepaksorn, P., Zheng, B., Koizumi, A., & Harada, K. H. (2016). Spatial and temporal trends in perfluorooctanoic and perfluorohexanoic acid in well, surface, and tap water around a fluoropolymer plant in Osaka, Japan. *Chemosphere, 164*, 603–610.
99. Kuroda, K., Murakami, M., Oguma, K., Takada, H., & Takizawa, S. (2014). Investigating sources and pathways of perfluoroalkyl acids (PFAAs) in aquifers in Tokyo using multiple tracers. *Science of the Total Environment, 488-489*, 51–60.
100. Duong, H. T., Kadokami, K., Shirasaka, H., Hidaka, R., Chau, H. T. C., Kong, L., Nguyen, T. Q., & Nguyen, T. T. (2015). Occurrence of perfluoroalkyl acids in environmental waters in Vietnam. *Chemosphere, 122*, 115–124.
101. Lam, N. H., Cho, C.-R., Kannan, K., & Cho, H.-S. (2017). A nationwide survey of perfluorinated alkyl substances in waters, sediment and biota collected from aquatic environment in Vietnam: Distributions and bioconcentration profiles. *Journal of Hazardous Materials, 323*, 116–127.
102. Hongkachok, C., Boontanon, S. K., Boontanon, N., Fujii, S., Tanaka, S., & Suzuki, Y. (2018). Levels of perfluorinated compounds (PFCs) in groundwater around improper municipal and industrial waste disposal sites in Thailand and health risk assessment. *Water Science and Technology, 2017*(2), 457–466.
103. Hongkachok, C., Boontanon, S. K., Boontanon, N., Sukeesan, S., Polprasert, C., & Fujii, S. (2023). Per-and polyfluoroalkyl substances in groundwater in Thailand: occurrence, source identification and spatial distribution. *International journal of Environmental Science and Technology, 20*(11), 12437–12448.
104. Yong, Z. Y., Kim, K. Y., & Oh, J.-E. (2021). The occurrence and distributions of per-and polyfluoroalkyl substances (PFAS) in groundwater after a PFAS leakage incident in 2018. *Environmental Pollution, 268*, 115395.
105. Park, S., Kim, D.-H., Yoon, J.-H., Kwon, J.-B., Choi, H., Shin, S.-K., Kim, M., & Kim, H.-K. (2023). Study on Pollution Characteristics of Perfluoroalkyl Substances (PFASs) in Shallow Groundwater. *Water, 15*(8), 1480.
106. Lin, Y.-C., Lai, W. W.-P., Tung, H.-H., & Lin, A. Y.-C. (2015). Occurrence of pharmaceuticals, hormones, and perfluorinated compounds in groundwater in Taiwan. *Environmental Monitoring and Assessment, 187*(5), 256.
107. Jin, Y. H., Liu, W., Sato, I., Nakayama, S. F., Sasaki, K., Saito, N., & Tsuda, S. (2009). PFOS and PFOA in environmental and tap water in China. *Chemosphere, 77*(5), 605–611.
108. Zhu, X., Jin, L., Yang, J., Wu, J., Zhang, B., Zhang, X., Yu, N., Wei, S., Wu, J., & Yu, H. (2017). Perfluoroalkyl acids in the water cycle from a freshwater river basin to coastal waters in eastern China. *Chemosphere, 168*, 390–398.
109. Liu, Y., Ma, L., Yang, Q., Li, G., & Zhang, F. (2018). Occurrence and spatial distribution of perfluorinated compounds in groundwater receiving reclaimed water through river bank infiltration. *Chemosphere, 211*, 1203–1211.
110. Sun, R., Wu, M., Tang, L., Li, J., Qian, Z., Han, T., & Xu, G. (2018). Perfluorinated compounds in surface waters of Shanghai, China: Source analysis and risk assessment. *Ecotoxicology and Environmental Safety, 149*, 88–95.
111. Liu, Z., Xu, C., Johnson, A. C., Sun, X., Wang, M., Xiong, J., Chen, C., Wan, X., Ding, X., & Ding, M. (2022). Exploring the source, migration and environmental risk of perfluoroalkyl acids and novel alternatives in groundwater beneath fluorochemical industries along the Yangtze River, China. *Science of The Total Environment, 827*, 154413.
112. Li, B.-B., Hu, L.-X., Yang, Y.-Y., Wang, T.-T., Liu, C., & Ying, G.-G. (2020). Contamination profiles and health risks of PFASs in groundwater of the Maozhou River basin. *Environmental Pollution, 260*, 113996.
113. Chen, H., Munoz, G., Duy, S. V., Zhang, L., Yao, Y., Zhao, Z., Yi, L., Liu, M., Sun, H., Liu, J., & Sauvé, S. (2020). Occurrence and Distribution of Per-and Polyfluoroalkyl Substances in Tianjin, China: The Contribution of Emerging and Unknown Analogues. *Environmental Science & Technology, 54*(22), 14254–14264.

114. Kaserzon, S. L., Vijayasarathy, S., Bräunig, J., Mueller, L., Hawker, D. W., Thomas, K. V., & Mueller, J. F. (2019). Calibration and validation of a novel passive sampling device for the time integrative monitoring of per-and polyfluoroalkyl substances (PFASs) and precursors in contaminated groundwater. *Journal of Hazardous Materials, 366*, 423–431.
115. Turner, B. D., Sloan, S. W., & Currell, G. R. (2019). Novel remediation of per-and polyfluoroalkyl substances (PFASs) from contaminated groundwater using Cannabis Sativa L. (hemp) protein powder. *Chemosphere, 229*, 22–31.
116. Szabo, D., Coggan, T. L., Robson, T. C., Currell, M., & Clarke, B. O. (2018). Investigating recycled water use as a diffuse source of per-and polyfluoroalkyl substances (PFASs) to groundwater in Melbourne, Australia. *Science of The Total Environment, 644*, 1409–1417.
117. Hepburn, E., Madden, C., Szabo, D., Coggan, T. L., Clarke, B., & Currell, M. (2019). Contamination of groundwater with per-and polyfluoroalkyl substances (PFAS) from legacy landfills in an urban re-development precinct. *Environmental Pollution, 248*, 101–113.
118. Currell, M., Northby, N., & Netherway, P. (2024). Examining changes in groundwater PFAS contamination from legacy landfills over a three-year period at Australia's largest urban renewal site. *Chemosphere, 352*, 141345.
119. Close, M., & Banasiak, L. (2023). *National Survey of Per-and Polyfluoroalkyl Substances (PFAS) in Groundwater 2022*. Institute of Environmental Science and Research Limited.
120. Zhao, Z., Li, J., Zhang, X., Wang, L., Wang, J., & Lin, T. (2022). Perfluoroalkyl and polyfluoroalkyl substances (PFASs) in groundwater: current understandings and challenges to overcome. *Environmental Science and Pollution Research, 29*(33), 49513–49533.

Chapter 6
Current Status of Contaminants in Developed Countries

Mahalaxmi Nagasubramanian, Amitha Unnikrishnan, C. T. Aravindakumar, and Usha K. Aravind

1 Introduction

Prior to the Industrial Revolution, the primary pollutants endangering the health of humans and ecosystems were naturally occurring infections such as bacteria, fungi, and viruses [1]. Industrialization drastically altered pollution patterns by bringing in new environmental contaminants such as particulate matter, industrial chemicals, and heavy metals [2]. Since the beginning of the Anthropocene, humans have increased their consumption of natural resources and created "new chemical compounds" to help humanity progress. Waste streams that cross planetary boundaries, upend natural ecosystems, and alter agricultural methods are the results, and they have sparked the evolution of viruses with wild-type characteristics. Metalloids and other hazardous substances are among the many naturally occurring chemicals that are regularly released into the environment by a wide range of human activities, including mining, mineral processing, energy production, construction, and agriculture. The growing threat of chemical pollution and its potentially catastrophic effect on humanity are discussed by Naidu et al. [3]. Chemicals or biological agents that are present in the environment are known as emerging contaminants (ECs) or contaminants of emerging concern (CECs) and are either recently discovered to be hazardous to humans or ecosystems or recently identified as synthetic or naturally

M. Nagasubramanian · C. T. Aravindakumar (✉)
Inter-University Instrumentation Centre (IUIC), Mahatma Gandhi University (MGU), Kottayam, Kerala, India
e-mail: cta@mgu.ac.in

A. Unnikrishnan · U. K. Aravind
School of Environmental Sciences, Cochin University of Science & Technology (CUSAT), Kochi, Kerala, India

M. K. Jindal et al. (eds.), *Emerging Contaminants in Water*, Emerging Contaminants and Associated Treatment Technologies,
https://doi.org/10.1007/978-3-031-89591-3_6

occurring chemicals. Our knowledge of the risks these contaminants represent is currently lacking. The growth of infections in an anthropogenic environment was examined by Lebarbenchon et al. [4].

Water quality and ecosystem health are significantly impacted by the worldwide presence of organic chemical pollutants, such as pesticides, medicines, and persistent organic pollutants (POPs) [5]. Globally, organic chemical pollutants such pesticides, medications, and persistent organic pollutants (POPs) have major impacts on ecosystem health and water quality [5]. Agriculture has historically employed pesticides, some of which are stable, durable, and impervious to degradation [6]. The pharmaceutical industry commonly employs multibatch operations and bulk production techniques, which lead to the improper release of medications into the environment. Industrial processes that contribute to pollution and jeopardize ecosystems include construction, chemical manufacturing, petrochemical refining, and the growth of telecommunications infrastructure. [7, 8]. The amounts of organic pollutants might vary significantly from those in the environment and can even be hazardous. Organisms that bioaccumulate toxins can serve as indicators of environmental contamination since they are vulnerable to toxic exposure. Consuming tainted foods, such as fish, can expose people to dangerously high concentrations of toxins, which is ultimately bad for their health [9, 10]. It is essential that everyone have access to safe, clean water as a fundamental human right since the risk increases when drinking water sources are contaminated with these toxins. The risk is further increased when drinking water sources that flow into rivers, lakes, and estuaries are contaminated [11, 12, 13]. According to [14], recent incidents such as the Norfolk Southern train crash in East Palestine, Ohio, have brought attention to the immediate and wide-ranging environmental repercussions of industrial disasters, highlighting the necessity of thorough study of their impact. Reviewing the available data on emerging contaminants (ECs) in developed nations and examining their position and destiny in the environment are the goals of this chapter. The illustration below highlights emerging contaminants (ECs) and how they enter water bodies (Fig. 6.1).

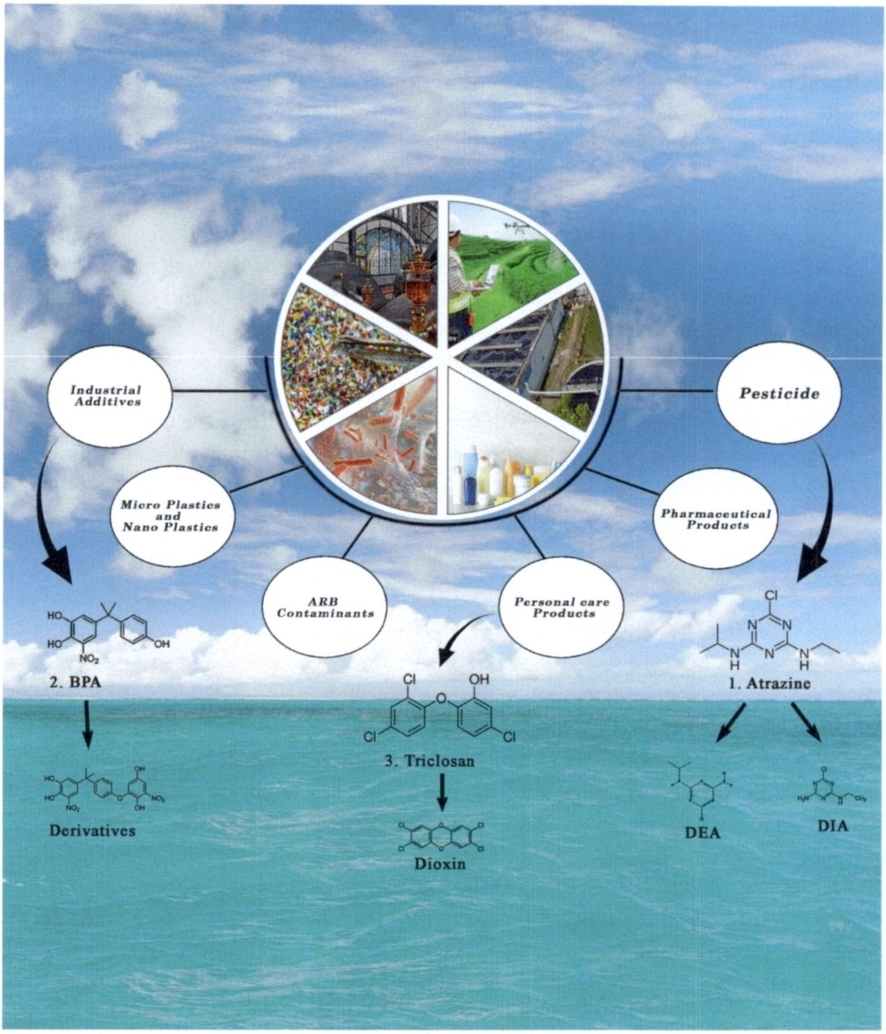

Fig. 6.1 The fate of emerging contaminant ECs in water bodies

2 The Present Situation of Emerging Contaminants in Developed Nations

In highly developed nations, the threat posed by emerging contaminants (ECs) is becoming more recognized and being addressed through enhanced regulation and monitoring protocols. Pollutants that commonly enter aquatic bodies through wastewater treatment plants include pharmaceuticals, personal care products, industrial chemicals including PFAS, biological pollutants, and microplastics [15]. Even though traditional water treatment facilities are usually insufficient for completely eradicating these compounds, nations like the United States and members of the European Union have created specific programs and updated directives to better monitor and manage ECs [16]. For instance, gathering data and setting limits for these substances are the main goals of the US EPA's Unregulated Contaminant Monitoring Rule (UCMR) and the EU's revised Drinking Water Directive [17]. Sweden is at the forefront of ECs through programs such as the Stockholm County Council, which promote pharmaceutical take-back programs and deploy cutting-edge treatment technology to limit the entry of drugs into the water supply [18]. Because of their stability and resistance to standard treatment methods, certain persistent pollutants—such as PFASs and some medications—are very difficult to eliminate. Even with these initiatives, many obstacles remain to be overcome to comprehend the long-term health impacts of low-level exposure and create affordable therapeutic options [19]. Overall, the growing recognition of ECs in developed countries is leading to more rigorous research, regulatory frameworks, and innovative solutions to protect the environment and public health. A comprehensive list of emerging contaminants found in developed countries is provided in Table 6.1.

2.1 *Recent Discoveries of New Contaminants*

New contaminants in developed countries highlight the challenge of managing chemical pollution. Despite established regulations, the list of chemicals continues to grow. Advanced analytical techniques reveal the presence of pharmaceuticals, personal care products, industrial chemicals, and microplastics. On the basis of REACH data (REACH, which stands for Registration, Evaluation, Authorization, and Restriction of Chemicals, is a European Union regulation), daily use products contain an estimated 30,000–50,000 industrial chemicals [55]. The complexity of these contaminants demands continuous vigilance and innovative solutions. Some emerging biocontaminants, such as algal blooms, are harmful (harmful algal blooms—HABs), posing an increasing threat to environmental and public health. European food safety is threatened by these algal blooms, which require legislation and monitoring to protect consumers from emerging toxins [56, 57]. Another threat is antibiotic-resistant bacteria, and antibiotic resistance is a growing global problem that threatens modern healthcare. Antibiotic use and bacterial resistance have increased as a result of the improper use of antibiotics in agricultural settings, such as growth boosters or nonspecific infection prevention. Higher rates of sickness and mortality can result from humans consuming or coming into contact with this resistance through food [58]. While traditionally viewed as a clinical issue, recent studies highlight the significant role of nonclinical environments in the spread of antibiotic resistance genes (ARGs). Antibiotic-resistant bacteria in the environment, which often spread through horizontal gene transfer (HGT), are a significant threat. High levels of ARGs near human activities are linked to antibiotic discharge and resistance genes from wastewater and agricultural runoff. Further research is needed to understand and mitigate this process [59].

Table 6.1 Comprehensive lists of emerging contaminants found in developed countries

Sl. No.	Class of emerging contaminants		Country	Reference
1.	Pesticides	Atrazine	Brazil, United States, Sweden, Spain	Machado et al. [20], Benotti et al. [21], Nystrom et al. [22], Troger et al. [23]
		Chlorpyrifos	Brazil, Sweden	Souza et al. [24], Mauriz et al. [25]
		Diuron	Brazil, Spain	Souza et al. [24], Troger et al. [23]
		Malathion	Brazil, Germany & Netherlands	Souza. et al. [24], Lundbergh et al. [26]
		Diazinon	United States, Sweden	Benotti et al. [21], Pihlstrom et al. [27]
		Linuron	United States, Spain	Benotti et al. [21], Troger et al. [23]
		Methoxychlor	United States	Benotti et al. [21]
		Octachlorostyrene	United States	Benotti et al. [21]
2.	Pharmaceuticals	Amitriptyline	Brazil, Sweden	Silveira et al. [28], Blum et al. [29]
		Enalapril	Brazil, United States	Silveira et al. [28], Snyder [30]
		Glibenclamide Haloperidol	Brazil, Spain	Silveira et al. [28], Troger et al. [23]
		Diclofenac	Malaysia, Spain	Wee et al. [31], Troger et al. [23]
		Triclosan	Malaysia, United States	Wee et al. [31], Halden & Paull [32]
		Ciprofloxacin	Malaysia, United states	Wee et al. [31], Watkinson [33]
		Propranolol	Malaysia, Germany	Wee et al. [31], Wilken et al. [34]
		Atenolol	United States, Spain	Troger et al. [23]
		Carbamazepine	United States	Benotti et al. [21]
		Ibuprofen	Canada	Hao et al. [35]
		Lincomycin	Canada	Hao et al. [35]
		Virginiamycin	Canada	Hao et al. [35]
		Naproxen	Canada	Hao et al. [35]
3.	Lifestyle compounds	Caffeine	Malaysia, Florida	Wee et al. [31], Gardinali & Zhao [36]
4.	Personal care products	Benzophenone-3	Spain, Australia	Gomez et al. [37], Malley et al. [38]
		Octocrylene	Spain	Gomez et al. [37]
		d-Limonene	California	Alvarez et al. [39]
		Galaxolide	California	Alvarez et al. [39]

(continued)

Table 6.1 (continued)

Sl. No.	Class of emerging contaminants		Country	Reference
5.	Industrial additives and byproducts (phthalates, bisphenols)	Dimethyl phthalate (DMP)	Portugal	Serodio & Nogueira [40]
		Diethyl phthalate (DEP)	Portugal, France	Serodio & Nogueira [40], Bach et al. [41]
		Di-n-butyl phthalate (DBP)	Portugal	Serodio & Nogueira [40]
		Butyl benzyl phthalate (BBP)	Portugal, France	Serodio & Nogueira [40], Bach et al. [41]
		Bis(2-ethylhexyl) phthalate (DEHP)	Portugal, France	Serodio & Nogueira [40], Bach et al. [41]
		diisobutylphthalate (DIBP)	France	Bach et al. [41]
		Bisphenol A	Malaysia, Canada, United States	Benotti et al. [21], Kleywegt et al. [42], Santhi et al. [43]
6.	Hormones and steroids	Testosterone	United States, Spain	Bexfield et al. [44], Troger et al. [23]
		Estrone (E1)	Australia, France	Green et al. [45], Jugan et al. [46]
		17β-estradiol (E2)	Australia, France	Green [45], Jugan et al. [46]
		17α-ethinylestradiol (EE2)	Australia, France	Green [45], Jugan et al. [46]
		Equilin	Canada	Kleywegt et al. [42]
7.	Surfactants	Per-Polyfluoroalkyl substances (PFAS)	Germany, New York, Canada, Australia	Sinclair [47], Domingo & Nadal [48], Kleywegt et al. [42], Gallen et al. [49]
		linear alkylbenzene sulfonates (LAS)	Spain, Sweden	Lara et al. [50], Larsson et al. [51]
		alkyl ethoxy sulfates (AES)	Spain	Lara et al. [50]
		nonylphenol ethoxylates (NPEOs)	Spain	Lara et al. [50]
		alcohol ethoxylates (AEOs)	Spain	Lara et al. [50]
8.	Flame retardants	Tris(2-butoxyethyl) phosphate (TBEP)	Spain	Esteban et al. [52]
		Tris(1,3-dichloropropan-2-yl) phosphate (TCPP)	Spain, United States	Benotti et al. [21], Esteban et al. [52]
		Tris(2-chloroethyl) phosphate (TCEP)	Spain, United States	Benotti et al. [21], Esteban et al. [52]
9.	Antibiotic resistance gene		New Mexico USA	Bhullar et al. [53]
10.	Microplastics and Nanoplastics	polyethylene terephthalate polystyrene	Central and Eastern European (CEE) surface waters	Bordós et al. [54]

The Stockholm Convention on Persistent Organic Pollutants (POPs) is a global treaty aimed at protecting human health and the environment from chemicals that remain in the environment for extended periods, accumulate in human and wildlife fatty tissue, and have harmful impacts [60, 61]. POPs can cause serious health effects, including cancer, birth defects, immune and reproductive system dysfunctions, increased susceptibility to disease, and damage to the central and peripheral nervous systems [62, 63]. The Convention, adopted in 2001 and enacted in 2004, requires parties to take measures to eliminate or reduce the release of POPs into the environment. Each party prohibits or eliminates the production and use of intentionally produced POPs, restricts the production and use of intentionally produced POPs, reduces or eliminates releases from unintentionally produced POPs, ensures safe and environmentally sound management of stockpiles and wastes containing POPs, and targets additional POPs [64, 65]. The convention has recently added several POPs to different annexes on the basis of their toxicity, as outlined in the new POP list mentioned in the table, along with the developed countries list and their detected values identified in the matrices (Table 6.2).

2.2 *Geographic and Temporal Distributions of Emerging Contaminants*

Emerging contaminants in developed countries are distributed primarily through industrial activities, agricultural practices, urban wastewater discharge, and high consumer product usage, despite advanced monitoring and regulatory efforts. For instance, it is widely recognized that "forever chemicals," such as perfluoroalkyl and polyfluoroalkyl substances (PFASs), pose a global threat due to prolonged exposure, which causes significant harm to both humans and aquatic ecosystems [92]. PFASs are particularly important in Australia, Canada, and North America, where high concentrations have been detected near sites of historical aqueous film-forming foams (AFFFs) used for firefighting, raising community concerns about potential long-term health impacts. In addition to their industrial use in textiles, paper, and packaging, the widespread use of foams containing PFASs at military installations, airports, and firefighting training sites exacerbates this contamination [93]. Regulatory shortcomings have allowed prolonged use and release of PFAS, and their persistence and mobility have led to widespread soil and water contamination. Heightened awareness and proactive monitoring have further underscored the extent of PFAS pollution across Australia [94–97].

Weissinger and colleagues investigated the ECs in the Colorado River from Colorado to the Gulf of California, including prescription medicines, illegal substances, and personal hygiene items. Using grab sampling, researchers have detected various ECs, including those of azithromycin, methamphetamine, and pseudoephedrine, with some showing temporal variation and estrogenic activity. Most contaminants in main surface water channels are near or below detection limits, highlighting the need for ongoing monitoring and assessment [98]. Another pollutant called brominated flame retardants (BFRs) is commonly used for thermal

Table 6.2 New POPs under the Stockholm Convention (held from 1–12 May 2023). Annex A (Elimination): The parties are required to take action to stop using and producing the substances specified in Annex A [65]. Annex B: Limitations In light of any applicable acceptable objectives and/or specific exemptions mentioned in the Annex, the parties should take appropriate steps to restrict the production and use of the substances listed under Annex B. [65]. Annex C (Unintentional production) Parties shall take action to minimize unintended releases of substances listed in Annex C (unintentional production), with the ultimate goal of ultimate elimination whenever practical. [65]

	Chemical name	Annexure	Matrices	River name	Range	Reference
1.	Alpha hexachlorocyclohexane	A	river water	Ebro River to Med. Sea, Spain	0.54–6 ng/L	Gómez-Gutiérrez et al. [66]
			Sea water	Bering-C. Seas, USA-Russia	2.28–2.7 ng/L	Strachan et al. [67]
			Sea water	Harbor, Sea-L Port-of-Spain	0.7–1.8 ng/L	Mohammed et al. [68]
2.	Beta hexachlorocyclohexane	A	ND	Across the globe	ND	Vijgen et al. [69]
3.	Chlordecone	A	wastewater treatment plant	Sludge was collected from the aeration tank of the wastewater treatment plant of Beaurade (Rennes, France)	ND	Cimetiere et al. [70]
4.	Decabromodiphenyl ether (commercial mixture, c-decaBDE)	A	Sediment	Lake and Marine Sediments in Sweden	0.48–11 ng/g	Ricklund et al. [71]
5.	Dechlorane Plus	A	River water	South Korea	ND	Kang et al. [72]
			River water	rivers in France	ND	Malak et al. [73]
6.	Dicofol	A	Drinking water	Tahtalı Dam Water or drinking water	Below the detection limit	Sazova [74]
7.	Hexabromobiphenyl	A				
8.	Hexabromocyclododecane (HBCDD)	A	Sediment from river	Ebro River basin (Spain)	1.6–1849 ng/g	Guerra et al. [75]

9.	Hexabromodiphenyl ether and heptabromodiphenyl ether (commercial octabromodiphenyl ether)	A	River water sediment	Diep River, Cape Town, South Africa Sediment of the Saginaw River, USA	2.60–4.29 ng/L 0.02–55.1 ng/g	Daso et al. [76]
			Sediment	Sediment of the Saginaw River, USA	0.02–55.1 ng/g	Yun et al. [77]
10.	Hexachlorobutadiene (HCBD)	A&C	Sediment	Coast of Southwestern Taiwan	0.6–6.8 ng/g	Lee & Fang [78]
11.	Lindane	A	Surface water	Surface water in western Canada	0.011–0.004 mg/L	Donald et al. [79]
12.	Methoxychlor	A	Drinking water	Drinking water were collected in the Warka and Grojec region, located in central Poland	9.0–24.9 μg/L 0.0423–0.9660 μg/L	Badach et al. [80]
13.	Pentachlorobenzene (PeCB)	A&C		subtropical North Atlantic		Lammel et al. [81]
14.	Pentachlorophenol and its salts and esters (PCP)	A	Lake water	Lake Rotorua catchment, New Zealand	0.04 $\mu g.L^{-1}$	Gifford et al. [82]
15.	Perfluorohexane sulfonic acid its salts and PFHxS-related compound	A		New Jersey drinking water system	ND	Quiñones et al. [83]
16.	Perfluorooctane sulfonic acid (PFOS), its salts and perfluorooctane sulfonyl fluoride (PFOS-F)	B	Lake water	Lake Ontario	2.6–22 ng/L.9	Furdui [84]
			wastewater treatment plants	California wastewater treatment plants	20–190 ng/L	Pontius [85]
			River water	Pearl River	0.90–99 ng/L	So et al. [86]
17.	Perfluorooctanoic acid (PFOA), its salts, and PFOA-related compounds	A	Lake water	Lake Ontario	4.1–38.1 ng/L	Furdui [84]
			wastewater treatment plants	California wastewater treatment plants	10–190 ng/L	Pontius [85]

(continued)

Table 6.2 (continued)

	Chemical name	Annexure	Matrices	River name	Range	Reference
18.	Polychlorinated naphthalenes (PCNs)	A&C	Groundwater	Groundwater Lobregat aquifer close to Barcelona in North East of Spain	0.0005 and 79.1 μg/L	Agunbiade et al. [87]
19.	Short-chained chlorinated paraffins	A	River water	Rivers in Taiwan	ND-12.6 mg/kg	Lee et al. [88]
20.	Technical endosulfan and its related isomers	A	Drinking water	Tahtalı Dam water or drinking water	0.002–0.004 μg/L Below the detection limit	Sazova [74]
21.	Tetrabromodiphenyl ether and pentabromodiphenyl ether (commercial pentabromodiphenyl ether)	A	River water	River Douro, Portugal	ND	Gama et al. [89]
			River water	Ebro River basin (NE), Spain	NA	Raldúa et al. [90]
22.	UV-328	A	Streams	water samples collected from Saitama Prefecture, Japan	102 ng/L	Kameda et al. [91]
			Heavily polluted river		701 ng/L	
			Moderately polluted river		152 ng/L	

insulation in constructions, textiles, electronic castings, and circuitry. Legacy BFRs peaked between the mid-1970s and the mid-2000s and have since stabilized, according to one study that looked at BFRs in the Rhône River in France. In the 1970s and 1980s, novel BFRs (nBFRs), which are still present at relatively low but stable levels, emerged. The importance of better understanding nBFR sources, distributions, and environmental effects before they reach dangerous levels has been emphasized by researchers. Despite the prohibition of several previous BFRs, nBFRs are still in use, which raises questions about their toxicity [99]. The European Food Safety Authority (EFSA) assesses the risks that brominated flame retardants (BFRs) pose to human health and the environment. To help ensure the safety of consumer goods, food, and environmental health in Europe, the EFSA evaluates scientific data and offers recommendations for guiding regulations. After being used to dispose of paper mill waste in the 1960s, toxic polychlorinated dibenzo-p-dioxins and dibenzofurans (PCCDDs/PCDFs) accumulated in Houston, Texas's San Jacinto River trash dumps [100, 101]. Owing to significant environmental damage, the site was uncovered in 2005 and is currently an EPA Superfund site [102].

Uranium, a heavy metal and radionuclide, has been found in groundwater over the past decade because of both natural and anthropogenic sources [103, 104]. Factors such as a decline in the groundwater table, nitrate pollution, and carbonates can accelerate the dissolution of uranium minerals, increasing the dissolved uranium concentration in groundwater [105]. Anthropogenic activities such as mining, processing, fertilizer manufacturing, and medical activities can also contribute to uranium contamination [106]. The chemical toxicity of uranium is six orders of magnitude greater than its radionuclide toxicity, and long-term exposure can lead to kidney failure, bone damage, and other health issues [107, 108]. The World Health Organization (WHO) has set a permissible concentration of 30 μg/L in drinking water because of the chemical toxicity of uranium [109]. However, countries have adopted different guidelines on the basis of socioeconomic conditions, health effects, species presence, and lack accurate data on radioactivity and chemical toxicity [110, 111, 112]. Current pollution practices and legacy industries are the sources of emerging contaminants (ECs), which affect their distribution and effects. These contaminants continue to seriously contaminate soil and water despite increased monitoring and regulatory actions.

3 Emerging Contaminants in Drinking Water: Guidelines

Emerging groundwater contaminants (EGCs) are newly identified substances in groundwater that originate from a variety of sources, such as biomedical waste, hospitals, homes, and municipalities, as well as industrial and agricultural waste. These chemicals were not tracked or detected previously. The number of regulated contaminants will gradually increase over the next several decades. Groundwater quality in Europe is now regulated by the Water Framework Directive and the Groundwater Daughter Directive [113, 114], whereas drinking water quality is governed by the Drinking Water Directive [115]. Pesticides are also regulated under the Plant Protection and Biocides Directive [116]. Some of these contaminants can have human or ecological health effects, and a better understanding of their fate in environmental systems is needed [117].

The European Drinking Water Directive sets limits for a small number of organic micropollutants, comprising aromatic hydrocarbons, chlorinated solvents, and disinfection byproducts [115]. The Priority Substances Directive establishes a number of priority substances, including benzene, octyl, and nonyl phenols; specified polyaromatic hydrocarbons (PAHs); di(2-ethylhexyl)phthalate; and a range of chlorinated hydrocarbons [118]. Revisions have included the emerging contaminants ibuprofen, diclofenac, α-ethinyloestradiol, β-estradiol, perfluorooctane sulfonate (PFOS), and perfluorooctanoic acid (PFOA) [115, 119]. The US EPA has derived statutory guideline values for approximately 90 contaminants in drinking water, 28 of which could be considered microorganic pollutants, excluding pesticides [117]. Among these, several pharmaceuticals and personal care products (PCPs) are now included, reflecting growing concerns over these substances [120].

The use of phthalates and BPA in various industrial, agricultural, and household applications underscores their significant presence in groundwater reservoirs, posing both national and global concerns. Groundwater contains significant amounts of BPA and phthalates from wastewater treatment plant (WWTP) effluents, leading to their transport into secondary environments such as surface water and soil via wastewater discharge, irrigation via either treated or untreated wastewater, and the use of sludge in farming. Phthalate concentrations in WWTP effluents vary from 0.07 to 1146.37 μg/L, whereas BPA values vary globally from 0.02 to 416 μg/L. [121]. The plasticizer bisphenol-A (BPA) in municipal wastewater can be harmful to environmental creatures, according to a study conducted in the southwestern region of the United Kingdom. BPA levels in wastewater ranged up to 100 μg L − 1, 10–37% greater than previously reported levels, and were greater than those reported at other sites. Additionally, even when 90% of BPA was removed during wastewater treatment, BPA was still present in the effluent [85]. Surface waters, which are more likely to contain higher concentrations of toxins from sources such as wastewater treatment work (WTW) discharges, have been the focus of the majority of emerging contaminant research [117].

4 Effects on Human Health and the Environmen: A Comprehensive Approach to Emerging Contaminants

To monitor and reduce the effects of emerging contaminants such as PFAS, dioxins, flame retardants, and phthalates on the environment and human health, industrialized nations are implementing strict regulations and cutting-edge technologies. Nevertheless, new pollutants continue to emerge, necessitating ongoing research and flexible solutions to address these evolving problems. The introduction of hazardous substances into the environment by these types of pollution places ecosystems and people in jeopardy. Because of their persistence and potential toxicological effects, emerging contaminants (ECs), which are not yet under control, have drawn attention. These toxins enter the ecosystem via a variety of channels, including industrial effluents and agricultural runoff, before reaching aquatic bodies and entering the food chain. Advanced detection and treatment technologies are needed to control ECs effectively.

Processes such as membrane bioreactors are instrumental in biodegradation and adsorption for the capture and removal of these contaminants. In addition, AOPs, including photocatalysts, aid in the breakdown of stable organic compounds. However, the production and disposal dynamics of EC-generating materials complicate monitoring efforts. Pathways and transformation products of ECs need to be understood, and modeling is needed for validation and understanding of their transport mechanisms. Some challenges exist notwithstanding the development of robust analytical strategies and remediation technologies to address byproduct toxicity and minimize secondary pollution. Hence, optimal data availability and validation are necessary for effective monitoring and appraisal of ECs. The effects of emerging pollutants on the environment and human health must be reduced by comprehensive strategies that incorporate cutting-edge detection technology, effective remediation techniques, and strict monitoring [122].

5 Conclusion and Outlook

The advent of the Industrial Revolution brought in new environmental pollutants such as heavy metals, industrial chemicals, and particulate matter, changing pollutant dynamics. Human activities such as mining, mineral processing, energy production, construction, and agriculture are continuously discharging dangerous substances into the environment. Emerging contaminants or contaminants of emerging concern (CECs) are generally said to be chemicals or biological agents found, regardless of their appearance, in environments that have either been recently discovered to be hazardous to humans or ecosystems or have only recently been identified as synthetic or naturally occurring chemicals [123].

The spread of organic chemical pollutants, including pharmaceuticals, pesticides, and persistent organic pollutants, at global scales has significantly affected water quality and ecosystem health [124]. Pesticide applications in agriculture, the pharmaceutical industry, and industrial operations contribute to pollution, which endangers water quality and ecosystems [125]. The concentrations of organic pollutants may differ considerably from ambient levels, causing the concentrations to be significantly hazardous [126]. The consumption of contaminated organisms may result in high dosages of toxic chemicals, ultimately resulting in health risks.

In developed nations, emerging contaminants (ECs) are increasingly acknowledged and, as a result, are being addressed by means of increased monitoring and regulation initiatives [127]. These contaminants, pharmaceuticals, personal care commodities, industrial chemicals, biological contaminants, and microplastics increasingly reach water bodies through wastewater treatment plants, industrial processes, and consumer products [128]. Countries such as the United States and the

EU have launched several programs and upgraded directives to monitor and manage ECs more effectively [129].

Newer contaminants in developed countries are adding to the challenge of managing chemical pollution [130]. However, established regulations do little to stem the increasing tide of chemicals entering the list, including pharmaceuticals, personal care products, industrial chemicals, and microplastics [131]. Thousands of industrial chemicals are contained in common products, requiring regular vigilance and innovative solutions. The intrusion of newly evolving biological agents such as algal blooms threatens environmental and public health, whereas antibiotic-resistant bacteria endanger modern health care. The Stockholm Convention on Persistent Organic Pollutants empowers our efforts to protect human health and the environment against persistent organic constituents that bioaccumulate in living beings and confer harmful effects thereafter [130]. Future research should have an increased focus on understanding the behavior and effects of these pollutants so that better mitigation strategies may be developed to provide sustainable environmental management for future generations.

References

1. Steffen, W., Richardson, K., Rockstrom, J., et al. (2015). Planetary boundaries: Guiding human development on a changing planet. *Science, 347*, 1259855.
2. Wang, F., Xiang, L., Leung, K. S. Y., Elsner, M., Zhang, Y., Guo, Y., & Tiedje, J. M. (2024). Emerging contaminants: A one health perspective. *The Innovation, 5*, 100612.
3. Naidu, R., Biswas, B., Willett, I. R., et al. (2021). Chemical pollution: A growing peril and potential catastrophic risk to humanity. *Environment International, 156*, 106616.
4. Lebarbenchon, C., Brown, S. P., Poulin, R., et al. (2008). Evolution of pathogens in a man-made world. *Molecular Ecology, 17*, 475–484.
5. Jalloul, G., Keniar, I., Tehrani, A., & Boyadjian, C. (2021). Antibiotics contaminated irrigation water: an overview on its impact on edible crops and visible light active titania as potential photocatalysts for irrigation water treatment. *Frontiers in Environmental Science, 9*, 767963.
6. Mukhopadhyay, A., Duttagupta, S., & Mukherjee, A. (2022). Emerging organic contaminants in global community drinking water sources and supply: A review of occurrence, processes and remediation. Journal of Environmental Chemical Engineering, 10(3), 107560.
7. Radelyuk, I., Tussupova, K., Klemeš, J. J., & Persson, K. M. (2021). Oil refinery and water pollution in the context of sustainable development: Developing and developed countries. Journal of Cleaner Production, 302, 126987.
8. González-González, R. B., Flores-Contreras, E. A., Parra-Saldívar, R., & Iqbal, H. M. (2022). Bio-removal of emerging pollutants by advanced bioremediation techniques. Environmental Research, 214, 113936.
9. Kapoor, D., & Singh, M. P. (2021). Heavy metal contamination in water and its possible sources. In Heavy metals in the environment (pp. 179-189). Elsevier.
10. Ray, S., & Shaju, S. T. (2023). Bioaccumulation of pesticides in fish resulting toxicities in humans through food chain and forensic aspects. Environmental Analysis, Health and Toxicology, 38, e2023017.
11. Li, P., Karunanidhi, D., Subramani, T., & Srinivasamoorthy, K. (2021). Sources and consequences of groundwater contamination. Archives of environmental contamination and toxicology, 80, 1-10.
12. Wanner, P. (2021). Plastic in agricultural soils–A global risk for groundwater systems and drinking water supplies?–A review. Chemosphere, 264, 128453.

13. Fida, M., Li, P., Wang, Y., Alam, S. K., & Nsabimana, A. (2023). Water contamination and human health risks in Pakistan: a review. Exposure and Health, 15(3), 619-639.
14. Gay, D. A., Blaydes, K., Schauer, J. J., & Shafer, M. (2024). Widespread impacts to precipitation of the East Palestine Ohio train accident. Environmental Research Letters, 19(7), 074022.
15. Cruz, S., & Barceló, D. (2015). *Personal care products in the aquatic environment*. Springer.
16. OECD. (2018). Summary report on the new comprehensive global database of per- and polyfluoroalkyl substances (PFASs). Organisation for Economic Co-operation and Development. https://www.oecd.org/content/dam/oecd/en/publications/reports/2018/05/summary-report-on-the-new-comprehensive-global-database-of-per-and-polyfluoroalkyl-substances-pfass_73d72705/1a14ad6c-en.pdf
17. Conroy-Ben, O., & Crowder, E. (2020). Unregulated and emerging contaminants in Tribal water. *Journal of Contemporary Water Research & Education, 169*(1), 92–100.
18. Kallaos, J., Wheeler, K., Wong, C., Zahller, M., & Kotchen, M. (2007). *Pharmaceuticals in wastewater streams: Disposal practices and policy options in Santa Barbara*. Donald Bren School of Environmental Science and Management.
19. Gavrilescu, M., Demnerová, K., Aamand, J., Agathos, S., & Fava, F. (2015). Emerging pollutants in the environment: present and future challenges in biomonitoring, ecological risks and bioremediation. *New Biotechnology, 32*(1), 147–156.
20. Machado, K. C., Grassi, M. T., Vidal, C., Pescara, I. C., Jardim, W. F., Fernandes, A. N., & Severo, F. J. (2016a). A preliminary nationwide survey of the presence of emerging contaminants in drinking and source waters in Brazil. *Science of the Total Environment, 572*, 138–146.
21. Benotti, M. J., Trenholm, R. A., Vanderford, B. J., Holady, J. C., Stanford, B. D., & Snyder, S. A. (2009). Pharmaceuticals and endocrine disrupting compounds in US drinking water. *Environmental Science & Technology, 43*(3), 597–603.
22. Nyström, B., Paulsson, M., Almgren, K., & Blank, H. (2000). Evaluation of the capacity for development of atrazine tolerance in periphyton from a Swedish freshwater site as determined by inhibition of photosynthesis and sulfolipid synthesis. *Environmental Toxicology and Chemistry: An International Journal, 19*(5), 1324–1331.
23. Tröger, R., Ren, H., Yin, D., Postigo, C., Nguyen, P. D., Baduel, C., & Wiberg, K. (2021). What's in the water?–Target and suspect screening of contaminants of emerging concern in raw water and drinking water from Europe and Asia. *Water Research, 198*, 117099.
24. de Souza, L. F. C. B., Montagner, C. C., Almeida, M. B., Kuroda, E. K., Vidal, C., & Freire, R. L. (2019). Determination of pesticides in the source and drinking waters in Londrina, Paraná, Brazil. *Semina: Ciências Agrárias, 40*(3), 1153–1164.
25. Mauriz, E., Calle, A., Lechuga, L. M., Quintana, J., Montoya, A., & Manclus, J. J. (2006). Real-time detection of chlorpyrifos at part per trillion levels in ground, surface and drinking water samples by a portable surface plasmon resonance immunosensor. *Analytica Chimica Acta, 561*(1–2), 40–47.
26. Lundbergh, I., Kreuger, J., & Johnson, A. (1995). Pesticides in surface waters: A review of pesticide residues in surface waters in Nordic Countries, Germany, and The Netherlands and problems related to pesticide contamination.
27. Pihlström, T., Hellström, A., & Axelsson, V. (1997). Gas chromatographic analysis of pesticides in water with off-line solid phase extraction. *Analytica Chimica Acta, 356*(2–3), 155–163.
28. Silveira, M. A. K., Caldas, S. S., Guilherme, J. R., Costa, F. P., Guimarães, B. D. S., Cerqueira, M. B., & Primel, E. G. (2013). Quantification of pharmaceuticals and personal care product residues in surface and drinking water samples by SPE and LC-ESI-MS/MS. *Journal of the Brazilian Chemical Society, 24*, 1385–1395.
29. Blum, K. M., Norström, S. H., Golovko, O., Grabic, R., Järhult, J. D., Koba, O., & Lindström, H. S. (2017). Removal of 30 active pharmaceutical ingredients in surface water under long-term artificial UV irradiation. *Chemosphere, 176*, 175–182.
30. Snyder, S. A. (2010). Occurrence of pharmaceuticals in US drinking water. In *Contaminants of emerging concern in the environment: ecological and human health considerations* (pp. 69–80). American Chemical Society.

31. Wee, S. Y., Haron, D. E. M., Aris, A. Z., Yusoff, F. M., & Praveena, S. M. (2020). Active pharmaceutical ingredients in Malaysian drinking water: consumption, exposure, and human health risk. *Environmental Geochemistry and Health, 42*, 3247–3261.
32. Halden, R. U., & Paull, D. H. (2005). Co-occurrence of triclocarban and triclosan in US water resources. *Environmental Science & Technology, 39*(6), 1420–1426.
33. Watkinson, A. J., Murby, E. J., Kolpin, D. W., & Costanzo, S. D. (2009). The occurrence of antibiotics in an urban watershed: from wastewater to drinking water. *Science of the Total Environment, 407*(8), 2711–2723.
34. Wilken, R. D., Ternes, T. A., & Heberer, T. (2000). Pharmaceuticals in sewage, surface and drinking water in Germany. In *Security of public water supplies* (pp. 227–240). Springer.
35. Hao, C., Lissemore, L., Nguyen, B., Kleywegt, S., Yang, P., & Solomon, K. (2006). Determination of pharmaceuticals in environmental waters by liquid chromatography/electrospray ionization/tandem mass spectrometry. *Analytical and Bioanalytical Chemistry, 384*, 505–513.
36. Gardinali, P. R., & Zhao, X. (2002). Trace determination of caffeine in surface water samples by liquid chromatography–atmospheric pressure chemical ionization–mass spectrometry (LC–APCI–MS). *Environment International, 28*(6), 521–528.
37. Gomez, M. J., Gomez-Ramos, M. M., Aguera, A., Mezcua, M., Herrera, S., & Fernandez-Alba, A. R. (2009). A new gas chromatography/mass spectrometry method for the simultaneous analysis of target and non-target organic contaminants in waters. *Journal of Chromatography A, 1216*(18), 4071–4082.
38. O'Malley, E., O'Brien, J. W., Verhagen, R., & Mueller, J. F. (2020). Annual release of selected UV filters via effluent from wastewater treatment plants in Australia. *Chemosphere, 247*, 125887.
39. Alvarez, D. A., Maruya, K. A., Dodder, N. G., Lao, W., Furlong, E. T., & Smalling, K. L. (2014). Occurrence of contaminants of emerging concern along the California coast (2009–10) using passive sampling devices. *Marine Pollution Bulletin, 81*(2), 347–354.
40. Serodio, P., & Nogueira, J. M. F. (2006). Considerations on ultra-trace analysis of phthalates in drinking water. *Water Research, 40*(13), 2572–2582.
41. Bach, C., Rosin, C., Munoz, J. F., & Dauchy, X. (2020). National screening study investigating nine phthalates and one adipate in raw and treated tap water in France. *Environmental Science and Pollution Research, 27*(29), 36476–36486.
42. Kleywegt, S., Raby, M., McGill, S., & Helm, P. (2020). The impact of risk management measures on the concentrations of per-and polyfluoroalkyl substances in source and treated drinking waters in Ontario, Canada. *Science of the Total Environment, 748*, 141195.
43. Santhi, V. A., Sakai, N., Ahmad, E. D., & Mustafa, A. M. (2012). Occurrence of bisphenol A in surface water, drinking water and plasma from Malaysia with exposure assessment from consumption of drinking water. *Science of the Total Environment, 427*, 332–338.
44. Bexfield, L. M., Toccalino, P. L., Belitz, K., Foreman, W. T., & Furlong, E. T. (2019). Hormones and pharmaceuticals in groundwater used as a source of drinking water across the United States. *Environmental Science & Technology, 53*(6), 2950–2960.
45. Green, C., Williams, R., Kanda, R., Churchley, J., He, Y., Thomas, S., & Jobling, S. (2013). Modeling of steroid estrogen contamination in UK and South Australian rivers predicts modest increases in concentrations in the future. *Environmental Science & Technology, 47*(13), 7224–7232.
46. Jugan, M. L., Oziol, L., Bimbot, M., Huteau, V., Tamisier-Karolak, S., Blondeau, J. P., & Lévi, Y. (2009). In vitro assessment of thyroid and estrogenic endocrine disruptors in wastewater treatment plants, rivers and drinking water supplies in the greater Paris area (France). *Science of the Total Environment, 407*(11), 3579–3587.
47. Sinclair, E., Mayack, D. T., Roblee, K., Yamashita, N., & Kannan, K. (2006). Occurrence of perfluoroalkyl surfactants in water, fish, and birds from New York State. *Archives of Environmental Contamination and Toxicology, 50*, 398–410.
48. Domingo, J. L., & Nadal, M. (2019). Human exposure to per-and polyfluoroalkyl substances (PFAS) through drinking water: A review of the recent scientific literature. *Environmental Research, 177*, 108648.

49. Gallen, C., Eaglesham, G., Drage, D., Nguyen, T. H., & Mueller, J. F. (2018). A mass estimate of perfluoroalkyl substance (PFAS) release from Australian wastewater treatment plants. *Chemosphere, 208*, 975–983.
50. Lara-Martín, P. A., Gómez-Parra, A., & González-Mazo, E. (2008). Sources, transport and reactivity of anionic and non-ionic surfactants in several aquatic ecosystems in SW Spain: a comparative study. *Environmental Pollution, 156*(1), 36–45.
51. Larsson, N., Otrembska, P., Villar, M., & Jönsson, J. Å. (2011). Liquid phase micro-extraction of linear alkylbenzene sulfonate anionic surfactants in aqueous samples. *Membranes, 1*(4), 299–313.
52. Esteban, S., Gorga, M., Gonzalez-Alonso, S., Petrovic, M., Barcelo, D., & Valcarcel, Y. (2014). Monitoring endocrine disrupting compounds and estrogenic activity in tap water from Central Spain. *Environmental Science and Pollution Research, 21*, 9297–9310.
53. Bhullar, K., Waglechner, N., Pawlowski, A., Koteva, K., Banks, E. D., Johnston, M. D., & Wright, G. D. (2012). Antibiotic resistance is prevalent in an isolated cave microbiome. *PLoS One, 7*(4), e34953.
54. Bordós, G., Urbányi, B., Micsinai, A., Kriszt, B., Palotai, Z., Szabó, I., & Szoboszlay, S. (2019). Identification of microplastics in fish ponds and natural freshwater environments of the Carpathian basin, Europe. *Chemosphere, 216*, 110–116.
55. Kumari, P. (2024). *Different approaches of quantitative structure retention relationship of small molecules in liquid chromatography (QSRR)* (Doctoral dissertation, Universite de Liege (Belgium)).
56. Karlson, B., Andersen, P., Arneborg, L., Cembella, A., Eikrem, W., John, U., & Suikkanen, S. (2021). Harmful algal blooms and their effects in coastal seas of Northern Europe. *Harmful Algae, 102*, 101989.
57. Otero, P., & Silva, M. (2022). Emerging marine biotoxins in European waters: potential risks and analytical challenges. *Marine Drugs, 20*(3), 199.
58. Economou, V., & Gousia, P. (2015). Agriculture and food animals as a source of antimicrobial-resistant bacteria. *Infection and Drug Resistance, 8*, 49–61.
59. Berglund, B. (2015). Environmental dissemination of antibiotic resistance genes and correlation to anthropogenic contamination with antibiotics. *Infection Ecology & Epidemiology, 5*(1), 28564.
60. Lallas, P. L. (2001). The Stockholm Convention on persistent organic pollutants. *American Journal of International Law, 95*(3), 692–708.
61. Secretariat of the Stockholm Convention. (2019). *All POPs listed in the Stockholm Convention.* http://chm.pops.int/TheConvention/ThePOPs/AllPOPs/tabid/2509/Default.aspx
62. Lee, H. K., & Pak, Y. K. (2018). Persistent organic pollutants, mitochondrial dysfunction, and metabolic syndrome. In *Mitochondrial dysfunction caused by drugs and environmental toxicants* (pp. 691–707). Wiley.
63. Pan, J., Liu, P., Yu, X., Zhang, Z., & Liu, J. (2024). The adverse role of endocrine disrupting chemicals in the reproductive system. *Frontiers in Endocrinology, 14*, 1324993.
64. Intosai, W. (2016). Inside INTOSAI. *International Journal of Government Auditing., 11*, 1.
65. Secretariat of the Stockholm Convention. (2023). *All POPs listed in the Stockholm Convention.* https://www.pops.int/TheConvention/ThePOPs/TheNewPOPs/tabid/2511/Default.aspx
66. Gomez-Gutierrez, A., Garnacho, E., Bayona, J. M., & Albaiges, J. (2007). Screening ecological risk assessment of persistent organic pollutants in Mediterranean sea sediments. *Environment International, 33*(7), 867–876.
67. Strachan, W. M. J., Burniston, D. A., Williamson, M., & Bohdanowicz, H. (2001). Spatial differences in persistent organochlorine pollutant concentrations between the Bering and Chukchi Seas (1993). *Marine Pollution Bulletin, 43*(1–6), 132–142.
68. Mohammed, A., Peterman, P., Echols, K., Feltz, K., Tegerdine, G., Manoo, A., & Orazio, C. (2011). Polychlorinated biphenyls (PCBs) and organochlorine pesticides (OCPs) in harbor sediments from Sea Lots, Port-of-Spain, Trinidad and Tobago. *Marine Pollution Bulletin, 62*(6), 1324–1332.

69. Vijgen, J., Abhilash, P. C., Li, Y. F. et al. (2011). Hexachlorocyclohexane (HCH) as new Stockholm Convention POPs—a global perspective on the management of Lindane and its waste isomers. *Environ Sci Pollut Res, 18,* 152–162. https://doi.org/10.1007/s11356-010-0417-9
70. Cimetiere, N., Giraudet, S., Papazoglou, M., Fallou, H., Amrane, A., & Le Cloirec, P. (2014). Analysis of chlordecone by LC/MS–MS in surface and wastewaters. *Journal of Environmental Chemical Engineering, 2*(2), 849–856.
71. Ricklund, N., Kierkegaard, A., & McLachlan, M. S. (2010). Levels and potential sources of decabromodiphenyl ethane (DBDPE) and decabromodiphenyl ether (DecaBDE) in lake and marine sediments in Sweden. *Environmental Science & Technology, 44*(6), 1987–1991.
72. Kang, J. H., Kim, J. C., Jin, G. Z., Park, H., Baek, S. Y., & Chang, Y. S. (2010). Detection of Dechlorane Plus in fish from urban-industrial rivers. *Chemosphere, 79*(8), 850–854.
73. Malak, I. A., Cariou, R., Vénisseau, A., Dervilly-Pinel, G., Jaber, F., Babut, M., & Le Bizec, B. (2018). Occurrence of Dechlorane Plus and related compounds in catfish (Silurus spp.) from rivers in France. *Chemosphere, 207,* 413–420.
74. Sazova, Y. (2004a). *Investigation of dicofol and endosulfan pesticide levels in Tahtalı Dam Water or drinking water*. Izmir Institute of Technology (Turkey).
75. Guerra, P., Eljarrat, E., & Barceló, D. (2010). Simultaneous determination of hexabromocyclododecane, tetrabromobisphenol A, and related compounds in sewage sludge and sediment samples from Ebro River basin (Spain). *Analytical and Bioanalytical Chemistry, 397,* 2817–2824.
76. Daso, A. P., Fatoki, O. S., & Odendaal, J. P. (2013). Occurrence of polybrominated diphenyl ethers (PBDEs) and 2, 2′, 4, 4′, 5, 5′-hexabromobiphenyl (BB-153) in water samples from the Diep River, Cape Town, South Africa. *Environmental Science and Pollution Research, 20,* 5168–5176.
77. Yun, S. H., Addink, R., McCabe, J. M., Ostaszewski, A., Mackenzie-Taylor, D., Taylor, A. B., & Kannan, K. (2008). Polybrominated diphenyl ethers and polybrominated biphenyls in sediment and floodplain soils of the Saginaw River watershed, Michigan, USA. *Archives of Environmental Contamination and Toxicology, 55,* 1–10.
78. Lee, C. L., & Fang, M. D. (1997). Sources and distribution of chlorobenzenes and hexachlorobutadiene in surficial sediments along the coast of southwestern Taiwan. *Chemosphere, 35*(9), 2039–2050.
79. Donald, D. B., Block, H., & Wood, J. (1997a). Role of ground water on hexachlorocyclohexane (lindane) detections in surface water in western Canada. *Environmental Toxicology and Chemistry: An International Journal, 16*(9), 1867–1872.
80. Badach, H., Nazimek, T., & Kaminska, I. A. (2007). Pesticide content in drinking water samples collected from orchard areas in central Poland. *Annals of Agricultural and Environmental Medicine, 14*(1), 109.
81. Lammel, G., Spitzy, A., Audy, O., Beckmann, S., Codling, G. P., Kretzschmann, L., & Stemmler, I. (2017). Organochlorine pesticides and polychlorinated biphenyls along an east-to-west gradient in subtropical North Atlantic surface water. *Environmental Science and Pollution Research, 24,* 11045–11052.
82. Gifford, J. S., Judd, M. C., McFarlane, P. N., & Anderson, S. M. (1995). Pentachlorophenol (PCP) in the New Zealand environment: assessment near contaminated sites and remote freshwater lakes. *Toxicological & Environmental Chemistry, 48*(1–2), 69–82.
83. Quiñones, O., & Snyder, S. A. (2009). Occurrence of perfluoroalkyl carboxylates and sulfonates in drinking water utilities and related waters from the United States. *Environmental Science & Technology, 43*(24), 9089–9095.
84. Furdui, V. I., Crozier, P. W., Reiner, E. J., & Mabury, S. A. (2008). Trace level determination of perfluorinated compounds in water by direct injection. *Chemosphere, 73*(1), S24–S30.
85. Pontius, F. (2019). Regulation of perfluorooctanoic acid (PFOA) and perfluorooctane sulfonic acid (PFOS) in drinking water: A comprehensive review. *Water, 11*(10), 2003.
86. So, M. K., Miyake, Y., Yeung, W. Y., Ho, Y. M., Taniyasu, S., Rostkowski, P., ... & Lam, P. K. S. (2007). Perfluorinated compounds in the Pearl river and Yangtze river of China. *Chemosphere, 68*(11), 2085–2095.
87. Agunbiade, I. V., Adeniji, A. O., Okoh, A. I., & Okoh, O. O. (2020). A review on occurrence and analytical procedures for the evaluation of polychlorinated naphthalenes in human and environmental matrices. *Environmental Pollutants and Bioavailability, 32*(1), 154–174.

88. Lee, C. C., Wu, Y. Y., Chen, C. S., & Tien, C. J. (2022). Spatiotemporal distribution and risk assessment of short-chain chlorinated paraffins in 30 major rivers in Taiwan. *Science of the Total Environment, 806*, 150969.
89. Gama, A. C., Sanatcumar, P., Viana, P., Barceló, D., & Bordado, J. C. (2006). The occurrence of polybrominated diphenyl ethers in river and coastal biota from Portugal. *Chemosphere, 64*(2), 306–310.
90. Raldua, D., Padros, F., Sole, M., Eljarrat, E., Barcelo, D., Riva, M. C., & Barata, C. (2008). First evidence of polybrominated diphenyl ether (flame retardants) effects in feral barbel from the Ebro River basin (NE, Spain). *Chemosphere, 73*(1), 56–64.
91. Kameda, Y., Kimura, K., & Miyazaki, M. (2011). Occurrence and profiles of organic sun-blocking agents in surface waters and sediments in Japanese rivers and lakes. *Environmental Pollution, 159*(6), 1570–1576.
92. Brase, R. A., Mullin, E. J., & Spink, D. C. (2021). Legacy and emerging per-and polyfluoroalkyl substances: analytical techniques, environmental fate, and health effects. *International Journal of Molecular Sciences, 22*, 995.
93. Kempisty, D. M., Xing, Y., & Racz, L. (Eds.). (2018). *Perfluoroalkyl substances in the environment: theory, practice, and innovation*. CRC Press.
94. Karásková, P., Venier, M., Melymuk, L., Bečanová, J., Vojta, Š., Prokeš, R., & Klánová, J. (2016). Perfluorinated alkyl substances (PFASs) in household dust in Central Europe and North America. *Environment International, 94*, 315–324.
95. Munoz, G., Liu, M., Duy, S. V., Liu, J., & Sauvé, S. (2023). Target and nontarget screening of PFAS in drinking water for a large-scale survey of urban and rural communities in Québec, Canada. *Water Research, 233*, 119750.
96. Sims, J. L., Stroski, K. M., Kim, S., Killeen, G., Ehalt, R., Simcik, M. F., & Brooks, B. W. (2022). Global occurrence and probabilistic environmental health hazard assessment of per-and polyfluoroalkyl substances (PFASs) in groundwater and surface waters. *Science of the Total Environment, 816*, 151535.
97. Toms, L. M. L., Bräunig, J., Vijayasarathy, S., Phillips, S., Hobson, P., Aylward, L. L., & Mueller, J. F. (2019). Per-and polyfluoroalkyl substances (PFAS) in Australia: Current levels and estimated population reference values for selected compounds. *International Journal of Hygiene and Environmental Health, 222*(3), 387–394.
98. Weissinger, R. H., Blackwell, B. R., Keteles, K., Battaglin, W. A., & Bradley, P. M. (2018). Bioactive contaminants of emerging concern in National Park waters of the northern Colorado Plateau, USA. *Science of the Total Environment, 636*, 910–918.
99. Vauclin, S., Mourier, B., Dendievel, A. M., Marchand, P., Vénisseau, A., Morereau, A., & Winiarski, T. (2021). Temporal trends of legacy and novel brominated flame retardants in sediments along the Rhône River corridor in France. *Chemosphere, 271*, 129889.
100. Tucker T. (2012). *Sourcing of dioxins from the San Jacinto waste pit superfund site into the Galveston Bay System*. Dissertation. University of Houston
101. Lakshmanan, D., Howell, N. L., Rifai, H. S., & Koenig, L. (2010). Spatial and temporal variation of polychlorinated biphenyls in the Houston Ship Channel. *Chemosphere, 80*(2), 100–112.
102. Iyer, R., Aggarwal, J., & Iken, B. (2016). A review of the Texas, USA San Jacinto Superfund site and the deposition of polychlorinated dibenzo-p-dioxins and dibenzofurans in the San Jacinto River and Houston Ship Channel. *Environmental Science and Pollution Research, 23*, 23321–23338.
103. Lapworth, D. J., Boving, T. B., Kreamer, D. K., Kebede, S., & Smedley, P. L. (2022). Groundwater quality: Global threats, opportunities and realising the potential of groundwater. *Science of the Total Environment, 811*, 152471.
104. Waseem, A., Ullah, H., Rauf, M. K., & Ahmad, I. (2015). Distribution of natural uranium in surface and groundwater resources: a review. *Critical Reviews in Environmental Science and Technology, 45*(22), 2391–2423.
105. Eisenbud, M., & Gesell, T. F. (1997). *Environmental radioactivity from natural, industrial and military sources: from natural, industrial and military sources*. Elsevier.
106. Gavrilescu, M., Pavel, L. V., & Cretescu, I. (2009). Characterization and remediation of soils contaminated with uranium. *Journal of Hazardous Materials, 163*(2–3), 475–510.

107. Cui, Q., Zhang, Z., Beiyuan, J., Cui, Y., Chen, L., Chen, H., & Fang, L. (2023). A critical review of uranium in the soil-plant system: Distribution, bioavailability, toxicity, and bioremediation strategies. *Critical Reviews in Environmental Science and Technology, 53*(3), 340–365.
108. Zamora, M. L. L., Zielinski, J. M., Moodie, G. B., Falcomer, R. A., Hunt, W. C., & Capello, K. (2009). Uranium in drinking water: renal effects of long-term ingestion by an aboriginal community. *Archives of Environmental & Occupational Health, 64*(4), 228–241.
109. World Health Organization. (2017). *A global overview of national regulations and standards for drinking-water quality*. WHO.
110. Jindal, M. K., Pandit, S. A., Karunakara, N., Chandrashekara, M. S., Kumara, S., Kumar, V., ... & Srinivasan, R. (2024). High uranium dose from the groundwater in a granitic terrain in the eastern part of Karnataka, India. *Journal of Radioanalytical and Nuclear Chemistry, 333*(6), 2905–2913. https://doi.org/10.1007/s10967-023-09053-6.
111. World Health Organization. (2021). *A global overview of national regulations and standards for drinking-water quality*. WHO.
112. Prat, O., Vercouter, T., Ansoborlo, E., Fichet, P., Perret, P., Kurttio, P., & Salonen, L. (2009). Uranium speciation in drinking water from drilled wells in southern Finland and its potential links to health effects. *Environmental Science & Technology, 43*(10), 3941–3946.
113. European Commission (EC). (2000). *Water framework directive*.
114. European Commission (EC). (2006). *Groundwater daughter directive*.
115. European Commission (EC). (2020). *Drinking water directive*.
116. European Commission (EC). (2009). *Plant protection and biocides directives*.
117. Stuart, M., Lapworth, D., Crane, E., & Hart, A. (2012). Review of risk from potential emerging contaminants in UK groundwater. *Science of the Total Environment, 416*, 1–21.
118. European Commission (EC). (2008). *Priority substances Directive*.
119. European Commission (EC). (2013). *Priority substances directive (revised)*.
120. US Environmental Protection Agency (US EPA). (2021). *National primary drinking water regulations*. US Environmental Protection Agency (US EPA).
121. Dueñas-Moreno, J., Mora, A., Cervantes-Avilés, P., & Mahlknecht, J. (2022). Groundwater contamination pathways of phthalates and bisphenol A: origin, characteristics, transport, and fate–a review. *Environment International, 170*, 107550.
122. Li, X., Shen, X., Jiang, W., Xi, Y., & Li, S. (2024). Comprehensive review of emerging contaminants: Detection technologies, environmental impact, and management strategies. *Ecotoxicology and Environmental Safety, 278*, 116420.
123. Brown, A., Smith, B., & Johnson, C. (2020). Emerging contaminants: A global perspective. *Environmental Science Journal, 15*(3), 102–115.
124. Green, D., Robinson, E., & Wilson, P. (2019). Impact of organic pollutants on water quality and ecosystems: A review. *Environmental Pollution Review, 28*(4), 321–335.
125. Johnson, L., & White, J. (2017). Chemical pollution from agricultural and industrial activities. *Journal of Environmental Management, 25*(2), 78–91.
126. Robinson, S., Wilson, M., & Brown, E. (2021). Concentration deviation of organic pollutants in environmental matrices. *Environmental Chemistry Journal, 18*(2), 145–158.
127. European Environment Agency. (2022). *Emerging contaminants in Europe: Monitoring and regulatory perspectives. EEA Report No. 2022/10*. European Environment Agency.
128. US Environmental Protection Agency. (2023). *Emerging contaminants: Pathways and impacts on water quality. EPA Report No. 2023/07*. US Environmental Protection Agency.
129. OECD. (2021). *Monitoring and managing emerging contaminants: A policy review. OECD Environment Policy Paper No. 2021/05*. OECD.
130. UNEP. (2020). *Stockholm Convention on Persistent Organic Pollutants: Protecting human health and the environment. UNEP Report No. 2020/11*. UNEP.
131. World Health Organization. (2022). *Chemicals in daily-use products: Implications for human health*. WHO Technical Report Series No. 2022/09.

Part IV
Red Line for Emerging Contaminants

Chapter 7
Highly Dangerous Contaminants and Their Potential Causes

Vaibhav Dagaji Aher, Buyinza Nicholas, Pavan Kumar Padarthi, Ramenani Hari Babu, and Sasikala Manickavasagam

1 Introduction

Highly dangerous contaminants threaten both ecological and human well-being because of their toxicity, long-lasting nature, and tendency to accumulate in living organisms. These include pharmaceuticals, persistent organic pollutants, radioactive materials, microbes, and chemicals, which can pose serious risks even in small amounts. One key characteristic is their persistence in the environment, leading to long-term exposure risks for ecosystems and human populations [1, 2]. Moreover, they can bioaccumulate within organisms and transfer through the food chain, impacting wildlife and humans, who rely on contaminated food sources. Prioritizing efforts to mitigate these contaminants, improve monitoring, and implement

V. D. Aher (✉)
Department of Pharmaceutical Medicine, Maharashtra University of Health Sciences, Nashik, Maharashtra, India

B. Nicholas
Department of Pharmaceutics and Pharmaceutical Technology, Kampala International University, Western Campus, Ishaka-Bushenyi, Uganda
e-mail: buyinza.nicholas@kiu.ac.ug

P. K. Padarthi
Academic Research and Development, St Martinus University, Willemstand, Curacao

R. H. Babu
MB School of Pharmaceutical Sciences, Mohan Babu University, Tirupati, Andhra Pradesh, India

S. Manickavasagam
Department of Pharmaceutical Analysis, Karpagam College of Pharmacy, Coimbatore, Tamil Nadu, India
e-mail: sasikala@kcp.edu.in

M. K. Jindal et al. (eds.), *Emerging Contaminants in Water*, Emerging Contaminants and Associated Treatment Technologies,
https://doi.org/10.1007/978-3-031-89591-3_7

practical restoration approaches is crucial for protecting environmental quality and public health [3].

Emerging contaminants (ECs) are unregulated pollutants that are increasingly recognized for their potential to harm human health, aquatic ecosystems, and the environment. These substances, which originate from sources such as industrial processes, pharmaceuticals, and consumer products, are defined by their persistence, bioaccumulative properties, and toxicity. Despite their low concentrations in water and wastewater—typically measured in ng/L to μg/L—ECs pose significant threats owing to their pervasive nature and the challenges associated with their removal via conventional treatment methods. The presence of ECs in the environment and their potential health risks have made ECs a critical focus of scientific research and environmental monitoring [4]. Among the most concerning ECs are pharmaceutical residues, which include antibiotics, synthetic hormones, and antidepressants. These substances, often excreted or improperly disposed of, accumulate in water systems because of the inability of conventional wastewater treatments to fully remove them. In particular, antibiotics exacerbate the global antimicrobial resistance crisis by promoting the development of resistant bacterial strains, whereas synthetic hormones disrupt the endocrine system in aquatic species, leading to reproductive abnormalities. Similarly, endocrine-disrupting chemicals (EDCs) mimic or block hormones, causing developmental and reproductive issues in wildlife and humans [5]. Persistent organic pollutants (POPs), including per- and polyfluoroalkyl substances (PFASs), dioxins, and certain pesticides, are another major concern. Owing to their stability and ability to travel long distances, POPs bioaccumulate in ecosystems, contaminating food chains and posing severe risks such as cancer, immune suppression, and developmental disorders [6].

Microplastics and nanoplastics, derived from the breakdown of larger plastic items or microbeads in cosmetics, are ubiquitous in the environment. These tiny particles are ingested by aquatic organisms, causing physical harm and acting as carriers for other toxic substances, such as heavy metals and POPs, amplifying their environmental impact [7]. Pesticides and herbicides, including neonicotinoids and glyphosate, also contribute to this problem. While glyphosate is linked to potential carcinogenic and endocrine-disrupting effects in humans, neonicotinoids have been implicated in the decline of pollinators such as bees, which threatens biodiversity and food security [8]. Volatile organic compounds (VOCs), such as benzene and trichloroethylene, are emitted from industrial processes and petroleum products, further exacerbating environmental degradation. VOCs contaminate air and water, contributing to respiratory illnesses, neurological disorders, and ecological harm [9].

Another emerging concern is the use of nanomaterials engineered for use in electronics, cosmetics, and medical applications. Their nanoscale size allows them to penetrate biological membranes, causing oxidative stress and potentially unknown health effects [10]. Similarly, antibiotic resistance genes (ARGs) are novel threats that spread antimicrobial resistance through wastewater and agricultural runoff. These genes persist in water and soil, transferring resistance to pathogens and undermining the efficacy of modern antibiotics [11]. Flame retardants, such as polybrominated diphenyl ethers (PBDEs), have been added to this growing list. Found in electronics and furniture, these chemicals are highly persistent and accumulate in ecosystems, interfering with thyroid hormone regulation and neurological development [12]. Collectively, these contaminants represent a significant challenge for environmental health and necessitate advanced monitoring, innovative treatment methods, and stringent regulations to mitigate their impacts.

2 Pharmaceutical Emerging Contaminants

Pharmaceuticals are among the most researched emerging contaminants (ECs). However, there is still limited knowledge about how antibiotics and nonantibiotic pharmaceuticals affect environmental microbiomes.

2.1 Antibiotics

Owing to constant exposure to human activities, aquatic environments have become key sites for the spread of antibiotic resistance. This resistance can be either intrinsic, naturally occurring and predictable, or acquired, arising unpredictably from genetic mutations or changes. In biofilms, bacteria gain an additional edge against antibiotics through a dual mechanism: a protective layer formed by the extracellular polymer matrix and physiological changes driven by genetic modifications. Horizontal gene transfer, where resistance genes move between species, further accelerates the process. Once resistance genes are acquired, they spread through vertical transmission during cell division, allowing resistant bacteria to multiply rapidly in aquatic systems [13]. The rise of antibiotic-resistant bacteria (ARBs) is associated with increased virulence, greater pathogenicity, and more frequent disease outbreaks, exposing humans to these pathogens through contaminated water. The overuse and misuse of antibiotics, often outside clinical settings, have significantly contributed to their presence in the environment. As these compounds enter ecosystems, they not only harm microbial communities but also exert selective pressure that favors the survival of ARB carrying resistance genes while suppressing vulnerable species. ARGs and ARBs are now considered environmental pollutants of concern, amplifying risks to both health and ecosystems [14].

Even small, subinhibitory concentrations of antibiotics, which are too low to inhibit bacterial growth or cause cell death, can promote resistance. At these levels, antibiotics act as chemical signals, influencing processes such as gene transcription, quorum sensing, biofilm formation, and microbial communication across species [15]. Several studies highlight these mechanisms. For example, sulfamethoxazole exposure has been shown to increase the expression of ARGs, such as the bla_d gene, in microbial communities such as Fusobacteria [16]. In drinking water systems, ciprofloxacin exposure was found to upregulate mexA (a gene encoding efflux pumps), and when combined with sulfadiazine, it significantly increased the abundance of the mexA and intI1 genes [17]. Another study revealed that *Escherichia coli* K12 exposed to fluoxetine developed resistance to antibiotics such as chloramphenicol (unclassified), amoxicillin (penicillin), and tetracycline, as well as multidrug resistance to fluoroquinolones, aminoglycosides, β-lactams, and other antibiotic classes [18].

2.2 Non-Antibiotic Pharmaceuticals

Interestingly, even nonantibiotic pharmaceuticals can contribute to antibiotic resistance. Common drugs such as ibuprofen, naproxen, diclofenac, gemfibrozil (used to lower cholesterol), and propranolol (a β-blocker) have been found to accelerate the spread of resistance at environmentally relevant concentrations. They do so by facilitating the uptake of exogenous ARGs [19]. These findings highlight a pressing issue: pharmaceuticals—whether antibiotics or not—can drive the spread of resistance genes, even at trace levels. The presence of multiclass ECs in the environment presents a serious challenge that requires urgent attention to protect public health and the balance of ecosystems.

2.3 Antidepressants

Pharmaceuticals, including antidepressants, persist in the environment due to their chemical stability, leading to widespread contamination of wastewater, surface water, and even drinking water. These compounds, which are excreted as either parent molecules or metabolites, enter aquatic systems where they pose risks to nontarget organisms and ecosystems.

Antidepressants, in particular, have been shown to accumulate in aquatic environments, disrupting the physiology and behavior of exposed organisms and allowing them to enter food chains through bioaccumulation. Studies have revealed their ability to interfere with serotonin-regulated systems in both vertebrates and invertebrates, affecting critical biological processes such as behavior, reproduction, morphology, gene expression, and survival. Sublethal effects, including altered swimming, feeding, and reproductive behaviors, highlight their significant ecological implications [20]. For example, zebrafish exposed to SSRIs and SNRIs exhibited reduced swimming activity, hypoactivity, and impaired feeding and reproductive behaviors. In goldfish, sertraline increased swimming activity but disrupted feeding patterns and shoaling tendencies, indicating complex and species-specific effects. Moreover, long-term studies have revealed transgenerational impacts, such as reduced cortisol responses in offspring of zebrafish exposed to fluoxetine, underscoring the potential for antidepressants to induce persistent physiological changes across generations [21]. The environmental impact of antidepressants is further compounded by their interactions with other pollutants, which often result in additive, synergistic, or antagonistic effects. For example, zebrafish exposed to a mixture of citalopram and tramadol exhibited reduced swimming activity, whereas Mediterranean mussels exposed to fluoxetine and propranolol presented altered oxidative stress markers [22, 23]. Similarly, the combination of fluoxetine and 17α-ethinylestradiol delayed reproduction in *Daphnia magna*, an effect not observed when these compounds were tested individually [24]. These findings emphasize the complexity of pharmaceutical mixtures in aquatic ecosystems and the need for comprehensive studies that consider such interactions. The bioaccumulation of antidepressants in aquatic organisms also raises concerns about potential human exposure through the food chain. Marine mussels (*Mytilus galloprovincialis*) along the Portuguese Atlantic coast have been found to accumulate SSRIs [25]. Similarly, commercially important fish species, such as *Salminus brasiliensis* and *Prochilodus lineatus*, have been shown to contain trace levels of antidepressants in muscle tissue [26]. Although current concentrations are considered safe for human consumption, these findings underscore the potential for antidepressants to bioaccumulate and highlight the need for further research into their long-term health implications. The persistence, bioactivity, and bioaccumulation potential of antidepressants in aquatic environments pose substantial ecological and health risks. Their ability to disrupt essential physiological processes in nontarget organisms, combined with their interaction with other pollutants, complicates the assessment of their full environmental impact. Continued research is critical to understand their long-term effects, develop effective mitigation strategies, and minimize their impact on ecosystems and human health. Adverse impacts of pharmaceuticals on health and the environment are shown in Fig. 7.1.

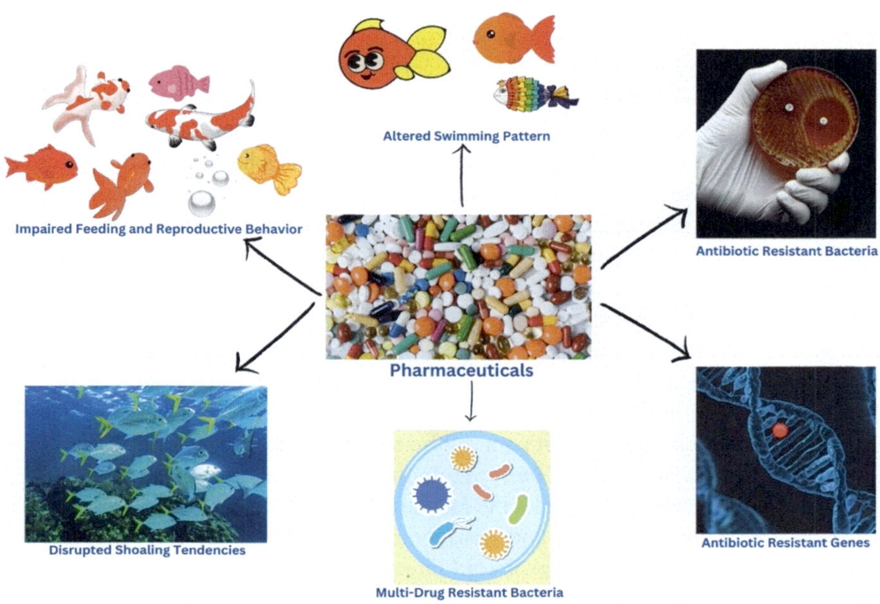

Fig. 7.1 Adverse impacts of pharmaceuticals on health and the environment

2.4 Hormones and Endocrine-Disrupting Chemicals

EDCs are substances, both natural and synthetic, that interfere with hormonal systems by mimicking, blocking, or altering natural hormone functions. These disruptions can lead to significant health problems, including developmental, reproductive, neurological, and immune disorders in humans and wildlife. Among the most concerning EDCs are pharmaceuticals such as steroidal hormones, veterinary growth hormones, and antibiotics [27]. As research continues, the list of harmful compounds is expanding, highlighting the urgency of addressing this issue.

One of the most pressing examples of EDCs is environmental contamination with estrogens. Naturally occurring estrogens, such as estrone, 17β-estradiol, 17α-estradiol, and estriol, are produced by humans and animals, whereas synthetic estrogens, such as 17α-ethinylestradiol, which is commonly used in contraceptives, contribute to environmental pollution [28]. Phytoestrogens from plants add another layer of complexity. These compounds find their way into ecosystems, disrupting reproductive and developmental processes across species [29]. For example, studies have shown that chronic exposure to ethynylestradiol can feminize male fish, reduce testis size, alter secondary sex characteristics, and skew sex ratios [30]. Amphibians are similarly affected, with exposure during key developmental stages impairing metamorphosis and disrupting sex determination [31]. The impacts of environmental estrogens are not limited to aquatic species. Mammalian studies reveal equally concerning outcomes. Research in rodents has shown that prenatal exposure to 17α-ethinylestradiol can interfere with placental development, increase the risk of oversized offspring, and alter maternal behaviors. In humans, these contaminants have been linked to an increased risk of breast cancer in women and infertility in men, illustrating their potential to harm both individuals and populations [32, 33].

Despite growing evidence of their dangers, much remains unknown about the prevalence and concentrations of these estrogens in various ecosystems. This knowledge gap complicates efforts to develop effective solutions. Furthermore, other emerging contaminants, such as bisphenol A (BPA), phthalates, and triclosan, exacerbate this problem, intensifying the disruption of endocrine systems and raising additional environmental and public health concerns [28]. Tackling these interconnected challenges requires stricter regulations, innovative mitigation strategies, and comprehensive research to safeguard both ecosystems and human well-being. BPA is a widely used industrial compound that is integral to the production of polycarbonate plastics and epoxy resins. These materials find applications in everyday items such as food and beverage containers, medical devices, safety equipment, and coatings for metal products such as food cans and water supply pipes. Despite its versatility, BPA has raised significant public health concerns because of its ability to interact with hormonal receptors, including androgen, estrogen, and thyroid receptors. This interaction disrupts the endocrine system, contributing to reproductive toxicity, tumor formation, and transgenerational effects [34]. Evidence from animal studies underscores the profound impact of BPA on reproductive health. In male mice, BPA exposure has been linked to reduced sperm motility, decreased sperm count, abnormal morphology, compromised membrane integrity, and impaired testicular development [35]. In females, exposure—particularly during gestation—has been associated with abnormal mammary gland development, such as increased epithelial volume and ductal hyperplasia, changes that may heighten the risk of mammary gland cancer. BPA also disrupts ovarian function, impairing follicle development, estradiol production, and germ cell nest breakdown [36, 37]. Furthermore, it triggers premature activation of primordial follicles through the PTEN/PI3K/AKT signaling pathway. Uterine functions are also affected, with BPA impairing implantation, altering steroid hormone signaling, and causing intrauterine growth restrictions [38]. In humans, the evidence is similarly troubling. Elevated BPA levels have been linked to impaired sperm quality, azoospermia, decreased antioxidant levels, and DNA damage in men [39]. In women, BPA exposure is correlated with diminished ovarian function, reduced oocyte retrieval during in vitro fertilization, and an increased risk of miscarriage. High BPA levels are also associated with polycystic ovary syndrome (PCOS), precocious puberty in girls, and reduced fecundability in women attempting to conceive [40]. Additionally, prenatal BPA exposure has been tied to low birth weight and restricted fetal growth, emphasizing its impact across generations [41]. While efforts have been made to replace BPA with structural analogs, many of these substitutes exhibit similar or even greater toxic effects. This highlights the urgent need for further research to elucidate the mechanisms underlying the toxicity of BPA and to develop safer, more sustainable alternatives [42]. Addressing these challenges is essential for safeguarding reproductive health and mitigating the broader environmental and health risks posed by BPA exposure. Phthalates are widely used chemicals because of their versatility as plasticizers, lubricants, and solvents across numerous industries. Phthalates are classified into high- and low-molecular-weight groups on the basis of their chemical structure and carbon chain length. High-molecular-weight phthalates, such as di(2-ethylhexyl) phthalate (DEHP), benzyl butyl phthalate (BBP), diisononyl phthalate (DiNP), and di-n-octyl phthalate, are widely used in products such as food storage containers, clothing, and healthcare devices, including IV tubing and transfusion bags [43]. In contrast, low-molecular-weight phthalates, including diethyl phthalate, dimethyl phthalate, dibutyl phthalate (DBP), and diisobutyl phthalate, are predominantly found in cosmetics, personal care products, pharmaceuticals, and adhesives [44].

Experimental studies have consistently demonstrated that phthalates negatively affect reproductive health in laboratory animals. In males, DEHP exposure has been shown to impair testosterone production, decrease sperm quality, and reduce fertility in CD-1 mice [45]. Similarly, DBP exposure in male Sprague–Dawley rats resulted in reduced sperm counts and a decline in Sertoli cell numbers across multiple generations [46]. In utero DEHP exposure further delays puberty onset and decreases sperm counts while increasing the prevalence of abnormal seminiferous tubules in subsequent generations of male mice [47].

In females, phthalates disrupt reproductive health through various mechanisms. Exposure to DEHP and its metabolite MEHP during critical developmental windows has been associated with premature reproductive aging, reduced oocyte quality, and impaired embryonic development in offspring [48]. Female mice exposed to DEHP in utero presented hormonal imbalances, including reduced estrogen levels and increased follicle-stimulating hormone (FSH) levels, alongside structural changes in ovarian tissue. These disruptions also extended to subsequent generations, with affected females experiencing reduced pregnancy rates, disrupted estrous cycles and altered litter sizes [49].

Phthalates also influence folliculogenesis. For example, DEHP exposure in adult mice accelerated follicle maturation, decreasing primordial follicle numbers while increasing the numbers of primary, preantral, and antral follicles. These changes were observed not only in directly exposed females but also across generations, suggesting transgenerational impacts on ovarian function and germ cell numbers [50].

The reproductive toxicity of phthalates is closely linked to their ability to disrupt endocrine signaling. In males, phthalates interfere with sex hormone production by altering the expression of steroidogenic genes, such as HSD3B, a key enzyme involved in converting hormone precursors into active steroids [51]. In females, phthalates such as DEHP downregulate genes responsible for androgen synthesis in ovarian thecal cells, such as Cyp17a1, Star, and Lhcgr [52]. These disruptions result in lower concentrations of key hormones, including progesterone, estradiol, and androstenedione, and structural abnormalities in ovarian follicles. Transgenerational studies further revealed that phthalates impact hormone levels and ovarian function across multiple generations [53]. Reproductive toxicity of phthalates in males and females is illustrated in Fig. 7.2.

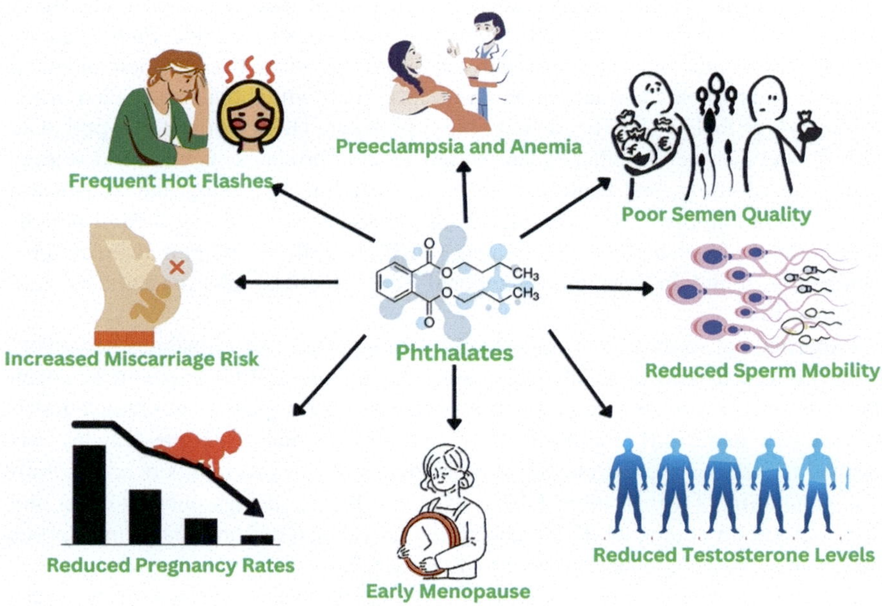

Fig. 7.2 Reproductive toxicity of phthalates in males and females

In humans, phthalates are recognized as endocrine disruptors with significant reproductive health consequences. In males, maternal phthalate exposure has been linked to reduced anogenital distance, a marker of altered androgen signaling [54, 55]. Higher urinary levels of DEHP and DiNP metabolites correlate with reduced sperm concentration, motility, and testosterone levels. DBP and BBP metabolites are similarly associated with poorer semen quality, increased sperm DNA damage, and heightened oxidative stress [56].

For women, phthalates are linked to a range of reproductive health issues. Chronic exposure has been associated with reduced pregnancy rates, increased miscarriage risk, and complications during pregnancy, such as pre-eclampsia and anaemia [57]. Elevated levels of urinary phthalate metabolites are associated with hormonal imbalances, including increased levels of FSH and progesterone. Furthermore, phthalates may contribute to adverse outcomes throughout a woman's life, including early menopause and more frequent hot flashes [58].

Given their widespread use and persistence in the environment, addressing the health risks posed by phthalates requires a more comprehensive understanding of their cumulative effects. Future research should prioritize the study of mixtures of phthalates that reflect real-world exposure scenarios. Investigating whether phthalate combinations produce synergistic effects could provide valuable insights into mitigating their impact on reproductive health. The development of such an evidence base is crucial for designing effective public health interventions and regulatory strategies.

3 Persistent Organic Pollutants (POPs)

Persistent organic pollutants (POPs) are synthetic substances that endure in the environment and traverse long distances through air and water currents. They encompass a variety of compounds, including agricultural insecticides, industrial chemicals, and residues from combustion processes [59]. POPs are notorious for their ability to accumulate in fatty tissues and magnify within the food chain, resulting in elevated levels in apex predators, including humans. Well-known examples of POPs include polychlorinated biphenyls (PCBs), dioxins, PFASs, and certain pesticides, such as dichlorodiphenyltrichloroethane (DDT) and lindane [60].

Owing to resistance to environmental degradation, POPs persist in the environment for extended periods, causing various diseases and health problems. These include cancer, gastrointestinal disorders, immune system dysfunction, birth defects, reproductive issues, weight loss, and neurological damage. Some POPs act as endocrine disruptors, altering the hormonal balance and impacting the reproductive and immune systems in both individuals and offspring [61, 62].

3.1 DDT

DDT, scientifically referred to as 4,4′-dichlorodiphenyltrichloroethane, exists in multiple isomeric forms, including 2,4'-DDT. Commercial DDT products primarily consist of 4,4'-DDT, with minor quantities of other closely related positional isomers, such as 4,4′- and 2,4′- dichlorodiphenyldichloroethanes (DDD) and 4,4′- and 2,4′- dichlorodiphenyldichloroethenes (DDE). These isomers are tasteless, nearly odorless solids, with a detection threshold in water of 0.35 mg/^{l} [63]. DDT, which acts as a broad-spectrum contact insecticide, faced bans in multiple nations during the 1970s due to environmental concerns. Although restrictions have been imposed in recent years, DDT is still used in some regions for disease vector control, notably against malaria. Designated a POP by the United Nations Environment Programme in 1997, DDT is persistent in the environment and resistant to complete biotransformation, although photochemical breakdown occurs [64, 65]. The International Agency for Research on Cancer (IARC) classifies commercial DDT as a carcinogen that does not directly affect genetic material. Instead, it serves as a potent promoter of liver tumors. While epidemiological studies have not consistently linked DDT exposure to various cancers, evidence suggests potential reproductive and developmental impairments in some species [66]. An analysis of exposure data and DDT concentrations in human tissues revealed declining mean concentrations globally, which coincided with restrictions on DDT usage. Although typical baseline exposure levels are generally not expected to pose significant health risks, efforts to minimize existing background exposure are crucial. Fetuses in gestation and newborns with rapidly maturing organ systems may be particularly vulnerable to DDT exposure [67].

3.2 Chlorinated Dioxin and Dioxin-Like Compounds

Chlorinated dioxins and dioxin-like compounds (DLCs) form a group of structurally and toxicologically related substances, including polychlorinated dibenzo-p-dioxins (PCDDs), polychlorinated dibenzofurans, and dioxin-like polychlorinated biphenyls (PCBs). These compounds, known for their toxic potency, pose significant health risks to humans and ecosystems [68, 69].

Decades of research indicate that the harmful health effects associated with DLC exposure are driven primarily by the activation of the aryl hydrocarbon receptor (AhR). This nuclear receptor, which regulates transcription, remains in a cytosolic complex in its inactive state. Upon binding with a ligand, AhR translocates to the nucleus, where it binds to DNA response elements, altering gene expression and triggering a range of biological and toxicological effects. The functional characteristics of AhR are highly conserved among vertebrates, although its activation by DLCs produces species-specific toxic and biological responses [70, 71].

Planar DLCs, defined by halogenated substitutions at specific positions, are highly relevant for evaluating human health risks because of their significant toxicological potency. Extensive research on biochemical and toxic responses to DLC exposure in experimental animals has demonstrated a range of effects, including enzyme induction, retinoid level alterations, substantial weight loss, thymus shrinkage, liver toxicity, immune system damage, hormonal disruption, reproductive and developmental issues, and cancer development [72].

In humans, prolonged exposure to tetrachlorodibenzo-p-dioxin (TCDD), the most toxic DLC, is linked to compromised immune function, developmental abnormalities in the nervous system, hormonal imbalances, reproductive impairments, and cancer-related outcomes [73].

Exposure to elevated levels of DLCs over a short duration can result in skin conditions such as chloracne, skin discoloration, and altered liver function [74]. Prolonged exposure is associated with immune system impairments [75], developmental issues in the nervous system [76], endocrine disruptions [77], and reproductive health problems [78, 79]. The IARC has classified TCDD as a confirmed human carcinogen. Nonetheless, TCDD does not directly harm DNA, and there is an exposure threshold below which the likelihood of cancer is considered minimal [80, 81]. Given the ubiquitous presence of DLCs in the environment, everyone experiences baseline exposure and consequently retains a certain level of dioxins in their bodies, known as the accumulated load. Typical baseline exposure levels are generally not expected to pose significant health risks to the average person [82, 83]. However, considering the potential toxicity of these compounds, it is imperative to undertake measures to minimize existing baseline exposure. The fetus during gestation is particularly vulnerable to DLC exposure, and newborns with rapidly maturing organ systems may also be at increased risk of certain effects [84, 85].

Assessing the health risks of DLC exposure remains challenging owing to exposure to complex mixtures, with certain congeners showing higher tissue retention [86]. The toxic equivalency concept was developed to account for the varying toxicities of DLC congeners, facilitating risk assessment of DLC mixtures [87]. The concept of toxic equivalency involves estimating the concentration equivalent to that of TCDD, enabling the assessment of potential adverse effects compared with established safety thresholds.

3.3 Lindane

Lindane, the ϒ isomer of 1,2,3,4,5,6-hexachlorocyclohexane (HCH), is a component of a technical mixture containing five principal isomers differing in their chlorine atom orientation around the cyclohexane ring. Among these compounds, ϒ-HCH has potent insecticidal properties [88, 89]. Chemical structures of the five principal isomers of hexachlorocyclohexane are shown in Fig. 7.3.

The environmental behavior of lindane varies depending on degradation pathways and environmental conditions [90]. Its estimated half-life in air ranges from 2

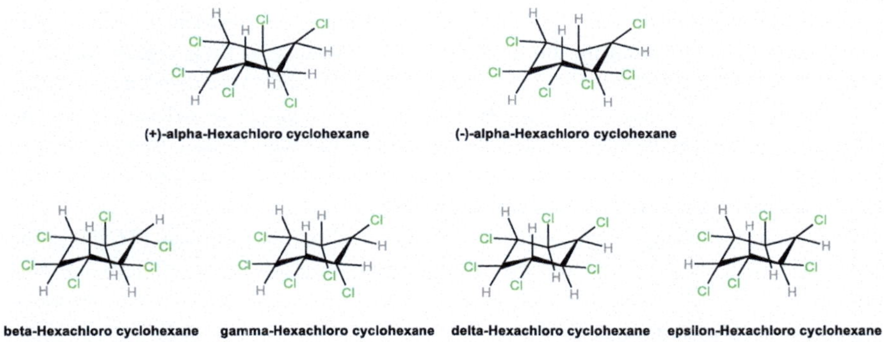

Fig. 7.3 Chemical structures of the five principal isomers of hexachlorocyclohexane

to 96 days, whereas in rivers and lakes, it ranges from 3 to 300 days. The half-lives of hydrolysis range from 92 to 3090 h, and degradation in soil can persist for 2–3 years. Factors such as climate and soil type influence lindane degradation [91]. Although hydrolysis is not a major degradation pathway in aquatic environments at neutral pH, lindane remains stable at pH 5 and 7, with degradation occurring at pH 9. Photolysis is not expected due to the absence of light-absorbing chromophores in lindane. Microbial degradation occurs slowly under aerobic conditions, with potential metabolites including pentachlorocyclohexene and isomeric trichlorobenzenes [92, 93].

Lindane acts as a GABA receptor antagonist in the brain, disrupting GABA neurotransmission and leading to hyperexcitation of nerve cells [94]. Additionally, it interacts with voltage-gated sodium channels, prolonging sodium influx into neurons and resulting in hyperexcitation [95]. While lindane primarily targets the insect nervous system, chronic exposure in humans can lead to neurological symptoms such as dizziness and convulsions [96]. Exposure to lindane can occur through various routes, including food intake, particularly from animal-derived products, and water containing the pesticide. Children are particularly vulnerable, with lindane detected in maternal adipose tissue, blood, umbilical cord blood, and breast milk [97]. Chronic exposure has been associated with hepatotoxic effects [98], immunosuppression [99], and potential carcinogenicity, leading to classification by the IARC as possibly carcinogenic to humans [100].

In summary, human exposure to lindane remains a significant concern because lindane is present in tissues and breast milk. Occupational exposure and direct exposure to pharmaceutical products containing lindane further increase concerns.

3.4 *Per- and Polyfluoroalkyl Substances*

PFASs represent a diverse group of fluorinated compounds widely used because of their chemical stability, water- and oil-responsive properties, and surfactant characteristics. However, their environmental persistence and potential harmful effects on ecosystems and human health have raised significant concerns. Human exposure to PFASs occurs primarily through drinking water, with additional sources including food, consumer products, and occupational settings [101]. Among the most extensively studied PFASs are perfluorooctanoic acid (PFOA) and perfluorooctane sulfonate (PFOS). PFOS, often used as a salt or a component of larger polymer structures, belongs to the perfluoroalkyl sulfonate family, which includes compounds capable of degrading PFOS through environmental or metabolic processes. While PFOS typically constitutes only a fraction of high-molecular-mass polymeric products, its persistence, toxicity, and bioaccumulation potential make it a considerable environmental and health concern [102]. PFOS is exceptionally resistant to environmental degradation, with no evidence of hydrolysis, photolysis, or biodegradation under typical conditions. The only known method for its destruction is high-temperature incineration under controlled conditions. Unlike many persistent organic pollutants that accumulate in fatty tissues, PFOS binds to plasma proteins and liver proteins, such as liver fatty acid-binding protein (L-FABP), disrupting normal physiological processes through unique bioaccumulation mechanisms [103].

The toxic effects of PFOS are multifaceted. It binds to plasma proteins such as albumin and lipoproteins, disrupting protein function and impacting cellular signaling pathways, including those involved in hormone regulation, leading to endocrine disruption. PFOS also induces oxidative damage by generating oxygen free radicals and inhibiting antioxidant systems, causing cellular damage and toxicity. Additionally, it impairs mitochondrial function, alters energy metabolism, modulates immune system activity, and disrupts lipid metabolism, increasing the risk of cardiovascular and metabolic disorders. Although its genotoxic potential remains unclear, evidence suggests that PFOS may cause DNA damage and mutations, necessitating further investigation into its mechanisms of toxicity [104–106].

PFOA, another persistent PFAS, is similarly resistant to degradation, with a half-life in aquatic environments exceeding 92 years and potentially reaching 235 years. Abiotic and biotic degradation processes have proven largely ineffective at breaking down PFOA under environmental conditions [107]. Photodegradation has minimal impact, whereas indirect photolysis, involving secondary reactions with sunlight-generated byproducts, extends its environmental persistence. PFOA contamination originates primarily from manufacturing processes, where wastewater discharge and air emissions contribute to its release into the environment [108].

Both PFOS and PFOA pose significant reproductive and developmental risks in animal models and humans. In male mice, PFOA exposure damages seminiferous tubules, reduces testosterone levels, and impairs sperm quality and fertility. Prenatal exposure to PFOA causes testicular damage, hormonal imbalances, and reduced offspring numbers. PFOS exposure has been shown to lower sperm counts, delay Leydig cell maturation and disrupt androgen production pathways [109]. Female mice exposed to PFOA exhibit delayed puberty, disrupted ovarian function, reduced steroid production, and adverse effects on pregnancy, including embryo resorption and impaired fetal growth. These reproductive effects extend to humans, with studies linking PFAS exposure to reduced sperm quality, hormonal imbalances, longer times to conception, and negative pregnancy outcomes, including reduced birth weight and delayed development [110, 111]. Emerging PFASs, such as GenX (perfluoro-2-propoxypropanoic acid), introduced as replacements for PFOA, raise additional concerns. Although marketed as safer alternatives, early studies suggest that they may exhibit similar or even greater toxicity [112]. Furthermore, most research has focused on individual PFAS compounds, whereas real-world exposure involves complex mixtures. These mixtures may have additive, synergistic, or antagonistic effects, complicating risk assessments and highlighting the need for studies on combined toxicity [113]. The persistence, bioaccumulation, and toxicity of PFASs demand urgent attention. Comprehensive research is essential to understand their long-term impacts, particularly for emerging compounds and mixtures. Regulatory measures to prevent emissions, limit exposure, and develop effective mitigation strategies are critical for safeguarding ecosystems and public health.

3.5 *Polybrominated Diphenyl Ethers*

PBDEs are a class of EDCs commonly used as flame retardants in industrial equipment and consumer products. Despite restrictions on certain PBDE mixtures under the 2009 and 2017 Stockholm Conventions, these persistent organic pollutants remain major environmental and public health concerns owing to their ability to bioaccumulate. Elevated levels of PBDEs are still found in marine mammals and endangered species, underscoring their persistent global impact [114].

From a chemical perspective, PBDEs consist of a diphenyl ether backbone in which up to ten hydrogen atoms are substituted by bromine, creating 209 possible congeners. These congeners typically exist as mixtures of brominated entities rather than singular compounds [115]. Unlike chemically bound additives, PBDEs are physically incorporated into products, making them prone to leaching and migration into the environment. This characteristic facilitates their dissemination into air, water, soil, and sediments, especially in areas of production, use, or disposal [116].

PBDEs are generally categorized into three primary mixtures—pentaBDEs, octaBDEs, and decaBDEs—each associated with specific industrial applications [117]. Once released, PBDEs accumulate in indoor air at concentrations significantly greater than those found outdoors. Contaminated soil and water further contribute to human exposure, with diet—particularly meat and dairy products—serving as a predominant source in certain regions [118].

The lipophilic nature of lower-brominated PBDEs enables them to persist in lipid-rich tissues, leading to bioaccumulation in aquatic organisms and subsequent transfer through the food chain to humans [119]. For example, pentaBDEs accumulate in fish and human tissues, whereas less soluble decaBDEs, such as BDE-209, have been detected in maternal and cord blood in industrial areas. This demonstrates their ability to infiltrate human systems despite their bulk and lower water solubility [120]. The toxicity of PBDEs often follows nonlinear, nonmonotonic dose–response patterns, where low and high doses may exert opposing effects, a phenomenon known as hormesis. For example, low PBDE doses may stimulate certain physiological parameters, whereas higher doses inhibit them. Evidence suggests that PBDE exposure can induce both genotoxic and epigenetic effects. While epigenetic changes are potentially reversible, they may contribute to long-term disease susceptibility [121]. Developmental exposure to PBDEs is particularly concerning, as the remodeling of tissues and immature defense mechanisms in developing organisms increase their vulnerability. Subtle epigenetic alterations during critical growth periods may not cause immediate effects but can increase disease risk later in life [122]. PBDEs and their hydroxylated metabolites disrupt thyroid hormone (TH) signaling by mimicking their structure and binding to transport proteins such as thyroxine-binding globulin and transthyretin. This mimicry interferes with normal TH function [123]. Additionally, PBDEs can disrupt nuclear thyroid hormone receptors and retinoic acid signaling, leading to downstream effects on metabolism and development. PBDEs also target steroid hormone receptors, such as estrogen, androgen, and progesterone receptors, impairing hormone binding and transcriptional activity [124].

Experimental studies have linked PBDE exposure to impaired glucose metabolism. For example, decaBDEs have been shown to increase fasting blood glucose levels and alter the expression of insulin receptors and glucose transporters in tissue-specific patterns influenced by dietary conditions. Chronic exposure to PBDEs may contribute to hyperinsulinemia and insulin resistance, implicating them in the development of metabolic disorders [125].

In terms of reproductive health, PBDE exposure has been associated with altered puberty timing, reduced semen quality, and increased risk of gestational diabetes mellitus (GDM). Notably, PBDE-154 has been identified as a significant contributor to GDM risk, whereas other congeners exhibit varied effects on male and female reproductive outcomes [126]. Prenatal exposure to PBDEs has also been linked to epigenetic changes, such as telomere elongation in newborn babies, which could increase cancer risk later in life [127].

Emerging evidence suggests that PBDEs may play a role in carcinogenesis (Fig. 7.4). For example, decaBDE exposure has been associated with increased cell proliferation in thyroid and colorectal cancer models [128]. PBDEs may promote cancer development through mechanisms such as activation of the PI3K/AKT signaling pathway and epigenetic modifications, including histone methylation and k-RAS activation [129]. While the US Environmental Protection Agency (EPA) has identified decaBDE as a potential carcinogen, further studies are needed to clarify the mechanisms and risks associated with other PBDE congeners [130].

The persistence and bioaccumulation of PBDEs, coupled with their ability to disrupt hormonal and metabolic systems, underscore their importance as environmental pollutants. Despite regulatory restrictions, their widespread presence in ecosystems and human tissues calls for continued research to elucidate their toxicological mechanisms and long-term health implications. Comprehensive investigations into PBDE mixtures and individual congeners are crucial for understanding their collective impact and informing future regulatory strategies.

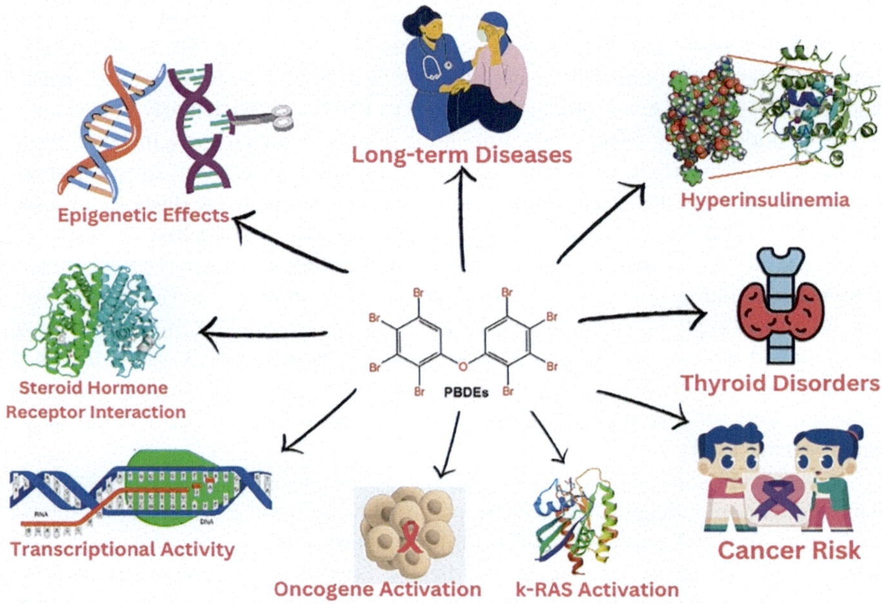

Fig. 7.4 Health impacts of PBDE exposure

4 Pesticides

Pesticides have revolutionized agriculture by significantly increasing food production and helping control vector-borne plant diseases. The rising global demand for food and agricultural productivity has driven the increasing use of pesticides, resulting in their pervasive presence in the environment [131]. One of the most concerning consequences of pesticide use is the contamination of drinking water, which is sourced from both surface water and groundwater. This chronic contamination, coupled with the intentional use of pesticides in water disinfection, results in long-term exposure for nontarget species, including humans, and has been linked to serious reproductive health problems [132]. The toxic effects of pesticides are not limited to humans. Restricted-use pesticides, some of the most hazardous, act as carcinogens and accumulate in nontarget organisms, including aquatic species, algae, and microbes. Malathion, for example, is known to cause genotoxic effects in humans [133]. Moreover, pesticides ingested through contaminated food interact with the gastrointestinal microbiome, leading to hormonal imbalances, digestive disorders, and an increased risk of lung cancer [134]. Each year, pesticide exposure through food is estimated to cause >20,000 deaths, underscoring the dire need for stricter regulation and innovative strategies to minimize their harmful effects [135].

Most pesticides are derived from chemical families such as carbamates, organochlorides, pyrethroids, and organophosphates, each of which serves specialized purposes, including promoting plant growth or repelling pests. Despite their agricultural benefits, only approximately 1% of applied pesticides reach their intended target. The remaining 99% disperse into the environment, degrade, or are transformed by microbial activity [136]. For example, farmers in the United States have spent >200 billion on neonicotinoids to increase pollination, yet these chemicals have been linked to honeybee colony collapse disorder, illustrating their unintended consequences [137].

The growing reliance on pesticides to meet food production demands has undoubtedly contributed to agricultural success, but their far-reaching consequences require immediate attention. To safeguard both human health and the environment, it is imperative to prioritize sustainable solutions that reduce pesticide use and mitigate pesticide presence in water and food systems.

4.1 Organochlorine Pesticides

Among various types of pesticides, organochlorides are particularly harmful. These persistent chemicals not only weaken biodiversity but also destroy bird habitats, harm pollinators, and endanger already vulnerable species. Their fat-solubility allows them to bioaccumulate in organisms and progressively increase in level as they ascend through the food chain. Once in the human body, organochlorides are stored in organs, where they can cause severe health problems, including neurodegenerative disorders such as Parkinson's disease. Long-term exposure to pesticides has also been linked to skin and eye conditions, nervous system damage, and millions of poisonings and deaths annually, particularly among agricultural workers.

Organochlorine (OC) pesticides, a group of synthetic chlorinated hydrocarbon derivatives, have been extensively used worldwide, particularly in agriculture and the chemical industry. Examples include DDT, lindane, dicofol, aldrin, dieldrin, chlordane, heptachlor, endosulfan, and toxaphene. These compounds share a common structural framework characterized by aliphatic or aromatic rings with chlorine substituents, which impart similar properties, such as environmental persistence, bioaccumulation, and high toxicity. Depending on the compound, their persistence can range from moderate (with a half-life of approximately 60 days) to extremely long (up to 10–15 years). This durability, while beneficial for pest control, has led to widespread environmental and health concerns [138].

Although banned in many developed nations, the use of OC pesticides continues, often through misuse and illegal practices. Their application, which is intended to target specific pests, frequently harms nontarget species. Birds have experienced population declines, behavioral disturbances, and habitat loss due to prolonged OC exposure. Notably, the accumulation of these compounds in seabird eggs serves as a stark indicator of environmental contamination [139].

The toxicity of organochlorines primarily arises from their effects on the central nervous system. Compounds such as endosulfan and lindane, which are classified as cyclodienes, act as GABA antagonists by interfering with calcium ion uptake and impairing the function of Ca- and Mg-ATPase, leading to an abnormal surge in neurotransmitter levels. Epidemiological studies have linked OC exposure to neurological disorders, including Parkinson's disease, as well as systemic conditions, such as hypertension, cardiovascular disease, and neuromuscular disorders. Moreover, organochlorines function as endocrine-disrupting chemicals (EDCs), interfering with hormonal pathways, disrupting steroid metabolism, and contributing to a range of endocrine and reproductive health problems [140–142].

Among the most concerning organochlorines is endosulfan, which exhibits high bioaccumulation and prolonged environmental persistence. This compound contaminates food chains, posing significant risks to human health. It primarily affects the central nervous system, with inhalation exposure being more acutely toxic than dermal exposure is and is readily absorbed through the gastrointestinal tract [143]. Beyond endosulfan, many organochlorines have been implicated in breast, prostate, stomach, and lung cancers [144]. Additionally, dioxins, a subset of OCs, have been detected in human ovarian follicular fluid, raising concerns about their role in conditions such as endometriosis and autoimmune diseases, including multiple sclerosis and eczema [79]. These compounds can act as xenoestrogens, disrupting thyroid hormone regulation through both inhibited synthesis and accelerated degradation, as demonstrated by compounds such as TCDD and methoxychlor [145].

The human health impacts of organochlorines are well documented, with numerous studies highlighting their detection in breast milk globally [146]. For example, in Croatia, p,p'-DDE, a metabolite of DDT, was the dominant organochlorine pesticide found in breast milk, whereas studies in Korea and Mexico reported the presence of chlordanes, aldrin, DDTs, and related compounds. This contamination underscores the potential health risks to infants owing to exposure through breastfeeding, reflecting the persistent and bioaccumulative nature of these chemicals [147, 148]. In addition to contamination, organochlorines are strongly associated with metabolic and endocrine disorders. Chronic exposure to these compounds has been linked to insulin resistance, type 2 diabetes, and gestational diabetes [149]. In the French Caribbean, for example, chlordecone exposure has been connected to hypertensive pregnancy disorders [150]. Similarly, research in Slovakia revealed elevated glucose, insulin, triglyceride, and cholesterol levels in individuals from polluted areas, alongside decreased testosterone in men [151]. These findings highlight the systemic impacts of organochlorine exposure, which extends from endocrine disruption to metabolic dysfunction. Studies also show that residues of β-HCH, HCB, and DDT can accumulate in maternal circulation, transfer across the placenta, and disrupt thyroid hormone balance in newborns, raising concerns about developmental risks [152].

The neurological and developmental impacts of organochlorines are equally alarming. For example, heptachlor has been shown to impair mitochondrial function, triggering neurotoxicity and potentially contributing to conditions such as Parkinson's disease. Even low-dose exposure has been associated with delayed psychomotor development in children [153, 154]. Additionally, OC exposure has been linked to vitamin D deficiency [155], gallstone disease [156], and neonatal thyroid hormone disruption [157]. Gender-specific endocrine effects [158, 159] have also been observed, with antiandrogenic impacts in males and estrogenic effects in females, further complicating the health outcomes of OC exposure [160]. Organochlorine pesticides pose significant environmental and health risks due to their persistence, bioaccumulation, and systemic toxicity. Their continued use highlights the tension between agricultural needs and public health, demanding a comprehensive approach to balance these priorities.

Metolachlor, a widely used chloroacetanilide herbicide, is one of the most commonly detected contaminants in drinking water worldwide. Structurally, it is an organochlorine compound derived from aniline, characterized by a 2-chloroacetamide group substituted with a (2-ethyl-6-methylphenyl)-N-(1-methoxypropan-2-yl) moiety. Its chemical profile places it within multiple classifications, including aromatic amides, ethers, benzenes, and organochlorines [161, 162].

The prevalence of metolachlor in water supplies has raised significant environmental and health concerns. Studies have reported its presence in over half of the tested water sources in regions such as Croatia [163], Spain [164], South Africa, India, and China [165], highlighting its global impact. Research on the effects of metolachlor has revealed disturbing findings. In prepubertal male rats, exposure led to hormonal imbalances, including increased levels of testosterone, estradiol, and FSH, while dihydrotestosterone levels were reduced. These changes were accompanied by abnormalities in the morphology of the seminiferous epithelium, suggesting potential reproductive toxicity [166]. Compared with those in areas with uncontaminated water, human populations exposed to metolachlor-contaminated water have also reported higher rates of intrauterine growth retardation and certain cancers, emphasizing the potential long-term health risks of herbicides [167].

Acute toxicity from metolachlor ingestion is primarily linked to the formation of methemoglobin, which inhibits oxygen transport to tissues. This disruption causes hypoxia and can lead to lactic acidosis, posing severe health threats to exposed individuals [168].

4.2 *Triazine*

Atrazine is among the most commonly identified triazine herbicide pollutants in both surface water and groundwater worldwide. The toxic effects of atrazine on reproductive health are well documented across animal and human studies [169]. In male rats, exposure disrupted the seminiferous epithelium, reduced sperm counts, increased the number of abnormal sperm, and lowered antioxidant capacity. Hormonal imbalances were also observed, including decreased serum testosterone and inhibin-B levels alongside elevated FSH and luteinizing hormone (LH) levels [170]. Female rats exposed to atrazine exhibited delayed vaginal opening, reduced ovarian and uterine weights, lower pubertal body weight, and irregular estrous cycles characterized by prolonged diestrus [171]. These findings are mirrored in human studies. Compared with urban residents, agricultural workers in Missouri exposed to atrazine experienced a 40% reduction in sperm count, with urinary atrazine levels strongly correlated with sperm quality [172]. Furthermore, communities consuming atrazine-contaminated water reported higher rates of intrauterine growth retardation [173]. The harmful effects of atrazine are largely attributed to its role as an endocrine disruptor, which interferes with the hypothalamic–pituitary–gonadal (HPG) axis [174].

Simazine, another widely used triazine herbicide, is associated with similar environmental and health concerns. In northern Italy, all tested surface and tap water samples contained simazine (16 ng/L), highlighting its ubiquity. Animal studies reveal its considerable reproductive toxicity [175]. Female offspring of mice exposed to simazine presented reduced body, ovarian, and uterine weights; shorter anogenital distances; and increased ovarian cell apoptosis [176]. Similarly, female rats showed delayed vaginal opening, disrupted estrous cycles, and reduced prolactin levels following exposure [177]. Male offspring exposed in utero presented reduced body, testicular, and epididymal weights; increased testicular apoptosis; and lower sperm concentrations. Additionally, simazine affects Leydig cell function, disrupting steroidogenesis by altering key regulatory genes such as Star, Hsd3b, and Cyp17a1 in a dose-dependent manner [178].

The pervasive presence and endocrine-disrupting effects of atrazine and simazine highlight their significant risks to both environmental and human health. These findings reinforce the urgent need for enhanced monitoring and stricter regulatory controls to limit contamination in water sources and mitigate their harmful impacts.

4.3 Neonicotinoids

Neonicotinoids, a widely used class of systemic insecticides, have become the cornerstone of pest control globally and are categorized into four distinct generations. The first generation includes imidacloprid, acetamiprid, thiacloprid, and nitenpyram, all of which share a chloropyridine ring and produce 6-chloronicotinic acid as a common metabolite. This metabolite serves as a reliable biomarker in biomonitoring studies. The second generation is represented by thiamethoxam, which is metabolized into the insecticidal compound clothianidin. Dinotefuran, a hallmark of the third generation, and the fourth-generation compounds sulfoxaflor and cycloxaprid further highlight the evolution of this class. However, their water-soluble and persistent nature has raised concerns about environmental contamination, especially in drinking water. Their persistence through water treatment processes underscores their potential threat to human health [179, 180].

Once absorbed, neonicotinoids are rapidly metabolized and eliminated within 48 h, preventing their accumulation in the body. However, their metabolic pathways, which are driven primarily by CYP450 enzymes, vary across species and often yield metabolites that are equally or more toxic than the parent compounds. For example, N-denitro-imidacloprid has a strong affinity for mammalian nicotinic acetylcholine receptors (nAChRs) [181]. These insecticides act by binding to postsynaptic nAChRs in the nervous system, promoting the influx of cations such as Ca^{2+}, Na^{+}, or K^{+}, depending on the receptor subtype. The structural differences in nAChRs between insects and mammals largely account for the high selectivity of neonicotinoids toward insects, coupled with relatively low toxicity in vertebrates [182]. In mammals, they primarily affect the central nAChR subtypes $\alpha4\beta2$ and $\alpha7$, with minimal impact on peripheral nAChRs [183]. Initially, believed to pose little risk to humans owing to their weak affinity for mammalian receptors and limited ability to cross the blood–brain barrier, emerging evidence has since refuted this perception. Studies in vitro, in vivo, and across ecological contexts have revealed that even sublethal doses can adversely affect mammalian nAChRs, drawing parallels to the action of nicotine [184, 185].

Acute exposure to high doses of neonicotinoids often results in gastrointestinal symptoms, including nausea, vomiting, abdominal pain, damage to the oral mucosa, and skin and eye irritation (Fig. 7.5). Activation of autonomic nervous system nAChRs can further trigger symptoms such as tachycardia, hypertension, excessive sweating, and dilated pupils [186]. Beyond these immediate effects, chronic exposure has been linked to neurodevelopmental and reproductive toxicity [187]. In male rats, imidacloprid exposure has been shown to reduce the sperm concentration, increase DNA fragmentation, lower testosterone levels, and impair the antioxidant defense system in testicular tissue [188]. Female rats exposed to imidacloprid experience reduced ovarian weight, altered levels of FSH and LH, and diminished progesterone production [189]. Similarly, clothianidin has been associated with reduced

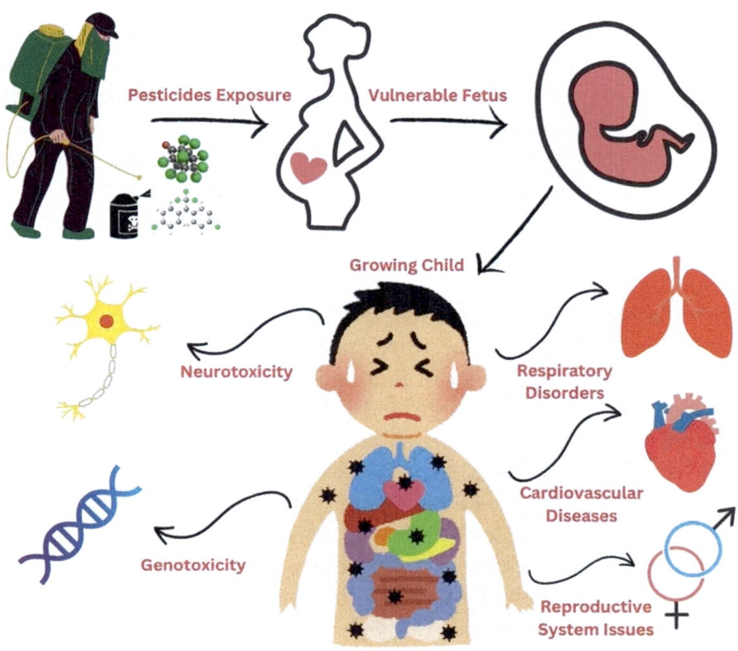

Fig. 7.5 Transgenerational effects of pesticide exposure on health and development

epididymal weight, changes in testicular fatty acid composition, and decreased antioxidant activity [190].

Despite the widespread presence of neonicotinoids in drinking water and their documented effects on human and environmental health, regulatory measures remain inadequate. Addressing this gap is essential to mitigate their long-term risks and ensure safer environmental and public health outcomes. Research into the mechanisms underlying pesticide toxicity is essential for developing targeted strategies to mitigate risks and protect human and environmental health.

5 Microplastics

Plastics are intricate synthetic materials with a carbon-based polymer backbone supplemented by additives that increase flexibility, durability, water resistance, flame retardation, and UV protection. Approximately 98% of plastics are derived from fossil fuels such as coal, oil, and natural gas. Many chemicals used in their production—including phthalates, bisphenols, PFAS, brominated flame retardants, and organophosphate flame retardants—are toxic. These substances, which include carcinogens, neurotoxicants, and endocrine disruptors, contribute to the harmful impacts of plastics on human health and the environment [191].

The pervasive presence of microplastics (MPs) and nanoplastics (NPs) in marine ecosystems exacerbates this problem. These particles have been detected in numerous marine species, many of which are consumed by humans. While larger microplastic particles (>10 μm) may not biomagnify, the hydrophobic chemicals they carry can bioaccumulate in marine organisms and be magnified through the food web. In contrast, smaller microplastics and nanoplastics (<10 μm) pose a greater risk owing to their ability to move freely within biological systems, although their long-term behavior and effects remain poorly understood [192].

Microplastics, defined as plastic fragments smaller than 5 mm, are categorized as primary or secondary. Primary microplastics are intentionally manufactured for use in cosmetics, biomedical products, and synthetic textiles, where they are released into the environment during production, use, or disposal. Secondary microplastics, on the other hand, result from the fragmentation of larger plastic debris through weathering processes. The impact of these particles depends on their size, shape, chemical composition, surface charge, and concentration. While microplastics typically range from 1–5 mm in size, nanoplastics, classified as 1–1000 nm in size, are particularly concerning owing to their increased mobility and surface area, which allow for unique interactions with biological systems [193].

Extensive research highlights the harmful effects of nanoplastics on cellular and organismal health. One of the most concerning outcomes is the generation of reactive oxygen species (ROS), which trigger oxidative stress, impair cellular functions, and disrupt mitochondrial activity. Although the exact mechanisms remain unclear, they may involve mechanical stretching of membranes or direct interactions with cellular components. These oxidative effects are often accompanied by inflammation, which is characterized by changes in immune gene expression and cytokine activation. The resulting physiological impacts include damage to the gastrointestinal tract, liver, reproductive system, and neurological functions in both vertebrates and invertebrates [194].

The small size of nanoplastics enables them to cross biological barriers, such as the intestinal lining and blood–brain barrier, and distribute throughout the body. For example, studies in zebrafish have shown that 50 nm particles can enter the bloodstream and disrupt dopamine metabolism, whereas larger particles, such as those at 100 nm, are less likely to infiltrate sensitive tissues. These findings suggest that smaller nanoplastics pose greater risks due to their ability to accumulate in critical organs over time. However, uncertainties persist regarding the specific particle sizes most likely to bioaccumulate and their long-term effects on human health [195]. In addition to physical toxicity, the chemical composition of plastics adds another layer of risk. Additives such as BPA, phthalates, and PBDEs can leach into the environment, whereas plastic surfaces readily adsorb environmental pollutants such as persistent organic pollutants (POPs). These include PCBs, PCDDs, and polycyclic aromatic hydrocarbons (PAHs) [196]. These chemicals are linked to a range of adverse health outcomes, including neurodevelopmental disorders, immune system dysfunction, reproductive impairments, metabolic disruptions, cardiovascular diseases, and cancers. Many of these effects are mediated through interactions with cellular and nuclear receptors that regulate hormonal activity, growth, reproduction, and cell proliferation [197]. The combined impacts of plastic particles and their associated chemicals remain underexplored. Laboratory studies often isolate the effects of either particles or chemicals, but real-world exposures involve complex interactions between the two. Understanding these synergistic effects is critical for accurately assessing the risks that plastics pose to ecosystems and human health [198]. Microplastics are now recognized as ubiquitous environmental pollutants that threaten ecological balance and public health. Their toxicological effects extend across terrestrial, aquatic, and atmospheric ecosystems, disrupting biological processes in plants, animals, and microorganisms. In aquatic ecosystems, microplastics have entered the human food chain via trophic transfer. Algae, as primary producers, are especially vulnerable, with studies showing disruptions in growth, photosynthesis, and oxidative processes when exposed to microplastics. Polystyrene (PS) particles, in particular, are among the most harmful. Furthermore, microplastics often interact with other pollutants, reducing their bioavailability to algae while introducing antagonistic effects. These disruptions cascade through the food chain, affecting higher organisms and threatening biodiversity [199].

Terrestrial ecosystems are not immune to these effects, although research in this area is limited. Microplastics have been shown to impair skeletal muscle regeneration in mice, reduce feeding efficiency in copepods, and infiltrate plant communities through soil [200]. Once in the soil, microplastics hinder plant growth by reducing biomass, interfering with nutrient uptake, and disrupting root and leaf development. While impacts vary depending on the type of plastic and plant species, the overall effects are overwhelmingly negative, warranting further investigation into their long-term ecological consequences [201].

Human exposure to microplastics occurs primarily through ingestion, inhalation, and dermal contact. Although definitive evidence is lacking, studies in other organisms suggest that microplastics may induce cytotoxicity by interacting with cellular structures. Laboratory experiments have shown that exposure to environmentally sourced plastic particles can trigger harmful cellular responses. Alarmingly, humans are estimated to ingest 39,000–52,000 microplastic particles annually, underscoring the urgent need for research into the potential health risks [202]. Microplastics also serve as carriers for various chemical contaminants, including pharmaceuticals, heavy metals, and personal care products. Additionally, they host bacteria and viruses, raising concerns about their role in spreading ARB and ARGs [203]. Three primary mechanisms facilitate the adsorption of contaminants onto microplastics: hydrophobic interactions, electrostatic interactions, and pore-filling interactions. Hydrophobic interactions, driven by the nonpolar nature of many polymers, enable plastics such as polyethylene (PE) and polypropylene (PP) to bind with contaminants. Electrostatic interactions, influenced by surface charges, determine sorption rates, whereas pore-filling interactions allow contaminants to embed within the micro- and nanopores of plastic particles. Weathering further enhances these interactions by increasing surface porosity [204]. The transfer of contaminants from microplastics to organisms has been well documented. For example, fluoranthene-contaminated microplastics elevated PAH levels in blue mussels, whereas cadmium-contaminated particles disrupted zebrafish embryonic development [205]. Microplastics also facilitate the spread of ARB and ARGs by providing a habitat for bacterial communities. Prolonged exposure to antibiotics on microplastic surfaces accelerates resistance development, enabling the spread of ARGs across aquatic, terrestrial and wastewater environments [206].

Microplastics represent a growing global crisis. The ability of these bacteria to bioaccumulate, disrupt biological systems, and act as vectors for contaminants highlights the urgent need for action. Comprehensive research into their mechanisms of toxicity, environmental interactions, and long-term effects is essential. Only through a holistic approach can we develop effective strategies to mitigate their far-reaching impacts on ecosystems, human health, and future generations.

6 Volatile Organic Compounds

Volatile organic compounds (VOCs) are carbon-based substances that easily evaporate at room temperature and are major contributors to indoor air pollution. Common VOCs found in products such as cleaning supplies, furniture, wallpaper, plastics, and plywood include benzene, toluene, xylene, and formaldehyde. While VOCs also occur naturally, emissions from human activities have significantly increased in recent decades because of their widespread use in residential and commercial products.

Based on their volatility, VOCs are categorized by the World Health Organization (WHO) into three groups: extremely volatile organic compounds (VVOCs), volatile organic compounds (VOCs), and semivolatile organic compounds (SVOCs). VVOCs, such as methyl chloride and propane, are so volatile that they are often present as gases, making them difficult to measure. VOCs, including hexanal, formaldehyde, and toluene, are more stable but still readily vaporize, whereas SVOCs, such as pesticides, are less volatile and are often associated with solid surfaces, liquids, or materials such as dust and construction products. As volatility increases, so does the likelihood of these compounds entering the environment, heightening concerns about their impact [207]. Although VOCs are essential in manufacturing and maintaining product stability, their environmental and health impacts are becoming increasingly evident. In the atmosphere, VOCs serve as precursors for ground-level ozone formation. Photochemical reactions involving ozone and primary pollutants release reactive compounds that aggravate air pollution and disrupt the physiology of plants. Biogenic VOCs, such as isoprenoids emitted by plants, add complexity to this process [208, 209]. However, human exposure to anthropogenic VOCs poses even greater concerns, as prolonged contact is associated with a range of adverse health effects [210]. Indoor environments are a primary source of VOC exposure, with significant consequences for skin and respiratory health. For example, studies have shown that airborne formaldehyde can impair skin barrier function, leading to increased transepidermal water loss and increased skin pH. Individuals with allergic dermatitis are particularly susceptible, with exposure exacerbating symptoms such as pruritus and inflammation [211]. Research also links elevated VOC levels in newly remodeled homes—such as those with PVC-based wallpaper—to higher rates of allergic dermatitis, particularly in children, where the risk can increase sevenfold when combined with food allergies [212].

VOCs are also a concern for respiratory health. Aromatic compounds such as benzene, xylene, and ethylbenzene, along with formaldehyde, have been implicated in conditions such as asthma. Epidemiological studies confirm that children living in homes with high VOC concentrations face a greater risk of developing asthma [213]. Additionally, renovation activities such as painting or decorating have been associated with an increased incidence of eczema and allergic symptoms in young children, likely due to the emission of VOCs. These activities may also increase the levels of inflammatory markers, such as interleukin-8, which further highlights the link between VOC exposure and immune responses [214].

In addition to homes, other environments, such as chlorinated swimming pools, also contribute to VOC-related health risks. Disinfection byproducts such as trihalomethanes and chloramines have been linked to respiratory conditions such as asthma, chronic bronchitis, and upper respiratory tract irritation. While some studies suggest a relationship between swimming pool attendance and these health issues, others find no significant association, underscoring the need for more robust research to clarify these findings [215]. Emerging evidence points to VOC exposure in neonatal incubators, where compounds such as 2-heptanone and n-butyl acetate have been detected at higher levels than in ambient air. Although the long-term effects on developing infants remain unclear, these findings highlight a critical area for further investigation [216]. Notably, VOC detection has potential in medical diagnostics. For example, patients with invasive aspergillosis (IA) exhale specific

VOCs, such as 2-pentylfuran, which can be detected with up to 81% accuracy. This noninvasive technique offers a promising tool for the early diagnosis of *Aspergillus* infections, although more studies are needed to increase its reliability [217]. In summary, while VOCs play a pivotal role in modern manufacturing and product development, their widespread presence raises significant environmental and health concerns. By exacerbating air pollution and impacting plant physiology to contribute to respiratory and skin-related illnesses, the risks associated with VOC exposure cannot be overlooked. As research continues, understanding and mitigating the effects of VOCs will be essential for balancing their utility with public health and environmental sustainability.

7 Radioactive Substances

Radioactive substances consist of unstable atoms that emit radiation as they decay, posing significant risks to ecosystems and human health. These substances can contaminate air, water, soil, and food supplies; increase cancer risk; cause genetic mutations; and disrupt ecosystems. Notable nuclear incidents such as the Chernobyl and Fukushima Daiichi disasters underscore the profound impacts of radioactive contamination on human health and the environment [218].

7.1 Ionizing Radiation and Radionuclides

Ionizing radiation comprises electromagnetic waves (e.g., ϒ- or X-rays) or particles (e.g., neutrons, β, or α particles) released during atomic decay. This process, known as radioactivity, occurs when atoms spontaneously disintegrate, emitting surplus energy as ionizing radiation. Elements undergoing disintegration and emitting ionizing radiation are termed radionuclides [219]. Radionuclides in drinking water may stem from natural or human-made sources. Naturally occurring radionuclides, which are abundant in rocks and soil, are prevalent in groundwater. Of particular concern for human radiation exposure from drinking water are radionuclides from thorium and uranium decay series, including radium-226, radium-228, polonium-210, lead-210, and radon [220]. Human-made radionuclides, such as cesium-134, cesium-137, strontium-90, iodine-131, tritium, and carbon-14, can infiltrate drinking water from various sources, albeit typically at low concentrations. While these contaminants often register below detectable limits through standard analytical techniques, they are generally considered to pose negligible risks to human health compared with microorganisms and chemicals [221].

7.2 *Health Impacts and Radiation Exposure*

The health hazards associated with radionuclides in drinking water are typically lower and not immediate than those posed by microorganisms and chemicals. The radiation dose from ingesting radionuclides is generally much lower than that from other common sources of radiation. The impact of radiation on tissues and organs depends on the absorbed dose, measured in grays (Gy). Potential damage varies on the basis of the type of radiation and tissue vulnerability. To assess potential harm, the effective dose, measured in sieverts (Sv), and the dose rate, indicated in microsieverts per hour (μSv/hour) or millisieverts per year (mSv/year), are crucial factors [222]. On average, individuals are exposed to approximately 0.3 mSv/year of radiation from naturally occurring radionuclides in their diet, with approximately 5% originating from drinking water. This dose constitutes ~10% of the total average annual radiation exposure from all natural sources, which includes cosmic rays, soil, radon, and dietary intake [223]. Radiation exposure can lead to acute effects such as skin redness, hair loss, or radiation burns at certain thresholds, typically approximately 1 Sv (1000 mSv). Conversely, low doses over prolonged periods diminish the risk of immediate harm, although long-term effects such as cataracts or cancer remain possible [224]. Certain demographics, such as those of children and adolescents, are more sensitive to radiation effects, highlighting the need for stringent safety measures and monitoring in radiation-exposed environments. Epidemiological studies have consistently shown elevated cancer risk at doses exceeding 100 mSv, suggesting increased risks even at lower doses [225].

Prenatal exposure to ionizing radiation presents specific risks, particularly concerning fetal brain development. Studies indicate potential fetal brain damage at exposure levels exceeding 100 mSv between weeks 8 and 15 of pregnancy and 200 mSv between weeks 16 and 25 [226].

7.3 *Uranium Metabolism and Elimination*

Uranium metabolism and elimination in the human body involve complex kinetics. Uranium primarily accumulates in the kidneys and skeleton, with minimal amounts in the liver, reflecting its localization in certain tissues. The overall elimination half-life of uranium in humans under normal conditions ranges from 180 to 360 days. Uranium interacts with proteins through diverse mechanisms, potentially inducing structural alterations and disrupting cellular processes [227].

7.3.1 Renal Effects of Uranium Exposure

Environmental uranium exposure may lead to renal effects, as indicated by studies in regions such as Nova Scotia and Saskatchewan, Canada. While no clear link between overt renal disease and uranium exposure has been established, increased urinary β-2-microglobulin and uranium concentrations in well water suggest potential early tubular defects.

Similarly, studies in Finland revealed modest alterations in proximal tubular function among individuals exposed to well water containing uranium [228]. These findings underscore the importance of continued research to understand the mechanisms underlying uranium exposure and renal health outcomes.

7.4 Radon Exposure and Health Effects

Radon, a naturally occurring noble gas, can easily migrate into various environments, including air, groundwater, and surface water. Ingestion of radon occurs when it dissolves in water, leading to intake through the stomach, small intestine, and large intestine. The toxicity of radon mainly stems from the alpha radiation emitted during its decay, which can cause DNA breakage and potentially lead to tumor formation. The repair of DNA damage from α particles involves mechanisms such as homologous recombination and nonhomologous end joining [229].

8 Microorganisms, Waterborne Diseases, and Host Interactions

Microorganisms, including bacteria, viruses, and parasites, are ubiquitous on Earth's surface and are commonly found in human sewage and animal waste. Consumption of water contaminated with these microorganisms can lead to water-related infectious diseases (WRID), which primarily cause gastrointestinal illnesses and other infections [230] (Fig. 7.6).

WRID transmission can occur through various pathways, including ingestion of contaminated water, inhalation of aerosols, and direct contact with contaminated surfaces during activities such as swimming or bathing [231].

The top five causes of drinking water microbial outbreaks from 1971 to 2020 are as follows: (1) *Legionella*, (2) *Giardia*, (3) *Norovirus*, (4) *Shigella*, and (5) *Campylobacter* [232].

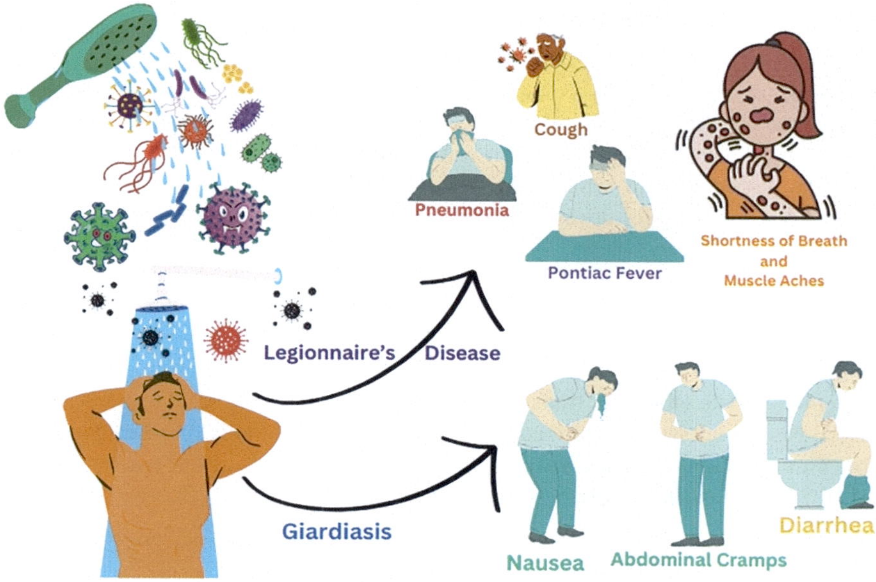

Fig. 7.6 Water-related infectious diseases: Legionnaires' disease and giardiasis

8.1 Legionnaires' Disease

Legionella pneumophila is responsible for two distinct illnesses: Legionnaires' disease and Pontiac fever. Legionnaires' disease manifests as severe pneumonia with symptoms such as cough, shortness of breath, fever, muscle aches, and headaches. Complications can be severe, leading to lung failure and death, particularly among high-risk individuals such as the elderly or those with weakened immune systems [233].

Pontiac fever, in contrast, presents with fever and muscle aches but does not involve pneumonia. Understanding Legionella–host interactions at the molecular level is crucial for identifying potential drug targets to combat Legionella infections [234].

8.2 Giardiasis

Giardia duodenalis, which is resilient to environmental conditions, can contaminate various sources, including water and food. Infection can lead to gastrointestinal symptoms such as diarrhea, stomach cramps, and nausea. Understanding *Giardia*–host interactions is essential for identifying drug targets and developing effective therapies against giardiasis [235].

8.3 Norovirus Infection

Norovirus, which is highly contagious and causes acute gastroenteritis, spreads through the ingestion of contaminated fecal matter or vomiting particles. Symptoms include diarrhea, vomiting, and fever, with complications such as dehydration posing risks, especially for vulnerable groups [236]. Understanding norovirus–host interactions is crucial for developing prevention and treatment strategies.

8.4 Shigellosis

Shigella bacteria, which are transmitted through contaminated food or water, cause shigellosis, which is characterized by symptoms such as bloody diarrhea and fever. The complications include reactive arthritis and hemolytic-uremic syndrome [237]. Understanding *Shigella*–host interactions is vital for effective treatment development.

8.5 **Campylobacter** *Infection*

Campylobacter bacteria, a leading cause of diarrheal illness globally, can lead to complications such as Guillain–Barré syndrome. Understanding *Campylobacter*–host interactions is essential for identifying effective drug targets and reducing the impact of campylobacteriosis on public health [238].

9 Conclusion

Addressing the challenges of highly dangerous contaminants requires a broad approach that integrates scientific advancements, strong regulations, and active public involvement. These substances have already caused significant harm to ecosystems and human health, but their impact can be reduced through effective and collaborative action. Cleaner production methods, improved waste management, and international cooperation are crucial steps toward minimizing their presence and persistence. By addressing key causes, such as unsustainable industrial practices, excessive agricultural inputs, and poor disposal methods, further damage can be prevented. Success depends on the combined efforts of science, policy, and communities working together. With a commitment to sustainability and environmental care, we can protect ecosystems, safeguard public health, and create a healthier future for generations to come.

References

1. Goutam Mukherjee, A., Ramesh Wanjari, U., Eladl, M. A., El-Sherbiny, M., Elsherbini, D. M. A., Sukumar, A., Kannampuzha, S., Ravichandran, M., Renu, K., Vellingiri, B., Kandasamy, S., & Valsala, G. A. (2022). Mixed contaminants: Occurrence, interactions, toxicity, detection, and remediation. *Molecules, 27*(8), 2577. https://doi.org/10.3390/molecules27082577
2. Chu, E. W., Karr, J. R. (2017). Environmental impact: Concept, consequences, measurement. reference module in life sciences:B978-0-12-809633-8.02380-3. https://doi.org/10.1016/B978-0-12-809633-8.02380-3.
3. Gunjyal, N., Rani, S., Asgari Lajayer, B., et al. (2023). A review of the effects of environmental hazards on humans, their remediation for sustainable development, and risk assessment. *Environmental Monitoring and Assessment, 195*, 795. https://doi.org/10.1007/s10661-023-11353-z
4. Samal, K., Saswat Mahapatra, M., & Ali, H. (2022). Pharmaceutical wastewater as Emerging Contaminants (EC): Treatment technologies, impact on environment and human health. *Energy Nexus, 6*, 100076., ISSN 2772-4271. https://doi.org/10.1016/j.nexus.2022.100076
5. Patel, M., Kumar, R., Kishor, K., Mlsna, T., Pittman Jr, C. U., & Mohan, D. (2019). Pharmaceuticals of emerging concern in aquatic systems: Chemistry, occurrence, effects, and removal methods. *Chemical Reviews, 119*(6), 3510–3673. https://doi.org/10.1021/acs.chemrev.8b00299
6. Guo, W., Archer, J., Moore, M., et al. (2021). Software-assisted pattern recognition of persistent organic pollutants in contaminated human and animal food. *Molecules, 26*(3), 685. https://doi.org/10.3390/molecules26030685
7. Jayavel, S., Govindaraju, B., Michael, J. R., et al. (2024). Impacts of micro and nanoplastics on human health. *Bulletin of the National Research Centre, 48*, 110. https://doi.org/10.1186/s42269-024-01268-1
8. Pathak, V. M., Verma, V. K., Rawat, B. S., Kaur, B., Babu, N., Sharma, A., Dewali, S., Yadav, M., Kumari, R., Singh, S., Mohapatra, A., Pandey, V., Rana, N., & Cunill, J. M. (2022). Current status of pesticide effects on environment, human health and it's eco-friendly management as bioremediation: A comprehensive review. *Frontiers in Microbiology, 13*, 962619. https://doi.org/10.3389/fmicb.2022.962619
9. Pan, Q., Liu, Q.-Y., Zheng, J., Li, Y.-H., Xiang, S., Sun, X.-J., & He, X.-S. (2023). Volatile and semi-volatile organic compounds in landfill gas: Composition characteristics and health risks. *Environment International, 174*, 107886., ISSN 0160-4120. https://doi.org/10.1016/j.envint.2023.107886
10. Tran, T.-K., Nguyen, M.-K., Lin, C., Hoang, T.-D., Nguyen, T.-C., Lone, A. M., Khedulkar, A. P., Gaballah, M. S., Jagpreet Singh, W., Jin Chung, D., & Nguyen, D. (2024). Review on fate, transport, toxicity and health risk of nanoparticles in natural ecosystems: Emerging challenges in the modern age and solutions toward a sustainable environment. *Science of The Total Environment, 912*, 169331., ISSN 0048-9697. https://doi.org/10.1016/j.scitotenv.2023.169331
11. Ahmed, W., Gyawali, P., Hamilton, K. A., Joshi, S., Aster, D., Donner, E., Simpson, S. L., & Symonds, E. M. (2021). Antibiotic resistance and sewage-associated marker genes in untreated sewage and a river characterized during baseflow and stormflow. *Frontiers in Microbiology, 12*, 632850. https://doi.org/10.3389/fmicb.2021.632850
12. Ding, G., Jing, Y., Chen, L., Wang, C., Zhou, Y., Yi, H., Shi, R., Zhang, Y., Chang Cui, Y., Gao, Y. T., & Liu, F. (2017). Polybrominated diphenyl ethers (PBDEs) and thyroid hormones in cord blood. *Environmental Pollution, 229*, 489–495., ISSN 0269-7491. https://doi.org/10.1016/j.envpol.2017.05.065
13. Gomes, I. B., Maillard, J. Y., Simões, L. C., et al. (2020). Emerging contaminants affect the microbiome of water systems—Strategies for their mitigation. *npj Clean Water, 3*, 39. https://doi.org/10.1038/s41545-020-00086-y

14. Sanganyado, E., & Gwenzi, W. (2019). Antibiotic resistance in drinking water systems: Occurrence, removal, and human health risks. *Science of the Total Environment, 669*, 785–797. https://doi.org/10.1016/j.scitotenv.2019.03.162
15. Ching, C., Orubu, E. S., Wirtz, V. J., & Zaman, M. H. (2020). Bacterial antibiotic resistance development and mutagenesis following exposure to subinhibitory concentrations of fluoroquinolones in vitro: A systematic review of the literature. *JAC-Antimicrobial Resistance, 2*, e030747. https://doi.org/10.1093/jacamr/dlaa068
16. Qiu, W., Sun, J., Fang, M., Luo, S., Tian, Y., Dong, P., Xu, B., & Zheng, C. (2019). Occurrence of antibiotics in the main rivers of Shenzhen, China: Association with antibiotic resistance genes and microbial community. *Science of the Total Environment, 653*, 334–341. https://doi.org/10.1016/j.scitotenv.2018.10.398
17. Wang, H., Hu, C., Shen, Y., Shi, B., Zhao, D., & Xing, X. (2019). Response of microorganisms in biofilm to sulfadiazine and ciprofloxacin in drinking water distribution systems. *Chemosphere, 218*, 197–204. https://doi.org/10.1016/j.chemosphere.2018.11.106
18. Jin, M., Lu, J., Chen, Z., Nguyen, S. H., Mao, L., Li, J., Yuan, Z., & Guo, J. (2018). Antidepressant fluoxetine induces multiple antibiotics resistance in Escherichia coli via ROS-mediated mutagenesis. *Environment International, 120*, 421–430. https://doi.org/10.1016/j.envint.2018.07.046
19. Wang, Y., Lu, J., Engelstädter, J., Zhang, S., Ding, P., Mao, L., Yuan, Z., Bond, P. L., & Guo, J. (2020). Non-antibiotic pharmaceuticals enhance the transmission of exogenous antibiotic resistance genes through bacterial transformation. *The ISME Journal, 14*, 2179–2196. https://doi.org/10.1038/s41396-020-0679-2
20. Gould, S. L., Winter, M. J., Norton, W. H. J., & Tyler, C. R. (2021). The potential for adverse effects in fish exposed to antidepressants in the aquatic environment. *Environmental Science & Technology, 55*(24), 16299–16312. https://doi.org/10.1021/acs.est.1c04724
21. Faria, M., Bellot, M., Soto, O., Prats, E., Montemurro, N., Manjarrés, D., Gómez-Canela, C., & Raldúa, D. (2022). Developmental exposure to sertraline impaired zebrafish behavioral and neurochemical profiles. *Frontiers in Physiology, 13*, 1040598. https://doi.org/10.3389/fphys.2022.1040598
22. Bachour, R. L., Golovko, O., Kellner, M., & Pohl, J. (2020). Behavioral effects of citalopram, tramadol, and binary mixture in zebrafish (Danio rerio) larvae. *Chemosphere, 238*, 124587. https://doi.org/10.1016/j.chemosphere.2019.124587
23. Franzellitti, S., Buratti, S., Du, B., et al. (2015). A multibiomarker approach to explore interactive effects of propranolol and fluoxetine in marine mussels. *Environmental Pollution, 205*, 60–69. https://doi.org/10.1016/j.envpol.2015.05.020
24. Luna, T. O., Plautz, S. C., & Salice, C. J. (2015). Chronic effects of 17α-ethinylestradiol, fluoxetine, and the mixture on individual and population-level end points in Daphnia magna. *Archives of Environmental Contamination and Toxicology, 68*(4), 603–611. https://doi.org/10.1007/s00244-014-0119-2
25. Silva, L. J. G., Pereira, A. M. P. T., Rodrigues, H., Meisel, L. M., Lino, C. M., & Pena, A. (2017). SSRIs antidepressants in marine mussels from Atlantic coastal areas and human risk assessment. *Science of the Total Environment, 603–604*, 118–125., ISSN 0048-9697,. https://doi.org/10.1016/j.scitotenv.2017.06.076
26. Rojo, M., Álvarez-Muñoz, D., Dománico, A., Foti, R., Rodriguez-Mozaz, S., Barceló, D., & Carriquiriborde, P. (2019). Human pharmaceuticals in three major fish species from the Uruguay River (South America) with different feeding habits. *Environmental Pollution, 252*(Part A), 146–154., ISSN 0269-7491. https://doi.org/10.1016/j.envpol.2019.05.099
27. Rhomberg, L. R., Seeley, M., Verslycke, T., & Goodman, J. E. (2014). Environmental Hormone Disruptors. In P. Wexler (Ed.), *Encyclopedia of Toxicology* (3rd ed., pp. 378–380., ISBN 9780123864550). Academic Press. https://doi.org/10.1016/B978-0-12-386454-3.01010-1
28. Torres, N. H., de Oliveira, G., Santos, S., Ferreira, L. F. R., Américo-Pinheiro, J. H. P., Eguiluz, K. I. B., & Salazar-Banda, G. R. (2021). Environmental aspects of hormones estriol, 17β-estradiol and 17α-ethinylestradiol: Electrochemical processes as next-generation tech-

nologies for their removal in water matrices. *Chemosphere, 267*, 128888., ISSN 0045-6535. https://doi.org/10.1016/j.chemosphere.2020.128888
29. Wyse, J., Latif, S., Gurusinghe, S., McCormick, J., Weston, L. A., & Stephen, C. P. (2022). Phytoestrogens: A review of their impacts on reproductive physiology and other effects upon grazing livestock. *Animals: an open access Journal from MDPI, 12*(19), 2709. https://doi.org/10.3390/ani12192709
30. Gross-Sorokin, M. Y., Roast, S. D., & Brighty, G. C. (2006). Assessment of feminization of male fish in English rivers by the environment Agency of England and Wales. *Environmental Health Perspectives, 114*(Suppl 1), 147–151. https://doi.org/10.1289/ehp.8068
31. Marlatt, V. L., Bayen, S., Castaneda-Cortès, D., Delbès, G., Grigorova, P., Langlois, V. S., Martyniuk, C. J., Metcalfe, C. D., Parent, L., Rwigemera, A., Thomson, P., & Van Der Kraak, G. (2022). Impacts of endocrine disrupting chemicals on reproduction in wildlife and humans. *Environmental Research, 208*, 112584., ISSN 0013-9351. https://doi.org/10.1016/j.envres.2021.112584
32. You, H. H., & Song, G. (2021). Review of endocrine disruptors on male and female reproductive systems. *Comparative Biochemistry and Physiology Part C: Toxicology & Pharmacology, 244*, 109002., ISSN 1532-0456. https://doi.org/10.1016/j.cbpc.2021.109002
33. Rumph, J. T., Stephens, V. R., Martin, J. L., Brown, L. K., Thomas, P. L., Cooley, A., Osteen, K. G., & Bruner-Tran, K. L. (2022). Uncovering evidence: Associations between environmental contaminants and disparities in women's health. *International Journal of Environmental Research and Public Health, 19*(3), 1257. https://doi.org/10.3390/ijerph19031257
34. Panagopoulos, P., Mavrogianni, D., Christodoulaki, C., Drakaki, E., Chrelias, G., Panagiotopoulos, D., Potiris, A., Drakakis, P., & Stavros, S. (2023). Effects of endocrine disrupting compounds on female fertility. *Best Practice & Research Clinical Obstetrics & Gynaecology, 88*, 102347., ISSN 1521-6934. https://doi.org/10.1016/j.bpobgyn.2023.102347
35. Mínguez-Alarcón, L., Hauser, R., & Gaskins, A. J. (2016). Effects of bisphenol a on male and couple reproductive health: A review. *Fertility and Sterility, 106*(4), 864–870. https://doi.org/10.1016/j.fertnstert.2016.07.1118
36. Fischer, C., Mamillapalli, R., Goetz, T. G., et al. (2016). Bisphenol A (BPA) exposure in utero leads to immunoregulatory cytokine dysregulation in the mouse mammary gland: A potential mechanism programming breast cancer risk. *Discover Oncology, 7*, 241–251. https://doi.org/10.1007/s12672-016-0254-5
37. Ma, Z., Parris, A. B., Howard, E. W., et al. (2020). In utero exposure to bisphenol a promotes mammary tumor risk in MMTV-Erbb2 transgenic mice through the induction of ER-erbB2 crosstalk. *International Journal of Molecular Sciences, 21*(9), 3095. Published 2020 Apr 28. https://doi.org/10.3390/ijms21093095
38. Hu, Y., Yuan, D. Z., Wu, Y., et al. (2018). Bisphenol a initiates excessive premature activation of primordial follicles in mouse ovaries via the PTEN signaling pathway. *Reproductive Sciences, 25*(4), 609–620. https://doi.org/10.1177/1933719117734700
39. Presunto, M., Mariana, M., Lorigo, M., & Cairrao, E. (2023). The effects of bisphenol a on human male infertility: A review of current epidemiological studies. *International Journal of Molecular Sciences, 24*(15), 12417. https://doi.org/10.3390/ijms241512417
40. Pivonello, C., Muscogiuri, G., Nardone, A., Garifalos, F., Provvisiero, D. P., Verde, N., de Angelis, C., Conforti, A., Piscopo, M., Auriemma, R. S., Colao, A., & Pivonello, R. (2020). Bisphenol A: An emerging threat to female fertility. *Reproductive Biology and Endocrinology: RB&E, 18*(1), 22. https://doi.org/10.1186/s12958-019-0558-8
41. Huang, Y. F., Chang, C. H., Chen, P. J., Lin, I. H., Tsai, Y. A., Chen, C. F., Wang, Y. C., Huang, W. Y., Tsai, M. S., & Chen, M. L. (2021). Prenatal bisphenol a exposure, DNA methylation, and low birth weight: A pilot study in Taiwan. *International Journal of Environmental Research and Public Health, 18*(11), 6144. https://doi.org/10.3390/ijerph18116144
42. Mustieles, V., D'Cruz, S. C., Couderq, S., Rodríguez-Carrillo, A., Fini, J.-B., Hofer, T., Steffensen, I.-L., Dirven, H., Barouki, R., Olea, N., Fernández, M. F., & David, A. (2020). Bisphenol A and its analogues: A comprehensive review to identify and prioritize effect

biomarkers for human biomonitoring. *Environment International, 144*, 105811., ISSN 0160-4120. https://doi.org/10.1016/j.envint.2020.105811
43. Casati, L., & Celotti, F. (2019). Chapter 17—Transgenerational epigenetics of endocrine-disrupting chemicals in mammals. In T. O. Tollefsbol (Ed.), *Translational epigenetics, transgenerational epigenetics (Second Edition)* (Vol. 13, pp. 369–388., ISSN 25425358, ISBN 9780128163634). Academic Press. https://doi.org/10.1016/B978-0-12-816363-4.00017-1
44. Wang, Y., Zhu, H., & Kannan, K. (2019). A review of biomonitoring of phthalate exposures. *Toxics, 7*(2), 21. https://doi.org/10.3390/toxics7020021
45. Hlisníková, H., Petrovičová, I., Kolena, B., Šidlovská, M., & Sirotkin, A. (2020). Effects and mechanisms of phthalates' action on reproductive processes and reproductive health: A literature review. *International Journal of Environmental Research and Public Health, 17*(18), 6811. https://doi.org/10.3390/ijerph17186811
46. Arzuaga, X., Walker, T., Yost, E. E., Radke, E. G., & Hotchkiss, A. K. (2020). Use of the adverse outcome pathway (AOP) framework to evaluate species concordance and human relevance of Dibutyl phthalate (DBP)-induced male reproductive toxicity. *Reproductive toxicology (Elmsford, N.Y.), 96*, 445–458. https://doi.org/10.1016/j.reprotox.2019.06.009
47. Brehm, E., & Flaws, J. A. (2019). Transgenerational effects of endocrine-disrupting chemicals on male and female reproduction. *Endocrinology, 160*(6), 1421–1435. https://doi.org/10.1210/en.2019-00034
48. Parra-Forero, L. Y., Veloz-Contreras, A., Vargas-Marín, S., Mojica-Villegas, M. A., Alfaro-Pedraza, E., Urióstegui-Acosta, M., & Hernández-Ochoa, I. (2019). Alterations in oocytes and early zygotes following oral exposure to di(2-ethylhexyl) phthalate in young adult female mice. *Reproductive Toxicology, 90*, 53–61., ISSN 0890-6238,. https://doi.org/10.1016/j.reprotox.2019.08.012
49. Rattan, S., Brehm, E., Gao, L., Niermann, S., & Flaws, J. A. (2018). Prenatal exposure to di(2-ethylhexyl) phthalate disrupts ovarian function in a transgenerational manner in female mice. *Biology of Reproduction, 98*(1), 130–145. https://doi.org/10.1093/biolre/iox154
50. Fletcher, E. J., Santacruz-Márquez, R., Mourikes, V. E., Neff, A. M., Laws, M. J., & Flaws, J. A. (2022). Effects of phthalate mixtures on ovarian folliculogenesis and steroidogenesis. *Toxics, 10*(5), 251. https://doi.org/10.3390/toxics10050251
51. Gray, L. E., Lambright, C. S., Conley, J. M., Evans, N., Furr, J. R., Hannas, B. R., Wilson, V. S., Sampson, H., & Foster, P. M. D. (2021). Genomic and hormonal biomarkers of phthalate-induced male rat reproductive developmental toxicity part II: A targeted RT-qPCR array approach that defines a unique adverse outcome pathway. *Toxicological Sciences, 182*(2), 195–214. https://doi.org/10.1093/toxsci/kfab053
52. Lai, F. N., Liu, J. C., Li, L., et al. (2017). Di (2-ethylhexyl) phthalate impairs steroidogenesis in ovarian follicular cells of prepuberal mice. *Archives of Toxicology, 91*(3), 1279–1292. https://doi.org/10.1007/s00204-016-1790-z
53. Zhou, C., Gao, L., & Flaws, J. A. (2017). Exposure to an environmentally relevant phthalate mixture causes transgenerational effects on female reproduction in mice. *Endocrinology, 158*(6), 1739–1754. https://doi.org/10.1210/en.2017-00100
54. Bornehag, C. G., Carlstedt, F., Jönsson, B. A., Lindh, C. H., Jensen, T. K., Bodin, A., Jonsson, C., Janson, S., & Swan, S. H. (2015). Prenatal phthalate exposures and anogenital distance in Swedish boys. *Environmental Health Perspectives, 123*(1), 101–107. https://doi.org/10.1289/ehp.1408163
55. Arbuckle, T. E., Agarwal, A., MacPherson, S. H., Fraser, W. D., Sathyanarayana, S., Ramsay, T., Dodds, L., Muckle, G., Fisher, M., Foster, W., Walker, M., & Monnier, P. (2018). Prenatal exposure to phthalates and phenols and infant endocrine-sensitive outcomes: The MIREC study. *Environment International, 120*, 572–583., ISSN 0160-4120. https://doi.org/10.1016/j.envint.2018.08.034
56. Radke, E. G., Braun, J. M., Meeker, J. D., & Cooper, G. S. (2018). Phthalate exposure and male reproductive outcomes: A systematic review of the human epidemiological evidence. *Environment International, 121*(Part 1), 764–793., ISSN 0160-4120. https://doi.org/10.1016/j.envint.2018.07.029

57. Nobles, C. J., Mendola, P., Kim, K., Pollack, A. Z., Mumford, S. L., Perkins, N. J., Silver, R. M., & Schisterman, E. F. (2023). Preconception phthalate exposure and women's reproductive health: Pregnancy, pregnancy loss, and underlying mechanisms. *Environmental Health Perspectives, 131*(12), 127013. https://doi.org/10.1289/EHP12287
58. Chiang, C., Pacyga, D. C., Strakovsky, R. S., Smith, R. L., James-Todd, T., Williams, P. L., Hauser, R., Meling, D. D., Li, Z., & Flaws, J. A. (2021). Urinary phthalate metabolite concentrations and serum hormone levels in pre- and perimenopausal women from the midlife women's health study. *Environment International, 156*, 106633. https://doi.org/10.1016/j.envint.2021.106633
59. Boulkhessaim, S., Gacem, A., Khan, S. H., Amari, A., Yadav, V. K., Harharah, H. N., Elkhaleefa, A. M., Yadav, K. K., Rather, S. U., Ahn, H. J., & Jeon, B. H. (2022). Emerging trends in the remediation of persistent organic pollutants using nanomaterials and related processes: A review. *Nanomaterials (Basel), 12*(13), 2148. https://doi.org/10.3390/nano12132148. PMID: 35807983; PMCID: PMC9268313.
60. Guo, W., Pan, B., Sakkiah, S., Yavas, G., Ge, W., Zou, W., Tong, W., & Hong, H. (2019). Persistent organic pollutants in food: Contamination sources, health effects and detection methods. *International Journal of Environmental Research and Public Health, 16*(22), 4361. https://doi.org/10.3390/ijerph16224361. PMID: 31717330; PMCID: PMC6888492.
61. Tran, N. Q. V., & Miyake, K. (2019). Chapter 23—Pharmacoepigenetics and toxicoepigenetics in neurodevelopmental disorders. In R. Cacabelos (Ed.), *Translational epigenetics, pharmacoepigenetics* (Vol. 10, pp. 711–719., ISSN 25425358, ISBN 9780128139394). Academic Press. https://doi.org/10.1016/B978-0-12-813939-4.00023-1
62. Alharbi, O. M. L., Basheer, A. A., Khattab, R. A., & Ali, I. (2018). Health and environmental effects of persistent organic pollutants. *Journal of Molecular Liquids, 263*, 442–453. https://doi.org/10.1016/j.molliq.2018.05.029
63. Kim, J., Sun, Q., Yue, Y., et al. (2016). 4,4'-dichlorodiphenyltrichloroethane (DDT) and 4,4′-dichlorodiphenyldichloroethylene (DDE) promote adipogenesis in 3T3-L1 adipocyte cell culture. *Pesticide Biochemistry and Physiology, 131*, 40–45. https://doi.org/10.1016/j.pestbp.2016.01.005
64. United States Environmental Protection Agency. DDT regulatory history: A brief survey (to 1975). https://www.epa.gov/archive/epa/aboutepa/ddt-regulatory-history-brief-survey-1975.html.
65. Sadasivaiah, S., Tozan, Y., & Breman, J. G. (2007). Dichlorodiphenyltrichloroethane (DDT) for indoor residual spraying in Africa: How can it be used for malaria control? In J. G. Breman, M. S. Alilio, & N. J. White (Eds.), *Defining and defeating the intolerable burden of Malaria III: Progress and perspectives: Supplement to Volume 77(6) of American Journal of Tropical Medicine and Hygiene*. American Society of Tropical Medicine and Hygiene. Available from: https://www.ncbi.nlm.nih.gov/books/NBK1724/
66. Harada, T., Takeda, M., Kojima, S., & Tomiyama, N. (2016). Toxicity and carcinogenicity of dichlorodiphenyltrichloroethane (DDT). *Toxicology Research, 32*(1), 21–33. https://doi.org/10.5487/TR.2016.32.1.021
67. Eskenazi, B., Chevrier, J., Rosas, L. G., Anderson, H. A., Bornman, M. S., Bouwman, H., Chen, A., Cohn, B. A., de Jager, C., Henshel, D. S., Leipzig, F., Leipzig, J. S., Lorenz, E. C., Snedeker, S. M., & Stapleton, D. (2009). The Pine River statement: Human health consequences of DDT use. *Environmental Health Perspectives, 117*(9), 1359–1367. https://doi.org/10.1289/ehp.11748
68. Young, R. A. (2014). Dioxins. In P. Wexler (Ed.), *Encyclopedia of toxicology (Third Edition)* (pp. 190–194., ISBN 9780123864550). Academic Press. https://doi.org/10.1016/B978-0-12-386454-3.00374-2
69. Elmore, S. A., & Boorman, G. A. (2013). Chapter 34—Environmental toxicologic pathology and human health. In W. M. Haschek, C. G. Rousseaux, & M. A. Wallig (Eds.), *Haschek and Rousseaux's Handbook of Toxicologic Pathology* (3rd ed., pp. 1029–1049., ISBN 9780124157590). Academic Press. https://doi.org/10.1016/B978-0-12-415759-0.00034-0

70. Avilla, M. N., Malecki, K. M. C., Hahn, M. E., Wilson, R. H., & Bradfield, C. A. (2020). The ah receptor: Adaptive metabolism, ligand diversity, and the Xenokine model. *Chemical Research in Toxicology, 33*(4), 860–879. https://doi.org/10.1021/acs.chemrestox.9b00476
71. Oesch-Bartlomowicz, B., Huelster, A., Wiss, O., et al. (2005). Aryl hydrocarbon receptor activation by cAMP vs. dioxin: Divergent signaling pathways. *Proceedings of the National Academy of Sciences of the United States of America, 102*(26), 9218–9223. https://doi.org/10.1073/pnas.0503488102
72. DeVito, M., Bokkers, B., van Duursen, M. B. M., et al. (2024). The 2022 World health organization reevaluation of human and mammalian toxic equivalency factors for polychlorinated dioxins, dibenzofurans and biphenyls. *Regulatory Toxicology and Pharmacology, 146*, 105525., ISSN 0273-2300. https://doi.org/10.1016/j.yrtph.2023.105525
73. Singh, N. P., Yang, X., Bam, M., Nagarkatti, M., & Nagarkatti, P. (2022). 2,3,7,8-Tetrachlorodibenzo-p-dioxin induces multigenerational alterations in the expression of microRNA in the thymus through epigenetic modifications. *PNAS Nexus, 2*(1), pgac290. https://doi.org/10.1093/pnasnexus/pgac290. Erratum in: PNAS Nexus. 2023 Jan 27;2(1):pgad016. doi: 10.1093/pnasnexus/pgad016.
74. Furue, M., & Tsuji, G. (2019). Chloracne and hyperpigmentation caused by exposure to hazardous aryl hydrocarbon receptor ligands. *International Journal of Environmental Research and Public Health, 16*(23), 4864. https://doi.org/10.3390/ijerph16234864
75. Committee to Review the Health Effects in Vietnam Veterans of Exposure to Herbicides (Tenth Biennial Update); Board on the Health of Select Populations; Institute of Medicine; National Academies of Sciences, Engineering, and Medicine. Veterans and Agent Orange: Update 2014. Washington (DC): National Academies Press (US); 2016 Mar 29. 7, Immune-System Disorders. Available from: https://www.ncbi.nlm.nih.gov/books/NBK356084/.
76. Kakeyama, M., & Tohyama, C. (2003). Developmental neurotoxicity of dioxin and its related compounds. *Industrial Health, 41*(3), 215–230. https://doi.org/10.2486/indhealth.41.215
77. Schjenken, J. E., Green, E. S., Overduin, T. S., Mah, C. Y., Russell, D. L., & Robertson, S. A. (2021). Endocrine disruptor compounds-a cause of impaired immune tolerance driving inflammatory disorders of pregnancy? *Frontiers in Endocrinology (Lausanne), 12*, 607539. https://doi.org/10.3389/fendo.2021.607539
78. Faiad, W., Soukkarieh, C., Murphy, D. J., & Hanano, A. (2022). Effects of dioxins on animal spermatogenesis: A state-of-the-art review. *Frontiers in Reproductive Health, 4*, 1009090. https://doi.org/10.3389/frph.2022.1009090
79. Aldeli, N., Murphy, D., & Hanano, A. (2024). Impact of dioxins on reproductive health in female mammals. *Frontiers in Toxicology, 6*, 1392257. https://doi.org/10.3389/ftox.2024.1392257
80. Turner, M. C., Cogliano, V., Guyton, K., Madia, F., Straif, K., Ward, E. M., & Schubauer-Berigan, M. K. (2023). Research recommendations for selected IARC-classified agents: Impact and lessons learned. *Environmental Health Perspectives, 131*(10), 105001. https://doi.org/10.1289/EHP12547
81. Boffetta, P., Mundt, K. A., Adami, H. O., Cole, P., & Mandel, J. S. (2011). TCDD and cancer: A critical review of epidemiologic studies. *Critical Reviews in Toxicology, 41*(7), 622–636. https://doi.org/10.3109/10408444.2011.560141
82. Tuyet-Hanh, T. T., Vu-Anh, L., Ngoc-Bich, N., & Tenkate, T. (2010). Environmental health risk assessment of dioxin exposure through foods in a dioxin hot spot-Bien Hoa City, Vietnam. *International Journal of Environmental Research and Public Health, 7*(5), 2395–2406. https://doi.org/10.3390/ijerph7052395
83. Wan, R., Wu, J., Guo, J., Qu, J., Li, L., & Tang, L. (2023). Baseline soil dioxin levels from sites where municipal solid waste incineration construction is planned throughout China: Characteristics, sources and risk assessment. *Sustainability, 15*(12), 9310. https://doi.org/10.3390/su15129310
84. Yim, G., Minatoya, M., Kioumourtzoglou, M.-A., Bellavia, A., Weisskopf, M., Ikeda-Araki, A., Miyashita, C., & Kishi, R. (2022). The associations of prenatal exposure to

dioxins and polychlorinated biphenyls with neurodevelopment at 6 Months of age: Multi-pollutant approaches. *Environmental Research, 209*, 112757., ISSN 0013-9351. https://doi.org/10.1016/j.envres.2022.112757

85. Long, M., Wielsøe, M., & Bonefeld-Jørgensen, E. C. (2022). Dioxin-like activity in pregnant women and indices of fetal growth: The ACCEPT birth cohort. *Toxics, 10*(1), 26. https://doi.org/10.3390/toxics10010026
86. Kortenkamp, A., Scholze, M., Ermler, S., Priskorn, L., Jørgensen, N., Andersson, A.-M., & Frederiksen, H. (2022). Combined exposures to bisphenols, polychlorinated dioxins, paracetamol, and phthalates as drivers of deteriorating semen quality. *Environment International, 165*, 107322. https://doi.org/10.1016/j.envint.2022.107322
87. Ring, C., Blanchette, A., Klaren, W. D., Fitch, S., Haws, L., Wheeler, M. W., DeVito, M., Walker, N., & Daniele Wikoff, A. (2023). multi-tiered hierarchical Bayesian approach to derive toxic equivalency factors for dioxin-like compounds. *Regulatory Toxicology and Pharmacology, 143*, 105464., ISSN 0273-2300. https://doi.org/10.1016/j.yrtph.2023.105464
88. ATSDR. (2005). *Toxicological profile for alpha-, beta-, gamma-, and delta-hexachlorocyclohexane*. Agency for Toxic Substances and Disease Registry. http://www.atsdr.cdc.gov/toxprofiles/tp43.html
89. National Center for Biotechnology Information. PubChem Compound Summary for CID 727,Hexachlorocyclohexane. Retrieved June 13, 2024, from https://pubchem.ncbi.nlm.nih.gov/compound/Hexachlorocyclohexane
90. Kumar, D., & Pannu, R. (2018). Perspectives of lindane (γ-hexachlorocyclohexane) biodegradation from the environment: A review. *Bioresources and Bioprocessing, 5*, 29. https://doi.org/10.1186/s40643-018-0213-9
91. Zhang, W., Lin, Z., Pang, S., Bhatt, P., & Chen, S. (2020). Insights into the biodegradation of Lindane (γ-Hexachlorocyclohexane) using a microbial system. *Frontiers in Microbiology, 11*, 522. https://doi.org/10.3389/fmicb.2020.00522
92. Nienow, A. M., Bezares-Cruz, J. C., Poyer, I. C., Hua, I., & Jafvert, C. T. (2008). Hydrogen peroxide-assisted UV photodegradation of Lindane. *Chemosphere, 72*(11), 1700–1705. https://doi.org/10.1016/j.chemosphere.2008.04.080
93. Kaur, I., Kumar Gaur, V., Rishi, S., et al. (2023). Deciphering the kinetics and pathway of lindane biodegradation by novel soil ascomycete fungi for its implication in bioremediation. *Bioresource Technology, 387*, 129581. https://doi.org/10.1016/j.biortech.2023.129581
94. Nagata, K., & Narahashi, T. (1995). Multiple actions of dieldrin and lindane on the GABAA receptor-chloride channel complex of rat dorsal root ganglion neurons. *Pesticide Science, 44*(1), 1–7. https://doi.org/10.1002/ps.2780440102
95. Jenson, L. J., Anderson, T. D., & Bloomquist, J. R. (2016). Insecticide sensitivity of native chloride and sodium channels in a mosquito cell line. *Pesticide Biochemistry and Physiology, 130*, 59–64. https://doi.org/10.1016/j.pestbp.2015.11.012
96. Cattabeni, F., Pastorello, M. C., & Eli, M. (1983). Convulsions induced by lindane and the involvement of the GABAergic system. In C. M. Chambers, P. L. Chambers, & S. Gitter (Eds.), *Toxicology in the use, misuse, and abuse of food, drugs, and chemicals. Archives of toxicology* (Vol. 6). Springer. https://doi.org/10.1007/978-3-642-69083-9_44
97. Qi, S. Y., Xu, X. L., Ma, W. Z., Deng, S. L., Lian, Z. X., & Yu, K. (2022). Effects of organochlorine pesticide residues in maternal body on infants. *Frontiers in Endocrinology (Lausanne), 13*, 890307. https://doi.org/10.3389/fendo.2022.890307
98. Videla, L. A., Barros, S. B., & Junqueira, V. B. (1990). Lindane-induced liver oxidative stress. *Free Radical Biology & Medicine, 9*(2), 169–179. https://doi.org/10.1016/0891-5849(90)90120-8
99. Dunier, M., Vergnet, C., Siwicki, A. K., & Verlhac, V. (1995). Effect of lindane exposure on rainbow trout (Oncorhynchus mykiss) immunity. IV. Prevention of nonspecific and specific immunosuppression by dietary vitamin C (ascorbate-2-polyphosphate). *Ecotoxicology and Environmental Safety, 30*(3), 259–268. https://doi.org/10.1006/eesa.1995.1029

100. Alsen, M., Vasan, V., Genden, E. M., Sinclair, C., & van Gerwen, M. (2022). Correlation between Lindane use and the incidence of thyroid cancer in the United States: An ecological study. *International Journal of Environmental Research and Public Health, 19*(20), 13158. https://doi.org/10.3390/ijerph192013158
101. Kancharla, S., Alexandridis, P., & Tsianou, M. (2022). Sequestration of per- and polyfluoroalkyl substances (PFAS) by adsorption: Surfactant and surface aspects. *Current Opinion in Colloid & Interface Science, 58*, 101571., ISSN 1359-0294. https://doi.org/10.1016/j.cocis.2022.101571
102. Meegoda, J. N., Bezerra de Souza, B., Casarini, M. M., & Kewalramani, J. A. (2022). A review of PFAS destruction technologies. *International Journal of Environmental Research and Public Health, 19*(24), 16397. https://doi.org/10.3390/ijerph192416397
103. Grgas, D., Petrina, A., Štefanac, T., Bešlo, D., & Landeka, D. T. (2023). A review: Per- and Polyfluoroalkyl substances-biological degradation. *Toxics, 11*(5), 446. https://doi.org/10.3390/toxics11050446
104. Fenton, S. E., Ducatman, A., Boobis, A., DeWitt, J. C., Lau, C., Ng, C., Smith, J. S., & Roberts, S. M. (2021). Per- and Polyfluoroalkyl substance toxicity and human health review: Current state of knowledge and strategies for informing future research. *Environmental Toxicology and Chemistry, 40*(3), 606–630. https://doi.org/10.1002/etc.4890
105. Lind, P. M., & Lind, L. (2020). Are persistent organic pollutants linked to lipid abnormalities, atherosclerosis and cardiovascular disease? A Review. *Journal of Lipid and Atherosclerosis, 9*(3), 334–348. https://doi.org/10.12997/jla.2020.9.3.334. Epub 2020 Aug 5. PMID: 33024729; PMCID: PMC7521972.
106. Zeng, Z., Song, B., Xiao, R., et al. (2019). Assessing the human health risks of perfluorooctane sulfonate by in vivo and in vitro studies. *Environment International, 126*, 598–610. https://doi.org/10.1016/j.envint.2019.03.002
107. Brunn, H., Arnold, G., Körner, W., et al. (2023). PFAS: Forever chemicals—Persistent, bioaccumulative and mobile. Reviewing the status and the need for their phase out and remediation of contaminated sites. *Environmental Sciences Europe, 35*, 20. https://doi.org/10.1186/s12302-023-00721-8
108. Chen, G., Liu, S., Shi, Q., Gan, J., Jin, B., Men, Y., & Liu, H. (2022). Hydrogen-polarized vacuum ultraviolet photolysis system for enhanced destruction of perfluoroalkyl substances. *Journal of Hazardous Materials Letters, 3*, 100072., ISSN 2666-9110. https://doi.org/10.1016/j.hazl.2022.100072
109. Panieri, E., Baralic, K., Djukic-Cosic, D., Buha Djordjevic, A., & Saso, L. (2022). PFAS molecules: A major concern for the human health and the environment. *Toxics, 10*(2), 44. https://doi.org/10.3390/toxics10020044
110. González-Alvarez, M. E., Antwi-Boasiako, C., & Keating, A. F. (2024). Effects of per- and Polyfluoroalkylated substances on female reproduction. *Toxics, 12*, 455. https://doi.org/10.3390/toxics12070455
111. Estefanía González-Alvarez, M., Severin, A., Sayadi, M., & Keating, A. F. (2022). PFOA-induced Ovotoxicity differs between lean and obese mice with impacts on ovarian reproductive and DNA damage sensing and repair proteins. *Toxicological sciences : an official journal of the Society of Toxicology, 190*(2), 173–188. https://doi.org/10.1093/toxsci/kfac104
112. Kancharla, S., Choudhary, A., Davis, R. T., Dong, D., Bedrov, D., Tsianou, M., & Alexandridis, P. (2022). GenX in water: Interactions and self-assembly. *Journal of Hazardous Materials, 428*, 128137., ISSN 0304-3894. https://doi.org/10.1016/j.jhazmat.2021.128137
113. Meng, L., Zhou, B., Liu, H., Chen, Y., Yuan, R., Chen, Z., Luo, S., & Chen, H. (2024). Advancing toxicity studies of per- and poly-fluoroalkyl substances (pfass) through machine learning: Models, mechanisms, and future directions. *Science of The Total Environment, 946*, 174201., ISSN 0048-9697. https://doi.org/10.1016/j.scitotenv.2024.174201
114. Zhu, L., Hajeb, P., Fauser, P., & Vorkamp, K. (2023). Endocrine disrupting chemicals in indoor dust: A review of temporal and spatial trends, and human exposure. *Science of The Total Environment, 874*, 162374., ISSN 0048-9697. https://doi.org/10.1016/j.scitotenv.2023.162374

115. Toxicological profile for polybrominated diphenyl ethers (PBDEs). Atlanta (GA): Agency for Toxic Substances and Disease Registry (US); 2017 Mar. 4, CHEMICAL AND PHYSICAL INFORMATION. Available from: https://www.ncbi.nlm.nih.gov/books/NBK592324/.
116. Wagner, S., & Schlummer, M. (2020). Legacy additives in a circular economy of plastics: Current dilemma, policy analysis, and emerging countermeasures. *Resources, Conservation and Recycling, 158*, 104800., ISSN 0921-3449. https://doi.org/10.1016/j.resconrec.2020.104800
117. Turner, A. (2022). PBDEs in the marine environment: Sources, pathways and the role of microplastics. *Environmental Pollution, 301*, 118943., ISSN 0269-7491. https://doi.org/10.1016/j.envpol.2022.118943
118. Marquès, M., Nadal, M., & Domingo, J. L. (2022). Human exposure to polybrominated diphenyl ethers (PBDEs) through the diet: An update of the scientific literature. *Food and Chemical Toxicology, 167*, 113322., ISSN 0278-6915. https://doi.org/10.1016/j.fct.2022.113322
119. EFSA Panel on Contaminants in the Food Chain (CONTAM), Schrenk, D., Bignami, M., Bodin, L., Chipman, J. K., Del Mazo, J., Grasl-Kraupp, B., Hogstrand, C., Ron Hoogenboom, L., Leblanc, J. C., Nebbia, C. S., Nielsen, E., Ntzani, E., Petersen, A., Sand, S., Schwerdtle, T., Wallace, H., Benford, D., Fürst, P., Hart, A., et al. (2024). Update of the risk assessment of polybrominated diphenyl ethers (PBDEs) in food. *EFSA Journal European Food Safety Authority, 22*(1), e8497. https://doi.org/10.2903/j.efsa.2024.8497
120. Ohoro, C. R., Adeniji, A. O., Okoh, A. I., & Okoh, O. O. (2021). Polybrominated diphenyl ethers in the environmental systems: A review. *Journal of Environmental Health Science & Engineering, 19*(1), 1229–1247. https://doi.org/10.1007/s40201-021-00656-3
121. Renzelli, V., Gallo, M., Morviducci, L., Marino, G., Ragni, A., Tuveri, E., Faggiano, A., Mazzilli, R., Natalicchio, A., Zatelli, M. C., Montagnani, M., Fogli, S., Giuffrida, D., Argentiero, A., Danesi, R., D'Oronzo, S., Gori, S., Franchina, T., Russo, A., Monami, M., et al. (2023). Polybrominated diphenyl ethers (PBDEs) and human health: effects on metabolism, diabetes and cancer. *Cancers, 15*(17), 4237. https://doi.org/10.3390/cancers15174237
122. Herbstman, J. B., & Mall, J. K. (2014). Developmental exposure to polybrominated diphenyl ethers and neurodevelopment. *Current Environmental Health Reports, 1*(2), 101–112. https://doi.org/10.1007/s40572-014-0010-3
123. Qin, W. P., Li, C. H., Guo, L. H., Ren, X. M., & Zhang, J. Q. (2019). Binding and activity of polybrominated diphenyl ether sulfates to thyroid hormone transport proteins and nuclear receptors. *Environmental Science. Processes & Impacts, 21*(6), 950–956. https://doi.org/10.1039/c9em00095j
124. Rehman, G., Kumari, N., Bano, F., & Tyagi, R. K. (2023). Thyroid hormone receptor beta: Relevance in human health and diseases. *Endocrine and Metabolic Science, 13*, 100144., ISSN 2666-3961. https://doi.org/10.1016/j.endmts.2023.100144
125. Kozlova, E. V., Chinthirla, B. D., Bishay, A. E., Pérez, P. A., Denys, M. E., Krum, J. M., DiPatrizio, N. V., & Currás-Collazo, M. C. (2023). Glucoregulatory disruption in male mice offspring induced by maternal transfer of endocrine disrupting brominated flame retardants in DE-71. *Frontiers in Endocrinology, 14*, 1049708. https://doi.org/10.3389/fendo.2023.1049708
126. Arowolo, O., Pilsner, J. R., Sergeyev, O., & Suvorov, A. (2022). Mechanisms of male reproductive toxicity of polybrominated diphenyl ethers. *International Journal of Molecular Sciences, 23*(22), 14229. https://doi.org/10.3390/ijms232214229
127. Reddam, A., Sjödin, A., Cowell, W., Jones, R., Wang, S., Perera, F., Herbstman, J. B., & Kupsco, A. (2023). Prenatal exposure to polybrominated diphenyl ethers and birth outcomes. *Environmental Research, 216*(Pt 4), 114830. https://doi.org/10.1016/j.envres.2022.114830
128. Wang, Y., Wang, X., Sui, S., & Liu, Z. (2023). Endocrine disrupting and carcinogenic effects of decabromodiphenyl ether. *Frontiers in Endocrinology, 14*, 1183815. https://doi.org/10.3389/fendo.2023.1183815
129. Anzalone, G., Moscato, M., Montalbano, A. M., Albano, G. D., Gagliardo, R., Marchese, R., Fucarino, A., Nigro, C. L., Drago, G., & Profita, M. (2021). PBDEs affect inflammatory and

oncosuppressive mechanisms via the EZH2 methyltransferase in airway epithelial cells. *Life Sciences, 282*, 119827., ISSN 0024-3205. https://doi.org/10.1016/j.lfs.2021.119827
130. Rao, S. K. P., Szabo, D. T., Stoker, T. E., & Fenton, S. E. (2011). Chapter 40—Brominated flame retardants. In R. C. Gupta (Ed.), *Reproductive and developmental toxicology* (pp. 523–541., ISBN 9780123820327). Academic Press. https://doi.org/10.1016/B978-0-12-382032-7.10040-2
131. Tudi, M., Daniel Ruan, H., Wang, L., Lyu, J., Sadler, R., Connell, D., Chu, C., & Phung, D. T. (2021). Agriculture development, pesticide application and its impact on the environment. *International Journal of Environmental Research and Public Health, 18*(3), 1112. https://doi.org/10.3390/ijerph18031112
132. Syafrudin, M., Kristanti, R. A., Yuniarto, A., Hadibarata, T., Rhee, J., Al-Onazi, W. A., Algarni, T. S., Almarri, A. H., & Al-Mohaimeed, A. M. (2021). Pesticides in drinking water-a review. *International Journal of Environmental Research and Public Health, 18*(2), 468. https://doi.org/10.3390/ijerph18020468
133. Intisar, A., Ramzan, A., Sawaira, T., Kareem, A. T., Hussain, N., Din, M. I., Bilal, M., & Iqbal, H. M. N. (2022). Occurrence, toxic effects, and mitigation of pesticides as emerging environmental pollutants using robust nanomaterials—A review. *Chemosphere, 293*, 133538., ISSN 0045-6535. https://doi.org/10.1016/j.chemosphere.2022.133538
134. Ali, A., & AlHussaini, K. I. (2024). Pesticides: Unintended impact on the hidden world of gut microbiota. *Metabolites, 14*(3), 155. https://doi.org/10.3390/metabo14030155
135. Moebus, S., & Boedeker, W. (2021). Case fatality as an indicator for the human toxicity of pesticides-a systematic scoping review on the availability and variability of severity indicators of pesticide poisoning. *International Journal of Environmental Research and Public Health, 18*(16), 8307. https://doi.org/10.3390/ijerph18168307
136. Guerrero Ramírez, J. R., Ibarra Muñoz, L. A., Balagurusamy, N., Frías Ramírez, J. E., Alfaro Hernández, L., & Carrillo Campos, J. (2023). Microbiology and biochemistry of pesticides biodegradation. *International Journal of Molecular Sciences, 24*(21), 15969. https://doi.org/10.3390/ijms242115969
137. Sabarwal, A., Kumar, K., & Singh, R. P. (2018). Hazardous effects of chemical pesticides on human health–cancer and other associated disorders. *Environmental Toxicology and Pharmacology, 63*, 103–114., ISSN 1382-6689. https://doi.org/10.1016/j.etap.2018.08.018
138. Jayaraj, R., Megha, P., & Sreedev, P. (2016). Organochlorine pesticides, their toxic effects on living organisms and their fate in the environment. *Interdisciplinary Toxicology, 9*(3–4), 90–100. https://doi.org/10.1515/intox-2016-0012
139. Tsygankov, V. Y., Lukyanova, O. N., & Boyarova, M. D. (2018). Organochlorine pesticide accumulation in seabirds and marine mammals from the Northwest Pacific. *Marine Pollution Bulletin, 128*, 208–213. https://doi.org/10.1016/j.marpolbul.2018.01.027
140. Martyniuk, C. J., Mehinto, A. C., & Denslow, N. D. (2020). Organochlorine pesticides: Agrochemicals with potent endocrine-disrupting properties in fish. *Molecular and Cellular Endocrinology, 507*, 110764. https://doi.org/10.1016/j.mce.2020.110764
141. Richardson, J. R., Fitsanakis, V., Westerink, R. H. S., & Kanthasamy, A. G. (2019). Neurotoxicity of pesticides. *Acta Neuropathologica, 138*(3), 343–362. https://doi.org/10.1007/s00401-019-02033-9
142. Garud, A., Pawar, S., Patil, M. S., Kale, S. R., & Patil, S. (2024). A scientific review of pesticides: Classification, toxicity, health effects, sustainability, and environmental impact. *Cureus, 16*(8), e67945. https://doi.org/10.7759/cureus.67945
143. Ray, S., & Shaju, S. T. (2023). Bioaccumulation of pesticides in fish resulting toxicities in humans through food chain and forensic aspects. *Environmental Analysis, Health and Toxicology, 38*(3), e2023017. https://doi.org/10.5620/eaht.2023017
144. Kaur, N., Swain, S. K., Banerjee, B. D., Sharma, T., & Krishnalata, T. (2019). Organochlorine pesticide exposure as a risk factor for breast cancer in young Indian women: A case-control study. *South Asian Journal of Cancer, 8*(4), 212–214. https://doi.org/10.4103/sajc.sajc_427_18

145. Iribarne-Durán, L. M., Castillero-Rosales, I., Peinado, F. M., Artacho-Cordón, F., Molina-Molina, J. M., Medianero, E., Nicolás-Delgado, S. I., Sánchez-Pinzón, L., Núñez-Samudio, V., Vela-Soria, F., Olea, N., & Alvarado-González, N. E. (2024). Placental concentrations of xenoestrogenic organochlorine pesticides and polychlorinated biphenyls and assessment of their xenoestrogenicity in the PA-MAMI mother-child cohort. *Environmental Research, 241*, 117622., ISSN 0013-9351. https://doi.org/10.1016/j.envres.2023.117622
146. Mekonen, S., Ambelu, A., Wondafrash, M., et al. (2021). Exposure of infants to organochlorine pesticides from breast milk consumption in southwestern Ethiopia. *Scientific Reports, 11*, 22053. https://doi.org/10.1038/s41598-021-01656-x
147. Klincic, D., Herceg Romanic, S., Matek Saric, M., Grzunov, J., & Dukic, B. (2014). Polychlorinated biphenyls and organochlorine pesticides in human milk samples from two regions in Croatia. *Environmental Toxicology and Pharmacology, 37*, 543–552. https://doi.org/10.1016/j.etap.2014.01.009
148. Chen, M. W., Santos, H. M., Que, D. E., Gou, Y. Y., Tayo, L. L., Hsu, Y. C., Chen, Y. B., Chen, F. A., Chao, H. R., & Huang, K. L. (2018). Association between organochlorine pesticide levels in breast milk and their effects on female reproduction in a Taiwanese population. *International Journal of Environmental Research and Public Health, 15*(5), 931. https://doi.org/10.3390/ijerph15050931
149. Lin, J. Y., & Yin, R. X. (2023). Exposure to endocrine-disrupting chemicals and type 2 diabetes mellitus in later life. *Exposure and Health, 15*, 199–229. https://doi.org/10.1007/s12403-022-00486-0
150. Saunders, L., Kadhel, P., Costet, N., et al. (2014). Hypertensive disorders of pregnancy and gestational diabetes mellitus among French Caribbean women chronically exposed to chlordecone. *Environment International, 68*, 171–176. https://doi.org/10.1016/j.envint.2014.03.024
151. Langer, P., Ukropec, J., Kocan, A., Drobna, B., Radikova, Z., Huckova, M., et al. (2014). Obesogenic and diabetogenic impact of high organochlorine levels (HCB, p,p′-DDE, PCBs) on inhabitants in the highly polluted Eastern Slovakia. *Endocrine Regulations, 48*, 17–24. https://doi.org/10.4149/endo_2014_01_17
152. Leemans, M., Couderq, S., Demeneix, B., & Fini, J. B. (2019). Pesticides with potential thyroid hormone-disrupting effects: A review of recent data. *Frontiers in Endocrinology, 10*, 743. https://doi.org/10.3389/fendo.2019.00743
153. Costa, L. G. (2015). The neurotoxicity of organochlorine and pyrethroid pesticides. *Handbook of Clinical Neurology, 131*, 135–148. https://doi.org/10.1016/B978-0-444-62627-1.00009-3
154. Hong, S., Hwang, J., Kim, J. Y., Shin, K. S., & Kang, S. J. (2014). Heptachlor induced nigral dopaminergic neuronal loss and parkinsonism-like movement deficits in mice. *Experimental & Molecular Medicine, 46*(2), e80. https://doi.org/10.1038/emm.2014.12
155. Brennan, E., Butler, A. E., Nandakumar, M., Drage, D. S., Sathyapalan, T., & Atkin, S. L. (2023). Association between organochlorine pesticides and vitamin D in female subjects. *Biomedicine, 11*(5), 1451. https://doi.org/10.3390/biomedicines11051451
156. Su, Y., Dai, Y., Lin, Y., Gao, X., Han, Y., & Zhao, B. (2012). Serum organochlorine pesticide residues and risk of gallstone disease: A case-control study in Xiamen. *Annals of Epidemiology, 22*(8), 592–597. https://doi.org/10.1016/j.annepidem.2012.05.002
157. Gheidarloo, M., Kelishadi, R., Hovsepian, S., Keikha, M., & Hashemipour, M. (2020). The association between prenatal exposure to organochlorine compounds and neonatal thyroid hormone levels: A systematic review. *Journal of Pediatric Endocrinology & Metabolism, 33*(1), 21–33. https://doi.org/10.1515/jpem-2019-0336
158. Hamid, E. R. A., Sharaf, N. E., Ahmed, H. H., Ahmed, A., & Mossa, A. T. H. (2020). In utero exposure to organochlorine pesticide residues and their potential impact on birth outcomes and fetal gender. *Environemental Science and Pollution Research, 27*, 33703–33711. https://doi.org/10.1007/s11356-020-09411-x
159. Richardson, J. R., Caudle, W. M., Wang, M. Z., Dean, E. D., Pennell, K. D., & Miller, G. W. (2008). Developmental heptachlor exposure increases susceptibility of dopamine neurons to N-Methyl-4-Phenyl-1,2,3,6-Tetrahydropyridine (MPTP)in a gender-specific manner. *Neurotoxicology, 29*, 855–863. https://doi.org/10.1016/j.neuro.2008.05.007

160. Kumar, R., & McEwan, I. J. (2024). Chapter 6—Endocrine disruptors: the enemy without. In R. Kumar & I. J. McEwan (Eds.), *Steroid hormone receptors in health and disease* (pp. 107–123., ISBN 9780323911412). Academic Press. https://doi.org/10.1016/B978-0-323-91141-2.00002-7
161. Silver, S. R., Bertke, S. J., Hines, C. J., Alavanja, M. C., Hoppin, J. A., Lubin, J. H., Rusiecki, J. A., Sandler, D. P., & Beane Freeman, L. E. (2015). Cancer incidence and metolachlor use in the agricultural health study: An update. *International Journal of Cancer, 137*(11), 2630–2643. https://doi.org/10.1002/ijc.29621
162. Badou-Jeremie Kouame, K., Savin, M. C., Willett, C. D., Bertucci, M. B., Butts, T. R., Grantz, E., & Roma-Burgos, N. (2022). S-metolachlor persistence in soil as influenced by within-season and inter-annual herbicide use. *Environmental Advances, 9*, 100318., ISSN 2666-7657. https://doi.org/10.1016/j.envadv.2022.100318
163. Fingler, S., Mendaš, G., Dvoršćak, M., Stipičević, S., Vasilić, Ž., & Drevenkar, V. (2017). Herbicide micropollutants in surface, ground and drinking waters within and near the area of Zagreb, Croatia. *Environmental Science and Pollution Research International, 24*(12), 11017–11030. https://doi.org/10.1007/s11356-016-7074-6
164. Marín-Benito, J. M., Herrero-Hernández, E., Ordax, J. M., Sánchez-Martín, M. J., & Rodríguez-Cruz, M. S. (2021). The role of two organic amendments to modify the environmental fate of S-metolachlor in agricultural soils. *Environmental Research, 195*, 110871., ISSN 0013-9351. https://doi.org/10.1016/j.envres.2021.110871
165. Mugudamani, I., Oke, S. A., Gumede, T. P., & Senbore, S. (2023). Herbicides in water sources: Communicating potential risks to the population of Mangaung metropolitan municipality, South Africa. *Toxics, 11*(6), 538. https://doi.org/10.3390/toxics11060538
166. Hamid, N., Junaid, M., & Pei, D.-S. (2021). Combined toxicity of endocrine-disrupting chemicals: A review. *Ecotoxicology and Environmental Safety, 215*, 112136., ISSN 0147-6513. https://doi.org/10.1016/j.ecoenv.2021.112136
167. Rusiecki, J. A., Hou, L., Lee, W. J., et al. (2006). Cancer incidence among pesticide applicators exposed to metolachlor in the Agricultural Health Study. *International Journal of Cancer, 118*(12), 3118–3123. https://doi.org/10.1002/ijc.21758
168. Suekane, A., Edamoto, M., Kubo, K., Abe, T., Nakatsutsumi, K., & Ameda, T. (2022). Metolachlor poisoning with lactic acidosis improved by thiamine administration: A case report. *The American Journal of Case Reports, 23*, e937873. https://doi.org/10.12659/AJCR.937873
169. Chang, J., Fang, W., Chen, L., Zhang, P., Zhang, G., Zhang, H., Liang, J., Wang, Q., & Ma, W. (2022). Toxicological effects, environmental behaviors and remediation technologies of herbicide atrazine in soil and sediment: A comprehensive review. *Chemosphere, 307, Part 3*, 136006., ISSN 0045-6535. https://doi.org/10.1016/j.chemosphere.2022.136006
170. Song, Y., Jia, Z. C., Chen, J. Y., Hu, J. X., & Zhang, L. S. (2014). Toxic effects of atrazine on reproductive system of male rats. *Biomedical and Environmental Sciences, 27*(4), 281–288. https://doi.org/10.3967/bes2014.050
171. Davis, L. K., Murr, A. S., Best, D. S., Fraites, M. J., Zorrilla, L. M., Narotsky, M. G., Stoker, T. E., Goldman, J. M., & Cooper, R. L. (2011). The effects of prenatal exposure to atrazine on pubertal and postnatal reproductive indices in the female rat. *Reproductive Toxicology, 32*, 43–51. https://doi.org/10.1016/j.reprotox.2011.04.004
172. Fucic, A., Duca, R. C., Galea, K. S., Maric, T., Garcia, K., Bloom, M. S., Andersen, H. R., & Vena, J. E. (2021). Reproductive health risks associated with occupational and environmental exposure to pesticides. *International Journal of Environmental Research and Public Health, 18*(12), 6576. https://doi.org/10.3390/ijerph18126576
173. Rinsky, J. L., Hopenhayn, C., Golla, V., Browning, S., & Bush, H. M. (2012). Atrazine exposure in public drinking water and preterm birth. *Public Health Reports*. (Washington, D.C. : 1974)*, 127*(1), 72–80. https://doi.org/10.1177/003335491212700108
174. Wirbisky, S. E., & Freeman, J. L. (2015). Atrazine exposure and reproductive dysfunction through the hypothalamus-pituitary-gonadal (HPG) Axis. *Toxics, 3*(4), 414–450. https://doi.org/10.3390/toxics3040414

175. Grasselli, F., Bussolati, S., Ramoni, R., Grolli, S., & Basini, G. (2018). Simazine, a triazine herbicide, disrupts swine granulosa cell functions. *Animal Reproduction, 15*(1), 3–11. https://doi.org/10.21451/1984-3143-2017-AR960
176. Park, S., Kim, S., Jin, H., Lee, K., & Bae, J. (2014). Impaired development of female mouse offspring maternally exposed to simazine. *Environmental Toxicology and Pharmacology, 38*(3), 845–851, ISSN 1382-6689. https://doi.org/10.1016/j.etap.2014.10.008
177. Zorrilla, L. M., Gibson, E. K., & Stoker, T. E. (2010). The effects of simazine, a chlorotriazine herbicide, on pubertal development in the female Wistar rat. *Reproductive Toxicology, 29*(4), 393–400., ISSN 0890-6238. https://doi.org/10.1016/j.reprotox.2010.03.010
178. Gonsioroski, A., Mourikes, V. E., & Flaws, J. A. (2020). Endocrine disruptors in water and their effects on the reproductive system. *International Journal of Molecular Sciences, 21*(6), 1929. https://doi.org/10.3390/ijms21061929
179. Simon-Delso, N., Amaral-Rogers, V., Belzunces, L. P., et al. (2015). Systemic insecticides (neonicotinoids and fipronil): Trends, uses, mode of action and metabolites. *Environmental Science and Pollution Research, 22*, 5–34. https://doi.org/10.1007/s11356-014-3470-y
180. Randhawa, J. S. (2024). Microbial-assisted remediation approach for neonicotinoids from polluted environment. *Bulletin of the National Research Centre, 48*, 70. https://doi.org/10.1186/s42269-024-01227-w
181. Wrobel, S. A., Bury, D., Hayen, H., Koch, H. M., Brüning, T., & Käfferlein, H. U. (2022). Human metabolism and urinary excretion of seven neonicotinoids and neonicotinoid-like compounds after controlled oral dosages. *Archives of Toxicology, 96*(1), 121–134. https://doi.org/10.1007/s00204-021-03159-0
182. Houchat, J.-N., Cartereau, A., Le Mauff, A., Taillebois, E., & Thany, S. H. (2020). An overview on the effect of neonicotinoid insecticides on mammalian cholinergic functions through the activation of neuronal nicotinic acetylcholine receptors. *International Journal of Environmental Research and Public Health, 17*, 3222. https://doi.org/10.3390/ijerph17093222
183. Cimino, A. M., Boyles, A. L., Thayer, K. A., & Perry, M. J. (2017). Effects of neonicotinoid pesticide exposure on human health: A systematic review. *Environmental Health Perspectives, 125*(2), 155–162. https://doi.org/10.1289/EHP515
184. Costas-Ferreira, C., & Faro, L. R. F. (2021). Neurotoxic effects of neonicotinoids on mammals: What is there beyond the activation of nicotinic acetylcholine receptors?-a systematic review. *International Journal of Molecular Sciences, 22*(16), 8413. https://doi.org/10.3390/ijms22168413
185. Pelkonen, O., Abass, K., Parra Morte, J. M., Panzarea, M., Testai, E., Rudaz, S., Louisse, J., Gundert-Remy, U., Wolterink, G., Jean-Lou, C. M. D., Coecke, S., & Bernasconi, C. (2023). Metabolites in the regulatory risk assessment of pesticides in the EU. *Frontiers in Toxicology, 5*, 1304885. https://doi.org/10.3389/ftox.2023.1304885
186. Selvam, V., & Srinivasan, S. (2019). Neonicotinoid poisoning and management. *Indian Journal of Critical Care Medicine: Peer-reviewed, Official Publication of Indian Society of Critical Care Medicine, 23*(Suppl 4), S260–S262. https://doi.org/10.5005/jp-journals-10071-23308
187. Mariani, A., Comolli, D., Fanelli, R., Forloni, G., & De Paola, M. (2024). Neonicotinoid pesticides affect developing neurons in experimental mouse models and in human induced pluripotent stem cell (iPSC)-derived neural cultures and organoids. *Cells, 13*(15), 1295. https://doi.org/10.3390/cells13151295
188. Tariba Lovaković, B., Kašuba, V., Sekovanić, A., Orct, T., Jančec, A., & Pizent, A. (2021). Effects of sub-chronic exposure to Imidacloprid on reproductive organs of adult male rats: Antioxidant state, DNA damage, and levels of essential elements. *Antioxidants (Basel, Switzerland), 10*(12), 1965. https://doi.org/10.3390/antiox10121965
189. Mourikes, V. E., Santacruz Márquez, R., Deviney, A., Neff, A. M., Laws, M. J., & Flaws, J. A. (2023). Imidacloprid and its bioactive metabolite, Desnitro-Imidacloprid, differentially affect ovarian antral follicle growth, morphology, and hormone synthesis in vitro. *Toxics, 11*, 349. https://doi.org/10.3390/toxics11040349

190. Bal, R., Türk, G., Yılmaz, Ö., et al. (2012). Effects of clothianidin exposure on sperm quality, testicular apoptosis and fatty acid composition in developing male rats. *Cell Biology and Toxicology, 28*, 187–200. https://doi.org/10.1007/s10565-012-9215-0
191. Landrigan, P. J., Raps, H., Cropper, M., Bald, C., Brunner, M., Canonizado, E. M., Charles, D., Chiles, T. C., Donohue, M. J., Enck, J., Fenichel, P., Fleming, L. E., Ferrier-Pages, C., Fordham, R., Gozt, A., Griffin, C., Hahn, M. E., Haryanto, B., Hixson, R., Ianelli, H., et al. (2023). The Minderoo-Monaco commission on plastics and human health. *Annals of Global Health, 89*(1), 23. https://doi.org/10.5334/aogh.4056
192. Ali, N., Katsouli, J., Marczylo, E. L., Gant, T. W., Wright, S., Bernardino, J., & de la Serna. (2024). The potential impacts of micro-and-nano plastics on various organ systems in humans. *eBioMedicine, 99*, 104901., ISSN 2352-3964. https://doi.org/10.1016/j.ebiom.2023.104901
193. Ray, S. S., Lee, H. K., Huyen, D. T. T., Chen, S.-S., & Kwon, Y.-N. (2022). Microplastics waste in environment: A perspective on recycling issues from PPE kits and face masks during the COVID-19 pandemic. *Environmental Technology & Innovation, 26*, 102290., ISSN 2352-1864. https://doi.org/10.1016/j.eti.2022.102290
194. Lee, S. E., Yi, Y., Moon, S., Yoon, H., & Park, Y. S. (2022). Impact of micro- and Nanoplastics on mitochondria. *Metabolites, 12*(10), 897. https://doi.org/10.3390/metabo12100897
195. Buzenchi Proca, T. M., Solcan, C., & Solcan, G. (2024). Neurotoxicity of some environmental pollutants to zebrafish. *Life (Basel, Switzerland), 14*(5), 640. https://doi.org/10.3390/life14050640
196. Gündoğdu, S., Bour, A., Köşker, A. R., Walther, B. A., Napierska, D., Mihai, F.-C., Syberg, K., Hansen, S. F., & Walker, T. R. (2024). Review of microplastics and chemical risk posed by plastic packaging on the marine environment to inform the Global Plastics Treaty. *Science of The Total Environment, 946*, 174000., ISSN 0048-9697. https://doi.org/10.1016/j.scitotenv.2024.174000
197. Belmaker, I., Anca, E. D., Rubin, L. P., Magen-Molho, H., Miodovnik, A., & van der Hal, N. (2024). Adverse health effects of exposure to plastic, microplastics and their additives: Environmental, legal and policy implications for Israel. *Israel Journal of Health Policy Research, 13*(1), 44. https://doi.org/10.1186/s13584-024-00628-6
198. Haleem, N., Kumar, P., Zhang, C., Jamal, Y., Hua, G., Yao, B., & Yang, X. (2024). Microplastics and associated chemicals in drinking water: A review of their occurrence and human health implications. *Science of The Total Environment, 912*, 169594., ISSN 0048-9697. https://doi.org/10.1016/j.scitotenv.2023.169594
199. Ali, N., Khan, M. H., Ali, M., Sidra, S. A., Khan, A., Nabi, G., Ali, F., Bououdina, M., & Kyzas, G. Z. (2024). Insight into microplastics in the aquatic ecosystem: Properties, sources, threats and mitigation strategies. *Science of The Total Environment, 913*, 169489., ISSN 0048-9697. https://doi.org/10.1016/j.scitotenv.2023.169489
200. Zolotova, N., Kosyreva, A., Dzhalilova, D., Fokichev, N., & Makarova, O. (2022). Harmful effects of the microplastic pollution on animal health: A literature review. *PeerJ, 10*, e13503. https://doi.org/10.7717/peerj.13503
201. Jia, L., Liu, L., Zhang, Y., Fu, W., Liu, X., Wang, Q., Tanveer, M., & Huang, L. (2023). Microplastic stress in plants: Effects on plant growth and their remediations. *Frontiers in Plant Science, 14*, 1226484. https://doi.org/10.3389/fpls.2023.1226484
202. Lee, Y., Cho, J., Sohn, J., & Kim, C. (2023). Health effects of microplastic exposures: Current issues and perspectives in South Korea. *Yonsei Medical Journal, 64*(5), 301–308. https://doi.org/10.3349/ymj.2023.0048
203. Gaur, V. K., Sirohi, R., Bhat, M. I., Gautam, K., Sharma, P., Srivastava, J. K., & Ashok Pandey. (2023). A review on the effect of micro- and nano-plastics pollution on the emergence of antimicrobial resistance. *Chemosphere, 311*(Part 1), 136877., ISSN 0045-6535. https://doi.org/10.1016/j.chemosphere.2022.136877
204. Joo, S. H., Liang, Y., Kim, M., Byun, J., & Choi, H. (2021). Microplastics with adsorbed contaminants. *Mechanisms and Treatment, Environmental Challenges, 3*, 100042., ISSN 2667-0100. https://doi.org/10.1016/j.envc.2021.100042

205. Stapleton, M. J., & Hai, F. I. (2023). Microplastics as an emerging contaminant of concern to our environment: A brief overview of the sources and implications. *Bioengineered, 14*(1), 2244754. https://doi.org/10.1080/21655979.2023.2244754
206. Kaur, K., Reddy, S., Barathe, P., Oak, U., Shriram, V., Kharat, S. S., Govarthanan, M., & Kumar, V. (2022). Microplastic-associated pathogens and antimicrobial resistance in environment. *Chemosphere, 291*(Part 2), 133005., ISSN 0045-6535. https://doi.org/10.1016/j.chemosphere.2021.133005
207. David, E., & Niculescu, V. C. (2021). Volatile organic compounds (VOCs) as environmental pollutants: Occurrence and mitigation using nanomaterials. *International Journal of Environmental Research and Public Health, 18*(24), 13147. https://doi.org/10.3390/ijerph182413147
208. Pozzer, A. C., Gómez, P. A., & Weiss, J. (2022). Volatile organic compounds in aquatic ecosystems—Detection, origin, significance and applications. *Science of The Total Environment, 838*(Part 2), 156155., ISSN 0048-9697. https://doi.org/10.1016/j.scitotenv.2022.156155
209. Ma, W., Feng, Z., Zhan, J., Liu, Y., Liu, P., Liu, C., Ma, Q., Yang, K., Wang, Y., He, H., Kulmala, M., Mu, Y., & Liu, J. (2022). Influence of photochemical loss of volatile organic compounds on understanding ozone formation mechanism. *Atmospheric Chemistry and Physics, 22*, 4841–4851. https://doi.org/10.5194/acp-22-4841-2022
210. Xiong, Y., Du, K., & Huang, Y. (2024). One-third of global population at cancer risk due to elevated volatile organic compounds levels. *npj Climate and Atmospheric Science, 7*, 54. https://doi.org/10.1038/s41612-024-00598-1
211. Kim, Y. M., Kim, J., Ha, S. C., & Ahn, K. (2021). Harmful effect of indoor formaldehyde on atopic dermatitis in children: A longitudinal study. *Allergy, Asthma & Immunology Research, 13*(3), 468–478. https://doi.org/10.4168/aair.2021.13.3.468
212. Tang, K. T., Chen, Y. S., Lee, M. F., Chen, T. T., Lai, C. C., Lin, C. C., & Chen, Y. H. (2024). Exposure to volatile organic compounds may contribute to atopic dermatitis in adults. *Biomedicine, 12*(7), 1419. https://doi.org/10.3390/biomedicines12071419
213. Subramanian, A., & Khatri, S. B. (2019). The exposome and asthma. *Clinics in Chest Medicine, 40*(1), 107–123., ISSN 0272-5231, ISBN 9780323655361. https://doi.org/10.1016/j.ccm.2018.10.017
214. Saeedi, M., Malekmohammadi, B., & Tajalli, S. (2024). Interaction of benzene, toluene, ethylbenzene, and xylene with human's body: Insights into characteristics, sources and health risks. *Journal of Hazardous Materials Advances, 16*, 100459., ISSN 2772-4166. https://doi.org/10.1016/j.hazadv.2024.100459
215. Abilleira, E., Goñi-Irigoyen, F., Aurrekoetxea, J. J., Cortés, M. A., Ayerdi, M., & Ibarluzea, J. (2023). Swimming pool water disinfection by-products profiles and association patterns. *Heliyon, 9*(2), e13673., ISSN 2405-8440. https://doi.org/10.1016/j.heliyon.2023.e13673
216. Course, C., et al. (2021). Volatile organic compounds as disease predictors in newborn infants: A systematic review. *Journal of Breath Research, 15*, 024002. https://doi.org/10.1088/1752-7163/abe283
217. Li, Z. T., Zeng, P. Y., Chen, Z. M., Guan, W. J., Wang, T., Lin, Y., Li, S. Q., Zhang, Z. J., Zhan, Y. Q., Wang, M. D., Tan, G. B., Li, X., & Ye, F. (2021). Exhaled volatile organic compounds for identifying patients with chronic pulmonary aspergillosis. *Frontiers in Medicine, 8*, 720119. https://doi.org/10.3389/fmed.2021.720119
218. Sharma, S., & Bhattacharya, A. (2017). Drinking water contamination and treatment techniques. *Applied Water Science, 7*, 1043–1067. https://doi.org/10.1007/s13201-016-0455-7
219. Averyn, V. S. (2021). Short refresher of radiobiology. In I. Naletoski, A. G. Luckins, & G. Viljoen (Eds.), *Nuclear and radiological emergencies in animal production systems, preparedness, response and recovery*. Springer. https://doi.org/10.1007/978-3-662-63021-1_2
220. Guidelines for drinking-water quality: Fourth edition incorporating the first and second addenda [internet]. Geneva: World Health Organization; 2022. 9, Radiological aspects. Available from: https://www.ncbi.nlm.nih.gov/books/NBK579448/.
221. Berthiaume, A. (2023). Radionuclide contamination in Canada: A scoping review. *Heliyon., 9*(6), e16602. https://doi.org/10.1016/j.heliyon.2023.e16602

222. Ferrari, C., Manenti, G., & Malizia, A. (2023). Sievert or gray: Dose quantities and protection levels in emergency exposure. *Sensors (Basel), 23*(4), 1918. https://doi.org/10.3390/s23041918
223. Dingwall, S., Mills, C. E., Phan, N., Taylor, K., & Boreham, D. R. (2011). Human health and the biological effects of tritium in drinking water: Prudent policy through science - addressing the ODWAC new recommendation. *Dose-Response: A Publication of International Hormesis Society, 9*(1), 6–31. https://doi.org/10.2203/dose-response.10-048.Boreham
224. Toxicological profile for ionizing radiation. Atlanta (GA): Agency for Toxic Substances and Disease Registry (US); 1999 Sep. 3, Summary of health effects of ionizing radiation. Available from: https://www.ncbi.nlm.nih.gov/ books/NBK597567/.
225. Kutanzi, K. R., Lumen, A., Koturbash, I., & Miousse, I. R. (2016). Pediatric exposures to ionizing radiation: Carcinogenic considerations. *International Journal of Environmental Research and Public Health, 13*(11), 1057. https://doi.org/10.3390/ijerph13111057
226. Kumar, R., & De Jesus, O. (2024). Radiation effects on the fetus. [Updated 2023 Aug 23]. In *StatPearls [Internet]*. StatPearls Publishing. Available from: https://www.ncbi.nlm.nih.gov/books/NBK564358/
227. Keith, S., Faroon, O., Roney, N., et al. (2013). *Toxicological profile for uranium*. Agency for Toxic Substances and Disease Registry (US). Health effects. Available from: https://www.ncbi.nlm.nih.gov/books/NBK158798/
228. Guéguen, Y., & Frerejacques, M. (2022). Review of knowledge of uranium-induced kidney toxicity for the development of an adverse outcome pathway to renal impairment. *International Journal of Molecular Sciences, 23*(8), 4397. https://doi.org/10.3390/ijms23084397
229. Riudavets, M., Garcia de Herreros, M., Besse, B., & Mezquita, L. (2022). Radon and lung cancer: Current trends and future perspectives. *Cancers (Basel), 14*(13), 3142. https://doi.org/10.3390/cancers14133142
230. World Health Organization 2022. Introduction to waterrelated infectious diseases-Module1.1. https://www.who.int/docs/librariesprovider2/default-document-library/module_1.1_introducti on-wrid_ds_220831.pdf.
231. UN-Water. Summary progress update 2021: SDG 6—water and sanitation for all. https://www.unwater.org/sites/default/files/app/uploads/2021/12/SDG-6-Summary-Progress-Update-2021_Version-July-2021a.pdf.
232. Kunz, J. M., Lawinger, H., Miko, S., et al. (2024). Surveillance of waterborne disease outbreaks associated with drinking water — United States, 2015–2020. *MMWR Surveillance Summaries, 73*(SS-1), 1–23. https://doi.org/10.15585/mmwr.ss7301a1
233. Magana-Arachchi, D. N., Wanigatunge, R. P. (2020). Chapter 2—Ubiquitous waterborne pathogens, Editor(s): Majeti Narasimha Vara Prasad, Anna Grobelak, Waterborne Pathogens, Butterworth-Heinemann, 15–42, ISBN 9780128187838, https://doi.org/10.1016/B978-0-12-818783-8.00002-5.
234. Edelstein, P. H., & Roy, C. R. (2015). 234—Legionnaires' disease and pontiac fever. In J. E. Bennett, R. Dolin, & M. J. Blaser (Eds.), *Mandell, Douglas, and Bennett's Principles and Practice of Infectious Diseases* (8th ed., pp. 2633–2644.e6, ISBN 9781455748013). W.B. Saunders. https://doi.org/10.1016/B978-1-4557-4801-3.00234-4
235. Lalle, M., & Cacciò, S. M. (2023). Giardiasis from a one health perspective. In A. Sing (Ed.), *Zoonoses: Infections affecting humans and animals*. Springer. https://doi.org/10.1007/978-3-031-27164-9_33
236. Shah, M. P., & Hall, A. J. (2018). Norovirus illnesses in children and adolescents. *Infectious Disease Clinics of North America, 32*(1), 103–118. https://doi.org/10.1016/j.idc.2017.11.004
237. Bennish, M. L., & Khan, W. A. (2013). 44—Shigellosis. In A. J. Magill, D. R. Hill, T. Solomon, & E. T. Ryan (Eds.), *Hunter's Tropical Medicine and Emerging Infectious Disease* (9th ed., pp. 454–461., ISBN 9781416043904). W.B. Saunders. https://doi.org/10.1016/B978-1-4160-4390-4.00044-8
238. Finsterer, J. (2022). Triggers of Guillain-Barré syndrome: Campylobacter jejuni predominates. *International Journal of Molecular Sciences, 23*(22), 14222. https://doi.org/10.3390/ijms232214222

Part V
Effects of Emerging Contaminants

Chapter 8
Effects of Emerging Contaminants on Plants, Seed, and Crop Production

R. Suganthi, R. Poornima, S. Priyadharsini, C. Sivaranjani, and R. Sabitha devi

1 Introduction

The agricultural sector is contaminated by various compounds, such as nutrients, sediments, agrochemicals, etc. The occurrence of emerging contaminants in water threatens both flora and fauna owing to its persistence and perturbs the functioning and physiology of the receptors in the environment. The continuous use of contaminated irrigation water results in the accumulation of toxic substances (Fig. 8.1), viz. toxic heavy metals such as mercury, cadmium, chromium, lead, nickel, arsenic, antibiotics, microplastics, hydrocarbons, and veterinary drug residues, which affect soil and plant physiology, health, etc. [1]. Other contaminants, such as nano- and microplastics, alter the physical structure of soil, resulting in reduced soil water and nutrient reduction and, under certain conditions, bioaccumulation [2, 3]. Various

R. Suganthi (✉)
Department of Environmental Sciences, JSA College of Agriculture and Technology, Cuddalore, Tamil Nadu, India

R. Poornima
Department of Environmental Sciences, Vanavarayar Institute of Agriculture, Pollachi, Coimbatore, Tamil Nadu, India

S. Priyadharsini
Department of Environmental Sciences, Tamil Nadu Agricultural University, Coimbatore, Tamil Nadu, India

C. Sivaranjani
Department of Soil Science & Agricultural Chemistry, Institute of Agriculture, Kumulur, Tamil Nadu, India

R. Sabitha devi
Department of Environmental Sciences, Don Bosco College of Agriculture and Technology, Sagayathottam, Tamil Nadu, India

M. K. Jindal et al. (eds.), *Emerging Contaminants in Water*, Emerging Contaminants and Associated Treatment Technologies,
https://doi.org/10.1007/978-3-031-89591-3_8

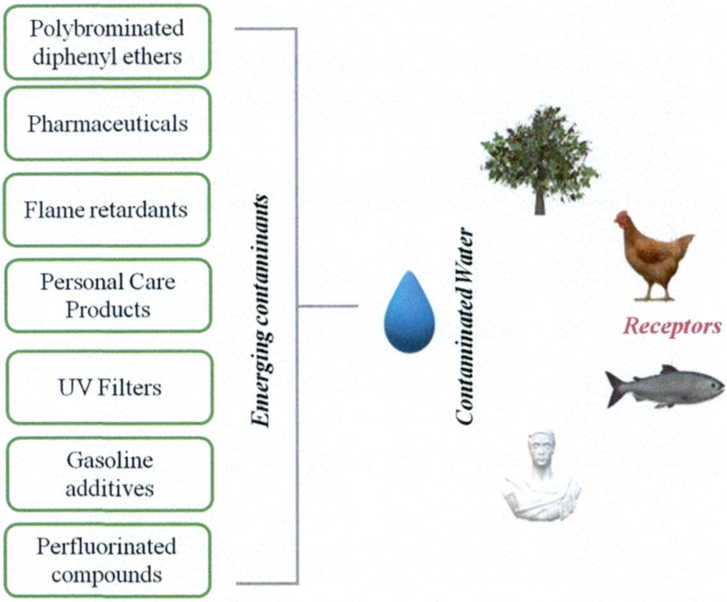

Fig. 8.1 Emerging contaminants in the environment

synthetic additives, such as plasticizers, flame retardants, stabilizers, chain extenders, and nucleating agents, are employed in biodegradable plastics [4]. Plasticizers are used to increase the elasticity of plastics, which tend to reside in soil for a longer period of time during degradation [5]. In terms of seed germination, microplastics such as polyethylene tend to negatively affect the growth of agricultural crops such as black grams, tomatoes [6], paddies [7], wheat [8], and lettuce [9]. Polystyrene, polyethylene, and polypropylene microplastics diminish Amaranthus morphological growth by creating oxidative stress and free radical formation [10]. The structural diversity of pharmaceutical and personal care products governs their mobility and entry into groundwater by leaching [11].

Petroleum, which is the major energy source, is widely used and contaminates the environment during extraction, processing, and transportation by various means, posing a sole threat to surface water and ground water [12, 13]. Conventional treatment methods are known to eliminate major pollutants from wastewater but not these emerging contaminants. Scientific methodologies must be designed to detect and eliminate these contaminants without affecting the environment. Phytoremediation has been recognized as a sustainable solution for the remediation of contaminated water and soil systems. Root exudates alter the rhizosphere pH, thereby increasing metal solubility and availability [14] via sorption and regulating cellular movement within root cells. The translocation of heavy metals from roots to aerial parts occurs mainly via apoplastic and symplastic pathways, which are formed by passive diffusion and later by active transport across the plasma membrane, which is ultimately sequestered in extracellular compartments (cell walls) or plant vacuoles and inhibits the accumulation of free metal ions in the cytosol [15].

2 ECs in Plant Systems

Nano- and microplastics, per- and polyfluoroalkyl substances (PFASs), endocrine-disrupting chemicals (EDCs), antibiotics, hormones, and pain relievers are recognized as potential environmental pollutants in agriculture that pose a threat to biotic and abiotic environments, which are instigated from diverse sources, such as agriculture and industry. To increase productivity, novel formulations have been developed to restrict pest and weed growth [16]. Long-term irrigation with contaminated water, organic and inorganic amendments, and other agri inputs affects agroecosystems [17, 18]. In addition, pharmaceutical and veterinary drug residues from livestock can be transported over long ranges, resulting in contamination of aquatic ecosystems, soil biota and agricultural productivity [19–21]. These emerging contaminants undergo various metabolic reactions within the plant system, aiding in phytoremediation (Fig. 8.2).

The bioaccumulate diminish emerging contaminants in agricultural systems through water or other sources diminishes the quality of the product after it affects its growth [22, 23] and reproductive and developmental processes [24, 25]. PFAS compounds are a threat to agricultural crops and the food chain because of the presence of aliphatic perfluorocarbon in the soil and water of adjacent agricultural areas owing to human activities [26, 27]. In addition, globalization and international trade pave the way for the entry of new, more virulent plant pathogens, which together affect agricultural productivity [28]. Soil structure, water, and nutrient retention are

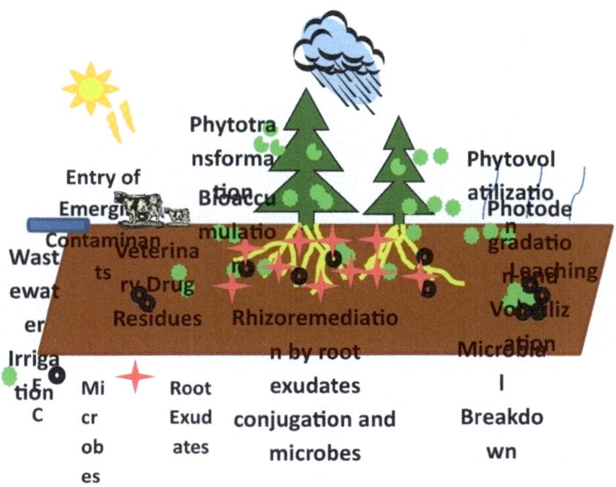

Fig. 8.2 Fate of emerging contaminants in agroecosystems

caused by the effects of nano- and microplastics on crop growth, which can lead to food safety and human health concerns [2, 3, 29, 30]. Biotic and abiotic components govern the bioaccumulation and biomagnification of emerging contaminants in plant tissues. The former factor includes plant genotype, physiological state, and soil fauna, and the latter considers the chemical properties of ECs and environmental conditions. Studies indicate that the ability of crops to take up and accumulate ECs decreases in the order of leafy vegetables, root vegetables, cereals, fodder crops, and fruit vegetables. However, further research is needed, as ECs have the potential to disturb plant physiological and molecular functions, significantly decreasing crop productivity. The presence of ECs in soil–plant systems present a challenge to the agricultural sector, necessitating ongoing research, monitoring, and regulatory efforts to address these issues.

3 Plant Physiology

Pharmaceutical compounds and their metabolites are toxic to the environment even at low concentrations after disposal. Nontarget organisms are largely affected by these pharmaceuticals and personal care products, which are metabolized in the environment. At contaminated sites, plants tend to consume these compounds, which undergo biotransformation, which may result in less toxic products or more toxic metabolites. Enzymes, especially those associated with drug metabolism or endogenous metabolism, may be responsible for the aforementioned biotransformation [31]. Plants have the capacity to nullify the effects of antibiotics or other pharma compounds by modifying their physiological process and functions though enzyme synthesis and drug transporter proteins. Detoxification of these compounds within the plant system occurs in three steps, as illustrated in Fig. 8.3. The metabolism of drugs within the plant system may result in toxic as well as nontoxic compounds, which affect the health of animals as well as humans.

The benzimidazole group, which includes flubendazole, thiabendazole, fenbendazole, etc., is often used in veterinary drugs against parasites and contributes to the largest group of anthelmintic drugs. They are commonly used against gastrointestinal parasites such as giardia, roundworms, and hookworms. They are also used in cancer treatment and fungal infection in humans. For example, fenbendazole in a plant system is oxidized into various metabolites, such as sulfones, sulfides, hydroxylated compounds, hydrolyzed compounds, and acetylated compounds [32]. Similarly, fenbendazole easily accumulated in soybean, and biotransformation resulted in 16 metabolites. These compounds did not affect plant yield, whereas they reduced the production of polyphenols and other antioxidant enzymes, which are found in leaves, pods, and beans [31]. In other cases, such as albendazole, it is oxidized into 18 metabolites by ribwort plants, in which a substantial portion of the metabolites are unstable and can be reverted to their toxic parent compounds. Cattle excreta are often used as fertilizer for crop production as well as pasture cultivation.

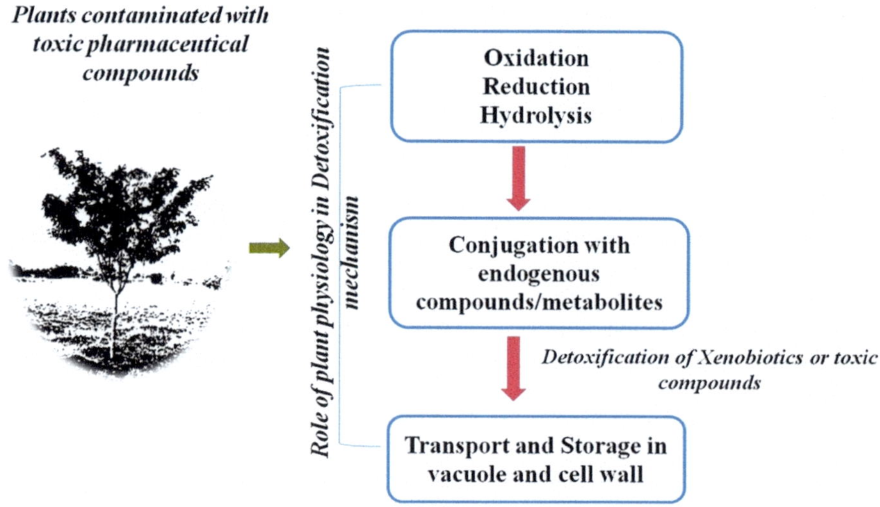

Fig. 8.3 Detoxification of xenobiotics within the plant system

This practice results in the circulation of compounds in the environment, which could affect herbivores and result in drug resistance over a period of time [33].

Owing to their numerous applications, heavy metals in the environment threaten the survival of living flora and fauna. Environmental pollution by heavy metals discharge to the environment by multiple human activities such as mining, energy and fuel production, electroplating, wastewater sludge treatment, and agriculture is one of the major environmental problems worldwide. Heavy metal pollution is a global concern these days due to intense urbanization, the adoption of luxurious lifestyles, industrialization, etc. A broad class of trace elements with significant biological and industrial applications are referred to as "heavy metals" or trace metals. Initially, heavy metals were naturally occurring components of soil, but as a result of human activity, their presence within the natural environment has increased. The issue of excessive concentrations of heavy metals, including Pb, Zn, Cr, Cu, Cd, Hg, and As, in soils is a global one. Compared with the control treatment, arsenic contamination increased water use efficiency by 30% and reduced gaseous exchange, i.e., intercellular carbon dioxide, by 29%. It increases the 5-methylcytosine level, which results in DNA hypermethylation. In addition, plant anomalies, such as cortex parenchyma, xylem poles, and the rhizodermis, have been observed [34].

Chlorophyll plays a major role in photosynthesis within the plant system. Mercury disrupts the normal functioning of leaves and the plant system, affecting the green economy by damaging the leaf cell wall. According to Raj et al. [35], a reduction in chlorophyll content was observed in soil treated with a high dosage (1000 mg Hg kg^{-1}). This reveals the lack of Hg uptake efficiency by *B. juncea* in the treatment of severely polluted soil. After poisons are absorbed, chlorophyll molecules breakdown or become less concentrated. Chlorophyll a, chlorophyll b, and total chlorophyll were substantially different ($p < 0.0001$) from those of the control. Gametophytes grown on mercury salts significantly reduce their chlorophyll content [36]. A low chlorophyll content was observed in the plants treated with 20 mg/kg, which also resulted in a reduction in photosynthetic activity. The highest chlorophyll content was detected in the control plants (26.5 in *Brassica juncea* and 6.5 in *Nephrolepis exaltata*). As stated by Suganthi et al. [37], other physiological characteristics also showed an inverse relationship with mercury concentration.

These toxic metals and their residues have harmful effects on the light and dark phases of photosynthesis, which results in disruption of the physiological functions of the plant, such as its photosynthetic pigments, stomatal conductance, and rates of transpiration [38]. Among other factors, stomatal conductance and CO_2 concentrations significantly influence photosynthesis. In controlled laboratory settings, Hg absorption by *Epipremnum* aureum can range from 170 to 1341 ng m^{-2} $day^{-1,}$ which results in distinct daily absorption patterns along with increased stomatal conductance [39].

The carcinogenic and teratogenic properties of hydrocarbons render them critical pollutants against human health and the environment. Hydrocarbons are major sources of energy, resulting in contamination of the majority of both aquatic and terrestrial ecosystems. Different physical, chemical, and mechanical techniques have been employed for the remediation of hydrocarbon-contaminated soil. Various methodologies have been utilized to address the contamination of soil environments by hydrocarbons, including physical, chemical, and mechanical approaches for the treatment or removal of pollutants.

A field study was performed at Shandong Agricultural University to study the effects of emerging contaminants such as polyvinylchloride and polyethylene on the soybean *Glycine max* [L.] merr.) (c.v. Zhonghuang 37) [40]. The research findings indicated that exposure to MPs, PVC, or PE alone or in combination, led to a notable increase in lipid peroxidation, ranging from 21.8% to 97.9%. Additionally, a significant decline in various parameters related to photosynthetic efficiency, including the net photosynthetic rate, PSII activity, energy uptake, capture, transport, and dissipation, was observed, ranging from 11.5% to 22.4% compared with those of the control group. Compared with any other treatment, PVC had a significant negative effect on soybeans.

4 Plant Biochemistry

In general, heavy metals damage plant tissues by direct and indirect means of generating reactive oxygen species (ROS), leading to oxidative stress. Plants possess their own detoxification mechanism to combat this oxidative damage through enzymatic and nonenzymatic antioxidants. Elevated levels of metals or other toxic compounds can be detrimental to plants, causing visible damage and physiological malfunctions. Heavy metals disrupt the opening and closing of stomata by interacting with water channel proteins and disrupt mitochondrial function by provoking oxidative stress through the generation of reactive oxygen species, ultimately resulting in disturbances in cellular and lipid metabolism in plants [41]. Exposure of tomato (*Lycopersicon esculentum*), cucumber (cucumerem), or alfalfa (Lucerne) to heavy metals led to an imbalance of ROS, heightened lipid peroxidation, suppressed activation of the antioxidant system, and diminished photosynthetic capability, as indicated by various studies. Malondialdehyde (MDA) serves as a marker of lipid peroxidation, contributing to membrane degradation and leakage. The combined impacts of mercury exposure hinder plant development, as evidenced by findings such as reduced plant height and biomass in alfalfa (Lucerne). The growth of plants under mercury-induced stress is believed to be dependent on antioxidant systems, lipid peroxidation byproducts, and photosynthetic pigments, as discussed earlier [42]. The antioxidant capacity of the plant increased in a dose-dependent manner, thereby demonstrating a notable increase in the levels of antioxidants such as proline, catalase, and peroxidase.

Proline plays a decisive role in heavy metal detoxification because of its inherent ability to shield cells from harm by scavenging free radicals or reactive oxygen species (ROS). It is produced within the plant system when exposed to excess metals in the growth medium [43], which leads to oxidative stress, resulting in the generation of superoxide radicals (O_2), hydroxyl radicals (OH), and hydrogen peroxide (H_2O_2) free radicals. Overall, an increase in heavy metals disrupts the equilibrium of the antioxidant protection system. When Hg^{2+} enters plant root cells, it binds with thiol (–SH) groups with a relatively high affinity, causing a reduction in GSH levels in the shoots and an increase in H_2O_2 production. The collaborative action of the CAT and SOD enzymes facilitates the decomposition of H_2O_2 into O_2 and H_2O, which is not toxic to plants. Consequently, the surplus H_2O_2 in pakchoi can substantially diminish the activities of CAT and SOD [37, 44].

To neutralize the aftermath of oxidative stress induced by mercury exposure, the plant synthesizes antioxidant enzymes to nullify the consequences within the plant system. SOD activity, which serves as a primary cellular defense mechanism, disintegrates free radicals, namely, superoxide radical (O^{2-}), into hydrogen peroxide and oxygen molecules. The ascorbate peroxidase enzyme hinders increased hydrogen peroxide accumulation, ultimately resulting in the transformation of water molecules. Both the catalase and ascorbate peroxidase activities were in line with previous findings [45]. Superoxide dismutase and catalase activities were markedly decreased by Hg-induced stress in garlic without any deviation in peroxidase activity. Si application enhances SOD activity in plants subjected to mercury stress. Si application did not have a profound effect on the activities of enzymes such as catalase and peroxidase in the leaves [46].

In the case of microplastics, plants maintain a dynamic equilibrium between the generation and elimination of ROS under normal circumstances. In response to external stressors, plants typically rely on their intrinsic antioxidant enzyme systems (SOD, POD, and CAT) and antioxidants to combat ROS, thus bolstering their resilience to stress. Nevertheless, when environmental pressures surpass the regulatory capacity of the antioxidant system, the defensive capabilities of plants may diminish or dissipate. These findings indicate that amaranth shoots are resilient against oxidative stress induced by MPs through an increase in POD activity. At a concentration of 100 mg/L, the generation of ROS under PS and PP stress surpassed the tolerance threshold, consequently inhibiting the expression of SOD and POD in amaranth. The enzymatic activity of SOD, along with its isoenzymes, facilitates the conversion of superoxide radicals to H_2O_2, which is subsequently decomposed by CAT, thereby preventing the premature conversion of ROS to H_2O_2. This study revealed that the activity of CAT remained unaffected by MPs. MDA serves as a breakdown product of polyunsaturated fatty acids, which are highly reactive compounds formed during lipid peroxidation under conditions associated with oxidative stress. MDA is typically positively correlated with ROS levels. This investigation revealed that the decline in MDA levels under stressful conditions serves as a crucial indicator of plant stress tolerance. However, lower concentrations of PS (50 μm) increased the MDA content in amaranth shoots, indicating the activation of amaranth's antioxidant defense system by MPs. In situations where antioxidant enzymes prove inadequate in swiftly and effectively counteracting the excessive accumulation of ROS, increased lipid peroxidation ensues, disrupting the normal metabolic processes of amaranth. The induction of ROS production within cells is triggered when plants encounter stress. H_2O_2 and O^{2-} represent the primary ROS species. Upon root injury, the accumulation of ROS, membrane peroxidation, and subsequent root cell impairment may transpire, leading to the inhibition of root elongation. The extent of ROS production is significantly influenced by the plant's response to stressors, including abiotic stress, UV-B radiation, and metal and material toxicity. This observation elucidates the phenomenon of ROS amplification subsequent to exposure to high concentrations of MPs. Notably, the ROS levels did not substantially change after treatment with low concentrations of MPs, confirming the minimal toxicity of the three types of MPs in comparison to heavy metals.

The adverse impacts of TiO_2 nanoparticles on the stress resistance of young mulberry plants were observed, leading to a notable increase in malondialdehyde levels in the treatment groups exposed to concentrations of 100, 200, 400, and 800 mg/L by 87.70%, 91.36%, 96.57%, and 192.19%, respectively, in comparison with those in the control group. Analysis of the transcriptomic data indicated that exposure to TiO_2 NPs predominantly influenced the expression of genes associated with energy production and transportation, protein metabolism, and stress response mechanisms. Moreover, the metabolomics findings revealed significant variations in 42 metabolites in mulberry plants, with 26 metabolites exhibiting upregulation and 16 metabolites showing downregulation, particularly affecting metabolic pathways such as secondary metabolite synthesis, the citric acid cycle, and the tricarboxylic acid cycle. This unfavorable impact was found to inhibit the germination and growth of mulberry saplings [47].

5 Seed Germination

Seed germination and seedling growth are clearly influenced by microplastics, which are reliant on factors such as microplastic concentration, microplastic type, and plant species. The polyethylene group of microplastics tends to decrease seed germination rates and root growth in black grams and tomatoes. In addition to seed germination and root elongation, nutrient transportation is impeded in pea seedlings, worsening oxidative stress and disrupting antioxidant systems [48]. Different phases of seed germination in *Daucus carota* are influenced by the microplastic structure along with films, and fibers affect the soil water content and rate of seed germination [49]. Moreover, seed viability was hindered by low-density polyethylene microplastics, in contrast to the increased biomass accumulation in water spinach. Antioxidant activity and soluble sugar content are significantly influenced [50], which together represent the major factors influencing seed germination.

It affects the functioning of micro- and macroorganisms present in agroecosystems. Polystyrene microplastics were recorded in the intercellular layer of carrot roots [51] and in the vascular system of rice seedlings [52]. Microplastics hinder micropyle growth (MPs), delaying the germination and root growth of garden cress (*Lepidium sativum* L.) [53], reducing the net photosynthetic rate and reducing shoot elongation in rice and lentils [54], whereas polystyrene nanoplastics (PSNPs) have no substantial effect on wheat seed germination [55]. Microplastic accumulation in seed capsule pores corresponds to a delay in germination [53]. In contrast, Amaranthus shoot length was increased by large polystyrene particles, which demand more nutrients for plant growth and development. In addition to the accumulation of microplastics affecting crops, their physical presence causes mechanical damage and impedes root epidermis and shoot growth [56].

For example, our findings are in line with those of a prior study conducted by [57], which revealed that in an outdoor experiment lasting approximately 4 months, soil containing 250–1000 mm LDPE microplastics at a concentration of 25,000 mg kg^{-1} dry soil had no appreciable effect on the biomass of shoots, roots, or fruits of *Phaseolus vulgaris* L. On the other hand, Pignattelli et al. [58] investigated the effects of 184 mg kg^{-1} polyethylene microplastics on *Lepidium sativum* or garden cress for a period of 21 days. They discovered that polyethylene microplastic-containing soil had considerably greater germination inhibition (7.1%) than the control (0%) and a 38% reduction in shoot biomass. Nevertheless, neither the shoot height nor the number of leaves was significantly affected.

Case Studies

Uptake and Degradation of Phthalates in Plants

Phthalates are widely used in the manufacturing of plastic products and are widely recognized as endocrine disruptors as well as mutagens. Fourteen phthalates are widely used in commercial plastic production, of which the usage of six is entirely restricted by the United States Environmental Protection Agency (USEPA) [59]. The uptake, translocation, and degradation of phthalates occur via various pathways. Since it is organic in nature, it enters the plant root system via the vascular system and vapor phase of the soil. Phthalates are taken by plants through nutrient solution from the soil, which includes symplastic and apoplastic pathways (Fig. 8.4). The vascular system aids in the translocation of phthalates from the soil to the roots and aerial parts of terrestrial plants through the xylem and phloem. The fate of di-n-butyl phthalate was studied in maize, wheat, and crust, which resulted in a maximum deposition of 14.26 mg kg^{-1} dw [60]. After plant uptake, di-n-butyl phthalate undergoes hydrolysis, resulting in the generation of monoesters followed by conjugation and compartmentalization. These monoesters are further broken down into nontoxic products. Carboxyl esterases play a major role in the detoxification of parent compounds into less toxic products [61]. In addition to other enzymes, such as cytochrome P450 monooxygenase, peroxidase is involved in oxidative metabolism, and glutathione-s-transferase and UDP glycosyl transferases are involved in the conjugation of these phthalates (Fig. 8.4). These conjugates are further sequestered in the form of lignin, which is subsequently detoxified [61, 62].

The effects of dimethyl phthalate on cucumber germination, the antioxidant system and the chloroplast structure were studied. There was no significant difference in the germination rate, but a delay in germination time was noted when the concentration of phthalates was greater than 50 mg L^{-1} [63, 64]. Lateral root growth and number were reduced by treatment with 30 mg/L. Phthalates are responsible for reducing chlorophyll contents and increasing the generation of free radicals such as hydrogen peroxide and the stress enzyme proline. To prevent damage, catalase and peroxidase activity significantly differed between 30 mg/L and 50 mg/L cucumber seedlings. Transmission electron microscopy confirmed the retrogressive effects of

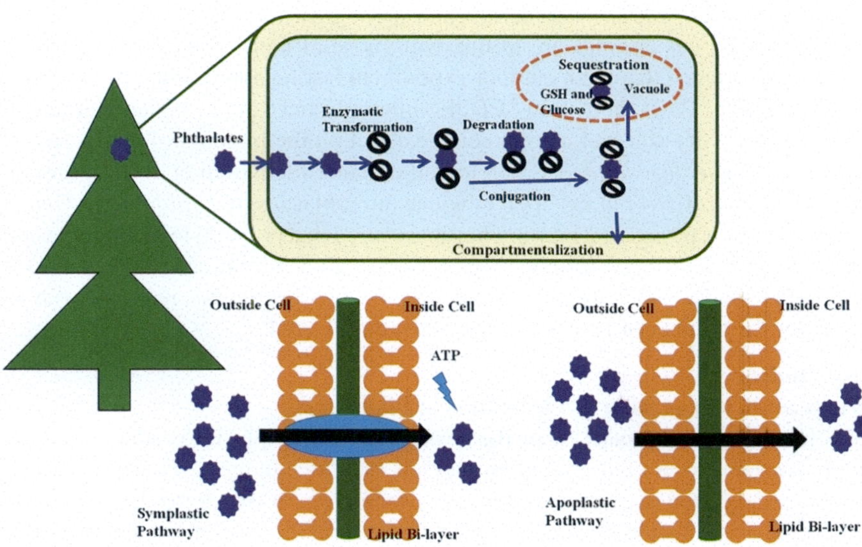

Fig. 8.4 Uptake and sequestration mechanism of phthalates by plants

phthalates on chloroplast structure, such as the degradation of starch grains, disordered lamellae, and damaged membrane structure when the concentration of phthalates exceeded 30 mg/L [63, 65].

Numerous studies have confirmed the bioaccumulation of phthalates even in edible plants. Di-n-butyl phthalate, di-2-ethylhexyl phthalate and their corresponding primary metabolites, mono-n-butyl phthalate, and mono-2-ethylhexyl phthalate, were assessed in strawberry, lettuce, and carrot to understand their accumulation, transformation, and degradation patterns. The bioconcentration factor exhibited a maximum value of 4.78, in the order carrot> strawberry>lettuce, and greater accumulation was observed in the roots than in the shoot systems of carrot and strawberry. A relatively high lipid content was detected in the roots, indicating the uptake of hydrophobic compounds, while comparing the treated phthalates and their metabolites, rapid hydrolysis of di-n-butyl phthalate to its monoester in carrot cell culture was detected. Owing to their efficient and rapid metabolism of phthalates to their monoesters, phthalates must be carefully considered when studying their impact on humans, animals, and other ecosystem components [61, 66].

Effects of Nanoparticles on Crop Growth and Function

Owing to their multifaceted characteristics, nanoparticles are widely used in various sectors, such as modern agriculture and industry [67]. NPs such as gold, copper, titanium, cerium oxide, copper oxide, and multiwalled carbon nanotubes are exclusively employed in agriculture because of their ability to withstand biotic and abiotic stresses [68]. In certain cases, it enters agricultural fields incidentally via biosolid application. Evident findings have confirmed the uptake and translocation of engineered nanoparticles in plants. It has been reported that plants can take up air-dispersed nanoparticles through stomatal openings and direct penetration [69]. Translocation occurs with the help of the vascular system and differs with each nanoparticle. Some nanoparticles are taken directly by the plant through roots and then translocated and transformed in the shoot system, indirectly through the formation of metal ions and chelation with plant metabolites [70, 71]. If nanoparticles enter the soil, they undergo different chemical modifications, such as ionization, oxidation, and sulfidation, which govern the potential to damage the growth and function of the plant system. It has detrimental effects on germination, seedling growth, and other physiological parameters. Weathered copper oxide nanoparticles were tested for their effects on spinach and foliar health; gas exchange; chlorophyll content; carotenoid, catalase, and ascorbate peroxidase enzyme activity; and copper bioaccumulation. It was concluded that the weathering process is responsible for the reduction in pigment production and foliar health [72]. Transcriptomic and metabolomic approaches were employed to study the effects of titanium oxide nanoparticles on protein metabolism, energy generation, and transport in mulberry plants. This analysis revealed 42 metabolites, of which 26 favor growth, whereas 16 metabolites are downregulated, which is not suitable for germination and growth [1].

6 Conclusion

The development and utilization of newer products result in the concentration of emerging contaminants in the environment. The sources of these contaminants, their fates, and their interactions with the biotic and abiotic components of ecosystems determine their impacts on the environment. Careful consideration of the possible risks stemming from these contaminants should be given in collaboration with various subject *domains,* i.e., biotechnology and nanotechnology, which might facilitate an enhanced understanding and remediation of these compounds. While much progress has been made in this area, there is still much to learn about the occurrence, sources, and hazards of ECs, as well as the need for full-scale implementation of efficient technologies to meet the requirements of practical application. To overcome these obstacles, research efforts must be directed toward the development of strong remediation strategies built on sustainable foundations, strategies that should integrate multiple removal processes to assure regulatory compliance and establish consistency between the characteristics of pollutants.

References

1. Naeem, M., Gill, R., Gill, S. S., Singh, K., Sofo, A., & Tuteja, N. (2023). Emerging contaminants and their effect on agricultural crops. *Frontiers in Plant Science, 14*, 1296252.
2. Jia, L., Liu, L., Zhang, Y., Fu, W., Liu, X., Wang, Q., Tanveer, M., & Huang, L. (2023). Microplastic stress in plants: Effects on plant growth and their remediations. *Frontiers in Plant Science, 14*, 1226484.
3. Quilliam, R. S., Pow, C. J., Shilla, D. J., Mwesiga, J. J., Shilla, D. A., & Woodford, L. (2023). Microplastics in agriculture–a potential novel mechanism for the delivery of human pathogens onto crops. *Frontiers in Plant Science, 14*, 1152419.
4. Cao, X., Liang, Y., Jiang, J., Mo, A., & He, D. (2023). Organic additives in agricultural plastics and their impacts on soil ecosystems: Compared with conventional and biodegradable plastics. *Trends in Analytical Chemistry, 117212*.
5. Billings, A., Jones, K. C., Pereira, M. G., & Spurgeon, D. J. (2021). Plasticisers in the terrestrial environment: Sources, occurrence and fate. *Environmental Chemistry, 18*(3), 111–130.
6. Sahasa, R. G. K., Dhevagi, P., Poornima, R., Ramya, A., Moorthy, P. S., Alagirisamy, B., & Karthikeyan, S. (2023). Effect of polyethylene microplastics on seed germination of Blackgram (*Vigna mungo* L.) and tomato (*Solanum lycopersicum* L.). *Environmental Advances, 11*, 100349.
7. Chen, S., Feng, Y., Han, L., Li, D., Feng, Y., Jeyakumar, P., Sun, H., Shi, W., & Wang, H. (2022). Responses of rice (*Oryza sativa* L.) plant growth, grain yield and quality, and soil properties to the microplastic occurrence in paddy soil. *Journal of Soils and Sediments, 22*(8), 2174–2183.
8. Zhang, K., Wang, M., Li, Y., Zhang, X., Xiao, K., Ma, C., Zhang, X., Zhang, H., & Chen, Y. (2024). Wheat (*Triticum aestivum* L.) seedlings performance mainly affected by soil nitrate nitrogen under the stress of polyvinyl chloride microplastics. *Scientific Reports, 14*(1), 4962.
9. Li, Z., Li, Q., Li, R., Zhao, Y., Geng, J., & Wang, G. (2020). Physiological response of lettuce to microplastic pollution. *Environmental Science and Pollution Research., 27*, 30306–30314.
10. Wang, J., Li, J., Liu, W., Zeb, A., Wang, Q., Zheng, Z., Shi, R., Lian, Y., & Liu, L. (2023). Three typical microplastics affect the germination and growth of amaranth (*Amaranthus mangostanus* L.) seedlings. *Plant Physiology and Biochemistry, 194*, 589–599.

11. Corada-Fernández, C., Jiménez-Martínez, J., Candela, L., González-Mazo, E., & Lara-Martín, P. A. (2015). Occurrence and spatial distribution of emerging contaminants in the unsaturated zone. Case study: Guadalete River basin (Cadiz, Spain). *Chemosphere, 119*, S131–S137.
12. Balasubramaniam, A., Boyle, A. R., & Voulvoulis, N. (2007). Improving petroleum contaminated land remediation decision making through the MCA weighting process. *Chemosphere, 66*, 791–798.
13. Nadim, F., Hoag, G. E., Liu, S., Carley, R. J., & Zack, P. (2000). Detection and remediation of soil aquifer systems contaminated with petroleum products: An overview. *Journal of Petroleum Science and Engineering, 26*, 169–178.
14. Dalvi, A. A., & Bhalerao, S. A. (2013). Response of plants towards heavy metaltoxicity: An overview of avoidance, tolerance and uptake mechanism. *Annals of Plant Science, 2*, 362–368.
15. Thakur, S., Singh, L., Wahid, Z. A., Siddiqui, M. F., Atnaw, S. M., & Din, M. F. M. (2016). Plant-driven removal of heavy metals from soil: Uptake, translocation, tolerance mechanism, challenges, and future perspectives. *Environmental Monitoring and Assessment, 188*, 206. https://doi.org/10.1007/s10661-016-5211-9
16. Rajak, B. K., Rani, P., Mandal, P., Chhokar, R. S., Singh, N., & Singh, D. V. (2023). Emerging possibilities in the advancement of herbicides to combat acetyl-CoA carboxylase inhibitor resistance. *Frontiers in Agronomy, 5*, 1218824.
17. Sun, N., Thompson, R. B., Xu, J., Liao, S., Suo, L., Peng, Y., et al. (2021). Arsenic and cadmium accumulation in soil as affected by continuous organic fertilizer application: Implications for clean production. *Agronomy, 11*(11), 2272.
18. Mohanty, B., & Das, A. (2023). Heavy metals in agricultural cultivated products irrigated with wastewater in India: A review. *AQUA Water Infrastructure Ecosystems and Society, 72*(6), 851–867.
19. Charuaud, L., Jarde, E., Jaffrezic, A., Thomas, M.-L., & Bot, B. L. (2019). Veterinary pharmaceutical residues from natural water to tap water: Sales, occurrence and fate. *Journal of Hazardous Materials, 361*, 169–186.
20. Nightingale, J., Carter, L., Sinclair, C. J., Rooney, P., & Kay, P. (2023). Influence of manure application method on veterinary medicine losses to water. *Journal of Environmental Management, 334*, 117361.
21. Khan, A. H. A., & Barros, R. (2023). Pharmaceuticals in water: Risks to aquatic life and remediation strategies. *Hydrobiology, 2*(2), 395–409.
22. Zhang, H., Li, X., Yang, Q., Sun, L., Yang, X., Zhou, M., et al. (2017). Plant growth, antibiotic uptake, and prevalence of antibiotic resistance in an endophytic system of pakchoi under antibiotic exposure. *International Journal of Environmental Research and Public Health, 14*(11), 1336.
23. Li, L., Li, T., Liu, Y., Li, L., Huang, X., & Xie, J. (2023). Effects of antibiotics stress on root development, seedling growth, antioxidant status and abscisic acid level in wheat (*Triticum aestivum L.*). *Ecotoxicology and Environmental Safety, 252*, 114621.
24. Pridemore, C. M. (2021). The effects of endocrine disrupting chemicals on plants. Theses, Dissertations and Capstones. Available at: https://mds.marshall.edu/etd/1412.
25. Thacharodi, A., Hassan, S., Hegde, T. A., Thacharodi, D. D., Brindhadevi, K., & Pugazhendhi, A. (2023). Water a major source of endocrine-disrupting chemicals: An overview on the occurrence, implications on human health and bioremediation strategies. *Environmental Research, 231*, 116097.
26. Adu, O., Ma, X., & Sharma, V. K. (2023). Bioavailability, phytotoxicity and plant uptake of per-and polyfluoroalkyl substances (PFAS): A review. *Journal of Hazardous Materials, 447*, 130805.
27. Nassazzi, W., Wu, T.-C., Jass, J., Lai, F. L., & Ahrens, L. (2023). Phytoextraction of per- and polyfluoroalkyl substances (PFAS) and the influence of supplements on the performance of short–rotation crops. *Environmental Pollution, 333*, 122038.
28. Singh, B. K., Delgado-Baquerizo, M., Egidi, E., Guirado, E., Leach, J. E., Liu, H., et al. (2023). Climate change impacts on plant pathogens, food security and paths forward. *Nature Reviews. Microbiology, 21*, 640–656.

29. Zantis, L. J., Borchi, C., Vijver, M. G., Peijnenburg, W., Lonardo, S. D., & Bosker, T. (2023). Nano- and microplastics commonly cause adverse impacts on plants at environmentally relevant levels: A systematic review. *The Science of the Total Environment, 867*, 161211.
30. Iqbal, B., Zhao, T., Yin, W., Zhao, X., Xie, Q., Khan, K. Y., et al. (2023). Impacts of soil microplastics on crops: A review. *Applied Soil Ecology, 181*, 104680.
31. Podlipná, R. (2022). Benzimidazoles and plants: Uptake, transformation and effect. *Toxics, 10*, 135. https://doi.org/10.3390/toxics10030135
32. Stuchlikova, L. R., Skalova, L., Szotakova, B., Syslova, E., Vokral, I., Vanek, T., & Podlipna, R. (2018). Biotransformation of flubendazole and fenbendazole and their effects in the ribwort plantain (*Plantago lanceolata*). *Ecotoxicology and Environmental Safety, 147*, 681–687.
33. Navratilova, M., Stuchlikova, L. R., Matouskova, P., Ambroz, M., Lamka, J., Vokral, I., Szotakova, B., & Skalova, L. (2021). Proof of the environmental circulation of veterinary drug albendazole in real farm conditions. *Environmental Pollution, 286*, 117590.
34. Popov, M., Kudrna, J., Lhotská, M., Hniliˇcka, F., Tunklová, B., Zemanová, V., Kubeš, J., Vachová, P., Ceská, J., Praus, L., et al. (2023). Arsenic ˇ soil contamination and its effects on 5-Methylcytosine levels in onions and arsenic distribution and speciation. *Toxics, 11*, 237. https://doi.org/10.3390/toxics11030237
35. Raj, D., Kumar, A., & Maiti, S. K. (2020). *Brassica juncea(L.) Czern.*(Indian mustard): A putative plant species to facilitate the phytoremediation of mercury contaminated soils. *International Journal of Phytoremediation, 22*(7), 733–744.
36. Thekkeveedu, R., & Hegde, S. (2021). In vitro spore germination and phytoremediation of Hg and Pb using gametophytes of *Pityrogramma calomelanos*. *International Journal of Phytoremediation, 23*(3), 307–315.
37. Suganthi, R., Avudainayagam, S., Davamani, V., Banu, K. S. P., Chandrasekhar, C. N., & Sivakumar, U. (2021). Indian mustard and Boston fern exhibits growth tolerance to increased dose of soil spiked mercury. *Environment and Ecology, 39*, 3.
38. Sheetal, K., Singh, S., Anand, A., & Prasad, S. (2016). Heavy metal accumulation and effects on growth, biomass and physiological processes in mustard. *Indian Journal of Plant Physiology, 21*(2), 219–223.
39. Naharro, R., Esbrí, J. M., Amorós, J. A., & Higueras, P. L. (2020). Experimental assessment of the daily exchange of atmospheric mercury in *Epipremnum aureum*. *Environmental Geochemistry and Health, 42*(10), 3185–3198.
40. Li, B., Huang, S., Wang, H., Liu, M., Xue, S., Tang, D., et al. (2021). Effects of plastic particles on germination and growth of soybean (*Glycine max*): A pot experiment under field condition. *Environmental Pollution, 272*, 116418.
41. Nagajyoti, P. C., Lee, K. D., & Sreekanth, T. (2010). Heavy metals, occurrence and toxicity for plants: A review. *Environmental Chemistry Letters, 8*(3), 199–216.
42. Tran, T. A. T., Zhou, F., Yang, W., Wang, M., Dinh, Q. T., Wang, D., & Liang, D. (2018). Detoxification of mercury in soil by selenite and related mechanisms. *Ecotoxicology and Environmental Safety, 159*, 77–84.
43. Mohanty, P., & Matysik, J. (2001). Effect of proline on the production of singlet oxygen. *Amino Acids, 21*(2), 195–200.
44. Sharma, P., Jha, A. B., Dubey, R. S., & Pessarakli, M. (2012). Reactive oxygen species, oxidative damage, and antioxidative defense mechanism in plants under stressful conditions. *Journal of Botany, 2012*.
45. Ansari, M. K. A., Ahmad, A., Umar, S., & Iqbal, M. (2009). Mercury-induced changes in growth variables and antioxidative enzyme activities in Indian mustard. *Journal of Plant Interactions, 4*(2), 131–136.
46. Hu, Y., Wang, Y., Liang, Y., Guo, J., Gong, H., & Xu, Z. (2020). Silicon alleviates mercury toxicity in garlic plants. *Journal of Plant Nutrition, 43*(16), 2508–2517.
47. Yu, D., Lu, Q., Wei, Y., Hou, D., Yin, X., Cai, K., Qiu, C., & Xu, K. (2023). Combined analysis of transcriptomics and metabolomics on the cumulative effect of nano-titanium dioxide on mulberry seedlings. *Frontiers in Plant Science, 14*, 1175012. https://doi.org/10.3389/fpls.2023.1175012

48. Mengen, K., Yi, L., Hao, W., Yu-Jong, W., Dongqing, G., & Xue, B. (2023). Physiological toxicity and antioxidant mechanism of photoaging microplastics on Pisum sativum L. Seedlings. *Toxics*. https://doi.org/10.3390/toxics11030242
49. Yudi, M., Lozano, P., Caesaria, U., & Matthias, C. R. (2022). Microplastics of different shapes increase seed germination synchrony while only films and fibers affect seed germination velocity. *Frontiers in Environmental Science*. https://doi.org/10.3389/fenvs.2022.1017349
50. Ying, Z., Wen-Ting, J., Xunyue, L., Gu, Z., Ronggui, T., Haibo, Z., Yanjiang, C. (2023). [Effects of low-density polyethylene microplastics on the growth and physiology characteristics of Ipomoea aquatica Forsk]. https://doi.org/10.13227/j.hjkx.202207205.
51. Dong, Y., Gao, M., Qiu, W., & Song, Z. (2021). Uptake of microplastics by carrots in presence of As (III): Combined toxic effects. *Journal of Hazardous Materials, 411*, 125055.
52. Liu, Y., Guo, R., Zhang, S., Sun, Y., & Wang, F. (2022). Uptake and translocation of nano/microplastics by rice seedlings: Evidence from a hydroponic experiment. *Journal of Hazardous Materials, 421*, 126700.
53. Bosker, T., Bouwman, L. J., Brun, N. R., Behrens, P., & Vijver, M. G. (2019). Microplastics accumulate on pores in seed capsule and delay germination and root growth of the terrestrial vascular plant Lepidium sativum. *Chemosphere, 226*, 774–781.
54. De Silva, Y. S. K., Rajagopalan, U. M., Kadono, H., & Li, D. (2022). Effects of microplastics on lentil (Lens culinaris) seed germination and seedling growth. *Chemosphere, 303*, 135162.
55. Lian, J., Wu, J., Xiong, H., Zeb, A., Yang, T., Su, X., Su, L., & Liu, W. (2020). Impact of polystyrene nanoplastics (PSNPs) on seed germination and seedling growth of wheat (Triticum aestivum L.). *Journal of Hazardous Materials, 385*, 121620.
56. Maity, S., & Pramanick, K. (2020). Perspectives and challenges of micro/nanoplastics-induced toxicity with special reference to phytotoxicity. *Global Change Biology, 26*(6), 3241–3250.
57. Meng, F., Yang, X., Riksen, M., Xu, M., & Geissen, V. (2021). Response of common bean (Phaseolus vulgaris L.) growth to soil contaminated with microplastics. *Science of the Total Environment, 755*, 142516.
58. Pignattelli, S., Broccoli, A., & Renzi, M. (2020). Physiological responses of garden cress (L. sativum) to different types of microplastics. *Science of the Total Environment, 727*, 138609.
59. United States Environmental Protection Agency (USEPA). (2013). Electronic code of federal regulations, Title 40-protection of environment, Part-423-team electric power generating point source category. Appendix A to Part 423–126, priority pollutants. https://www.ecfr.gov/cgi-bin/text
60. Zorníková, G., Jarošová, A., & Hřivna, L. (2014). Distribution of phthalic acid esters in agricultural plants and soil. *Acta Universitatis Agriculturae et Silviculturae Mendelianae Brunensis, 59*, 233–238.
61. Sun, J., Wu, X., & Gan, J. (2015). Uptake and metabolism of phthalate esters by edible plants. *Environmental Science & Technology, 49*, 8471–8478.
62. Sharma, R., & Kaur, R. (2020). Physiological and metabolic alterations induced by phthalates in plants: Possible mechanisms of their uptake and degradation. *Environmental Sustainability, 3*(4), 391–404. https://doi.org/10.1007/s42398-020-00141-x
63. Zhang, Y., Zhang, H., Sun, X., Wang, L., Du, N., Tao, Y., et al. (2016). Effect of dimethyl phthalate (DMP) on germination, antioxidant system, and chloroplast ultrastructure in *Cucumis sativus* L. *Environmental Science and Pollution Research, 23*, 1183–1192.
64. Ma, T. T., Christie, P., Teng, Y., & Luo, Y. M. (2013). Rape (*Brassica chinensis* L.) seed germination, seedling growth, and physiology in soil polluted with di-n-butyl phthalate and bis(2-ethylhexyl) phthalate. *Environmental Science and Pollution Research, 20*, 5289–5298.
65. Gomes, M. P., Duarte, D. M., Carneiro, M. M., Barreto, L. C., Carvalho, M., Soares, A. M., Guilherme, L. R., & Garcia, Q. S. (2013). Zinc tolerance modulation in Myracrodruon urundeuva plants. *Plant Physiology and Biochemistry, 67*, 1–6.
66. Singh, B. N., Hidangmayum, A., Singh, A., Guru, A., Yashu, B. R., & Singh, G. S. (2020). Effect of emerging contaminants on crops and mechanism of toxicity. *Sustainable Agriculture Reviews, 40*, 217–241.

67. Zain, M., Yasmeen, H., Yadav, S. S., Amir, S., Bilal, M., Shahid, A., et al. (2022). Applications of nanotechnology in biological systems and medicine in nanotechnology for hematology, blood transfusion, and artificial blood. Eds. A. Denizli, T. A. Nguyen, M. Rajan, M. F. Alam and K. Rahman (Elsevier), 215–235.
68. Pullagurala, V. L. R., Rawat, S., Adisa, I. O., Hernandez-Viezcas, J. A., Peralta-Videa, J. R., & Gardea-Torresdey, J. L. (2018). Plant uptake and translocation of contaminants of emerging concern in soil. *Science of the Total Environment, 636*, 1585–1596.
69. Raliya, R., Franke, C., Chavalmane, S., Nair, R., Reed, N., & Biswas, P. (2016). Quantitative understanding of nanoparticle uptake in watermelon plants. *Frontiers in Plant Science, 7*, 1288.
70. Servin, A. D., Castillo-Michel, H., Hernandez-Viezcas, J. A., Diaz, B. C., Peralta-Videa, J. R., & Gardea-Torresdey, J. L. (2012). Synchrotron micro-XRF and micro-XANES confirmation of the uptake and translocation of TiO2 nanoparticles in cucumber (*Cucumis sativus*) plants. *Environmental Science & Technology, 46*(14), 7637–7643.
71. Gui, X., Zhang, Z., Liu, S., Ma, Y., Zhang, P., He, X., et al. (2015). Fate and phytotoxicity of CeO2 nanoparticles on lettuce cultured in the potting soil environment. *PLoS One, 10*(8), e0134261.
72. Rawat, S., Cota-Ruiz, K., Dou, H., Pullagurala, V. L., Zuverza-Mena, N., White, J. C., et al. (2021). Soil-weathered CuO nanoparticles compromise foliar health and pigment production in spinach (*Spinacia oleracea*). *Environmental Science & Technology, 55*(20), 13504–13512.

Chapter 9
Health Effects of Emerging Contaminants: Effects on Humans Via Ingestion of Contaminated Water

C. M. Archana, A. Kanakalakshmi, K. Nithya, E. Kaarunya, and K. Renugadevi

1 Introduction

Among the five elements in the world, one of the most vital resources is water. Pure water influences the quality and balance of life of all living organisms. Currently, obtaining pure water is a major task in any geographical region, even in forestry places, due to the presence of dangerous contaminations. Contaminated water leads to improper sanitation, which can affect regular activity and hygienic practices, thus causing drastic socioeconomic issues. Every year, millions of individuals who lack access to a suitable water system may become ill from waterborne illnesses.

The depth of knowledge and updation are important to rule out the entry of toxins, pathogens, and chemical substances into water systems, particularly water, which is used by humans. Currently, under climate change, sea level variation can lead to a high prevalence of waterborne pathogens. This information is important for evaluating and monitoring the exposure of contaminants in water systems to address the safety precautions taken by various diseases.

C. M. Archana · A. Kanakalakshmi · E. Kaarunya
Department of Microbiology, Thiagarajar College, Madurai, India

K. Renugadevi (✉)
Department of Biotechnology, Thiagarajar College, Madurai, India

K. Nithya
Department of Environmental Science, Madurai Kamaraj University, Madurai, India

M. K. Jindal et al. (eds.), *Emerging Contaminants in Water*, Emerging Contaminants and Associated Treatment Technologies,
https://doi.org/10.1007/978-3-031-89591-3_9

2 Types of Water Contamination

2.1 Physical Contamination

Contaminants that alter the color, odor, or pH of water are referred to as physical contaminants. Although residential water consumption is a crucial risk to human health, such contamination of water does not immediately affect human health. The following are issues with contamination that impact physical properties: turbidity, color contamination, corrosiveness (pH), and disagreeable scents. Metals, dye pollution, soil particles, and water blooms contribute to the color contamination of bodies of water. Metals and bacteria that create musty odors are sources of water's disagreeable taste and odor. Water treatment must entirely eliminate any colored or foul-smelling materials from water because people become upset if their drinking water or other household usage water is colored or smells terrible. The growth of ecosystem-susceptible creatures is impacted by the pH of the water body, which is too close to neutral, and the growth of plants and agricultural products is hindered by excessively acidic soil or agricultural water. Therefore, one of the most important markers of environmental deterioration is pH. Because of the presence of suspended particles, physical contaminants alter the physical appearance of water. These could be soil, clay, organic materials, or silt. They might have an impact on color, fragrance, and turbidity [1, 2]. One major form of physical pollution is suspended solids (SS), which includes dust, contaminants, silt, clay, algae, and undissolved particles larger than 2 μm. Suspended particles in water can cause a number of diseases and toxins that are detrimental to human health, in addition to producing color and turbidity. Suspended solids are often easily removed via filtration or sedimentation [3, 4]. Pregnant women should be especially concerned about high levels of total dissolved solids, as they can increase the risk of hypertension, which can lead to (pre)eclampsia, cardiovascular disease, and other pregnancy-related issues [5].

In addition to having an impact on taste, moderate TDS levels might limit the amount of water available or force a community to switch to another source of water, both of which can lead to an unfavorable influence on health [6]. The primary causes of color contamination in aquatic bodies are metals, dye contamination, soil atoms, and the growth of eutrophication-induced water blooms. The color contamination of a body of water is problematic since it ruins the surrounding environment. When people use it for consumption or other home functions, they are also troubled by its hue. The introduction of dye-containing wastewater results in color pollution. The effluent from dye manufacturers comprises significant amounts of inorganic salts, organic acids, surfactants, water-soluble polymer complexes, and other dyes, including basic, acid, and direct dyes. In addition to contaminating color, dye wastewater also has other issues because many dyes contain harmful and/or persistent materials. Rainwater carries soil into rivers, reservoirs, lakes, and marshes, causing brown turbidity to emerge in these bodies of water after it rains. Material is the reason why water bodies in areas where laterite, a low specific gravity material, has been deposited are always murky. By preventing swimming, for example, turbidity substantially detracts from the hydrophilic purpose of water bodies and damages the landscape. Turbid water causes damage to treatment plants by clogging their treatment systems when it is used as an edible source. Thermal pollution is a typical physical manifestation of water contamination. This contains waste, foam, floating debris, and warm water from cooling towers. When warm water is released into a naturally occurring body of water, it can have an impact on ecosystems downstream and change their ecological equilibrium. It is necessary to avoid the use of excessive chemical treatments, mainly chlorine. Pathogens can build immunity to defend themselves against chlorine and its byproducts [7]. Moreover, reducing water pollution can help maintain source cleanliness and provide a long-term sustainable solution. It is essential to increase the public's understanding of how safety and quality are perceived in water. This applies to both developed and developing nations.

2.2 *Chemical Contamination*

Over a billion people worldwide are impacted by chemical contamination of water sources, which puts them at risk for negative health effects and water scarcity. Increased health effects, such as malignancies, cognitive decline, neurological disorders, hypertension, bone abnormalities, endocrine disorders, and cardiovascular disease, are a result of widespread exposure to chemical contamination around the world [8]. Numerous safe water methods for dealing with chemical pollution have been shown to be effective in full-scale or pilot-scale trials meant for application in environments with limited resources. Chemical pollutants can include inorganic or organic chemicals, as well as synthetic or natural elements. These pollutants can be manmade or naturally occurring. Chemical pollutants fall into two primary categories: organic and inorganic. Examples of organic contaminants include hormones, pesticides, herbicides, and fungicides from surrounding agricultural and horticultural enterprises, as well as oil and gasoline spills from concrete areas and roads [9, 10].

High rates of morbidity and mortality can result from drinking water contaminated with heavy metals, which can cause cardiovascular problems, neurological damage, multiple sclerosis, Parkinson's disease, muscular dystrophy, Alzheimer's disease, kidney problems, an increased risk of cancerous conditions, and a higher prevalence of diabetes. In addition to causing a number of diseases, such as cardiovascular disorders, neuronal damage, multiple sclerosis, Parkinson's disease, muscular dystrophy, Alzheimer's disease, renal injuries, cancer risk, and a high prevalence of diabetes, heavy metals can contaminate drinking water. By changing the endocrine system, causing catastrophic illnesses such as skin cancer, and modifying metabolic anomalies even in subsequent generations, cosmetic treatments containing metal elements can occasionally change the entire human physiology.

In addition to existing as individual molecules dissolved in water, suspended particles containing organic pollutants are also known to exist. Phosphorus (P), potassium (K), and nitrogen (N) are examples of inorganic pollutants. Increases in these basic chemicals in streams are almost always the consequence of land use practices such as fertilizer runoff or direct industrial discharge. There is a wide range of these chemical concentrations and how they enter waterways. Metals and metal particles are also considered inorganic pollutants [11]. These are found in the runoff of stormwater from urban expansion, and they tend to collect in areas of land that are low lying or in drainage systems. Eventually, many of these sources of contamination leak into rivers with little to no preceding chemical removal treatment. On the other hand, wastewater from the mining, energy, wastewater treatment, forest processing, and meat and dairy processing industries may contain specific metals such as mercury, copper, chrome, and zinc, as well as inorganic chemical contaminants such as bleach and curing agents [1]. Chemical pollutants can also include insecticides, medications, personal hygiene items, and dissolved organic materials [12]. These elements disintegrate when water seeps through the soil layers [13]. A waterway's bacterial population and oxygen consumption may both be accelerated by certain pollutants. Eutrophication can occur in rivers with relatively low amounts of nutrient contamination. The inorganic fertilizers phosphate and nitrate, which promote the growth of algae, are mostly responsible for eutrophication. After that, the algae die, leaving behind more organic debris and low oxygen levels [14].

Human health is significantly harmed by environmental chemical pollution. Climate change and greater industrialization are predicted to cause this influence to significantly increase throughout the decades that followed [15]. Waste substances dumped into streams cause biochemical reactions that deplete oxygen and breakdown organic materials. This leads to a decrease in dissolved oxygen (DO) levels and the local extinction of fish and invertebrate species. Fish "breathe" oxygen through their gills; they can perish if there is a decrease in oxygen availability (anoxia) in the water column. Native populations that are less tolerant of low dissolved oxygen levels may be replaced with fish that can handle them, such as the introduced species gambusia. Higher turbidity and lower water clarity impair fish vision, making it more difficult for them to catch prey and recognize predators [16]. Fish may exhibit physiological and behavioral changes due to multiple exposures to sublethal doses of certain contaminants. The population may be affected in the long run by these changes, which include reduced reproductive success; nest and brood desertion; reduced immunity to illnesses, cancers, and lesions; harm to the central nervous system; and a greater incapacity to evade predators. Certain pollutants, such as mercury, have the potential to bioaccumulate in animal tissues and make their way into the bodies of people who eat fish. Therefore, to reduce health risks, treatment procedures that eliminate chemical pollution, as well as alternate tactics that mix or replace polluted sources with chemically safe water sources, are necessary.

2.3 Radiological Contamination

Overdoses of radiation can alter human physiology, including chromosomal disruption, cancer, leukemia, nuclear mutations, cataracts, osteonecrosis, acute radiation syndrome, cutaneous radiation injury, or radiation sickness. Cancer, leukemia, genetic mutations, osteonecrosis, cataracts, chromosomal disruption, lung disorders, genitals, and bone marrow death are among the other changes in human physiology. Radioactivity in minute amounts is typically present in all drinking water. Although radioactive water pollution is a relatively new phenomenon, it poses a major hazard to both human health and the environment [17]. The phenomenon in which elements or waves are spontaneously emitted from the unbalanced nucleus of some materials is known as radioactivity. Radioactive emissions come in three different forms: beta, gamma, and alpha. Beta particles are negatively charged electrons, alpha particles are positively charged He atoms, and gamma rays are neutral electromagnetic radiations and particles. The crust of a planet contains naturally occurring radioactive materials. Groundwater contamination results from naturally occurring materials (NORM) seeping into the aquifer from sol sediments. Nuclear weapon research, nuclear disasters, nuclear power plants, and the disposal of radioactive waste are important anthropogenic sources of radioactive pollution [18, 19]. Anthropogenic sources of radioactive pollution include the study of nuclear weapons, nuclear disasters, nuclear power plants, and the disposal of radioactive waste; minor sources include the use of radioisotopes in scientific laboratories and industry; human exposure to radioactive elements occurs from the use of these elements in nuclear weapons, MRIs, X-rays, and other medical equipment; surface water pollution results from the disposal of these radioactive wastes; and groundwater is more likely than surface water to be contaminated owing to exposure to radioactive materials discovered in rocks. Magma can occasionally emit radioactive gases into the surrounding air. Radioactive pollution is also caused by the deposition of these radioactive gases in water bodies. Wet and dry cosmogenic radionuclide air deposition contributes radioactive nuclei to surface water. Surface and subsurface waters contain a variety of radionuclides, commonly ^{3}H, ^{14}C, ^{40}K, ^{210}Pb, ^{210}Po, ^{222}Rn, ^{226}Ra, ^{228}Ra, ^{232}Th, and $^{234,235,238}U$. Water resources are contaminated by three NORM series: actinium, uranium, and thorium. One of the distinctly radiotoxic elements present in water systems is radium, a member of the NORM series, which can cross the threshold of groundwater by (a) aquifer rock dissolution, (b) ^{238}U and ^{232}Th decay, or (c) desorption mechanisms.

The natural stores of naturally occurring uranium are oceans and seas. When uranyl carbonate ions are present, they can be found. The higher salinity of marine water is thought to contain a sizable amount of uranium. By emitting radioactive waste, nuclear power facilities located along the shore increase the radiological pollutants in marine water. These powerhouses also use polluted water for coolant purposes. The sea environment is contaminated by radioactivity from nuclear submarines. There have been reports of radioactive pollution from submarine sinkings and accidents. Examples of such nuclear mishaps are the Colorado Rocky Flats facility, Fukushima, and the Chernobyl nuclear accident [20, 21]. The measurements of radioactivity are either in Curie or Becquerel (an SI unit). Gray measures the amount of energy absorbed per unit mass, whereas Sievert measures the amount of radiation received by human tissues. Although some radiation is present in all types of water, extended radiation exposure is harmful to human health [22]. The presence of radioactivity can be measured via a gross alpha test for evaluation in drinking water.

The World Health Organization established standards for the quality of drinking water, including an annual maximum allowable reference dosage level of 0.1 microsieverts. In addition to having an impact on the environment, radioactive elements present a risk to human health when the environment is breathed, injected, or exposed. Radiation is absorbed by human tissues through contaminated food and drink, which poses major health hazards. Acute radiation syndrome or cutaneous radiation injury can be caused by high radiation exposure. Radiation exposure results in a number of physiological diseases in humans, such as chromosomal disruption, cancer, leukemia, genetic mutations, osteonecrosis, and cataracts [23]. Genetic effects: Ionizing radiation causes mutations in germ cells, which are male and female sperm and egg cells [24]. These mutations change the structural makeup of germ cell DNA, which is subsequently passed on to progeny [25]. Severe mental illness and early mortality are two consequences of hereditary illnesses. The concentration and makeup of these radioactive elements differ from location to location, primarily on the basis of the radiochemical makeup of the rock layers and soil that the raw water may have gone through. Water has been shown to contain a wide variety of manufactured and natural radionuclides, but only a small number of them and their decay products account for the majority of the radioactivity. These include the following radiation emitters with low linear energy transfer (LET) that are available at different levels: rubidium-87 (^{87}Rb), carbon-14 (^{14}C), tritium (^{3}H), and potassium-40 (^{40}K). Furthermore, there may also be high-LET, alpha-emitting radionuclides such as polonium-210 (^{210}Po), uranium (U), thorium (Th), radon-220 (^{220}Rn), radon-222 (^{222}Rn), and daughters of radium-228 (^{228}Ra) [26]. The planet was formed with some radioactive elements and their decay products, which are the source of some of the naturally occurring radioactivity found in drinking water. Other radionuclides are continuously created by cosmic ray bombardment. Tritium is produced when cosmic rays combine with atmospheric nitrogen and oxygen. After that, it undergoes oxidation to become tritiated water, which enters the hydrosphere. Water resources can be protected against anthropogenic sources of radioactive contamination by taking preventative and precautionary measures. Remedial procedures that are efficient in treating radioactively polluted water include granule carbon adsorption, ion exchange, reverse osmosis, and aeration [27]. Currently, a safe water supply also requires appropriate radioactive contaminant analysis and monitoring.

2.4 Biological Contamination

Life on Earth is fundamentally dependent on water. Water cross-contamination is a worldwide issue. These could be manmade or geological [8]. Elevated levels of pollutants in drinking water rarely have an immediate adverse impact on health. It depends, of course, on the person's sensitivity and how the body is touched. Two million tons of industrial, agricultural, and sewage waste are discharged into the world's waterways every day, according to a 2003 UN assessment [28]. Pathogenic bacteria, viruses, parasites, and parasitic worms are the main biological pollutants. Salmonellosis, cholera, dysentery, typhoid, giardiasis, *E. coli* infection, infectious hepatitis, leptospirosis, and diarrhea can all be caused by these factors. Through contaminated water sources, a number of diseases with common symptoms have spread to people through the fecal–oral route. An estimated 1.7 billion children suffer from diarrhea each year, and 525,000 of them die as a result. In addition, diarrhea kills approximately one million adults annually. In addition, diarrhea kills almost one million adults annually. Microorganisms such as bacteria, viruses, and parasites such as worms and small protozoa are among the microbiological pollutants. Each of these can result in different issues in water [29, 30]. Algae are tiny, unicellular creatures that depend on phosphorus (or other nutrients) in the water and are rather prevalent. Most nutrients come from indoor-trial pollutants or from home runoff. Excessive algae development can not only affect the taste and odor of water but also clog filters and result in unwanted slime growth on the carriers. These blue–green algae, which include *Anabaena, Aphanizomenon*, and *Microcystis*, can occasionally release toxins that harm the skin, liver, and nervous system [31, 32].

Moreover, bacteria are single-celled tiny organisms. Numerous harmful microorganisms exist, and water can infect them [33]. They may cause gastroenteritis, cholera, diarrhea, and typhoid. While not hazardous, certain nonpathogenic bacteria (such as sulfur and crenothrix iron bacteria) might lead to problems with taste and odor [34, 35]. Protozoans are tiny, single-celled organisms that function in a similar way. *Giardia* and *Cryptosporidium* are two examples of protozoans that are frequently discovered in rivers, lakes, and streams that have been contaminated by animal waste or that receive effluent from sewage treatment facilities. These could result in headaches, exhaustion, dehydration, nausea, diarrhea, and cramping in the stomach. The tiniest living things that can spread illness and cause infection are viruses. There are frequent reports of polio- and hepatitis-causing viruses in tainted water. Unsurprisingly, human and animal waste can disperse these living organisms. Testing the color, turbidity, taste, and odor of water can help quickly identify some impurities. To determine whether or not water is contaminated, the majority, however, are difficult to identify and need to be tested. As a result, pollutants may stain in addition to having a disagreeable taste or odor and having detrimental effects on health.

Water quality standards are dependent on the intended usage, but generally, fecal coliform counts are employed to validate the water quality when humans come into direct contact with it. Lakes, reservoirs, and ground water can spread infectious diseases if dangerous bacteria are present or if the water purification process leaves some bacteria behind. A number of people died from cholera and dysentery, two common waterborne infectious diseases, in the late nineteenth and early twentieth centuries. The spread of these waterborne infectious diseases has been halted by modern water systems and water sterilization initiatives, but in 1991, a significant cholera outbreak claimed many lives in Africa and Central and South America. Declining living conditions and subpar sewage and waterworks treatment systems are thought to be the causes. *Shigella* causes dysentery, whereas *Vibrio cholerae* is the bacterium that causes cholera. Among the microorganisms known to cause infections transmitted by water are *Salmonella*, *Staphylococcus*, *Campylobacter*, *Clostridium*, and the pathogenic colon *Bacillus*. All of these harmful bacteria cause intestinal infections in humans, leak out of the body through excrement, are treated, and then end up in lakes, rivers, and reservoirs.

In Japan, the incidence of waterborne infectious illnesses caused by fecal coliforms such as *Escherichia coli* O157 has increased in recent years [36]. *Escherichia coli* O157 is a pathogenic colon bacillus that causes hemorrhagic colitis, hemolytic uremic syndrome, thrombotic thrombocytopenic purpura, and, in rare cases, death. This pathogenic colon bacteria also produces verotoxin. It is vitally important to prevent infection by such dangerous germs. According to Schriewer et al. [37], the indicator organism *E. coli* may typically be detected with adequate accuracy generally by PCR-based detection. Overall, the understanding of major issues resulting from interactions among different populations can be improved by developing standard epidemiological data for diseased populations, advancing technology to more precisely identify causal factors, and increasing the knowledge of disease dynamics [38–41]. Since the prevalence of pathogen contamination worldwide is a serious concern, it is imperative that people become more aware of the main pathogen sources and their significant consequences for water supplies. Many investigations on pathogen contamination have been carried out in laboratory settings; however, to improve our comprehension of pathogen interactions in the environment, field research should receive greater attention.

The assessment of pathogen contamination at the watershed scale is expected to benefit from the creation of new models and the enhancement of current modeling techniques that are frequently employed to forecast the levels of water-borne pathogens [42]. Combining expertise from several disciplines (such as hydrology, microbiology, and ecology) can assist in the development of long-term plans to enhance water quality and improve our understanding of pollution levels and their sources.

2.4.1 PFAS

Compounds referred to as per- and polyfluoroalkyl substances (PFASs) have existed since the 1940s. They work on anything from putting out fires to dyeing and waterproofing clothing, carpets, and food packaging [43]. Aqueous film-forming foam (AFFF), a firefighting foam, and consumer goods such as apparel, food packaging, cookware, cosmetics, and carpet all contain perfluoroalkyl substances (PFASs). Individuals come into contact with consumer products that contain perfluoroalkyl and polyfluoroalkyl substances (PFAS), breathe in dust that contains PFAS, and ingest contaminated food and water [44, 45]. With a half-life of 3.5–7.3 years (geometric mean) [46], PFASs primarily affect blood serum proteins, the kidney, and the liver. They also attach to proteins and can disrupt other metabolic processes [47]. PFAS-contaminated landfills and sites where PFAS-based firefighting foam (AFFF) has been used, including airports, military installations, chemical plants, and aboveground petroleum storage tanks, are additional possible sources of pollution in addition to industrial and manufacturing facilities [48]. Because they are made to last for a long time, PFASs are difficult to remove and do not degrade quickly. PFASs can be released into the air by incinerator facilities, contributing to the PFAS cycle in the environment. Vegetables grown on agricultural fields treated with contaminated biosolids from wastewater treatment facilities have been found to contain PFAS [49]. Certain PFASs are still present in the environment and have the potential to contaminate surface and groundwater near the sites where they are used or produced, even though their production has stopped owing to concerns about the environment and human health, despite the lack of knowledge on their potential effects on human health and the environment. Studies on people and animals exposed to perfluorooctanoic acid (PFOA) and perfluorooctanesulfonic acid (PFOS) have provided information about the health impacts of PFAS. Human blood has also been found to contain more recent PFAS that are still being produced. Increased cholesterol, altered liver enzymes, hormone interruption, an increased risk of thyroid disease, a decreased likelihood of pregnancy, high blood pressure or preeclampsia during pregnancy, slight reductions in newborn birth weights, a diminished response to vaccination in children, and an increased risk of kidney or testicular cancers are all possible outcomes of high levels of PFAS exposure in contaminated drinking water [50–52]. Furthermore, while PFASs are not known to cause death, their effects as endocrine disruptors are mostly linked to hormonal imbalances, decreased fertility, birth abnormalities, and changes in a variety of biochemical markers [53, 54]. Two examples of in-home treatment devices that can remove PFAS from drinking water include reverse osmosis and carbon filtration. Carbon filters can be placed at the point of entry into one's home or at the point of usage, such as a kitchen faucet or refrigerator. Filtered water pitchers can also be purchased with carbon filters. Reverse osmosis devices can also be mounted beneath a kitchen sink to purify water that is mostly used for cooking or drinking. Given that the effectiveness of treatment is mostly dependent on the quantity of PFAS contamination and ongoing system maintenance, in-home treatment may not always reduce PFAS below levels of concern [55]. It is imperative to halt industrial discharges and initiate a switch to firefighting systems devoid of per- and polyfluoroalkyl substances (PFASs) to address this situation. In the long run, maintaining public health requires stopping continuous PFAS leaks.

2.4.2 Microplastics

Thompson et al. [56] first used the term "microplastic" (MP) in a formal sense in 2004 to draw attention to the rising issue of plastic pollution in the ocean. Since then, scientists, authorities, the general public, and the media have all become more aware of its existence in the environment. In 2018, Provencher et al. [57] reported that MPs are commonly characterized as plastic particles with an unknown lower limit of less than 5 mm in any dimension, notwithstanding the lack of an official definition or complete consensus among the writers. Primary and secondary MPs are typically divided into two categories [58]. Primary microplastics (MPs) are those that have previously been created at the microscale. Examples of these include the microspheres (less than 500 μm) used in various cosmetic items, mixes used in shotblasting and sandblasting, and MPs utilized in 3D printing and as medicinal vectors. The breakdown of larger plastic materials via mechanical or photooxidative mechanisms results in secondary MPs. Silva et al. [59] reported that MPs can also be categorized on the basis of their chemical composition and morphology, which is usually in the form of fibers, fragments, and spherical beads, which include polyethylene (PE), low-density PE (LDPE), PE terephthalate (PET), and polyacrylates (PAs), among others. MPs are a diverse class of materials that include a wide range of polymer types, particle sizes (ranging over 6 orders of magnitude), shapes (from spheres to fibers), and chemical formulations (thousands of different types). Certain PFASs are still present in the environment and have the potential to contaminate surface and groundwater near the sites where they are used or produced, even though their production has stopped owing to concerns about the environment and human health, despite the lack of knowledge on their potential effects on human health and the environment. Many groups of MPs are present on the basis of their characteristics. These materials are likely to be found in water [60]. Microplastics from water distribution systems can enter bottled and tap water. They can contribute to air pollution since they are found in dust particles [61]. Numerous studies have emphasized the need to monitor microplastics since they can migrate into the human body through water [62, 63]. Drinking water sources include raw water that reaches drinking water treatment plants (DWTPs) [64], groundwater [65], bottled water [66], tap water [67], public drinking water fountains [68, 69], and water kiosks [70]. This consumption of drinking water comes from different sources. In addition to the contamination of the aquifer that supplies the water, other potential sources of contamination for this matrix might include pipelines, filters, water delivery systems, and packaging materials [71]. Using more biodegradable materials is an additional tactic to address the harmful issues of plastics, MPs, and NPs in the environment. Renewable raw ingredients such as starch, cellulose, lignin, and bioethanol are used to make bioplastics [72]. An alternative bioengineering-based approach involves the identification of novel pathways for the biodegradation of traditional plastics, such as distinct kinds of fungi and bacteria, or the isolation of relevant enzymes to guarantee the hydrolysis of plastics by enzymes. Extracellular carboxylesterases have the ability to hydrolyze biodegradable polyesters [73]. Specialized bacteria can breakdown plastics. For example, *Ideonella sakaiensis* can breakdown PET [74], whereas the marine fungus *Zalerion maritimum* can breakdown PE [75].

The remediation of macro- and MPs at the laboratory scale is still in its infancy, despite seemingly promising alternatives. The development of methods for the in situ biodegradation of MPs through the use of local microflora or the inclusion of microorganisms is imperative. This allows for greater natural attenuation [76]. The creation of more intelligent and recyclable plastic materials, the investigation and isolation of degrading microbes to improve recycling procedures, and the tracking and elimination of MPs by cutting-edge wastewater treatments must all contribute to the decisions made. These mechanisms can be combined in ways that modulate and reduce MP contamination in water. In addition, to decrease the usage of plastic, governments, nongovernmental organizations, and scientists must constantly make press releases and provide general public information. These actions, along with new technology for recycling, remediation, and removal, will contribute to a decrease in the pollution of plastic and litter on Earth.

2.4.3 Nanoparticles

Pesticides, medications, individual maintenance items, plasticizers, perfumes, flicker retardants, hormones, siloxanes, algal pollutants, and various micronutrients, such as radionuclides and occasional earth rudiments, are just a few of the numerous environmental contaminants that are on the rise [77–79]. Nanoparticles (NPs) and nanomaterials (NMs) are two distinct categories of pollutants that fall under this category. "Material with any external dimension at the nanoscale or having an internal structure or surface structure at the nanoscale" is how the European Union (EU) and the International Organization for Standardization (ISO) describe nanometer materials (NMs) [80, 81]. Furthermore, according to the ISO and EU, NPs are defined as "nano-objects with all three external dimensions at the nanoscale"; here, the nanoscale represents the range of sizes between approximately 100 nm and larger [82]. Rainwater runoff or wind can release naturally occurring minerals (NMs) into the aquatic environment, either intentionally or unintentionally. This occurs in solid wastes and wastewater effluents [83, 84]. The ability of NMs to firmly adhere to and clump together with mineral surfaces is another effect of their small size, which also influences their mobility through porous media [85]. For example, NM mobility decreases in groundwater aquifers when they adhere to mineral surfaces [86]. Heavy clay soils tend to stabilize NMs and encourage increased dispersal since NMs can now travel farther before becoming trapped in the sand matrix. The sizes of colloids in aquatic systems range from 1 nm to 1 μm. Natural materials such as proteins, humic acids, fulvic acids, and peptides, as well as inorganic materials in colloidal form such as hydrous iron and manganese oxides, are examples of aquatic colloids. Owing to their tiny size and large surface area per unit mass, these materials are significant binding phases for both organic and inorganic pollutants [87].

The benthic organisms in freshwater were adversely impacted by the NP aggregates that settled to the bottom and gradually accumulated in the silt. NPs may accumulate in marine ecosystems between warm and cold currents [88]. In marine environments, biota may also recycle NPs [83]. As a result, species such as tuna that live in areas where warm and cold currents meet may be more vulnerable [88]. Nam et al. [89] reported that NPs settled in a simple microecosystem, resulting in a significant concentration of TiO_2 NPs in the sediment layer. Furthermore, Nam et al. [89] demonstrated that artificially produced nanoparticles (NPs), such as Ag and TiO_2 NPs, can pass through the feeding habits of aquatic organisms. Additionally, this study demonstrated that the attachment of NPs to the cell wall causes algal cells to concentrate the NPs. Therefore, the fate and movement of engineered nanomaterials (ENMs) in aquatic environments are influenced by their stability under such conditions. Large aggregates have rapidly deposited ENMs, with significantly limited transport and bioavailability [90]. On the other hand, tiny, evenly distributed aggregates of ENMs travel a greater distance and are more likely to come into contact with living things and pose a threat to them. Research on potential ENM forms that might occur in actual natural water systems is scarce [90]. However, ENMs frequently aggregate in water, which reduces their total surface area and, as a result, reduces their reactivity [90].

The significance of pH and natural organic matter on the fate and transport of ENMs in an aquatic environment was emphasized in an article by Lin et al. [90]. Research on how NMs affect different wildlife aquatic species, including "trout," where TiO_2 NP exposure leads to nullification of their activity, is currently ongoing [91]. Research on the biodegradability of NMs is still ongoing. For example, metals and metal oxides are not biodegradable, although some biodegradable fullerenes (such as C^{60} and C^{70}) have been shown to require many months to degrade. On-site remediation employs zero-valent iron nanoparticles (NPs), but little is known about their destiny and environmental transfer. Additionally, researching the destiny of NMs in the environment can be aided by modeling methods. According to Jeevanandam et al. [92], cigarette smoke, which contains NPs, can cause genetic changes, middle ear disease, pancreatic cancer, chronic respiratory illnesses, and worsened asthma. Conversely, biological NPs, also known as nanobacteria, are found in serum, blood, and organs and are ubiquitous in humans, animals, and other species. According to Buzea et al. [93], they may be the source of calcification disorders, which include ovarian and breast cancers, heart valve problems, aortic aneurysms, chronic prostatitis, renal stone development, and arterial plaque. Environmental samples and NM contaminants are assessed via several methods, such as size separation, particle counting, and morphological/chemical analysis. Aerosol fractionation is used to determine the size fractions of nanomaterials on the basis of their mobility in an electrical field. Additionally, by vaporizing and analyzing the resulting ions, aerosol mass spectrometers are employed in the chemical study of NMs dispersed in gases and liquids [94]. Expansion and condensation nucleus counters, with a 3 nm detection limit, can measure the density of NMs in gas suspensions via optical measurements [95]. Size exclusion chromatography, ultrafiltration, and field flow fractionation are methods that can be used to extract and size fractionate nanometer particles (NMs) from liquid environmental samples. Effective methods for eliminating NMs from the air include respirators and air filters [86]. There is less information available on the toxicity, risk, impact, and destiny of NMs. Therefore, more studies are necessary to close this knowledge gap. It is vital to reduce the risk of handling and utilizing these materials, minimize their discharge into the environment, and establish particular norms and standards for NMs. Creating a successful plan for recovering and recycling new NMs is essential for reducing any possible harm. If NMs are released into the environment, a life cycle analysis can help assess the actual environmental effects.

2.4.4 Pesticides

The use of pesticides and herbicides in agriculture is quite popular to increase output and guarantee enough high-quality, reasonably priced food for the expanding population, but alarms have been raised about the possible harmful effects on the environment, particularly water resources. The fundamental cause of the related environmental effects is the persistent and pervasive nature of several pesticides, which severely damage biodiversity [96]. Many of these herbicides are difficult to break down and remain in the environment for a long time, even if some can breakdown naturally in the soil [97]. Pesticides end up in water bodies because of runoff from industrial effluent and agricultural fields. Pesticides with low concentrations that accumulate in water can become more concentrated along the food chain and infiltrate aquatic organisms, which poses a risk to human health [96]. Prolonged exposure to pesticides through water consumption can mimic human hormones, compromising hormone balance, lowering immunity, causing issues with reproduction, increasing the risk of cancer, and lowering IQ, particularly in young children [98]. Individuals exposed to these chemicals frequently experience health problems such as cancer, fetal deformity, infertility, neurological disorders, and abnormalities of the hematopoietic and hormonal systems. Genotoxic, neurotoxic, and endocrine-disrupting activities are the fundamental mechanisms driving these consequences [99–101]. Pesticide usage in Asia is the highest in the world, with a maximum of 25 kg per hectare. Pesticides are used annually in approximately two million tons worldwide, with Europe accounting for 45%, the US accounting for 25%, and the rest of the world for 25%. Only 3.75% of pesticides are consumed in India. India uses only 0.5 kg/ha pesticides, compared with 12.0 kg/ha in Japan and 6.6 kg/ha in Korea [102]. India uses only 0.5 kg/ha pesticides, compared with 12.0 kg/ha in Japan and 6.6 kg/ha in Korea [102]. Herbicides constitute 47.5% of the pesticides used worldwide, followed by insecticides (29.5%), fungicides (17.5%), and miscellaneous substances (5.5%). Because they linger in the environment for a long period of time, OCPs, OPPs, and their derivatives are significant pesticides discovered in water from various sources. Toxic heavy metal and pesticide discharge causes major pollution of waterways. The use of pesticides in agriculture has been linked to autoimmune illnesses, including systemic lupus erythematosus [103].

According to Kim et al. [104], it has also been connected to a number of cancer types, respiratory conditions, and neurological illnesses. High quantities of heavy metals, which are used to make pesticides, can also be a result of the overuse of synthetic pesticides. This can affect the biochemistry and microbial activity of the soil, which can have a detrimental effect on plants [105]. Herbicides and pesticides enter water systems mostly through agricultural and urban runoff, where they seep into the soil or are discharged directly as contaminated wastewater [106]. High pesticide amounts have also been discovered in surface water, ground water, river water, and—interestingly—rainwater, according to a thorough investigation [102]. The Gomti River, one of the main tributaries of the Ganga in India, rises from a naturally occurring reservoir in an area that is both swampy and heavily forested. With a population of more than 3.5 million, Lucknow City is the state capital of

Uttar Pradesh and one of the main sources of drinking water for the city. Among the samples analyzed, 21 pesticides were found in river water and bed sediment collected from the Huangpu River. Pesticide residues ranged from 2.16 to 567.49 ng/L in the Gomti River water and from 0.92 to 813.59 ng/L in the bed sediments. Aged and weathered agricultural soils with recently used DDT in river catchments were proposed as sources of DDT contamination. The findings showed that the Gomti River's bed sediments are contaminated with DDT, lindane, endrin, and heptachlor epoxides and that this contamination may be a factor in the sediment toxicity that the river's freshwater environment experiences [107]. The phase transfer of pollutants is the basis for conventional pesticide treatment methods, which include coagulation–flocculation, adsorption, filtration, and sedimentation. Lopez-Loveira et al. [108] reported that these strategies can result in high operational costs and secondary pollutants, such as the formation of sludge. Advanced oxidation processes (AOPs) are clean solutions for treating water that contains biorefractory and recalcitrant contaminants such as pesticides. Its wide range of applications and thermodynamic viability have led to its adoption as a contemporary water purification technique [109]. The primary method of treating water with AOPs involves creating highly reactive hydroxyl radicals in situ, which indiscriminately oxidize a variety of resistant organic pollutants to completely mineralize them into carbon dioxide, water, and mineral salts. They can also convert pesticide compounds into more biodegradable forms [110]. The presence of pesticides in water has a negative impact on human health, and the severity of the impact varies depending on the solubility, half-life, adsorption capacity, and biodegradability of the pesticide components. Chemical pesticides will continue to constitute an essential part of pest management in the future. In urban and agricultural contexts, integrated pest management (IPM) is a great way to manage insects and pests while providing long-term, natural pest prevention. Therefore, to reduce the occurrence of pesticide chemicals, drinking water sources should undergo regular water quality monitoring [111]. Hence, biopesticides should be identified as alternatives to pesticides to minimize their negative environmental effects.

2.4.5 Personal Care

PCPs, or personal care products, are typically used for grooming, cleaning, personal hygiene, and beautifying products such as skin moisturizers, lipsticks, nail polishes, eye and facial makeup primers, shampoos, hair dyes, toothpaste, and deodorants. However, some may contain prescription drugs such as lip balms, mouthwashes, diaper ointments, protectants, and antiperspirants. PCPs include both classes of medications and cosmetics used for medicinal and cosmetic purposes [112]. Hence, they easily expose people because they fit the description of pharmaceuticals and/or cosmetics. Examples of such products are toothpaste, which strengthens teeth due to its fluoride concentration, and moisturizers, which calm the skin. These substances continuously and persistently discharge components that are both biologically active and inactive into the atmosphere, biosphere, and geosphere, and they have negative consequences for marine, wild, and human life [113]. Inhalation, cutaneous contact, ingestion, and absorption are the direct routes of exposure; environmental pollution and the use of other goods are the indirect pathways [114]. Numerous carcinogenic, mutagenic, and reproductive effects have been connected to PPCPs through studies on their exposure to people and animals [115, 116]. In addition to causing oxidative stress, histological and cytological alterations, and the suppression of hormones and enzymes in vertebrates, PPCPs have been shown to interfere with the hormonal system [117]. PCP environmental pollutants are found in drinking water, groundwater, surface water, wastewater solids, sediment and wastewater effluent, among other environmental compartments [118]. Evgenidou et al. [119] reported that sedimentation, adsorption, bioaccumulation, biodegradation, hydrolysis, and photolysis are the main environmental processes that contribute to particulate matter pollution in surface water. Elements that are both physical and chemical, such as suspended particles, pH, heavy metal concentrations, and soil permeability, all influence the environmental mobility of PPCPs [120]. The physical–chemical characteristics of PCP constituents, dilution factor, concentration, organic carbon content, and mode of action are some of the elements that affect the environmental distribution of PCP pollutants [121]. Serious health risks can arise from prolonged exposure to PPCPs, even at low concentrations (low to high μg/L). According to Keerthanan et al. [122], exposure to PPCPs can lead to the formation of ARGs, disturbance of the endocrine system, and other health effects. Therefore, addressing the identification, destiny, transformation, transfer, and management of petroleum products and petroleum-based waste products (PPCPs) in the environment is crucial. Additionally, because of the low occurrence concentrations of these contaminants and their difficulty in analyzing them, the removal of these contaminants in wastewater treatment plants (WWTPs) prior to their discharge into surface water has proven to be challenging [123]. Therefore, it is necessary to improve and modify current WWTPs to treat or eradicate these contaminants. Physical, chemical, or a combination of physicochemical processes, including ozonation, photocatalysis, UV irradiation, ultrasonication, flocculation, coagulation, sedimentation, biological processes, membrane filtration, chlorination, adsorption via activated carbon (AC), carbon nanotubes (CNTs), graphene oxide (GO), and others, are used in traditional wastewater treatment methods. Unfortunately, these methods are insufficient for removing PCPP pollutants from wastewater. Microalgae/fungal strains and anaerobic membrane bioreactors have emerged as two more promising techniques for removing PPCPs in recent years [120]. The problems posed by PPCPs have attracted increasing attention, as the world's water resources are being depleted daily because of their extremely harmful impacts, even at low concentrations. To lessen problems with location and social health. The development of a treatment process that prioritizes the safe and effective removal of ECs from wastewater is essential.

3 Effects of Water Contamination and Human Health

Water is an essential resource for human survival. Water source contamination has a significant effect on domestic and industrial growth, public health, and the survival of aquatic organisms, which provide food for humans. According to UNESCO's 2021 World Water Development Report, freshwater use has increased sixfold over the past century and has increased by approximately 1% annually since the 1980s. The increase in water consumption has resulted in significant problems with water quality. The rivers and oceans that sustain life are negatively impacted, which eventually affects human health and the advancement of sustainable social development. Urbanization, agriculture, and industrialization have all contributed to environmental pollution and degradation [124]. An estimated 80% of urban and industrial wastewater worldwide is released into the environment untreated, endangering both ecosystems and human health. There is a stronger link between pollution and health problems. Microorganisms that cause disease are referred to as pathogens, which can cause disease in people. [125]. In both developed and developing countries, floods and heavy rains are associated with harsh weather and the spread of numerous diseases. Ten percent of people are dependent on produce and food cultivated in tainted water. Numerous infectious illnesses that are transmitted through the fecal–oral route are associated with the contamination of water sources by feces [126]. The health hazards associated with contaminated water include a wide range of ailments, including cancer, respiratory issues, diarrheal problems, neurological disorders, and cardiovascular conditions. Nitrogenous chemicals are the cause of both blue baby syndrome and cancer [127] in water sources. Contaminated water has a major negative effect on the health of the fetus and increases the risk of low birth weight if pregnant women are exposed to pollutants [128]. Bad water threatens aquatic and human life by ruining crop harvests and contaminating our food. The food chain is disrupted by pollutants, and the fish respiratory system is impacted by heavy metals, particularly iron. The health of people may suffer significantly if they eat fish with iron blockage in their gills, which can have a serious negative impact on their health. Water containing metals can cause neurological disorders, liver cirrhosis, hair loss, and renal failure [129].

3.1 Bacterial Illness

The main cause of diarrhea is contaminated or untreated drinking water. Between 4% and 15% of diarrhea cases worldwide are caused by *Campylobacter jejuni*, which is characterized by fever, headache, nausea, and abdominal pain. Antibiotic use together with good hygiene habits can help prevent this illness. The tainted water is what causes the disease cholera caused by *Vibrio cholerae*. Watery diarrhea, nausea, vomiting, and dehydration, followed by renal failure, are the symptoms of this sickness. *Shigella* is the name for the bacterium that causes shigellosis. It affects the human digestive system and damages the intestinal lining. The digestive system is contaminated by salmonellosis. When polluted water contains *Salmonella* bacteria, the intestinal tract becomes inflamed, which frequently leads to death [130].

3.2 Viral Illness

Hepatitis is a liver-infecting virus caused by contaminated water. Hepatitis symptoms include jaundice, loss of appetite, exhaustion, pain and elevated fever. The long-term persistence of this condition increases the possibility of mortality. An inflammatory illness called encephalitis is transmitted by mosquito bites. For this illness, there is no vaccination available. Poliomyelitis virus is the cause of poliomyelitis. Sore throat, fever, nausea, constipation, diarrhea, and sometimes paralysis are all signs of poliomyelitis. Numerous viruses, such as rotaviruses, adenoviruses, calciviruses and Norwalk viruses, can cause gastroenteritis, resulting in fever, headache and vomiting. One to two days after infection, symptoms start to appear. For small children, elderly individuals, and those with impairments, illness can be dangerous [131].

3.3 Parasitic Illnesses

The parasite *Cryptosporidium parvum* is the source of the disease caused by cryptosporidiosis. Diarrhea, loss of or watery stools, stomach cramps, and an upset stomach are among the symptoms of this global illness. Human sickness and diarrhea are caused by the immune system-compromising, disinfection-resistant bacteria *Cryptosporidium. Entamoeba histolytica* is the cause of galloping amoebae, which damage the lining of the stomach. Watery diarrhea, chills, and fever are symptoms [132]. *Giardia lamblia* causes giardiasis. Intestinal lining cells can sustain damage. *Giardia* is immune to both disinfectants and chilly temperatures. It is also occasionally referred to as travelers' illness. Giardiasis symptoms include watery diarrhea, bloating, increased gas, and weight loss.

3.4 Skin Disorders

The maximum amount of arsenic toxins leads to skin disease, and the amount of arsenic in drinking water has a direct effect on the health of the local population. The amount of arsenic in biological samples (blood and hair) from individuals with skin disorders is correlated with drinking water contaminated with arsenic. There is a high prevalence of scabies in Bangladesh as a result of river pollution. Furthermore, skin cancer can also be caused by industrial water contamination [133].

3.5 Cancer

A controlled study of lung cancer patients and a frequency-matched hospital conducted in the northern United States revealed a substantial correlation between drinking water arsenic and lung cancer in Chile between 1994 and 1996. An elevated risk of liver cancer has also been linked to high levels of arsenic in drinking water, although this association was not statistically significant at exposure levels lower than 0.64 mg/L [134]. High levels of hexavalent chromium, which cause cancer, can be contracted by drinking water containing high levels of chromium. Cancer mortality is correlated with drinking water trihalomethane (THM) levels. The most robust and persistent correlation was found between the THM exposure index and bladder cancer mortality. Positive correlations were also observed with brain and non-Hodgkin's lymphoma, kidney cancer, and bladder cancer in men.

3.6 Child Health

Adult height loss is associated with childhood pollution exposure at critical developmental stages. In addition to causing hunger, illnesses that are directly related to water and sanitation can also lead to other infections, such as measles and pneumonia. Globally, more than one-third of children still experience stunting and malnutrition as a result of poor water and sanitation. According to a study conducted in rural India, children who lived in homes with tap water experienced a considerable reduction in the frequency and duration of illness [135]. Infant and child mortality are positively correlated with the level of industrial organic water pollution in less developed countries, and industrial water pollution has a significant role in both of these outcomes.

4 Prevalence and Impact of Health Consequences Due to exposure to Contaminated Water

One of the most important essentials for human survival and development is water. The Worldwide Water Development Report, which was published in 2021 by UNESCO, states that the amount of water used worldwide has increased sixfold. The quality of water was also negatively impacted by the increase in water usage. Increased agricultural production, urbanization, and industrialization all contribute to the contamination of entire water bodies [124]. The quality of drinking water has a significant impact on human health and is one of the key determinants. Waterborne infections have spread as a result of poor drinking water quality. Inadequate drinking water quality is linked to more than 50 diseases and accounts for 50% of infant mortality worldwide, according to the WHO research. By 2030, the Sustainable

Development Goal of 2017 aims to guarantee universal access to and sustainable management of water and sanitation (United Nations. The Millennium Development Goals Report. United Nations 2015:72. 978-92-1-101,320-7). Globally, 7.80 million people lack access to safe drinking water, whereas 2.5 billion people in the developing world lack adequate sanitation [136, 137]. People are at risk for health problems due to contaminated water and inadequate sanitation. Both developed and developing nations are at risk from emerging waterborne pathogens, which spread rapidly and affect large populations [138]. Contaminated drinking water containing dangerous microbes such bacteria, viruses, intestinal parasites, and protozoa can spread waterborne illnesses. Waterborne disease was the second leading cause of death in 1990, but by 2020, it was the ninth most significant cause of death, according to the Global Burden Disease Report [139]. Owing to contaminated drinking water and unsanitary sanitation practices, an estimated 829,000 individuals per year pass away from diarrheal illnesses, mostly cholera, dysentery, and typhoid fever [140]. To maintain a secure health status, adequate, safe, and easily accessible water supplies and adequate sanitation are crucial [141]. The stated that enhancing cleanliness, sanitation, and water supply could mitigate approximately 4% of the world's disease burden ([142] update). An estimated 37.7 million Indians are thought to be affected by watery illnesses each year; an estimated 1.5 million children may be away from diarrhea alone, and 73 million working days are lost as a result of waterborne illnesses [143]. Infections caused by water have an important impact on a country's economic growth while causing a high disease burden [144]. Across all age categories, the majority of Indians are ignorant of safe and hygienic behaviors. Consequently, the likelihood of contracting infectious infections increases. Bacterial infections are caused primarily by inadequate wastewater treatment procedures or fecal contamination of drinking water. The most prevalent of these are cholera, diarrhea, salmonellosis, schigellosis, and viral illnesses, including polio, encephalitis, hepatitis, and gastroenteritis, followed by parasitic illnesses such as schistosomiasis, galloping amoeba, cryptosporidiosis, and giardiasis. Diarrhea is the most prevalent illness caused by water pollution and is mostly spread by enteroviruses in aquatic environments.

4.1 Sources of Water Pollution

Industrialization, agriculture, natural disasters, and inadequate water supply and sewage treatment systems are the main causes of water pollution. Water pollution may result from the discharge of these pollutants into aquatic environments without sufficient treatment [145]. Wastewater from industrial activity has steadily increased as urbanization has accelerated [146]. Through drinking water, pesticides have detrimental effects on human health. A 10% increase in pesticide use was shown to be associated with a 1% increase in the medical disability index for individuals over 65, according to data from the Health Life Expectancy Longitudinal Survey [147]. The aquatic environment provides habitat for a variety of enteroviruses. Almost 100 harmful microbes identified in human and animal excrement can reach the ecosystem through drinking water, residential wells, sewage treatment plants, groundwater, estuaries, and the ocean [148]. The primary cause of waterborne infections in this area, such as viral hepatitis, diarrhea, dysentery, and gastroenteritis, is coliform bacterial water contamination [149]. Therefore, the main goal of water and sanitation health interventions is to prevent the transmission of diarrheal diseases from the environment to people [150]. An 82% reduction in infant mortality among Malaysian neonates has been attributed to easy access to clean water and hygienic conditions, especially among those who did not receive breastfeeding [151].

4.2 Water Pollution as the Main Source of Diarrhea

Diarrhea is the most prevalent ailment caused by water pollution and a leading source of illness and mortality for young children in low-income countries. According to Waddington et al. [150], diarrheal diseases account for 21% of the annual deaths of children under 5 years of age in developing countries. Consuming drinking water that has not been purified can result in diseases such as diarrhea [152]. Parasites present in nonpurifying drinking water can cause illnesses because of their presence [153]. People of all ages who drink purified water from water treatment facilities have been shown to experience fewer cases of diarrhea [154]. For example, a study carried out in the southern region of Brazil revealed a clear correlation between an increased risk of death from diarrhea and several factors, including crowded dwellings, poor housing conditions, and a lack of flush toilets. According to Victora et al. [155], the likelihood of infant mortality due to diarrhea was shown to be 4.8 times greater in families lacking piped water than in those with piped water. Diarrheal diseases such as cholera claim the lives of 1.8 million people annually; most of these deaths are from children under 5 years of age and are concentrated in developing countries. Eighty-eight percent of diarrheal infections are caused by poor sanitation, poor hygiene, and a lack of access to clean water [156]. One important source of these compounds is exposure to tainted food and water. Malnutrition and a weakened immune system can result from diarrhea in infants and young children, which increases the risk of chronic and recurring diarrhea [157]. According to a study performed in rural India, children who lived in homes with tap water suffered drastically fewer sicknesses and for shorter periods of time [158]. In conclusion, water contamination is a significant cause of illnesses in children. In 2016, air, water, and soil pollution killed 940,000 children worldwide, with two-thirds of those deaths occurring in underdeveloped or middle-income nations [159].

According to an analysis of more than 100 papers, water pollution significantly harms people's health overall, even if the consequences may vary depending on a person's country, geography, age, and gender. Water contamination causes a wide range of human diseases, such as cancer, skin disorders, diarrhea, and other pediatric issues. Thus, drinking water intervention is needed to prevent enteroviruses from spreading from the environment to humans. Second, there is a chance of developing skin conditions if one uses or is exposed to highly contaminated water. The primary pathogenic factors of skin illnesses are heavy metals in drinking water and an abundance of germs in saltwater. Third, there are three ways in which water contamination can endanger human health: the water source, treatment, and delivery system. The country should consider a few parameters: water quality at the point of use is a priority, and measures to enhance it include chlorination, safe storage, and the supply of clean, treated water [149, 160]. Countries should carry out epidemiological research independently to create health-promoting bathing water quality guidelines appropriate for their unique circumstances to lessen the effects of water pollution on skin ailments [161]. The purity of water sources, the scientific nature of water treatment, and the effectiveness of drinking water monitoring are important for reducing the incidence of cancer caused by water pollution. Every civilization also needs to prevent and regulate the sources of pollution that come from consumption, production, and transportation [159]. To increase public awareness, it is crucial to implement health education. Environmental education should be incorporated; public health awareness should be increased; and locals should be informed about sanitary water through newspapers, periodicals, television, the internet, and other media. Farmers should not use excessive amounts of agricultural chemicals that harm water supplies. Similarly, to address public concern, the government sector needs to improve water intervention management and implement various strategies to enhance water quality and lessen the negative health effects of water pollution.

5 Heavy metal Accumulation and Disease

Environmental contamination is aggravated as an outcome of the Industrial Revolution and by anthropogenic activities. Significant pollutants, such as compounds with atomic numbers >20 and densities >5 g/cm^3, are a class of heavy metals, and metalloids pose serious clinical issues. Similarly, heavy metals can be categorized into important and nonessential heavy metals. The essential metals are copper, chromium, cobalt, iron, manganese, molybdenum, selenium, and zinc, whereas the nonessential metals include arsenic, cadmium, mercury, nickel, and vanadium [162, 163]. Typically, nonessential heavy metals do not have any biological importance and have adverse effects on both the environment and human beings. However, both heavy metals are likely to be toxic when they are present at high concentrations; they are further transported and partitioned in cells, destroy macromolecules and interrupt cellular functions in the human body.

The International Agency for Research on Cancer (IARC) has grouped heavy metals on the basis of their level of toxicity or deleterious effects. Group I includes arsenic, lead, nickel, cadmium, and chromium, which are likely to be carcinogens. Usually, the group I category affects glucose metabolism and increases the risk of developing diabetes and cardiovascular diseases [164–167]. Metals with fewer risk effects fall into the group II category, which includes arsenic, bismuth, indium, thallium, and antimony, and group III includes compounds that play important roles in our body system, such as biochemical processes, and become toxic only when the concentration is exceeded [168]. Humans are exposed to heavy metals through several means, such as mining, e-waste disposal, the use of fertilizers and pesticides, and other occupational exposures. Cadmium, lead, chromium, and arsenic are among the most toxic, and their contents are discussed here.

Typically, toxic heavy metal exposure elevates lipid peroxidation and induces oxidative stress. Metals such as lead, cadmium, arsenic, and mercury are involved in the process of generating reactive oxygen species (ROS), which are composed of nitric oxide, superoxide radicals, and hydrogen peroxide [169, 170]. The toxicity of cadmium can be visualized even at very low concentrations. It is involved in the DNA repair mechanism by producing ROS and results in the apoptosis process. Additionally, it also binds to the mitochondria and hinders oxidative phosphorylation and cellular respiration [171]. However, metallothionein is a protein that functions as a free radical scavenger. Cadmium toxicity is resistant to metallothionein, whereas cells that cannot harvest metallothionein are sensitive [172]. Authors have also discussed acute and chronic types of exposure and their effects on bodily mechanisms. Typically, acute exposure to cadmium activates DNA hypomethylation and prevents DNA methyltransferase 1, but the opposite conditions are observed under chronic exposure; under these conditions, methyltransferase 1 is activated, resulting in hypermethylation [173, 174]. Lead is not a biodegradable compound, and the main source of lead is lead-based paint; other sources might include tap water, soil, and dust particles. Their entry into the human body involves ingesting, breathing, and absorbing through the skin. Sometimes, dietary intake, such as contaminated vegetables, fruits and other foodstuffs, tends to lead the human body. Short-term exposure to lead influences headache, loss of appetite, renal dysfunction, insomnia and arthritis [175]. Furthermore, lead is specifically known for its ability to produce reactive oxygen species (ROS), and oxidative stress can ruin nucleic acids and proteins and lead to lipid peroxidation [176]. In the human body, the organs that specifically absorb lead are the kidney at very high concentrations, followed by the brain, heart and liver [177]. In the human body, the mechanism of toxicity starts with the replacement of calcium ions, which helps lead to interactions with protein molecules and interferes with sulfhydryl enzymes and inhibits their functions. Research by Flora et al. [178] revealed that when a rat model is exposed to lead compounds, delta-aminolaevulinic acid dehydratase activity is deleterious, and a decrease in the glutathione level in the blood results in a reduction in the blood hemoglobin level. However, there was a steady increase in brain reactive oxygen species. Finally, it starts damaging overall organs, such as the brain, heart, kidney, and reproductive organs [179]. Chromium is a chief micronutrient that is an essential heavy metal that is specifically needed for protein, carbohydrate, and lipid metabolism [180]. Chromium compounds usually exist in two forms: trivalent and hexavalent chromium (Cr(VI)). Trivalent chromium (Cr(III)) was found to be safe, less absorbed, and highly desirable for human metabolism, whereas the hexavalent form is toxic and has been found to be a carcinogen. It is very difficult for Cr(III) to enter cells, whereas Cr(VI) can easily enter cell membranes through sulfate transporters [181, 182]. However, intercellular Cr(VI) is converted to Cr(III) and is involved in covalent interactions with macrophages. In an in vitro study, the immunotoxicities of several metals, such as lead, cadmium, nickel and chromium, were analyzed in natural killer (NK) cells. These metals have no effect on the functions of NK cells [183]. However, when chromium enters the system, it initially attaches to immune cells and is focused on damaging the DNA. In research conducted by Thomas et al. [184], exposure to these chromium compounds led to delayed immunity, whereas increased levels of mRNA expression of IL-4, IL-6, and IFN-Q were detected. The metallic compound arsenic enters the body of a person by consuming tainted foods, air, water, and other occupational exposures [185]. The mode of entry starts with ingestion, inhalation, and absorption through the dermal layers. Acute and long-term arsenic exposure leads to a condition called arsenicosis [186]. This exposure can lead to complications in the nervous system and the development of symmetrical sensorimotor neuropathy because the sensory nerves are more sensitive to arsenic. Studies have also confirmed that arsenic exposure leads to impairments in learning and memory [187]. Furthermore, arsenic toxicity can also damage the renal system by increasing the serum creatinine level and proteinuria. The cause of these complications is increased reactive oxygen species (ROS) activity, and oxidative stress increases the expression of heme oxygenase-1 and the pathways that regulate transcription factors [188]. Arsenic compounds also affect the male and female reproductive systems. In males, decreased testosterone, necrosis, and apoptosis were found due to dysfunction of gonad synthesis [189]. In females, birth defects such as late fetal and infant mortality are greater [190]. Factors affecting embryogenesis include alterations in cell proliferation and abnormal DNA methylation patterns [191]. Thus, researchers must focus on proper environmentally friendly and cost-effective remediation techniques so that heavy metal exposure can be mapped quickly.

5.1 Ingesting microplastics and Metal Ions in Human Physiology

The individual toxicity of microplastics and heavy metals has been widely studied, but very few literature sources are available on the connections between microplastics and heavy metals and their related health hazards. However, both microplastics and heavy metals exist in the environment for longer periods of time. Typically, microplastics have active sites with greater surface areas, which enables them to attract a wide variety of pollutants, such as heavy metals, antibiotics and several other organic substances, from the environment. Once contaminants are attached to microplastics, they are transferred among food webs and start accumulating in humans and other organisms [192–194]. The combined effects of microplastics and heavy metals are becoming a global concern, and researchers are lagging in their understanding of the exact mechanism of interaction and its negative effects. Heavy metals are typically used as pigments and stabilizers in plastic products. The average concentration of metals usually ranges from 1% to 10% [195]. Numerous studies have recognized the adsorption and desorption of heavy metals to focus on the bioaccessibility of microplastics in releasing heavy metals into the human system via in vitro dynamic models. Chen et al. [196] focused on the bioaccessibility of microplastics and heavy metals via an in vitro model for assessing their implications for human health. Heavy metals As(V), Cr(VI), Cd(II), and Pb(II) are also microplastics, namely, polyvinyl chloride (PVC), used in this study. They focused on adsorption and desorption studies. The amount of heavy metals adsorbed was high during the first 1–6 h, after which it reached a constant equilibrium state because adsorption took place on the external surface of the PVC [197]. The accessibility of the abovementioned heavy metals to the PVC was checked via a soluble bioavailability research consortium (SBRC) assay in gastric juice and the small intestine, which revealed that the level of Pb(II) was greater in the gastric phase than were the other heavy metals. Furthermore, via FTIR, they reported that purposefully active groups present on the surface of PVC are not responsible for capturing heavy metals and identified carcinogenic risk factors upon oral ingestion of Cr(VI) and As(V). Another study conducted by Liao and Yang [198] explored heavy metal-contaminated microplastics. Chromium was chosen as the heavy metal, and four types of nondegradable microplastics, such as polypropylene (PP), polystyrene (PS), polyethylene (PE), polyvinyl chloride (PVC), and degradable plastic, such as polylactic (PLA), were used. The whole digestive in vitro method was used in this study. Degradable PLA was found to have the highest value of 19.9% in the stomach, 15.6% in the small intestine, and 3.9% in the large intestine because of surface decomposition by the acids secreted in the human intestine and digestive enzymes. Similarly, the hazard quotient was 1.0, which is noncarcinogenic, with daily values ranging from 0.50 to 1.18 μg/day. In addition, Godoy et al. [199] used a dynamic in vitro stimulator to determine the bioaccessibility of Cr and Pb with polypropylene and polyethylene and reported that the heavy metal Cr was released earlier under gastric conditions, whereas Pb was released more slowly, and when the pH reached 6.5 in the duodenum, it started precipitating. Similarly, 23.11% of the Cr and 23.17% of the Pb in the microplastic cross the tubular membrane and start intestinal absorption on the basis of the type and properties of the polymer. Thus, the toxicity of heavy metal and microplastic combinations in humans is based on several factors, such as age, sex, dosage level, and exposure. However, if the concentration increases, it may damage the tissue, cells, and estrogenic responses.

6 Microbial Pollution of Drinking Water Current Status and Disease Outbreaks

Rivers, lakes, or ground waters are typically the foundations of drinking water, depending on the availability of water resources that have been redirected for human consumption through distribution networks. Microbiological contamination may still occur even when the water has previously been disinfected [38]. People who come into contact with tainted water may suffer from *E. coli* infections, diarrhea, giardiasis, cholera, typhoid fever, salmonellosis, hepatitis A, and polio. Admittance to sufficient water infrastructure and a clean water supply is still not guaranteed for the entire world's population. The need for clean drinking water has increased in recent years, but the supply of high-quality water resources has decreased as a result of human activities, including industrialization, urbanization, farming, raising livestock, and the consequences of climate change. The likelihood of human pathogen contamination and the spread of watery illnesses has increased due to combinations of these factors.

6.1 Types of Biological Contaminants in Drinking Water Sources

6.1.1 Pathogenic Bacteria

The most common harmful bacteria that can be detected in water are *Salmonella typhi*, *Vibrio cholerae*, and *Escherichia coli*. causes cholera, typhoid, and diarrhea, among other waterborne illnesses [200, 201]. According to Rossolini et al. [202], enterotoxigenic *E. coli* (ETEC, also known as O148) and enterohemorrhagic *E. coli* (EHEC, also known as O157), which cause 100 million instances of diarrhea and thousands of deaths annually, are the bacterial enteropathogens most frequently discovered in young kids under the age of five. *Vibrio cholerae* is a common species of *Vibrio* that can spread illness through water. It is responsible for waterborne diseases. Cholera and gastroenteritis are caused by *V. cholerae* O1 and *V. cholerae* O139 [201, 203]. There are between 1.3 million and four million cholera cases annually, as well as between 21,000 and 14,300 cholera-related deaths. With 473,000 cases reported to the WHO in 2022—more than twice as many as those recorded in 2021—cholera continues to pose a concern to public health worldwide. Preliminary data for 2023 show even higher numbers, with over 700,000 cases reported. The disease cholera can be prevented and treated by inadequate waste management, hygienic conditions, and access to safe water [204]. Another gram-negative, rod-shaped bacterium, *Salmonella*, can spread to humans when they drink water containing animal excrement. In addition to gastrointestinal illnesses, it can cause typhoid and paratyphoid fever [205]. *Shigella* is a gram-negative rod-shaped bacterium that is the oldest pathogen specific to humans. It causes shigellosis, also known as bacillary dysentery, in people. *Shigella dysenteriae*, *S. flexneri*, and *S. sonnei* are the most common *Shigella* species, accounting for the global outbreak of dysentery sickness [201, 206].

6.1.2 Virus

Waterborne infections are linked to more than 150 enteric viruses. Numerous disease outbreaks are caused by the predominance of bacteria in drinking water: bacteriophages [207], adenoviruses, rotaviruses, enteroviruses, astrosviruses, noroviruses, and hantaviruses cause various disease outbreaks [208]. Human adenoviruses (HAdVs) are double-stranded DNA viruses with icosahedral capsid structures that are nonenveloped, nonsegmented, and nonsegmented. From 2007 to 2019, HAdV was responsible for approximately 10.8% of all gastroenteritis cases in children under 5 years of age in Sub-Saharan Africa [209]. A noteworthy and important waterborne pathogen is rotavirus (RV), which is a member of the Reoviridae family [210]. Approximately 258 million cases of diarrhea in children under 5 years of age worldwide have been connected to rotavirus infection [211]. A spherical, nonenveloped virus with a positive-sense single-stranded RNA genome is called an enterovirus (EV). EV has caused sporadic outbreaks of sickness that have affected millions of people globally. Enterovirus EV-D68 is the most commonly reported type; it accounted for 56% of all recorded types between 2014 and 2016 and 68% of the detected types in 2014 [212, 213]. Diarrhea caused by an enteric virus is now the second most common cause of death for children under 5 years of age worldwide [214]. Small nonenveloped single-stranded RNA viruses are known as noroviruses. Immunocompromised individuals and young children are typically the most affected by gastroenteritis due to noroviruses, which can lead to chronic illnesses. After being identified as the cause of gastroenteritis in Asia, a new strain of norovirus called GII.17 has been detected globally [215]. Hepatitis E virus (HEV) is the cause of hepatitis E disease, a liver disorder that is common in Sub-Saharan African and Asian countries with poor sanitation and personal hygiene, limited admittance to harmless drinking water, and a fecal–oral transmission route that is primarily through contaminated water [216].

6.1.3 Parasites

It has been estimated that parasitic waterborne diseases account for 842,000 deaths each year. Giardiasis, a major threat in developing countries, is caused by *Giardia intestinalis*. It has been reported that 33% of the population globally is affected each year. According to Dunn and Juergens [217], it is spread primarily via contaminated water or fecal contamination to people through the oral route. Amoebic dysentery, the third leading cause of death worldwide from parasitic infections, is caused *by Entamoeba histolytica.* According to reports, the parasite kills close to 100,000 individuals annually [218]. When there is no access to clean, safe water, it is common in nations with low socioeconomic status and poor public health [219]. Climate change and socioeconomic factors play vital roles in microbial contamination, and monitoring the concentrations of pathogens in drinking water sources to predict future human health risks is crucial [220, 221].

References

1. Manual, Centre for Affordable Water and Sanitation Technology (CAWST) (2009). Introduction to drinking water quality testing.
2. Binnie, C., Kimber, M., & Smethurst, G. (2002). *Basic water treatment*. Royal Society of Chemistry.
3. WHO (2017). Guidelines for drinking water quality, 4th edition.
4. CDC A. W. (2020). Centers for disease control and prevention.
5. Shammi, M., Rahman, M.M., Bondad, S.E. and Bodrud-Doza, M., 2019. Impacts of salinity intrusion in community health: A review of experiences on drinking water sodium from coastal areas of Bangladesh. In Healthcare (7, 1, 50). MDPI.
6. Lei, M., Zhang, L., Lei, J., Zong, L., Li, J., Wu, Z., & Wang, Z. (2015). *Overview of emerging contaminants and associated human health effects*. BioMed Research International.
7. Shah, A., Arjunan, A., Baroutaji, A., & Zakharova, J. (2023). A review of physicochemical and biological contaminants in drinking water and their impacts on human health. *Water Science and Engineering, 16*, 333.
8. Fawell, J., & Nieuwenhuijsen, M. J. (2003). Contaminants in drinking water: Environmental pollution and health. *British Medical Bulletin, 68*(1), 199–208.
9. Bhattacharya, A. (2006). Remediation of pesticide-polluted waters through membranes. *Separation and Purification Reviews, 35*(1), 1–38.
10. Damalas, C. A., & Eleftherohorinos, I. G. (2011). Pesticide exposure, safety issues, and risk assessment indicators. *International Journal of Environmental Research and Public Health, 8*(5), 1402–1419.
11. Ahluwalia, S. S., & Goyal, D. (2007). Microbial and plant derived biomass for removal of heavy metals from wastewater. *Bioresource Technology, 98*(12), 2243–2257.
12. Palansooriya, K. N., Yang, Y., Tsang, Y. F., Sarkar, B., Hou, D., Cao, X., Meers, E., Rinklebe, J., Kim, K. H., & Ok, Y. S. (2020). Occurrence of contaminants in drinking water sources and the potential of biochar for water quality improvement: A review. *Critical Reviews in Environmental Science and Technology, 50*(6), 549–611.
13. Howe, K. J., Hand, D. W., Crittenden, J. C., Trussell, R. R., & Tchobanoglous, G. (2012). *Principles of water treatment*. John Wiley & Sons.
14. Lipczynska-Kochany, E. (2018). Effect of climate change on humic substances and associated impacts on the quality of surface water and groundwater: A review. *Science of the Total Environment, 640*, 1548–1565.
15. Allaire, M., Wu, H., & Lall, U. (2018). National trends in drinking water quality violations. *Proceedings of the National Academy of Sciences, 115*(9), 2078–2083.
16. Pekcan-Hekim, Z. (2007). Effects of turbidity on feeding and distribution of fish. Department of Biological and Environmental Sciences University of Helsinki Finland, Academic dissertation, ISBN 978-952-92-2127-1, Yliopistopaino, Helsinki 2007.
17. Singh, N., Kamboj, S., Siwal, S. S., Srivastav, A. L., & Naresh, R. K. (2024). Toxic, nontoxic, and essential elements in drinking water: Sources and associated health issues in rural Asia. In *Water resources Management for Rural Development* (pp. 171–190). Elsevier.
18. World Health Organization (2011). Nuclear accidents and radioactive contamination of foods. *Retrieved July, 8*, p. 2017.
19. Batchelor, A. L., Phillips, R. J., & Searle, A. G. (1969). The ineffectiveness of chronic irradiation with neutrons and gamma rays in inducing mutations in female mice. *The British Journal of Radiology, 42*(498), 448–451.
20. Berry, R. J. (1987). Early radiation casualties-what Chernobyl has taught us. *Journal of the Society for Radiological Protection, 7*(2), 55.
21. Parmentier, N., & Nenot, J. C. (1989). Radiation damage aspects of the Chernobyl accident. *Atmospheric Environment, 23*, 771–775.
22. Rajkhowa, S., Sarma, J., & Das, A. R. (2021). Radiological contaminants in water: Pollution, health risk, and treatment. In *Contamination of Water* (pp. 217–236). Academic Press.

23. Cadet, J., Douki, T., Gasparutto, D., & Ravanat, J. L. (2005). Radiation-induced damage to cellular DNA: Measurement and biological role. *Radiation Physics and Chemistry, 72*(2–3, 293), –299.
24. Brewen, J. G., Preston, R. J., & Gengozian, N. (1975). Analysis of X-ray-induced chromosomal translocations in human and marmoset spermatogonial stem cells. *Nature, 253*(5491), 468–470.
25. Shaw, P., Duncan, A., Vouyouka, A., & Ozsvath, K. (2011). Radiation exposure and pregnancy. *Journal of Vascular Surgery, 53*(1), 28S–34S.
26. Lucas, H. F., & Krause, D. P. (1960). Preliminary survey of radium 226 and radium 228 (MsThI) contents of drinking Water. *Radiology, 74*(1), 114–114.
27. Korostynska, O., Mason, A., Ikezawa, S., & Al-Shamma'a, A. I. (2014). Modern methods of monitoring radiological contamination of water reservoirs. *Sensing Technology: Current Status and Future Trends, I*, 309–324.
28. Paquin, M., & Cosgrove, C. (2016). *The United Nations world water development report 2016: Water and jobs*. UNESCO for UN-Water.
29. Daschner, F. D., Rüden, H., Simon, R., & Clotten, J. (1996). Microbiological contamination of drinking water in a commercial household water filter system. *European Journal of Clinical Microbiology and Infectious Diseases, 15*, 233–237.
30. Ashbolt, N. J. (2004). Microbial contamination of drinking water and disease outcomes in developing regions. *Toxicology, 198*(1–3), 229–238.
31. Hitzfeld, B. C., Höger, S. J., & Dietrich, D. R. (2000). Cyanobacterial toxins: Removal during drinking water treatment, and human risk assessment. *Environmental Health Perspectives, 108*(suppl 1), 113–122.
32. Rao, P. V., Gupta, N., Bhaskar, A. S., & Jayaraj, R. (2002). Toxins and bioactive compounds from cyanobacteria and their implications on human health. *Journal of Environmental Biology, 23*(3), 215–224.
33. Inamori, Y., & Fujimoto, N. (2009). Water quality and standards—Vol. II, microbial/biological contamination of water. *Encyclopaedia of Life support systems (EOLSS).*
34. Nwachcuku, N., & Gerba, C. P. (2004). Emerging waterborne pathogens: Can we kill them all? *Current Opinion in Biotechnology, 15*(3), 175–180.
35. Rusin, P. A., Rose, J. B., Haas, C. N. and Gerba, C. P. (1997). Risk assessment of opportunistic bacterial pathogens in drinking water. *Reviews of Environmental contamination and toxicology: Continuation of residue reviews*, pp.57–83.
36. Nwabor, O. F., Nnamonu, E. I., Martins, P. E., & Ani, O. C. (2016). Water and waterborne diseases: A review. *International Journal of Tropical Disease and Health, 12*(4), 1–14.
37. Schriewer, A., Miller, W. A., Byrne, B. A., Miller, M. A., Oates, S., Conrad, P. A., Hardin, D., Yang, H. H., Chouicha, N., Melli, A., & Jessup, D. (2010). Presence of Bacteroidales as a predictor of pathogens in surface waters of the Central California coast. *Applied and Environmental Microbiology, 76*(17), 5802–5814.
38. Sharma, S., & Bhattacharya, A. J. A. W. S. (2017). Drinking water contamination and treatment techniques. *Applied Water Science, 7*(3), 1043–1067.
39. Harvell, C. D., Kim, K., Burkholder, J. M., Colwell, R. R., Epstein, P. R., Grimes, D. J., Hofmann, E. E., Lipp, E. K., Osterhaus, A. D. M. E., Overstreet, R. M., & Porter, J. W. (1999). Emerging marine diseases—climate links and anthropogenic factors. *Science, 285*(5433), 1505–1510.
40. Daszak, P., Cunningham, A. A., & Hyatt, A. D. (2000). Emerging infectious diseases of wildlife—threats to biodiversity and human health. *Science, 287*(5452), 443–449.
41. Harvell, C. D., Mitchell, C. E., Ward, J. R., Altizer, S., Dobson, A. P., Ostfeld, R. S., & Samuel, M. D. (2002). Climate warming and disease risks for terrestrial and marine biota. *Science, 296*(5576), 2158–2162.
42. Pandey, P. K., Kass, P. H., Soupir, M. L., Biswas, S., & Singh, V. P. (2014). Contamination of water resources by pathogenic bacteria. *AMB Express, 4*, 1–16.

43. Banzhaf, S., Filipovic, M., Lewis, J., Sparrenbom, C. J., & Barthel, R. (2017). A review of contamination of surface-, ground-, and drinking water in Sweden by perfluoroalkyl and polyfluoroalkyl substances (PFASs). *Ambio, 46*(3), 335–346.
44. Vestergren, R., Berger, U., Glynn, A., & Cousins, I. T. (2012). Dietary exposure to perfluoroalkyl acids for the Swedish population in 1999, 2005 and 2010. *Environment International, 49*, 120–127.
45. Bjermo, H., Darnerud, P. O., Pearson, M., Barbieri, H. E., Lindroos, A. K., Nälsén, C., Lindh, C. H., Jönsson, B. A., & Glynn, A. (2013). Serum concentrations of perfluorinated alkyl acids and their associations with diet and personal characteristics among S wedish adults. *Molecular Nutrition & Food Research, 57*(12), 2206–2215.
46. Olsen, G. W., Burris, J. M., Ehresman, D. J., Froehlich, J. W., Seacat, A. M., Butenhoff, J. L., & Zobel, L. R. (2007). Half-life of serum elimination of perfluorooctanesulfonate, perfluorohexanesulfonate, and perfluorooctanoate in retired fluorochemical production workers. *Environmental Health Perspectives, 115*(9), 1298–1305.
47. Han, X., Snow, T. A., Kemper, R. A., & Jepson, G. W. (2003). Binding of perfluorooctanoic acid to rat and human plasma proteins. *Chemical Research in Toxicology, 16*(6), 775–781.
48. Scher, D. P., Kelly, J. E., Huset, C. A., Barry, K. M., Hoffbeck, R. W., Yingling, V. L., & Messing, R. B. (2018). Occurrence of perfluoroalkyl substances (PFAS) in garden produce at homes with a history of PFAS-contaminated drinking water. *Chemosphere, 196*, 548–555.
49. Brown, J. B., Conder, J. M., Arblaster, J. A., & Higgins, C. P. (2020). Assessing human health risks from per-and polyfluoroalkyl substance (PFAS)-impacted vegetable consumption: A tiered modeling approach. *Environmental Science & Technology, 54*(23), 15202–15214.
50. Sunderland, E. M., Hu, X. C., Dassuncao, C., Tokranov, A. K., Wagner, C. C., & Allen, J. G. (2019). A review of the pathways of human exposure to poly-and perfluoroalkyl substances (PFASs) and present understanding of health effects. *Journal of Exposure Science & Environmental Epidemiology, 29*(2), 131–147.
51. Breysse, P. N. (2018). *Toxicological profile for perfluoroalkyls*. Agency for Toxic Substances & Disease Registry.
52. Lewis, R. C., Johns, L. E., & Meeker, J. D. (2015). Serum biomarkers of exposure to perfluoroalkyl substances in relation to serum testosterone and measures of thyroid function among adults and adolescents from NHANES 2011–2012. *International Journal of Environmental Research and Public Health, 12*(6), 6098–6114.
53. Webster, G. (2010). Potential human health effects of perfluorinated chemicals (PFCs). *Blood, 4*(6), 1–10.
54. Darrow, L. A., Stein, C. R., & Steenland, K. (2013). Serum perfluorooctanoic acid and perfluorooctane sulfonate concentrations in relation to birth outcomes in the mid-Ohio Valley, 2005–2010. *Environmental Health Perspectives, 121*(10), 1207–1213.
55. Stoiber, T., Evans, S., Temkin, A. M., Andrews, D. Q., & Naidenko, O. V. (2020). PFAS in drinking water: An emergent water quality threat. *Water Solutions, 1*(40), e49.
56. Thompson, R. C., Olsen, Y., Mitchell, R. P., Davis, A., Rowland, S. J., John, A. W., McGonigle, D., & Russell, A. E. (2004). Lost at sea: Where is all the plastic? *Science, 304*(5672), 838–838.
57. Provencher, J. F., Vermaire, J. C., Avery-Gomm, S., Braune, B. M., & Mallory, M. L. (2018). Garbage in guano? Microplastic debris found in faecal precursors of seabirds known to ingest plastics. *Science of the Total Environment, 644*, 1477–1484.
58. Pico, Y., Alfarhan, A., & Barcelo, D. (2019). Nano-and microplastic analysis: Focus on their occurrence in freshwater ecosystems and remediation technologies. *TrAC Trends in Analytical Chemistry, 113*, 409–425.
59. Silva, A. B., Bastos, A. S., Justino, C. I., da Costa, J. P., Duarte, A. C., & Rocha-Santos, T. A. (2018). Microplastics in the environment: Challenges in analytical chemistry-a review. *Analytica Chimica Acta, 1017*, 1–19.
60. Jiang, J. Q. (2018). Occurrence of microplastics and its pollution in the environment: A review. *Sustainable Production and Consumption, 13*, 16–23.

61. Bhardwaj, L. K., Rath, P., Yadav, P., & Gupta, U. (2024). Microplastic contamination, an emerging threat to the freshwater environment: A systematic review. *Environmental Systems Research, 13*(1), 1–17.
62. Sun, J., Dai, X., Wang, Q., Van Loosdrecht, M. C., & Ni, B. J. (2019). Microplastics in wastewater treatment plants: Detection, occurrence and removal. *Water Research, 152*, 21–37.
63. Viaroli, S., Lancia, M., & Re, V. (2022). Microplastics contamination of groundwater: Current evidence and future perspectives. A review. *Science of the Total Environment, 824*, 153851.
64. Wang, Z., Lin, T., & Chen, W. (2020). Occurrence and removal of microplastics in an advanced drinking water treatment plant (ADWTP). *Science of the Total Environment, 700*, 134520.
65. Mintenig, S. M., Löder, M. G., Primpke, S., & Gerdts, G. (2019). Low numbers of microplastics detected in drinking water from ground water sources. *Science of the Total Environment, 648*, 631–635.
66. Winkler, A., Santo, N., Ortenzi, M. A., Bolzoni, E., Bacchetta, R., & Tremolada, P. (2019). Does mechanical stress cause microplastic release from plastic water bottles? *Water Research, 166*, 115082.
67. Tong, H., Jiang, Q., Hu, X., & Zhong, X. (2020). Occurrence and identification of microplastics in tap water from China. *Chemosphere, 252*, 126493.
68. Shruti, V. C., Pérez-Guevara, F., & Kutralam-Muniasamy, G. (2020). Metro station free drinking water fountain-a potential "microplastics hotspot" for human consumption. *Environmental Pollution, 261*, 114227.
69. Danopoulos, E., Twiddy, M., & Rotchell, J. M. (2020). Microplastic contamination of drinking water: A systematic review. *PLoS One, 15*(7), e0236838.
70. Shruti, V. C., Kutralam-Muniasamy, G., Pérez-Guevara, F., Roy, P. D., & Elizalde-Martínez, I. (2022). Free, but not microplastic-free, drinking water from outdoor refill kiosks: A challenge and a wake-up call for urban management. *Environmental Pollution, 309*, 119800.
71. Eerkes-Medrano, D., Leslie, H. A., & Quinn, B. (2019). Microplastics in drinking water: A review and assessment. *Current Opinion in Environmental Science & Health, 7*, 69–75.
72. Paço, A., Jacinto, J., da Costa, J. P., Santos, P. S., Vitorino, R., Duarte, A. C., & Rocha-Santos, T. (2019). Biotechnological tools for the effective management of plastics in the environment. *Critical Reviews in Environmental Science and Technology, 49*(5), 410–441.
73. Zumstein, M. T., Rechsteiner, D., Roduner, N., Perz, V., Ribitsch, D., Guebitz, G. M., Kohler, H. P. E., McNeill, K., & Sander, M. (2017). Enzymatic hydrolysis of polyester thin films at the nanoscale: Effects of polyester structure and enzyme active-site accessibility. *Environmental Science & Technology, 51*(13), 7476–7485.
74. Yoshida, S., Hiraga, K., Takehana, T., Taniguchi, I., Yamaji, H., Maeda, Y., Toyohara, K., Miyamoto, K., Kimura, Y., & Oda, K. (2016). A bacterium that degrades and assimilates poly (ethylene terephthalate). *Science, 351*(6278), 1196–1199.
75. Paço, A., Duarte, K., da Costa, J. P., Santos, P. S., Pereira, R., Pereira, M. E., Freitas, A. C., Duarte, A. C., & Rocha-Santos, T. A. (2017). Biodegradation of polyethylene microplastics by the marine fungus Zalerion maritimum. *Science of the Total Environment, 586*, 10–15.
76. Picó, Y., & Barceló, D. (2019). Analysis and prevention of microplastics pollution in water: Current perspectives and future directions. *ACS Omega, 4*(4), 6709–6719.
77. Gwenzi, W., Mangori, L., Danha, C., Chaukura, N., Dunjana, N., & Sanganyado, E. (2018). Sources, behaviour, and environmental and human health risks of high-technology rare earth elements as emerging contaminants. *Science of the Total Environment, 636*, 299–313.
78. Sengupta, S., Roy, U., Chowdhary, S., & Das, P. (2021). New bioremediation technologies to remove heavy metals and radionuclides. Removal of emerging contaminants through microbial processes, pp.23–45.
79. Priya, A. K., Gnanasekaran, L., Rajendran, S., Qin, J., & Vasseghian, Y. (2022). Occurrences and removal of pharmaceutical and personal care products from aquatic systems using advanced treatment-a review. *Environmental Research, 204*, 112298.
80. ISO (2010). International Organization for Standardization. Nanotechnologies vocabulary part 1: core terms. ISO/TS 80004-1.

81. Rauscher, H., Mech, A., Gibson, N., Gilliland, D., Held, A., Kestens, V., Koeber, R., Linsinger, T. P. J., & Stefaniak, E. A. (2019). *Identification of nanomaterials through measurements*. Publications Office of the European Union.
82. ISO (2008). International Organization for Standardization. Technical specification: Nanotechnologies terminology and definitions for nanoobjects nanoparticle, nanofibre and nanoplate. ISO/TS 80004-2:2008.
83. Klaine, S. J., Alvarez, P. J., Batley, G. E., Fernandes, T. F., Handy, R. D., Lyon, D. Y., Mahendra, S., McLaughlin, M. J., & Lead, J. R. (2008). Nanomaterials in the environment: Behavior, fate, bioavailability, and effects. *Environmental Toxicology and Chemistry: An International Journal, 27*(9), 1825–1851.
84. Lowry, G. V., Hotze, E. M., Bernhardt, E. S., Dionysiou, D. D., Pedersen, J. A., Wiesner, M. R., & Xing, B. (2010). Environmental occurrences, behavior, fate, and ecological effects of nanomaterials: An introduction to the special series. *Journal of Environmental Quality, 39*(6), 1867–1874.
85. Tourinho, P. S., Van Gestel, C. A., Lofts, S., Svendsen, C., Soares, A. M., & Loureiro, S. (2012). Metal-based nanoparticles in soil: Fate, behavior, and effects on soil invertebrates. *Environmental Toxicology and Chemistry, 31*(8), 1679–1692.
86. Wiesner, M. R., Lowry, G. V., Alvarez, P., Dionysiou, D., & Biswas, P. (2006). Assessing the risks of manufactured nanomaterials. *Environmental Science and Technology, 40*(14), 4336–4345.
87. Zhang, M., Yang, J., Tang, L., Pan, X., & Zhang, D. (2020). What occurs in colloidal gas aphron-induced separation of titanium dioxide nanoparticles? Particle fate analysis by tracking technologies. *Science of the Total Environment, 716*, 137104.
88. Hyseni, S. (2016). Toxicological effects of nanomaterials on aqueous and terrestrial ecosystems. *Center for Development and Strategy, 2016*, 1.
89. Nam, D. H., Lee, B. C., Eom, I. C., Kim, P., & Yeo, M. K. (2014). Uptake and bioaccumulation of titanium-and silver-nanoparticles in aquatic ecosystems. *Molecular & Cellular Toxicology, 10*, 9–17.
90. Lin, D., Tian, X., Wu, F., & Xing, B. (2010). Fate and transport of engineered nanomaterials in the environment. *Journal of Environmental Quality, 39*(6), 1896–1908.
91. Federici, G., Shaw, B. J., & Handy, R. D. (2007). Toxicity of titanium dioxide nanoparticles to rainbow trout (Oncorhynchus mykiss): Gill injury, oxidative stress, and other physiological effects. *Aquatic Toxicology, 84*(4), 415–430.
92. Jeevanandam, J., Barhoum, A., Chan, Y. S., Dufresne, A., & Danquah, M. K. (2018). Review on nanoparticles and nanostructured materials: History, sources, toxicity and regulations. *Beilstein Journal of Nanotechnology, 9*(1), 1050–1074.
93. Buzea, C., Pacheco, I. I., & Robbie, K. (2007). Nanomaterials and nanoparticles: Sources and toxicity. *Biointerphases, 2*(4), MR17–MR71.
94. EPA, U.S.E.P.A. (2010). Emerging contaminants nanomaterials. https://archive.epa.gov/regio n9/mediacenter/web/pdf/emerging_contaminant_nanomaterials.pdf.
95. Saghafifar, H., Kürten, A., Curtius, J., von der Weiden, S. L., Hassanzadeh, S., & Borrmann, S. (2009). Characterization of a modified expansion condensation particle counter for detection of nanometer-sized particles. *Aerosol Science and Technology, 43*(8), 767–780.
96. Sharma, A., Kumar, V., Shahzad, B., Tanveer, M., Sidhu, G. P. S., Handa, N., Kohli, S. K., Yadav, P., Bali, A. S., Parihar, R. D., & Dar, O. I. (2019). Worldwide pesticide usage and its impacts on ecosystem. *SN Applied Sciences, 1*, 1–16.
97. Youssef, L., Younes, G., & Al-Oweini, R. (2019). Photocatalytic degradation of atrazine by heteropolyoxotungstates. *Journal of Taibah University for Science, 13*(1), 274–279.
98. Yadav, I. C., Devi, N. L., Syed, J. H., Cheng, Z., Li, J., Zhang, G., & Jones, K. C. (2015). Current status of persistent organic pesticides residues in air, water, and soil, and their possible effect on neighboring countries: A comprehensive review of India. *Science of the Total Environment, 511*, 123–137.
99. Monneret, C. (2017). What is an endocrine disruptor? *Comptes Rendus. Biologies, 340*(9–10), 403–405.

100. Gundogan, K., Donmez-Altuntas, H., Hamurcu, Z., Akbudak, I. H., Sungur, M., Bitgen, N., Baskol, G., & Bayram, F. (2018). Evaluation of chromosomal DNA damage, cytotoxicity, cytostasis, oxidative DNA damage and their relationship with endocrine hormones in patients with acute organophosphate poisoning. *Mutation Research/Genetic Toxicology and Environmental Mutagenesis, 825*, 1–7.
101. Jokanović, M. (2018). Neurotoxic effects of organophosphorus pesticides and possible association with neurodegenerative diseases in man: A review. *Toxicology, 410*, 125–131.
102. Agarwal, A., Prajapati, R., Singh, O. P., Raza, S. K., & Thakur, L. K. (2015). Pesticide residue in water—A challenging task in India. *Environmental Monitoring and Assessment, 187*, 1–21.
103. Parks, C. G., Costenbader, K. H., Long, S., Hofmann, J. N., Beane, F. L., & Sandler, D. P. (2022). Pesticide use and risk of systemic autoimmune diseases in the agricultural health study. *Environmental Research, 209*, 112862.
104. Kim, K. H., Kabir, E., & Jahan, S. A. (2017). Exposure to pesticides and the associated human health effects. *Science of the Total Environment, 575*, 525–535.
105. Muturi, E. J., Donthu, R. K., Fields, C. J., Moise, I. K., & Kim, C. H. (2017). Effect of pesticides on microbial communities in container aquatic habitats. *Scientific Reports, 7*(1), 44565.
106. Ikehata, K., & Gamal El-Din, M. (2005). Aqueous pesticide degradation by ozonation and ozone-based advanced oxidation processes: A review (part I). *Ozone: Science and Engineering, 27*(2), 83–114.
107. Malik, A., Ojha, P., & Singh, K. P. (2009). Levels and distribution of persistent organochlorine pesticide residues in water and sediments of Gomti River (India)—A tributary of the Ganges River. *Environmental Monitoring and Assessment, 148*, 421–435.
108. López-Loveira, E., Ariganello, F., Medina, M. S., Centrón, D., Candal, R., & Curutchet, G. (2017). Degradation alternatives for a commercial fungicide in water: Biological, photo-Fenton, and coupled biological photo-Fenton processes. *Environmental Science and Pollution Research, 24*, 25634–25644.
109. Zapata, A., Malato, S., Sánchez-Pérez, J. A., Oller, I., & Maldonado, M. I. (2010). Scale-up strategy for a combined solar photo-Fenton/biological system for remediation of pesticide-contaminated water. *Catalysis Today, 151*(1–2), 100–106.
110. Ribeiro, A. R., Nunes, O. C., Pereira, M. F., & Silva, A. M. (2015). An overview on the advanced oxidation processes applied for the treatment of water pollutants defined in the recently launched directive 2013/39/EU. *Environment International, 75*, 33–51.
111. Schulz, R. (2004). Field studies on exposure, effects, and risk mitigation of aquatic nonpoint-source insecticide pollution: A review. *Journal of Environmental Quality, 33*(2), 419–448.
112. Dey, S., Bano, F., & Malik, A. (2019). Pharmaceuticals and personal care product (PPCP) contamination—A global discharge inventory. In *Pharmaceuticals and personal care products: Waste management and treatment technology* (pp. 1–26). Butterworth-Heinemann.
113. Fernandes, A. S., Bragança, I., & Homem, V. (2023). Personal care products in soil-plant and hydroponic systems: Uptake, translocation, and accumulation. *Science of the Total Environment, 912*, 168894.
114. Heudorf, U., Mersch-Sundermann, V., & Angerer, J. (2007). Phthalates: Toxicology and exposure. *International Journal of Hygiene and Environmental Health, 210*(5), 623–634.
115. Catanese, M. C., Suvorov, A., & Vandenberg, L. N. (2015). Beyond a means of exposure: A new view of the mother in toxicology research. *Toxicology Research, 4*(3), 592–612.
116. Kiyama, R., & Wada-Kiyama, Y. (2015). Estrogenic endocrine disruptors: Molecular mechanisms of action. *Environment International, 83*, 11–40.
117. Chaturvedi, P., Shukla, P., Giri, B. S., Chowdhary, P., Chandra, R., Gupta, P., & Pandey, A. (2021). Prevalence and hazardous impact of pharmaceutical and personal care products and antibiotics in environment: A review on emerging contaminants. *Environmental Research, 194*, 110664.

118. Hopkins, Z. R., & Blaney, L. (2016). An aggregate analysis of personal care products in the environment: Identifying the distribution of environmentally-relevant concentrations. *Environment International, 92*, 301–316.
119. Evgenidou, E. N., Konstantinou, I. K., & Lambropoulou, D. A. (2015). Occurrence and removal of transformation products of PPCPs and illicit drugs in wastewaters: A review. *Science of the Total Environment, 505*, 905–926.
120. Shahid, M. K., Kashif, A., Fuwad, A., & Choi, Y. (2021). Current advances in treatment technologies for removal of emerging contaminants from water–a critical review. *Coordination Chemistry Reviews, 442*, 213993.
121. Vimalkumar, K., Seethappan, S., & Pugazhendhi, A. (2019). Fate of Triclocarban (TCC) in aquatic and terrestrial systems and human exposure. *Chemosphere, 230*, 201–209.
122. Keerthanan, S., Jayasinghe, C., Biswas, J. K., & Vithanage, M. (2021). Pharmaceutical and personal care products (PPCPs) in the environment: Plant uptake, translocation, bioaccumulation, and human health risks. *Critical Reviews in Environmental Science and Technology, 51*(12), 1221–1258.
123. Rehman, M. U., Nisar, B., Yatoo, A. M., Sehar, N., Tomar, R., Tariq, L., Ali, S., Ali, A., Rashid, S. M., Ahmad, S. B., & Aldossari, R. M. (2024). After effects of pharmaceuticals and personal care products (PPCPs) on the biosphere and their counteractive ways. *Separation and Purification Technology, 342*, 126921.
124. Xu, X., Yang, H., & Li, C. (2022). Theoretical model and actual characteristics of air pollution affecting health cost: A review. *Ijerph, 19*, 3532. https://doi.org/10.3390/ijerph19063532
125. Halder, J. N., & Islam, M. N. (2015). Water pollution and its impact on the human health. *Journal of Environment and Human, 2*(1), 36–46.
126. Nel, L. H., & Markotter, W. (2009). New and emerging waterborne infectious diseases. *Encyclopaedia of Life Support System, 1*, 1–10.
127. Ullah, S., Javed, M. W., Shafique, M., et al. (2014). An integrated approach for quality assessment of drinking water using GIS: A case study of lower Dir. *Journal of Himalayan Earth Sciences, 47*(2), 163–174.
128. Currie, J., Joshua, G. Z., Katherine, M., et al. (2013). Something in the water: Contaminated drinking water and infant health. *Canadian journal of economics., 46*(3), 791–810.
129. Chowdhury, S., Annabelle, K., Klaus, F. Z. (2015). Arsenic contamination of drinking water and mental health. ZEF Discussion Papers on Development Policy, No. 222, University of Bonn, Center for Development Research (ZEF), Bonn. ISSN: 1436-9931 Published by: Zentrum für Entwicklungsforschung (ZEF), Center for Development Research Walter-Flex-Straße 3 D – 53113 Bonn Germany.
130. Kamble, S. M. (2014). Water pollution and public health issues in Kolhapur city in Maharashtra. *International Journal of Scientific and Research Publications, 4*(1), 1–6.
131. Ahmad, S. M., Yusafzai, F., Bari, T., et al. (2014). Assessment of heavy metals in surface water of River Panjkora Dir Lower, KPK Pakistan. *Journal of Biodiversity and Environmental Sciences, 5*, 144–152.
132. Corcoran, E., Nellemann, C., Baker, E., et al. (2010). *Sick water? The central role of wastewater management in sustainable development. A rapid response assessment.* United Nations Environment Programme.
133. Arif, A., Malik, M. F., Liaqat, S., Aslam, A., Mumtaz, K., & Afzal, A. (2020). 3. Water pollution and industries. *Pure Applied Biology (PAB), 9*(4), 2214–2224. https://doi.org/10.19045/bspab.2020.90237
134. Lin, H.-J., Sung, T.-I., Chen, C.-Y., & Guo, H.-R. (2013). Arsenic levels in drinking Water and mortality of liver cancer in Taiwan. *Journal of Hazardous Materials, 262*, 1132–1138. https://doi.org/10.1016/j.jhazmat.2012.12.049
135. Zaveri, E. D., Russ, J. D., Desbureaux, S. G., Damania, R., Rodella, A. S., De Souza, R. P., & G. (2023). *The nitrogen legacy: The long-term effects of water pollution on human capital.* World Bank Policy Research Working Paper.
136. Agingu, J. B. (2020). Levels and differentials of occurrence of water borne diseases at Moi University. *Kenya Academic Research Institute, 11*, 10–18.

137. Bidhuri, S., Taqi, M., & Khan, M. M. A. (2018). Water-borne disease: Link between human health and water use in the Mithepur and Jaitpur area of the NCT of Delhi. *Journal of Public Health, 26*, 119–126. https://doi.org/10.1007/s10389-017-0835-y
138. Sharma, S., Sachdeva, P., & Virdi, J. S. (2003). Emerging water-borne pathogens. *Applied Microbiology and Biotechnology, 61*(5–6), 424–428. https://doi.org/10.1007/s00253-003-1302-y. Epub 2003 Apr 9.
139. Lopez, A. D., Murray, C. J., eds. (1996). 2020. The global burden of disease: a comprehensive assessment of mortality and disability from diseases, injuries, and risk factors in 1990 and projected to; summary. Harvard School of Public Health.
140. Prüss-Ustün, A., Bartram, J., Clasen, T., Colford, J. M., Jr., Cumming, O., Curtis, V., et al. (2014). Burden of disease from inadequate water, sanitation and hygiene in low-and middle-income settings: A retrospective analysis of data from 145 countries. *Tropical Medicine and International Health, 19*, 894–905.
141. Szálkai, K. W.-B. (2019). Diseases BT. In S. Romaniuk, M. Thapa, & P. Marton (Eds.), *The Palgrave encyclopedia of global security studies* (pp. 1–7). Springer International Publishing. https://doi.org/10.1007/978-3-319-74336-3_562-1
142. WHO/UNICEF. (2015). *Joint Water supply, & sanitation monitoring Programme. Progress on sanitation and drinking water: 2015 update and MDG assessment*. World Health Organization.
143. UNICEF (2019). Clean drinking water ensuring survival and improved outcomes across all outcomes for every child.
144. Pathak, H. (2015). Effect of Water Borne Diseases on Indian economy: A cost-benefit analysis. *Analele Univ Din Ordea, Ser Geogr., 2015*(1), 74–78.
145. Chowdhary, P., Bharagava, R. N., Mishra, S., & Khan, N. (2020). Role of industries in water scarcity and its adverse effects on environment and human health. Environmental concerns and sustainable development, 235–256. https://doi.org/10.1007/978-981-13-5889-0_12.
146. Wu, H., Gai, Z., Guo, Y., Li, Y., Hao, Y., & Lu, Z. N. (2020). Does environmental pollution inhibit urbanization in China? A new perspective through residents' medical and health costs. *Environmental Research, 182*, 109128–109128.9. https://doi.org/10.1016/j.envres.2020.109128
147. Lai, W. (2017). Pesticide use and health outcomes: Evidence from agricultural water pollution in China. *Journal of Environmental Economics and Management, 86*, 93–120. https://doi.org/10.1016/j.jeem.2017.05.006
148. Fong, T.-T., & Lipp, E. K. (2005). Enteric viruses of humans and animals in aquatic environments: Health risks, detection, and potential Water quality assessment tools. *Microbiology and Molecular Biology Reviews, 69*(2), 357–371. https://doi.org/10.1128/mmbr.69.2.357-371.2005
149. Khan, S., Shahnaz, M., Jehan, N., Rehman, S., Shah, M. T., & Din, I. (2013). Drinking water quality and human health risk in Charsadda District, Pakistan. *Journal of Cleaner Production, 60*, 93–101. https://doi.org/10.1016/j.jclepro.2012.02.016
150. Waddington, H., Snilstveit, B., White, H., & Fewtrell, L. (2009). *Water, sanitation and hygiene interventions to combat childhood diarrhoea in developing countries*. New Delhi India Global Development Network International Initiative for Impact Evaluation.
151. Esrey, S. A., Potash, J. B., Roberts, L., & Shiff, C. (1991). Effects of improved Water supply and sanitation on ascariasis, Diarrhoea, Dracunculiasis, hookworm infection, schistosomiasis, and trachoma. *Bulletin of the World Health Organization, 69*(5), 609.
152. Ahmed, S., & Ismail, S. (2018). Water pollution and its sources, effects and management: A case study of Delhi. *International Journal of Current Advanced Research, 7*(2), 10436–10442. https://doi.org/10.24327/ijcar.2018.10442.1768
153. Ansari, Z. Z., & Akhmatov, S. V. (2020). Impacts of water pollution on human health: A case study of Delhi. January 2020 https://doi.org/10.17223/978-5-94621-954-9-2020-39. Conference: Prirodopol'zovanie i ohrana prirody: Ohrana pamjatnikov prirody, biologicheskogo i landshaftnogo raznoobrazija Tomskogo Priob'ja i drugih regionov Rossii.

154. Clasen, T. F., Alexander, K. T., Sinclair, D., Boisson, S., Peletz, R., Chang, H. H., et al. (2015). Interventions to improve Water quality for preventing Diarrhoea. *Cochrane Database of Systematic Reviews, 10*, CD004794. https://doi.org/10.1002/14651858.CD004794.pub3
155. Victora, C. G., Smith, P. G., Vaughan, J. P., Nobre, L. C., Lombard, C., Teixeira, A. M. B., et al. (1988). Water supply, sanitation and housing in relation to the risk of infant mortality from Diarrhoea. *International Journal of Epidemiology, 17*(3), 651–654. https://doi.org/10.1093/ije/17.3.651
156. Team, S. H. (2004). Water, sanitation and hygiene links to health: Facts and figures: World Health Organization.
157. Marino, D. D. (2007). Water and food safety in the developing world: Global implications for health and nutrition of infants and young children. *Journal of American Dietetic Association, 107*(11), 1930–1934. https://doi.org/10.1016/j.jada.2007.08.013
158. Jalan, J., & Ravallion, M. (2003). Does piped water reduce diarrhea for children in rural India? *Journal of Econometrics, 112*(1), 153–173. https://doi.org/10.1016/s0304-4076(02)00158-6
159. Landrigan, P. J., Fuller, R., Fisher, S., Suk, W. A., Sly, P., Chiles, T. C., et al. (2018). Pollution and children's health. *Scientific and Total Environment, 650*(Pt 2), 2389–2394. https://doi.org/10.1016/j.scitotenv.2018.09.375
160. Gundry, S., Wright, J., & Conroy, R. (2004). A systematic review of the health outcomes related to household water quality in developing countries. *Journal of Water and Health, 2*(1), 1–13. https://doi.org/10.2166/wh.2004.0001
161. Cheung, W. H. S., Chang, K. C. K., Hung, R. P. S., & Kleevens, J. W. L. (1990). Health effects of beach Water pollution in Hong Kong. *Epidemiology and Infection, 105*(1), 139–162. https://doi.org/10.1017/s0950268800047737
162. Rengel, Z. (2004). Heavy metals as essential nutrients. In M. N. V. Prasad (Ed.), *Heavy metal stress in plants* (pp. 271–294). Springer.
163. Bánfalvi, G. (2011). Heavy metals, trace elements and their cellular effects. In G. Banfalvi (Ed.), *Cellular effects of heavy metals* (pp. 3–28). Springer.
164. Sevim, C., Dogan, E., & Comakli, S. (2020). Cardiovascular disease and toxic metals. *Current Opinion in Toxicology, 19*, 88–92.
165. Cai, J., Li, Y., Liu, S., Liu, Q., Min, X., Zhang, J., Wei, Y., Mo, X., Lin, Y., Tang, X., Mai, T., Mo, C., Luo, T., Huang, S., Lu, H., Zhang, Z., & Qin, J. (2022). Associations between multiple heavy metals exposure and glycated hemoglobin in a Chinese population. *Chemosphere, 287*(Pt 2), 132159.
166. Yang, A. M., Lo, K., Zheng, T. Z., Yang, J. L., Bai, Y. N., Feng, Y. Q., Cheng, N., & Liu, S. M. (2020). Environmental heavy metals and cardiovascular diseases: Status and future direction. *Chronic Diseases and Translational Medicine, 6*(4), 251–259.
167. Haidar, Z., Fatema, K., Shoily, S. S., & Sajib, A. A. (2023). Disease-associated metabolic pathways affected by heavy metals and metalloid. *Toxicology Reports, 10*, 554–570.
168. Odobašić, A., Šestan, I., & Begić, S. (2019). *Biosensors for determination of heavy metals in waters. Biosensors for environmental monitoring*. IntechOpen.
169. Sharma, B., Singh, S., & Siddiqi, N. J. (2014). Biomedical implications of heavy metals induced imbalances in redox systems. *BioMed Research International*, 640754.
170. Solenkova, N. V., Newman, J. D., Berger, J. S., Thurston, G., Hochman, J. S., & Lamas, G. A. (2014). Metal pollutants and cardiovascular disease: Mechanisms and consequences of exposure. *American Heart Journal, 168*(6), 812–822.
171. Rafati Rahimzadeh, M., Rafati Rahimzadeh, M., Kazemi, S., & Moghadamnia, A. A. (2017). Cadmium toxicity and treatment: An update. *Caspian Journal of Internal Medicine, 8*(3), 135–145.
172. Han, Y. L., Sheng, Z., Liu, G. D., Long, L. L., Wang, Y. F., Yang, W. X., & Zhu, J. Q. (2015). Cloning, characterization and cadmium inducibility of metallothionein in the testes of the mudskipper Boleophthalmus pectinirostris. *Ecotoxicology and Environmental Safety, 119*, 1–8.

173. Takiguchi, M., Achanzar, W. E., Qu, W., Li, G., & Waalkes, M. P. (2003). Effects of cadmium on DNA-(Cytosine-5) methyltransferase activity and DNA methylation status during cadmium-induced cellular transformation. *Experimental Cell Research, 286*(2), 355–365.
174. Suzuki, M., Takeda, S., Teraoka-Nishitani, N., Yamagata, A., Tanaka, T., Sasaki, M., Yasuda, N., Oda, M., Okano, T., Yamahira, K., Nakamura, Y., Kobayashi, T., Kino, K., Miyazawa, H., Waalkes, M. P., & Takiguchi, M. (2017). Cadmium-induced malignant transformation of rat liver cells: Potential key role and regulatory mechanism of altered apolipoprotein E expression in enhanced invasiveness. *Toxicology, 382*, 16–23.
175. Zaynab, M., Al-Yahyai, R., Ameen, A., Sharif, Y., Ali, L., Fatima, M., Khan, K., & Li, S. (2022). Health and environmental effects of heavy metals. *Journal of King Saud University—Science, 34*(1), 101653.
176. Peters, J. L., Kubzansky, L. D., Ikeda, A., Fang, S. C., Sparrow, D., Weisskopf, M. G., Wright, R. O., Vokonas, P., Hu, H., & Schwartz, J. (2012). Lead concentrations in relation to multiple biomarkers of cardiovascular disease: The normative aging study. *Environmental Health Perspectives, 120*(3), 361–366.
177. Perrelli, M., Wu, R., Liu, D. J., Lucchini, R. G., Del Bosque-Plata, L., Vergare, M. J., Akhter, M. P., Ott, J., & Gragnoli, C. (2022). Heavy metals as risk factors for human diseases—a Bayesian network approach. *European Review for Medical and Pharmacological Sciences, 26*(24), 9275–9310.
178. Flora, S. J., Saxena, G., Gautam, P., Kaur, P., & Gill, K. D. (2007). Response of lead-induced oxidative stress and alterations in biogenic amines in different rat brain regions to combined administration of DMSA and MiADMSA. *Chemico-Biological Interactions, 170*(3), 209–220.
179. Ahamed, M., & Siddiqui, M. K. (2007). Low level lead exposure and oxidative stress: Current opinions. *Clinica Chimica Acta; International Journal of Clinical Chemistry, 383*(1–2), 57–64.
180. Zaeemzadeh, N., Jahanian Sadatmahalleh, S., Ziaei, S., Kazemnejad, A., Movahedinejad, M., Mottaghi, A., & Mohamadzadeh, N. (2021). Comparison of dietary micronutrient intake in PCOS patients with and without metabolic syndrome. *Journal of Ovarian Research, 14*(1), 10.
181. Zhang, X. H., Zhang, X., Wang, X. C., Jin, L. F., Yang, Z. P., Jiang, C. X., Chen, Q., Ren, X. B., Cao, J. Z., Wang, Q., & Zhu, Y. M. (2011). Chronic occupational exposure to hexavalent chromium causes DNA damage in electroplating workers. *BMC Public Health, 11*, 224.
182. Henkler, F., Brinkmann, J., & Luch, A. (2010). The role of oxidative stress in carcinogenesis induced by metals and xenobiotics. *Cancers, 2*(2), 376–396.
183. Yücesoy, B., Mirshahidi, S., Yücesoy, C., & Karakaya, A. (1999). In vitro effects of various metals on natural killer cell activity in cultured human lymphocytes. *Immunopharmacology and Immunotoxicology, 21*(3), 599–607.
184. Thomas, P., Summer, B., Sander, C. A., Przybilla, B., Thomas, M., & Naumann, T. (2000). Intolerance of osteosynthesis material: Evidence of dichromate contact allergy with concomitant oligoclonal T-cell infiltrate and TH1-type cytokine expression in the peri-implantar tissue. *Allergy, 55*(10), 969–972.
185. Abdul, K. S., Jayasinghe, S. S., Chandana, E. P., Jayasumana, C., & De Silva, P. M. (2015). Arsenic and human health effects: A review. *Environmental Toxicology and Pharmacology, 40*(3), 828–846.
186. McCarty, K. M., Hanh, H. T., & Kim, K. W. (2011). Arsenic geochemistry and human health in South East Asia. *Reviews on Environmental Health, 26*(1), 71–78.
187. Vahidnia, A., van der Voet, G. B., & de Wolff, F. A. (2007). Arsenic neurotoxicity—a review. *Human & Experimental Toxicology, 26*(10), 823–832.
188. Sasaki, A., Oshima, Y., & Fujimura, A. (2007). An approach to elucidate potential mechanism of renal toxicity of arsenic trioxide. *Experimental Hematology, 35*(2), 252–262.
189. Shen, H., Xu, W., Zhang, J., Chen, M., Martin, F. L., Xia, Y., Liu, L., Dong, S., & Zhu, Y. G. (2013). Urinary metabolic biomarkers link oxidative stress indicators associated with

general arsenic exposure to male infertility in a han chinese population. *Environmental Science & Technology, 47*(15), 8843–8851.
190. Hopenhayn-Rich, C., Browning, S. R., Hertz-Picciotto, I., Ferreccio, C., Peralta, C., & Gibb, H. (2000). Chronic arsenic exposure and risk of infant mortality in two areas of Chile. *Environmental Health Perspectives, 108*(7), 667–673.
191. Li, D., Lu, C., Wang, J., Hu, W., Cao, Z., Sun, D., Xia, H., & Ma, X. (2009). Developmental mechanisms of arsenite toxicity in zebrafish (Danio rerio) embryos. *Aquatic Toxicology (Amsterdam, Netherlands), 91*(3), 229–237.
192. Liu, P., Lu, K., Li, J., Wu, X., Qian, L., Wang, M., & Gao, S. (2020). Effect of aging on adsorption behavior of polystyrene microplastics for pharmaceuticals: Adsorption mechanism and role of aging intermediates. *Journal of Hazardous Materials, 384*, 121193.
193. Groh, K. J., Backhaus, T., Carney-Almroth, B., Geueke, B., Inostroza, P. A., Lennquist, A., Leslie, H. A., Maffini, M., Slunge, D., Trasande, L., Warhurst, A. M., & Muncke, J. (2019). Overview of known plastic packaging-associated chemicals and their hazards. *The Science of the Total Environment, 651*(Pt 2), 3253–3268.
194. Zhou, Y., Liu, X., & Wang, J. (2020). Ecotoxicological effects of microplastics and cadmium on the earthworm Eisenia foetida. *Journal of Hazardous Materials, 392*, 122273.
195. Town, R. M., van Leeuwen, H. P., & Blust, R. (2018). Biochemodynamic features of metal ions bound by micro- and Nano-plastics in aquatic media. *Frontiers in Chemistry, 6*, 627.
196. Chen, X. J., Ma, J. J., Yu, R. L., Hu, G. R., & Yan, Y. (2022). Bioaccessibility of microplastic-associated heavy metals using an in vitro digestion model and its implications for human health risk assessment. *Environmental Science and Pollution Research International, 29*(51), 76983–76991.
197. Gao, F., Li, J., Sun, C., Zhang, L., Jiang, F., Cao, W., & Zheng, L. (2019). Study on the capability and characteristics of heavy metals enriched on microplastics in marine environment. *Marine Pollution Bulletin, 144*, 61–67.
198. Liao, Y. L., & Yang, J. Y. (2020). Microplastic serves as a potential vector for Cr in an in-vitro human digestive model. *The Science of the Total Environment, 703*, 134805.
199. Godoy, V., Martínez-Férez, A., Martín-Lara, M. Á., Vellido-Pérez, J. A., Calero, M., & Blázquez, G. (2020). Microplastics as vectors of chromium and Lead during dynamic simulation of the human gastrointestinal tract. *Sustainability, 12*(11), 4792.
200. Ali, D., Almarzoug, M. H., Al Ali, H., Samdani, M. S., Hussain, S. A., & Alarifi, S. (2020). Fish as bio indicators to determine the effects of pollution in river by using the micronucleus and alkaline single cell gel electrophoresis assay. *Journal of King Saud University-Science, 32*(6), 2880–2885.
201. Cabral, J. P. (2010). Water microbiology. Bacterial pathogens and water. *International Journal of Environmental Research and Public Health, 7*(10), 3657–3703.
202. Rossolini, G. I. A. N., Arena, F., & Giani, T. (2016). Mechanisms of antibacterial resistance. https://doi.org/10.1016/B978-0-7020-6285-8.00138-6 In book: Infectious Diseases. pp.1181–1196.e1.
203. Cooper, S. (2001). Helical growth and the curved shape of vibrio cholerae. *FEMS Microbiology Letters, 198*(2), 123–124.
204. Ali, M., Nelson, A. R., Lopez, A. L., & Sack, D. A. (2015). Updated global burden of cholera in endemic countries. *PLoS Neglected Tropical Diseases, 9*(6), e0003832.
205. Kristanti, R. A., Hadibarata, T., Syafrudin, M., Yılmaz, M., & Abdullah, S. (2022). Microbiological contaminants in drinking water: Current status and challenges. *Water, Air, & Soil Pollution, 233*(8), 299.
206. Kotloff, K. L., Riddle, M. S., Platts-Mills, J. A., Pavlinac, P., & Zaidi, A. K. (2018). Shigellosis. *The Lancet, 391*(10122), 801–812.
207. Radin, D. (2014). New trends in food-and waterborne viral outbreaks. *Archives of Biological Sciences, 66*(1), 1–9.

208. Chen, L., Deng, Y., Dong, S., Wang, H., Li, P., Zhang, H., & Chu, W. (2021). The occurrence and control of waterborne viruses in drinking water treatment: A review. *Chemosphere, 281*, 130728.
209. Oppong, T. B., Yang, H., Amponsem-Boateng, C., Kyere, E. D., Abdulai, T., Duan, G., & Opolot, G. (2020). Enteric pathogens associated with gastroenteritis among children under 5 years in sub-Saharan Africa: A systematic review and meta-analysis. *Epidemiology & Infection, 148*, e64.
210. Atabakhsh, P., Kargar, M., & Doosti, A. (2021). Detection and evaluation of rotavirus surveillance methods as viral indicator in the aquatic environments. *Brazilian Journal of Microbiology, 52*(2), 811–820.
211. Troeger, C., Khalil, I. A., Rao, P. C., Cao, S., Blacker, B. F., Ahmed, T., Armah, G., Bines, J. E., Brewer, T. G., Colombara, D. V., & Kang, G. (2018). Rotavirus vaccination and the global burden of rotavirus diarrhea among children younger than 5 years. *JAMA Pediatrics, 172*(10), 958–965.
212. Cassidy, H., Poelman, R., Knoester, M., Van Leer-Buter, C. C., & Niesters, H. G. (2018). Enterovirus D68–the new polio? *Frontiers in Microbiology, 9*, 2677.
213. Puenpa, J., Wanlapakorn, N., Vongpunsawad, S., & Poovorawan, Y. (2019). The history of enterovirus A71 outbreaks and molecular epidemiology in the Asia-Pacific region. *Journal of Biomedical Science, 26*, 1–11.
214. Murray, C., & Newby, H. (2012). Data resource profile: United Nations children's fund (UNICEF). *International Journal of Epidemiology, 41*(6), 1595–1601.
215. Zhou, X., Kong, D. G., Li, J., Pang, B. B., Zhao, Y., Zhou, J. B., Zhang, T., Xu, J. Q., Kobayashi, N., & Wang, Y. H. (2019). An outbreak of gastroenteritis associated with GII. 17 norovirus-contaminated secondary water supply system in Wuhan, China, 2017. *Food and Environmental Virology, 11*, 126–137.
216. Smith, D. B., & Simmonds, P. (2018). Classification and genomic diversity of enterically transmitted hepatitis viruses. *Cold Spring Harbor Perspectives in Medicine, 8*(9), a031880.
217. Dunn, N., & Juergens, A. L. (2024). Giardiasis. In *StatPearls*. StatPearls Publishing.
218. Ghosh, S., Padalia, J., & Moonah, S. (2019). Tissue destruction caused by Entamoeba histolytica parasite: Cell death, inflammation, invasion, and the gut microbiome. *Current Clinical Microbiology Reports, 6*, 51–57.
219. Chou, A., & Austin, R. L. (2023). Entamoeba histolytica infection. In *StatPearls*. StatPearls Publishing.
220. Islam, M. M., Iqbal, M. S., Leemans, R., & Hofstra, N. (2018). Modelling the impact of future socio-economic and climate change scenarios on river microbial water quality. *International Journal of Hygiene and Environmental Health, 221*(2), 283–292.
221. Whitehead, P. G., Barbour, E., Futter, M. N., Sarkar, S., Rodda, H., Caesar, J., Butterfield, D., Jin, L., Sinha, R., Nicholls, R., & Salehin, M. (2015). Impacts of climate change and socio-economic scenarios on flow and water quality of the Ganges, Brahmaputra and Meghna (GBM) river systems: Low flow and flood statistics. *Environmental Science: Processes & Impacts, 17*(6), 1057–1069.

Chapter 10
Effects on Humans Via the Ingestion of Contaminated Water: A Radiological Aspect

Ankush Verma, Amit Pratap Singh Chouhan, Vandana Singh, Pankaj Verma, Lekha Singh, Neha Kumari, Amarjeet, Manisha Kharaliya, and Umang Sain

1 Introduction

Water contaminated by radioactivity is harmful to human health because water consists of radioactive materials. Radioactive wastes, deposits of natural origin, and nuclear accidents or industrial processes form several sources of these chemicals

A. Verma (✉)
Department of Radiology, Sharda School of Allied Health Sciences, Sharda University, Greater Noida, U.P, India
e-mail: ankush.verma@sharda.ac.in

A. P. S. Chouhan
Department of Radiology, Sharda School of Allied Health Sciences, Galgotias University, Greater Noida, U.P, India
e-mail: amit.chouhan@galgotiasuniversity.edu.in

V. Singh
Department of Microbiology, Sharda School of Allied Health Sciences, Sharda University, Greater Noida, U.P, India
e-mail: vandana.singh@sharda.ac.in

P. Verma
Water Resources Engineering, School of Civil & Environmental Engineering, IIT Mandi, Mandi, Himachal Pradesh, India

L. Singh
Department of Allied Health Sciences in Kailash Institute of Nursing and Parametric Sciences, Greater Noida, U.P, India

N. Kumari · Amarjeet · M. Kharaliya
Department of Anthropology, University of Delhi, Delhi, India

U. Sain
Department of Radiology, Noida International University, Noida, U.P, India
e-mail: umang.sain@niu.edu.in

M. K. Jindal et al. (eds.), *Emerging Contaminants in Water*, Emerging Contaminants and Associated Treatment Technologies,
https://doi.org/10.1007/978-3-031-89591-3_10

[1]. The appearance of radioactive materials in drinking water can result in the leakage of harmful radiation into a human body through drinking water, which can lead to significant health effects. The radionuclides present in water can include iodine-131, cesium-137, strontium-90, and radium-226 isotopes. All of them emit different types of radiation, including alpha, beta, and gamma radiation, that have specific features and influence biological tissues in different ways [2]. Although it is a lightly penetrating ionization radiation, the ingested alpha particles cause gross cellular damage to internal organs such as the bones, liver, and lungs. Wherever radionuclides penetrate, beta particles may penetrate deeper into soft body tissues and threaten the inner structures of bones. Gamma rays have large penetrations, and thus, they interfere with a larger body content, consequently increasing the severe risk even in negligible amounts in the environment. Ingestion of radially contaminated water can cause acute and chronic health effects. Acute exposure to such levels of radiation is always related to acute health problems, including radiation sickness and nausea, vomiting, and exhaustion. Chronic exposure has been found to increase the lifetime risk of cancer, genetic defects and other dire health conditions. Vulnerable sections include children, pregnant women, and those whose health is compromised due to disease [3]. Since the pathways and impacts of radioactive pollutants in water are directly related to public health, this aspect must be taken quite seriously, especially in relation to stringent monitoring and regulatory frameworks that respond effectively to these problems [4]. These impurities are due to natural geological formations, nuclear accidents such as the Chernobyl disaster and the Fukushima Daiichi nuclear disaster, industrial processes, and inappropriate nuclear waste disposal [5]. Such water with radioactive elements radiates ionizing radiation in water, which is dangerous if it is consumed inside and produces unfavorable health effects. One of the most complex public health issues with considerable implications for forensic investigations is the ingestion of contaminated water. Contamination by biological, chemical, or physical contaminants can lead to acute illness to chronic disease; hence, a holistic approach toward forensic investigation is needed to trace out cause and accountability. Forensic investigations for determining the cause and consequences of health effects related to contamination through the ingestion of water constitute extremely challenging, multidisciplinary work. This would be an integration of clinical results, epidemiological data, environmental science, and advanced analytical tools in establishing cause-and-effect relationships between aquatic pollutants and health consequences. This comprehensive method enhances the understanding and reduction of contaminated water effects by holding accountable responsible entities who can prevent future incidents from occurring. Knowing the causes and impacts of radioactive contaminants in water can protect public health, but very strict monitoring of water pollution and the development of appropriate response mechanisms to these problems are needed [6]. Public education or enlightening the public on the dangers possibly presented by radiologically contaminated water enhances preparedness and resilience in members of the community. Waters should be safe and free from contaminants to ensure safety from radioactive hazards and maintenance of ecological balance as far as human health is concerned.

2 Movement of the Transport of Radiological Contamination Into Water Bodies

Radiological contamination of water sources arises from different sources. The natural sources are, however, limited to the uranium, thorium, and radium that can enter ground water as well as surface water through geologic formations. In addition, industrial operations, such as mining and processing of radioactive material, may cause contamination [7]. The failure of appropriate disposal of radioactive waste from medical, industrial, and scientific institutions also results in the presence of radionuclides in water supplies. Nuclear weapon testing and subsequent fallout have historically caused extensive radioactive pollution. These actions result in the release of radionuclides that can endure in the environment for prolonged durations, presenting enduring threats to water quality and human health.

2.1 Natural Radionuclides

Natural radionuclides are radioactive isotopes that exist inherently in the environment and are commonly found in soil, rocks, water, and the atmosphere. It undergoes decay into other radionuclides, such as radium and radon, which may further increase ambient radioactivity. Thorium-232 (Th-232) is a notable natural radionuclide that has a half-life of 14 billion years and is extensively found in the Earth's crust. It undergoes decay into radium-228, ultimately resulting in stable lead. Potassium-40 (K-40), which has a half-life of 1.25 billion years, is a naturally occurring isotope of potassium found in all potassium-bearing minerals and is vital for living beings [8]. Natural radionuclides contribute to background radiation levels and can infiltrate water sources via geological processes, erosion, and leaching, resulting in significant health problems when they accumulate in drinking water. Their presence in the environment underscores the importance of monitoring and managing natural radiation to protect public health.

2.1.1 Examples of Naturally Occurring Radionuclides in Water

Radionuclides that occur naturally are present in several water sources, such as groundwater, surface water, and drinking water. Radionuclides infiltrate aquatic systems via natural mechanisms such as rock weathering, leaching, and erosion. Examples of naturally occurring radionuclides frequently present in water are shown in Table 10.1.

Table 10.1 Examples of radionuclides and their sources

Radionuclide	Characteristics	Sources in water
Uranium-238 (U-238)	Heavy metal with radioactive properties; half-life of ~4.5 billion years	Leaches into water from rocks and soil containing uranium minerals; commonly found in groundwater
Radium-226 (Ra-226)	Decay product of uranium-238; half-life of 1600 years	Found in groundwater as it leaches from uranium and thorium ores; often detected in higher concentrations in deep wells and aquifers
Radon-222 (Rn-222)	Radioactive gas; half-life of 3.8 days; formed from the decay of radium-226	Dissolves in groundwater, especially in areas with high uranium and radium content in geological formations; can enter homes through water use
Thorium-232 (Th-232)	Radioactive metal; half-life of 14 billion years	Enters water systems through the weathering and erosion of thorium-containing minerals
Potassium-40 (K-40)	Naturally occurring isotope of potassium; half-life of 1.25 billion years	Present in small amounts in all potassium-containing minerals; typically found in trace amounts in both groundwater and surface water
Lead-210 (Pb-210)	Part of the uranium-238 decay series; half-life of 22.3 years	Accumulates in water bodies from the decay of radon-222; often found in sediments
Radionuclide	Characteristics	Sources in water

2.2 *Human-Made Radionuclides*

Human-made radionuclides, or anthropogenic radionuclides, are radioactive isotopes released into the environment due to diverse human activities, such as nuclear power generation, nuclear weapon testing, and medical applications [8]. Radionuclides, including cesium-137 (Cs-137), strontium-90 (Sr-90), and iodine-131 (I-131), are generated during nuclear fission processes. Cesium-137 has a half-life of approximately 30 years and emits beta particles and gamma radiation. It can be produced during nuclear accidents and fall out from weapons testing, presenting considerable health hazards when it accumulates in muscle tissues. Strontium-90, which has a half-life of approximately 29 years, resembles calcium and accumulates in bones and teeth, hence increasing the risk of bone cancer and leukemia. I-131, which has a half-life of approximately 8 days, is utilized in medical diagnostics and therapies; nevertheless, it may also be emitted during nuclear incidents, leading to accumulation in the thyroid gland and an increased risk of thyroid cancer [9]. These radionuclides can contaminate water sources through atmospheric deposition, industrial discharge, and improper disposal of radioactive waste, leading to both acute and chronic health effects upon ingestion. Ensuring stringent monitoring, regulatory measures, and effective response strategies is crucial to mitigate the radiological risks posed by human-made radionuclides in the environment.

2.2.1 Accidents and Disasters

Incidents and disasters associated with radioactive materials can easily contaminate human health through the use of polluted water. When radioactive materials leak into the environment, they seep into rivers, lakes, and underground waters, causing severe damage to human health. The intake of radionuclides such as iodine-131, cesium-137, and strontium-90 with water causes acute and long-term health effects [10]. Acute exposure involves acute symptoms such as nausea, vomiting, and diarrhea, but in cases of acute exposure, health effects are quite likely to include cancers, genetic mutations, and organ damage to the liver or kidneys. This depends on the type and number of radionuclides ingested, the period of exposure, and the ages or types of health conditions of the involved people. Radioactive incidents and disasters would never pose public health risks if there were untainted sources of water, with disaster response techniques applied in time.

2.2.2 Industrial Processes

Industrial processes involving radioactive elements cause negative health impacts on human life through the use of polluted water. The extraction, processing, and manufacturing of nuclear materials can release radionuclides into aquatic systems, causing pollution [11]. For example, radionuclides such as uranium, thorium, and radon decay products in drinking water supplies. Chronic exposure to low levels of radioactive pollutants can increase the likelihood of disease being caused by cancer, especially stomach, liver and bone cancer. Continuous internal radiation exposure. Exposure to some radionuclides for longer periods can cause renal dysfunction and cardiovascular diseases. The effect is most pronounced in communities living near heavy industries because of high exposure levels. Neonates, children, and pregnant women are more susceptible to damage from radiological contamination. This is because they have developing bodies and metabolic activity. For effective monitoring, industrial releases must be regulated to avoid radiological pollution through remediation sources of water and safeguard public health against the potential future impacts of industrial activities [12].

3 Radionuclide Incorporation and Bioaccumulation

Radionuclide incorporation and bioaccumulation through the ingestion of contaminated water may profoundly impact human health once radionuclides penetrate water sources, whereby they can be taken up by aquatic organisms, leading to bioaccumulation in the food chain. Humans who consume radioactive fallout contaminate water and aquatic organisms, including fish and shellfish, which ingest these radionuclides, which may accumulate in their bodies over time [12]. Radionuclides include cesium-137 and strontium-90, the most dangerous of which are plutonium-239 because of their long half-lives and tendency to concentrate. In particular, cesium-137 accumulates in soft tissues, strontium-90 in bones, and plutonium-239 in the liver and bones. This bioaccumulation leads to protracted internal exposure to ionizing radiation, which can cause cancers, genetic mutations, and damage to organs. The danger is more virulent in susceptible groups, such as children and pregnant women, who are more radiation sensitive. It is against a whole set of stringent regulations and through public health interventions. Bioaccumulation of radionuclides and safe provision of drinking water and food.

3.1 How Radionuclides Enter the Body Through Ingestion

The main source of exposure to radionuclides is through the ingestion of contaminated food and water. They present significant health risks due to their radioactive properties. When radioactive materials are released into the environment from sources such as nuclear accidents, industrial processes, or improper radioactive waste disposal, they can contaminate water bodies, soil, and plants. Contamination continues into food chains and leads to the consumption of crops that have taken in radionuclides [13]. Humans ingest these radionuclides by consuming contaminated water, agricultural produce, dairy products, meat, and seafood. Once ingested, radionuclides can be absorbed through the gastrointestinal tract into the bloodstream, from which they are distributed to various tissues and organs depending on their chemical nature and biological affinity.

4 Health Effects of Ingesting Contaminated Water

Ingesting radiologically contaminated water can lead to a range of serious health effects, both acute and chronic, due to the presence of harmful radionuclides. Acute exposure to high levels of radioactive contamination can cause immediate symptoms such as nausea, vomiting, diarrhea, and acute radiation syndrome (ARS), which include fatigue, fever, and even life-threatening damage to the bone marrow and gastrointestinal tract. Chronic exposure, even at relatively low levels, poses significant long-term health risks. The accumulation of radionuclides in the body can lead to the development of cancers, particularly in the thyroid (from iodine-131), bones (from strontium-90), and soft tissues (from cesium-137). Specific radionuclides have distinct affinities for certain tissues, causing localized damage. For example, iodine-131 is concentrated in the thyroid gland, increasing the risk of thyroid cancer and other thyroid disorders [14]. Strontium-90 mimics calcium and accumulates in bones and teeth, increasing the risk of bone cancer and leukemia. Cesium-137, which spreads throughout soft tissues, can cause a range of cancers and tissue damage. The long-term ingestion of radiologically contaminated water can lead to genetic mutations, which may not only affect the individual exposed but also be passed on to future generations, potentially causing congenital disabilities and other hereditary conditions. Noncancer health effects include damage to the kidneys, liver, and cardiovascular system due to the toxic effects of certain radionuclides on these organs. The vulnerability of specific populations, such as children, pregnant women, and individuals with compromised immune systems, exacerbates health risks. Children are particularly susceptible due to their developing bodies and relatively high rates of cell division, which increase their sensitivity to radiation-induced damage.

4.1 Deterministic Effects (Acute Radiation Sickness)

Deterministic effects, also known as nonstochastic effects, are health effects that increase in severity with higher doses of radiation and typically have a threshold below which they do not occur. When humans ingest contaminated water containing radioactive material, these deterministic effects can occur after internal exposure to ionizing radiation [15]. Deterministic effect intensities depend on the dose and increase with internal radiation exposure contamination that causes more hazardous health effects [16]. Short-term effects from high alcohol consumption can present as radiation disease, which presents with symptoms such as nausea, vomiting, diarrhea, and, in severe cases, hemorrhage and death due to destruction of the bone marrow and gastrointestinal lining. The long-term effects include cataracts, fibrosis, and organ atrophy. Bone marrow destruction leads to aplastic anemia, whereas damage to the gut lining can lead to chronic digestive disorders and increase susceptibility to infections. Most deterministically acting agents have an inflection dose below which effects are improbable. This threshold varies with each organ or tissue depending on the radiosensitivity of the cells. The dose that results in radiation-induced cataracts in the ocular lens is much lower than that for gastrointestinal effects. The latency period is the interval between exposure and the appearance of symptoms and varies with the radiation dose and the type of tissue involved.

4.1.1 Symptoms of Acute Radiation Sickness

Acute radiation sickness, which is sometimes called radiation poisoning or radiation toxicity, is a severe condition caused by an excessive dose of ionizing radiation within a short duration. This is a result of the use of water containing radioactive materials. When radioactive isotopes such as cesium-137, iodine-131, or strontium-90 are ingested, they enter the bloodstream and irradiate body tissues and organs [17]. The onset and severity of symptoms associated with ARS depend on the level of exposure and the specific isotopes ingested. Initially, individuals may experience a prodromal phase, which usually occurs within hours to days of exposure and includes nausea, vomiting, diarrhea, and fatigue. Most of these symptoms are generally associated with injury to the rapidly dividing cells of the gastrointestinal tract and bone marrow. As the condition progresses, the latent phase may follow, where symptoms temporarily subside, resulting in an incorrect impression of recovery. This stage is an illusion because deep inner damage continues to manifest. The next phase, the manifest illness phase, is characterized by more severe and life-threatening symptoms. Hematopoietic syndrome, a result of bone marrow damage, leads to dramatic decreases in white blood cells, red blood cells, and platelets, increasing susceptibility to infections, anemia, and hemorrhaging. Owing to extensive damage to the lining of the intestines, gastrointestinal syndrome results in severe nausea, vomiting, diarrhea, and dehydration, which can be fatal if not managed promptly. Cardiovascular and neurological symptoms may also emerge, particularly at higher doses, including dizziness, headaches, decreased blood pressure, and, in extreme cases, seizures and coma due to damage to blood vessels and the brain. Cutaneous radiation syndrome can occur if the skin is also exposed, resulting in burns, blisters, and ulcers that are slow to heal.

The long-term effects of and recovery from ARS depend on the dose and medical intervention. Survivors of high-dose exposure may face chronic health issues such as an increased risk of cancers, particularly leukemias and thyroid cancer, due to the incorporation of radioactive isotopes into the body's cells most affected [18].

The management of ARS requires immediate medical intervention, including the administration of antiemetics to control nausea, antibiotics to prevent infections, blood transfusions to treat anemia, and fluids to prevent dehydration. In cases of internal contamination, treatments such as potassium iodide can block radioactive iodine uptake by the thyroid, whereas agents such as Prussian blue can bind cesium and thallium, promoting their excretion. Additionally, colony-stimulating factors can stimulate bone marrow recovery, and advanced care may include stem cell transplants for severe bone marrow damage.

4.2 *Stochastic Effects (Increased Cancer Risk)*

Stochastic effects (Fig. 10.1), which increase the risk of cancer, are significant concerns when humans ingest water contaminated with radioactive substances [19]. Unlike deterministic effects, stochastic effects do not have a threshold dose and can occur at any level of radiation exposure, with the probability of occurrence increasing with dose but the severity of the effect remaining independent of dose. When radioactive isotopes such as cesium-137, iodine-131, and strontium-90 enter the body through contaminated water, they irradiate internal tissues and organs over time. This chronic internal exposure can cause damage to the DNA within cells, leading to mutations that may initiate carcinogenesis.

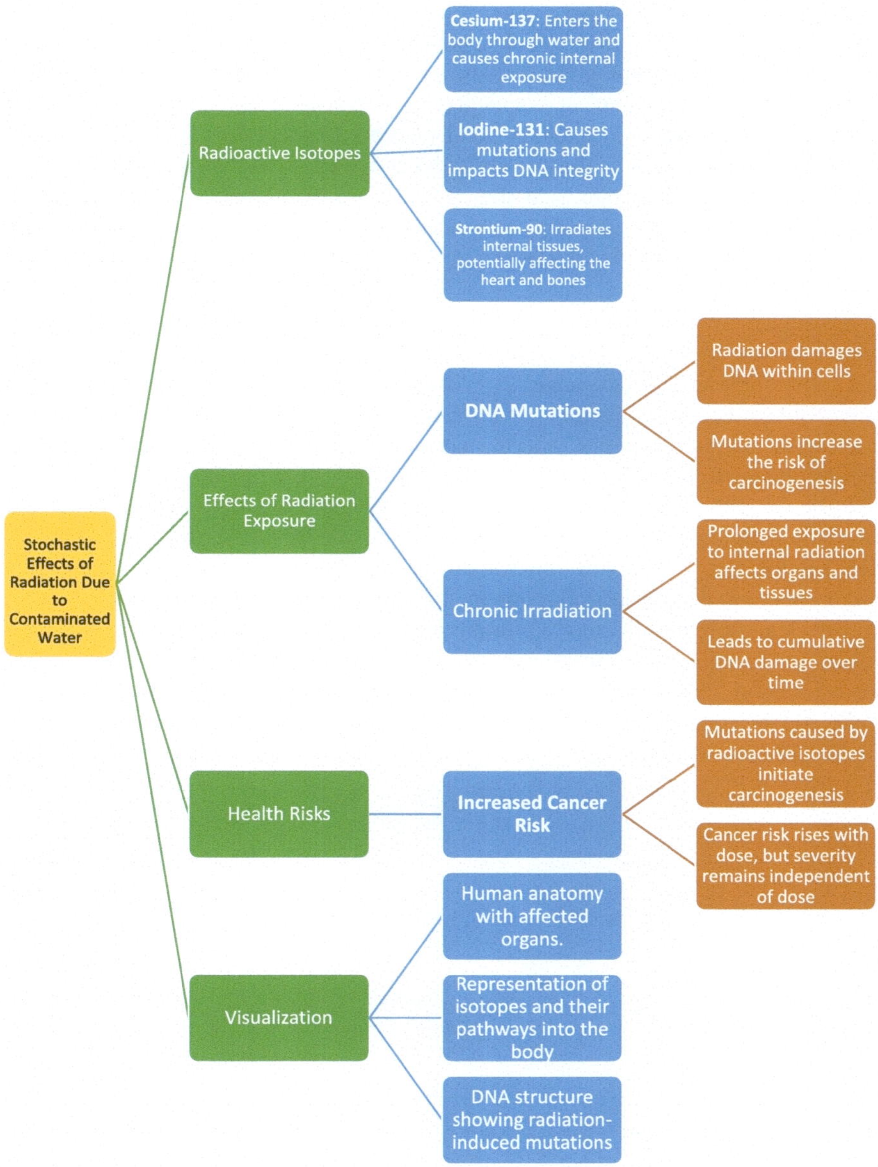

Fig. 10.1 Presentation of stochastic effects

The latency period for these cancers can range from several years to decades, making it challenging to directly link specific cases of cancer to past radiation exposure. However, epidemiological studies, such as those conducted on survivors of atomic bombings and nuclear accidents, have consistently shown a correlation between internal radiation exposure and increased cancer incidence. The risk is affected by factors such as the specific radioactive isotope, the cumulative dosage absorbed, the ingestion rate, and the age and health of the individual exposed. Children and fetuses are especially vulnerable because of their quickly proliferating cells and extended anticipated lifespan, which offers a broader opportunity for cancer progression.

The mechanism of stochastic effects involves the ionization of cellular molecules, resulting in direct DNA damage or the generation of reactive oxygen species that cause indirect DNA damage [20]. Inaccurate healing of damage may lead to irreversible genetic alterations. These mutations may activate oncogenes or inactivate tumor suppressor genes, altering normal cellular regulation and resulting in uncontrolled cellular proliferation, a characteristic feature of cancer.

Public health strategies to alleviate these concerns involve guaranteeing the safety of drinking water via stringent monitoring and treatment to eliminate radioactive pollutants. In cases of contamination, supplying alternative water sources, communicating danger information, and providing medical examinations for early cancer detection in affected communities are essential measures. Regulatory thresholds for allowable concentrations of radioactive isotopes in drinking water are established through a comprehensive study aimed at reducing cancer risk while considering the cumulative effects of low-dose exposure. A flowchart showing the effects of ingestion of contaminated water on humans is illustrated in Fig. 10.2.

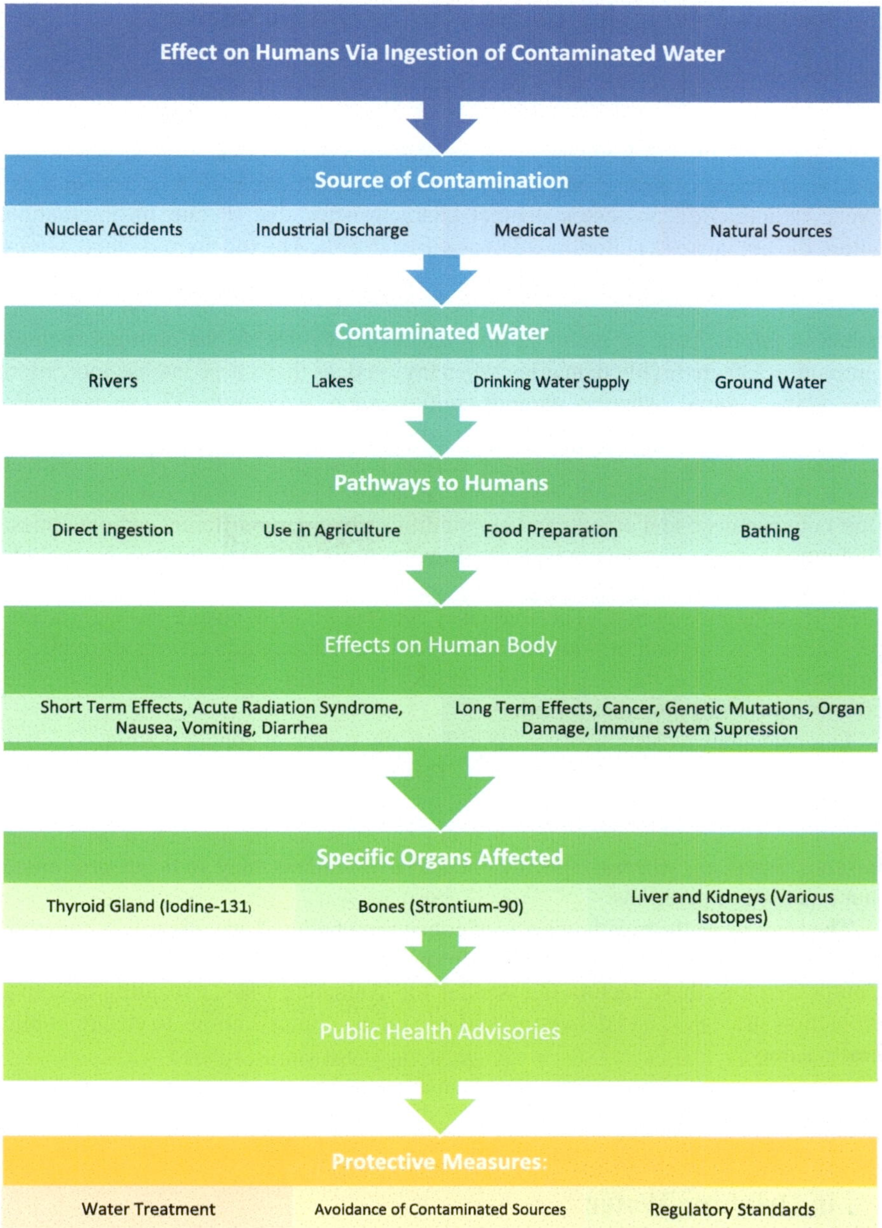

Fig. 10.2 Flowchart showing the effects of ingestion of contaminated water on humans

5 Factors Influencing the Severity of Health Effects

The severity of health effects from consuming water contaminated with radionuclides is determined by various factors, including the type and quantity of radionuclide, the duration and frequency of exposure, individual physiological variations, and the presence of synergistic or antagonistic interactions with other substances. Every radionuclide possesses distinct characteristics that dictate its interaction within the organism [21]. Iodine-131 is swiftly absorbed by the thyroid gland, strontium-90 accumulates in bones, and cesium-137 disperses uniformly throughout soft tissues. The radioactive decay properties, such as the half-life and type of emitted radiation (alpha, beta, or gamma), also play crucial roles. Alpha emitters such as plutonium-239 are highly damaging when ingested, as they cause intense ionization over a short range, whereas gamma emitters such as cesium-137 can penetrate deeper into tissues, causing widespread damage [22].

The dose, defined as the amount of radiation absorbed by the body, is a critical factor. Higher doses generally result in more severe health effects, and the relationship between dose and effect can vary significantly among different radionuclides. Prolonged low-dose exposure may result in enduring health complications, including cancer, but acute high-dose exposure can induce quick and severe consequences, such as radiation sickness. The frequency and length of exposure are significant; persistent consumption of minimal quantities of a radionuclide can result in bioaccumulation, wherein the material accumulates in the body over time, increasing the effective dose and potential for injury.

Forensic analysis also involves identifying the source of contamination, which is crucial for legal and remedial measures. This may entail tracking contamination in industrial effluents, agricultural runoff, or insufficient wastewater management. The detection of particular pesticides in water samples can be associated with local farming practices, whereas specific industrial pollutants may indicate proximate manufacturing operations.

The severity of the health impacts of consuming polluted water involves complex interactions among radiological, biological, and environmental elements. Comprehending these factors is essential for evaluating risks, executing effective protection strategies, and delivering suitable medical interventions to ensure public health safety.

6 Regulatory Standards for Radionuclides in Drinking Water

Rules for the regulation of radionuclides in drinking water are crucial to protect public health from the harmful effects of radiological contamination. These standards are set on the basis of extensive scientific studies, risk assessments, and recommendations from international and national sources. Health and safety agencies.

The World Health Organization (WHO) and the United States, among others, have imposed strict regulations on limiting radionuclide concentrations in water used for drinking to reduce the risk of cancer, and other irradiation-related EPAs have set maximum contaminant levels for radionuclides. The concentration of radium-226 and radium-228 was limited to 5 picocuries per liter (pCi/L), and the total standard was 15 pCi/L for gross alpha particles, excluding radon and uranium. The imposed limits are such that the potential dose from drinking water is below a lifetime exposure threshold regarded as acceptable, typically at 4 millirem according to the EPA (mrem) per year. Specific radionuclides that carry serious health risks are recommended. Iodine-131 is associated mainly with nuclear incidents. It has specific guideline levels to protect against harmful beta and gamma radiation. Other isotopes, such as cesium-137 and strontium-90, which have much longer half-lives and greater potential for important internal exposure, are controlled, and their concentrations in drinking water should be kept as low as reasonably achievable (ALARA). These standards are dynamic; they are reviewed and updated periodically on the basis of new scientific findings and advances in detection and remediation technology [23]. Public education and communication are integral parts of the holistic approach to risk reduction. Associated with radionuclides in drinking water. Governments should see that citizens are better informed of potential hazards, the importance of regulatory restrictions imposed, and measures undertaken to safeguard water quality. It is ensured through strict regulatory standards and an informed public: risks associated with radionuclide-contaminated water can be significantly reduced, thereby safeguarding public health and maintaining confidence in the safety of drinking water supplies.

6.1 International Guidelines

The numbers are derived from risk assessments that consider dose–response relationship epidemiological data, together with the health impact of long-term exposure. The GVs are designed to ensure that any health effect, for instance, cancer, occurs at a likelihood as small as reasonably achievable, considering operational and technological factor constraints [24]. The IAEA issues standards of radiation protection, and safety guidelines encompass advice for managing risks from radioactive contamination of potable water sources. The IAEA's Safety Standards Series Safety Guide No. WS-G-2.3 presents principles and procedures for evaluating and controlling radioactive contamination in drinking water sources. These requirements are the basic necessity of launching full-fledged monitoring programs, using efficient treatment practices, and engaging the public in the stewardship of radioactive hazards. International alliances and collaborations enhance the sharing of best practices; technically, competencies, skills, and capability building; and high-level implementation in an international context. Regional bodies, such as the European Union (EU) and the Pan American Health Organization (PAHO), play the most critical role in preparing and enforcing regulatory regimes for radionuclides in drinking water in their respective jurisdictions. Adherence to international standards is critical to maintain the consistency and effectiveness of efforts to protect public health from the dangers of radioactive contamination in drinking water. International guidelines enhance the water supply system's resilience and protect the health and well-being of communities around the world through advocacy for standardization and promotion collaboration among governments.

7 Treatment for Radiation Sickness Caused by Contaminated Water

The treatment of radiation sickness from contaminated drinking water is a multifaceted approach toward symptom management, a reduction in radiation exposure, and support for the body's healing processes, which vary with the dose of radiation and address particular radionuclides and the overall health condition of a person. In the case of acute radiation sickness (ARS) caused by drinking contaminated water, prompt medical intervention is crucial to mitigate symptom progression and improve outcomes. The main aim of treatment is to decrease gastrointestinal symptoms such as nausea, vomiting, and diarrhea, which are often secondary manifestations of radiation-induced damage to the gastrointestinal tract [25]. Antiemetic agents are used for the prevention of nausea and vomiting, whereas antidiarrheal agents are prescribed to control diarrhea and thus dehydration. IV fluids and electrolyte solutions include hydration and electrolyte balance, especially in the setting of extreme dehydration. Supportive care is a part of treatment that addresses issues related to radiation but is indispensable. Antibiotics may be used for prophylaxis or treatment of infections due to a compromised immune system, and mucosal membrane blood transfusions or hematopoietic growth factors may require the management of other hematological complications, such as anemia [26]. Leukopenia and hemorrhagic manifestations may arise due to interference with the bone marrow. Radiation treatment, a multidisciplinary and comprehensive approach, is therefore needed to address waterborne sickness, including medical treatment, supportive care, and psychosocial support. The exposure of healthcare providers to radiation can improve quality of life through symptom control, minimize radiation exposure, and reinforce the body's own healing procedures.

7.1 Supportive Care

Supportive therapy includes the treatment of victims exposed to radiation from contaminated water. This approach focuses on alleviating symptoms, preventing complications, and activating the body's natural healing response. The ultimate goal of supportive therapy is to relieve gastrointestinal symptoms such as nausea and vomiting. Diarrhea, which commonly results from irradiation damage to the gastrointestinal system, is treated with antiemetic drugs to prevent nausea and vomiting. Antidiarrheal agents are used to control diarrhea and prevent dehydration. Intravenous fluids and electrolyte solutions are used to maintain hydration and electrolyte balance, especially in severe dehydration. Supportive treatment, which includes bone marrow suppression and cytopenias, can be used to monitor and manage the hematological problems associated with radiation exposure. Blood transfusions may be prescribed to treat anemia though hematopoietic growth factors such as granulocytes. Colony-stimulating factor (G-CSF) and erythropoietin may be used to stimulate the production of white blood cells and red blood cells, respectively. Platelet transfusion is carried out to control thrombocytopenia and prevent bleeding complications: preventive measures for infection that could result from compromised immune status, an infection management system, and mucosal membrane malfunction. Antibiotics may be instituted either prophylactically or therapeutically to prevent or treat bacterial infections. Antifungal agents are needed for fungal therapy, especially in immunocompromised patients. Psychological support and counseling form part of supportive care that helps people overcome the psychoreactive effects of stress and anxiety caused by radiation exposure and possible long-term consequences. Radiation safety recommendations, which include any precautionary measures and available support services, can improve people's and their families' understanding during the recovery process. Nutritional care is important for patients who are irradiated, as proper nutrition improves immune function and facilitates tissue repair and healing. Dietary supplements and dietary counseling can be provided to establish the appropriate intake of nutrients, particularly in cases of loss of appetite or gastrointestinal disorders that discourage food intake. Thus, supportive care becomes an essential factor in decreasing the radiation exposure burden due to the application of polluted water. Supportive treatment enhances and improves the quality of life of individuals exposed to radiation by treating symptoms, relieving complications and activating the body's natural healing system.

8 Conclusion

Those people who consume contaminated water containing radionuclides are exposed to a dose that can lead to acute radiation syndrome and a lifetime of cancer-inducing radiation. Radionuclides, including iodine-131, strontium-90, cesium-137, and uranium, pass through water and food chains. It may cause cancer and hematological, gastrointestinal, and renal disorders by affecting specific organs and tissues. The severity of such effects is affected by numerous factors, including the type and degree of the radionuclide, length of time exposed, susceptibility of the individual, and effect of other materials present. To contain radioactive pollution of drinking water, it is necessary to adopt a wide range of measures that include laws and regulations, routine surveillance, prompt action, and large-scale health education of the population. International recommendations and legislation at the regional and national level are essential for the standardization, evaluation and correction of the preventive aspects of population health. Treatment of radiation syndrome resulting from the use of contaminated water consists of supportive measures aimed at alleviating symptoms, decreasing radiation exposure and promoting repair processes in the body. The whole treatment regimen incorporates psychological support, nutritional therapies, and education on radiation safety as auxiliary components. Rigorous legal requirements, increased monitoring and surveillance, and better public awareness and involvement are all useful in reducing the likelihood of contamination of drinking water sources with radionuclides and ultimately improving and enhancing health and wellbeing worldwide. Continuous inquiry, improvements in technologies and universal cooperation are fundamental to solving existing problems and assessing prospective radiological catastrophes. By managing coordinated efforts at the local, national and global levels, we can reduce the health effects of Palestinians as a result of radioactive pollution and provide clean and safe drinking water for all. In addition to providing suggestions for regulatory approaches and treatment techniques, community outreach and preparedness are critical elements in addressing the radiological aspects of polluted water. Mass education programs meant to sensitize the populace toward radionuclide sources and risks and emphasize the importance of active participation in mitigation and countermeasures can prepare the public to do things that would be good for their communities. This can encourage authorities to use other water sources in instances of pollution, improve water treatment and water supply systems, and encourage involvement in local monitoring and reporting programs. For effective prevention, preparedness, and response to radiological risks, radiological health within regulatory bodies, water companies, nongovernmental organizations and academic institutions need to work together in an intersectoral manner. This collaboration can enhance the sharing of experiences, resources and good practices and therefore improve overall risk assessment, management and response actions. By integrating the capabilities and resources of different stakeholders, communities are able to be more proactive in prevention, response, and recovery efforts related to nuclear incidents; hence, the impact on health and the environment is lower. To resolve the radioactive dimensions of contaminated water, a wide-ranging and well-integrated strategy that incorporates regulatory, scientific, technological, and community-based elements is needed. We may avert risks from radioactive pollution through emphasis on preventative measures, preparedness, and cooperation, hence safeguarding the health and welfare of both current and future populations.

References

1. Rajkhowa, S., Sarma, J., & Das, A. R. (2021). Radiological contaminants in water: Pollution, health risk, and treatment. In *Contamination of water* (pp. 217–236). Academic Press.
2. Campbell, K. M., & Kane, T. J. (2021). Radionuclides in surface water and groundwater. In *Handbook of water purity and quality* (pp. 199–230). Academic Press.
3. Ritter, K. S., Sibley, P., Hall, K., Keen, P., GevanMattu, B. L., & L. (2002). Sources, pathways, and relative risks of contaminants in surface water and groundwater: A perspective prepared for the Walkerton inquiry. *Journal of Toxicology and Environmental Health Part A, 65*(1), 1–142.

4. McKinley, I. G., Grogan, H. A., & McKinley, L. E. (2011). Fukushima: Overview of relevant international experience. *Genshiryoku Backend Kenkyu, 18*(2), 89–99.
5. Villanueva, C. M., Kogevinas, M., Cordier, S., Templeton, M. R., Vermeulen, R., Nuckols, J. R., & Levallois, P. (2014). Assessing exposure and health consequences of chemicals in drinking water: Current state of knowledge and research needs. *Environmental Health Perspectives, 122*(3), 213–221.
6. Steinhauser, G., Brandl, A., & Johnson, T. E. (2014). Comparison of the Chernobyl and Fukushima nuclear accidents: A review of the environmental impacts. *Science of the Total Environment, 470*, 800–817.
7. Mas, J. L., García-León, M., García-Tenorio, R., & Bolívar, J. P. (2006). Radionuclide concentrations in water. In *Radionuclide concentrations in food and the environment* (pp. 59–112). CRC Press.
8. Al-Khawlany, A. H., Khan, A. R., & Pathan, J. M. (2018). Review on studies in natural background radiation. *Radiation Protection and Environment, 41*(4), 215–222.
9. Braverman, E. R., Blum, K., Loeffke, B., Baker, R., Kreuk, F., Yang, S. P., & Hurley, J. R. (2015). Managing terrorism or accidental nuclear errors, preparing for iodine-131 emergencies: A comprehensive review. In *Anti-aging therapeutics* (Vol. XVII). Academic Press.
10. Madhav, S., Ahamad, A., Singh, A. K., Kushawaha, J., Chauhan, J. S., Sharma, S., & Singh, P. (2020). Water pollutants: Sources and impact on the environment and human health. In *Sensors in water pollutants monitoring: Role of material* (pp. 43–62). Springer.
11. Li, C., Dos Reis, A. A., Ansari, A., Bertelli, L., Carr, Z., Dainiak, N., & Zhang, J. (2022). Public health response and medical management of internal contamination in past radiological or nuclear incidents: A narrative review. *Environment International, 163*, 107222.
12. Neff, J. M. (2002). *Bioaccumulation in marine organisms: Effect of contaminants from oil well produced water*. Elsevier.
13. Adeola, A. O., Iwuozor, K. O., Akpomie, K. G., Adegoke, K. A., Oyedotun, K. O., Ighalo, J. O., & Conradie, J. (2023). Advances in the management of radioactive wastes and radionuclide contamination in environmental compartments: A review. *Environmental Geochemistry and Health, 45*(6), 2663–2689.
14. Sajjan, R., & Bomanji, J. (2014). Radionuclide scanning in the diagnosis and treatment of endocrine disorders. In *Endocrinology in clinical practice* (p. 139). CRC Press.
15. Hamada, N., & Fujimichi, Y. (2014). Classification of radiation effects for dose limitation purposes: History, current situation and future prospects. *Journal of Radiation Research, 55*(4), 629–640.
16. Kamiya, K., Ozasa, K., Akiba, S., Niwa, O., Kodama, K., Takamura, N., & Wakeford, R. (2015). Long-term effects of radiation exposure on health. *The Lancet, 386*(9992), 469–478.
17. Knizhnikov, V. A., & Barkhudarov, R. M. (1975). Radioactive contamination of the environment and the possible consequences. *Atomic Energy Review, 13*(2), 171–214.
18. Phipps, A. J., Bergmann, J. N., Albrecht, M. T., Singh, V. K., & Homer, M. J. (2022). Model for evaluating antimicrobial therapy to prevent life-threatening bacterial infections following exposure to a medically significant radiation dose. *Antimicrobial Agents and Chemotherapy, 66*(10), e00546–e00522.
19. Leier, M., Suursoo, S., & Kiisk, M. (2021). Cost-benefit analysis of health risks caused by indicative dose from drinking water consumption. *Journal of Environmental Radioactivity, 231*, 106546.
20. Friedland, W., & Kundrat, P. (2018). Stochastic multi-scale modeling of biological effects induced by ionizing radiation. In *A guide to outcome modeling in radiotherapy and oncology* (pp. 147–180). CRC Press.
21. Avwiri, G. O., Ononugbo, C. P., & Olasoji, J. M. (2021). Radionuclide transfer factors of staple foods and its health risks in Niger Delta region of Nigeria. *International Journal of Innovation Environmental Studies Research, 9*(1), 21–32.
22. Ahmed, M. F., Alam, L., Mohamed, C. A. R., Mokhtar, M. B., & Ta, G. C. (2018). Health risk of polonium 210 ingestion via drinking water: An experience of Malaysia. *International Journal of Environmental Research and Public Health, 15*(10), 2056.

23. Hossain, F. (2020). Natural and anthropogenic radionuclides in water and wastewater: Sources, treatments and recoveries. *Journal of Environmental Radioactivity, 225*, 106423.
24. Pintilie-Nicolov, V., Georgescu, P. L., Iticescu, C., Moraru, D. I., & Pintilie, A. G. (2021). The assessment of the annual effective dose due to ingestion of radionuclides from drinking water consumption: Calculation methods. *Journal of Radioanalytical and Nuclear Chemistry, 327*, 49–58.
25. Annamalai, S. K., Arunachalam, K. D., & Selvaraj, R. (2017). Natural radionuclide dose and lifetime cancer risk due to ingestion of fish and water from fresh water reservoirs near the proposed uranium mining site. *Environmental Science and Pollution Research, 24*, 15427–15443.
26. Yadav, R., Ali, M., Kumar, A., & Pandey, B. N. (2017). Mechanism of carcinogenesis after exposure of actinide radionuclides: Emerging concepts and missing links. *Journal of Radiation and Cancer Research, 8*(1), 20–34.

Chapter 11
Effects of Rare Earth Element (REE) Contamination in Water on Human Health Through Ingestion

Soumya Kar, Abinash Sahoo, and Sandeep Samantaray

1 Introduction

With the world's growing metrics being influenced significantly by the unconstrained increase in transport, industry, urbanization, and agriculture, growth in consumer demands and living ethics has magnified pollution. Recent discoveries of emerging contaminants (ECs), a broad range of composites characterized by perceived or veridical threats to public health and the environment, include groups of uncontrolled pollutants from primary sources, including pharmaceuticals, pesticides, and personal care products; flame retardants; disinfection byproducts; endocrine-disrupting chemicals; antibiotic-resistant bacteria; antibiotic resistance genes; man-made nanomaterials; and other substances and additives that exist in groundwater, surface water, and soil [1]. These chemicals increasingly enter the environment as nonbiodegradable substances and hazardous wastes. Furthermore, there is insufficient documentation and a lack of robust epidemic information on their behavior and fate in the global environment, including information concerning human exposure, serum and tissue concentrations, and threats to ecological health [2]. These pollutants, which are harmful in very small amounts, are often mixed with other pollutants and released as breakdown products into the worldwide natural ecosystem. ECs now persist across the global ecosystem and have a prolonged lifespan, as they are characteristically bioaccumulative and bioactive, potentially

S. Kar · A. Sahoo
Department of Civil Engineering, Odisha University of Technology and Research, Bhubaneswar, Odisha, India

S. Samantaray (✉)
Department of Civil Engineering, National Institute of Technology, Srinagar, India
e-mail: sandeep@nitsri.ac.in

M. K. Jindal et al. (eds.), *Emerging Contaminants in Water*, Emerging Contaminants and Associated Treatment Technologies,
https://doi.org/10.1007/978-3-031-89591-3_11

contributing to more than nine million deaths resulting from adulterated water, food, consumer products, or workplaces.

The lack of defined laws and quality standards addressing the acceptable amounts of ECs is one of the main obstacles to controlling their mobility. ECs have greater potential to harm ecosystems because they are not typically analyzed in the environment. In many nations, there is significant concern over the presence of ECs comprising endocrine-disrupting chemicals (EDCs), pesticides, and pharmaceuticals and personal care products (PPCPs) in aquatic systems. The human hormone balance is disrupted by the bioaccumulation of PPCPs in aquatic environments, which can lead to a variety of negative consequences, such as decreased fertility, reproductive impairments, and a high risk of breast and testosterone cancers. Among various ECs, major contaminants found in the environment include triazines, chloroacetanilides, carbamates, organophosphates, dichlorodiphenyltrichloroethane (DDT), and polychlorinated biphenyls (PCBs). Consequently, further research is necessary to characterize the ubiquitous ECs in natural ecosystems.

The field of technological advancements has made significant progress in recent years, resulting in the emergence of new pollutants or contaminants, which are found in air, water, and soil. These emerging contaminants have both natural and anthropogenic origins and include microplastics, pesticides, endocrine-disrupting chemicals, heavy metals, flame retardants, surfactants, industrial products, and organic pollutants that persist over time. Because of their potential ecological and human risks, emerging contaminants (ECs) have attracted considerable public and research attention in recent years. The sources of these contaminants are the same as those traditionally recognized and can be categorized as point or nonpoint sources. These contaminants enter the environment through industrial processes and emissions, agriculture, wastewater, and effluent disposal. Emerging contaminants have a broad range of environmental effects, affect soil quality and freshwater sources, and enter the food chain through crop uptake, bioaccumulation, and biomagnification. Consequently, continuous exposure to REEs leads to health risks for humans and other living organisms [3–5].

Lanthanide group components, or REEs, are examples of ECs, despite being frequently disregarded. The last several years have seen a sharp rise in the production of REEs, coinciding with new developments in nanotechnology that are finding greater use in the fields of electronics, medicine, transportation, agriculture, building, and catalysis. Like most ECs, these are (1) unregulated in the environment at the moment [6], (2) not routinely checked in public health and environmental programs, (3) detectable with very low recognition analytical equipment as micropollutants, and (4) poorly understood mechanisms of environmental and human toxicity [7]. REEs are a class of metals that include yttrium, scandium, and 14 lanthanide components. Owing to their significant contributions to both technological advancement and traditional industry development, these metals have been referred to as "industrial vitamins" and "treasurys" of innovative materials [8, 9].

Three categories are used by the Australian Industry Commission to categorize REEs, mostly in relation to mineral extraction terminology: light REEs, Eu, Sm, and Gd; medium REEs, Nd, Ce, Pm, La, and Pr; and heavy REEs, Yb, Lu, Sc, Yb,

Ho, Dy, Tm, and Er. The US Geological Survey (USGS), which follows the IUPAC categorization scheme and classifies Pr, Ce, La, Pr, Sm, Nd, Gd, and Eu as LREEs and Dy, Tb, Ho, Tm, Er, Lu, and Yb as HREEs, is likewise deserving of recognition [10]. REEs are typically found in perovskites, carbonatites, pegmatites, and granites. Phosphominerals, such as apatite, bastnaesite, monazite, and others, are typically added. For high-tech industrial uses, REEs are considered essential resource commodities. REEs are referred to as "vitamins of modern industry" because they are now incredibly significant constituents of all high-tech instruments. For many years, human activity has been the primary source of REEs in the environment, ultimately polluting the entire planet. REEs may be released into the environment during the course of extraction, mining, and processing of REE ores. Moreover, REEs are broadly utilized in medical technologies, agriculture, and industry because of their outstanding chemical and physical qualities. REEs have been thoroughly examined by Massari and Ruberti for use as feed additives and fertilizers.

2 REE Sources in the Environment

REEs are mostly derived from geogenic sources in the environment, but they are also increasingly derived from high-tech equipment, agriculture, and medicines. In actuality, REEs have an extensive range of uses, including in permanent magnets (Nd, Pr, Ho, Dy), batteries (Nd, Ce, La), agents (Gd), and the production of cancer treatment medications [11, 12]. According to Tepe et al. (2014), they are used in agricultural products, information storage, flints, lighters, fluorescent lights, high-temperature, high-tech superconductors, and military defense systems [13]. They are essential to currently switch from conventional sources of energy to clean energy machineries, electric cars, and energy-efficient lighting since they are indispensable in developing clean energy [14]. By 2030, the demand for rare earth ores from green technologies is expected to reach 51.9 kilotons, with Nd and Dy being the most in demand. In addition to examining the application of REEs in aquaculture, forestry, animal husbandry, and agriculture, Cai et al. (2015) investigated how adding REE-enriched yeast to a meal improved the quality and digestion of meat [15].

REEs are released into the environment via mineral processing and mining operations, industrial activities that employ REEs as effluents or wastewater, and disposal of industrial products (e.g., landfills) and consumer products [7]. REEs have been obtained from acid mine drainage (Gwenzi et al., 2018), solid waste, and wastewater from mining and mineral processing facilities [7, 16–18]. According to Sloof's (1995) research, soils in mine-affected areas had concentrations of REEs that were at least 100 times greater than the baseline levels found in Earth's crust. According to Li et al. (2013), there was contamination in the soils of an REE processing plant within a 7 km radius, and concentrations decreased as one moved farther from the tailings [11]. According to Tan et al. (2015), micropollutants are released into the environment by solid wastes, effluents/wastewaters, and air emissions from industrial activities (such as petroleum purification and healthcare facilities) that use REEs [19]. Moreover, point sources of REEs are urban wastewater treatment systems that receive different streams of municipal and industrial wastewater [20, 21]. According to Thomsen (2017), one of the major sources of anthropogenic Gd found in aquatic systems is wastewater treatment plants [22]. REEs in wastewater are further distributed by hydrological processes (such as runoff and leaching) and wastewater irrigation in both aquatic and terrestrial systems.

Because landfills are frequently used to dispose of waste electrical and electronic equipment (WEEEs) or postconsumer REE items, they serve as REE reservoirs [23]. For example, REEs, especially Eu, are present in computer monitors and television screens up to 72%, or 0.76%, on a weight basis [24]. Since Eu is present mostly as a liquid, it can permeate the soil [19]. Recycling facilities for WEEEs and other REE-containing wastes, such as fluorescent bulbs that are widely used in China, Vietnam, India, Ghana, Nigeria, and Pakistan, are additional sources of REEs [25, 26]. REE-containing fertilizers and animal feeds also add to anthropogenic REEs [27–31].

REE is utilized as a fertilizer supplement in China to increase crop quality and production [32]. The surface soil layer of China contains 100–200 mg/kg lanthanides, which could lead to bioaccumulation [33, 34]. Because of this widespread application, the global intake of REEs was 540 kt in 2020 and is still growing at a rate of 6% annually. Economic indicators that impact cutoff grades in mining metallurgical operations include market prices for the primary product and its byproduct, expenses linked with mining and processing, and the distribution of metallurgical recovery tonnage grades [35]. As a result, determining the ore cutoff grade is exceedingly challenging. One important carbonatite deposit with considerable quantities of REEs is the Saranu deposit in Rajasthan, India. The deposit is located in carbonatite dikes that are approximately 10 cm wide and contain $\geq$5.5% REO (rare earth oxides). There have been reports of LREE minerals from Gujarat's Amba Dongar carbonate complex, including bastnäsite, parisite, and synchysite [36]. Carbonatite sills, dikes, and plugs hosting REE concentrations were found at Kamthai (near the Saranu deposit), Barmer region, Rajasthan, via grid channel geochemical sampling techniques and geological mapping [37]. The deposits are estimated to be up to 84 m deep, with the highest LREE value being 17.31% and the mean grade being 3.33%. These deposits have possible REE resources of 4.91 Mt. In a later investigation of these carbonatites, Bhushan (2015) suggested that mantle "hot spot" activity contemporaneous with Deccan volcanism was responsible for the deposit's formation [38].

Because REEs are relatively new pollutants, nothing is known about how they behave in the environment in terms of toxicity, bioaccumulation, and bioavailability. Hence, governments typically do not restrict the environmental discharge of REEs. Furthermore, the establishment of control standards is hampered by diffuse REE inputs from water runoff from hard-standing places and air, particularly in estuaries, which receive distinct inputs from their watersheds [39]. According to the literature, three European rules that set REE threshold concentrations only apply to treated water and waste from mining activities (Directive/EU 0621/; EIA directive/EU1452/; Directive/EU 1359/) [40, 41]. Therefore, REEs in aquatic environments need to be viewed as pervasive pollutants. Its presence has already been documented in rivers, estuaries, ocean fluids, suspended particles, and sediments [42, 43]. Trace amounts of REEs have been found in coastal locations [44].

Different paths may be used by REEs to interact with the resident biota in the marine environment. They have the ability to bioaccumulate in entities and then be transported further up the trophic chain, where they may cause biodilution or biomagnification at higher trophic levels. Because of their distinct electrical structures, REEs and their composites are extremely reactive. REEs react easily with most nonmetals (such as H, O, and S). Together with a variety of inorganic (such as carbonates and phosphates) and organic (such as fulvic and humic acids) ligands, REEs can also form ionic REE–ligand complexes. The coordination numbers of REE-ligand complexes are frequently greater than six and occasionally as high as 12 [7].

The concentrations of REEs in soil continue to rise because of their use in agriculture as fertilizers to increase crop growth and productivity [45]. Mineral fertilizers, sometimes known as phosphate fertilizers, and soil conditioners typically comprise micronutrients such as Fe and Si; micronutrients such as Mg, Ca, N, S, and P; and REEs. Higher REE values were discovered in rice than in corn, suggesting that rice tends to accumulate more REEs than corn does. Concerns about the environment are growing that excessive REE application in agriculture could cause these components to infiltrate the food chain via plant absorption and harm human health [10]. According to Hu et al. (2004), China was the first nation to employ REEs as fertilizers in agricultural production and as performance developers in animal feed [46]. As of 2007, China has applied 54,100 tons of REEs to agriculture overall. REEs can enter the environment through the use of phosphate fertilizers made from phosphate minerals containing REEs, in addition to being purposefully provided as feed additives and fertilizers [47]. Phosphate fertilizer application in Canada began to rise gradually in 1995 and reached approximately 750,000 tons annually from 2011 to 2012 [48]. It was projected that the use of phosphate fertilizers caused the soil in Brazil to absorb 12,000 tons of cedar every year in 2014.

REE contamination levels in soil, water, and air have increased due to the significant and widespread rise in REE mining and applications, which has also had an impact on the food chain [7, 49]. One significant method by which individuals are exposed to REEs that may have negative impacts on their bodies is through food. There are not as many toxicological studies on REEs as there are for other common metals [50]. Nonetheless, the scant toxicological evidence from research on animals and humans indicates that REEs may cause tissue-centric bioaccumulation upon entering the human body, leading to harm to the immune system, neurological system, and reproductive system [51–54]. Moreover, there is a shortage of toxicological data concerning other REEs, which are limited primarily to Ce and La in relation to REE-associated health impacts.

3 Projection of REE Intake and Cumulative Risk Assessment

Equation (11.1) was used to compute the daily average intake of REE. The daily projected total REE intake was obtained by multiplying the mean consumption data of each food classification by the associated mean absorptions of REEs for that food category.

$$\mathrm{EDI} = \frac{\sum_{j=1}^{n} Cj * Fj}{BW} \tag{11.1}$$

where *BW* represents the average body weight of each subgroup or the general population (kg); *n* represents the number of food classifications utilized for deriving the EDI for each subgroup or the general population; BW represents the mean body weight of each subgroup or the general population; *EDI* represents the estimated daily dietary REE (μg kg^{-1}); *Fj* represents the daily average consumption of food classification j (g day^{-1}); and *Cj* represents the average REE concentration in the corresponding food classification *j* (mg/kg). Assuming that these users were vastly exposed to two food classifications and a mediocre level for the remaining classifications, the exposure evaluation model of dietary REE exposure was completed in three phases in accordance with EFSA management for the usage of a brief database [55]. Two food classifications that have the greatest nutritional assistance to consumers were added to the mean exposure levels of other food classifications for the common population to determine high exposure (the 95th percentile, or P95).

The hazard index (HI) and hazard quotient (HQ) of different compounds were used Eq. (11.2) to evaluate cumulative exposure to REEs.

$$\mathrm{HI} = \sum_{i=1}^{n} HQi = \sum_{i=1}^{n} \frac{EDIi}{RVi} \tag{11.2}$$

where *n* represents the number of REEs utilized to estimate *HI* for each subgroup or general population, *EDIi* represents the daily dietary intake of *REEi*, *HQi* represents the hazard quotient of REE i, *RVi* represents the health-based guideline value of *REEi*, and *HI* represents the hazard index. Cautious consideration should be given to adverse health impacts if $HI > 1$, whereas $HI \leq 1$ recommends that risk is tolerable. The maximum amount of REEs allowed in food has not yet been established by the European Union, Codex Alimentarius Commission, New Zealand, Australia, the United States, Canada, or Japan. From 1991 to 2017, the China Food Safety Standard (GB 2762–2012) established upper limits for food-related elements. REE limits were set at 2.0 mg kg^{-1} for rice, corn, wheat, and tea and 0.7 mg kg^{-1} for fruits and vegetables [56].

3.1 *Toxicity Effects of REEs*

3.1.1 Human Beings

The literature has established the influence of REE mining on the environment and people living nearby. Studies have revealed significantly higher urine REE concentrations in people residing near the Baiyun Obo mining region in Inner Mongolia, China, than in those living outside the mining region, which can disturb protein levels in the serum that take part in different biological processes [57, 58]. REE investigations through hair samples revealed increased concentrations in the scalps of people residing near mining locations in Inner Mongolia and in hair samples from children living in Liao and Shang villages in comparison with those from reference areas [59, 60].

Separate studies have revealed that blood and hair samples from people residing in large-scale mining sites are enriched with REEs, primarily LREEs, in Fujian Province, China. Excess Ce in dust from mining regions and possible health complications through inhalation were established in a prior study [61]. Kowalczyk et al. performed an extensive study for estimating the exposure of the adult Dane population to REEs resulting from tea consumption, followed by individual elemental risk assessment [62]. According to the risk assessment, consumers who consume tea have a low risk of being exposed to all analyzed substances, and even high consumers should not expect any negative consequences.

In addition to mining, people are exposed to REEs from the environment. Although children from Uganda had much higher REE concentrations in their primary teeth than children from the UK did, Moroccans had higher levels of REEs in their blood than Canarians did, primarily because of improper handling of e-waste [63, 64]. According to Guo et al., resident blood from an e-waste location has greater levels of Ce and La, which are linked to higher levels of thyroid-stimulating hormone (TSH) [65]. Studies have proven that Madrid's high traffic volume is correlated with higher Ce levels in women's serum [66]. Zaichick et al. reported that accumulated REE concentrations in rib bone tissue were age dependent [67]. REEs are associated with a lower risk of cancer, as established through studies of soil samples and subsequent human health risk evaluations by Da Silva Ferreira et al. [68]. Furthermore, the noncarcinogenic presence of REEs in street dust in industrialized areas is attributed less to anthropogenic pollution and largely to heavy local soil [69]. Conflicting results regarding REE and associated risks of neural tube defects (NTDs) have been reported [70, 71]. While a previous study suggested that higher REEs affect pregnancies, they are not related to the risk of NTD; a later study reported that the risk for NTDs increases with REE concentration, specifically La. The idea that REEs contribute to the development of anemia has been highlighted by Henríquez-Hernández et al. (2017) [72], as REEs were observed to be greater in the blood of sub-Saharan migrants who had anemia, regardless of Fe levels. A French community residing close to a site where Ce was harvested showed an increase in mortality from bladder and lung cancer, according to De Vathaire et al.

(1998) [73]. Similarly, deposits of monazite elements, such as Ce, have been observed by Kutty et al. (1996) to increase the incidence of endomyocardial fibrosis (EMF) [74]. Infants that are exposed to REEs in utero, particularly La and Nd, have a greater chance of developing orofacial clefts [75]. The TSH levels of infants born to pregnant women exposed to Ce and Yb were found to be lower [76]. By accumulating in the cerebral cortex, long-term environmental ingestion of trace amounts of REEs can lead to subclinical harm [77]. Relative to normal brain tissues, the brain tumor tissues of patients with astrocytomas contain significantly more REEs, particularly La, Ce, and Gd [78]. A host of ailments, including dyspepsia, abdominal distension, diarrhea, weakness, anorexia, and exhaustion, affect people who live close to high-REE background locations. Blood serum has increased amounts of IgM and decreased levels of total serum-glutamic pyruvic transaminase (SGPT), albumin, protein, and globulin [79].

According to assessments, one of the main causes of REEs in indoor respirable particles is tobacco smoking. Owing to tobacco smoke, cerium (Ce) and La levels considerably rose in smokers' homes as well as in hospitality establishments such as pubs, restaurants, dance clubs, offices, and cafeterias. These findings suggest that tobacco smoke poses a risk to people's respiratory systems [80, 81]. Additionally, higher indoor Ce and La levels are linked to indoor smoking and respiratory problems in children [82]. REE dissolution in PM10 and inhalation bioaccessibility are greater in nonsmoking homes, although smokers' households have higher levels of REEs in PM10 [83]. One possible source of increased consumption of REEs is e-cigarettes. According to Badea et al. (2018), those who use e-cigarettes have greater serum levels of REEs than do people who use traditional cigarettes [84]. Additionally, people who use e-cigarettes longer have higher blood serum levels of Ce and Er. According to Rodushkin and Axelsson (2000), individuals who wear porcelain bridges may also be at risk of consuming REEs because of greater ytterbium levels in their teeth [85]. For many years, magnetic resonance angiography (MRA) and magnetic resonance imaging (MRI) have employed gadolinium-based contrast agents (GBCAs). Patients experienced no adverse effects from these agents throughout the initial studies on their safety for human health. A few studies have used a variety of cytotoxicological tests to examine the possible toxicity of GBCAs to human health. Edward et al. (2010) investigated how various GBCAs affect human skin fibroblasts. A few of them promoted the synthesis of hyaluronan and proliferated cells, but none of the GBCAs had any effect on the creation of collagen [86].

According to Wei et al. (2013), the concentration of REEs, specifically light REEs such as Ce, La, Nd, and Pr, in the hair of miners was significantly greater than that in nonminers in both the mining region and the control region [59]. The lung tissue of smelter workers who passed away also presented elevated La levels [87]. Decreased bone mineral density in the femoral neck, lumbar vertebrae, intertrochanter, and greater trochanter was linked to elevated REE levels in miners' hair, which were also linked to Fe and Ca in hair [75]. According to these studies, the working conditions of miners may expose them to elevated quantities of REEs. Inhaling dust containing REEs can lead to rare occupational sickness, known as rare earth pneumoconiosis [88]. Early research revealed that employees engaged in movie projection were subjected to REEs, which were released into the air by arc carbon lamp dust and fumes, which contained a sizable quantity of REEs [89]. Pneumoconiosis was observed in photoengravers who had been exposed to dust and fumes from carbon arc lamps for many years. Clinical analyses revealed a noticeably high concentration of REEs in the lymph node and pulmonary biopsies, indicating a possible link between occupational exposure to REE dust and pneumoconiosis. Therefore, it is necessary to establish air exposure threshold limit values of REEs to prevent workers from being exposed to these pollutants at work [90, 91].

Importantly, DNA oxidative destruction and elevated blood Cr concentrations in chromate-exposed workers have been linked to elevated REE levels in the blood [92]. More Eu is found in the urine and blood of e-waste recyclers; however, this is only true for urine in terms of La [93]. According to Roncati et al. (2018), a woman who worked in the ceramics sector had her triple-negative breast cancer cells ingest REEs [94]. Gd and La have been mentioned as potential mutagenesis agents for human primary peripheral lymphocytes [95]. In a thorough investigation carried out by Yang et al. (2022), food samples were combined into 11 major food classifications and then further separated into subclassifications for evaluating dietary exposure to REEs in the Chinese population. Grain, vegetable, fruit, meat, fish, dairy, eggs, beans, thallus, tea, and nuts constituted the primary diet groups. In the dietary samples, 100% of the REEs were detected [96]. The various dietary categories had average concentrations of total REEs ranging from 0.04 to 1.41 mg/kg^{-1}. Tea had the highest mean concentration of all REEs (1.41 mg kg^{-1}), followed by eggs (0.29 mg kg^{-1}), thalli (1.14 mg kg^{-1}), and aquatic goods (0.28 mg kg^{-1}). Meat, beans, grains, vegetables, fruits, and milk all had mean values of 0.14 mg kg^{-1} or less.

LREEs are easier for plants to absorb because of their greater mobility in soils; as a result, they are found in greater concentrations in plant tissues than HREEs are. In addition, the superior absorption of LREEs was encouraged because HREEs form considerably more stable complexes in the soil solution. The average ratio of all LREEs to all HREEs was 3.01. Even while the exposure did not surpass the tADI, toddlers (2–6 years old) had a higher food intake of REEs than other sex–age groups did, suggesting a greater potential health risk for them. Toddlers' greater food intake may have resulted from their greater food consumption relative to their body weight compared with other subgroup populations. The dietary exposures to Sc, La, Y, Nd, and Ce were greater than those to other REEs, whereas the dietary exposures to other REEs differed depending on the location of China. Nigerian surface and groundwater bodies are known to be contaminated by small-scale gold mining and artisanal activities [97, 98]. In the vicinity of the mining region, samples from surface water, bottom sediments, groundwater, fish, and floating plants were examined for REEs. For both groundwater (0.81 ± 0.63 μg L^{-1}) and surface water (1.00 ± 0.65 μg L^{-1}), Scandinavia had the highest REE concentrations. In groundwater, as opposed to surface water, the majority of REEs are present in relatively high concentrations. The most prevalent REE found in the bottom sediments was Eu

(98.60 ± 94.32 μg/kg), followed by Ce (31.20 ± 64.12 μg/kg) and La (16.57 ± 28.78 μg/kg). The total REEs in the fish samples were 31.91 ± 35.80 μg kg^{-1} in the gut, 18.60 ± 19.47 μg kg^{-1} in the muscle, and 15.78 ± 22.45 μg/kg in the gills [99]. Exposure to REEs through the intake of polluted fish results in the bioaccumulation of REEs in the human body, initiating significant human health effects [100].

The consumption of contaminated vegetables is a significant pathway for metal uptake in humans [101–103]. Galhardi et al. (2020) conducted a study in two tropical uranium ore fields on the risks associated with REEs located in Brazil [104]. Researchers have evaluated the metal concentrations in 50 vegetable samples (soybean and corn) and their equivalent agricultural soils in both U-rich coal mining areas and U-mining areas. The results revealed that the REE concentrations for both areas, ranging from La to Lu, were greater than those proposed by the Brazilian Environmental Agency. Nonetheless, daily estimated REE intakes were found to be considerably lower than the intake dose considered harmful to human health (100–110 μg/kg/day) [105]. Brouziotis et al. (2022) conducted a structured review examining the influence of REEs on public health [106]. Their analysis revealed that more than one-third of research pertains to cytotoxicity. REE mining has been revealed to have adverse effects on human health, as these components must end up in the hair, blood, or urine of mining workers who are directly exposed to them and residents living near mining areas. Furthermore, people can be exposed to REEs on a daily basis through other means, such as inappropriate management of e-waste or increased traffic volume, as REEs are commonly utilized to develop additives in diesel fuels and as technological devices.

A cross-sectional investigation conducted by Li et al. (2021) revealed that a high level of lanthanum (La) in the serum of women who undergo in vitro fertilization-embryo transfer can lead to a 30% increase in failure of clinical pregnancy and an adverse correlation related to the number of better-quality oocytes [107]. Wang et al. (2018) established a correlation between increased REEs and the risk of hypertension and its concentration in the hair of a housewife with indoor air pollution [108]. A study showed that Ce nanoparticles could help reduce the production of oxidative and reactive oxygen species (ROS) that damage DNA in the BEAS-2B cell line [109]. In contrast, researchers have suggested that La and Gd may be mutagenic elements for human primary peripheral lymphocytes [95]. Gadolinium oxide nanoparticles have been found to be toxic to human umbilical vein endothelial cells, causing ROS production, lipid peroxidation, autophagic modulation, and mitochondrial dysfunction through necrosis and apoptosis [110]. Moreover, Y_2O_3 nanoparticles have been found to induce necrosis and apoptosis in human embryonic kidney cells by increasing mitochondrial membrane permeability, increasing cellular ROS levels, and promoting DNA destruction [111].

Several studies have highlighted the damage caused by Gd-based contrast agents to the human nephrological system [112]. La has been linked to dysfunctional neurological behavior, leading to lower IQ levels and impaired learning in children. Nd has been reported to affect the gastrointestinal structure, causing nausea and vomiting [7]. The ingestion of polluted water and food, direct intake via occupational exposure, and medical administration are the main ways in which REEs enter the human body, resulting in potential risks [59]. The impacts of three LREEs and HREEs on the biomass and germination of five plant species were assessed [113]. These plant species included two crops and three native, wild Canadian species, which were grown in contaminated soils with increasing REE concentrations (chloride forms of neodymium, praseodymium, terbium, samarium, erbium, and dysprosium). All the plant species tested in this study accumulated and absorbed REEs from soils.

Lanthanide uptake was inversely related to the pH of the soil and positively correlated with the soil concentration. Other studies have also suggested that the pH of soil plays a crucial role in the bioavailability of different REEs and their discharge into soil. It is likely that in the worst-case situation, where the soil becomes acidic, plants growing at the site are more vulnerable to REE toxic contamination [114–116]. Prior to the stabilization of soil concentrations to natural levels, REE soil contamination has reportedly spread up to seven kilometers, caused by tailings from an REE processing factory in China. In particular, the amounts of praseodymium (Pr) and Nd in relation to the source were 310 and 85 mg kg^{-1} at 2.1 km, 1279 and 373 mg kg^{-1} at 1.3 km, and 2266 and 650 mg/kg at 0.8 km. High concentrations of REEs—500–700 $mgkg^{-1}$ La, 800–900 $mgkg^{-1}$ cerium, 100 $mgkg^{-1}$ Pr, and 400 mg kg^{-1} Nd—are found in soils in contaminated areas close to industrial locations in the Netherlands. For the majority of REEs, including samarium (Sm), pr, and nd, concentrations in Chinese mining sites are up to 200 times higher than those in the Earth's crust [117].

3.1.2 Aquatic Organisms

Although there are several known sources of REEs, the majority of them unmistakably point to anthropogenic industrial sources. REEs can originate from wind-driven movement to open waters as well as from processes of rock and soil erosion [118]. Waste electronic and electrical equipment (WEEE), or postconsumer REE items, is a fantastic source of REEs for oceans once they reach landfills [23]. Up to 72% Eu can be found in the liquid stage in some of these devices, such as computer screens [24]. Therefore, recycling facilities for WEEE and other wastes containing REEs, including fluorescent bulbs, poses a significant risk and are widespread in a number of nations, including Ghana, Vietnam, Nigeria, Ghana, India, and China [25]. The release of wastewater polluted with REEs is one of the primary sources of REEs for aquatic ecosystems [119], which poses a serious risk to coastal environments through runoff and leaching. Animal feed and fertilizers are two more significant sources of REEs. Because they are frequently used to make phosphate fertilizers, phosphoric rocks with significant loads of REEs have the potential to contaminate estuaries [120]. Depending on how the environment changes, two key factors in estimating the quantity of metal accumulated in aquatic species are bioaccessibility and bioavailability. Owing to their widespread use in a variety of industries and absence of safety limitations, REEs have been shown to have a wide range of effects on aquatic life forms.

Lanthanide speciation, which is impacted by the physicochemical factors of alkalinity, pH value, and ionic strength, determines the distribution and bioavailability of lanthanides in marine settings. In river or estuarine waters, lanthanides are separated into two phases: the solution phase and the suspended particles and colloids. Extensive observations have indicated that the distribution and movement of REEs in colloids and particles are caused mostly by the affinity of lanthanides for these particles. The majority of biota contain REEs, and the accumulation pattern typically complies with the Oddo–Harkin rule, which indicates that components with even atomic numbers are more profuse than those with odd atomic numbers [32, 47, 121, 122].

The bioaccumulation and bioavailability of REEs are closely related in estuarine aquatic ecosystems. According to Zhang and Shan (2001), the more accessible pool is typically exchangeable and soluble in water [27]. REE speciation, which is impacted by physicochemical factors, pH, ionic strength, alkalinity, and the presence of various inorganic and organic complexing agents, determines REE bioavailability [123]. Little is known about the bioavailability of REEs, despite several investigations on their fractionation and distribution in estuarine systems. Numerous effects of la have been reported, mainly on the neurological, excretory, and smooth muscle systems [124]. Research has revealed that total REEs are not present in the muscle tissue of fish, ringed seals, or terrestrial herbivores but are more abundant in liver tissues. The findings demonstrated that the cytotoxicity patterns of each REE varied among species, indicating that the toxicity of different REEs scales differently [125]. It has been demonstrated that hydra species are susceptible to the harmful effects of heavy metals and REEs, which have been related to metal contamination in previous studies [126]. After a 21-day study on three marine microcrustaceans (*Thamnocephalus platyurus*, *Heterocypris incongruens*, and *Daphnia magna*), the toxicity of five lanthanides (La, Pr, Ce, Gd, and Nd) was subsequently determined. These findings indicate that lanthanides are hazardous to crustaceans in chronic bioassays. Their research also recommended treating lanthanides as a homogeneous class with a comparable toxicity mechanism on the basis of a long-term D. magna study [127].

Because lanthanum and alkaline earth elements have comparable chemical compositions, their cytotoxic effects can be characterized. Significant alterations were observed in an investigation to examine the acute toxicity of La^{3+} on the liver and gill of rare minnows over 21 days, confirming the toxic effects of La^{3+} on these organs [128]. In a study where glass eels were exposed to LaC13 for 14 days, fast La buildup was observed after 3 days, and even after continuous exposure, absorption decreased. According to the findings of Paoli et al. (2014), bioaccumulation occurs via both extracellular and intracellular mechanisms, intensifying as the concentration of Ce increases and ultimately resulting in a decrease in lichen viability [129]. To clarify the harmful effects, other lanthanides, including gadolinium (Gd), have also been employed. According to Rogosnitzky and Branch (2016), it is a common paramagnetic contrast agent for MRI that has an adverse effect on people's health [130]. Martino et al. (2017) examined the effects of gadolinium on the embryonic development of four sea urchin species: *A. lixula* and *P. lividus* from Europe and

Centrostephanus rodgersii and *Haliotis tuberculata* from Australia [131]. This work revealed dose-dependent skeletal injury and varying gadolinium sensitivities. To understand the process of neodymium (Nd) biouptake by the green alga Chlamydomonas reinhardtii, Yang and Wilkinson (2018) carried out a study [132]. The physiological, ultrastructural, and biochemical responses of *Spirodela polyrhiza* to 0–60 μM *Spirodela polyrhiza* were tested over 20 days. The results revealed that the impact of praseodymium on this species was concentration-dependent. According to Xu et al. (2016), a notable decline in photosystem II activity was accompanied by a notable increase in cell death, suggesting the occurrence of Pr-induced oxidative stress [133].

The biological impact of Ho^{3+} on Halobacterium was examined via the microcalorimetric method. The growth rate constants and maximum heat production were reduced by Ho^{3+}. It was determined that the addition of Ho^{3+} altered the metabolic process of *H. halobium* R1 development by comparing the findings of transmission electron microscopy (TEM) and thermogenic curves [134]. Recycling and other industrialized activities have led to increased accumulation of LREEs in oysters from suspended particles in China's Pearl River Delta [135]. Jiao et al. (2021) measured REE levels in *Oratosquilla oratoria* mantis shrimp taken from Shandong Province, China. REEs were found to be below dangerous limits [136].

3.1.3 REE Removal Technologies

The amount of RE metals added to water as a result of industrial activity is increasing significantly and is starting to raise concerns internationally. Consequently, the retrieval of REEs is a substantial concern that has to be given the attention it deserves. As a result, methods and procedures for the retrieval of metallic ions from diverse sources have advanced. According to the concentration of valuable or poisonous metals, the solution composition, the process economics, and the target metals, the recovery of REEs might include a variety of different procedures [137]. The retrieval of REEs from solid waste or minerals can be achieved through two methods: hydrometallurgy (also known as bioleaching or chemical leaching), which is the most popular technique for extracting individual REEs from mineral concentrate, or pyrometallurgy, an energy-intensive procedure that utilizes high temperatures to chemically transform input materials and separate them to extract valuable metals [138, 139].

3.1.4 Chemical Precipitation

It is common practice to recover metals by treating them with hydroxides, carbonates, or sulfides [140]. An efficient and economical method for removing RE metals from aqueous solutions is precipitation. In this procedure, coagulants such as iron salts, lime, or alum are added, the pH is adjusted, a chemical precipitant is added, and flocculation occurs. Sedimentation or filtering can then be used to remove the precipitates from the water. It is a well-liked technique for extracting metals from polluted wastewater because of its easy, practical, safe operation and low equipment requirements. The enormous amount of chemicals needed for reducing metals to tolerable levels before disposal, however, is significant. Other associated shortcomings include excessive sludge production, which requires extra treatment, poor settling, slow metal precipitation kinetics, long-term environmental effects, and increased cost of sludge disposal [141].

An increase in the HCl and NaOH concentrations, liquid-to-solid ratio, conversion temperature, and time increased the REE leachability, according to findings from a previous study on the effects of process factors [142]. When sodium hydroxide and hydrochloric acid were used to leach the remaining solids, 99.4% lanthanum, 100% gadolinium oxide, and 100% yttrium were recovered from the spent optical glass, which included 43.22% lanthanum oxide, 9.38% yttrium oxide, and 4.50% gadolinium oxide.

3.1.5 Electrocoagulation

EC is a promising method for removing contaminants from water or wastewater without the need for the addition of additional chemicals or regeneration involving the electrochemical production of destabilization agents that help in charge neutralization, which leads to the removal of pollutants. EC has been extensively utilized for water or wastewater treatment and can be performed via various electrodes, such as iron, steel, aluminum, zinc, magnesium, or a mixture of them. These electrodes generate metallic ions that undergo hydrolysis near the anode. This process produces a series of stimulated intermediates that help destabilize the finely discrete elements present in water or wastewater that need treatment [143]. Compact treatment facilities, relatively low costs, high particulate removal effectiveness, and the potential for total automation are some of the benefits of EC, but its main drawback is high operational expenses. Murthy and Parmar (2011) investigated the effectiveness of aluminum and stainless steel electrodes in eliminating strontium. The removal efficiencies were assessed in terms of temperature, initial strontium concentration, distance between electrodes, current density, EC time, solution pH, and distance between electrodes [144].

3.1.6 Ion Flotation

Ion flotation is an effective method utilized to remove metal ions from dilute aqueous solutions. This separation technology involves the use of surfactants (surface active agents) to make the metallic species hydrophobic and, subsequently, with the help of air bubbles, to remove hydrophobic species. The ion flotation process has a number of benefits for handling diluted wastes, such as low energy needs, minimal metal residue concentration, quick operation, minimal space requirements, and adaptability in the use of different metals at different concentrations. Furthermore, this procedure lowers processing costs by yielding tiny quantities of sludge that are greatly improved with the desired components. However, the ionic strength of a solution has a major effect on how effective this process is [145, 146].

3.1.7 Membrane Filtration

This technique is called molecular sieving, which separates particles on the basis of their size. Membranes with different pore sizes, such as microfiltration, ultrafiltration, ED, RO, and NF membranes, utilize size exclusion to remove heavy metals. The effectiveness of membrane filtration is mainly determined by the pore size of the membrane, with smaller pore sizes attaining greater effectiveness. However, smaller pore sizes can lead to limited diffusion of the feed, which requires high pressure for improving diffusion through membranes, leading to increased energy consumption and operational costs [147]. To remove samarium, an REE, from an aqueous solution, a polyelectrolyte-enhanced ultrafiltration procedure was used with poly(acrylic acid) (PAA) with an average molecular weight of 100,000 Da utilizing a tangential cell system. The results indicated that the permeate flux increased linearly with increasing transmembrane pressure and with increasing PAA concentration [148].

3.1.8 Electrodialysis

Electrodialysis (ED) is a process that uses a special membrane to separate ionized species in a solution. This technique is useful for effectively separating charged particles in a solution. Compared with other membrane separation techniques, electrodialysis is economically expensive because of the use of ion-exchange membranes, particularly bipolar membranes. Additionally, the electrodes used in ED can be expensive, and the membranes have a relatively short lifespan when operating in high-density electric fields. The formation of metal hydroxides during this process can clog the membrane, causing damage to the stack. Therefore, careful operation, clean feed, periodic maintenance, and preventative measures are necessary to ensure the longevity and effectiveness of the process.

3.1.9 Reverse Osmosis

RO is a procedure used to remove heavy metals from wastewater by straining them through a semipermeable membrane. The membrane allows purified fluid to pass through while blocking contaminants [149]. The efficacy of heavy metal deduction by RO is affected mainly by pressure rather than pH. When the pressure is high, the metal removal efficiency is high, and consequently, the energy consumption is high. Compared with other membrane filtration approaches, RO is more effective for removing metals from inorganic solutions, with a rejection percentage of over 97%

for metal concentrations fluctuating between 21 and 200 mgL^{-1}. A high water flux rate, strong salt rejection, resilience to biological attack, chemical stability, mechanical robustness, and temperature tolerance are some of the other benefits of RO. The pore size of RO membranes can reach 10–4 μm. The membrane's tiny pores, however, increase the likelihood of fouling from suspended solids or oxidized substances such as chlorine oxides, which increases operating expenses. Gadolinium (Gd) was successfully detected at extremely low amounts by Lawrence et al. (2010), who also created purified recycled water from wastewater. The main removal step occurred across the RO membrane when anthropogenic Gd was decreased from 0.39 nmol/kg to 0.59 pmol/kg at a reduction rate of 99.85%, but coagulation and microfiltration resulted in very little removal [150].

3.1.10 Nanofiltration

The smallest membrane pore size employed is approximately 1°A. NF membranes require pore diameters. Compared with RO, this method has the benefits of low functional pressure and strong denial of organic molecules compared with the ultrafiltration (UF) method [151]. The capacity of NF membranes to extract ions from water is a significant characteristic; the separation of these ionic species is highly dependent on pore size and membrane charge. Murthy and Choudhary (2011) investigated the use of commercial NF to treat water, including neodymium (REE). According to their experimental findings, the rate of Nd(III) rejection increased as the employed pressure and feed flow rate increased and decreased as the feed solution concentration increased [152]. Several authors have explored the NF potential in the recovery of REEs from rare earth wastewater, acidic mine water, and coal fly ash leachate [153–156].

3.1.11 Biosorption

The chemical species that attach themselves to biopolymers is known as biosorption. Moreover, biological materials have the capacity to absorb heavy metals from wastewater via metabolically or physicochemically mediated absorption routes. Through the action of functional groups or ligands on the outer surface of a cell, biosorption allows live, dead, or their derivatives to complex with metal ions. This procedure is a biotechnological invention that offers an economical way to extract RE metals from aqueous solutions. Numerous biomaterials have been identified as possible adsorbents for the recovery of REEs, including activated bacteria, algae, carbon, fungi, yeasts, and resins [157–160].

One of the main benefits of biosorption over traditional treatment techniques is its low cost. Additionally, it produces minimal chemical or biological sludge, has a high metal-binding ability, and is highly efficient. Biosorption also does not require any additional nutrients and allows for the regeneration of biosorbents. The original motivations for developing biosorption in industrial processes were the rapid kinetics of desorption and adsorption, as well as the lack of generation of secondary residues. The process of metal accumulation at biosorption locations is complex and involves multiple mechanisms. Metal binding can occur through surface complexation, ion exchange, precipitation, and electrostatic interactions, either individually or in combination. Several factors affect biosorption processes, including temperature, pH, initial metal concentration, sorbent dosage, contact period, and agitation rate. While Das et al. (2014) studied the potential of biosorbents prepared from fish scales and neem sawdust to recover lanthanum from aqueous solutions, Palmieri et al. (2000) studied the biosorption potential of microalgae, yeast, and Penicillium to remediate neodymium from aqueous media [161, 162].

3.1.12 Reducing REE Fluxes

Various approaches can be used to reduce the amount of REEs that are introduced into the environment, particularly in water systems. Empirical examples include the manufacturing of products with lower REE requirements, such as power generators and magnets, or longer-lasting devices, such as wind turbines, which last more than 10 years [163]. The improvement of mineral extraction technology and the encouragement of e-waste recycling from used cell phones, catalysts, and magnets are further solutions [164]. Another way to lower REE flux is through trans-materialization, which involves the replacement of Ga-doped sintered Nd-Fe-B magnets with Dy in Nd, Dy–Fe–B magnets [165].

3.1.13 Bioavailability of REEs

It has been proposed that nonessential cationic metal species can penetrate membranes through processes known as molecular or ionic mimicry [166]. For example, the ionic binding capacity and dimension of La^{3+} are similar to those of calcium ions. The high charge-to-volume ratio of La^{3+} makes it more likely to attach to protein sites, where it can take the place of Mg^{2+}, Fe^{3+}, Ca^{2+}, and Mn^{2+} [167]. In marine primary producers, bioaccumulation follows pathways derived from terrestrial plants: it occurs concurrently with elevated glutamic oxaloacetic transaminase activity [168].

The shared ability of REEs to remove calcium from locations where it binds, leading to enzyme inhibition or other biochemical dysfunctions, is a crucial toxicological feature [169]. Additional possible impacts on human health include altered bone structure [67] and dysfunctional neurological problems, such as children's lower IQ, which are mostly linked to La. Furthermore, several REEs are linked to fibrotic tissue damage and genotoxicity (Fig. 11.1) [7]. Numerous negative consequences on the liver, kidney, and lungs, in addition to the neurological systems of the animals, were documented in the case of La. The impacts included histopathological alterations, disruption of the homeostasis of vital components and enzymes, and oxidative stress (Fig. 11.2) [62].

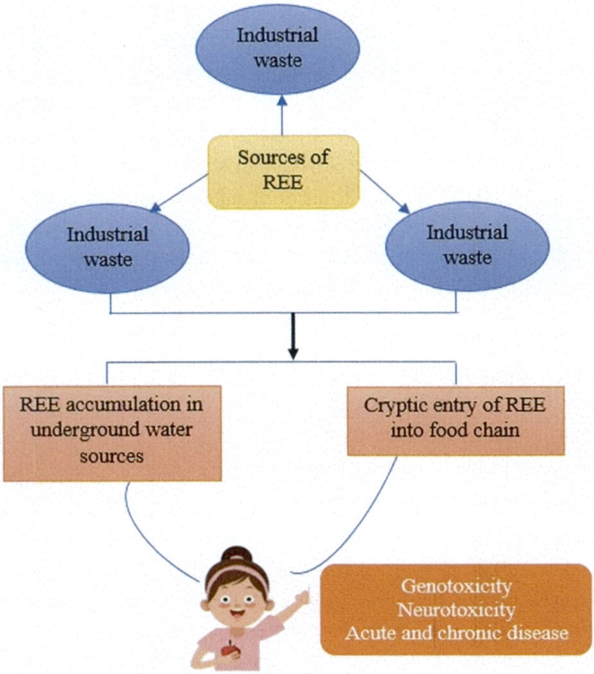

Fig. 11.1 Various sources of REEs in water and their buildup affect human health

Numerous variables, including water hardness, pH, dissolved organic carbon (DOC), alkalinity, Ca^{2+}, Mg^{2+}, HCO_{3-}, and CO_3^{2-} , affect the bioavailability of La. Each of these characteristics alters the activity of bioavailable species, which in turn alters how the organism's membranes interact with it. According to published research, there is little possibility for biomagnification in the bioaccumulation of REEs in the marine environment, and trophic dilution frequently occurs [170, 171]. In contrast, Reindl and Falkowska (2021) reported that HREEs were mostly present in the muscles of important clupeids from the Iberian Peninsula, Northeast Atlantic, and Gdansk Bay, South Baltic Sea [172].

A different study examined the food web of neotropical mangrove habitats and trophic transfer. According to Souza et al. (2021), the authors found substantial indications of biomagnification among plankton, oysters, shrimp, fish, and crabs for few REEs and biodilution for others, depending on varying degrees of pollution [40]. The study also revealed that Nb was connected to airborne particulate matter and that Ce, Y, and La inputs were connected to solid waste from steel mills. According to some findings, La is hazardous because it bioaccumulates in the nervous systems and brains of marine species [173]. The majority of the literature refers primarily to LREEs and La, and there is a lack of understanding regarding the toxicological pathways of REEs in marine environments.

pH-dependent REE speciation and subsequent toxicity play another unidentified role in the health effects of REEs. Mining regions can release industrial

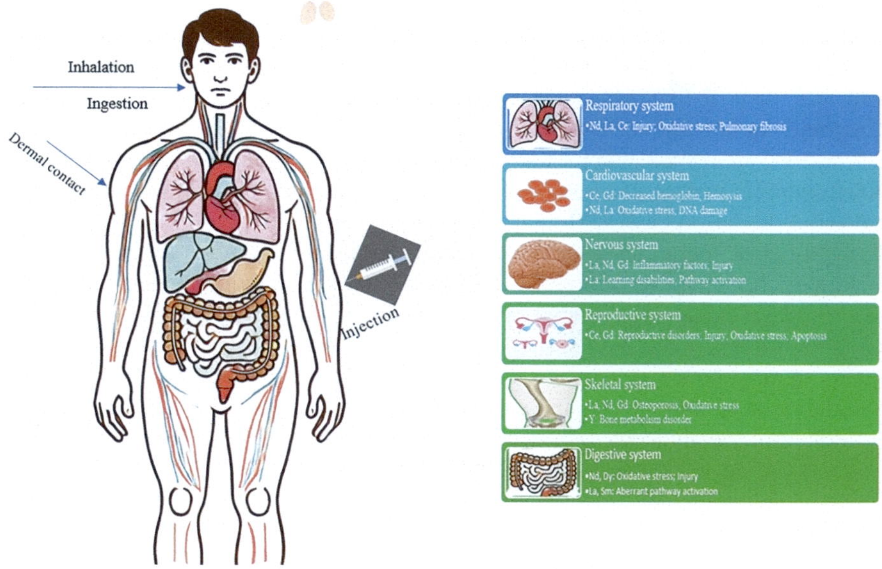

Fig. 11.2 Effects of rare earth elements on human health

effluents into estuaries that contain both acidic pollutants and REE byproducts. On the basis of the scant material that is currently available, it appears that estuary acidification increases the toxicity of REEs. The trophodynamics of metals in the Arctic marine food web were studied by Campbell et al. (2005), who described the REE biodilution process in the food web [174]. With respect to accumulation in fish, birds, and seals, scientists have discovered that La is highly concentrated in algae and zooplankton. The authors concluded that pelagic plankton could readily bioconcentrate these components from seawater with little trophic transmission.

4 Conclusions

The globe has experienced the "rare earth" problem as a result of China's monopoly, which accounts for 80% of the expanding REE market, and export prohibitions. The crisis spurred global exploration for REE reserves, resulting in a significant increase in the amount of REEs (98 Mt) contained in mineral resources outside of China, which are located in Canada (38 Mt), Africa (10.3 Mt), Greenland (39 Mt), and Greenland (39 Mt). India (6.9 Mt), Russia (12 Mt), Brazil (22 Mt), Vietnam (22 Mt), and other countries are projected to have 120 Mt. of REE deposits in the world today (13 Mt). The world's supplies of RM and REE are currently sufficient to meet the needs of the various sectors. However, it seems that societal and environmental concerns are the primary limitations rather than technological ones.

This work focused on REEs, which are currently regarded as emergent pollutants from high-tech industries, and examined their possible sources, bioavailabilities, health impacts, and remediation strategies. A variety of sources, such as pharmaceutical and medical facilities, mineral and mining processing, waste electrical and electronic equipment, REE recycling facilities, fertilizers, animal feed, and high-tech sectors, are responsible for the release of REEs. The main mechanisms responsible for the environmental mobilization and transportation of REEs are hydrological and wind-driven processes. Humans can come into contact with infected objects during medical operations, swallow contaminated food or water, or inhale contaminated particles. However, a broad variety of organisms, including plants, microbes, and terrestrial and aquatic species, are susceptible to the small- and long-term ecotoxicological impacts of REEs. Reduced root function and growth, plant nutritional quality in the shoots, oxidative stress, animal neurotoxicity and genotoxicity, bioaccumulation at trophic levels, and acute and chronic toxicity to soil organisms and bacteria have specific ecotoxicological impacts. There are few epidemiological and ecotoxicological data connecting environmental exposure to REEs to harmful human health outcomes, aside from the risks to human health posed by their medical applications.

For the purpose of tracking REE matrices, sampling and analysis methods must be improved. This will necessitate the development of sophisticated techniques. Global databases, reference standards, and safety restrictions will help manage pollution in the environment, prohibit unrestrained exploitation of REEs, and open new research paths.

In this still-developing field of study, there are a number of unanswered questions regarding REEs. These questions include (1) quantitative information about possible sources of REEs, such as unengineered landfills; (2) source separation data; (3) the behavior and fate of REEs in a variety of environmental conditions, including tropical ones with particular soils; (4) human epidemiology and aquatic ecotoxicology; and (5) the impact of different low-cost water and wastewater treatment techniques on REE removal. To better understand human health and the environmental dangers connected with REEs, advanced analytical amenities, research funds, and experience are necessary.

References

1. Gomes, A. R., Justino, C., Rocha-Santos, T., Freitas, A. C., Duarte, A. C., & Pereira, R. (2017). Review of the ecotoxicological effects of emerging contaminants to soil biota. *Journal of Environmental Science and Health, Part A, 52*(10), 992–1007.
2. Kavanaugh, M. C. (2003, January). Unregulated and emerging chemical contaminants: Technical and institutional challenges. In *WEFTEC 2003* (pp. 1–19). Water Environment Federation.
3. Arciszewska, Ż., Gama, S., Leśniewska, B., Malejko, J., Nalewajko-Sieliwoniuk, E., Zambrzycka-Szelewa, E., & Godlewska-Żyłkiewicz, B. (2022). The translocation pathways

of rare earth elements from the environment to the food chain and their impact on human health. *Process Safety and Environmental Protection, 168*, 205–223.
4. Liu, Q., Shi, H., An, Y., Ma, J., Zhao, W., Qu, Y., Chen, H., Liu, L., & Wu, F. (2023). Source, environmental behavior and potential health risk of rare earth elements in Beijing urban park soils. *Journal of Hazardous Materials, 445*, 130451.
5. Guimarães, D., Praamsma, M. L., & Parsons, P. J. (2016). Evaluation of a new optic-enabled portable X-ray fluorescence spectrometry instrument for measuring toxic metals/metalloids in consumer goods and cultural products. *Spectrochimica Acta Part B: Atomic Spectroscopy, 122*, 192–202.
6. Kulaksız, S., & Bau, M. (2013). Anthropogenic dissolved and colloid/nanoparticle-bound samarium, lanthanum and gadolinium in the Rhine River and the impending destruction of the natural rare earth element distribution in rivers. *Earth and Planetary Science Letters, 362*, 43–50.
7. Gwenzi, W., Mangori, L., Danha, C., Chaukura, N., Dunjana, N., & Sanganyado, E. (2018). Sources, behaviour, and environmental and human health risks of high-technology rare earth elements as emerging contaminants. *Science of the Total Environment, 636*, 299–313.
8. Du, X., & Graedel, T. E. (2011). Uncovering the global life cycles of the REEs. *Scientific Reports, 1*(1), 145.
9. Wang, L., Huang, X., & Zhou, Q. (2009). Protective effect of rare earth against oxidative stress under ultraviolet-B radiation. *Biological Trace Element Research, 128*(1), 82–93.
10. Balaram, V. (2019). REEs: A review of applications, occurrence, exploration, analysis, recycling, and environmental impact. *Geoscience Frontiers, 10*(4), 1285–1303.
11. Li, X., Chen, Z., Chen, Z., & Zhang, Y. (2013). A human health risk assessment of REEs in soil and vegetables from a mining area in Fujian Province, Southeast China. *Chemosphere, 93*(6), 1240–1246.
12. Paulick, H., & Machacek, E. (2017). The global REE exploration boom: An analysis of resources outside of China and discussion of development perspectives. *Resources Policy, 52*, 134–153.
13. Tepe, N., Romero, M., & Bau, M. (2014). High-technology metals as emerging contaminants: Strong increase of anthropogenic gadolinium levels in tap water of Berlin, Germany, from 2009 to 2012. *Applied Geochemistry, 45*, 191–197.
14. Alonso, E., Sherman, A. M., Wallington, T. J., Everson, M. P., Field, F. R., Roth, R., & Kirchain, R. E. (2012). Evaluating REE availability: A case with revolutionary demand from clean technologies. *Environmental Science & Technology, 46*(6), 3406–3414.
15. Cai, L., Park, Y. S., Seong, S. I., Yoo, S. W., & Kim, I. H. (2015). Effects of REEs-enriched yeast on growth performance, nutrient digestibility, meat quality, relative organ weight, and excreta microflora in broiler chickens. *Livestock Science, 172*, 43–49.
16. Chakhmouradian, A. R., & Wall, F. (2012). REEs: Minerals, mines, magnets (and more). *Elements, 8*(5), 333–340.
17. Haque, N., Hughes, A., Lim, S., & Vernon, C. (2014). REEs: Overview of mining, mineralogy, uses, sustainability and environmental impact. *Resources, 3*(4), 614–635.
18. Navarro, J., & Zhao, F. (2014). Life-cycle assessment of the production of rare-earth elements for energy applications: A review. *Frontiers in Energy Research, 2*, 45.
19. Tan, Q., Li, J., & Zeng, X. (2015). REEs recovery from waste fluorescent lamps: A review. *Critical Reviews in Environmental Science and Technology, 45*(7), 749–776.
20. Klaver, G., Verheul, M., Bakker, I., Petelet-Giraud, E., & Négrel, P. (2014). Anthropogenic REE in rivers: Gadolinium and lanthanum. Partitioning between the dissolved and particulate phases in the Rhine River and spatial propagation through the Rhine-Meuse Delta (The Netherlands). *Applied Geochemistry, 47*, 186–197.
21. Lawrence, M. G., Ort, C., & Keller, J. (2009). Detection of anthropogenic gadolinium in treated wastewater in south East Queensland, Australia. *Water Research, 43*(14), 3534–3540.
22. Thomsen, H. S. (2017). Are the increasing amounts of gadolinium in surface and tap water dangerous? *Acta Radiologica, 58*(3), 259–263.

23. Gutiérrez-Gutiérrez, S. C., Coulon, F., Jiang, Y., & Wagland, S. (2015). REEs and critical metal content of extracted landfilled material and potential recovery opportunities. *Waste Management, 42*, 128–136.
24. Resende, L. V., & Morais, C. A. (2010). Study of the recovery of REEs from computer monitor scraps–Leaching experiments. *Minerals Engineering, 23*(3), 277–280.
25. Ikhlayel, M. (2018). An integrated approach to establish e-waste management systems for developing countries. *Journal of Cleaner Production, 170*, 119–130.
26. Sthiannopkao, S., & Wong, M. H. (2013). Handling e-waste in developed and developing countries: Initiatives, practices, and consequences. *Science of the Total Environment, 463*, 1147–1153.
27. Zhang, S., & Shan, X. Q. (2001). Speciation of REEs in soil and accumulation by wheat with rare earth fertilizer application. *Environmental Pollution, 112*(3), 395–405.
28. Golev, A., Scott, M., Erskine, P. D., Ali, S. H., & Ballantyne, G. R. (2014). Rare earths supply chains: Current status, constraints and opportunities. *Resources Policy, 41*, 52–59.
29. He, M. L., Wehr, U., & Rambeck, W. A. (2010). Effect of low doses of dietary REEs on growth performance of broilers. *Journal of Animal Physiology and Animal Nutrition, 94*(1), 86–92.
30. Otero, N., Vitoria, L., Soler, A., & Canals, A. (2005). Fertiliser characterisation: Major, trace and REEs. *Applied Geochemistry, 20*(8), 1473–1488.
31. Pagano, G., Aliberti, F., Guida, M., Oral, R., Siciliano, A., Trifuoggi, M., & Tommasi, F. (2015). REEs in human and animal health: State of art and research priorities. *Environmental Research, 142*, 215–220.
32. Migaszewski, Z. M., & Gałuszka, A. (2015). The characteristics, occurrence, and geochemical behavior of REEs in the environment: A review. *Critical Reviews in Environmental Science and Technology, 45*(5), 429–471.
33. Liang, T., Zhang, S., Wang, L., Kung, H. T., Wang, Y., Hu, A., & Ding, S. (2005). Environmental biogeochemical behaviors of REEs in soil–plant systems. *Environmental Geochemistry and Health, 27*, 301–311.
34. Weber, R. J., & Reisman, D. J. (2012). REEs: A review of production, processing, recycling, and associated environmental issues. *US EPA Region, 8*, 189–200.
35. Nieto, A., & Zhang, K. Y. (2013). Cutoff grade economic strategy for byproduct mineral commodity operation: Rare earth case study. *Mining Technology, 122*(3), 166–171.
36. Doroshkevich, A. G., Viladkar, S. G., Ripp, G. S., & Burtseva, M. V. (2009). Hydrothermal REE mineralization in the Amba Dongar carbonatite complex, Gujarat, India. *The Canadian Mineralogist, 47*(5), 1105–1116.
37. Bhushan, S. K., & Kumar, A. (2013). First carbonatite hosted REE deposit from India. *Journal of the Geological Society of India, 81*(1), 41–60.
38. Bhushan, S. K. (2015). Geology of the Kamthai rare earth deposit. *Journal of the Geological Society of India, 85*(5), 537–546.
39. MacMillan, G. A., Chételat, J., Heath, J. P., Mickpegak, R., & Amyot, M. (2017). Rare earth elements in freshwater, marine, and terrestrial ecosystems in the eastern Canadian Arctic. *Environmental Science: Processes & Impacts, 19*(10), 1336–1345.
40. Souza, I. C., Morozesk, M., Azevedo, V. C., Mendes, V. A., Duarte, I. D., Rocha, L. D., & Fernandes, M. N. (2021). Trophic transfer of emerging metallic contaminants in a neotropical mangrove ecosystem food web. *Journal of Hazardous Materials, 408*, 124424.
41. Lortholarie, M., Zalouk-Vergnoux, A., Couderc, M., Kamari, A., François, Y., Herrenknecht, C., & Poirier, L. (2020). REE bioaccumulation in the yellow and silver European eel (Anguilla anguilla): A case study in the Loire estuary (France). *Science of the Total Environment, 719*, 134938.
42. Négrel, P., Guerrot, C., Cocherie, A., Azaroual, M., Brach, M., & Fouillac, C. (2000). REEs, neodymium and strontium isotopic systematics in mineral waters: Evidence from the Massif Central, France. *Applied Geochemistry, 15*(9), 1345–1367.

43. Johannesson, K. H., Palmore, C. D., Fackrell, J., Prouty, N. G., Swarzenski, P. W., Chevis, D. A., & Burdige, D. J. (2017). REE behavior during groundwater–seawater mixing along the Kona coast of Hawaii. *Geochimica et Cosmochimica Acta, 198*, 229–258.
44. Trifuoggi, M., Donadio, C., Ferrara, L., Stanislao, C., Toscanesi, M., & Arienzo, M. (2018). Levels of pollution of REEs in the surface sediments from the Gulf of Pozzuoli (Campania, Italy). *Marine Pollution Bulletin, 136*, 374–384.
45. Tyler, G. (2004). REEs in soil and plant systems-A review. *Plant and Soil, 267*, 191–206.
46. Hu, Z., Richter, H., Sparovek, G., & Schnug, E. (2004). Physiological and biochemical effects of REEs on plants and their agricultural significance: A review. *Journal of Plant Nutrition, 27*(1), 183–220.
47. Ramos, S. J., Dinali, G. S., Oliveira, C., Martins, G. C., Moreira, C. G., Siqueira, J. O., & Guilherme, L. R. (2016). REEs in the soil environment. *Current Pollution Reports, 2*, 28–50.
48. Yin, X., Martineau, C., Demers, I., Basiliko, N., & Fenton, N. J. (2021). The potential environmental risks associated with the development of REE production in Canada. *Environmental Reviews, 29*(3), 354–377.
49. Adeel, M., Lee, J. Y., Zain, M., Rizwan, M., Nawab, A., Ahmad, M. A., & Xing, B. (2019). Cryptic footprints of REEs on natural resources and living organisms. *Environment International, 127*, 785–800.
50. Pagano, G., Thomas, P. J., Di Nunzio, A., & Trifuoggi, M. (2019). Human exposures to REEs: Present knowledge and research prospects. *Environmental Research, 171*, 493–500.
51. He, X., Zhang, Z., Zhang, H., Zhao, Y., & Chai, Z. (2008). Neurotoxicological evaluation of long-term lanthanum chloride exposure in rats. *Toxicological Sciences, 103*(2), 354–361.
52. Liu, Q., Yang, J., Wu, S., Di, L., & Cai, Y. (2012). Effect of lanthanum chloride on reproductive function and antioxidative kinase in mice. *Chinese Journal of Integrative Medicine, 25*, 277–278.
53. Zheng, L., Yang, J., Liu, Q., Yu, F., Wu, S., Jin, C., & Cai, Y. (2013). Lanthanum chloride impairs spatial learning and memory and downregulates NF-κB signalling pathway in rats. *Archives of Toxicology, 87*, 2105–2117.
54. Cheng, J., Cheng, Z., Hu, R., Cui, Y., Cai, J., Li, N., & Hong, F. (2014). Immune dysfunction and liver damage of mice following exposure to lanthanoids. *Environmental Toxicology, 29*(1), 64–73.
55. EFSA (European Food Safety Authority). (2008). *Guidance document for the use of the concise European Food Consumption Database in Exposure Assessment*. EFSA (European Food Safety Authority).
56. Feng LiPing, F. L., Liu Ya, L. Y., Lai JinFeng, L. J., & Wang Yin, W. Y. (2018). *Analysis of chemical pollutants and harmful factors in 195 food products in Luzhou city in 2017*. CABI.
57. Liu, H., Wang, J., Yang, Z., & Wang, K. (2015). Serum proteomic analysis based on iTRAQ in miners exposed to soil containing REEs. *Biological Trace Element Research, 167*, 200–208.
58. Liang, Q., Yin, H., Li, J., Zhang, L., Hou, R., & Wang, S. (2018). Investigation of REEs in urine and drinking water of children in mining area. *Medicine, 97*(40), e12717.
59. Wei, B., Li, Y., Li, H., Yu, J., Ye, B., & Liang, T. (2013). REEs in human hair from a mining area of China. *Ecotoxicology and Environmental Safety, 96*, 118–123.
60. Tong, S. L., Zhu, W. Z., Gao, Z. H., Meng, Y. X., Peng, R. L., & Lu, G. C. (2004). Distribution characteristics of REEs in children's scalp hair from a rare earths mining area in southern China. *Journal of Environmental Science and Health, Part A, 39*(9), 2517–2532.
61. Tian, S., Liang, T., & Li, K. (2019). Fine road dust contamination in a mining area presents a likely air pollution hotspot and threat to human health. *Environment International, 128*, 201–209.
62. Kowalczyk, E., Givelet, L., Amlund, H., Sloth, J. J., & Hansen, M. (2022). Risk assessment of REEs, antimony, barium, boron, lithium, tellurium, thallium and vanadium in teas. *EFSA Journal, 20*, e200410.
63. Brown, C. J., Chenery, S. R., Smith, B., Mason, C., Tomkins, A., Roberts, G. J., & Tiberindwa, J. V. (2004). Environmental influences on the trace element content of teeth—Implications for disease and nutritional status. *Archives of Oral Biology, 49*(9), 705–717.

64. Henríquez-Hernández, L. A., González-Antuña, A., Boada, L. D., Carranza, C., Pérez-Arellano, J. L., Almeida-González, M., & Luzardo, O. P. (2018). Pattern of blood concentrations of 47 elements in two populations from the same geographical area but with different geological origin and lifestyles: Canary Islands (Spain) vs. Morocco. *Science of the Total Environment, 636*, 709–716.
65. Guo, C., Wei, Y., Yan, L., Li, Z., Qian, Y., Liu, H., & Wang, J. (2020). REEs exposure and the alteration of the hormones in the hypothalamic-pituitary-thyroid (HPT) axis of the residents in an e-waste site: A cross-sectional study. *Chemosphere, 252*, 126488.
66. Höllriegl, V., González-Estecha, M., Trasobares, E. M., Giussani, A., Oeh, U., Herraiz, M. A., & Michalke, B. (2010). Measurement of cerium in human breast milk and blood samples. *Journal of Trace Elements in Medicine and Biology, 24*(3), 193–199.
67. Zaichick, S., Zaichick, V., Karandashev, V., & Nosenko, S. (2011). Accumulation of REEs in human bone within the lifespan. *Metallomics, 3*(2), 186–194.
68. da Silva Ferreira, M., Fontes, M. P. F., Lima, M. T. W. D. C., Cordeiro, S. G., Wyatt, N. L. P., Lima, H. N., & Fendorf, S. (2022). Human health risk assessment and geochemical mobility of REEs in Amazon soils. *Science of the Total Environment, 806*, 151191.
69. Sun, G., Li, Z., Liu, T., Chen, J., Wu, T., & Feng, X. (2017). REEs in street dust and associated health risk in a municipal industrial base of Central China. *Environmental Geochemistry and Health, 39*, 1469–1486.
70. Huo, W., Zhu, Y., Li, Z., Pang, Y., Wang, B., & Li, Z. (2017). A pilot study on the association between REEs in maternal hair and the risk ofneural tube defects in North China. *Environmental Pollution, 226*, 89–93. https://doi.org/10.1016/j.envpol.2017.03.046
71. Wei, J., Wang, C., Yin, S., Pi, X., Jin, L., Li, Z., & Ren, A. (2020). Concentrations of REEs in maternal serum during pregnancy and risk for fetal neural tube defects. *Environment International, 137*, 105542.
72. Henríquez-Hernández, L. A., Boada, L. D., Carranza, C., Pérez-Arellano, J. L., González-Antuña, A., Camacho, M., & Luzardo, O. P. (2017). Blood levels of toxic metals and REEs commonly found in e-waste may exert subtle effects on hemoglobin concentration in sub-Saharan immigrants. *Environment International, 109*, 20–28.
73. de Vathaire, F., de Vathaire, C. C., Ropers, J., & Mollié, A. (1998). Cancer mortality in the commune of Pargny Sur Saulx in France. *Journal of Radiological Protection, 18*(1), 23.
74. Kutty, V. R., Abraham, S., & Kartha, C. C. (1996). Geographical distribution of endomyocardial fibrosis in South Kerala. *International Journal of Epidemiology, 25*(6), 1202–1207.
75. Liu, L., Wang, L., Ni, W., Pan, Y., Chen, Y., Xie, Q., & Ren, A. (2021). REEs in umbilical cord and risk for orofacial clefts. *Ecotoxicology and Environmental Safety, 207*, 111284.
76. Liu, Y., Wu, M., Zhang, L., Bi, J., Song, L., Wang, L., & Wang, Y. (2019). Prenatal exposure of REEs cerium and ytterbium and neonatal thyroid stimulating hormone levels: Findings from a birth cohort study. *Environment International, 133*, 105222.
77. Zhu, W., Xu, S., Shao, P., Zhang, H., Wu, D., Yang, W., & Feng, J. (1997). Bioelectrical activity of the central nervous system among populations in a REE area. *Biological Trace Element Research, 57*, 71–77.
78. Zhuang, G., Zhou, Y., Lu, H., Lu, W., Zhou, M., Wang, Y., & Tan, M. (1996). Concentration of REEs, As, and Th in human brain and brain tumors, determined by neutron activation analysis. *Biological Trace Element Research, 53*, 45–49.
79. Zhu, W., Xu, S., Shao, P., Zhang, H., Wu, D., Yang, W., & Feng, L. (2005). Investigation on liver function among population in high background of rare earth area in South China. *Biological Trace Element Research, 104*, 1–7.
80. Böhlandt, A., Schierl, R., Diemer, J., Koch, C., Bolte, G., Kiranoglu, M., & Nowak, D. (2012). High concentrations of cadmium, cerium and lanthanum in indoor air due to environmental tobacco smoke. *Science of the Total Environment, 414*, 738–741.
81. Slezakova, K., Pereira, M. C., & Alvim-Ferraz, M. C. (2009). Influence of tobacco smoke on the elemental composition of indoor particles of different sizes. *Atmospheric Environment, 43*(3), 486–493.

82. Drago, G., Perrino, C., Canepari, S., Ruggieri, S., L'Abbate, L., Longo, V., & RESPIRA Collaborative Project Group. (2018). Relationship between domestic smoking and metals and REEs concentration in indoor PM2. 5. *Environmental Research, 165*, 71–80.
83. Kastury, F., Ritch, S., Rasmussen, P. E., & Juhasz, A. L. (2020). Influence of household smoking habits on inhalation bioaccessibility of trace elements and light REEs in Canadian house dust. *Environmental Pollution, 262*, 114132.
84. Badea, M., Luzardo, O. P., González-Antuña, A., Zumbado, M., Rogozea, L., Floroian, L., & Henríquez-Hernández, L. A. (2018). Body burden of toxic metals and REEs in non-smokers, cigarette smokers and electronic cigarette users. *Environmental Research, 166*, 269–275.
85. Rodushkin, I., & Axelsson, M. D. (2000). Application of double focusing sector field ICP-MS for multielemental characterization of human hair and nails. Part II. A study of the inhabitants of northern Sweden. *Science of the Total Environment, 262*(1–2), 21–36.
86. Edward, M., Quinn, J. A., Burden, A. D., Newton, B. B., & Jardine, A. G. (2010). Effect of different classes of gadolinium-based contrast agents on control and nephrogenic systemic fibrosis–derived fibroblast proliferation. *Radiology, 256*(3), 735–743.
87. Gerhardsson, L., & Nordberg, G. F. (1993). Lung cancer in smelter workers—Interactions of metals as indicated by tissue levels. *Scandinavian Journal of Work, Environment & Health, 19 Suppl 1*, 90–94.
88. Yoon, H. K., Moon, H. S., Park, S. H., Song, J. S., Lim, Y., & Kohyama, N. (2005). Dendriform pulmonary ossification in patient with rare earth pneumoconiosis. *Thorax, 60*(8), 701–703.
89. Waring, P. M., & Watling, R. J. (1990). Rare earth deposits in a deceased movie projectionist: A new case of rare earth pneumoconiosis? *Medical Journal of Australia, 153*(11–12), 726–730.
90. Vocaturo, G., Colombo, F., Zanoni, M., Rodi, F., Sabbioni, E., & Pietra, R. (1983). Human exposure to heavy metals: Rare earth pneumoconiosis in occupational workers. *Chest, 83*(5), 780–783.
91. Sabbioni, E., Pietra, R., Gaglione, P., Vocaturo, G., Colombo, F., Zanoni, M., & Rodi, F. (1982). Long-term occupational risk of rare-earth pneumoconiosis A case report as investigated by neutron activation analysis. *Science of the Total Environment, 26*(1), 19–32.
92. Bai, Y., Long, C., Hu, G., Zhou, D., Gao, X., Chen, Z., Wang, T., Yu, S., Han, Y., & Yan, L. (2019). Association of blood chromium and REEs with the risk of DNA damage in chromate exposed population. *Environmental Toxicology and Pharmacology, 72*, 103237. https://doi.org/10.1016/j.etap.2019.103237
93. Takyi, S. A., Basu, N., Arko-Mensah, J., Dwomoh, D., Houessionon, K. G., & Fobil, J. N. (2021). Biomonitoring of metals in blood and urine of electronic waste (E-waste) recyclers at Agbogbloshie, Ghana. *Chemosphere, 280*, 130677.
94. Roncati, L., Gatti, A. M., Barbolini, G., Piscioli, F., Pusiol, T., & Maiorana, A. (2018). In vivo uptake of rare earth metals by triple-negative breast cancer cells. *Pathology & Oncology Research, 24*, 161–165.
95. Yongxing, W., Xiaorong, W., & Zichun, H. (2000). Genotoxicity of lanthanum (III) and gadolinium (III) in human peripheral blood lymphocytes. *Bulletin of Environmental Contamination & Toxicology, 64*(4), 611.
96. Yang, D., Sui, H., Mao, W., Wang, Y., Yang, D., Zhang, L., & Song, Y. (2022). Dietary exposure assessment of REEs in the Chinese population. *International Journal of Environmental Research and Public Health, 19*(23), 15583.
97. Galadima, A., Garba, Z. N., Leke, L., Almustapha, M. N., & Adam, I. K. (2011). Domestic water pollution among local communities in Nigeria-causes and consequences. *European Journal of Scientific Research, 52*(4), 592–603.
98. Ayantobo, O. O., Awomeso, J. A., Oluwasanya, G. O., Bada, B. S., & Taiwo, A. M. (2014). Gold mining in Igun-Ijesha, Southwest Nigeria: Impacts and implications for water quality. *American Journal of Environmental Sciences, 10*(3), 289.
99. Awomeso, J. A., Taiwo, A. M., Dada, E. O., & Ayantobo, O. O. (2017). Human health and environmental assessments of small-scale and artisanal mining activities in the Gold City of Ijeshaland, southwestern Nigeria. *Environmental Systems Research, 6*, 1–11.

100. Chester-Paul, K. (2013, October). A review of environmental and health concerns of REEs (ree): Cautionary notes on an increasingly important geologic resource. In *2013 GSA Annual Meeting in Denver.*
101. Baraud, F., & Leleyter, L. (2012). Prediction of phytoavailability of trace metals to plants: Comparison between chemical extractions and soil-grown radish. *Comptes Rendus Geoscience, 344*(8), 385–395.
102. Xu, D., Zhou, P., Zhan, J., Gao, Y., Dou, C., & Sun, Q. (2013). Assessment of trace metal bioavailability in garden soils and health risks via consumption of vegetables in the vicinity of Tongling mining area, China. *Ecotoxicology and Environmental Safety, 90*, 103–111.
103. Zhang, J., Li, H., Zhou, Y., Dou, L., Cai, L., Mo, L., & You, J. (2018). Bioavailability and soil-to-crop transfer of heavy metals in farmland soils: A case study in the Pearl River Delta, South China. *Environmental Pollution, 235*, 710–719.
104. Galhardi, J. A., de Mello, J. W., & Wilkinson, K. J. (2020). Environmental and health risk assessment of agricultural areas adjacent to uranium ore fields in Brazil. *Environmental Geochemistry and Health, 42*(11), 3965–3981.
105. Weifang, Z., Suqin, X., Pingping, S., Hui, Z., Jia, F., Donglin, W., & Wenjiao, Y. (1997). Investigation on intake allowance of rare earth--A study on bio-effect of rare earth in South Jiangxi. *Zhongguo Huanjing Kexue, 17*(1), 63–66.
106. Bruoziotis, A. A., Giarra, A., Libralato, G., Pagano, G., Guida, M., & Trifuoggi, M. (2022). Toxicity of REEs: An overview on human health impact. *Frontiers in Environmental Science, 10*, 948041.
107. Li, M., Zhuang, L., Zhang, G., Lan, C., Yan, L., Liang, R., & Wang, B. (2021). Association between exposure of light REEs and outcomes of in vitro fertilization-embryo transfer in North China. *Science of the Total Environment, 762*, 143106.
108. Wang, B., Zhu, Y., Pang, Y., Xie, J., Hao, Y., Yan, H., Li, Z., & Ye, R. (2018). Indoor air pollution affects hypertension risk in rural women in northern China by interfering with the uptake of metal elements: A preliminary cross-sectional study. *Environmental Pollution, 240*, 267–272.
109. Rubio, L., Annangi, B., Vila, L., Hernández, A., & Marcos, R. (2016). Antioxidant and anti-genotoxic properties of cerium oxide nanoparticles in a pulmonary-like cell system. *Archives of Toxicology, 90*, 269–278.
110. Akhtar, M. J., Ahamed, M., & Alhadlaq, H. (2020). Gadolinium oxide nanoparticles induce toxicity in human endothelial HUVECs via lipid peroxidation, mitochondrial dysfunction and autophagy modulation. *Nanomaterials, 10*(9), 1675.
111. Selvaraj, V., Bodapati, S., Murray, E., Rice, K. M., Winston, N., Shokuhfar, T., & Blough, E. (2014). Cytotoxicity and genotoxicity caused by yttrium oxide nanoparticles in HEK293 cells. *International Journal of Nanomedicine, 9*, 1379–1391.
112. Ranga, A., Agarwal, Y., & Garg, K. (2017). Gadolinium based contrast agents in current practice: Risks of accumulation and toxicity in patients with normal renal function. *Indian Journal Of Radiology And Imaging, 27*(02), 141–147.
113. Carpenter, D., Boutin, C., Allison, J. E., Parsons, J. L., & Ellis, D. M. (2015). Uptake and effects of six rare earth elements (REEs) on selected native and crop species growing in contaminated soils. *PLoS One, 10*(6), e0129936.
114. Cao, X., Chen, Y., Wang, X., & Deng, X. (2001). Effects of redox potential and pH value on the release of REEs from soil. *Chemosphere, 44*(4), 655–661.
115. Fu, F., Akagi, T., Yabuki, S., & Iwaki, M. (2001). The variation of REE (rare earth elements) patterns in soil-grown plants: A new proxy for the source of rare earth elements and silicon in plants. *Plant and Soil, 235*, 53–64.
116. von Tucher, S., & Schmidhalter, U. (2005). Lanthanum uptake from soil and nutrient solution and its effects on plant growth. *Journal of Plant Nutrition and Soil Science, 168*(4), 574–580.
117. Liang, T., Li, K., & Wang, L. (2014). State of REEs in different environmental components in mining areas of China. *Environmental Monitoring and Assessment, 186*, 1499–1513.

118. Cao, S., Duan, X., Ma, Y., Zhao, X., Qin, Y., Liu, Y., & Wei, F. (2017). Health benefit from decreasing exposure to heavy metals and metalloid after strict pollution control measures near a typical river basin area in China. *Chemosphere, 184*, 866–878.
119. Kim, T., Kim, H., & Kim, G. (2020). Tracing river water versus wastewater sources of trace elements using REEs in the Nakdong River estuarine waters. *Marine Pollution Bulletin, 160*, 111589.
120. Arienzo, M., Ferrara, L., Trifuoggi, M., & Toscanesi, M. (2022). Advances in the fate of rare earth elements, REE, in transitional environments: Coasts and estuaries. *Water, 14*(3), 401.
121. Weltje, L., Heidenreich, H., Zhu, W., Wolterbeek, H. T., Korhammer, S., de Goeij, J. J., & Markert, B. (2002). Lanthanide concentrations in freshwater plants and molluscs, related to those in surface water, pore water and sediment. A case study in The Netherlands. *Science of the Total Environment, 286*(1–3), 191–214.
122. Binnemans, K., Jones, P. T., Blanpain, B., Van Gerven, T., Yang, Y., Walton, A., & Buchert, M. (2013). Recycling of rare earths: A critical review. *Journal of Cleaner Production, 51*, 1–22.
123. Moermond, C. T., Tijink, J., van Wezel, A. P., & Koelmans, A. A. (2001). Distribution, speciation, and bioavailability of lanthanides in the Rhine-Meuse estuary, The Netherlands. *Environmental Toxicology and Chemistry: An International Journal, 20*(9), 1916–1926.
124. Pagano, G., Guida, M., Tommasi, F., & Oral, R. (2015). Health effects and toxicity mechanisms of REEs—Knowledge gaps and research prospects. *Ecotoxicology and Environmental Safety, 115*, 40–48.
125. Sholkovitz, E. R. (1995). The aquatic chemistry of REEs in rivers and estuaries. *Aquatic Geochemistry, 1*, 1–34.
126. Marmolejo-Rodriguez, A. J., Prego, R., Meyer-Willerer, A., Shumilin, E., & Sapozhnikov, D. (2007). REEs in iron oxy– hydroxide rich sediments from the Marabasco River-estuary system (pacific coast of Mexico). REE affinity with iron and aluminium. *Journal of Geochemical Exploration, 94*(1–3), 43–51.
127. Blinova, I., Lukjanova, A., Muna, M., Vija, H., & Kahru, A. (2018). Evaluation of the potential hazard of lanthanides to freshwater microcrustaceans. *Science of the Total Environment, 642*, 1100–1107.
128. Hua, D., Wang, J., Yu, D., & Liu, J. (2017). Lanthanum exerts acute toxicity and histopathological changes in gill and liver tissue of rare minnow (Gobiocypris rarus). *Ecotoxicology, 26*, 1207–1215.
129. Paoli, L., Fiorini, E., Munzi, S., Sorbo, S., Basile, A., & Loppi, S. (2014). Uptake and acute toxicity of cerium in the lichen Xanthoria parietina. *Ecotoxicology and Environmental Safety, 104*, 379–385.
130. Rogosnitzky, M., & Branch, S. (2016). Gadolinium-based contrast agent toxicity: A review of known and proposed mechanisms. *Biometals, 29*(3), 365–376.
131. Martino, C., Bonaventura, R., Byrne, M., Roccheri, M., & Matranga, V. (2017). Effects of exposure to gadolinium on the development of geographically and phylogenetically distant sea urchins species. *Marine Environmental Research, 128*, 98–106.
132. Yang, G., & Wilkinson, K. J. (2018). Biouptake of a rare earth metal (Nd) by Chlamydomonas reinhardtii–bioavailability of small organic complexes and role of hardness ions. *Environmental Pollution, 243*, 263–269.
133. Xu, T., Su, C., Hu, D., Li, F., Lu, Q., Zhang, T., & Xu, Q. (2016). Molecular distribution and toxicity assessment of praseodymium by Spirodela polyrrhiza. *Journal of Hazardous Materials, 312*, 132–140.
134. Peng, L., Yi, L., Xi, L., & Ping, S. (2008). Microcalorimetric studies of the biological effect of holmium (iii) on halobacterium halobium r1 growth. *Biological Trace Element Research, 121*, 80–86.
135. Ma, L., Dang, D. H., Wang, W., Evans, R. D., & Wang, W. X. (2019). REEs in the Pearl River Delta of China: Potential impacts of the REE industry on water, suspended particles and oysters. *Environmental Pollution, 244*, 190–201.

136. Jiao, Y., Yang, L., Kong, Z., Shao, L., Wang, G., Ren, X., & Liu, Y. (2021). Evaluation of trace metals and REEs in mantis shrimp Oratosquilla oratoria collected from Shandong Province, China, and its potential risks to human health. *Marine Pollution Bulletin, 162*, 111815.
137. Tolba, A. A., Mohamady, S. I., Hussin, S. S., Akashi, T., Sakai, Y., Galhoum, A. A., & Guibal, E. (2017). Synthesis and characterization of poly (carboxymethyl)-cellulose for enhanced La (III) sorption. *Carbohydrate Polymers, 157*, 1809–1820.
138. Qu, Y., & Lian, B. (2013). Bioleaching of rare earth and radioactive elements from red mud using Penicillium tricolor RM-10. *Bioresource Technology, 136*, 16–23.
139. Yoon, H. S., Kim, C. J., Chung, K. W., Lee, J. Y., Shin, S. M., Lee, S. J., & Yoo, S. J. (2014). Leaching kinetics of neodymium in sulfuric acid of REEs (REE) slag concentrated by pyrometallurgy from magnetite ore. *Korean Journal of Chemical Engineering, 31*, 1766–1772.
140. Ku, Y., & Jung, I. L. (2001). Photocatalytic reduction of Cr (VI) in aqueous solutions by UV irradiation with the presence of titanium dioxide. *Water Research, 35*(1), 135–142.
141. Barakat, M. A. (2011). New trends in removing heavy metals from industrial wastewater. *Arabian Journal of Chemistry, 4*(4), 361–377.
142. Jiang, Y., Shibayama, A., Liu, K., & Fujita, T. (2005). A hydrometallurgical process for extraction of lanthanum, yttrium and gadolinium from spent optical glass. *Hydrometallurgy, 76*(1–2), 1–9.
143. Chen, X., Chen, G., & Yue, P. L. (2002). Investigation on the electrolysis voltage of electrocoagulation. *Chemical Engineering Science, 57*(13), 2449–2455.
144. Murthy, Z. V. P., & Parmar, S. (2011). Removal of strontium by electrocoagulation using stainless steel and aluminum electrodes. *Desalination, 282*, 63–67.
145. Polat, H., & Erdogan, D. (2007). Heavy metal removal from waste waters by ion flotation. *Journal of Hazardous Materials, 148*(1–2), 267–273.
146. Kurniawan, T. A., Chan, G. Y., Lo, W. H., & Babel, S. (2006). Physico–chemical treatment techniques for wastewater laden with heavy metals. *Chemical Engineering Journal, 118*(1–2), 83–98.
147. Vigneswaran, S., Ngo, H. H., Chaudhary, D. S., & Hung, Y. T. (2005). Physicochemical treatment processes for water reuse. In *Physicochemical treatment processes* (pp. 635–676). Humana Press.
148. Hammami, M., Ennigrou, D. J., Naifer, K. H., & Ferid, M. (2015). Retention of samarium ions from aqueous solutions by poly (acrylic acid)-enhanced ultrafiltration. *Desalination and Water Treatment, 56*(10), 2715–2722.
149. Fu, F., & Wang, Q. (2011). Removal of heavy metal ions from wastewaters: A review. *Journal of Environmental Management, 92*(3), 407–418.
150. Lawrence, M. G., Keller, J., & Poussade, Y. (2010). Removal of magnetic resonance imaging contrast agents through advanced water treatment plants. *Water Science and Technology, 61*(3), 685–692.
151. Yang, X. J., Livingston, A. G., & Dos Santos, L. F. (2001). Experimental observations of nanofiltration with organic solvents. *Journal of Membrane Science, 190*(1), 45–55.
152. Murthy, Z. V. P., & Choudhary, A. (2011). Application of nanofiltration to treat REE (neodymium) containing water. *Journal of Rare Earths, 29*(10), 974–978.
153. Mutlu, B. K., Cantoni, B., Turolla, A., Antonelli, M., Hsu-Kim, H., & Wiesner, M. R. (2018). Application of nanofiltration for REEs recovery from coal fly ash leachate: Performance and cost evaluation. *Chemical Engineering Journal, 349*, 309–317.
154. López, J., Reig, M., Gibert, O., & Cortina, J. L. (2019). Integration of nanofiltration membranes in recovery options of REEs from acidic mine waters. *Journal of Cleaner Production, 210*, 1249–1260.
155. Fonseka, C., Ryu, S., Naidu, G., Kandasamy, J., & Vigneswaran, S. (2022). Recovery of water and valuable metals using low pressure nanofiltration and sequential adsorption from acid mine drainage. *Environmental Technology & Innovation, 28*, 102753.
156. Zhao, Z., Feng, S., Xiao, C., Luo, J., Song, W., Wan, Y., & Li, S. (2022). Exploring ions selectivity of nanofiltration membranes for rare earth wastewater treatment. *Separation and Purification Technology, 289*, 120748.

157. Das, N., & Das, D. (2013). Recovery of rare earth metals through biosorption: An overview. *Journal of Rare Earths, 31*(10), 933–943.
158. Gupta, N. K., Gupta, A., Ramteke, P., Sahoo, H., & Sengupta, A. (2019). Biosorption-a green method for the preconcentration of REEs (REEs) from waste solutions: A review. *Journal of Molecular Liquids, 274*, 148–164.
159. Breuker, A., Ritter, S. F., & Schippers, A. (2020). Biosorption of REEs by different microorganisms in acidic solutions. *Metals, 10*(7), 954.
160. Giese, E. C. (2020). Biosorption as green technology for the recovery and separation of REEs. *World Journal of Microbiology and Biotechnology, 36*(4), 52.
161. Das, D., Varshini, C. J. S., & Das, N. (2014). Recovery of lanthanum (III) from aqueous solution using biosorbents of plant and animal origin: Batch and column studies. *Minerals Engineering, 69*, 40–56.
162. Palmieri, M. C., Garcia, O., Jr., & Melnikov, P. (2000). Neodymium biosorption from acidic solutions in batch system. *Process Biochemistry, 36*(5), 441–444.
163. Stegen, K. S. (2015). Heavy rare earths, permanent magnets, and renewable energies: An imminent crisis. *Energy Policy, 79*, 1–8.
164. Trapasso, G., Chiesa, S., Freitas, R., & Pereira, E. (2021). What do we know about the ecotoxicological implications of the REE gadolinium in aquatic ecosystems? *Science of the Total Environment, 781*, 146273.
165. Hono, K., & Sepehri-Amin, H. (2018). Prospect for HRE-free high coercivity Nd-Fe-B permanent magnets. *Scripta Materialia, 151*, 6–13.
166. Bridges, C. C., & Zalups, R. K. (2005). Molecular and ionic mimicry and the transport of toxic metals. *Toxicology and Applied Pharmacology, 204*(3), 274–308.
167. Evans, C. H. (1983). Interesting and useful biochemical properties of lanthanides. *Trends in Biochemical Sciences, 8*(12), 445–449.
168. Khan, A. M., Behkami, S., Yusoff, I., Zain, S. B. M., Bakar, N. K. A., Bakar, A. F. A., & Alias, Y. (2017). Geochemical characteristics of REEs in different types of soil: A chemometric approach. *Chemosphere, 184*, 673–678.
169. Pałasz, A., & Czekaj, P. (2000). Toxicological and cytophysiological aspects of lanthanides action. *Acta Biochimica Polonica, 47*(4), 1107–1114.
170. Amyot, M., Clayden, M. G., MacMillan, G. A., Perron, T., & Arscott-Gauvin, A. (2017). Fate and trophic transfer of REEs in temperate lake food webs. *Environmental Science & Technology, 51*(11), 6009–6017.
171. Bonnail, E., Pérez-López, R., Sarmiento, A. M., Nieto, J. M., & DelValls, T. Á. (2017). A novel approach for acid mine drainage pollution biomonitoring using REEs bioaccumulated in the freshwater clam Corbicula fluminea. *Journal of Hazardous Materials, 338*, 466–471.
172. Reindl, A. R., & Falkowska, L. (2021). Trace elements in the muscle, ova and seminal fluid of key clupeid representatives from the Gdansk Bay (South Baltic Sea) and Iberian Peninsula (north-East Atlantic). *Journal of Trace Elements in Medicine and Biology, 68*, 126803.
173. Jing-Xi, L. I., Zheng, L., Cheng-Jun, S. U. N., Jiang, F. H., Xiao-Fei, Y. I. N., Jun-Hui, C. H. E. N., & Xiao-Ru, W. A. N. G. (2016). Study on ecological and chemical properties of REEs in tropical marine organisms. *Chinese Journal of Analytical Chemistry, 44*(10), 1539–1546.
174. Campbell, L. M., Norstrom, R. J., Hobson, K. A., Muir, D. C., Backus, S., & Fisk, A. T. (2005). Mercury and other trace elements in a pelagic Arctic marine food web (Northwater Polynya, Baffin Bay). *Science of the Total Environment, 351*, 247–263.

Chapter 12
Effects of Emerging Contaminants on Wildlife

Rwitabrata Mallick, Kuldip Dwivedi, and Swapnil Rai

1 Introduction to Emerging Contaminants

Emerging contaminants, also known as contaminants of emerging concern (CECs), are chemicals and materials that have been found in the environment but are not currently monitored or controlled and may constitute a threat to human health and ecosystems. These pollutants come from a variety of chemical categories and sources, including pharmaceuticals, personal care items, industrial chemicals, agricultural runoff, and microplastics [1].

2 Characteristics of Emerging Contaminants

These pollutants are frequently newly identified as a result of advancements in detection methods and growing awareness of their potential consequences. They are found in a variety of environmental media, including water, soil, air, and biota, frequently in low amounts but with potentially serious consequences. They come from a wide range of sources, including domestic goods, industrial operations, agriculture, and medical procedures. Many developing pollutants have traits that may endanger human health and the environment, such as persistence, bioaccumulation, and toxicity [2]. Emerging contaminants are classified as organic or inorganic micropollutants, such as pesticides, personal care products (PCPs), pharmaceuticals, synthetic organic dyes, polycyclic aromatic hydrocarbons (PAHs), heavy metal

R. Mallick (✉) · K. Dwivedi · S. Rai
Department of Environmental Science, Amity School of Life Science, Amity University
Madhya Pradesh, Gwalior, India

M. K. Jindal et al. (eds.), *Emerging Contaminants in Water*, Emerging
Contaminants and Associated Treatment Technologies,
https://doi.org/10.1007/978-3-031-89591-3_12

ions, plasticizers, perfluorinated compounds, flame retardants, surfactants, etc., which are produced by human activities such as domestic, healthcare, agricultural, and industrial pathways. These compounds are a source of concern due to their physical and chemical features, as they are widely dispersed in the environment and are toxic to humans and wildlife (Fig. 12.1). These contaminants are difficult to detect and are derived from a variety of activities and sources. Its presence in low quantities causes chronic toxicity, endocrine disruption, and the spread of disease resistance [3].

Emerging pollutants have the potential to harm ecosystem health by interfering with the reproductive, developmental, and behavioral processes of wildlife. This can result in decreased biodiversity and altered ecological dynamics [4]. Exposure to developing toxins in drinking water, food, and air can cause hormonal imbalances, developmental difficulties, immune system suppression, and an increased risk of cancer, etc. (Fig. 12.2). Some pollutants can accumulate in creatures and

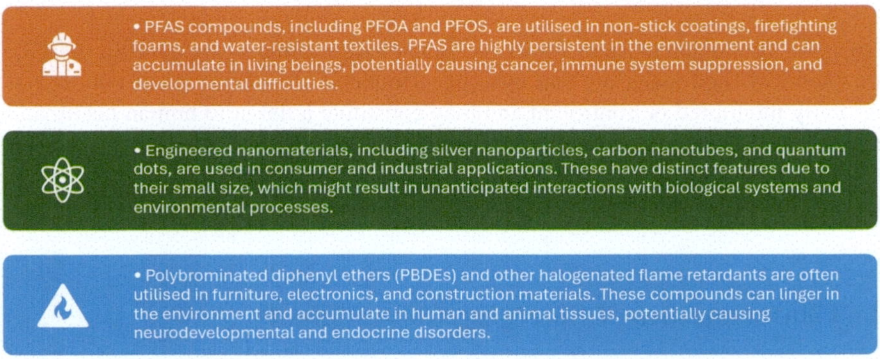

Fig. 12.1 Emerging contaminants toxic to humans and wildlife

Fig. 12.2 Effects of contaminants on wildlife

move up the food chain, resulting in larger amounts of top predators, including humans [5].

Many developing pollutants exist at low concentrations and require advanced analytical techniques for detection and monitoring. Current regulatory frameworks frequently fail to address numerous new pollutants, resulting in gaps in management and control [6]. The scarcity of toxicological data and complicated interactions between multiple compounds and environmental conditions make it difficult to assess the risks presented by the development of pollutants. There is a need for enhanced public knowledge and education about the safe use and disposal of products that may contribute to emerging contaminant pollution [7].

3 Toxicity and Bioaccumulation in Biodiversity/Wildlife

The toxicity and bioaccumulation of contaminants pose serious threats to wildlife and biodiversity. Understanding these processes is crucial for assessing environmental risks and developing strategies to mitigate their impacts. Toxicity refers to the harmful effects of substances on living organisms. The severity of toxicity depends on the nature of the contaminant, its concentration, the duration of exposure, and the sensitivity of the affected species. Various types of toxicity have been observed in wildlife. Pesticides such as organophosphates can cause acute toxicity, resulting in mortality in birds and fish. Oil spills can also cause immediate health crises in marine life. In this context, long-term exposure to low amounts of toxins, such as polychlorinated biphenyls (PCBs), can cause liver damage and reproductive difficulties in mammals. Sublethal effects include behavioral changes, physiological changes, and reproductive issues. Pharmaceuticals and other contaminants can change eating, mating, and predator avoidance behaviors. Heavy metals, such as lead, can cause brain damage and decrease cognitive function. Endocrine-disrupting chemicals (EDCs) can cause changes in sex ratios, decreased fertility, and developmental abnormalities. Because of the bioaccumulation of developing poisons, organisms absorb pollutants from their surroundings at higher concentrations than in the surrounding environment. Lipophilic (fat-loving) substances such as DDT and PCBs bioaccumulate because they break down in fat cells [8]. Temperature, pH, and other factors can influence bioaccumulation rates. Pollutants are stored and metabolized differently depending on the organism's species, age, sex, and health status. In contrast, biomagnification increases the concentration of a contaminant as it moves up the food chain [9]. Mercury enters aquatic habitats via industrial discharge, mining, and atmospheric deposition. Mercury accumulates in fish and other aquatic species, causing brain damage, reproductive failure, and even death in birds and mammals that feed on contaminated fish [10]. DDT has a significant effect on bird populations. It is mostly produced by agricultural operations and accumulates in insects, which are then ingested by birds. Once in the food chain, it causes substantial harm to birds [11]. In some bird species, such as bald eagles, eggshell thinning has resulted in poorer hatching success and population decline. PCBs in marine animals can be sourced from industrial waste and improper disposal of PCB-containing items, which then accumulate in fish and marine invertebrates. PCBs can impair immunological function, cause reproductive failure, and increase disease susceptibility in marine mammals such as seals and whales [12]. Microplastics in marine life have been a problem for decades and are generated primarily by the breakdown of large amounts of plastic waste and microbeads in personal care products. Furthermore, they are consumed by plankton, fish, and marine invertebrates, resulting in physical injury, decreased feeding, and the possible transfer of adsorbed hazardous compounds through the food chain [13].

3.1 Mitigation Strategies

Regulations can be put in place and enforced to limit the release of dangerous substances into the environment. This can be performed by limiting the use of particularly hazardous substances, such as some pesticides and persistent organic pollutants, as well as developing and implementing technologies for cleaning industrial effluents and wastewater. To prevent environmental pollution, hazardous waste should be disposed of and recycled safely. Regular monitoring of contaminant levels in various ecosystems can be used to detect and manage pollution problems, and this can be augmented by research into the long-term effects of toxins and the creation of safer alternatives. Public education can also be used to increase knowledge about the sources and effects of harmful substances, as well as to promote environmentally friendly practices by involving local populations in conservation and pollution control activities [14].

4 Endocrine Disruption and Its Impact

Endocrine-disrupting chemicals (EDCs) are substances that can disturb an organism's endocrine (hormonal) system. These changes can have a wide range of health and environmental effects on both wildlife and people. EDCs are chemicals that mimic, inhibit, or interfere with hormones found in the body's endocrine system. Hormones are necessary for regulating a wide range of biological functions, including growth, reproduction, and metabolism. Some common examples of endocrine disrupting chemicals include bisphenol A (BPA), which is found in plastics and epoxy resins; phthalates, which are used as plasticizers in various products; pesticides, such as DDT and atrazine; polychlorinated biphenyls (PCBs), which are industrial chemicals used in electrical equipment; and pharmaceuticals, such as hormones and drugs, which enter the environment through waste [15]. EDCs can mimic natural hormones, bind to hormone receptors, and activate them inappropriately. They can bind to hormone receptors without activating them, blocking the binding of natural hormones. EDCs can affect the synthesis, release, transport, and elimination of hormones, leading to imbalances.

5 Impact on Wildlife

EDCs can impair the immune system of wildlife, increasing their susceptibility to illnesses and infections and potentially reducing population levels. They can interact with growth hormones, causing stunted growth and delayed maturation in many species. They have an impact on humans, both directly and indirectly. EDC exposure has been associated with low sperm quality, irregular menstrual periods, and other reproductive problems. Prenatal exposure can cause developmental issues such as congenital abnormalities and neurological impairments. EDCs have been linked to an increased risk of hormone-sensitive malignancies, including breast, prostate, and ovarian cancer. Some EDCs are associated with obesity and metabolic syndrome because they disrupt metabolism and fat storage. EDCs can affect thyroid function, resulting in hypothyroidism or hyperthyroidism, and exposure to EDCs, especially during important developmental stages, can cause behavioral and cognitive issues in children, such as attention-deficit/hyperactivity disorder (ADHD) and low IQ [16].

EDCs can cause feminization or masculinization, resulting in intersex disorders in which people exhibit both male and female features. This may reduce reproductive success and population viability. Exposure to EDCs such as DDT has been related to eggshell weakening, which decreases reproductive success. EDCs can also affect typical mating habits. They can also impact marine species, causing changes in reproductive organs, lower fertility, and skewed sex ratios in progeny. Foraging efficiency and predator avoidance are two examples of behavioral modifications that might have an impact on survival. Hormonal imbalances can affect migratory and navigational ability in birds and fish [17].

To reduce the impact of known EDCs on consumer products and industrial processes, bans and restrictions on their usage can be implemented. There are methods for developing safety standards and guidelines for acceptable levels of EDCs in the environment and consumer goods. However, improved approaches can be developed and implemented to identify EDCs in the environment and assess their effects on animal and human health. Long-term ecological and epidemiological research can be undertaken to better understand the consequences of EDCs. It is critical to educate the public about EDCs and promote the use of nontoxic products. Initiatives can be taken to promote the proper disposal of medications and other items containing EDCs, preventing them from entering the environment. The development of safer chemicals and alternatives that do not disturb the endocrine system can be encouraged by creating ways to remove or neutralize EDCs in contaminated environments, such as enhanced water treatment procedures [18].

6 Genotoxicity and Mutagenicity Among Wildlife

Genotoxicity and mutagenicity are two important elements influencing wildlife. The ability of chemical substances to harm genetic information within a cell, resulting in mutations that can lead to cancer. Genotoxic chemicals can result in DNA mutations or chromosomal abnormalities. Mutagens are classified into three groups: physical mutagens, chemical mutagens, and biological mutagens. Mutagenicity refers to a substance's ability to cause genetic mutations or alterations in the DNA sequence. Mutagenic compounds particularly alter the structure or quantity of genetic material. The impact of these two factors on wildlife is considerable because the majority of them are toxic and mutagenic agents, primarily chemicals from industrial processes, agricultural runoff (pesticides and herbicides), heavy metals, and oil spills. Ultraviolet (UV) light and ionizing radiation from radioactive materials can significantly affect wildlife movement and habitat changes. Local climate change and thermal impact can combine with these two elements to produce genotoxic and mutagenic effects [19].

Certain naturally occurring chemicals, such as mycotoxins, have the potential to cause genotoxicity. It may also include human-caused activities such as waste dumping, plastic contamination, and the entry of medications into streams. These two factors can damage DNA by directly interacting with it, resulting in strand breakage, crosslinking, or adduct formation. It is also important for generating structural alterations in chromosomes, such as deletions, duplications, inversions, and translocations [20]. Wild-life mutations can take the form of point mutations, insertions, or deletions in the DNA sequence. It has a significant effect on the health of wildlife populations, particularly reproductive health. Mutagenic compounds can affect gamete production or function, whereas genotoxic drugs can result in developmental abnormalities or embryonic death. During the investigation, it was discovered that the buildup of genetic damage can cause accelerated aging or lower survival rates. The current study also revealed that neurotoxicity changed activities such as foraging, mating, and predator avoidance [21]. The reduction in genetic diversity among wildlife is caused by selective pressures on individuals with genetic mutations. High mortality rates and low reproductive success rates can also cause population reductions or local extinctions. Changes in the population dynamics of a species can have a wide-ranging impact on the food web. Genetic damage in keystone species can have a cascading effect on ecosystem structure and function [22].

Industrial pollutants such as polychlorinated biphenyls (PCBs) and heavy metals are related to DNA damage and decreased reproductive success (Fig. 12.3). Pesticides and UV radiation have been found to induce developmental defects and DNA damage in amphibian populations. The incidence of chromosomal abnormalities and mutations is increased in birds exposed to organochlorine insecticides and radiation. Rodent populations in polluted areas have greater mutation rates and genotoxic effects. Heavy metals and other contaminants can have genotoxic consequences, affecting both health and reproductive success. Biomonitoring can be an effective approach for assessing environmental health by using species that are

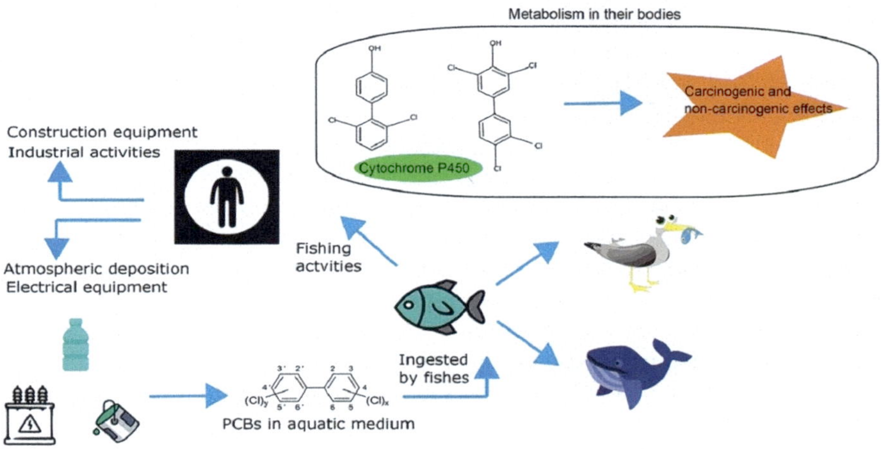

Fig. 12.3 Toxicity of polychlorinated biphenyls in aquatic environments

sensitive to genotoxic and mutagenic chemicals. Techniques such as the Comet assay, micronucleus test, and Ames test are executed to detect DNA damage and mutations among organisms [23].

One option to combat the harmful influence on animals is to implement stronger emissions and waste disposal rules. The remediation of contaminated areas is critical for reducing exposure to genotoxic chemicals. Habitat conservation and restoration are essential for maintaining healthy wildlife populations. Ongoing studies have aimed to understand the processes and effects of genotoxic and mutagenic substances. The current study demonstrates that educating the public and policymakers on the necessity of safeguarding animals from genetic harm can accelerate the influence on wildlife to a substantial level [24]. There have been instances of ecological damage in many ecosystems. Ecological disruption is defined as major changes in the structure, composition, and function of ecosystems induced by natural or manmade sources. These changes have the potential to reduce biodiversity, alter ecosystem services, and weaken the resilience of ecological communities. Many key factors contribute to ecological disruption, such as the conversion of forests to agricultural or urban areas, which causes the loss of habitat for many species; the expansion of cities and infrastructure fragment habitats; the reduction in natural areas; and the conversion of natural habitats to farmland, which disrupts local ecosystems [25]. Chemical contamination is another key element affecting wildlife. Pesticides, heavy metals, and industrial pollution can harm wildlife and ruin ecosystems. Plastic garbage accumulates in land and marine habitats, causing physical and chemical harm to creatures. Excessive nutrients from agricultural runoff cause eutrophication and death in aquatic systems. Rising temperatures affect species distributions and phenologies [26]. An increased frequency of extreme weather events disturbs ecosystems and may result in habitat loss. Rising sea levels pose a hazard to coastal habitats such as marshes and mangroves. The introduction of alien species can also be an issue because they outcompete, prey on, or transmit illnesses to native species, resulting in decreases in or extinction. Invasive species can change the physical and biological structure of ecosystems, deplete fish populations, and disrupt marine food webs. Hunting and poaching are again two significant factors that reduce populations of targeted species and can cause cascading effects on ecosystems; in addition, unsustainable harvesting practices degrade forest ecosystems.

There is evidence that direct species extinction diminishes biodiversity, whereas population size declines reduce genetic variation, decreasing the ability of species to adapt to change. As a result, resources such as food, fresh water, and raw materials have become scarcer. Disruptions in temperature management, water purification, and pollination services can also result in the loss of recreational, aesthetic, and spiritual qualities in natural settings. The current study revealed that adjustments in species dominance and community composition can cause altered interactions and ecosystem processes and that changes at one trophic level can cascade through the food web, influencing many species and interactions [27].

Habitat fragmentation can limit habitat connectivity and exacerbate edge effects, resulting in additional degradation. Deforestation, overgrazing, and pollution all have the potential to reduce soil fertility and structure. Increased sea temperatures and acidification lead to coral bleaching, lowering biodiversity and ecosystem complexity. The removal of essential species, such as herbivorous fishes, causes algae overgrowth, which further stresses coral reefs [28]. Large-scale deforestation and agricultural conversion limit habitats for a variety of species while also contributing to climate change. Increased fire frequency and intensity undermined the ecological balance and recovery mechanisms.

Draining wetlands for agriculture or urban development destroys biodiversity while disrupting water control. Runoff from pesticides and nutrients causes eutrophication and the extinction of aquatic organisms. Reserves and protected areas should be created to preserve crucial habitats and biodiversity. Effective management plans for preserving and restoring ecological health. Reforestation, wetland restoration, and other approaches can be used to restore degraded environments. Reintroducing locally extinct species can help to restore ecological services. Sustainable agricultural methods should be implemented to reduce habitat loss and pollution. Sustainable fishing techniques should be promoted to avoid overfishing and preserve marine biodiversity. Improving laws and policies to protect habitats and decrease pollution. Participating in international initiatives to combat climate change and biodiversity loss. Raising public awareness about the importance of biodiversity and ecosystem services. Local communities should be engaged in conservation and sustainable resource management [29].

7 Behavioral and Physiological Effects Among Wildlife

Environmental changes can affect the quantity and distribution of food resources, forcing species to adapt their foraging techniques. For example, habitat damage may cause animals to travel longer distances or spend more time hunting for food.

Increased predation pressure can alter foraging behavior. Prey species may forage at different times of day or in safer but less optimal environments. Urbanization and human activity can disturb natural feeding habits. Animals may graze more at night to avoid the presence of humans. Climate change can alter the timing of mating seasons. Some birds, for example, lay eggs earlier in the year as temperatures rise [30]. Pollution and habitat degradation can affect physical features critical for mate selection, resulting in changes in mating behavior and success rates. Environmental stressors can influence the quantity and quality of parental care. For example, exposure to pollution may impair parents' capacity to care for their children.

Stressors such as pollution, habitat damage, and human disturbance can increase corticosteroid levels in wildlife, causing chronic stress. Chronic stress can disrupt immunological function, reproduction, and growth. Stress can increase heart rate and change metabolic rates, compromising overall health and energy balance. Endocrine-disrupting chemicals (EDCs) in the presence of pollutants can disrupt the hormonal balance, resulting in reproductive difficulties such as lower fertility, aberrant development, and altered sex ratios [31]. Environmental contaminants can degrade the quality of sperm and eggs, lowering reproductive success and contributing to population decreases. Climate change can have an impact on people when they migrate. Some species may migrate early or later than expected, affecting their survival and reproductive success. Habitat degradation and fragmentation might drive animals to change their migration paths, resulting in increased mortality or worse reproductive success [32]. Human activities such as construction and deforestation can displace animals from their native habitats, causing changes in their migration patterns and behavior. Environmental change has the potential to have an impact on societal systems. For example, food scarcity may result in smaller group numbers or altered social hierarchies. Noise pollution from urban environments can interfere with communication among animals, such as birds and marine mammals, affecting their ability to find mates, avoid predators, and coordinate group activities (Fig. 12.4).

Pollutants and persistent stress can impair immune function, increasing the vulnerability of wildlife to illnesses and infections. Heavy metals and herbicides, for example, can impair the immune systems of amphibians and birds. Changes in

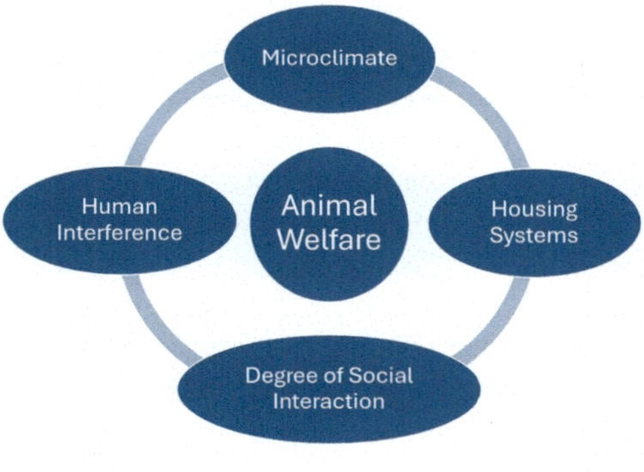

Fig. 12.4 Critical points of animal welfare

habitat and climate can aid in the spread of diseases and parasites, increasing disease prevalence in wildlife populations. Exposure to pollution and environmental contaminants during critical developmental times might result in deformities in wildlife. For example, amphibians exposed to pesticides frequently experience malformations and developmental delays. Nutritional deficits caused by habitat loss and reduced food availability can delay development rates and negatively impact animal health [33]. Climate change can expose species to temperatures outside their normal range, causing thermal stress. At warmer temperatures, for example, marine organisms such as corals and fish may experience bleaching and decreased physiological function. Animals may adapt their behavior to cope with temperature fluctuations, such as seeking shade, altering activity patterns, or relocating to cooler places. Reduced sea ice due to climate change compels polar bears to spend more time on land, increasing human–wildlife conflicts and shifting their diet from seals to land-based food sources [34]. Longer fasting periods and increased energy expenditure in pursuit of food result in weight loss, lower reproductive success, and increased death. Pollution and habitat degradation cause changes in frog calling behavior, which affects mating attractiveness and reproduction. Pesticides and UV radiation induce developmental defects, immunological suppression, and increased susceptibility to infections such as chytridiomycosis.

Noise pollution caused by shipping and industrial activity impairs whale and dolphin communication, navigation, and foraging behavior. Marine mammals absorb contaminants such as PCBs and heavy metals, which cause immunological suppression, reproductive failure, and increased mortality rates [35]. Urbanization and light pollution affect migration patterns, nesting timing, and feeding behavior in birds such as European robins and American warblers. Pesticides and pollutants affect endocrine function, resulting in decreased fertility, aberrant development, and compromised immunological responses.

8 Synergistic and Cumulative Effects on Wildlife

Synergistic effects arise when the combined impact of several stressors on animals exceeds the sum of their individual impacts. This interaction may amplify the negative consequences for wildlife more than intended. On the other hand, cumulative effects occur when various stressors accumulate over time, resulting in compounding effects on species and ecosystems. These stressors may be of the same or distinct types, but their cumulative impact increases over time [36]. Increased CO_2 levels cause ocean acidification, which, combined with rising temperatures, has a profound impact on marine life, particularly for calcifying species such as corals and shellfish. The combined stress of acidity and temperature increase is more severe than that of either alone. Pollutants such as polychlorinated biphenyls (PCBs) can impair the immune system of amphibians, increasing their vulnerability to the negative effects of increasing UV exposure. The fragmentation of habitats renders ecosystems more vulnerable to invading species. Native species, which are already stressed by habitat loss, may be unable to compete effectively with invasives, resulting in more severe reductions [37]. Overfishing depletes fish populations, and when combined with habitat degradation, such as coral reef damage, the ability of fish populations to rebound is greatly reduced. This combined stressor has a disproportionately negative influence on the entire marine ecosystem. During the investigation, it was discovered that coral reefs are exposed to greater temperatures, which eventually leads to bleaching, and that pollutants are more susceptible to disease. The combination of temperature stress and pathogen exposure causes more extensive reef damage than each stressor alone. Pesticide exposure weakens the immune system of amphibians, increasing their susceptibility to infections such as chytrid fungi. This synergistic effect has resulted in major decreases in amphibian populations worldwide [38]. Continuous exposure to contaminants such as heavy metals and persistent organic pollutants causes bioaccumulation in wildlife, which results in chronic health problems, reproductive failure, and increased mortality over time. Repeated nutrient runoff from agriculture causes eutrophication, hypoxic conditions, and dead zones in aquatic systems, severely harming aquatic life over time. Climate change causes species to alter their ranges, and when combined with habitat loss, new acceptable habitats may be fragmented or nonexistent, resulting in population decreases [39]. Changes in the timing of biological activities, such as breeding or migration, combined with habitat loss, can cause mismatches between species and their food supplies, compromising survival and reproduction. Continuously overfishing depletes fish stocks while also affecting the entire marine food chain over time, resulting in long-term ecological imbalances. Sustained hunting pressure can reduce population sizes such that recovery becomes difficult, especially when this pressure is combined with other stressors, such as habitat loss and climate change [40].

The persistent loss of sea ice caused by climate change has led polar bears to spend more time on land, where they have less access to their major prey, seals. Reduced hunting success combined with greater energy expenditure influences health and reproductive performance. With long-term exposure to contaminants such as PCBs and mercury, which bioaccumulate in the food chain, strains are polar. Bear populations and reduces their overall fitness and longevity. The cumulative impact of hydropower dams, which impede migratory paths, along with overfishing, decreases salmon populations and influences spawning success. Over time, the cumulative pressures cause considerable population decreases. Conservation techniques that address numerous stressors at the same time should be implemented to establish protected zones that prioritize habitat preservation and pollution management. Adaptive management approaches that can respond to new knowledge and changing conditions can be used to mitigate cumulative and synergistic effects effectively [41].

The solutions are as follows:

- Stricter regulations on pollutants, greenhouse gas emissions, and land use should be enforced to reduce the sources of multiple stressors.
- International agreements should be promoted to address transboundary issues such as climate change and ocean pollution.

- Long-term ecological studies should be conducted to understand the cumulative impacts of multiple stressors and develop effective mitigation strategies.
- Comprehensive monitoring programs should be implemented to track the health of ecosystems and wildlife populations and identify early signs of stress and disruption.
- Restoring degraded habitats to improve resilience against multiple stressors.
- Conservation plans that prioritize areas and species most vulnerable to cumulative and synergistic effects should be developed.

9 Conclusion

Emerging pollutants represent a substantial and growing threat to environmental and public health. To address this issue, a complete approach is needed, including improved detection and monitoring, the development of regulatory frameworks, appropriate waste management techniques, and increased public awareness. Understanding and limiting the effects of these toxins will protect ecosystems and human health for future generations. The toxicity and bioaccumulation of pollutants in wildlife and biodiversity present substantial environmental challenges. Addressing these concerns involves a combination of regulatory measures, modern pollution control technologies, ongoing research, and public education. By reducing the impact of harmful compounds, we can help conserve ecosystems and ensure the health and sustainability of wildlife populations.

Endocrine-disrupting chemicals pose a substantial threat to both animal and human health because of their capacity to disrupt hormonal systems. The consequences are numerous and far-reaching, impacting reproduction, development, behavior, immunological function, and general health. Addressing these difficulties will necessitate collaborative efforts in regulation, research, public education, and the development of safer alternatives. By reducing EDC release and exposure, we can conserve ecosystems and promote public health. Genotoxicity and mutagenicity are major dangers to animals, influencing individual health, population dynamics, and ecological stability. Understanding the causes and mechanisms of genetic damage is critical for developing effective monitoring and mitigation techniques.

Protecting wildlife from genotoxic and mutagenic chemicals is critical for sustaining biodiversity and maintaining healthy ecosystems. Ecological disruption poses an enormous threat to habitats and the organisms that rely on them. Addressing these issues requires a holistic approach that includes conservation activities, sustainable behaviors, and effective policy implementation. Protecting and restoring ecosystems allows us to lessen the effects of ecological disruption while also ensuring the health and resilience of our natural surroundings for future generations. The behavioral and physiological effects of environmental changes and human activities on wildlife are extensive and diverse. Understanding these implications is critical for creating successful conservation strategies and reducing the detrimental effects on wildlife populations and habitats. Protecting habitats, decreasing pollution, and managing climate change are critical measures for ensuring the health and resilience of species worldwide. Synergistic and cumulative effects pose substantial dangers to species and ecosystems, frequently magnifying the effects of individual stressors. Understanding these intricate connections is critical for creating successful conservation measures and reducing detrimental effects on biodiversity. By implementing integrated management approaches, stringent regulations, and comprehensive monitoring, we can better protect wildlife and ensure the resilience of ecosystems in the face of multiple environmental challenges.

References

1. Abdelbasir, S. M., & Shalan, A. E. (2019). An overview of nanomaterials for industrial wastewater treatment. *Korean Journal of Chemical Engineering, 36*(8), 1209–1225. https://doi.org/10.1007/s11814-019-0306-y
2. Al-Wafi, R., Ahmed, M. K., & Mansour, S. F. (2020). Tuning the synthetic conditions of graphene oxide/magnetite/hydroxyapatite/cellulose acetate nanofibrous membranes for removing Cr (VI), se (IV) and methylene blue from aqueous solutions. *Journal of Water Process Engineering, 38*, 101543. https://doi.org/10.1016/j.jwpe.2020.101543
3. Borrego, B., Lorenzo, G., Mota-Morales, J. D., Almanza-Reyes, H., Mateos, F., López-Gil, E., et al. (2016). Potential application of silver nanoparticles to control the infectivity of Rift Valley fever virus in vitro and in vivo. *Nanomedicine: Nanotechnology, Biology and Medicine, 12*(5), 1185–1192. https://doi.org/10.1016/j.nano.2016.01.021
4. Calderon, A. G., Duan, H., Seo, K. Y., Macintosh, C., Astals, S., Li, K., et al. (2021). The origin of waste activated sludge affects the enhancement of anaerobic digestion by free nitrous acid pre-treatment. *Science of the Total Environment, 795*, 148831. https://doi.org/10.1016/j.scitotenv.2021.148831
5. Chinnaiyan, P., Thampi, S. G., Kumar, M., & Mini, K. (2018). Pharmaceutical products as emerging contaminant in water: Relevance for developing nations and identification of critical compounds for Indian environment. *Environmental Monitoring and Assessment, 190*(5), 1–13. https://doi.org/10.1007/s10661-018-6672-9
6. Crini, G., & Lichtfouse, E. (2019). Advantages and disadvantages of techniques used for wastewater treatment. *Environmental Chemistry Letters, 17*(1), 145–155. https://doi.org/10.1007/s10311-018-0785-9
7. de Oliveira, M., Frihling, B. E. F., Velasques, J., Magalhães Filho, F. J. C., Cavalheri, P. S., & Migliolo, L. (2020). Pharmaceuticals residues and xenobiotics contaminants: Occurrence, analytical techniques and sustainable alternatives for wastewater treatment. *Science of the Total Environment, 705*, 135568. https://doi.org/10.1016/j.scitotenv.2019.135568
8. Dhangar, K., & Kumar, M. (2020). Tricks and tracks in removal of emerging contaminants from the wastewater through hybrid treatment systems: A review. *Science of the Total Environment, 738*, 140320. https://doi.org/10.1016/j.scitotenv.2020.140320
9. Donkadokula, N. Y., Kola, A. K., Naz, I., & Saroj, D. (2020). A review on advanced physico-chemical and biological textile dye wastewater treatment techniques. *Reviews in Environmental Science and Bio/technology, 19*(3), 1–18. https://doi.org/10.1007/s11157-020-09543-z
10. Fagan, R., McCormack, D. E., Dionysiou, D. D., & Pillai, S. C. (2016). A review of solar and visible light active TiO_2 photocatalysis for treating bacteria, cyanotoxins and contaminants of emerging concern. *Materials Science in Semiconductor Processing, 42*, 2–14. https://doi.org/10.1016/j.mssp.2015.07.052
11. Gogoi, A., Mazumder, P., Tyagi, V. K., Chaminda, G. T., An, A. K., & Kumar, M. (2018). Occurrence and fate of emerging contaminants in water environment: A review. *Groundwater for Sustainable Development, 6*, 169–180. https://doi.org/10.1016/j.gsd.2017.12.009

12. Hemavathy, R. V., Kumar, P. S., Kanmani, K., & Jahnavi, N. (2020). Adsorptive separation of cu (II) ions from aqueous medium using thermally/chemically treated Cassia fistula based biochar. *Journal of Cleaner Production, 249*, 119390. https://doi.org/10.1016/j.jclepro.2019.119390
13. Ngoubeyou, P. S., Wolkersdorfer, C., Ndibewu, P. P., & Augustyn, W. (2022). Toxicity of polychlorinated biphenyls in aquatic environments–A review. *Aquatic Toxicology, 251*, 106284. https://ars.els-cdn.com/content/image/1-s2.0-S0166445X22002107-ga1.jpg
14. Teresa Capucchio, M., Colombino, E., Tarantola, M., Biagini, D., Giovanni Alborali, L., Marco Maisano, A., Scali, F., Raspa, F., Valle, E., Biasato, I., & Schiavone, A. (2019). The disturbed habitat and its effects on the animal population. In *The disturbed habitat and its effects on the animal population* (pp. 1–21). IntechOpen. https://cdnintech.com/media/chapter/65961/1512345123/media/F1.png
15. Cabral, H., Fonseca, V., Sousa, T., & Costa Leal, M. (2019). Synergistic effects of climate change and marine pollution: An overlooked interaction in coastal and estuarine areas. *International Journal of Environmental Research and Public Health, 16*(15), 2737. https://www.mdpi.com/ijerph/ijerph-16-02737/article_deploy/html/images/ijerph-16-02737-g002-550.jpg
16. Vlachogianni, T., Fiotakis, K., Loridas, S., & Valavanidis, A. (2013). Environmental pollution by endocrine disrupting chemicals. In *Adverse developmental, reproductive and immune effects in the wildlife and in human health*. WEB SITE, Dpt of Chemistry, Univ. of Athens. www.chem.uoa.gr. 1.1–54. https://www.researchgate.net/profile/Athanasios-Valavanidis/publication/237078264/figure/fig4/AS:341331940528143@1458391284995/EDCs-affect-fish-by-arious-endocrine-mechanisms_W640.jpg
17. Juliano, C., & Magrini, G. A. (2017). Cosmetic ingredients as emerging pollutants of environmental and health concern. A mini-review. *Cosmetics, 4*(2), 11. https://doi.org/10.3390/cosmetics4020011
18. Karimi-Maleh, H., Ranjbari, S., Tanhaei, B., Ayati, A., Orooji, Y., Alizadeh, M., et al. (2021). Novel 1-butyl-3-methylimidazolium bromide impregnated chitosan hydrogel beads nanostructure as an efficient nanobio-adsorbent for cationic dye removal: Kinetic study. *Environmental Research, 195*, 110809. https://doi.org/10.1016/j.envres.2021.110809
19. Kim, E., Jung, C., Han, J., Her, N., Park, C. M., Jang, M., et al. (2016). Sorptive removal of selected emerging contaminants using biochar in aqueous solution. *Journal of Industrial and Engineering Chemistry, 36*, 364–371. https://doi.org/10.1016/j.jiec.2016.03.004
20. Li, J., Zhou, Q., Liu, Y., & Lei, M. (2017). Recyclable nanoscale zero-valent iron-based magnetic polydopamine coated nanomaterials for the adsorption and removal of phenanthrene and anthracene. *Science and Technology of Advanced Materials, 18*(1), 3–16. https://doi.org/10.1080/14686996.2016.1246941
21. Lu, H., Wang, J., Stoller, M., Wang, T., Bao, Y., & Hao, H. (2016). An overview of nanomaterials for water and wastewater treatment. *Advances in Materials Science and Engineering, 2016*, 4964828. https://doi.org/10.1155/2016/4964828
22. Mohamed Noor, M. H., Wong, S., Ngadi, N., Mohammed Inuwa, I., & Opotu, L. A. (2021). Assessing the effectiveness of magnetic nanoparticles coagulation/flocculation in water treatment: A systematic literature review. *International Journal of Environmental Science and Technology, 28*(22), 1–22. https://doi.org/10.1007/s13762-021-03369-0
23. Munthe, J., Brorström-Lundén, E., Rahmberg, M., Posthuma, L., Altenburger, R., Brack, W., et al. (2017). An expanded conceptual framework for solution-focused management of chemical pollution in European waters. *Environmental Sciences Europe, 29*(1), 1–16.
24. Nasrollahzadeh, M., Baran, T., Baran, N. Y., Sajjadi, M., Tahsili, M. R., & Shokouhimehr, M. (2020). Pd nanocatalyst stabilized on amine-modified zeolite: Antibacterial and catalytic activities for environmental pollution remediation in aqueous medium. *Separation and Purification Technology, 239*, 116542. https://doi.org/10.1016/j.seppur.2020.116542
25. Nasrollahzadeh, M., Sajjadi, M., Iravani, S., & Varma, R. S. (2021). Carbon-based sustainable nanomaterials for water treatment: State-of-art and future perspectives. *Chemosphere, 263*, 128005. https://doi.org/10.1016/j.chemosphere.2020.128005

26. Ngueagni, P. T., Woumfo, E. D., Kumar, P. S., Siéwé, M., Vieillard, J., Brun, N., et al. (2020). Adsorption of cu (II) ions by modified horn core: Effect of temperature on adsorbent preparation and extended application in river water. *Journal of Molecular Liquids, 298*, 112023. https://doi.org/10.1016/j.molliq.2019.112023
27. Ouni, L., Ramazani, A., & Fardood, S. T. (2019). An overview of carbon nanotubes role in heavy metals removal from wastewater. *Frontiers of Chemical Science and Engineering, 13*(2), 1–22. https://doi.org/10.1007/s11705-018-1765-0
28. Poonia, T., Singh, N., & Garg, M. C. (2021). Contamination of arsenic, chromium and fluoride in the Indian groundwater: A review, meta-analysis and cancer risk assessment. *International Journal of Environmental Science and Technology, 18*(9), 1–12. https://doi.org/10.1007/s13762-020-03043-x
29. Rajabi, M., Mahanpoor, K., & Moradi, O. (2017). Removal of dye molecules from aqueous solution by carbon nanotubes and carbon nanotube functional groups: Critical review. *RSC Advances, 7*(74), 47083–47090. https://doi.org/10.1039/c7ra09377b
30. Rajan, A., Sreedharan, S., & Babu, V. (2016). Solvent extraction and adsorption technique for the treatment of pesticide effluent. *Civil Engineering and Urban Planning: An International Journal (CiVEJ), 3*(2), 155–165. https://doi.org/10.5121/civej.2016.3214
31. Raouf, M. E. A., Maysour, N. E., & Farag, R. K. (2019). Wastewater treatment methodologies, review article. *International Journal of Environment and Agricultural Science, 3*(1), 18.
32. Rathi, B. S., Kumar, P. S., & Show, P. L. (2021). A review on effective removal of emerging contaminants from aquatic systems: Current trends and scope for further research. *Journal of Hazardous Materials, 409*, 124413. https://doi.org/10.1016/j.jhazmat.2020.124413
33. Richardson, S. D., & Kimura, S. Y. (2017). Emerging environmental contaminants: Challenges facing our next generation and potential engineering solutions. *Environmental Technology & Innovation, 8*, 40–56. https://doi.org/10.1016/j.eti.2017.04.002
34. Rout, P. R., Zhang, T. C., Bhunia, P., & Surampalli, R. Y. (2021). Treatment technologies for emerging contaminants in wastewater treatment plants: A review. *Science of the Total Environment, 753*, 141990. https://doi.org/10.1016/j.scitotenv.2020.141990
35. Taheran, M., Naghdi, M., Brar, S. K., Verma, M., & Surampalli, R. Y. (2018). Emerging contaminants: Here today, there tomorrow! *Environmental Nanotechnology, Monitoring & Management, 10*, 122–126. https://doi.org/10.1016/j.enmm.2018.05.010
36. Tijani, J. O., Fatoba, O. O., Babajide, O. O., & Petrik, L. F. (2016). Pharmaceuticals, endocrine disruptors, personal care products, nanomaterials and perfluorinated pollutants: A review. *Environmental Chemistry Letters, 14*(1), 27–49. https://doi.org/10.1007/s10311-015-0537-z
37. Tran, N. H., Reinhard, M., & Gin, K. Y. H. (2018). Occurrence and fate of emerging contaminants in municipal wastewater treatment plants from different geographical regions-a review. *Water Research, 133*, 182–207. https://doi.org/10.1016/j.watres.2017.12.029
38. Ungureanu, N., Vlăduț, V., & Voicu, G. (2020). Water scarcity and wastewater reuse in crop irrigation. *Sustainability, 12*(21), 9055. https://doi.org/10.3390/su12219055
39. Varsha, M., Senthil Kumar, P., & Senthil, R. B. (2022). A review on recent trends in the removal of emerging contaminants from aquatic environment using low-cost adsorbents. *Chemosphere, 287*, 132270. https://doi.org/10.1016/j.chemosphere.2021.132270
40. Vieira, W. T., de Farias, M. B., Spaolonzi, M. P., da Silva, M. G. C., & Vieira, M. G. A. (2020). Removal of endocrine disruptors in waters by adsorption, membrane filtration and biodegradation. A review. *Environmental Chemistry Letters, 18*(4), 1113–1143. https://doi.org/10.1007/s10311-020-01000-1
41. Yaashikaa, P. R., Kumar, P. S., Varjani, S. J., & Saravanan, A. (2019). Advances in production and application of biochar from lignocellulosic feedstocks for remediation of environmental pollutants. *Bioresource Technology, 292*, 122030. https://doi.org/10.1016/j.biortech.2019.122030

Part VI
The Fate of Microplastics, Plasticizers, and PFAS: Emerging Contaminants in Aquatic Ecosystems

Chapter 13
Occurrence, Distribution, and Removal of Microplastics from Aquatic Ecosystems

Sameer Shafi Ganaie and Suparna Mukherji

1 Introduction

Plastics are among the most extensively used materials globally due to their versatility and ease of molding, which makes them suitable for a wide array of applications. They are predominantly used in packaging, such that packaging accounts for 39.9% of their usage. Other significant sectors include construction and infrastructure (19.7%), the automotive sector (10%), and electronics manufacturing (6.2%), as well as household products, sports equipment, recreational items (4.2%) and agriculture (3.3%). The remaining 16.7% is spread across various other sectors. The global production of plastics has steadily increased, rising from 348 billion kilograms in 2017 to 359 billion kilograms in 2018 [1]. This surge in production has consequently led to an increase in plastic waste, posing significant environmental challenges. The most commonly used types of plastics include polyethylene (PE), polypropylene (PP), polyethylene terephthalate (PET), polystyrene (PS), and polyvinyl chloride (PVC). Notably, PE and PP constitute more than half of the global plastic output, highlighting their widespread use in various industries [2]. Owing to their properties, such as durability, flexibility, and resistance to moisture, PE & PP are used extensively in applications ranging from packaging materials to automotive components. However, their resilience also makes them resistant to degradation and increases the generation of plastic waste. Therefore, effective measures for these materials are necessary to mitigate their detrimental environmental effects.

Plastic has become the most significant source of environmental contamination because of its widespread use, increasing demand, significant interference with the

S. S. Ganaie · S. Mukherji (✉)
Environmental Science & Engineering Department, Indian Institute of Technology Bombay, Mumbai, Maharashtra, India
e-mail: 23D1262@iitb.ac.in; mitras@iitb.ac.in

M. K. Jindal et al. (eds.), *Emerging Contaminants in Water*, Emerging Contaminants and Associated Treatment Technologies,
https://doi.org/10.1007/978-3-031-89591-3_13

food web, and impact on air, water, and soil [3]. Plastic substances in the environment are exposed to solar and UV radiation and are subjected to abrasion in rivers and streams and temperature changes. These forces cause the physical deterioration of these plastics into smaller fragments and contribute to the partial chemical deterioration of these polymers. This process also releases various compounds or additives associated with plastics, such as heat stabilizers, plasticizers, dyes, catalysts, or plastic monomers. Thus, microplastics (MPs) and nanoplastics (NPs), referred to as micro/nanoplastics (MNPs), build up in the ecosystem and cause harm to diverse life forms [4].

The widespread presence of MPs in the ocean, lakes, rivers, soils, sediments, and atmosphere is a concern not only for other organisms in ecosystems but also for humans. These MPs accumulate and propagate through different species within the ecosystem, creating a complex web of interactions. Owing to their small size, aquatic organisms can ingest them. There is increasing concern regarding the bioconcentration of MPs since marine animals play a crucial role in the food web and can potentially affect human health [5]. The bioaccumulation of MPs has been linked with critical adverse effects in various organisms, such as death, physical impairments, and numerous molecular changes [6]. The consumption of water and seafood or the inhalation of airborne MPs constitute the main route of MP exposure in humans. As a result, there is an accumulation of MPs, which can pose potential risks to human health [7]. The toxicity of MPs to various species can be influenced by their physicochemical characteristics, such as their size and shape, color, specific surface area, surface charge, dose, and duration of exposure. These variables can also significantly impact their environmental transport and bioavailability [8]. Owing to their surface charge and unique morphologies, MPs may serve as vectors for other environmental pollutants, such as heavy metals and polynuclear aromatic hydrocarbons (PAHs). Such associations could lead to synergistic effects on the receiving environment, in which the combined impact on the environment is greater than the sum of their individual effects [9]. A thorough analysis of MPs and other contaminants is crucial for understanding their environmental impacts. This enables large-scale comparisons in terms of abundance, distribution, and chemical composition over time in diverse locations.

Plastic particles ranging from <5 mm down to 100 nm are called MPs. Over time, these particles can further fragment into nanoplastics, with sizes ranging from <100 to 1 nm. On the basis of their origin, MPs are categorized into primary and secondary MPs [10]. Primary MPs are released directly into the environment and originate from sources such as plastic waste generated during manufacturing processes. Secondary MPs are formed through large-scale plastic disintegration or macroplastic degradation during aging processes involving physical, chemical, and biological interactions, such as tire wear and discharge from wastewater treatment facilities [11], as shown in Fig. 13.1.

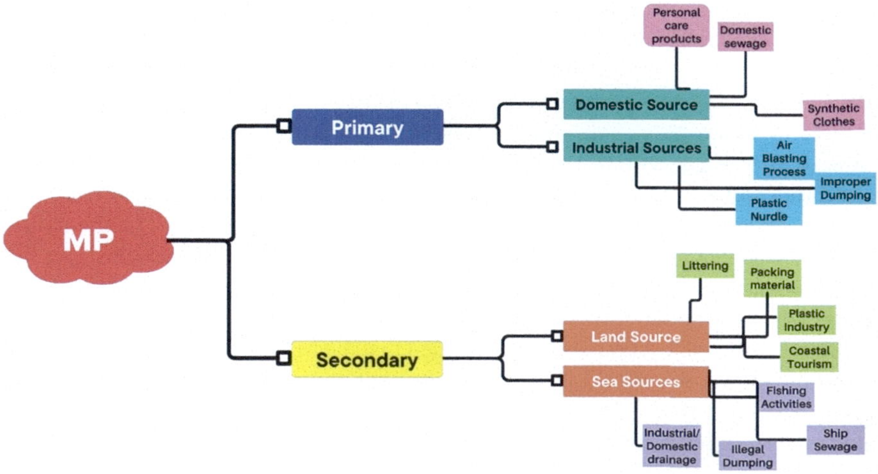

Fig. 13.1 Sources and types of MPs

2 Occurrence and Distribution of MPs

MPs contaminate various environmental compartments, including aquatic systems (both marine and freshwater), terrestrial ecosystems, and the atmosphere. These compartments are linked through various source–pathway–sink interactions, which affect the transport and accumulation of MPs [12]. MPs find their way into rivers and oceans via three primary pathways—soil, water, and air—with water being the predominant route for MP distribution [13]. Groundwater, a vital resource in the hydrological cycle, is also subject to microplastic contamination. Landfills, which are used for waste disposal, and agricultural lands, where various wastes and wastewater sludge are utilized, serve as sinks for primary and secondary MPs. These particles traverse through soil layers due to factors such as rainwater and eventually seep into groundwater after percolating through the soil [14]. Notably, most plastics produced in aquatic systems originate from production, use, and improper disposal on land due to various anthropogenic activities. As a result, terrestrial habitats are considered large reservoirs of MPs, potentially providing various exposure pathways for organisms within terrestrial ecosystems [15]. The recent discoveries of MPs in urban, suburban, and rural atmospheres highlight the atmosphere as a key reservoir and source of pollution. Airborne MPs pose a threat to the biosphere, travel considerable distances and accumulate in various terrestrial and aquatic environments. However, the atmosphere has received the least attention among all environmental compartments in terms of the occurrence and spatial distribution of MPs [16]. To determine the extent of MP pollution in aquatic ecosystems and develop remediation measures, it is important to identify the various routes of contamination.

Several factors affect the occurrence and distribution of MPs in aquatic ecosystems. The land usage patterns in the proximity of these aquatic systems and the population density are considered key determinants of the quantity and types of MPs found in such environments. A study was conducted in Tijuana, northwestern Mexico, in which seven rain events were sampled at five sites on or near a low-traffic density street with different surrounding land use patterns. Three sites were residential, one was predominantly commercial, and the last one was industrial. The study revealed the highest median MP concentration of 191 particles/L at the industrial site, whereas the second highest MP concentration was 158 particles/L at the residential site [17]. Rainfall events significantly influence the distribution of MPs. The seasonal variation in MPs was studied in the surface waters of Manipal Lake in southwest Karnataka. These findings revealed that the MP concentration was greater during the monsoon season, at 0.423 MPs/L, than it was during the post-monsoon period, at 0.117 MPs/L. The pollution load index (PLI) data suggested that the elevated microplastic levels during the monsoon were attributed to an increased influx of particles from surrounding areas [18]. Furthermore, changes in hydrodynamics during rainfall events can resuspend microplastic particles from sediments, contributing to an increase in microplastics in surface waters [19].

Furthermore, various intrinsic properties of MPs, such as their small surface area and low density, allow them to travel long distances through wind and water currents. In freshwater systems, the density of MPs influences their buoyancy and, consequently, their vertical distribution. Low-density MPs, such as PE, are usually found near the surface, whereas high-density MPs, such as PVC, tend to settle at greater depths. As a result, high-density microplastics often accumulate within benthic organisms. Furthermore, changes in particle density are caused by the adsorption of co-contaminants, such as heavy metals, or by biofouling. The shape and texture of MPs may further impact their transport behavior. For example, compared with spherical MPs, fiber-shaped MPs are more prone to biofouling, which causes them to sink and remain in surface waters for less time [20]. The overall abundance and physicochemical characteristics of MPs in these environmental systems vary significantly across different study areas. A few studies are summarized in Table 13.1.

Table 13.1 Occurrence & distribution of MPs

Environmental system	Area	Sample type	MP Type	Morphology	Concentration	Reference
	Melbourne city, Australia	Stormwater	PET, PA, HDPE, PP, PS	Beads, films, fragments	17–26 particles/liter	[21]
Aquatic	Vembanad, Kerala, India	Lake water	HDPE, LDPE, PS, & PP	Fibers, pellets, fragments, foam, films	96–496 items/m^2	[22]
	Helsinki, Finland	Stormwater	PE, PP, PET, PS	Fibers, films, fragments	29 particles/liter	[23]
	Guangzhou, China	River water	PE, PP, PET	Fiber, film, & fragments	379–7924 items/m^3	[24]
	Chennai, India	Groundwater	PP and PS	Pellets, foam, fragments, fibers	2–80 particles/liter	[25]
	Naivasha, Kenya	Lake water	PE, PP, polyester	Fragments, fiber, and films,	0.4 ± 0.1particles/m^2	[26]
	Tamil Nadu India	Groundwater	Nylon, PP, & polyester	Foam, fiber, film, & pellet	4.2 particles/liter	[14]
	San Francisco, USA	Stormwater	PE, PP, PS, PC, PVC, PET, ABS, PAN, nylon	Films, fibers, fragments, beads	1.1–24.6 particles/L	[27]
	Netravathi, India	River water	PE & PET	Fibers, films, & fragments	288 items/m^3	[28]
	Antuã, Portugal	River water	PE & PP	Foams, fibers, & fragments	64.5–729 items/m^3	[29]

(continued)

Table 13.1 (continued)

Environmental system	Area	Sample type	MP Type	Morphology	Concentration	Reference
Terrestrial	Brisbane, Australia	River sediments	PE, PA, PP	Films, fragments, and fibers	10–520 items/kg	[30]
	Ganga River, India	River sediments	PET and PE	Films and fibers	99.27–409.86 items/kg	[31]
	Mumbai, India	Beach sediments	PE, PET, PVC, PP, PS	Fibers, granules, & films	220 ± 50 items/kg	[32]
	Andaman & Nicobar Islands, India	Sediments	PP and PVC	Fragment, fibers, & spheres	414.35 ± 87.4 particles/kg	[15]
	Bhopal, India	Soil samples	PE, PP, PVC, & PET.	Fragments & filaments	180 ± 13 to 2.5 ± 0.7 particles/kg	[33]
	Hauts-DE France	Beach sediments	PE, PP, PS, PET, PVC, ABS	Fibers, fragments, pellets, films, & foams	23.4–69.3 items/kg	[34]
Atmospheric	East Indian Ocean	Dust	PET, PP, PAN-AA, & PR	Fibers, fragment	0.4 ± 0.6 items/100 m^3	[16]
	Patna, India	Indore dust	PET & PC	N. A	2000 & 20 μg/g	[35]

[a]*PE* polyethylene, *PP* polypropylene, *PET* polyethylene terephthalate, *PAN-AA* polyacrylonitrile-coacrylic acid, *PR* phenoxy resin, *PC* polycarbonate, *HDPE* high-density polyethylene, *LDPE* low-density polyethylene, *PVC* polyvinyl chloride, *PS* polystyrene, *ABS* acrylonitrile butadiene styrene, *PA* poly amide

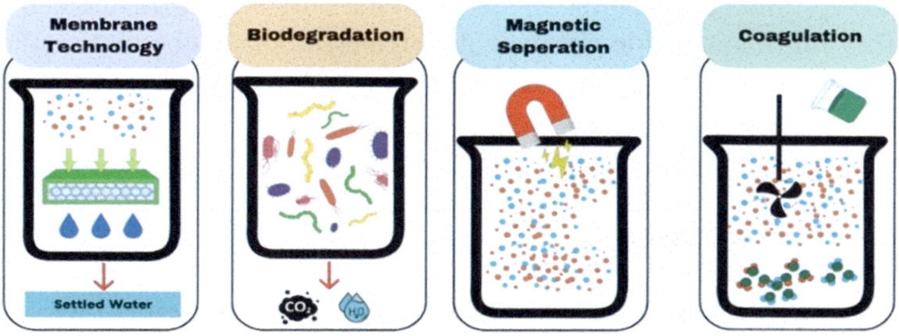

Fig. 13.2 Treatment methods for MPs

3 Removal of MPs

Preventing MPs from entering various environmental matrices is crucial. This involves strategies such as reducing sources, embracing a circular economy to minimize waste, and raising awareness through education. However, if MPs are present, various treatment technologies are needed to capture and remove them from the aquatic environment. These methods include physical processes, biological methods, and chemical treatments, as shown in Fig. 13.2.

Physical sorption and filtration involve the use of materials that can adsorb or filter out MPs from the environment. These materials can range from chemically synthesized sponge materials and graphene-based materials to biochar-loaded materials. For instance, metal organic frameworks and different biopolymers have been explored for their potential to remove MPs from water and wastewater. Magnetic separation involves the use of magnetic seeds and an external magnetic field to extract MPs. Biodegradation involves the use of organisms such as algae, fungi, and bacteria that can ingest or degrade MPs. Chemical treatments through coagulation and flocculation can cause the aggregation of MPs and generate flocs that can settle out of the water column. Electrocoagulation can generate coagulants in situ without the addition of chemicals [36]. A range of these various physical, chemical, and biological processes can accelerate the degradation and removal of MPs; however, enhancing these processes in large-scale aquatic systems is challenging.

3.1 *Bioremediation*

Microorganisms have a remarkable ability to adapt to diverse environmental conditions and degrade a wide range of substances, including MPs. Leveraging microorganisms for MP degradation offers a viable and eco-friendly approach to accelerate biodegradation processes without negatively impacting the environment. Microorganisms such as bacteria and fungi can form biofilms on plastic surfaces and release specific enzymes that break down the polymeric structure and its side chains. This enzymatic action results in depolymerization, generating smaller molecules such as oligomers, dimers, and monomers, along with byproducts such as CO_2 and water [37].

3.1.1 Bacterial Degradation

A significant amount of research has focused on the use of bacteria for the remediation of MPs. Bacteria naturally possess the ability to degrade plastics by metabolizing long-chain fatty acids. Enzymes such as oxidoreductases, lipases, esterases, and cutinases play crucial roles in the degradation of MPs [38]. A variety of studies, as illustrated in Table 13.2, have demonstrated the degradation of MPs, such as PE, PP, and PS, via both pure cultures and bacterial consortia. These studies attempted to understand the process and measure the extent of degradation under controlled laboratory conditions. In a previous study, two bacterial strains were isolated from the gut of a waxworm, and the MP degradation rates were reported to be 6.1 ± 0.3% by *Enterobacter asburiae* YT1 and 10.7 ± 0.2% by *Bacillus* sp. YP1 after 60 days of incubation [39]. Similarly, another study reported that *Microbacterium paraoxydans* and *Pseudomonas aeruginosa* could biodegrade low-density polyethylene (LDPE), resulting in gravimetric losses of 61% and 50.5%, respectively, after 2 months [40]. In another study, two bacterial strains, *Bacillus cereus* and *Sporosarcina globispora,* were identified from mangrove sediments in peninsular Malaysia because of their ability to degrade PP MPs. Gravimetric weight losses of 12% and 11%, respectively, were observed after 40 days of incubation [41]. The use of pure bacterial strains to study MP degradation allows the investigation of metabolic pathways and the evaluation of environmental factors influencing degradation. This method also enables close monitoring of the entire degradation process by bacteria, including changes in MPs. However, various studies have demonstrated that relying on a single strain for polymer biodegradation may be insufficient and can even inhibit microbial growth through the production of toxic end products [42].

A more effective approach involves the use of multiple bacteria, in which polymer degradation products from one bacterium serve as substrates for the growth of other bacteria. This approach can potentially increase the efficiency of microplastic degradation. A consortium of thermophilic bacterial strains, *Aneurinibacillus aneurinilyticus*, *Brevibacillus agri*, *Brevibacillus* sp., and *Brevibacillus brevis*, was used to degrade LDPE, HDPE, and PP films and pellets. After 140 days at 50 °C, the consortium achieved weight reductions of approximately 58.21 ± 2%, 46.6 ± 3%, and 56.3 ± 2%, respectively [43]. Another study used a consortium of mesophilic bacteria *Bacillus* sp. and *Paenibacillus* sp. from municipal landfill sediments and reported a 14.7% reduction in PE MPs after 60 days [44].

Table 13.2 Degradation of MPs by various microbes

Culture	Plastic	Microbial species	Degradation Time	Gravimetric Weight loss (%)	References
Pure bacterial strain	PE	*Enterobacter asburiae* YT1 & *Bacillus* sp. YP1	60 days	6.1 & 10.7	[39]
		M. paraoxydans & *P. aeruginosa*	2 months	61% & 50.5%	[40]
	PP	*Bacillus* sp. and *Rhodococcus* sp.	40 days	4% & 6.4%	[41]
		Pseudomonas sp. ADL15 & *Rhodococcus* sp. ADL36	40 days	17.3% & 7.3%	[52]
		Bacillus cereus & *Sporosarcina globispora*	40 days	12% and 11%	[53]
Bacterial consortium	PE	*Bacillus* sp. and *Paenibacillus* sp.	60 days	14.7%	[44]
		Aneurinibacillus aneurinilyticus; *Brevibacillus agri*; *Brevibacillus* sp. & *Brevibacillus brevis*	140 days	45.7% & 37.2%	[43]
	PP	*Aneurinibacillus aneurinilyticus*; *Brevibacillus agri; Brevibacillus* sp. & *Brevibacillus brevis*	140 days	44.2	[43]
	PS	*Thermus* sp., *Bacillus* sp., and *Geobacillus* sp.	56 days	7.3%	[54]
Fungus	PE	*Alternaria alternata*	120 days	95%	[45]
		Zalerion maritimum	14 days	43%	[46]
	PP	*Aspergillus niger* & *Aspergillus fumigatus*	45 days	71.1% & 53.1%	[47]
Algae	PP	*Spirulina* sp.	112 days	NA	[50]
	PS	*Scenedesmus abundans*	6 days	84%	[49]
	PET	*Spirulina* sp.	112 days	NA	[50]

3.1.2 Fungal Degradation

Owing to their widespread distribution and reproductive capabilities, fungi play a significant role in degrading MPs. Fungi are widely distributed and are capable of reproduction. They can degrade polymers such as high-density polyethylene (HDPE) by producing an oxidase referred to as laccase. This enzyme oxidatively cleaves amorphous regions and targets carbonyl regions in polymer chains [36]. Various fungal strains, such as *Zalerion maritimum*, *Aspergillus niger*, and *Aspergillus fumigatus*, have been found to degrade MPs, using them as their sole carbon source (Table 13.2). In one study, significant degradation of PE by the fungal strain *Alternaria alternata was reported.* After an incubation period of 120 days, the molecular weight of the PE film was reduced by 95%. Scanning electron microscopy (SEM) analysis revealed efficient degradation of the PE film, with holes up to

0.5 µm in diameter, some nearly piercing the film. After 28 days, X-ray diffraction (XRD) analysis revealed a significant reduction in relative crystallinity, from 62.79% to 52.02% [45]. A study demonstrated the potential of a marine fungal strain, *Zalerion maritimum*, to degrade PE plastic pellets. Within just 14 days of incubation, there was a 43% reduction in the mass of the pellets [46]. In another study, the abilities of various fungal strains, including *Aspergillus niger*, *Aspergillus nidulans, Penicillium* sp., *Mucor* sp., *Rhizopus* sp., *Fusarium* sp., *Microsporum canis,* and *Aspergillus fumigatus,* to degrade PP microplastics were analyzed. Among these strains, *Aspergillus niger* and *Aspergillus fumigatus* showed the highest efficiency of degradation, reducing the weight by 71.1% and 53.1%, respectively, after 45 days of incubation [47].

Fungi possess a significant advantage over bacteria in the degradation of MPs. Unlike bacteria, which often rely on specific enzymes for substrate degradation, fungi secrete a broad spectrum of enzymes that can act on nonspecific substrates. This gives fungi a high potential for effective plastic and MP degradation [37]. The use of a mixed fungal consortium, similar to a mixed bacterial consortium, offers greater benefits for MP degradation. In one study, various fungal species were evaluated for their capacity to degrade PE. These fungi included *Penicillium simplicum*, *Curvularia lunata, Alternaria alternata,* and *Fusarium* sp. These fungi had relatively low PE degradation efficiencies. However, when these four fungi were combined as a mixed culture, their synergistic action increased the PE degradation rate to 27% within 90 days, demonstrating the efficacy of the mixed fungal consortium in MP degradation [48].

3.1.3 Algal Removal

Algae are distinct from other microbial groups. They perform photosynthesis and utilize atmospheric CO_2 as their main source of carbon and energy. Hence, their metabolic processes may not be effective for the mineralization of MPs [38]. However, certain studies have demonstrated that some algae can eliminate MPs via surface colonization and absorption. The extracellular polymeric substances secreted by microalgae aid in the aggregation of MPs with extracellular polymers during colonization and absorption. This process facilitates automatic precipitation, eliminates the need for energy-intensive filtration or centrifugation, and is both repeatable and easily scalable [2]. Moreover, extracellular polymeric substances enhance MP adsorption on solid surfaces, reducing the presence of free and suspended MPs [49].

A study examined the interaction between the microalga *Spirulina* sp. and two types of MPs, PP and PET. Energy dispersive X-ray (EDX) analysis revealed a reduction in the carbon content of PP by 36.7% and in that of PET by 48.61% after 112 days of incubation. The tensile strengths of PP and PET were also evaluated before and after, and the results served as indicators of deterioration. This value decreased by 0.1977 MPa/day for PP and 0.9939 MPa/day for PET [50].

3.1.4 Mechanism of Microbial Degradation

Microorganisms break down MPs through a series of biochemical reactions, as shown in Fig. 13.3. Given that MPs are polymers, their degradation pathways mirror those of other polymers. Microorganisms depolymerize MPs, and extracellular enzymes catalyze the breakdown of biodegraded plastics into smaller components. This process involves breaking down large polymer chains into shorter chains or smaller molecules, including oligomers, dimers, and monomers. Inside the cells, these monomers undergo further transformations, with aerobic conditions leading to mineralization into CO_2 and H_2O. However, under anaerobic conditions, the production of CO_2, H_2O, and CH_4 occurs, and this process also leads to biomass generation [37]. As illustrated in Fig. 13.3, the biodegradation of MPs typically occurs in four stages: biofilm formation on the MP surface, bio-fragmentation, assimilation, and, finally, mineralization [2].

In the initial stage of biodegradation, microorganisms attach to MP particles, forming biofilms. These biofilms are complex communities of various microorganisms, such as bacteria, fungi, and algae. They develop on the surfaces of MP particles and use them as the sole carbon source necessary for their growth and reproduction. The biofilms are held together by a slimy extracellular matrix that the microorganisms produce themselves [2]. Generally, MPs are hydrophobic polymers. To enhance interactions with microorganisms, the surfaces of MPs require microbial attachment. Before microbial attachment, a continuous film with hydrophilic functional groups covers the surface to improve polymer–microorganism interactions [51].

Biofilms provide various benefits to these microorganisms. They facilitate the horizontal transfer of genetic material, allowing beneficial genes to be exchanged

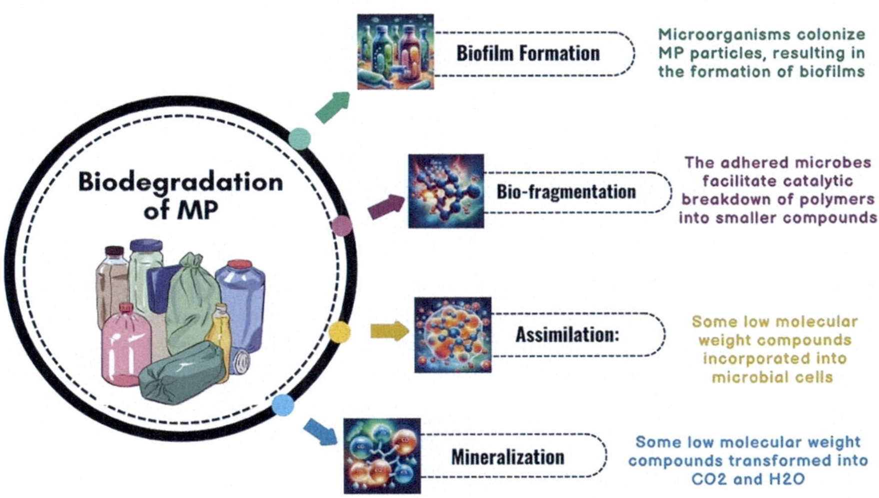

Fig. 13.3 Illustration of the different steps involved in the biodegradation of MPs

among different microorganisms. Additionally, biofilms aid in the absorption of essential nutrients, providing microorganisms with a competitive advantage. They also act as barriers against potentially toxic substances in the environment and prevent dehydration, thereby increasing the survival of these microbes [55].

After adhering to the surface of MPs, microorganisms facilitate the breakdown of polymers into lower-molecular-weight substances, including oligomers, dimers, and monomers. This degradation is initiated by free radicals and aided by extracellular enzymes. During this stage, enzymes such as oxidoreductases and hydrolases play important roles in hydrolyzing the polymeric structure of MPs. These enzymes effectively weaken the carbon bonds within the plastic particles, leading to their fragmentation. Compounds with lower molecular weights produced during hydrolysis are then incorporated into microbial cells in the assimilation stage, forming part of the growth and maintenance of the microbial cells through both active and passive uptake pathways [2]. Inside the cell, these low-molecular-weight compounds may undergo enzymatic reactions, which result in their complete degradation and transformation into oxidative metabolites, such as CO_2, N_2, CH_4, and H_2O. Through metabolic pathways such as the TCA cycle, these metabolites are further processed to generate ATP to provide energy for cell synthesis and other functions [56].

Biodegradation is sustainable and scalable and is vital for the effective and long-term mitigation of MP pollution. Currently, large-scale MP biodegradation faces challenges due to slow degradation rates. Ongoing research aims to address limitations such as lower efficiency and longer timeframes, which require improved biotechnological strategies. Optimizing microbial degradation through continued research may pave the way for more efficient and widespread applications.

Various abiotic pretreatments, such as UV irradiation and thermal treatment, accelerate MP biodegradation by oxidizing and breaking down their crystallized structure. This step can increase the particle surface area and expose more MPs to enzymes, thus facilitating effective biodegradation. For example, a study used an isolate of *Penicillium citrinum* for MP degradation. The isolate was selected on the basis of rapid colony growth in the screening medium. *P. citrinum* exhibited 38.82 ± 1.08% weight loss when the untreated LDPE polymer was tested in biodegradation studies. The use of LDPE pretreated with 99% (v/v) nitric acid at 80 °C for 10 days resulted in a significant increase in biodegradation of 47.22 ± 2.04% after 90 days [57]. Moreover, genetic engineering and protein engineering can be utilized to increase the activity of microbial enzymes and thus increase the efficiency of biodegradation. In one study, the genetically engineered thermophilic anaerobic bacterium *Clostridium thermocellum* was used to secrete a PET-degrading enzyme, lignocellulolytic cutinase (LCC). This enzyme, capable of functioning at high temperatures up to 70 °C, was originally sourced from a plant compost metagenome. When the engineered bacterium was applied to the PET films at 60 °C, substantial PET degradation occurred. Over a 14-day period, approximately 60% of the initial PET film mass (approximately 50 mg) was transformed into soluble monomers [58]. This approach outperforms traditional biodegradation methods that utilize mesophilic bacteria or microalgae.

3.2 Adsorption

Adsorption, a crucial physicochemical process governed by van der Waals forces or ion exchange interactions, enables adsorbate particles to adhere to the surface of adsorbents, facilitating their removal from water systems. A variety of removal processes utilize these adsorption mechanisms to eliminate the presence of MPs in aquatic ecosystems. These methods often employ chemically engineered sponges, graphene-based composites, or biochar, all of which are highly efficient at removing MPs from natural waters. A sustainable and biodegradable green sponge material was created through the chemical crosslinking of plant proteins. Owing to the presence of multiple active side chains on its amino acid residues, this sponge exhibited enhanced adsorption capabilities for MPs, reaching a removal efficiency of 81.2%. Moreover, this protein sponge exhibited properties such as reusability and biodegradability, making it an eco-friendly solution for tackling MP contamination [20]. In one study, three-dimensional reduced graphene oxide (3D RGO) was explored for its ability to enhance the adsorption of polystyrene microplastics in water. Under the optimized conditions of pH 6, NaCl concentration of 600 mg/L, adsorption time of 120 min, and temperature of 26 °C, 3D RGO demonstrated a maximum adsorption capacity, reaching 617.28 mg/g for PS-MPs [59]. The use of biochar for MP removal has shown considerable promise in various environmental settings, particularly in aquatic systems. For example, activated carbon and biochar have been effectively utilized to remove MPs from stormwater. Steam-activated biochar at 800 °C emerged as the most potent adsorbent for removing MPs. Furthermore, activated biochar, with a relatively low surface area of 200–600 m^2/g, has shown the ability to trap larger MPs [60].

3.3 Magnetic Separation

Magnetic extraction, also known as magnetic separation, is a sophisticated technique that utilizes the power of a magnetic field in conjunction with an acidic solution to accelerate the separation process. This technique employs magnetic seeds, which are typically made of iron nanoparticles, and an external magnetic force to efficiently separate the microplastics that attach to these seeds. It is effective at removing MPs from various water sources, including ultrapure water, seawater, rivers, and sewage [61]. An innovative magnetic technique was introduced for microplastic removal, utilizing hydrophobic iron nanoparticles to extract even small-sized MPs from various water sources. These nanoparticles were selected because of their ferromagnetic nature, cost efficiency, and high specific surface area. To increase their hydrophobicity and improve the adhesion of the MPs, a coating of hexadecyltrimethoxysilane was applied to the nanoparticles. A recovery of 92% for MPs (PE and PS) smaller than 20 μm was demonstrated. Moreover, a recovery of 93% was achieved for MPs (PE, PET, PS, PU, PVC, and PP) larger than 1 mm. This magnetic method was further applied for separating microplastics between 200 μm and 1 mm in size from freshwater, yielding an 84% recovery rate [62]. However, the iron-based nanoparticles used were nonbiodegradable and not reusable. Thus, iron nanoparticles may cause secondary pollution concerns. Furthermore, the presence of soil particles and lipophilic substances can reduce the effectiveness of plastic removal by damaging the nanoparticles.

3.4 Membrane Filtration

Membrane filtration serves as a physical barrier to remove plastic particles from water, even though it was not originally designed for plastic removal. Membranes offer several advantages, including efficient removal of microplastics, consistent effluent quality, and ease of operation [63]. In particular, when other tertiary treatment methods, such as rapid sand filtration, disk filtration, and dissolved air flotation, achieve 97%, 98.5%, and 95% removal efficiency of secondary effluent, respectively, membrane filtration can achieve 99.9% removal efficiency. The membranes, classified by pore size, include microfiltration (MF), ultrafiltration (UF), nanofiltration (NF), and reverse osmosis (RO) membranes. MF, UF, and NF are effective at removing particles larger than 0.08–2 μm, 0.005–0.02 μm, and 0.002 μm, respectively. RO is predominantly utilized for desalination purposes [64].

However, the membrane filtration process may be adversely affected by membrane fouling. During wastewater treatment, large molecules and particles, including micro- and nanoplastics, physically and chemically interact with the membrane materials, resulting in their accumulation both on the membrane surface and within the membrane pores. This undesired buildup constricts the pores of the membrane, thereby causing membrane clogging and reduced water permeation. Biofouling may also cause a severe reduction in treated water throughput. The consequences of this fouling include reduced water flow, increased transmembrane pressure, elevated energy consumption, higher maintenance costs, and longer operating times [65].

The membrane bioreactor (MBR) is an intricate system that combines biological catalysts, such as bacteria or enzymes, with membrane filtration. It operates via a film-based mechanism and is vital in addressing water pollution caused by MPs. Compared with conventional methods in wastewater treatment (WWT) plants, MBR technology has proven significantly more effective in the removal of MPs. They can remove more than 90% of MPs from primary treated wastewater. However, the efficiency of MBRs in degrading and removing pollutants is heavily influenced by various factors, including the physicochemical characteristics of the contaminants and key operational parameters, such as the feeding rate, hydraulic retention time, and pollutant concentration [66]. In one study, the effectiveness of various tertiary treatment technologies in removing MPs from wastewater treatment plant effluents was investigated. These methods include MBR, disk filtering, rapid sand filtration, and dissolved air flotation. The results revealed that the MBR exhibited an impressive removal efficiency of 99.9%, reducing the MP concentration from 6.9 to 0.005 particles/L. Additionally, the study demonstrated that MBR, rapid sand filtration, and dissolved air flotation effectively removed MPs of all sizes, including the smallest particles ranging from 20 to 100 μm. Notably, the removal efficiency was consistent regardless of the shape of the MPs, indicating efficient removal of textile fibers, which are prevalent in both the influent and effluent during the treatment process [67].

3.5 Electrocoagulation

Electrocoagulation (EC) presents itself as an environmentally friendly option for removing microplastics from the environment, integrating coagulation, floatation, and electrochemical reactions. Notably, this method avoids the use of chemicals, making it a sustainable choice. During EC, coagulants are generated through the application of electricity. Generally, metal electrodes, which often contain Fe^{2+} and/or Al^{3+} ions, react with the OH^- ions produced during electrolysis. This reaction leads to the formation of metal hydroxide coagulants, which trap colloids and are subsequently separated from the solution. The MPs existing as solid particles become trapped in the metal hydroxide precipitates and are removed through the sweep floc mechanism. The flocs containing MP particles can then be efficiently collected and removed from water or wastewater [68].

In one study, an Al-based electrocoagulation approach was applied at a pH of 7.5, NaCl concentration of 2 g/L, and current density = 11 A/m^2. This method effectively removes approximately 90% of plastic microbeads from synthetic wastewater [69]. Another study assessed the effectiveness of electrocoagulation in removing microplastics from both synthetic and real wastewater effluents from a municipal wastewater treatment plant. In synthetic wastewater, at current densities of 2.88 and 8.07 mA/cm^2, pH values of 4 and 7, and electrolysis times ranging from 0 to 90 min, a maximum removal efficiency of over 99% was achieved in all cases. The optimal conditions for removing microplastics from synthetic wastewater were an initial pH of 4, a current density of 2.88 mA/cm^2, and an electrolysis duration of 90 min. When these conditions were applied to real wastewater effluent, a maximum removal efficiency of 96.5% was achieved. This reduction was due to the presence of other competing colloidal particles in real wastewater, which consumed the available metal hydroxide ions needed for effective coagulation [70].

4 Conclusion

The widespread issue of MP pollution requires a multifaceted approach that addresses both its prevalence in the environment and the associated health concerns. Among the various treatment methods, bioremediation of MPs via various microorganisms, such as bacteria, fungi, and algae, is promising. However, the slow degradation rates pose challenges for large-scale implementation. The selected bacterial strains and reconstituted consortia that have shown good potential should be explored further for bioaugmentation. Compared with bacteria, fungi, which can secrete a wide array of nonspecific enzymes, typically have greater potential for MP degradation. In contrast, algae can remove MPs from water through an indirect mechanism.

Among various physicochemical methods, adsorption via chemically synthesized sponges, graphene-based substances, and biochar has resulted in high removal rates of MPs through adsorption. However, challenges such as secondary pollution concerns and the limited reusability of these adsorbents need to be addressed. The use of magnetic fields and nanoparticles to extract MPs from various aquatic systems also holds promise. However, concerns regarding the biodegradability and reusability of magnetic nanoparticles must be explored before their widespread application. Membrane filtration, while highly effective at removing MPs, may be adversely affected by membrane fouling, which can reduce treatment efficiency and increase maintenance costs. MBR, which integrates microbial degradation with membrane filtration, has shown excellent MP removal efficiency. However, their performance is heavily influenced by their operational parameters and pollutant properties. Electrocoagulation generates coagulants in situ that can destabilize colloids and promote floc formation, in which MPs can also be trapped and removed from water through settling. Further research is needed to develop effective approaches for mitigating MP pollution and for protecting aquatic ecosystems for future generations.

References

1. Plastic Europe. (2018). Plastics-the facts 2018: An analysis of European plastic production, demand and waste data. *Plastic Europe*.
2. Amobonye, A., Bhagwat, P., Singh, S., & Pillai, S. (2021). Plastic biodegradation: Frontline microbes and their enzymes. *Science of the Total Environment, 759*(143536), 143536.
3. Puri, H. S., Mishra, D. S., & Jindal, V. K. (2019). Plastic waste management-issues, solutions and case studies. *Ministry of Housing and Urban Affairs Government of India.*, 105–113.
4. Ogunola, O. S., Onada, O. A., & Falaye, A. E. (2018). Mitigation measures to avert the impacts of plastics and microplastics in the marine environment (a review). *Environmental Science and Pollution Research International, 25*(10), 9293–9310.
5. Sul, J. A. I., & Costa, M. F. (2014). The present and future of microplastic pollution in the marine environment. *Environmental Pollution, 185*, 352–364.
6. Bhattacharya, A., & Khare, S. K. (2020). Ecological and toxicological manifestations of microplastics: Current scenario, research gaps, and possible alleviation measures. *Journal of Environmental Science and Health, Part C:Toxicology and Carcinogenesis, 38*(1), 1–20.
7. Yee, M. S. L., Hii, L. W., Looi, C. K., Lim, W. M., Wong, S. F., Kok, Y. Y., et al. (2021). Impact of microplastics and nano plastics on human health. *Nanomaterials, 11*(2), 496.
8. Wright, S. L., Thompson, R. C., & Galloway, T. S. (2013). The physical impacts of microplastics on marine organisms: A review. *Environmental Pollution., 178*, 483–492.

9. Huang, W., Song, B., Liang, J., Niu, Q., Zeng, G., Shen, M., et al. (2021). Microplastics and associated contaminants in the aquatic environment: A review on their ecotoxicological effects, trophic transfer, and potential impacts to human health. *Journal of Hazardous Materials, 405*(124187), 124187.
10. Rai, P. K., Lee, J., Brown, R. J., & Kim, K. H. (2021). Micro-and nano-plastic pollution: Behavior, microbial ecology, and remediation technologies. *Journal of Cleaner Production., 291*, 125240.
11. Xu, B., Liu, F., Cryder, Z., Huang, D., Lu, Z., He, Y., et al. (2020). Microplastics in the soil environment: Occurrence, risks, interactions, and fate–a review. *Critical Reviews in Environmental Science and Technology., 50*(21), 2175–2222.
12. Zhang, Y., Kang, S., Allen, S., Allen, D., Gao, T., & Sillanpää, M. (2020). Atmospheric microplastics: A review on current status and perspectives. *Earth-Science Reviews., 203*, 103118.
13. Shen, M., Zeng, G., Zhang, Y., Wen, X., Song, B., & Tang, W. (2019). Can biotechnology strategies effectively manage environmental (micro) plastics? *Science of the Total Environment., 697*, 134200.
14. Selvam, S., Jesuraja, K., Venkatramanan, S., Roy, P. D., & Kumari, V. J. (2021). Hazardous microplastic characteristics and its role as a vector of heavy metal in groundwater and surface water of coastal South India. *Journal of Hazardous Materials., 402*, 123786.
15. Patchaiyappan, A., Ahmed, S. Z., Dowarah, K., Jayakumar, S., & Devipriya, S. P. (2020). Occurrence, distribution and composition of microplastics in the sediments of South Andaman beaches. *Marine Pollution Bulletin, 156*(111227), 111227.
16. Wang, X., Li, C., Liu, K., Zhu, L., Song, Z., & Li, D. (2020). Atmospheric microplastic over the South China Sea and East Indian Ocean: Abundance, distribution and source. *Journal of Hazardous Materials., 389*, 121846.
17. De Jesus Piñon-Colin, T., Rodriguez-Jimenez, R., Rogel-Hernandez, E., Alvarez-Andrade, A., & Wakida, F. T. (2020). *Microplastics in stormwater runoff in a semiarid region* (Vol. 704). Tijuana.
18. Warrier, A. K., Kulkarni, B., Amrutha, K., Jayaram, D., Valsan, G., & Agarwal, P. (2022). Seasonal variations in the abundance and distribution of microplastic particles in the surface waters of a southern Indian Lake. *Chemosphere, 300*, 134556.
19. Elizalde-Velázquez, G. A., & Gómez-Oliván, L. M. (2021). Microplastics in aquatic environments: A review on occurrence, distribution, toxic effects, and implications for human health. *Science of the Total Environment, 780*(146551), 146551.
20. Pan, Y., Gao, S. H., Ge, C., Gao, Q., Huang, S., Kang, Y., et al. (2023). Removing microplastics from aquatic environments: A critical review. *Environmental Science and Ecotechnology, 13*(100222), 100222.
21. Monira, S., Roychand, R., Bhuiyan, M. A., Hai, F. I., & Pramanik, B. K. (2022). Identification, classification and quantification of microplastics in road dust and stormwater. *Chemosphere, 299*, 134389.
22. Sruthy, S., & Ramasamy, E. V. (2017). Microplastic pollution in Vembanad Lake, Kerala, India: The first report of microplastics in lake and estuarine sediments in India. *Environmental pollution., 222*, 315–322.
23. Pankkonen P. Urban stormwater microplastics—Characteristics and removal using a developed filtration system.
24. Lin, L., Zuo, L. Z., Peng, J. P., Cai, L. Q., Fok, L., Yan, Y., et al. (2018). Occurrence and distribution of microplastics in an urban river: A case study in the Pearl River along Guangzhou City, China. *Science of the Total Environment, 644*, 375–381.
25. Natesan, U., Vaikunth, R., Kumar, P., Ruthra, R., & Srinivasalu, S. (2021). Spatial distribution of microplastic concentration around landfill sites and its potential risk on groundwater. *Chemosphere, 277*, 130263.
26. Migwi, F. K., Ogunah, J. A., & Kiratu, J. M. (2020). Occurrence and spatial distribution of microplastics in the surface waters of Lake Naivasha, Kenya. *Environmental Toxicology and Chemistry, 39*(4), 765–774.

27. Werbowski, L. M., Gilbreath, A. N., Munno, K., Zhu, X., Grbic, J., Wu, T., et al. (2021). Urban Stormwater runoff: A major pathway for anthropogenic particles, black rubbery fragments, and other types of microplastics to urban receiving waters. *ACS EST Water., 1*(6), 1420–1428.
28. Amrutha, K., & Warrier, A. K. (2020). The first report on the source-to-sink characterization of microplastic pollution from a riverine environment in tropical India. *Science of the Total Environment, 739*(140377), 140377.
29. Rodrigues, M. O., Abrantes, N., Gonçalves, F. J. M., Nogueira, H., Marques, J. C., & Gonçalves, A. M. M. (2018). Spatial and temporal distribution of microplastics in water and sediments of a freshwater system (Antuã River, Portugal). *Science of the Total Environment., 633*, 1549–1559.
30. He, B., Goonetilleke, A., Ayoko, G. A., & Rintoul, L. (2020). Abundance, distribution patterns, and identification of microplastics in Brisbane River sediments, Australia. *Science of the Total Environment., 700*, 134467.
31. Sarkar, D. J., Sarkar, S. D., Das, B. K., Manna, R. K., Behera, B. K., & Samanta, S. (2019). Spatial distribution of meso and microplastics in the sediments of river ganga at eastern India. *Science of the Total Environment., 694*, 133712.
32. Tiwari, M., Rathod, T. D., Ajmal, P. Y., Bhangare, R. C., & Sahu, S. K. (2019). Distribution and characterization of microplastics in beach sand from three different Indian coastal environments. *Marine Pollution Bulletin., 140*, 262–273.
33. Singh, S., Chakma, S., Alawa, B., Kalyanasundaram, M., & Diwan, V. (2023). Identification, characterization, and implications of microplastics in soil–A case study of Bhopal, Central India. *Journal of Hazardous Materials Advances., 9*, 100225.
34. Doyen, P., Hermabessiere, L., Dehaut, A., Himber, C., Decodts, M., Degraeve, T., et al. (2019). Occurrence and identification of microplastics in beach sediments from the Hauts-de-France region. *Environmental Science and Pollution Research International, 26*(27), 28010–28021.
35. Zhang, J., Wang, L., & Kannan, K. (2020). Microplastics in house dust from 12 countries and associated human exposure. *Environment International., 134*, 105314.
36. Gao, W., Zhang, Y., Mo, A., Jiang, J., Liang, Y., Cao, X., et al. (2022). Removal of microplastics in water: Technology progress and green strategies. *Green Analytical Chemistry., 3*(100042), 100042.
37. Liu, L., Xu, M., Ye, Y., & Zhang, B. (2022). On the degradation of (micro) plastics: Degradation methods, influencing factors, environmental impacts. *Science of the Total Environment., 806*, 151312.
38. Li, S., Yang, Y., Yang, S., Zheng, H., Zheng, Y., Jun, M., et al. (2023). Recent advances in biodegradation of emerging contaminants-microplastics (MPs): Feasibility, mechanism, and future prospects. *Chemosphere, 331*, 138776.
39. Yang, J., Yang, Y., Wu, W. M., Zhao, J., & Jiang, L. (2014). Evidence of polyethylene biodegradation by bacterial strains from the guts of plastic-eating waxworms. *Environmental Science and Technology., 48*(23), 13776–13784.
40. Rajandas, H., Parimannan, S., Sathasivam, K., Ravichandran, M., & Yin, L. S. (2012). A novel FTIR-ATR spectroscopy-based technique for the estimation of low-density polyethylene biodegradation. *Polymer Testing., 31*(8), 1094–1099.
41. Auta, H. S., Emenike, C. U., & Fauziah, S. H. (2017). Screening of Bacillus strains isolated from mangrove ecosystems in Peninsular Malaysia for microplastic degradation. *Environmental Pollution, 231*, 1552–1559.
42. Lin, Z., Jin, T., Zou, T., Xu, L., Xi, B., Xu, D., et al. (2022). Current progress on plastic/microplastic degradation: Fact influences and mechanism. *Environmental Pollution, 304*(119159), 119159.
43. Skariyachan, S., Patil, A. A., Shankar, A., Manjunath, M., Bachappanavar, N., & Kiran, S. (2018). Enhanced polymer degradation of polyethylene and polypropylene by novel thermophilic consortia of Brevibacillus sps. and Aneurinibacillus sp. screened from waste management landfills and sewage treatment plants. *Polymer Degradation and Stability., 149*, 52–68.

44. Park, S. Y., & Kim, C. G. (2019). Biodegradation of micro-polyethylene particles by bacterial colonization of a mixed microbial consortium isolated from a landfill site. *Chemosphere, 222*, 527–533.
45. Gao, R., Liu, R., & Sun, C. (2022). A marine fungus Alternaria alternata FB1 efficiently degrades polyethylene. *Journal of Hazardous Materials, 431*(128617), 128617.
46. Paço, A., Duarte, K., da Costa, J. P., Santos, P. S. M., Pereira, R., Pereira, M. E., et al. (2017). Biodegradation of polyethylene microplastics by the marine fungus Zalerion maritimum. *Science of the Total Environment, 586*, 10–15.
47. Williams, J. O., & Osahon, N. T. (2021). Assessment of microplastic degrading potential of fungal isolates from an estuary in rivers state, Nigeria. *South Asian Journal of Research in Microbiology., 9*(2), 11–19.
48. Sowmya, H. V., Ramalingappa, B., Nayanashree, G., Thippeswamy, B., & Krishnappa, M. (2015). Polyethylene degradation by fungal consortium. *International Journal of Environmental Research., 9*(3), 823–830.
49. Cheng, Y. R., & Wang, H. Y. (2022). Highly effective removal of microplastics by microalgae Scenedesmus abundans. *Chemical Engineering Journal, 435*(135079), 135079.
50. Khoironi, A., Anggoro, S., & Sudarno, S. (2019). Evaluation of the interaction among microalgae Spirulina sp, plastics polyethylene terephthalate and polypropylene in freshwater environment. *Inż Ekol., 20*(6), 161–173.
51. Schwibbert, K., Menzel, F., Epperlein, N., Bonse, J., & Krüger, J. (2019). Bacterial adhesion on femtosecond laser-modified polyethylene. *Materials., 12*(19), 3107.
52. Habib, S., Iruthayam, A., Abd Shukor, M. Y., Alias, S. A., Smykla, J., & Yasid, N. A. (2020). Biodeterioration of untreated polypropylene microplastic particles by Antarctic bacteria. *Polymers (Basel)., 12*(11), 2616.
53. Auta, H. S., Emenike, C. U., Jayanthi, B., & Fauziah, S. H. (2018). Growth kinetics and biodeterioration of polypropylene microplastics by Bacillus sp. and Rhodococcus sp. isolated from mangrove sediment. *Marine Pollution Bulletin, 127*, 15–21.
54. Chen, Z., Zhao, W., Xing, R., Xie, S., Yang, X., Cui, P., et al. (2020). Enhanced in situ biodegradation of microplastics in sewage sludge using hyperthermophilic composting technology. *Journal of Hazardous Materials, 384*(121271), 121271.
55. Ghosh, S., Qureshi, A., & Purohit, H. J. (2019). Microbial degradation of plastics: Biofilms and degradation pathways. In *Contaminants in agriculture and environment: Health risks and remediation* (pp. 184–199). Agro Environ Media—Agriculture and Ennvironmental Science Academy.
56. Kothawale, S. S., Kumar, L., & Singh, S. P. (2023). Role of organisms and their enzymes in the biodegradation of microplastics and nanoplastics: A review. *Environmental Research, 232*(116281), 116281.
57. Khan, S., Ali, S. A., & Ali, A. S. (2023). Biodegradation of low density polyethylene (LDPE) by mesophilic fungus `Penicillium citrinum'isolated from soils of plastic waste dump yard, Bhopal, India. *Environmental Technology, 44*(15), 2300–2314.
58. Yan, F., Wei, R., Cui, Q., Bornscheuer, U. T., & Liu, Y. J. (2021). Thermophilic whole-cell degradation of polyethylene terephthalate using engineered Clostridium thermocellum. *Microbial Biotechnology., 14*(2), 374–385.
59. Yuan, F., Yue, L., Zhao, H., & Wu, H. (2020). Study on the adsorption of polystyrene microplastics by three-dimensional reduced graphene oxide. *Water Science and Technology., 81*(10), 2163–2175.
60. Dad, F. P., Khan, W. U. D., Kirkham, M. B., Bolan, N., & Tanveer, M. (2023). Microplastics: A review of their impacts on different life forms and their removal methods. *Environmental Science and Pollution Research International, 30*(37), 86632–86655.
61. Rhein, F., Nirschl, H., & Kaegi, R. (2022). Separation of microplastic particles from sewage sludge extracts using magnetic seeded filtration. *Water Research, X, 17*, 100155.
62. Grbic, J., Nguyen, B., Guo, E., You, J. B., Sinton, D., & Rochman, C. M. (2019). Magnetic extraction of microplastics from environmental samples. *Environmental Science & Technology Letters, 6*(2), 68–72.

63. Park, H. B., Kamcev, J., Robeson, L. M., Elimelech, M., & Freeman, B. D. (2017). Maximizing the right stuff: The trade-off between membrane permeability and selectivity. *Science, 356*(6343).
64. Dey, T. K., Uddin, M. E., & Jamal, M. (2021). Detection and removal of microplastics in wastewater: Evolution and impact. *Environmental Science and Pollution Research International, 28*(14), 16925–16947.
65. Iorhemen, O. T., Hamza, R. A., & Tay, J. H. (2016). Membrane bioreactor (MBR) technology for wastewater treatment and reclamation: Membrane fouling. *Membranes, 6*(2), 33.
66. Ali, S. S., Elsamahy, T., Al-Tohamy, R., & Sun, J. (2024). A critical review of microplastics in aquatic ecosystems: Degradation mechanisms and removing strategies. *Environmental Science and Ecotechnology, 21*(100427), 100427.
67. Talvitie, J., Mikola, A., Koistinen, A., & Setälä, O. (2017). Solutions to microplastic pollution–removal of microplastics from wastewater effluent with advanced wastewater treatment technologies. *Water Research, 123*, 401–407.
68. Adegoke, K. A., Adu, F. A., Oyebamiji, A. K., Bamisaye, A., Adigun, R. A., Olasoji, S. O., et al. (2023). Microplastics toxicity, detection, and removal from water/wastewater. *Marine Pollution Bulletin, 187*(114546), 114546.
69. Perren, W., Wojtasik, A., & Cai, Q. (2018). Removal of microbeads from wastewater using electrocoagulation. *ACS Omega, 3*(3), 3357–3364.
70. Elkhatib, D., Oyanedel-Craver, V., & Carissimi, E. (2021). Electrocoagulation applied for the removal of microplastics from wastewater treatment facilities. *Separation and Purification Technology, 276*(118877), 118877.

Chapter 14
Occurrence and Migration of Plasticizers in River Water: A Review and Bibliometric Analysis

Deep Chakraborty, Apurba Koley, Shromana Dey, Nitu Gupta, Arindam Ghosh, Swapnil Rai, Paromita Chakraborty, Srinivasan Balachandran, and Krishnendu Mukhopadhyay

1 Introduction

Plasticizers, mainly phthalate esters, are widely used in different plastic products because of their flexibility attributes [1]. These substances can be easily detected in almost all ecosystems, such as marine and terrestrial ecosystems, as they tend to leach from plastic materials [2]. To increase the transparency, strength, flexibility, and durability of materials, many synthetic chemical additives, namely, phthalate esters (PAEs), are used. PAEs exhibit diverse physical and chemical properties that enable their multidisciplinary application [3]. Despite these advantages, phthalate

Deep Chakraborty and Apurba Koley contributed equally with all other contributors.

D. Chakraborty (✉) · K. Mukhopadhyay
Department of Environmental Health Engineering, Sri Ramachandra Faculty of Public Health, Sri Ramachandra Institute of Higher Education and Research (Deemed to be University), Porur, Chennai, India
e-mail: deepchakraborty@sriramachandra.edu.in

A. Koley · S. Dey · S. Balachandran
Bioenergy Laboratory, Department of Environmental Studies, Siksha-Bhavana (Institute of Science), Visva-Bharati, Santiniketan, West Bengal, India

N. Gupta
Department of Environmental Science, Tezpur University, Tezpur, Assam, India

A. Ghosh · S. Rai
Department of Environmental Science, Amity School of Life Science, Amity University Madhya Pradesh, Gwalior, India

P. Chakraborty
Centre for Research in Environment, Sustainability Advocacy and Climate Change (REACH), SRM Institute of Science and Technology, Kattankulathur, Tamil Nadu, India

M. K. Jindal et al. (eds.), *Emerging Contaminants in Water*, Emerging Contaminants and Associated Treatment Technologies,
https://doi.org/10.1007/978-3-031-89591-3_14

esters such as di(2-ethylhexyl) phthalate (DEHP) are considered endocrine-disrupting chemicals (EDCs). With minimal dosage, this chemical can disrupt the endocrine systems of aquatic organisms and mammals [2]. Other concerns related to these chemicals include reproductive disorders, birth defects, reduced testosterone levels, the proliferation of testicular Leydig cells, compromised semen quality, lowered serum estradiol levels, thyroid dysfunction, disruptions in the hormonal and endocrine systems, extended estrus and ovulation periods, impaired growth in children, asthma, and an increased risk of breast cancer [4–6].

Generally, the process of plasticization involves reducing intermolecular interactions between polymer molecules, thus decreasing rigidity and enhancing attributes such as flexibility and durability. During this process, the plasticizer is subjected to heating and amalgamated with different polymers, forming a homogenous material that can be transformed into different shapes, sizes, and products [7–9]. The results of the present study indicate greater amounts of DBP, DEHP, and DIBP in both surface sediments and pore water. However, the distribution of phthalate esters (PAEs) varies between urban and nonurban regions. The increased levels of DEHP and DBP in surface sediments pose a significant ecological risk. Similarly, DEHP, DBP, and DIBP represent ecological threats in pore water [10]. Therefore, phthalate esters can be considered major pollutants across various surface water bodies, including rivers, lakes, canals, glaciers, sediments, and even bottled water [7].

Recently, it has been widely explored that all major rivers across the world are contaminated with different classes of plastics. Plastic menace nurtures the hazards associated with plasticizers—chemical substances that are added to plastics for numerous purposes, such as strength, plasticity, durability, and flexibility. Numerous studies have suggested that plastic pollution in river water indicates that one of the most crucial implications of plastics is that these plasticizers are slowly released or degraded in river water over time due to a mix of physicochemical factors in the river [11].

Phthalates in river ecosystems are becoming a serious concern daily because of the ever-growing burden of plastics. Researchers have shown that these chemicals not only contribute to serious health issues [1] but also affect entire river and marine ecosystems in a state of ecological fragility [12]. There is growing concern over the impact of phenolic estrogens in river systems because they present significant environmental risks due to their widespread distribution and threat to biodiversity [13, 14]. In particular, bisphenol-A (BPA) and PAEs can disrupt ecological structures in marine and freshwater ecosystems, increasing their vulnerability. These compounds have a significant impact on various trophic levels, engulfing the entire ecological pyramid in danger. Species such as fish, crustaceans, and amphibians are severely impacted by butyl benzyl phthalate and BPA [15].

Considering the existing research on plasticizers reported to date, a compilation and comprehension of the literature on river water is crucial. This chapter focused on plasticizers, their properties, analytical methods, occurrence, and fate within riverine environments. Furthermore, this chapter contains a bibliometric analysis to examine research trends, key sources, and international collaborations related to plasticizers in river water. The analysis will reveal current research trends, the topic's relevance, and the various methodologies employed over time. The goal is to pinpoint major differences in the suggested approaches and understand how different types of plasticizers occur in river water and their impacts.

2 Characteristics and Properties of the Plasticizers

There are two primary methods for softening polymers to achieve plasticization. First, to increase flexibility, internal plasticization entails chemically altering the polymer or monomer. Second, external plasticization, which is more common due to its lower cost and versatility, involves the addition of a plasticizing agent to the polymer (Table 14.1). This approach allows manufacturers greater flexibility in creating formulations for various products [18].

Table 14.1 Characteristics and properties of plasticizers [16, 17]

Name of the plasticizers	Character	Highly recommended for these polymers	Used in
Abietates	–	–	Caulks, adhesives, and varnishes
Adipates	Melting point: 70 °C Boiling point: 509 °C	CAB, PVC, CN, CP, PS, PVAC	Water pipe, tank liners, nail care, food wrap, cable, and wire
Alkyl sulfonates	Pour point: −32 °C	CR, IIR, NR, PVC, SBR	Waterbed, goof fusion properties, high resistance to hydrolysis
Amides and amines	Pour point: - 2–3 °C	Starch[18,19] and acrylics[20]	Acrylic adhesives and tapes, biodegradable articles
Azelates	Boiling point: 278 °C	PS, PVC, cellulosics	Synthetic lubricants, food film
Benzoates	Pour point: −40 °C Boiling point: 323 °C	EVA, PVA, PU, PVC, PVAc	Caullks, sealants, melt adhesives, gaskets, binding agents
Bioplasticizers	Pour point: −20 °C	PHB, PLA, PVC, chitosan	Medical adhesive
Biodegradable plasticizers	–	PLA,	Model articles, film
Chlorinated paraffins	–	PU, PVC	Flexible hose, paints, sealants, cable insulation
Citrates	Melting point: 294 °C	CN, PVDC, PVC, PVB, PU	Bottle cap, cling, film, lacquers, cigarette filters
Cyclohexane dicarboxylate	–	PVC	Plastisols
Cyclohexane dicarboxylic acid, diisononyl ester	Boiling point: 250 °C	PVC	Sports and leisure products, toys

(continued)

Table 14.1 (continued)

Name of the plasticizers	Character	Highly recommended for these polymers	Used in
Energetic plasticizers (bis(2,2-dinitropropyl) formal)	–	Hydrocarbon resin	Explosives, pyrotechnics
Epoxides		PVA, PVB, EC	Food, pipes, blood bags, upholstery, foils, films, cables
Esters of C10–30 Dicarboxylic Acids		PVC	–
Ether-ester plasticizers	Boiling point: 344 °C	EVA, PU, CR, CTA, acrylics	Glues, automotive, floor tiles, paint
Glutarates	Boiling point: −280 °C	Acrylics, SBR, CP, CN, NBR, PVC	Coating, cosmetics
Hydrocarbon oils (paraffinic oils)	Boiling point: Above 250, dec	PVC, EP, SBR, EPDM	Medical, road marking paints
Isobutyrates	Boiling point: 288 (decomp)	CTA, CAB, CN, CA	Toys, printing inks, beverage, hair care products
Maleates	–	–	Textiles, latex paint
Oleates	–	PVC	Flooring, sheet, films, inks
Pentaerythritol derivatives	–	PEI, PVC	Plastisols, thin walls
Phosphates	Boiling point: 420 °C	SBR, PU, PP, PF, PA, PE, PVC, PVA, PS	Wire, cable, tubing, sealants, lacquers, sheet goods, printed inks
Phthalate-free plasticizers	Boiling point: 375 °C	PVC	Sports and leisure, toys, food containers
Phthalates	–	CN, CA, PVAc, EC, PMMA, PS	Nail care, food wrapper, sheeting, tubing, carpet backing, foam
Esters	Pour point: −1 °C Boiling point: 398.9 °C	CAB, PVC-PVA, EC, NBR	Upholstery, footwear, print roll, automotive parts, o-rings, tank liners
Polybutenes	–	PP, PE, SBR, PS, polyterpene	Cables, films, coating, sealants
Sebacates	Boiling point: 345 °C	PVB, CN, EC, PVC, PMMA	Pharmaceutical coating, floorings, aircraft and car interior, paints, food wrap
Sulfonamides (p-toluenesulfonamide)	Boiling point: 350 °C	PC, PA11, PA6, PA12, urea	Automotive, films, fishing, medical device
Superplasticizers and plasticizers for concrete (β-naphthalene sulfonate)	Boiling point: 105 °C	–	Concrete industry
Tri- and pyromellitates	Boiling point: 430 °C	CAB, PS, PVC, EC, PMMA, CN	Upholstery, cable and wire, film, artificial leather

Phthalate esters, particularly DEHP, DIDP, DINP, and DITDP, are the most widely used plasticizers globally. DEHP is the most prevalent phthalate among the more than 30 different phthalates available. Phthalates have been widely employed in PVC, thermoplastic ester molding compounds, and other vinyl chloride copolymers for more than 60 years. Phthalates constitute 92% of the plasticizers produced worldwide, with DEHP alone accounting for 51% of phthalate usage. Phthalates are favored for their combination of desirable plasticizer properties, including minimal interaction with resins at room temperature, good fusion properties, satisfactory cable insulation, high elasticity with adequate cold strength, low volatility under ambient conditions, and cost-effectiveness [19, 20]. The polarizable benzene ring in phthalates enhances their similarity with that of PVC, providing significant flexibility to the polymer chains. However, as the length of the disubstituted alkyl ester chains increases, the compatibility decreases. Phthalates with shorter alkyl chains are easier to formulate because of their faster diffusion rates, but they are also more volatile. Increasing branching reduces the plasticizing efficiency, particularly when the main chain is reduced or when the branches are nearer the polar group. This reduction in effectiveness is accompanied by an increase in viscosity, highlighting the close relationship between viscosity and plasticizer efficiency. Although terephthalates, oligoesters of o-phthalic acids, and solid phthalate esters such as dicyclohexyl phthalate and diphenyl phthalate possess some plasticizing properties, their use is limited due to high costs [21].

3 Methods Used for Determination and Quality Control of Plasticizers

Numerous technological obstacles have been solved as the plasticizer business has developed to improve formulations, handle end-use concerns, and satisfy new demands. Among the main barriers are the volatility or UV sensitivity of traditional plasticizers, which causes them to evaporate or degrade; leaching into polar and nonpolar liquids; migration to other polymeric materials or environments; inadequate lubrication at extremely low temperatures; and possible carcinogenic effects on living things. Numerous plasticizer testing methods are standardized at both the national and international levels. Current efforts toward harmonizing testing procedures are reflected in the documents provided by the International Organization for Standardization and ASTM International, which outline the primary analytical methods [21]. The methods are summarized in Table 14.2.

A standardized marking system for plastics, as per ASTM standards, indicates the amounts of additives but does not include a specific symbol for plasticizers. In highly crosslinked thermoplastic vulcanizates, the "other" category can be used to denote the plasticizer concentration if necessary. The standard classification allows for a comprehensive marking system that includes composition, properties, and volatile loss, which is indicated with a numerical value after the symbol "R" [32, 34, 40, 42].

Table 14.2 Methods for occurrence and quality analysis of plasticizers [16, 17]

Parameter	Methods used	Reference
Color	• ASTM method of plasticizer testing • Transparent liquid color has been determined by Gardner color scale • Saybolt chromometer which is designed to measure the color of petroleum products can also be used for mineral oils	[22–24]
Compatibility	• The evaluation of compatibility of plasticizers is generally conducted under humid conditions as well. In this assessment, samples measuring 25 cm^2 and 0.75 mm in thickness are suspended above water within sealed containers at temperatures of either 60 or 80 °C • Plasticizer spew is assessed by bending samples 180°, which can reveal decreased compatibility under compressive stress	[25, 26]
Purity	• Plasticizer purity is assessed by measuring ester content through titration as per protocol • The pureness of monomeric plasticizers can be determined using GC-FID or TCD. Capable of withstanding temperatures between 250 and 350 °C, are used for adequate component separation	[27, 28]
Specification	• The ISO standard 135 provides specifications for test methods applicable to phthalate plasticizers. This standard covers the measurement of water content, color, density, and other physical parameters	[29]
Specific gravity	• The specific gravity of plasticizers can be determined using several methods, including hydrometers, digital density meters, displacement methods, and pycnometers	[27, 30–33]
Staining	• Antioxidants migrating from rubber can stain PVC immediately or under UV exposure. Rubber specimens are pressed against PVC specimens with a 10 g/cm^2 pressure and then heated in a 70 °C open-air oven for 20 h	[34]
Residual contamination	• For quantitative measurement of the less volatile and semivolatile, a procedure was developed for plasticizers and residual detergents on cleaning materials that could contaminate substrates • The same study was done using a different method that involved ultrasonic extraction, materials being held in a test fluid in an ultrasonic bath for a specific amount of time and a regulated temperature. Operating circumstances and bath type have a big influence on the outcome. The extracted material can be analyzed using methods like gravimetric after solvent evaporation, liquid chromatography, or total carbon analysis	[32, 35, 36]
Saybolt viscosity	• The efflux time of 60 ml samples is recorded through a calibrated orifice in the Saybolt viscosity measurement, which is used for petroleum liquids between 21 and 99 degrees centigrade. The liquid flows into a 60 ml graduated flask while the viscometer is submerged in a thermostatic bath. The result is expressed in Saybolt viscosity units by multiplying the recorded duration by a correction factor derived from a viscosity standard	[37, 38]

(continued)

Table 14.2 (continued)

Parameter	Methods used	Reference
Relative volatility	• Plasticizers relatively volatile are determined by placing them on a rotating table in circulating air at over 105 or 155 °C. After 24 h later, samples are weighed to calculate the weight loss as percentage or average rate. Temperature control and airflow control are critical and comparing different plasticizers in the same experiment helps minimize experimental variability • Activated carbon process can be used for loss of volatiles • When observing volatile loss from calendared, extruded, and cast films and sheets, the direct activated carbon approach is employed. A graph showing the maximum volatile loss as a function of film thickness is also included in the standard	[32, 37–39]
Concentration of water	• Karl Fischer titration can detect free and bound water in solid and liquid samples, making it suitable for plasticizer testing	[27, 39]
Viscosity of plastisols and organosols	• Organosol and plastisol viscosity is determined using an extrusion viscometer operating at high shear rates. A sample is placed in the extrusion chamber after being preconditioned to the right temperature. Viscosity is computed using the recorded extrusion pressure and flow rate through the orifice	[35, 36]
Refractive index	• The refractive index of plasticizers is determined using a refractometer with a sodium spectral line of 589.3 nm at 25 °C. This method was initially designed for electrical insulating liquids	[27, 40]
Powder-mix time	• In general-purpose PVCs, mixing time and reaction to various plasticizers are measured by a torque rheometer. According to the standard, powder-mix time falls with process temperature and rises with plasticizer viscosity	[41]

4 Occurrence of Plasticizers in River Water

Plasticizers, particularly BPA and phthalates, enter river systems from a variety of sources. Key pathways include the leaching of phthalates from urban plastic waste into drainage networks and the degradation of plastic containers and bottles used in household activities such as washing, cooking, and drinking, which release phthalates into the soil, water, and air. The most common pipes used in water supply and distribution networks are made of polyvinyl chloride (PVC), which is a significant source of phthalates that eventually enter rivers [43]. Rivers are also contaminated by plastic waste products, such as bags, packaging materials, and other discarded items, and these plastics contain macroplastics and microplastics loaded with organic plastic additives such as phthalic acid esters (PAEs). Industrial activities, biomedical equipment, and pharmaceutical products can further exacerbate this issue by discharging phthalate-rich wastewater into rivers. The manufacturing and disposal of kid toys involves the release of harmful plastic fragments such as phthalates, which can be released into aquatic environments. Furthermore, polymer production units, which use high concentrations of BPA and phthalates, significantly

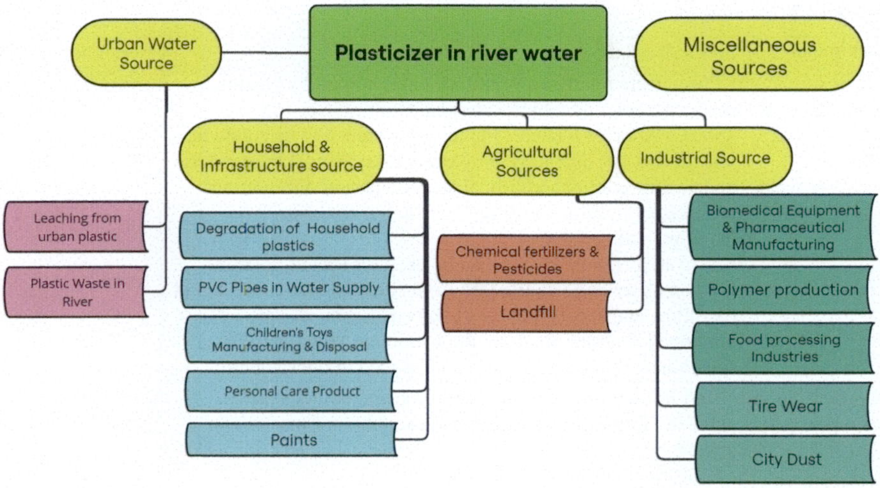

Fig. 14.1 Sources of plasticizers in river water

contribute to river pollution [11]. The food processing industry can be another source of river contamination by plasticizers (Chakraborty et al., 2021). Agricultural runoff containing chemical fertilizers, pesticides, herbicides, and fungicides serves as an additional pathway for phthalates to enter rivers (Fig. 14.1).

Some studies have explored the impact of seasonal changes that affect plasticizer levels in coastal waters and sediments. Rainfall can dilute plasticizer concentrations but also increase their transport from rivers to estuaries through higher flow, sediment resuspension, and runoff. Therefore, rainfall was considered to be one of the key factors. Storms may increase sewage discharge into rivers, and increased river discharge can disperse contaminated sediments away from the coast, which can potentially reduce estuarine plasticizer levels. Seasonal human activities such as industrial production, tourism, and shipping also influence plasticizer occurrence throughout the year [44–46].

5 Migration of Plasticizers in River Water

Various studies have shown that pollutants containing plasticizers, especially phthalates and bisphenol A, are capable of entering river systems via different transport mechanisms, such as advection and dispersion-diffusion, as well as reaction processes, such as adsorption, abiotic transformation, and biodegradation. Phthalates are capable of bioaccumulating at higher trophic levels through complex biochemical pathways. Therefore, these chemical additives pose significant ecological threats to aquatic life. The fate of plasticizers is largely governed by their hydrophobicity, which influences their partitioning between sediment and water in estuarine environments. The "estuarine filter" effect explains why an increase in salinity enhances the sorption of contaminants to sediments, playing a critical role in this process [47].

5.1 *Plasticizers Into the Environment*

Phthalates are ubiquitous environmental contaminants due to their persistence and widespread distribution. Primary sources of phthalates are associated with wastewater treatment plants. However, these compounds are inefficient at removing them from effluent and landfills. Therefore, phthalate-containing plastics can leach into groundwater and surface water, causing pollution. Various studies have reported phthalate concentrations in different ecosystems, such as surface water, marine environments, and soil. Detection of di(2-ethylhexyl) phthalate (DEHP) in river sediments at a concentration of 110 mg/L. The persistence of phthalates in these environments is due to their long relative half-lives, which range from several hours to months [48].

5.2 *Persistence of Plasticizers in Sediments and Water*

The study of plasticizer degradation in sediments and water has focused entirely on phthalates. Plasticizers receive attention because of their persistent environment. The persistence of plasticizers in the environment is strongly influenced by their physiochemical properties, especially the $\log K_{OW}$ values. Plasticizers with higher $\log K_{OW}$ values tend to bind more effectively to sediments and decrease their susceptibility to biodegradation [49]. Various environmental factors, such as temperature, pH, UV light exposure, dissolved organic matter, and oxygen levels, play crucial roles in determining how plasticizers degrade and persist in sediment and water environments. The presence of indigenous microbial communities within these environments also influences the rate of plasticizer degradation. These factors have been thoroughly scrutinized in soil ecosystems but have been less explored in estuarine and coastal ecosystems [50].

Biodegradation is the main process involved in the breakdown of plasticizers in estuarine sediments. The relative degradation rates of plasticizers can be influenced by the chain length and molecular weight. Longer-chain plasticizers are more resistant to degradation because of their greater hydrophobicity. Therefore, they are more susceptible to photolysis, hydrolysis, or biodegradation. In a laboratory study on the aerobic degradation of DnBP and DEHP in mangrove sediments, the half-lives were found to be 1.6–2.9 days for DnBP (C4 chains) and 5–8.3 days for DEHP (C8 chains) [44, 51].

5.3 Uptake by Organisms

Plasticizers in the environment can be degraded or taken up by organisms via a significant pathway. Ecotoxicological studies have shown that phthalate plasticizers can exhibit toxicity and endocrine-disrupting effects. These compounds have adverse effects on the metabolism and reproduction of marine species such as fish and mussels. When determining the danger of phthalates in estuarine and coastal habitats, it is essential to consider how these compounds are absorbed and accumulated by marine creatures. Phthalates are found in many marine animals, including fish, mollusks, crabs, and turtles, and plants. They also accumulate in adipose and muscle tissues, which are the main storage locations. High-molecular-weight phthalates such as DnBP and DEHP are commonly found at relatively high concentrations in plants and the fat and muscle tissues of cetaceans, mollusks, and turtles. However, phthalate profiles in crustaceans and fish are more evenly distributed across different molecular weights. These differences in phthalate profiles may be attributed to factors such as feeding behavior, life history, tissue composition, or species-specific metabolic capacities [44].

5.4 Migration into the Foods

Food components, especially fats, which migrate into plastics such as PE or PP can significantly increase the mobility of plastic components, thereby increasing their migration into the food contained within them [52]. In food packaging, plasticizers can migrate from the wrapping material into the food, becoming indirect food additives. However, testing with actual foods is challenging because of the difficulty in quantifying small amounts of migrating substances and the vast variety of foods, making comprehensive testing nearly impossible [53]. Glass jars with metal twist closures are widely used for a variety of items, including jams, pickles, spaghetti sauce, and fish. These lids use plasticized PVC gaskets to seal against the glass rim. Migration can also be significant in products with lower fat contents, such as infant food (1–2% fat), especially if the food is heated after filling. In contrast, less than 3 mg/kg ESBO was detected in aqueous foods such as jam, honey, asparagus, and pickles, as well as in foods that do not contact the lid, such as mustard or firm mayonnaise [54.

5.5 Impact of Plasticizer Migration

Plastics include chemicals that can be released during manufacturing, usage, disposal, and recycling, contaminating a variety of items and the environment. These chemicals are resistant to chemical, biological, and photolytic degradation. This

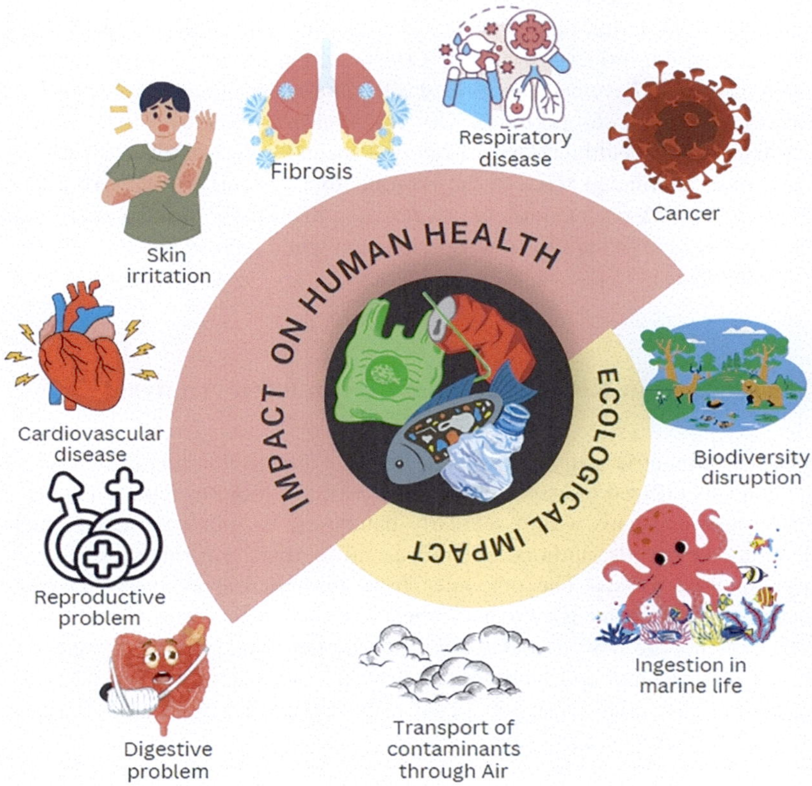

Fig. 14.2 Impact of plasticizer migration on ecology and human health

allows them to persist for extended periods and readily accumulate in living organisms [55]. The impacts of plasticizer migration on ecology and human health are described in Fig. 14.2.

Phthalate esters are considered endocrine-disrupting compounds (EDCs) that cause serious health issues when exposed for long periods. These include reproductive problems, birth defects, reduced testosterone, altered semen quality, thyroid disruption, hormonal imbalances, and an increased risk of conditions such as breast cancer and asthma. Phthalates can also affect the cellular targets associated with aryl hydrocarbon, androgen, and insulin receptors [3]. Upon entering the body, ortho-phthalate diesters can be quickly metabolized into phthalate monoesters, which are excreted in body fluids. These monoesters, which have one ester and one carboxylic acid group on the aromatic ring, are common degradation products. Additionally, these compounds have gained attention for further biomonitoring and epidemiological studies [48].

The harmful effects of BPA in water can affect both vertebrates and invertebrates. BPA has been linked to inflammatory bowel disorders and other digestive problems in people. As an endocrine disruptor, BPA has been found in body fluids such as urine, blood, and plasma at levels exceeding safe limits. Its presence and endocrine-disrupting effects have also been studied in marine and aquatic ecosystems [44–46]. PAEs have been confirmed as hazardous pollutants that contribute to indoor, water, air, and soil pollution. Growing concerns exist over their impact on human and animal health. Phthalate esters, especially di(2-ethylhexyl) phthalate (DEHP), pose significant risks to the reproductive system, particularly affecting sperm production. A recent study in rats revealed that DEHP reduces sperm motility, viability, nuclear maturation, and sperm count while increasing DNA damage in the sperm nucleus [47, 48].

6 Plasticizers in River Water: A Bibliometric Analysis

Currently, bibliometric analysis has gained crucial attention in literature. It is a statistical analysis utilized to assess publication patterns, monitor research areas, and identify trends and gaps within scholarly databases. Various parameters, such as publications, keywords, authors, countries, and journals, are important for determining literature trends. The outcomes from bibliometric analysis can provide researchers with a better idea to plan their research. This chapter's bibliometric analysis explains the risks associated with plasticizers and highlights the need for additional in vivo data and environmental toxicological studies [49, 50]. In this study, bibliometric analysis is conducted to investigate research trends related to plasticizers in river water.

6.1 Methodology

The literature search was conducted from the Web of Science (WoS) database [51, 52] via the advanced search option with the keywords "TS = (Plasticizers)" and "TS = (River Water)." A total of 404 articles were exported from WoS during the period 1991–2004. These articles present the contributions of 1774 authors from different countries. The dataset comprises detailed information on the annual growth rate, average citations per year, author collaborations, abstracts, and types of documents obtained, as summarized in Fig. 14.3.

The articles were analyzed for keyword co-occurrences via VOSviewer (version: 1.6.20), a software tool designed for visualizing scientific literature. VOSviewer, developed by the Centre for Science and Technology Studies at Leiden University in the Netherlands, was employed to facilitate this analysis [53–56]. The research trends, most relevant sources, and country collaborations have been analyzed via the Biblioshiny package [69, 70] of RStudio (Version 2023.06.1 Build 524) [71].

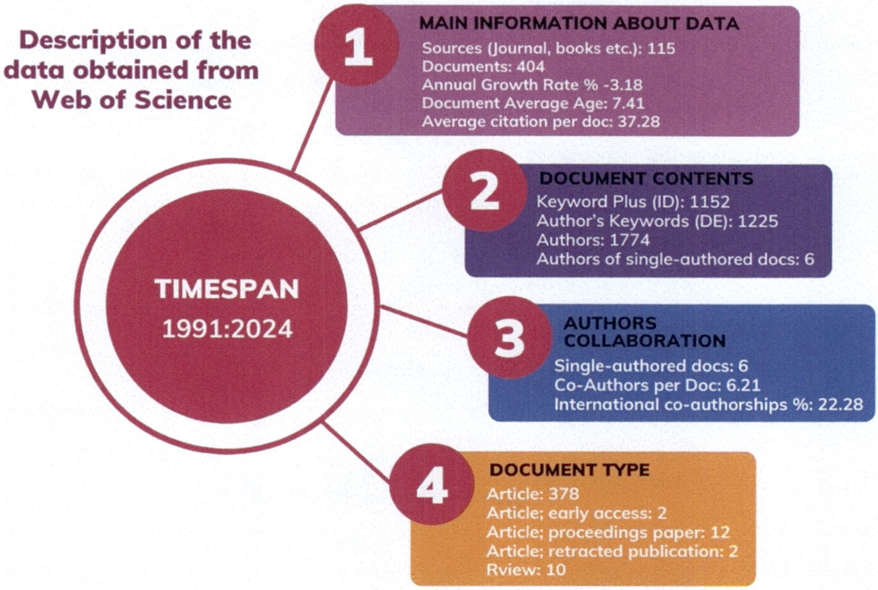

Fig. 14.3 Description of the data obtained from the Web of Science database via the keywords "Plasticizers" and "River Water"

6.2 *Co-occurrence Network*

In this bibliographic study, 404 articles were sourced from the Web of Science via combinations of keywords such as plasticizer and river water terms. The keywords were analyzed via VOSviewer software. Co-occurring keywords, selected up to five times, were used to create a fractionalization network map, illustrating the most frequently searched terms within these articles, as shown in Fig. 14.4. The analysis revealed three distinct clusters in the keyword co-occurrence data. Each cluster and its associated keywords are represented by unique sizes and colors, with straight lines indicating their connections. This keyword network analysis revealed that these articles focused predominantly on the following five clusters or themes: Cluster 1 (red color) indicates that plasticizers in river sediment and surface water pose significant risks to drinking water safety, contributing to human exposure through contamination. Assessing the toxicity of these chemicals is crucial for understanding potential health impacts and mitigating risks. Cluster 2 (green color), related to plasticizers, particularly phthalate esters, enters rivers and soils, where their degradation and biodegradation processes produce metabolites that can pollute groundwater. These acid ester compounds, often associated with polycyclic aromatic hydrocarbons (PAHs), persist in the environment, leading to long-term

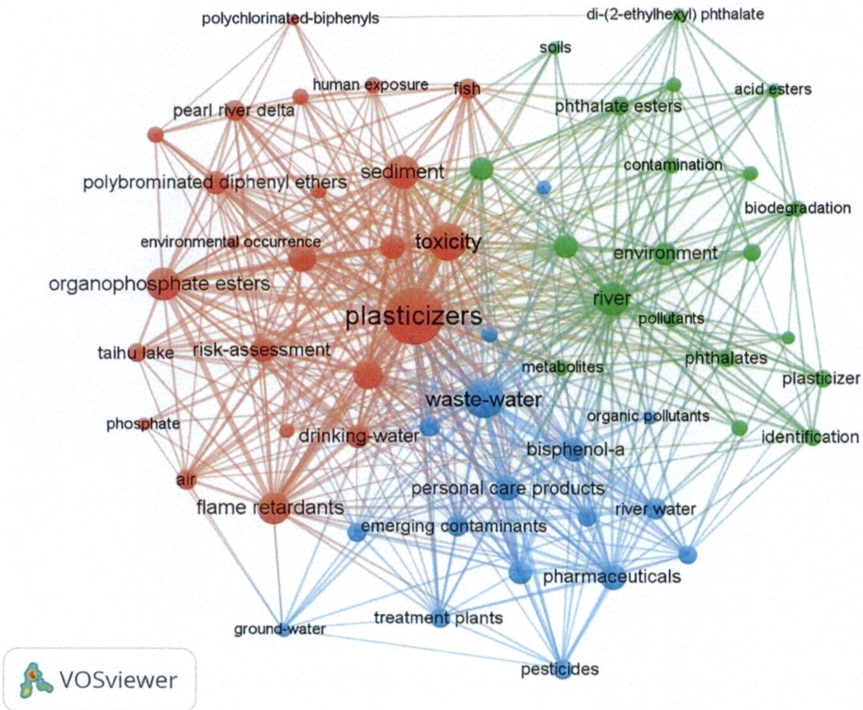

Fig. 14.4 Co-occurrence network map

contamination concerns. Cluster 3 (blue color) indicates that plasticizers such as phthalates, along with organic pollutants such as bisphenol, pesticides, pharmaceuticals, and personal care products, are emerging contaminants in river water. These substances challenge conventional water treatment processes, necessitating advanced methods to ensure safe drinking water and protection of aquatic ecosystems.

6.3 Annual Scientific Production

The data on plasticizer publications reveal a generally increasing trend from 2000 to 2023, with notable peaks in recent years. From 2000 to 2010, the publication count remained relatively low, fluctuating between two and nine records per year. A steady increase started approximately 2011, reaching a significant rise from 2016 onward (Fig. 14.5). Initially, the number of publications was relatively low, but there was a marked rise starting in approximately 2018, indicating a growing interest or new

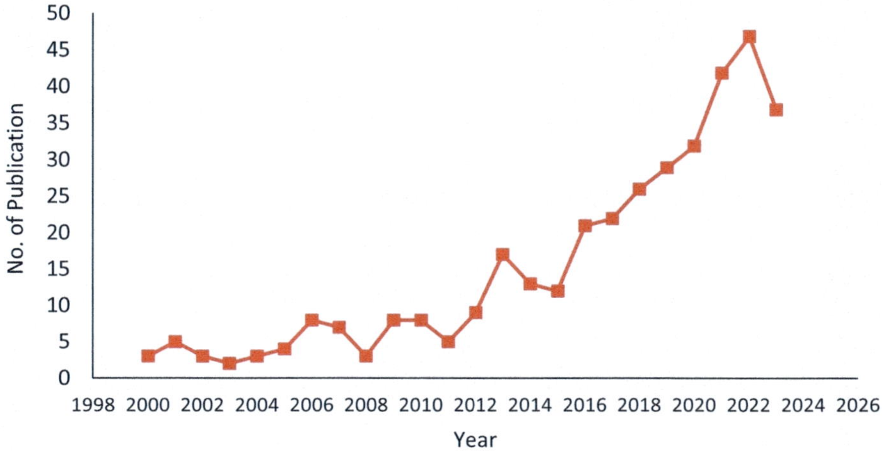

Fig. 14.5 Annual publication counts on plasticizers based on the Web of Science database

developments in the field. The decline observed in 2023 compared with 2022 might be attributed to factors such as publication cycles or temporary shifts in research focus. This decrease could be short-term, with the potential for a rebound if new trends, regulatory changes, or emerging topics drive renewed interest in plasticizers. Such fluctuations are common in research fields and often reflect broader trends or evolving scientific and regulatory landscapes.

6.4 Most Relevant Sources

Various publishers have contributed to the literature on plasticizers and river water, with publications spanning a range of document types, including articles, early access papers, proceedings papers, and reviews. Among these, *Science of the Total Environment* has the highest number of publications, with 57 articles. This is followed by *Chemosphere* and *Environmental Pollution*, each with 26 articles, and *Environmental Science and Pollution Research*, with 20 articles (Fig. 14.6).

Additionally, journals such as the *Journal of Chromatography*, *Journal of Hazardous Materials*, and *Marine Pollution Bulletin* each published 15 articles on the topic during the examined period. Other significant contributions come from *Water Research* with 14 articles, *Environmental Science & Technology* with 11, and *Environment International* with 10 publications. Furthermore, *Environmental Research*, *Analytica Chimica Acta*, *Environmental Toxicology and Chemistry*, and *Environmental Monitoring and Assessment* also show a notable number of publications, highlighting the broad interest and ongoing research efforts in this field across multiple prominent journals.

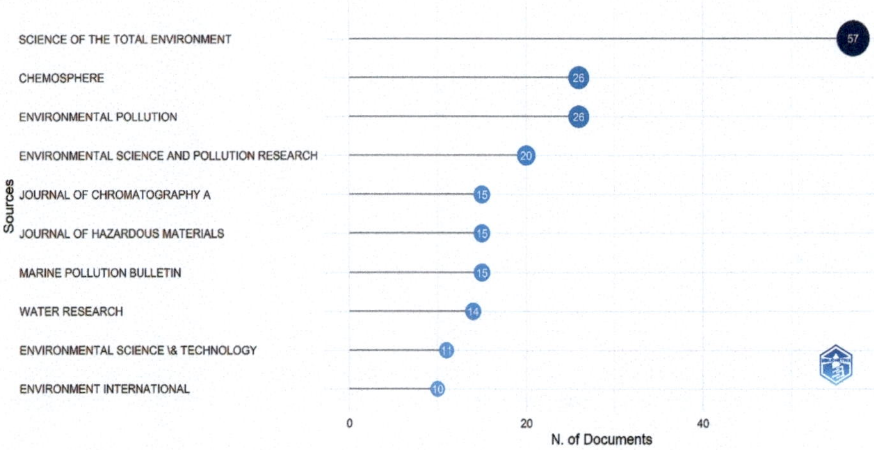

Fig. 14.6 Most relevant sources of documents published on the topics of plasticizers and river water

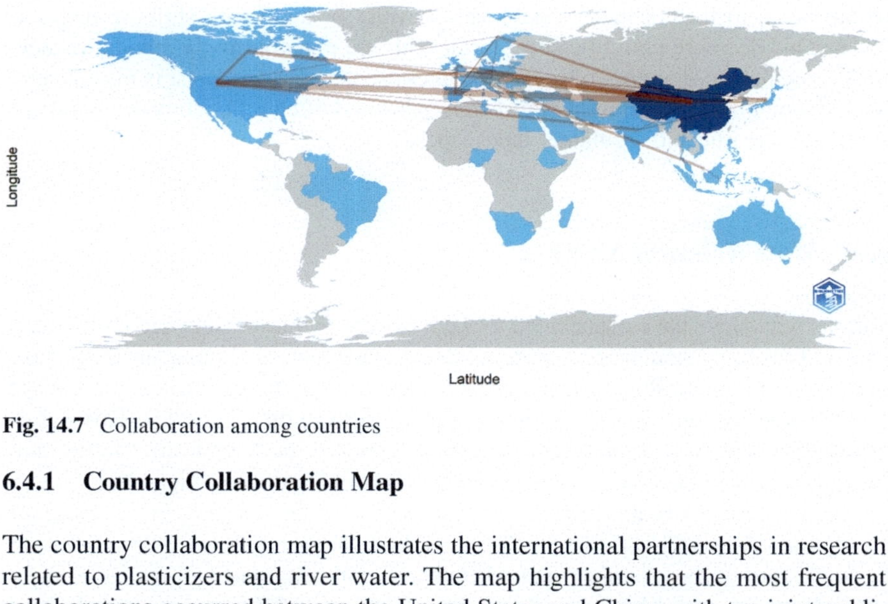

Fig. 14.7 Collaboration among countries

6.4.1 Country Collaboration Map

The country collaboration map illustrates the international partnerships in research related to plasticizers and river water. The map highlights that the most frequent collaborations occurred between the United States and China, with ten joint publications. This is followed by collaboration between China and Germany, with seven publications; China and Japan, with five; and China and Canada, with four. Additionally, the map shows that there were three collaborative publications each between the United States and India, the United States and Canada, and Spain and the United Kingdom. These data underscore the prominent role of China in collaborative research on this topic, often partnering with major research hubs such as the United States, India, and other European countries (Fig. 14.7).

7 Conclusion

The predominant transport pathways of various plasticizers into river systems must be investigated, with greater emphasis on their ecotoxicological implications for the aquatic life of riverine ecosystems. Ecotoxicological assessments carried out in recent years have confirmed that proper disposal mechanisms at various levels in municipal and solid waste management systems are lacking. Owing diligence toward plastic waste processing, units where many of these phthalates are generated must be provided. Owing to their high carcinogenicity and ability to affect many biochemical pathways in the human body, plasticizers such as phthalates are highly dangerous. Phthalates such as DHP and DEHP ultimately make their way into the human body through fishes (a process called biomagnification), eventually reaching carcinogenic concentrations. According to this review, there is an urgent need to implement stringent standards for different plasticizers, as these pollutants are ubiquitous in nature. There is also a gap in proper policy for the use of different plasticizers in different industrial segments. The proper mechanism of action of plasticizers inside the animal and human body is still unclear. Thus, there is considerable scope of scientific studies on the biochemistry of plasticizers in the future. Finally, there is an urgent need for source identification of different plasticizers, which would be possible with the collaboration of national and international policy makers.

References

1. Mondal, R., Chakraborty, D., & Majumdar, D. (2022). Phthalate esters from packaged milk and associated human health risk: A Monte Carlo probabilistic simulation approach. *Mapan, 37*(2), 409–419. https://doi.org/10.1007/s12647-022-00531-y
2. Sirivithayapakorn, S., Thuyviang, K., & Jansak, P. (2015). Distribution and ecological risk assessment of plasticizers in the Chao Phraya Delta, Gulf of Thailand. *ASEAN Engineering Journal, 4*(1), 73–80.
3. Weizhen, Z., Xiaowei, Z., Peng, G., Ning, W., Zini, L., Jian, H., & Zheng, Z. (2020). Distribution and risk assessment of phthalates in water and sediment of the Pearl River Delta. *Environmental Science and Pollution Research, 27*, 12550–12565.
4. Casals-Casas, C., & Desvergne, B. (2011). Endocrine disruptors: From endocrine to metabolic disruption. *Annual Review of Physiology, 73*(1), 135–162.
5. Foster, P. M., Mylchreest, E., Gaido, K. W., & Sar, M. (2001). Effects of phthalate esters on the developing reproductive tract of male rats. *APMIS, 109*(S103), S272–S2S7.
6. Lovekamp-Swan, T., & Davis, B. J. (2003). Mechanisms of phthalate ester toxicity in the female reproductive system. *Environmental Health Perspectives, 111*(2), 139–145.
7. Aslam, H. M., Qadir, A., Ahmad, A., Aslam, M., Mumtaz, M., & Ahmad, S. R. (2022). Impacts of plasticizers on riverine ecological integrity in context to sustainability challenges. In *Microplastic pollution: Environmental occurrence and treatment technologies* (pp. 323–346). Springer International Publishing.
8. Bocqué, M., Voirin, C., Lapinte, V., Caillol, S., & Robin, J. J. (2016). Petro-based and bio-based plasticizers: Chemical structures to plasticizing properties. *Journal of Polymer Science Part A: Polymer Chemistry, 54*(1), 11–33.

9. Muobom, S. S., Umar, A. M., Soongseok, Y., & Brolin, A. P. (2020). A review on plasticizers and eco-friendly bioplasticizers: Biomass sources and market. International Journal of Engineering Research, 9(5), 1138–1144.
10. Liu, Y., He, Y., Zhang, J., Cai, C., Breider, F., Tao, S., & Liu, W. (2020). Distribution, partitioning behavior, and ecological risk assessment of phthalate esters in sediment particle-pore water systems from the main stream of the Haihe River, Northern China. *Science of the Total Environment, 745*, 141131.
11. Mukhopadhyay, M., Sampath, S., Muñoz-Arnanz, J., Jiménez, B., & Chakraborty, P. (2020). Plasticizers and bisphenol A in Adyar and Cooum riverine sediments, India: Occurrences, sources and risk assessment. *Environmental Geochemistry and Health, 42*(9), 2789–2802. https://doi.org/10.1007/s10653-020-00516-3
12. Vignesh, E. R., Gireeshkumar, T. R., Arya, K. S., Nair, M. M., Rakesh, P. S., Jayadev, B. S., & Shirin, P. P. A. (2025). Phthalic acid esters as an ecological hazard to the coral reef ecosystems: A case study from the coral reef waters of the Lakshadweep Archipelago, Arabian Sea. *Marine Pollution Bulletin, 215*, 117866. https://doi.org/10.1016/j.marpolbul.2025.117866
13. Chae, Y., & An, Y. J. (2017). Effects of micro-and nanoplastics on aquatic ecosystems: Current research trends and perspectives. *Marine Pollution Bulletin, 124*(2), 624–632.
14. Guzzetti, E., Sureda, A., Tejada, S., & Faggio, C. (2018). Microplastic in marine organism: Environmental and toxicological effects. *Environmental Toxicology and Pharmacology, 64*, 164–171.
15. Oehlmann, J., Schulte-Oehlmann, U., Kloas, W., Jagnytsch, O., Lutz, I., Kusk, K. O., Wollenberger, L., Santos, E. M., Paull, G. C., Van Look, K. J., & Tyler, C. R. (2009). A critical analysis of the biological impacts of plasticizers on wildlife. *Philosophical Transactions of the Royal Society B: Biological Sciences, 364*(1526), 2047–2062.
16. Wypych, G. (2017). Plasticizer types. In *Handbook of plasticizers* (3rd ed., pp. 1–7).
17. Wypych, G. (2017). Typical methods of quality control of plasticizers. In *Handbook of plasticizers* (pp. 85–109). ChemTec Publishing.
18. SUSlick, K. S. (1998). Kirk-Othmer encyclopedia of chemical technology. *J. Wiley & Sons: New York, 26*, 517–541.
19. Gachter, R., & Muller, H. (Eds.). (1990). *Plastics additives handbook* (3rd ed.). Carl Hanser.
20. Murphy, J. (Ed.). (2001). *Additives for plastics handbook*. Elsevier.
21. Rahman, M., & Brazel, C. S. (2004). The plasticizer market: An assessment of traditional plasticizers and research trends to meet new challenges. *Progress in Polymer Science, 29*(12), 1223–1248.
22. ASTM. (2018). D6864-11 Standard specification for color and appearance retention of solid colored plastic siding products.
23. ASTM. (2010). D1544-04 Test method for color of transparent liquids (Gardner Color Scale).
24. ASTM. (2003). D156-15 Standard test method for saybolt color of petroleum products (saybolt chromometer method).
25. ASTM. (2019). D2383-14 Standard practice for testing plasticizer compatibility in poly(vinyl chloride) (PVC) compounds under humid conditions.
26. ASTM. (2022). D3291-11 Standard practice for compatibility of plasticizers in poly (vinyl chloride) plastics under compression.
27. ASTM. (2018). D891-09 Test methods for specific gravity, apparent, of liquid industrial chemicals.
28. ASTM. (2021). D3465-14 Standard test method for purity of monomeric plasticizers by gas chromatography
29. ISO 1385-1:1977. (1977). Phthalate esters for industrial use—Methods of test—Part I: General. Part II: Measurement of color after heat treatment (Diallyl phthalate excluded). Part II: Determination of ash. Part IV: Determination of acidity to phenolphthalein.
30. ASTM. (2020). D792-20 Standard test methods for density and specific gravity (relative density) of plastics by displacement.

31. ASTM. (2019). D4659-19 Standard test methods for polyurethane raw materials: Determination of specific gravity of isocyanates.
32. ASTM. (2016). D1593-13 Standard specification for nonrigid vinyl chloride plastic film and sheeting.
33. ASTM. (2023). D287-12b Test method for API gravity of crude petroleum and petroleum products (hydrometer method).
34. ASTM. (2000). D2151-95 Standard practice for staining of poly(vinyl chloride) compositions by rubber compounding ingredients (withdrawn).
35. ASTM. (2009). D1823-95 Standard test method for apparent viscosity of plastisols and organosols at.
36. ASTM. (2009). G136-03 Standard practice for determination of solution residual contaminants in materials by ultrasonic extraction.
37. ASTM. (2013). D2288-91 Standard test method for weight loss of plasticizers on heating.
38. ASTM. (2013). D88-07 Standard test method for saybolt viscosity.
39. ASTM. (2019). D1203-10 Standard test methods for volatile loss from plastics using activated carbon methods.
40. ASTM. (2005). D1807-00 Test methods for refractive index and specific optical dispersion of electrical insulating liquids.
41. ASTM. (2012). D2396-94 Standard test methods for powder-mix time of poly(vinyl chloride) (PVC) resins using a torque rheometer.
42. ASTM. (2017). D5024-95a Standard test method for plastics: dynamic mechanical properties: In compression.
43. Selvaraj, K. K., Mubarakali, H., Rathinam, M., Harikumar, L., Sampath, S., Shanmugam, G., & Ramaswamy, B. R. (2016). Cumulative exposure and dietary risk assessment of phthalates in bottled water and bovine milk samples: A preliminary case study in Tamil Nadu, India. *Human and Ecological Risk Assessment: An International Journal, 22*(5), 1166–1182. https://doi.org/10.1080/10807039.2016.1146984
44. Billings, A., Jones, K. C., Pereira, M. G., & Spurgeon, D. J. (2023). Emerging and legacy plasticisers in coastal and estuarine environments: A review. *Science of the Total Environment, 12*, 168462.
45. Paluselli, A., Aminot, Y., Galgani, F., Net, S., & Sempere, R. (2018). Occurrence of phthalate acid esters (PAEs) in the northwestern Mediterranean Sea and the Rhone River. *Progress in Oceanography, 163*, 221–231.
46. Zhang, Z. M., Yang, G. P., Zhang, H. H., Shi, X. Z., Zou, Y. W., & Zhang, J. (2020). Phthalic acid esters in the sea-surface microlayer, seawater and sediments of the East China Sea: Spatiotemporal variation and ecological risk assessment. *Environmental Pollution, 259*(113802).
47. Mukhopadhyay, M., & Chakraborty, P. (2021). Plasticizers and bisphenol A: Emerging organic pollutants along the lower stretch of River Ganga, north-east coast of the Bay of Bengal. *Environmental Pollution, 276*, 116697. https://doi.org/10.1016/j.envpol.2021.116697
48. Mathieu-Denoncourt, J., Wallace, S. J., de Solla, S. R., & Langlois, V. S. (2015). Plasticizer endocrine disruption: Highlighting developmental and reproductive effects in mammals and non-mammalian aquatic species. *General and Comparative Endocrinology, 219*, 74–88.
49. Mohammadian, S., Ghanemi, K., & Nikpour, Y. (2016). Competitive adsorption of phthalate esters on marine surface sediments: Kinetic, thermodynamic, and environmental considerations. *Environmental Science and Pollution Research, 23*, 24991–25002.
50. Billings, A., Jones, K. C., Pereira, M. G., & Spurgeon, D. J. (2021). Plasticisers in the terrestrial environment: Sources, occurrence and fate. *Environmental Chemistry, 18*(3), 111–130.
51. Yuan, S. Y., Huang, I. C., & Chang, B. V. (2010). Biodegradation of dibutyl phthalate and di-(2-ethylhexyl) phthalate and microbial community changes in mangrove sediment. *Journal of Hazardous Materials, 184*(1–3), 826–831.
52. Arvanitoyannis, I. S., & Bosnea, L. (2004). Migration of substances from food packaging materials to foods. *Critical Reviews in Food Science and Nutrition, 44*(2), 63–76.

53. Till, D. E., Reid, R. C., Schwartz, P. S., Sidman, K. R., Valentine, J. R., & Whelan, R. H. (1982). Plasticizer migration from polyvinyl chloride film to solvents and foods. *Food and Chemical Toxicology, 20*(1), 95–104.
54. Fankhauser-Noti, A., & Grob, K. (2006). Migration of plasticizers from PVC gaskets of lids for glass jars into oily foods: Amount of gasket material in food contact, proportion of plasticizer migrating into food and compliance testing by simulation. *Trends in Food Science & Technology., 17*(3), 105–112.
55. Groh, K. J., Backhaus, T., Carney-Almroth, B., Geueke, B., Inostroza, P. A., Lennquist, A., Leslie, H. A., Maffini, M., Slunge, D., Trasande, L., & Warhurst, A. M. (2019). Overview of known plastic packaging-associated chemicals and their hazards. *Science of the Total Environment, 651*, 3253–3268.
56. Beg, M. A., & Sheikh, I. A. (2020). Endocrine disruption: Molecular interactions of environmental bisphenol contaminants with thyroid hormone receptor and thyroxine-binding globulin. *Toxicology and Industrial Health, 36*(5), 322–335.
57. DeLuca, J. A., Allred, K. F., Menon, R., Riordan, R., Weeks, B. R., Jayaraman, A., & Allred, C. D. (2018). Bisphenol-A alters microbiota metabolites derived from aromatic amino acids and worsens disease activity during colitis. *Experimental Biology and Medicine, 243*(10), 864–875.
58. Liu, J., Zhang, L., Lu, G., Jiang, R., Yan, Z., & Li, Y. (2021). Occurrence, toxicity and ecological risk of Bisphenol A analogues in aquatic environment—A review. *Ecotoxicology and Environmental Safety, 208*, 111481.
59. Karimpour, Z., Hasanzadeh, S., & Razi, M. (2020). The ameliorative effects of achillea millefolium inflorescence hydro alcoholic extract on sperm parameters in male rats exposed to diethyl-hexyl–phthalate. *Journal of Ilam University of Medical Sciences, 28*(2), 38–49.
60. Li, X., Zhang, W., Lv, J., Liu, W., Sun, S., Guo, C., & Xu, J. (2021). Distribution, source apportionment, and health risk assessment of phthalate esters in indoor dust samples across China. *Environmental Sciences Europe, 33*, 1–4.
61. Bueno, I. M., & Teixeira, J. E. (2024). Waste plastic in asphalt mixtures via the dry method: A bibliometric analysis. *Sustainability, 16*(11), 4675.
62. Uddin, M., Kormoker, T., Uddin, M. S., Idris, A. M., Ampah, J. D., Haque, M. K., Hossain, M. Y., Siddique, M. A., Islam, M. S., Islam, A. R., & Amran, M. I. (2024). Micro (nano) plastics as emerging pollutants in global aquatic and terrestrial ecosystems: A bibliometric analysis. *Environmental Forensics, 14*, 1–33.
63. Gupta, N., Banerjee, S., Koley, A., Basu, A., Gogoi, N., Hoque, R. R., Mandal, N. C., & Balachandran, S. (2024). Fungal strategies for the remediation of polycyclic aromatic hydrocarbons. In *Bioremediation for sustainable environmental cleanup* (pp. 86–108). CRC Press.
64. Koley, A., GhoshThakur, R., Das, K., Gupta, N., Banerjee, A., Show, B. K., Ghosh, A., Chaudhury, S., Hazra, A. K., Nahar, G., & Ross, A. B. (2024). Growth dynamics and nutrient removal from biogas slurry using water hyacinth. *Sustainability, 16*(11), 4450.
65. Bagdi, T., Ghosh, S., Sarkar, A., Hazra, A. K., Balachandran, S., & Chaudhury, S. (2023). Evaluation of research progress and trends on gender and renewable energy: A bibliometric analysis. *Journal of Cleaner Production, 8*, 138654.
66. Koley, A., Bray, D., Banerjee, S., Sarhar, S., Thahur, R. G., Hazra, A. K., Mandal, N. C., Chaudhury, S., Ross, A. B., Camargo-Valero, M. A., & Balachandran, S. (2022). Water hyacinth (Eichhornia crassipes) a sustainable strategy for heavy metals removal from contaminated waterbodies. In *Bioremediation of toxic metal(loid)s* (pp. 95–114). CRC Press.
67. Koley, A., Mukhopadhyay, P., Gupta, N., Singh, A., Ghosh, A., Show, B. K., GhoshThakur, R., Chaudhury, S., Hazra, A. K., & Balachandran, S. (2023). Biogas production potential of aquatic weeds as the next-generation feedstock for bioenergy production: A review. *Environmental Science and Pollution Research, 30*(52), 111802–111832.
68. Van Eck, N. J., Waltman, L., Dekker, R., & Van Den Berg, J. (2010). A comparison of two techniques for bibliometric mapping: Multidimensional scaling and VOS. *Journal of the American Society for Information Science and Technology, 61*(12), 2405–2416.

69. Aria, M., & Cuccurullo, C. (2017). Bibliometrix: An R-tool for comprehensive science mapping analysis. *Journal of Informetrics, 11*(4), 959–975.
70. Gupta, N., Banerjee, S., Koley, A., Bharali, P., GhoshThakur, R., Hoque, R. R., & Balachandran, S. (2024). Strategies for remediation of polycyclic aromatic hydrocarbons in contaminated soil: A systematic review and bibliometric analysis. *Applied Soil Ecology, 204*, 105688.
71. Gupta, N., Koley, A., Banerjee, S., Ghosh, A., Hoque, R. R., & Balachandran, S. (2024). Nanomaterial-mediated strategies for enhancing bioremediation of polycyclic aromatic hydrocarbons: A systematic review. *Hybrid Advances, 100315.*

Chapter 15
Occurrence, Distribution, and Removal of PFAS in Aquatic Ecosystem

Pooja Dhiman, Arun Lal Srivstva, and Ajay Kumar

Abbreviations

ADONA	Dodecafluoro-3H-4,8-dioxanonanoate
AFFF	Aqueous film-forming foams
BOD	Biological oxygen demand
CEC	Contaminant of emerging concern
Cl-PFESA	Chlorinated polyfluoroalkyl ether sulfonic acid
COD	Chemical oxygen demand
DOC	Dissolved organic carbon
EU	European Union
F-53B	6:2 chlorinated polyfluoroalkyl ether sulfonate
FOSA/PFOSA	Perfluorooctane sulfonamide
FOSE	Perfluorooctane sulfonamido ethanol
FT	Fluorotelomer
FTOH	Fluorotelomer alcohol
GenX/HFPO-DA	Hexafluoropropylene oxide dimer acid
HPLC	High-performance liquid chromatography
NMeFBSE	N-methyl perfluorobutane sulfonamidoethanol
NOx	Nitrogen oxide
PFAAs	Perfluoroalkyl acids
PFASs	Per- and polyfluoroalkyl substances
PFBA	Perfluorobutanoic acid
PFBS	Perfluorobutane sulfonic acid
PFC	Per-fluorinated compounds

P. Dhiman · A. Kumar (✉)
Department of Chemistry, School of Basic and Applied Sciences, Maharaja Agrasen University, Atal Shiksha Kunj, Baddi, Solan, HP, India

A. L. Srivstva
Chitkara University School of Engineering and Technology, Chitkara University, Rajpura, Himachal Pradesh, India

M. K. Jindal et al. (eds.), *Emerging Contaminants in Water*, Emerging Contaminants and Associated Treatment Technologies,
https://doi.org/10.1007/978-3-031-89591-3_15

PFCA	Perfluoroalkyl carboxylic acids
PFDA	Per-fluoro-decanoic acid
PFESA	Polyfluoroalkyl ether sulfonic acids
PFHpA	Per-fluoro-heptanoic acid
PFHxA	Perfluorohexanoic acid
PFHxS	Perfluorohexane sulfonic acid
PFNA	Perfluorononanoic acid
PFOA	Perfluorooctanoic acid
PFOS	Perfluorooctane sulfonic acid
PFPeA	Perfluoropentanoic acid
PFPEs	Polyfluoroalkyl ethers
PFSAs	Perfluoroalkane sulfonic acids
PFUnDA	Perfluoroundecanoic acid
POP	Persistent organic pollutants
PreFOS	Perfluorooctane sulfonate precursors
SOC	Soil organic carbon-water partitioning coefficient
TFA	Tri-fluoro-acetic acid
USEPA	United States (US) Environmental Protection Agency
WWTP	Wastewater treatment plant

1 Introduction

Per- and polyfluoroalkyl substances (PFASs) constitute one of the critical groups of emerging waterborne pollutants because exposure of living cells to PFASs results in undesirable effects on health [1]. The chemical inertness, shorter length of the C–F bond, low molecular polarity and greater energy associated with the C–F bond (536 kJ/mol), surface-borne properties, and amphiphilicity and stability of PFASs toward heat make them favorable for customer- and industrial-based products for their utilization. This is the rationale; such compounds are so routinely consumed throughout the world [2]. The reason for their ubiquitous presence is that they are persistent in nature and are being deployed in industrial applications. Their detection has been confirmed even in snow, lakes, groundwater, rivers, biosolids, landfill leachate, rainwater surface water bodies, oceans, municipal sewage-treated drinking water, and several other environment-related matrices, including the biota of water bodies. Perfluorooctanoic acid (PFOA) and perfluorooctane sulfonic acid (PFOS) most commonly occur, are distributed, and have been identified in numerous environment-based analyses [3]. In the late 1950s, point sources of occurrence and potential toxicity of PFOA and PFOS were identified for the first time in 1990. Therefore, the occurrence and scattering of PFASs in aqueous media generally vary over time. The impact of structure and chain length on PFAS remediation mechanisms has been observed and reported; however, investigations on the scattering of PFASs in the global aquatic ecosystem have not been conducted [4]. Viticoski et al. [5] critically analyzed the occurrence and scattering of PFASs [6] in drinking water, groundwater, and marine water; however, the occurrence and potential toxicological consequences of PFASs in surface water have also been reviewed but in an uncomprehensive manner. An analysis of the occurrence and distribution patterns of PFASs in surface water matrices is needed to emphasize their outcomes with respect to their source and occurrence, in addition to their consequential toxic impacts at the global level. The present literature contains a review of PFASs associated with surface water parameters reported in previous years in the water matrices of all seven reported continents, with the following key aspects considered: (a) PFAS sources and detection; (b) their geographic patterns; and (c) their potential toxicological effects on all their consumers. The guidelines proposed by the advisory bodies restrict the deployment of PFOS throughout the world, and it has been hypothesized that a shift in the utilization of PFOS in mass production and packaging industries from Western nations to Eastern Asian countries will soon be observed globally. A critically reviewed literature primarily emphasized the historic and geography-based outcomes from PFASs of surface waters, and their toxic impacts pertained to PFASs consequently outcome as a result of exposure to surface waters. The effects of PFAS toxicity on plants and biota and their toxic impacts pertaining to sediment exposure have not been reviewed in the present literature. However, the primary focus of the present literature is to elucidate the state of occurrence, distribution, and remediation techniques of PFASs [7]. In the present literature, usage, regulations, PFAS features, distribution, occurrence in sea and coastal water, PFAS toxicity, the effects of PFAS on aquatic ecosystems, and PFAS remediation techniques have been discussed in detail.

2 Broad Overview of the Usage and Regulation of PFASs

PFASs have been reported as a diverse category of aliphatic compounds with one or more perfluoroalkyl moieties (C_nF_{2n+1}-). Polyfluoroalkyl and perfluoroalkyl substances are classified on the basis of being partially fluorinated (poly) or completely fluorinated (per) or alkyl chains attached with functional groups. These alkyl chains are tightened during fluorination; consequently, the resulting compounds become comparatively compact, which is facilitated by chemical structural features with thicker layers of trifluoromethyl groups [8]. PFASs exhibit both lipophobic (lipid, grease repellent) and hydrophobic (water-phobic) behaviors [9]. Therefore, their widespread deployment as grease and water repellents is quite common. The rationale behind the occurrence and dissemination of PFASs in aquatic ecosystems is their deployment not only in numerous commercial and industrial applications, such as metallic coatings, surfactants, and water repellents for textiles, leather, and firefighting materials such as foams, but also in cosmetic-based products [10]. PFASs can be categorized into polymeric PFASs, including fluoro-poly-ether- and fluoro-polymer side-chain fluorination-based polymers, and nonpolymeric PFASs, such as fluorotelomers (FTs), perfluoroalkyl acids (PFAAs), and perfluoroalkyl ethers (PFPEs). Perfluoroalkane sulfonic acids (PFSAs) and perfluoro-carboxylates

(PFCAs) are the most prominent categories of PFAAs [11]. Considering the length of the carbon chain, PFASs can be categorized broadly as short or long-chain PFASs; however, the number of carbons in the structure of PFASs bifurcating into short- or long-chain compounds has not yet been clearly defined. For example, according to the European Agency, working in the directions of safety and health of employees at the workplace, under regulation 1907/2006, short-chain PFAS molecules are referred to as having fewer than seven carbons in the structure of PFCAs and fewer than six carbons for PFSAs, whereas the US EPA refers to PFCAs with fewer than eight carbons and PFSAs with fewer than six carbon atoms in the short-chain PFAS. Compared with long-chain PFASs, short-chain PFASs also exhibit lower hydrophobicity, lower bioaccumulation capabilities, and relatively lower acute toxicity. For decades, long-chain PFASs have been favored over short-chain PFASs in the mass production and packaging industries; their bioaccumulation potential, particularly for PFOA and PFOS in aquatic animals and fish, and their toxicity, because of their exposure to humans, have restricted their utilization in the mass production of long-chain PFASs and have favored their replacement with shorter-chain PFASs since the last decade by producers in factories [12]. On the other hand, plausible adverse health effects associated with short-chain PFASs in potable water are also a matter of brain storage. Owing to the high toxicity-related issues associated with both long- and short-chain PFASs, which are known to occur in the environment, the European Commission has limited all plausible deployments of PFASs, with the goal of ruling out PFASs before 2030 (Lerner, 2019). Initially, the agenda of ruling out PFOS started with 3 M, a PFAS mass production factory running in the United States in 2000, and the target was also achieved within 2 years of its start. PFASs can also be released into natural water bodies during their manufacturing, utilization, and release from industries as well as households, and they are considered direct (primary) sources of PFAS release in the environment. However, there are numerous PFAS precursors, such as fluorotelomer alcohols (FTOHs), which have the potential to convert biologically and abiotically into ionic PFCAs or PFSAs and thus become an indirect (secondary) source of release of PFAS in the environment. Municipal facilities such as landfills and wastewater treatment plants behave as secondary points of contact with PFASs because these compounds are highly resistant to traditional treatment strategies. Owing to their persistent chemically stable features, they mostly remain untreated by traditional water treatment strategies in wastewater treatment plants. As an outcome, PFASs produced from industrial and domestic sources, as well as PFASs originating from their respective precursors in treatment plants, are most likely present in the effluent of these WWTPs. Untreated sewage and effluent from WWTPs are directly discharged into water bodies, and in other scenarios, the majority of customers' cosmetic products with PFASs in their composition also vanish in landfills [13]. The leachate of landfills is generally treated at WWTPs [14]; that is why they both act as potential point sources of PFAS in aquatic ecosystems. Lewis et al. [15] expressed PFAS concentrations of up to 6950 and 93,100 ng L^{-1} in the wastewater treatment plant influences and landfill leachate, respectively. An average PFAS concentration of 110 ng was recorded for the effluent discharged from 19 WWTPs in Australia

(Oz). L^{-1} PFAS is 500 times greater than the recommended limit of 0.23 ng L^{-1} as per the norms of Australia's (Oz) National Environmental Management Plan (NEMP). This makes implementation exercises for the sustainability of water resources such as environmental discharge and its recycling challenges await their address and resolution. Owing to the widespread occurrence and distribution of PFASs and their correlated toxicity level, the regulatory body's advisory guidelines related to health have been listed for recommendation and implementation throughout the world for recreational and potable water. The United States' Environmental Protection Agency (EPA) has set up health advisory restrictions very cautiously for drinkable water for PFOA and PFOS (1 and 150 μg L^{-1}, respectively), before the previous decade [16]. According to the United States' EPA (2009), the health advisory level is ~0.4 μg L^{-1} for PFOA as well as 0.2 μg L^{-1} for PFOS, both of which were reduced to 70 ng. L^{-1} for both PFOA and PFOS in potable water in late 2016.

3 PFAS Features

The chemical and physical properties of PFASs depend upon their molecular formula and chemical structure. Long PFASs contain 6–20 carbon atoms, whereas short-chain PFASs have been reported to contain 4–7 carbon atoms in their structure. For example, eight-carbon PFASs can be categorized into four types: (a) short-chain PFASs, (b) oxygen bridge-bonded PFASs, (c) cyclic structured PFASs, and (d) branched PFASs. Figure 15.1 is an illustration of the structure of short- and long-chain PFASs. Familiarization with the chemical and physical features of PFASs is mandatory to better understand their fate and dissemination in aquatic surroundings. The C–F bond has been observed as the strongest covalent bond in organic materials, which is the reason for the thermal stability of PFASs (ITRC, 2017). Additionally, PFASs exhibit mutually lipophilic and hydrophobic features,

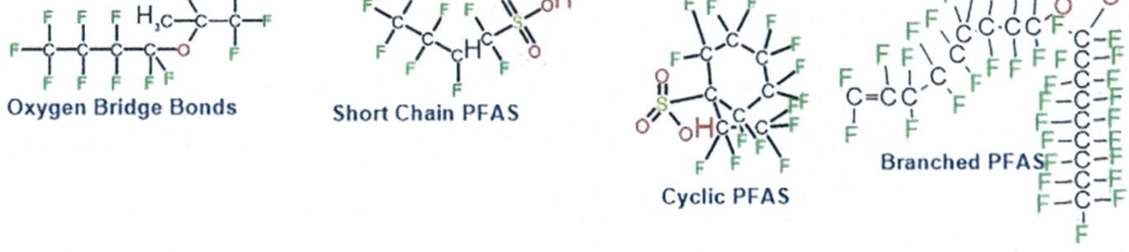

Fig. 15.1 Structures of short- and long-chain PFASs

which are related primarily to the poor polarizability of the compositional fluorine. Moreover, the functional groups (termini) connected to fluoroalkyl chains are the reason for the chemical stability of PFASs. The increasing substitution of hydrogen (H) atoms by fluorine atoms with a corresponding increase in the length of the carbon chains is the reason for the greater chemical inertness of PFAS. However, their solubility in water increases as the carbon chain length decreases [17].

PFSAs and PFCAs feature relatively lower vapor pressures, making the stripping of air an unfavorable technique for the remediation of PFASs [18]. The reported increase in the length of the C–F bond makes PFASs more lipophilic in nature, as confirmed by the partition coefficients of the octanol/water partition (Kow). At the observed environmental pH, PFASs have been reported to be anionic species on the basis of their relatively lower acid dissociation constant (pKa). However, other characteristics, such as resistance to thermal, chemical, and biological degradation in addition to redox stability, incorporate unique characteristics into PFASs, making them favorable for numerous industrial uses [19].

4 Occurrence and Distribution of PFASs in Water

In aquatic ecosystems, the detection of PFASs has even been reported at relatively low quantities ranging from pg L^{-1} to μg L^{-1} [20], and long-chain molecules, such as perfluorooctanoic acid (PFOA) and perfluorooctane sulfonic acid (PFOS), are among the most easily detected PFASs worldwide (ITRC, 2017). The common presence of PFASs in the water cycle of urban regions, such as river water (RW), wastewater (WW), lake water (LW), stormwater, potable water (PW), and groundwater (GW), has been demonstrated by numerous analyses [21]. The sources of surface water pollution by PFAS are illustrated in Fig. 15.2.

The quantities of PFAS remaining in wastewater treatment plants (WWTPs) even after their treatment and, consequently, their release into the aquatic environment are a matter of concern for the plausible occurrence of such PFAS in the supply of water deployed for drinking purposes. A recent study reported that the plausible effects of PFASs on groundwater (GW) are related to the deployment of treated wastewater (WW) for irrigation [22]. The prevalence and transport of PFAS in water have still not been fully characterized because of the limited data concerning the remediation effectiveness of typical water treatment facilities and the related transformation of PFAS precursors [23]. Although the utilization of PFOA and PFOS has decreased owing to their adverse impacts on health, the total concentrations of PFASs discharged into the environment have not decreased, as long-chain PFASs have been replaced by short and ultrashort PFASs [24]. The short and ultrashort PFASs are potentially less bioaccumulated and bioconcentrated through trophic levels than long-chain PFASs because these are environmentally persistent [25]; OECD/UNEP Global. Notably, the current PFAS regulatory guidelines for drinkable water in Western countries may not be useful for citizens of the Asian continent. The ingestion pattern of drinkable water in Asia remarkably differs from

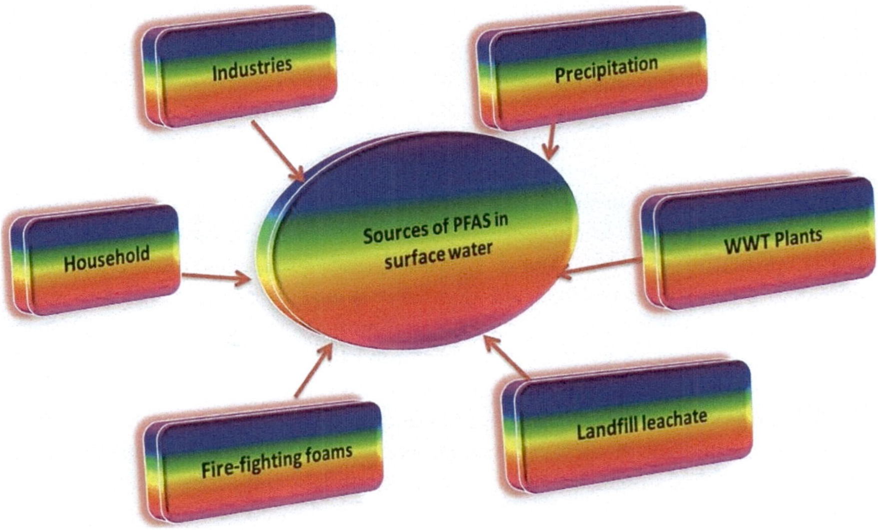

Fig. 15.2 Different sources of PFAS contamination in surface water

that in Western nations, specifically with respect to the type and nature of the point source of water. For example, in South and Southeast Asian nations, groundwater is commonly ingested by trillions of people [26]. This consumption trend was later followed by surface water, particularly in the lower-income rural population, with restricted access to facilities for water treatment [27]. For this reason, the possibility of PFAS leaching from industrial areas to drinkable water resources poses more serious risks to Asian citizens than to Western nationals. In Western countries, current regulatory guidelines as well as threshold values for PFAS offer protection to drinkable water by more encompassing PFAS regulation, according to which environmental freshwater sources are more favorable for Asian countries for protection of the Asian population. The limited information on the biological effect of the acceptable concentration is also an inherent concern for determining the safety level for exposure to PFASs. These findings were expressed in the United States with variable provisional guidelines in the range of 13–1000 ng L^{-1} from every state that may not mandatorily obey the US EPA advisory levels as ~70 ng L^{-1} for both PFOS and PFOA (U.S. EPA, 2016). Additionally, a number of mixtures of PFASs exist in aquatic environments, and the combined effects of such PFASs in biological environments have been poorly understood. Moreover, the impacts of individual PFASs and mixtures of numerous PFASs with other prevailing chemicals in the surrounding environment may behave synergistically, resulting in harmful effects on humans [28]. For this reason, approaches guided by bioassays, such as in vivo and in vitro models, are capable of resolving some of the disadvantages of individual PFASs while taking pure quantitative approaches into consideration. Owing to limited matrix knowledge concerning the relevant physiological concentrations of PFASs and their associated biological consequences, the modern provisional guidelines can offer only minimal protection against health risks associated with exposure to PFASs. In Asia, most of the investigations reported PFHpA/PFOA ratios below 1 (0.04–0.88), indicating the impact of point sources of PFAS. In Europe, this ratio ranges from 0.05 to 2.62 and expresses the impact of both atmospheric and urban mineralization. In North America, near urban areas, the ratios ranging from 0.04 to 0.75 reflect the impact of point sources of PFAS. The ratios were less than 1 for investigations in Africa in the Pra and Kakum Rivers and Lake Victoria in Ghana and Uganda in 2015 [3]. Shockingly, higher levels of PFOS and PFOA have been reported primarily in North America, with a few cases also observed in Asia. Atmospheric deposition has been the dominant source of PFAS prevalence on the African continent, as no reports of PFAS contamination were recorded in Africa until October 2023. The levels of contamination in Africa are relatively high compared with those in Asia or North America, where reported cases have been more common. Previously, many PFAS contaminations were concentrated in point sources. The majority of the observations recorded that the maximum range for PFOS and PFOA as a mixture was more than 70 ng L^{-1}. Importantly, in addition to the utilization of PFHpA/PFOA ratios to locate points of contamination, numerous other ratios were also reported in this decade, and these ratios were utilized to better understand the sources of contamination with PFAS in surface waters. These ratios have been reported to include the ratios of PFOS/PFOA and PFOA/PFNA [29]. Greater ratios (>1) of PFOA/PFNA have been used as indicators of direct emission from industrial production processes for precursor mineralization. Cheng et al. [30] suggested the ratio of the PFAS with (n-1) to PFAS with n, where "n" has been used to express the number of carbon atoms, to impose atmospheric deposition, instead of merely deploying PFHpA/PFOA ratios. Researchers have recommended the use of multivariate statistical tools, including hierarchical clustering of PFAS profiles and principal component analysis, in association with matrices on the hydrologic proximity of the most likely contamination sources [31]. Few factors, such as transportation pathways, including hydrologic distance and the flow direction of rivers, considerably influence the determination of point sources of contamination; however, PFAS transport through the atmosphere is difficult to identify through modern techniques. The dominance of short-chain PFASs such as PFHxA, PFPeA, and PFHpA has been reported in numerous investigations carried out in the current decade. The relatively low concentrations of PFOS and relatively high concentrations of PFHxA suggested that the incorporation of PFOS on the 2009 POP list was quite effective in Eastern China. Similar observations were reported in a study by [32] in the Yangtze River estuary and Jiulong River estuary in China, in which short-chain PFASs, such as PFBS and PFHxA, were found to be prevalent. ODCS and BRAC [33] performed a statistical analysis of their research outcomes and reported that short-chain PFASs, particularly those with more than four carbon atoms, had more diverse point sources than other PFASs did, indicating that these may have been used in industrial applications, whereas others might have been due to their long-range atmospheric deposition or atmospheric transport. PFBS and PFHxA were used as substituents owing to their comparatively lower toxicity than PFOS and PFOA. Other examples of short-chain common PFASs deployed as alternatives include PFHxS, which has been deployed in products for sale in Europe since 2005, and PFBS, a plausible mineralization product of N-methyl perfluorobutane sulfonamidoethanol (NMeFBSE) [34].

4.1 Seas and Coastal Waters

The greater number of detections and measurements of PFAS in coastal waters, such as estuaries and gulfs, does not translate into widespread geographic distribution. The majority of the investigation findings, including 80% for PFCAs and 85% for PFSAs, are for the coastal seas of China, Korea, Western Europe, and Japan. The elevated levels were due to PFOA, PFBA, and PFPeA. The Baltic Sea and North Sea also reported greater median concentrations of PFSAs and PFCAs than did the majority of other coastal regions, especially for PFHxS, PFBS, and PFOS. The greatest concentrations of ΣPFSAs, along with PFOS, were detected in coastal waters and urban-influenced estuaries in Western Europe, such as in the northern Adriatic, Rhine River Delta, and in the area surrounding Elbe. In the United States [35], the western coast, including San Francisco Bay and Puget Sound, also had relatively elevated ΣPFSA. Compared with other seas of the world, the South China Sea, such as the Mediterranean and offshore sites of Hong Kong, generally has lower median PFCA and PFSA quantities. A relatively larger number of samples for the three different time periods allowed qualitative comparisons of medians for 2000–2009 with those for 2015–2019 for 14 different types of PFAS in the seven different coastal regions. The Yellow/Bohai Seas, South China Sea, and East China Sea presented the greatest alterations in measured data from 2000–2009 and 2015–2019 [36]. These apparent elevations were particularly high for PFBA and PFPeA in Chinese coastal waters. Greater variations were also observed for PFHxA in the Yellow Sea, Bohai Sea, East China Sea, the Great Lakes, and South China Sea. The elevated values of some ratios are due to considerations of even nondetectable levels in samples obtained and analyzed in the previous decade and may also reflect diverse recorded limits rather than alterations in emissions. The ratios for 2015–2019/2000–2009 were <1 because lower median concentrations were recorded for most PFCAs and PFOS in the North Sea. The ratios for PFBS and PFHxS values were >1 in the Baltic Sea because of greater medians in 2015–2019; at the same time, PFOS was constant. The different routes of exposure of PFAS to human health are illustrated in Fig. 15.3. In the North American Great Lakes, the median concentrations of PFHxA, PFOA, and PFHpA increased by 3.0-, 1.4-, and 16-fold, respectively, between 2000 and 2009 in addition to 2015 and 2019. Greater alterations were also recorded for PFBA because of the nondetectable concentrations reported from 2000 to 2009 [37]. The median PFOS concentration was also greater in the Great Lakes from 2015 to 2019, whereas the PFBS and PFHxS concentrations were lower. Higher concentrations of PFOS, PFHxA, PFPeA, PFOA, and PFHpA were detected in the mixed urban and nonurban influenced sites than in the open lake sites, with differences of <2 times. However, a comparison of only

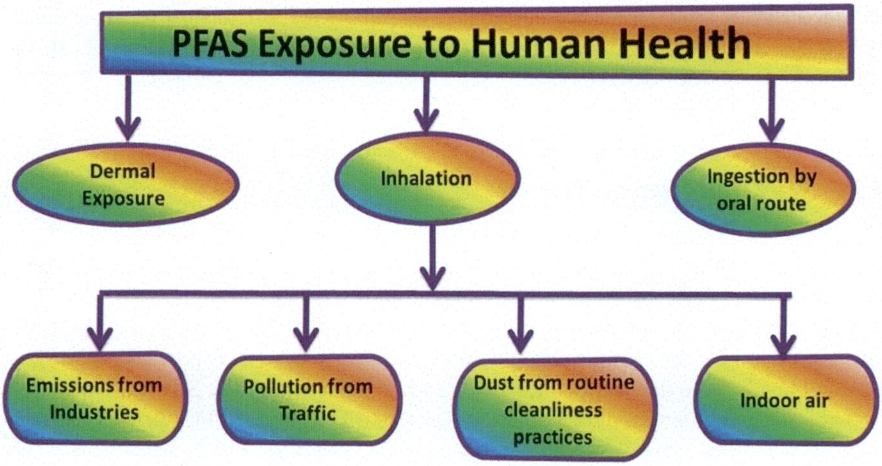

Fig. 15.3 Different routes of exposure of PFASs to human health

nonurban sites in the Great Lakes was problematic owing to a greater percentage of outcomes that were reported to be lower than MDLs (2000–2009, 76%; 2010–2014, 60%; and 2015–2019, 51%), especially in Lakes Superior, Huron, and Michigan [38]. All required and previously recorded data were used to calculate annual medians of $\Sigma C7 - C12$ PFCAs and ΣPFSAs for the Great Lakes and six coastal sea regions for the assessment of temporal trends. A remarkable decrease in PFSAs ($P = 0.014$) and PFOS ($P = 0.048$) in the Yellow Sea and Bohai Sea was observed from 2012 to 2018. A pattern of decrease was also confirmed for $\Sigma C7 - C12$ PFCAs ($P = 0.066$) as well as PFOA ($P = 0.053$) after a significantly elevated median value for 2017 was ignored. Distinctive trends over similar time periods have not been confirmed for the South China Sea and East China Seas [36].

5 PFAS Toxicity

PFASs are generally assimilated in the living body only after consumption by the mouth; these PFASs cannot be metabolized by human metabolic mechanisms, which is why they are detected in blood serum and human tissues at general concentrations of ng mL^{-1}. Specifically, long-chain PFASs are absorbed for their accumulation in the liver. The magnitude of the half-life of PFOA as well as PFOS in humans generally ranges from 2.3 to 5.4 years [39]. Epidemiological analyses have reported that the presence of PFOS and PFOA in the human body is associated with a greater probability of thyroid disease, ulcerative colitis, high cholesterol, pregnancy-induced hypertension, kidney cancer, and testicular cancer. Keeping the recommendations of the WHO and precautionary principles in mind, the revised Drinking Water Directive (98/83/EC, Directive on potable water quality) was

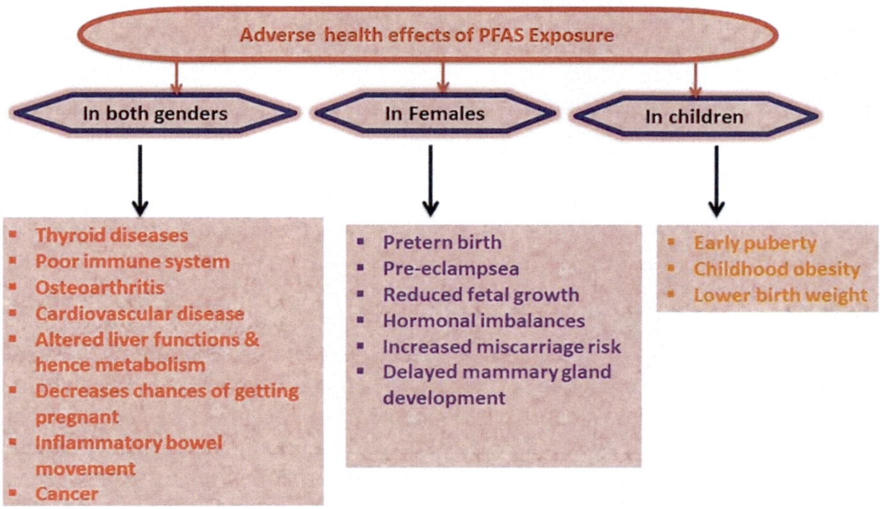

Fig. 15.4 The adverse health effects of PFAS exposure on the human body

published on February 1, 2018. The published report mentioned the group of per- and polyfluoroalkyl substances (PFASs) in the category of regulated chemicals. The adverse health effects of PFASs on the human body are depicted in Fig. 15.4.

The proposal has recommended that the total amount of PFAS be 0.5 μg. $L^{-1,}$ and the concentration of each PFAS was 0.1 μg L^{-1}. For PFASs, drinking water quality standards were set by the Veneto region of Italy in 2017 (Veneto Legislation 1590/201), with a restrictive value of 90 ng L^{-1} for PFOA as well as PFOS (with PFOS <30 ng L^{-1}) and a restrictive value of 300 ng L^{-1} for PFASs such as PFPeA, PFBA, PFBS, PFHxA, PFHxS, PFHpA, PFNA, PFUnA, PFDoA, and PFDeA [40].

6 Impact of PFASs on Aquatic Ecosystems

The negative impact of these compounds on the biology of the aquatic environment has prompted researchers to investigate their occurrence, distribution, and remediation techniques. However, their deployment has been limited by recent legislative regulations. However, greater concern has to be paid by researchers to short-chain PFASs, owing to their deployment as a substitute for long-chain PFASs and their remarkable activity in aquatic ecosystems. Therefore, short-chain PFASs and microplastics are easily identified in aqueous environments [41].

In the last decade, a remarkable increase in the levels of PFAS in China has been reported. However, whether neighboring Asian countries, particularly Southeast Asia and South Asia, also demonstrate this pattern has yet to be investigated because of the lack of PFAS monitoring analysis, regulatory guidelines, and initiatives in these regions of the world. Furthermore, more research on PFASs and monitoring initiatives in symbiosis with proposed policy draft initiatives can further yield improved results through a subsequent decrease in the levels of PFASs [42]. These countries will definitely take the required initiatives after drafting and implementing appropriate monitoring and regulatory initiatives to limit the increase in the extent of contamination caused by PFAS in the specified regions. Because Asian countries have been continuously competing and contributing to the economy at the global scale, each nation should continue to be cautious about the remediation of environmental pollution in the form of the outcome of modern progress in the industrial sector, which should ensure that its progress in terms of the economy is unacceptable at the cost of health and the well-being of its citizens [43].

In-depth evaluations of the patterns of occurrence and distribution of PFASs in surface water have been conducted globally. The concentration of PFAS, specifically PFOS and PFOA, was reported to be greater than 70 ng L^{-1}, as per present US EPA advisory guidelines. Several researchers have reported that approximately 4.5 million consumers from the Pearl River in China, the primary drinking water source, had concentrations of PFOS <2 ng L^{-1} well below the permissible limit. Another researcher reported that in 2010, greater concentrations of PFOS and PFOA, 26.5 and 0.11 ng, respectively, were detected. L^{-1} well below its identification limit of 320.5 ng L^{-1}, respectively, in the collected matrix from the Pearl River Delta [44]. As per previously published reports, the optimum limit for mixing PFOS and PFOA has remarkably been <70 ng. L^{-1}, which has demonstrated severe health hazards to aquatic faunal ecosystems [45]. Our present investigation revealed that the concentration of PFAS in Asian countries was elevated in comparison with the recommended optimal acceptable concentration by exposure, as determined by Sweden's National Food Agency. Several simultaneous attempts have been made by eminent researchers in this area to identify PFAS-associated risk factors, emphasizing their impact on the life of aquatic flora, fauna, and biota, especially when they are brought in contact with PFAS-contaminated surface water. Risk assessments of aquatic ecosystems have reported that PFASs potentially affect the microbiota of the guts of aquatic microorganisms [46]. The consumption of PFAS-contaminated water by aquatic birds has led to poor reproduction rates in birds [47]. In a few aquatic animals, the exposure of their bodies to PFASs weakens their immune system. When considerable amounts of PFAS bioaccumulate in the bodies of aquatic animals, such situations lead to endocrine disruption, affecting the regulation of reproductive and hormonal functions. PFASs not only lead to developmental toxicity in aquatic animals [48]. For example, oxidative compression in gill tissues as well as the hepatopancreas in pearl mussels has been reported to be caused by exposure to PFAS. Additionally, the exposure of zebra fish embryos to PFAS has resulted in swimming differences, startle responses, and burst activity, which are evidence of developmental toxicity in aquatic animals [49].

Advisory guidelines pertaining to PFAS quantities in water matrices worldwide are summarized in Table 15.1. This regulation was proposed and implemented after enormous pollution of groundwater and surface water occurred because of industrial wastewater discharge, consequently resulting in the bioaccumulation of PFAS

Table 15.1 Advisory guidelines regarding PFAS concentrations in water matrices globally

Continent	Country	Advisory department	Standard	Category of water	Contaminant	Reference
Europe (EU)	Germany (GE)	GMH, abbreviation for German Ministry of Health	100 ng L^{-1}	–	Mixture of PFOS and PFOA	[51]
Europe (EU)	Italy (IT)	INHS, abbreviation of Italian National Health Service	500, 30, and 500 ng L^{-1}	Drinkable water	Mixture of PFOA, PFOS, PFAS, and others	[52]
Europe (EU)	Sweden (SE)	NFA, which stands for National Food Agency	90 ng L^{-1}	Drinkable water	Mixture of PFOS, PFOA, PFHxA, PFHxS PFBS, 6:2 FTS, PFHpA, PFPeA, PFBA, PFNA, PFDA	[53]
Australia (Oz)	Australia (Oz)	NEMP, which stands for National Environmental Management Plan	1900, 0.23 ng L^{-1}	Marine water and fresh water	Mixture of PFOA and PFOS	[3]
Australia (Oz)	Australia (Oz)	NHMRC, which stands for National Health and Medical Research Council	2, 10 μg L^{-1}	Water for entertainment	Mixture of PFHxS, PFOA, and PFOS	[54]
North America (NA)	Canada	HC which stands for Health Canada	600 and 200 ng L^{-1}	Drinkable water	Mixture of PFOS and PFOA	[55]
North America (NA)	United States (USA)	USEPA, known as United States (US) Environmental Protection Agency	70 ng L^{-1}	Drinkable water	Combination of PFOS and PFOA	[56]
North America (NA)	United States (USA)	MDH which stands for Minnesota Department of Health	35, 27 ng L^{-1}	Drinkable water	Both PFOA and PFOS	[56]

(continued)

Table 15.1 (continued)

Continent	Country	Advisory department	Standard	Category of water	Contaminant	Reference
North America (NA)	California, United States (USA)	CSWRCB, which stands for California State Water Resources Control Board	14, 13 ng L^{-1}	Drinkable water	Both PFOA and PFOS	[57]
North America (NA)	United States (USA)	VDEC, which stands for Vermont Department of Environmental Conservation	20 ng L^{-1}	Drinkable water	Sum of PFOA, PFOS, PFHxS, PFHpA, and PFNA	[58]
North America (NA)	New Jersey, United States (USA)	NJDEP, which stands for new Jersey Department of Environmental Protection	14, 13 ng L^{-1}	Drinkable water	PFOA, PFOS	[59]
North America (NA)	Hampshire, United States (USA)	NHDES, which stands for new Hampshire Department of Environmental Services	12, 11, 18, 15 ng L^{-1}	Drinkable water	Combined presence of PFOA, PFOA, PFOS, PFNA, PFHxS	[60]
North America (NA)	Wisconsin, United States (USA)	WDH, which stands for Wisconsin Department of Health	2 ng L^{-1}	Drinkable water	Mixture of PFOA and PFOS	[61]

in the serum of living beings. Therefore, 175 types of PFASs as emerging contaminants have already been identified and regulated in Italy. Numerous countries obey the advisory restrictions established by the US EPA, and several nations as well as some states within the United States obey these guidelines pertaining to PFAS quantities more stringently [50].

7 PFAS Remediation Techniques

Keeping the chemical features of PFAS, such as solubility and greater mobility, traditional treatment processes for drinking water, such as filtration, sedimentation, and coagulation, are not efficient for removing PFAS compounds from aqueous ecosystems [62]. Water treatment technologies, such as reverse osmosis, granular activated carbon, adsorption, and ion exchange, are used to remediate PFASs via high-pressure membrane filtration. The selection of an appropriate water treatment technique to remove PFASs requires the investigation of each individual water treatment system. Factors such as the type of PFAS, including short- and long-chain and sulfonate or carboxylate functional groups; the initial concentration of PFAS; co-occurring contaminants; and water chemistry, including salinity and pH, affect the remediation of PFAS from aqueous bodies [63]. Traditional and conventional water treatment processes are ineffective for the remediation and degradation of the majority of PFASs owing to the high degree of straightening of the covalently bonded C–F bonds in their chains, their highly hydrophilic nature, and their relatively low concentration in aqueous ecosystems [64]. Both aerobic and anaerobic biological treatments are capable of cleaving only C–C bonds, and they proceed to form shorter-chain PFASs. In the absence of the implementation of advanced oxidation treatment techniques, biological wastewater treatment plants impact the water bodies receiving water. The detected concentration of PFAS in treated water is often greater than that in untreated water [65]. This behavior supports the recalcitrant nature of PFAS toward biological treatment methods and, furthermore, their formation through the biodegradation of precursor compounds [66]. According to data originating from monitoring and regulatory investigations, PFASs with long chains of carbon atoms, such as PFOS, tend to be stored in sludge, although perfluorobutane sulfonic acid (PFBS) has been detected primarily in the effluents of wastewater treatment plants (WWTPs) and in potable water treatment plants (PWTPs) [67]. Typically, in water treatment plants, water disinfection and sterilization by nascent chlorine or UV irradiation are quite ineffective and inefficient [68]. Compared with other microcontaminants, PFASs are remarkably resistant to oxidation by hydroxyl radicals and ozone gas owing to the electron withdrawing of their functional groups and their strong C–F bonds [69]. Additionally, advanced oxidation processes (AOPs) generally result in the partial mineralization of PFASs, with the generation of shorter-carbon-chain PFASs, such as fluorinated alkyl chains [12]. Anion exchange (AE), reverse osmosis membrane processes, and high-pressure nanofiltration and adsorption are quite effective for substantial PFAS remediation [70]. Prime concerns inherent to PFAS remediation through these elaborate strategies impact the water matrix, such as organic matter and competing anions, due to fast breakthrough, frequent regeneration, and disposal of relatively high PFAS concentrations [71]. Adsorption is generally an established approach for PFAS remediation, both as a step in water treatment plants and as a single technique of point-of-use deployment [72]. Additionally, it is more economical than elevated-pressure membrane processes are [73]. As mentioned earlier, PFASs act as anions at optimum pH values to make strong base AE resins favorable for effective remediation. However, the efficiency of the AE process largely depends on the characteristic features of the resins, including their functional groups, porosity, and polymer matrix [74]. To date, adsorption analysis has resolved single long-chain PFASs such as PFOS and PFOA, although short-chain and long-term remediation demands further research [66]. The different PFAS removal techniques from aqueous environments are shown in Fig. 15.5.

Moreover, numerous PFASs present in aqueous ecosystems are capable of impacting adsorption efficiency, and additional investigations must consider their

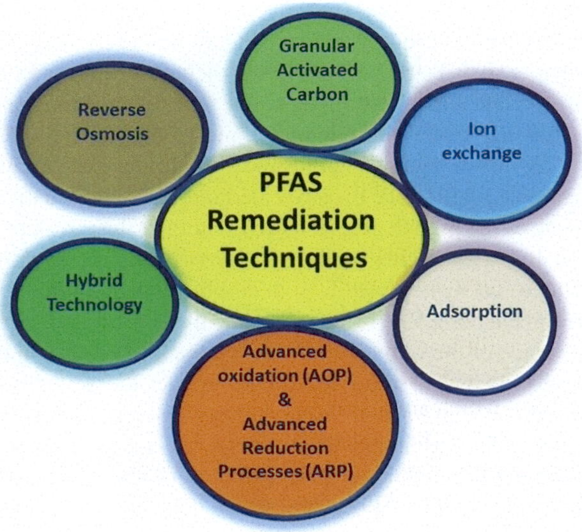

Fig. 15.5 Different PFAS removal techniques from aquatic environments

potential capabilities [72]. However, the costs of water treatment differ greatly and are reasonably governed by the type of technology chosen, plant size, capacity of the water treatment system, and volume of water treated on a routine basis. Owing to their lower cost, granular activated carbon (GAC) filters are generally the primary choice for PFAS remediation. Overall, cost investments are prime factors that restrict the installation of PFAS remediation technologies and even water evaluation for PFAS.

7.1 Granular Activated Carbon (GAC) Treatment

GAC filtration is common for point-of-use water filters both at domestic and business enterprises and at water treatment plants. The factors affecting the efficiency of these types of water treatment techniques include the contact time, type of carbonaceous material, and amount of natural organic carbonaceous material in the water, which can potentially decrease the ability of GAC to remediate PFAS. Investigations have shown that fluorinated sulfonates are adsorbed more quickly than fluoroalkyl acids by activated carbon; PFASs with long-chain carbon atoms are more readily adsorbed than short-chain PFASs [75]. For PFASs, compounds with shorter chains and longer GAC media contact times with more frequent media alterations are needed. Furthermore, the source of GAC may impact the effectiveness of PFAS remediation. Bituminous coal-derived GAC has been more effective and efficient than GAC synthesized from other materials, including the GAC derived from coconut shells [76]. The remediation efficacy also greatly depends on the duration for which the GAC has been in service before replenishment or reclaiming [77], and timely replacement of GAC filters is also needed. After the adsorption of contaminants by activated carbon reaches its maximum capacity, contaminant "breakthrough" generally takes place. The previously adsorbed contaminants can be discharged to the flow-through water. The remediation of short-chain and long-chain perfluoroalkyl acids and sulfonates through granular activated carbons: A comparison-based column study [78]. Researchers from Carbon Corporation, an enterprise that produces GAC, reported that coal-derived reagglomerated GAC targeted and met cost-effective PFAS remediation with an operational cost of \$0.12/1000 gallons treated and an estimated construction cost of \$275,000 per million gallons per day of treated water. The operational costs drop to \$0.08/1000 gallons when carbon is regenerated periodically [79]. In a simulated small-scale column investigation, GAC is efficient enough to decrease the quantities of both short-chain and long-chain PFAS, and its treatment is estimated to be efficient for the initial 1 or 2 years before it requires replenishment of carbonaceous material [80]. All PFAS remediation treatments encounter challenges for the disposal of PFAS residues, such as exhausted GAC media, which need to be reclaimed or translocated to an incinerator or a landfill for combustion at temperatures >1000 °C. Regeneration by thermal means involves heating at higher temperatures, similar to incineration, which permits GAC media to be redeployed. However, during this process, fluorinated chemicals are discharged into the air [81]. Thermal incineration/regeneration and incomplete incineration of PFAS-contaminated water release fluorinated chemicals into the atmosphere, then returning to water bodies through rain. More investigational analysis and community involvement in addition to government actions are needed to confirm the appropriate disposal practices for PFAS residuals. It not only transfers environmental contamination concerns from one medium to another but also wastes considerable capital and operational costs [82].

7.2 *Reverse Osmosis (RO) Treatment*

Membrane separation is a potentially efficient technique for the remediation of different types of shorter- and longer-chain PFASs, including [83]. Reverse osmosis generally results in better remediation of contaminants than does nanofiltration. Membrane separation technology can also involve symbiotic mixing with GAC for greater remediation rates and longer membrane lifetimes. Compared with other PFAS remediation techniques, membrane filtration generally requires more energy [84]. However, water treated by reverse osmosis is more corrosive for plumbing fixtures and distributive pipes. Furthermore, since drinking demineralized water may lead to numerous nutritional deficiencies, water treated by reverse osmosis needs to be demineralized. Like GAC, for the membrane to "reject water," disposal solutions must subsequently be identified to concentrate the PFAS source. The discharge of "drained water" has been reported to be equivalent to the amount of PFAS contaminants returning to aquatic ecosystems downstream, polluting other water sources [85].

7.3 Ion Exchange

Anion exchange treatment, a recently implementable technique for the mitigation of PFAS, can also be efficient. Like anion exchange and GAC, this method yields better and satisfactory results in removing PFASs with longer chains of carbon atoms than in removing PFASs with shorter chains [86]. The initial setup cost for installations of anion exchange treatment plants is generally greater than that of GAC. In contrast, anion exchange bed materials require less frequent replenishment than GAC does. The matrix water compositional parameters influence the effectiveness and efficiency of the anion exchange resin technique for the mitigation of PFAS, owing to the possible availability of competing anions. Single-use resin columns may be satisfactorily effective, and further in-depth research is needed on reactivated resins [87]. The spent and exhausted anion exchange bed resin needs to be incinerated or transferred to a landfill, similar to GAC media. Anion exchange chambers can also be deployed in association with the inclusion of GAC, and the symbiosis of anion exchange resin columns with nanofiltration technology offers far better mitigation-related efficiencies than anion exchange alone and GAC [88].

7.4 Adsorption

Adsorption technology is generally well established, widespread, and deployed ex situ PFAS remediation strategies in water by deploying activated carbon either in granular or powdered form, ion exchange-based resins or polymers, and few applications by deploying numerous naturally occurring adsorbents. Because these strategies concentrate PFASs from the liquid form to the solid form, a favorable strategy is needed to either treat or discard exhausted adsorbents [89]. The adsorption process involves transportation, safe storage, and efficient degradation of adsorbed PFAS either by reclamation or by discarding the exhausted adsorbent. The remarkable limitation of the adsorption process is its initial greater operational and maintenance cost, specifically, high consumption of energy for reclamation or replacement led by off-site waste treatment as well as its management [90].

7.5 Advanced Oxidation (AOP) and Advanced Reduction (ARP) Process

"Advanced oxidation processes" were set up by Glaze and his team in 1987. His report indicated that AOPs generate the most powerful oxidant, i.e., OH free radicals, which are different from fluorine and degrade organic water contaminants into nonhazardous small molecules or are completely mineralized to H_2O and CO_2 [19]. OH free radicals break carbon–nitrogen, carbon–carbon (C–C), and (C–S)

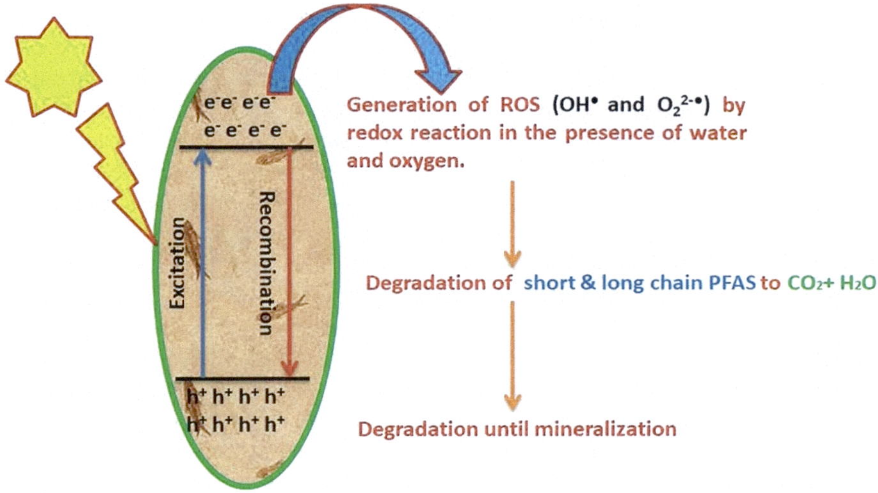

Fig. 15.6 Demonstration of the degradation of short- and long-chain PFAS by the advanced oxidation process (AOP)

carbon–sulfur as well as other chemical bonds. The generation of the required free radicals has been reported to take place via hydrogen peroxide (H_2O_2), ozone (O_3), catalysts (noble metals or transition metal ion-based photocatalysts), potassium persulfate (KPS), and illumination by light (either visible light or ultraviolet light), electron beams, or ultrasound. AOPs, including electrochemical oxidation, ultrasonication, and plasma processes, have the potential to generate hydroxyl free radicals ($OH^\bullet$) and attack target pollutants to degrade them into smaller nonhazardous molecules or their compositional parent elements [91]. A demonstration of the degradation of short- and long-chain PFASs by the advanced oxidation process (AOP) is shown in Fig. 15.6. A comprehensive report has been published by the Southern Nevada Water Authority on an in situ chemical oxidation strategy in which persulfate is subsequently used, followed by PFASs adsorbed at the surface of GAC. Unfortunately, promising outcomes have not been achieved for the destruction of PFOS and PFOA and their adsorption on the surface of GAC. A previously published document suggested that more advanced investigations explore other more effective, efficient, and cost-effective treatment techniques [92].

7.6 *Hybrid Technology*

The hybrid method of water treatment assembly is regarded as a symbiosis or conjunction of more than two independent individual treatment techniques, including diverse adsorption, biological, and membrane processes; AOPs; and ion exchange. Compared with single treatment processes, hybrid treatment assemblies yield more improved results and ameliorate the whole treatment of PFASs as micro-pollutants. Loganathan et al. carried out their research by deploying a hybrid membrane–ozone system comprising a unique ceramic-based membrane contactor for ozonization without bubbles of recently developed micro-pollutants [93]. The outcomes of their investigations revealed that the foul odor of membranes is the major setback; however, membranes are generally deployed for improved techniques for the treatment of naturally occurring wastewater. To resolve such concerns, numerous techniques have been explored. Among these, hybrid processes have been reported to be associated with more than more membrane filtration, and other membrane processes that are symbiotic with relevant water treatment techniques have always been in demand. Sorn et al. suggested that hybrid sorption systems of GAC as well as IX provide plausible substitutions to improve the removal of short-chain PFASs and overcome the existing competition between long-chain PFASs and NOM [94]. Medina et al. investigated the ozone–sonolysis hybrid technique to suggest that ozone attacks organic contaminants either by direct electrophilic attack or through indirect radically active chain reactions at varying pH values of the matrix [95]. Thermal degradation and mineralization of ozone in the subsidiary bubbles generate additional $OH^{\bullet}$ in the hybrid process, leading to far improved degradation of targeted pollutants. According to investigations carried by Mojiri et al., the ultrasonication with AOPs has demonstrated much improved performance-based results in comparison to those approaches, which have been carried out individually in numerous studies [91]. According to the investigational analysis by Ilić et al., several laboratory-based experimental hybrid ultrasound-based approaches presented increased possibilities for the breakdown of micro-pollutants in the matrix [96]. However, researchers have highlighted concerns associated with economic feasibility and implementation in large-scale applications. Few plausible technological substitution approaches are already in practice for their consideration in treatment systems for improvements and optimization. This offers an opportunity for researchers in research and innovation to engineer collectively in deploying improved economic practical implementations for the remediation of PFAS [90].

8 Addressable Sources of PFAS Contamination

With the objective of protecting water transportation from contamination by PFAS, identification, detection, avoidance, and elimination of the existing and plausible origins of contamination by PFAS are urgently needed. Pollution caused by contamination by PFAS in groundwater surrounding airports and military bases has been observed in multiple nations. Careful solution-oriented cautious steps have already been taken to resolve this concern, including the upgradation of the latest fluorine-free firefighting foam [97], and very recently, the US congressional mandate banned PFAS in foams for firefighters in 2024. Municipal committees run treatment plants to treat wastewater, which generally collect industrial PFAS-contaminated wastewater, and can implement the required technology for the conversion of larger and heavier molecular weight PFASs into small, lightweight, and more mobile perfluoroalkyl acids. An investigation in Australia (Oz) examined effluent and influent water samples from wastewater treatment plants collected from 19 different locations and reported that the amount of one particular type of PFAS, specifically perfluorocarboxylic acids, was greater in the effluent than in the influent, suggesting that larger PFAS precursor molecules are generally mineralized into smaller and lighter PFAS molecules during the wastewater treatment process [98]. Contamination of wastewater effluent with PFASs poses a risk to water treatment and recycling processes, including the recharge of natural water aquifers with treated demineralized water [99]. Furthermore, in wastewater effluent, PFASs are also converted into biosolid forms of sludge [100], which was later deployed for enriching agricultural soil. In food crops, PFAS contamination in the soil acts as a point of origin for PFAS and is a reason for PFAS pollution not only in aquatic ecosystems but also in aquatic organisms [101]. The widespread awareness of PFAS contamination has increased; reputed enterprises have started producing PFAS-free consumer-required products. Numerous apparel design and mass production companies have taken the required initiatives to replace PFAS usage in their textile treatments and fabrication processes [102]. IKEA, a global quality furniture manufacturer and retailer, has taken a bold step of complete phaseout of PFAS in all the required fabrications and has finished the final products since 2016. The majority of the local retailers in the United States had declared and advertised that their carpets as well as rugs no longer contain PFAS [103]. Such initiatives in the long run announce the reduced PFAS release in the years to come.

9 Future Scope

The present analysis revealed that several surface water samples were severely polluted with PFOA and PFOS, with concentrations greater than the advisory guidelines of 70 ng. L^{-1} for drinkable water; exposure of the human body through PFAS ingestion via drinkable surface water and its long-term exposure impacts have thus far not been evaluated, so observational analysis in this direction truly needs to be performed. Because rainwater and snowfall are associated with the skin exposure of living beings for a reasonable period and because vapor form and particulate matter containing PFAS can be generated as an outcome of emissions from industrial point origins and their corresponding routes to cloud-based droplets and the resulting rain and snowfall toxicity-related evaluations focused primarily on the impact of rainwater on living beings, research in this direction has not yet been conducted. Analytical evaluations of this contract can be conducted. The present study covered only a limited number of cases and distributions of PFAS in Antarctica, South America, and Africa. A more in-depth study in this regard in these regions is yet to be conducted. PFOA has been observed to be dominant throughout the world, and its subsequent impacts after its incorporation into the POP list of surface water have yet to be analyzed. An analysis of the impacts of long-term exposure as well as its imparted toxicity as an outcome of the consumption of surface water contaminated with short-chain PFASs, including PFBS and PFHxA as substituents for long-chain PFASs, should be performed in humans. The impacts of toxicity on the health of living beings owing to their long-term exposure to fluorinated substituents, including ADONA and GenX, via either dermal contact or ingestion have yet to be investigated. The most favorable nonhazardous, safe, and eco-friendly substituent of PFOA still needs to be determined. According to previous studies, the discharge of WWTPs has been documented as the origin point of PFASs in surface waters. Recently, evaluations reported that the secondary water treatment process, named the activated sludge process, has demonstrated promising outcomes in reducing PFOS concentrations, and more in-depth analysis should be conducted to verify the research outcomes. In addition, a practically more adaptable, cost-effective, and sustainable screening medium should be further developed and upgraded for the mitigation of PFAS to treat agricultural release, storm runoff water, and waste effluent when different water treatment processes are applied.

10 Conclusions

The elevated concentrations of PFAS in water bodies have recently caused alarming alerts not only in Western nations but also throughout the globe. The extent of PFAS contamination has been reported to be greater in surface water than in potable water, tap water, and groundwater. In the initial 2 years, GAC was efficient enough to decrease the concentrations of both short-chain and long-chain PFAS in a simulated small-scale column test. Reverse osmosis yielded satisfactory results in the removal of PFHxS, PFOA, PFOS, PFNA, PFBS, and PFHxA. Compared with individual anion exchange and GAC technology, the symbiotic combination of anion exchange resin-based columns with nanofiltration technology resulted in far better removal efficiencies for PFAS. The adsorption process addresses practical issues related to transportation, safe storage, and efficient degradation of adsorbed PFAS either by reclamation or by discarding the exhausted adsorbent. AOPs, including ultrasonication, plasma processes, and electrochemical oxidation, have the potential to generate hydroxyl free radicals ($OH^{\bullet}$) to degrade target short- or long-chain PFAS pollutants into smaller nonhazardous molecules. Compared with individual technologies, the combination of two hybrid technologies has led to more improved results. For example, ultrasonication with AOPs had far better results than did individual technology outcomes.

References

1. Sinclair, G. M., Long, S. M., & Jones, O. A. H. (2020). What are the effects of PFAS exposure at environmentally relevant concentrations? *Chemosphere, 258*, 127340.

2. Rodgers, K. M., Swartz, C. H., Occhialini, J., Bassignani, P., McCurdy, M., & Schaider, L. A. (2022). How well do product labels indicate the presence of PFAS in consumer items used by children and adolescents? *Environmental Science & Technology, 56*(10), 6294–6304.
3. Podder, A., Sadmani, A. H. M. A., Reinhart, D., Chang, N.-B., & Goel, R. (2021). Per and poly-fluoroalkyl substances (PFAS) as a contaminant of emerging concern in surface water: A transboundary review of their occurrences and toxicity effects. *Journal of Hazardous Materials, 419*, 126361.
4. Kurwadkar, S., et al. (2022). Per-and polyfluoroalkyl substances in water and wastewater: A critical review of their global occurrence and distribution. *Science of the Total Environment, 809*, 151003.
5. Viticoski, R. L., et al. (2022). Spatial distribution and mass transport of perfluoroalkyl substances (PFAS) in surface water: A statewide evaluation of PFAS occurrence and fate in Alabama. *Science of the Total Environment, 836*, 155524.
6. Sims, J. L., et al. (2022). Global occurrence and probabilistic environmental health hazard assessment of per-and polyfluoroalkyl substances (PFASs) in groundwater and surface waters. *Science of the Total Environment, 816*, 151535.
7. Melo, T. M., et al. (2022). Ecotoxicological effects of per-and polyfluoroalkyl substances (PFAS) and of a new PFAS adsorbing organoclay to immobilize PFAS in soils on earthworms and plants. *Journal of Hazardous Materials, 433*, 128771.
8. Guo, H., et al. (2024). Roles of varying carbon chains and functional groups of legacy and emerging per−/polyfluoroalkyl substances in adsorption on metal-organic framework: Insights into mechanism and adsorption prediction. *Environmental Research*, 118679.
9. Kirkwood, K. I. (2023). *Advancing analytical techniques to investigate per-and Polyfluoroalkyl substances (PFAS) exposure and metabolic impacts*. North Carolina State University.
10. Gaines, L. G. T. (2023). Historical and current usage of per-and polyfluoroalkyl substances (PFAS): A literature review. *American Journal of Industrial Medicine, 66*(5), 353–378.
11. Dalmijn, J., Glüge, J., Scheringer, M., & Cousins, I. T. (2024). *Emission inventory of PFASs and other fluorinated organic substances for the fluoropolymer production industry in Europe* (Vol. 26, p. 269). Processes & Impacts.
12. Li, F., et al. (2020). Short-chain per-and polyfluoroalkyl substances in aquatic systems: Occurrence, impacts and treatment. *Chemical Engineering Journal, 380*, 122506.
13. Harris, K. J., Munoz, G., Woo, V., Sauvé, S., & Rand, A. A. (2022). Targeted and suspect screening of per-and polyfluoroalkyl substances in cosmetics and personal care products. *Environmental Science & Technology, 56*(20), 14594–14604.
14. Masoner, J. R., et al. (2020). Landfill leachate contributes per-/poly-fluoroalkyl substances (PFAS) and pharmaceuticals to municipal wastewater. *Environmental Science: Water Research & Technology, 6*(5), 1300–1311.
15. Lewis, A. J., Yun, X., Spooner, D. E., Kurz, M. J., McKenzie, E. R., & Sales, C. M. (2022). Exposure pathways and bioaccumulation of per-and polyfluoroalkyl substances in freshwater aquatic ecosystems: Key considerations. *Science of the Total Environment, 822*, 153561.
16. Choi, G.-H., Lee, D.-Y., Song, A.-R., Moon, B.-Y., & Kim, J.-H. (2022). The dietary risk assessment of perfluorooctanoic acid (PFOA) and perfluorosulfonic acid (PFOS) in the root crops from the survey of the residue in agricultural soil and the crops. *Applied Biological Chemistry, 65*(1), 60.
17. Lin, B., et al. (2022). Self-healing performance of ethyl-cellulose based supramolecular gel coating highly loaded with different carbon chain length imidazoline inhibitors in NaCl corrosion medium. *Corrosion Science, 197*, 110084.
18. Mudlaff, M., Sosnowska, A., Gorb, L., Bulawska, N., Jagiello, K., & Puzyn, T. (2024). Environmental impact of PFAS: Filling data gaps using theoretical quantum chemistry and QSPR modeling. *Environment International, 185*, 108568.
19. Verma, S., Lee, T., Sahle-Demessie, E., Ateia, M., & Nadagouda, M. N. (2023). Recent advances on PFAS degradation via thermal and nonthermal methods. *Chemical Engineering Journal Advances, 13*, 100421.

20. Caniglia, J., Snow, D. D., Messer, T., & Bartelt-Hunt, S. (2022). Extraction, analysis, and occurrence of per-and polyfluoroalkyl substances (PFAS) in wastewater and after municipal biosolids land application to determine agricultural loading. *Frontiers in Water, 4*, 892451.
21. Viticoski, R. L. (2022). *Spatial distribution and mass transport of per-and polyfluoroalkyl substances (PFAS) in surface water systems* (Vol. 836, p. 155524).
22. Kavusi, E., et al. (2023). Critical review on phytoremediation of polyfluoroalkyl substances from environmental matrices: Need for global concern. *Environmental Research, 217*, 114844.
23. Tavasoli, E., Luek, J. L., Malley, J. P., & Mouser, P. J. (2021). Distribution and fate of per-and polyfluoroalkyl substances (PFAS) in wastewater treatment facilities. *Environmental Science: Processes & Impacts, 23*(6), 903–913.
24. Koulini, G. V., & Nambi, I. M. (2024). Occurrence of forever chemicals in Chennai waters, India. *Environmental Sciences Europe, 36*(1), 60.
25. Guelfo, J. L., et al. (2021). Environmental sources, chemistry, fate, and transport of per-and polyfluoroalkyl substances: State of the science, key knowledge gaps, and recommendations presented at the august 2019 SETAC focus topic meeting. *Environmental Toxicology and Chemistry, 40*(12), 3234–3260.
26. Tang, Z. W., Hamid, F. S., Yusoff, I., & Chan, V. (2023). A review of PFAS research in Asia and occurrence of PFOA and PFOS in groundwater, surface water and coastal water in Asia. *Groundwater for Sustainable Development*, 100947.
27. Liddie, J. M., Schaider, L. A., & Sunderland, E. M. (2023). Sociodemographic factors are associated with the abundance of PFAS sources and detection in US community water systems. *Environmental Science & Technology, 57*(21), 7902–7912.
28. East, A., Anderson, R. H., & Salice, C. J. (2021). Per-and polyfluoroalkyl substances (PFAS) in surface water near US Air Force bases: Prioritizing individual chemicals and mixtures for toxicity testing and risk assessment. *Environmental Toxicology and Chemistry, 40*(3), 871–882.
29. Chiu, W. A., et al. (2022). Bayesian estimation of human population toxicokinetics of PFOA, PFOS, PFHxS, and PFNA from studies of contaminated drinking water. *Environmental Health Perspectives, 130*(12), 127001.
30. Cheng, Z., et al. (2021). Interpretation of reductive PFAS defluorination with quantum chemical parameters. *Environmental Science & Technology Letters, 8*(8), 645–650.
31. Fiedler, H., Baabish, A., & Sadia, M. (2022). Multivariate analysis of abiotic and biota samples for three perfluoroalkane acids. *Frontiers in Analytical Science, 2*, 954915.
32. Xie, X., et al. (2024). Per-and polyfluoroalkyl substances in a subtropical river-mangrove estuary-bay system. *Journal of Hazardous Materials, 464*, 132937.
33. G. ODCS, & I. S. E. BRAC (2023). Preliminary assessment of per-and polyfluoroalkyl substances at Umatilla Chemical Depot, Oregon.
34. Riaz, R., et al. (2023). Spatial distribution, compositional profile, sources, ecological and human health risks of legacy and emerging per-and polyfluoroalkyl substances (PFASs) in freshwater reservoirs of Punjab, Pakistan. *Science of the Total Environment, 856*, 159144.
35. Muir, D., & Miaz, L. T. (2021). Spatial and temporal trends of perfluoroalkyl substances in global ocean and coastal waters. *Environmental Science & Technology, 55*(14), 9527–9537.
36. Zhong, H., et al. (2021). Legacy and emerging per-and polyfluoroalkyl substances (PFAS) in sediments from the East China Sea and the Yellow Sea: Occurrence, source apportionment and environmental risk assessment. *Chemosphere, 282*, 131042.
37. Goodrow, S. M., Ruppel, B., Lippincott, R. L., Post, G. B., & Procopio, N. A. (2020). Investigation of levels of perfluoroalkyl substances in surface water, sediment and fish tissue in New Jersey, USA. *Science of the Total Environment, 729*, 138839.
38. Codling, G., et al. (2018). Current and historical concentrations of poly and perfluorinated compounds in sediments of the northern Great Lakes–Superior, Huron, and Michigan. *Environmental Pollution, 236*, 373–381.

39. Batzella, E., et al. (2024). Determinants of PFOA serum half-life after end of exposure: A longitudinal study on highly exposed subjects in the Veneto region. *Environmental Health Perspectives, 132*(2), 27002.
40. Ingelido, A. M., et al. (2020). Serum concentrations of perfluorinated alkyl substances in farmers living in areas affected by water contamination in the Veneto region (northern Italy). *Environment International, 136*, 105435.
41. Zhang, C., et al. (2024). Quantification of long–chain, short–chain, and ultrashort–chain liquid chromatography–amenable PFASs in water: Evaluation of approaches and tradeoffs for AFFF–impacted water. *Journal of Hazardous Materials, 466*, 133591.
42. Bilela, L. L., Matijošytė, I., Krutkevičius, J., Alexandrino, D. A., Safarik, I., Burlakovs, J., ... & Carvalho, M. F. (2023). Impact of per-and polyfluorinated alkyl substances (PFAS) on the marine environment: Raising awareness, challenges, legislation, and mitigation approaches under the One Health concept. Marine pollution bulletin, 194, 115309.
43. Donley, N., Cox, C., Bennett, K., Temkin, A. M., Andrews, D. Q., & Naidenko, O. V. (2024). Forever pesticides: A growing source of PFAS contamination in the environment. *Environmental Health Perspectives, 132*(7), 75003.
44. Liu, L.-S., et al. (2023). Per-and polyfluoroalkyl substances in ambient fine particulate matter in the Pearl River Delta, China: Levels, distribution and health implications. *Environmental Pollution, 334*, 122138.
45. Heidari, H., Abbas, T., Ok, Y. S., Tsang, D. C. W., Bhatnagar, A., & Khan, E. (2021). GenX is not always a better fluorinated organic compound than PFOA: A critical review on aqueous phase treatability by adsorption and its associated cost. *Water Research, 205*, 117683.
46. Peritore, A. F., Gugliandolo, E., Cuzzocrea, S., Crupi, R., & Britti, D. (2023). Current review of increasing animal health threat of per-and polyfluoroalkyl substances (pfas): Harms, limitations, and alternatives to manage their toxicity. *International Journal of Molecular Sciences, 24*(14), 11707.
47. Sebastiano, M., et al. (2023). Physiological effects of PFAS exposure in seabird chicks: A multi-species study of thyroid hormone triiodothyronine, body condition and telomere length in South Western France. *Science of the Total Environment, 901*, 165920.
48. Gaballah, S., et al. (2020). Evaluation of developmental toxicity, developmental neurotoxicity, and tissue dose in zebrafish exposed to GenX and other PFAS. *Environmental Health Perspectives, 128*(4), 47005.
49. Hawkey, A. B., Mead, M., Natarajan, S., Gondal, A., Jarrett, O., & Levin, E. D. (2023). Embryonic exposure to PFAS causes long-term, compound-specific behavioral alterations in zebrafish. *Neurotoxicology and Teratology, 97*, 107165.
50. Panieri, E., Baralic, K., Djukic-Cosic, D., Buha Djordjevic, A., & Saso, L. (2022). PFAS molecules: A major concern for the human health and the environment. *Toxics, 10*(2), 44.
51. von der Trenck, K. T., et al. (2018). Significance thresholds for the assessment of contaminated groundwater: Perfluorinated and polyfluorinated chemicals. *Environmental Sciences Europe, 30*, 1–20.
52. Mastrantonio, M., Bai, E., Uccelli, R., Cordiano, V., Screpanti, A., & Crosignani, P. (2018). Drinking water contamination from perfluoroalkyl substances (PFAS): An ecological mortality study in the Veneto region, Italy. *The European Journal of Public Health, 28*(1), 180–185.
53. Forsberg, J. (2022). PFAS in a Swedish wastewater treatment plant: An analysis of the effectiveness of major treatment steps on 33 PFAS.
54. Sharp, S., et al. (2021). Per-and polyfluoroalkyl substances in ducks and the relationship with concentrations in water, sediment, and soil. *Environmental Toxicology and Chemistry, 40*(3), 846–858.
55. Kleywegt, S., Raby, M., McGill, S., & Helm, P. (2020). The impact of risk management measures on the concentrations of per-and polyfluoroalkyl substances in source and treated drinking waters in Ontario, Canada. *Science of the Total Environment, 748*, 141195.

56. Garnick, L., Massarsky, A., Mushnick, A., Hamaji, C., Scott, P., & Monnot, A. (2021). An evaluation of health-based federal and state PFOA drinking water guidelines in the United States. *Science of the Total Environment, 761*, 144107.
57. Boston, C. M., Banacos, N., & Heiger-Bernays, W. (2019). Per-and polyfluoroalkyl substances: A national priority for safe drinking water. *Public Health Reports, 134*(2), 112–117.
58. Boone, J. S., et al. (2019). Per-and polyfluoroalkyl substances in source and treated drinking waters of the United States. *Science of the Total Environment, 653*, 359–369.
59. Yu, C. H., Weisel, C. P., Alimokhtari, S., Georgopoulos, P. G., & Fan, Z. T. (2021). Biomonitoring: A tool to assess PFNA body burdens and evaluate the effectiveness of drinking water intervention for communities in New Jersey. *International Journal of Hygiene and Environmental Health, 235*, 113757.
60. Moran, G. (2023). *Mass loading model of per-and Polyfluoroalkyl substances to New Hampshire's Great Bay with a focus on surface waters*. University of New Hampshire.
61. Silver, M., et al. (2023). Prevalence and source tracing of PFAS in shallow groundwater used for drinking water in Wisconsin, USA. *Environmental Science & Technology, 57*(45), 17415–17426.
62. Wanninayake, D. M. (2021). Comparison of currently available PFAS remediation technologies in water: A review. *Journal of Environmental Management, 283*, 111977.
63. Qi, L., Li, R., Wu, Y., Lin, X., & Chen, G. (2022). Effect of solution chemistry on the transport of short-chain and long-chain perfluoroalkyl carboxylic acids (PFCAs) in saturated porous media. *Chemosphere, 303*, 135160.
64. Yang, Y., et al. (2020). High-hydrophobic CF3 groups within PTFPMS membrane for enhancing the furfural pervaporation performance. *Separation and Purification Technology, 235*, 116144.
65. Crone, B. C., et al. (2019). Occurrence of per-and polyfluoroalkyl substances (PFAS) in source water and their treatment in drinking water. *Critical Reviews in Environmental Science and Technology, 49*(24), 2359–2396.
66. Zhang, Z., Sarkar, D., Biswas, J. K., & Datta, R. (2022). Biodegradation of per-and polyfluoroalkyl substances (PFAS): A review. *Bioresource Technology, 344*, 126223.
67. Helmer, R. W., Reeves, D. M., & Cassidy, D. P. (2022). Per-and Polyfluorinated alkyl substances (PFAS) cycling within Michigan: Contaminated sites, landfills and wastewater treatment plants. *Water Research, 210*, 117983.
68. Wang, D., Junker, A. L., Sillanpää, M., Jiang, Y., & Wei, Z. (2023). Photo-based advanced oxidation processes for zero pollution: Where are we now? *Engineering, 23*, 19–23.
69. Lorpaiboon, W., & Ho, J. (2023). High-level quantum chemical prediction of C–F bond dissociation energies of perfluoroalkyl substances. *The Journal of Physical Chemistry A, 127*(38), 7943–7953.
70. Kancharla, S. (2022). *Fluorinated surfactants: Self-organization and binding in aqueous media*. State University of New York at Buffalo.
71. Dixit, F., Barbeau, B., Lompe, K. M., Kheyrandish, A., & Mohseni, M. (2021). Performance of the HSDM to predict competitive uptake of PFAS, NOM and inorganic anions by suspended ion exchange processes. *Environmental Science: Water Research & Technology, 7*(8), 1417–1429.
72. Pauletto, P. S., & Bandosz, T. J. (2022). Activated carbon versus metal-organic frameworks: A review of their PFAS adsorption performance. *Journal of Hazardous Materials, 425*, 127810.
73. Lee, T., Speth, T. F., & Nadagouda, M. N. (2022). High-pressure membrane filtration processes for separation of per-and polyfluoroalkyl substances (PFAS). *Chemical Engineering Journal, 431*, 134023.
74. Wang, R., et al. (2022). A tunable porous β-cyclodextrin polymer platform to understand and improve anionic PFAS removal. *ACS Central Science, 8*(5), 663–669.
75. Zhang, Y. (2023). *Adsorption of Long and short per-and Polyfluoroalkyl substances (PFAS) onto granular activated carbon and porous organic polymers*. University of South Florida.

76. Rhakho, N., Yadav, S., Jinagi, M., Altaee, A., Saxena, M., Jadhav, A. H., & Samal, A. K. (2024). Mitigating PFAS contaminants in water: A comprehensive survey of remediation strategies. Journal of Environmental Chemical Engineering, 113425.
77. Abdelraheem, W. H. M. (2020). *Catalytic and photocatalytic removal of contaminants of emerging concerns (CECs) and per-/Polyfluoroalkyl substances (PFAS) from wastewater effluents for water reuse applications*. University of Cincinnati.
78. Westreich, P., Mimna, R., Brewer, J., & Forrester, F. (2018). The removal of short-chain and long-chain perfluoroalkyl acids and sulfonates via granular activated carbons: A comparative column study. *Remediation Journal, 29*(1), 19–26.
79. Sonmez Baghirzade, B., Zhang, Y., Reuther, J. F., Saleh, N. B., Venkatesan, A. K., & Apul, O. G. (2021). Thermal regeneration of spent granular activated carbon presents an opportunity to break the forever PFAS cycle. *Environmental Science & Technology, 55*(9), 5608–5619.
80. Borthakur, A., et al. (2022). Rechargeable stormwater biofilters: In situ regeneration of PFAS removal capacity by using a cationic polymer, polydiallyldimethylammonium chloride. *Journal of Cleaner Production, 375*, 134244.
81. Gagliano, E., Falciglia, P. P., Zaker, Y., Karanfil, T., & Roccaro, P. (2021). Microwave regeneration of granular activated carbon saturated with PFAS. *Water Research, 198*, 117121.
82. Boyer, T. H., Ellis, A., Fang, Y., Schaefer, C. E., Higgins, C. P., & Strathmann, T. J. (2021). Life cycle environmental impacts of regeneration options for anion exchange resin remediation of PFAS impacted water. *Water Research, 207*, 117798.
83. Mastropietro, T. F., Bruno, R., Pardo, E., & Armentano, D. (2021). Reverse osmosis and nanofiltration membranes for highly efficient PFASs removal: Overview, challenges and future perspectives. *Dalton Transactions, 50*(16), 5398–5410.
84. Srivastava, A., Singh, R., Rajput, V. D., Minkina, T., Agarwal, S., & Garg, M. C. (2022). A systematic approach towards optimization of brackish groundwater treatment using nanofiltration (NF) and reverse osmosis (RO) hybrid membrane filtration system. *Chemosphere, 303*, 135230.
85. Ketharani, J., et al. (2022). A comparative study of community reverse osmosis and nanofiltration systems for total hardness removal in groundwater. *Groundwater for Sustainable Development, 18*, 100800.
86. Zhang, M., Wang, W., Gong, T., Wu, Y., & Chen, G. (2024). Cutting-edge technologies and relevant reaction mechanism difference in treatment of long-and short-chain per-and polyfluoroalkyl substances: A review. *Chemosphere, 354*, 141692.
87. Ellis, A. C., et al. (2022). Pilot study comparison of regenerable and emerging single-use anion exchange resins for treatment of groundwater contaminated by per-and polyfluoroalkyl substances (PFASs). *Water Research, 223*, 119019.
88. Virpiranta, H., Leiviskä, T., Taskila, S., & Tanskanen, J. (2022). Bioregeneration of sulfate-laden anion exchange resin. *Water Research, 224*, 119110.
89. Baskar, A. V., et al. (2022). Recovery, regeneration and sustainable management of spent adsorbents from wastewater treatment streams: A review. *Science of the Total Environment, 822*, 153555.
90. Murray, C. C., et al. (2023). PFAS adsorbent selection: The role of adsorbent use rate, water quality, and cost. *Journal of Hazardous Materials, 454*, 131481.
91. Mojiri, A., Zhou, J. L., Ozaki, N., KarimiDermani, B., Razmi, E., & Kasmuri, N. (2023). Occurrence of per-and polyfluoroalkyl substances in aquatic environments and their removal by advanced oxidation processes. *Chemosphere*, 138666.
92. Soker, O., Hao, S., Trewyn, B. G., Higgins, C. P., & Strathmann, T. J. (2023). Application of hydrothermal alkaline treatment to spent granular activated carbon: Destruction of adsorbed PFASs and adsorbent regeneration. *Environmental Science & Technology Letters, 10*(5), 425–430.
93. Loganathan, P., Kandasamy, J., Ratnaweera, H., & Vigneswaran, S. (2024). Treatment trends and hybrid methods for the removal of poly-and Perfluoroalkyl substances from water—A review. *Applied Sciences, 14*(6), 2574.

94. Sorn, S., Hara-Yamamura, H., Vet, S., Xiao, M., Hoek, E. M. V., & Honda, R. (2023). Biological treatment of perfluorooctanesulfonic acid (PFOS) using microbial capsules of a polysulfone membrane. *Chemosphere, 329*, 138585.
95. Medina, R., Pannu, M. W., Grieco, S. A., Hwang, M., Pham, C., & Plumlee, M. H. (2022). Pilot-scale comparison of granular activated carbons, ion exchange, and alternative adsorbents for per-and polyfluoroalkyl substances removal. *AWWA Water Science, 4*(5), e1308.
96. Ilić, N., et al. (2023). Ultrasonic degradation of GenX (HFPO-DA)–performance comparison to PFOA and PFOS at high frequencies. *Chemical Engineering Journal, 472*, 144630.
97. Roy-Basu, A., Bharat, G. K., & Chakraborty, P. (2022). A review on distribution and removal techniques for Perfluoroalkyl and Polyfluoroalkyl substances (PFASs) in water and wastewater. *Legacy and Emerging Contaminants in Water and Wastewater*, 389–408.
98. Saawarn, B., Mahanty, B., Hait, S., & Hussain, S. (2022). Sources, occurrence, and treatment techniques of per-and polyfluoroalkyl substances in aqueous matrices: A comprehensive review. *Environmental Research, 214*, 114004.
99. Gravesen, C. R., Lee, L. S., Alukkal, C. R., Openiyi, E. O., & Judy, J. D. (2023). Per-and polyfluoroalkyl substances in water treatment residuals: Occurrence and desorption. *Journal of Environmental Quality, 54*, 31.
100. Sharma, N., et al. (2024). A comprehensive review on the need for integrated strategies and process modifications for Per-and polyfluoroalkyl substances (PFAS) removal: current insights and future prospects. *Case Studies in Chemical and Environmental Engineering*, 100623.
101. Ehsan, M. N., Riza, M., Pervez, M. N., Li, C.-W., Zorpas, A. A., & Naddeo, V. (2024). PFAS contamination in soil and sediment: Contribution of sources and environmental impacts on soil biota. *Case Studies in Chemical and Environmental Engineering*, 100643.
102. Jia, Y., Shan, C., Fu, W., Wei, S., & Pan, B. (2023). Occurrences and fates of per- and polyfluoralkyl substances in textile dyeing wastewater along full-scale treatment processes. *Water Research, 242*, 120289.
103. Pollack, J. B., Carey, I. Q., & Xu, V. Y. (2024). Regulation of products with PFAS. *The Environmental Law Reporter, 54*, 10148.

Chapter 16
Research Gaps and Future Directions for Addressing Microplastics and PFAS: Impacts in Aquatic Environments

Mohammad Yaasar Moosa, Mohanraj Thangarasu, Indra Sulania, Jeyanthinath Mayandi, and Vairathevar Sivasamy Vasantha

1 Introduction

Microplastics (MPs) and per- and polyfluoroalkyl substances (PFASs), which are colloquially recognized as forever chemicals and are also called anthropogenic substances, are omnipresent pollutants in our daily activities [1, 2]. The first synthetic plastics were synthesized by Baekeland in 1909. Since then, they have become ubiquitous in modern society because of their low production cost, durability, and versatility [3, 4]. Plastic manufacturing has skyrocketed from 2 million tons in 1950 to 400.3 million tons in 2022, with projections reaching 1480 million tons by 2050, nearly triple the weight of the current human population and outweighing the mass of fish [12]. China held the top spot, accounting for 32% of global plastic production, and North America followed closely behind in second place, holding a 17% share [5]. However, their widespread use is associated with significant environmental and health hazards. MPs (< 5 mm size), first identified by Thompson et al. in 2004, have become a major environmental concern and exhibit distinct physicochemical characteristics depending on their source, color, shape, polymer type, and density [6]. According to the OECD Global Plastic Outlook Database, 9% of plastic

M. Y. Moosa · M. Thangarasu · V. S. Vasantha (✉)
Department of Natural Products Chemistry, School of Chemistry, Madurai Kamaraj University, Madurai, Tamil Nadu, India

I. Sulania
Inter University Accelerator Centre, New Delhi, India
e-mail: indra@iuac.res.in

J. Mayandi
Department of Materials Science, School of Chemistry, Madurai Kamaraj University, Madurai, Tamil Nadu, India
e-mail: jeyanthinath.chem@mkuniversity.ac.in

M. K. Jindal et al. (eds.), *Emerging Contaminants in Water*, Emerging Contaminants and Associated Treatment Technologies,
https://doi.org/10.1007/978-3-031-89591-3_16

waste is recycled, 19% is incinerated, 50% ends up in landfills, and 22% is mismanaged [7]. Continuous massive production and rapid dispersal lead to larger amounts of plastic debris, and aquatic species become trapped and entangled in plastic webs. These materials break down into floatable and incessant micro- to nanoplastics through physical, chemical, and biological degradation in the environment. Packaging (44%) and building and construction (18%) are the most contributing sources [7–10]. The plastic debris in the environment are of different sizes as follows: size ≥25 mm (macroplastics), <5–25 mm (mesoplastics), 5–1 μm (microplastics), and 1–1000 nm (nanoplastics). Alarmingly, 1.5 million tons of primary MPs enter aquatic environments annually, and the most common types include PET, (Polyethylene Terapthalate) polyester, LDPE, HDPE, PVC, PP, PA, PS, ABS, and PTFE [10, 11]. MPs can penetrate all aquatic habitats via rivers, lakes, and sediments to open seas and deep oceans. Recently, Naixin Qian et al. estimated that the average liter of bottled drinking water contains approximately 25 million MPs, approximately 90% of which are nanoplastics. Furthermore, the hydrophobic nature of these particles facilitates the sorption of pollutants, which become sources and transport vectors of noxious chemical substances that ultimately accumulate in aquatic life.

In series, fluorinated polymeric chains have at least one completely fluorinated methyl (-CF_3) or methylene carbon (-CF_2-) group; in other words, a family of fluorine-swaddled carbon chains with abundant homologs, isomers, and terminal functional groups is associated with their persistence and toxicity [1]. Owing to its strong electronegativity and small atomic size, fluorine imparts properties such as chemical and electrical inertness; hydrophobicity; oleophobicity; lipophobicity; temperature and radiation resistance; repellency to oil, water, and stain; corrosion protection; and durability. The widespread application of fluoropolymers in commercial and industrial products, including nonstick coatings (Teflon®), sealants (Gore-Tex®), lubricants, food packaging, firefighting foams, cosmetics, medical devices, and textiles, poses significant environmental risks [13]. Globally, the cumulative emission of PFAS is estimated to be ~46,000 tons. PFASs with longer chains (>C_6) are more prevalent and environmentally persistent and tend to bioaccumulate.3b) Among the thousands of PFAS classes, perfluoroalkyl sulfonic acids ($C_nF_{2n+1}SO_3H$, PFSAs) and perfluoroalkyl carboxylic acids ($C_nF_{2n+1}COOH$, PFCAs) are resistant to degradation and have higher accumulation potential. PFOA and PFOS are the predominant final stable intermediates and well-studied PFAS compounds, and their average removal half-lives exceed 5.4 years and 3.8 years, respectively [14]. Gellrich et al. reported that short-chain PFASs (<C8) were prevalent in tap water samples, with a maximum concentration of 42.7 ng.L^{-1} and estimated that the average daily human intake of PFOS and PFOA ranges from 0.17 to 0.21 ng.kg^{-1} body weight/day [16]. In 2016, the USEPA set a lifetime drinking water health advisory level of PFAS in the range of 300–7000 ng.L^{-1} for PFASs with C_4–C_7 chain lengths and 70 ng.L^{-1} for PFOA and PFOS [17]. These ubiquitous and persistent pollutants discharged as landfills and transported by river and stormwater runoff, adsorbed by MPs via noncovalent interactions [18], enter the food web, increasing their detrimental impact on aquatic environments. Schematic

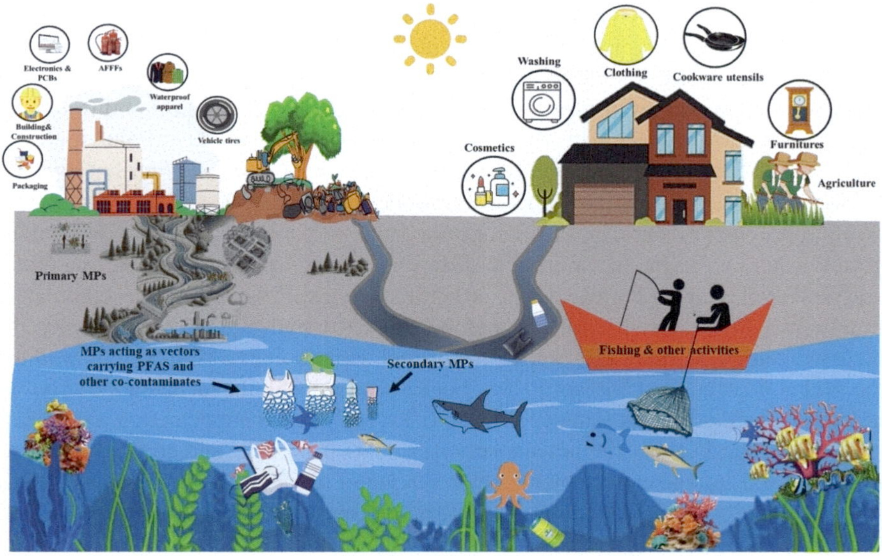

Fig. 16.1 Schematic overview of the sources and distributions of MPs and PFASs in aquatic environments

overview of the sources and distributions of MPs and PFASs in aquatic environments are illustrated in Fig. 16.1.

2 Environmental Footprint of MPs and PFASs

MPs originate from a variety of sources, which can be categorized into primary and secondary sources. Primary MPs are intentionally manufactured as small plastic particles and used in products such as plastic microbeads and fibers released from the textile and cosmetic industries, industrial abrasives, and plastic pellets (nurdles). They account for a significant portion of the MPs found in the environment and are estimated to constitute 15–31% of the total MPs in oceans [19]. Secondary MPs are released during the degradation of mismanaged plastic debris, including surface blasting/molding materials, plastic bags, bottles, fishing nets, and vehicle tires. They breakdown over time into smaller fragments due to physicochemical and biological stimuli and constitute 69–81% of the MPs in the ocean [20]. Recent research has indicated that landfill leachate is responsible for 80–90% of the MPs in aquatic environments [21]. These include construction materials, personal care products, and industrial wastes, which are distributed via sewage sludge and river runoff from urban areas to the ocean. These distributed MPs are inadvertently consumed by marine organisms, causing severe obstruction or damage to their digestive tracts. Additionally, additives such as AFFFs and plasticizers, along with cocontaminants such as PFAS, adhere to the MPs during distribution. These persistant contaminants accumulate and biomagnify, eventually entering the food chain through seafood, causing serious health threats to both seafood sources and predators.

PFAS contamination originates from industrial products, and firefighting foams often contain low levels of PFAS and are treated through WWTPs. Point sources, such as WWTPs and AFFF-based firefighting sites, are significant contributors, and PFASs are dispersed via atmospheric transport, surface runoff, and the breakdown of fluoropolymer precursors. PFAS concentrations vary widely in surface water and groundwater, with bioaccumulation differing among species, making accurate assessment of their ecological effects difficult. Among all PFAS, PFOA accumulates in relatively high concentrations in seafood sources. Challenges include the incomplete evaluation of many PFAS compounds and their ecological impacts, as well as the global distribution of novel PFAS. Future research should focus on identifying unknown precursors, assessing ecological risks, and improving detection methods.

Although earlier reports indicate serious threats from MPs and cocontaminants, i.e., PFASs in aquatic environments, there is a major research gap in the understanding of the sources and differentiation of different types of MPs and PFASs. Many of them found in the environment cannot be traced back to their original sources because of the multiple degradation states involved during weathering until they reach stable end products. For example, once plastic breaks down into smaller fragments, it becomes challenging to determine its original form or material. This complicates efforts to manage and mitigate such pollution. Understanding these sources is crucial for developing methods and policies for identification, treatment, production, and consumption [7]. Studies suggest that MPs can also be transported through air, although the sources of airborne MPs remain largely unidentified.

In response to the growing concern over MP pollution, numerous countries and international bodies have implemented policies and regulations to mitigate this issue. The USA led the Microbead-Free Waters Act of 2015, which insists prohibiting the use of plastic microbeads in personal care products. Members of the European Parliament advised the European Commission to establish an EU-wide ban on intentionally added MPs in cosmetics and detergents by 2020 [22, 23]. Moreover, the Fourth UN Environment Assembly in 2019 witnessed a global commitment to significantly reduce single-use plastics by 2030 [24]. Recently, environmental agencies in five European countries (Germany, Denmark, Sweden, the Netherlands, and Norway) submitted a Restriction Proposal to the EU on PFASs and fluorinated MPs. In that proposal, the ECHA planned to restrict the manufacturing of more than 12,000 PFASs across Europe [25].

3 Importance of Addressing Microplastics and PFAS

MP infiltration in aquatic livestock results in reduced food intake, developmental disorders, blockage of the gastrointestinal system, and bowel obstructions [26]. In vivo studies confirmed growth and reproductive impairment, oxidative stress, neurological disorders and apoptosis, DNA damage, gut microbiota modifications, and histopathological alterations in aquatic organisms. The bioaccumulation potency of MPs increases with decreasing size, posing a risk across the aquatic food web. Nearly 700 aquatic organisms worldwide, including fishes, sea turtles, penguins, and various crustaceans, are adversely affected by MPs [27].

With the support of MPs, the distribution of PFASs is becoming a growing global issue because of their environmental persistence, long half-lives, bioaccumulation potential, toxicity and environmental mobility [2]. Every single PFAS is transformed into recalcitrant stable end compounds, i.e., PFOS and PFOA [28]. All PFASs are endocrine-disrupting compounds, but there is still no clear exposure pathway for PFASs. The degradation of precursor polyfluoro substances into amphiphilic end products is a major cause of environmental distribution [29]. The prolonged health effects of PFAS pollutants are not yet well known but affect biodiversity via mutation and recombination in marine organisms such as mussels, oysters, copepods, freshwater turtles, and apex predators such as dolphins and polar bears and their seal prey. Mainly, they block the gut microbiome by mimicking food sources, destroy mucus layers, and damage gills during aquatic life [30]. The health effects of MPs and PFAS contaminated seafood consumption and their accumulation sites in the human 0body are presented in Fig. 16.2.

Despite the alarming statistics and growing body of research documented in numerous scientific publications, a significant knowledge gap persists in addressing MPs in aquatic environments. Public awareness remains limited, and effective solutions to address this issue are urgently needed. While the research community has actively investigated MPs, crucial knowledge gaps hinder our ability to effectively manage this environmental crisis. These gaps lie primarily in the areas of detection and remediation of MPs and associated cocontaminants within aquatic ecosystems. Addressing these critical areas is paramount in mitigating the adverse effects of MPs on both human health and the health of aquatic habitats.

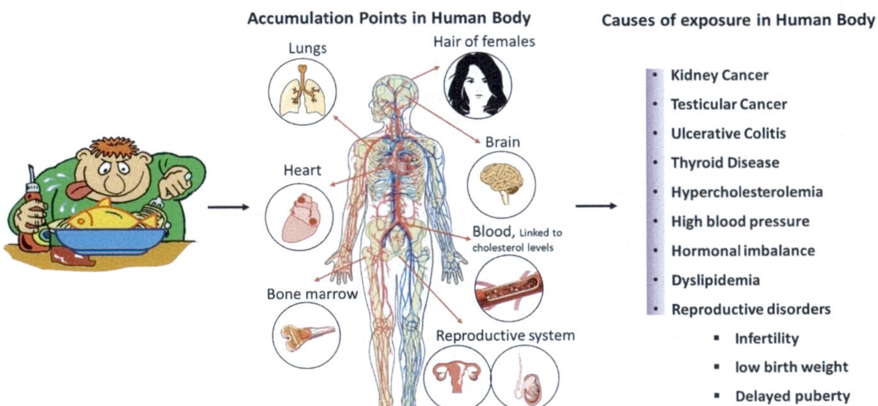

Fig. 16.2 Health effects of MPs and PFAS contaminated seafood consumption and their accumulation sites in the human body

4 Detection and Quantification of Microplastics and PFASs in Aquatic Environments

MPs in different environments require different detection techniques with multifaceted approaches because of their diverse sizes, shapes, colors, and polymeric compositions. Visual identification [31, 32], such as stereomicroscopic [33], morphological [34], and spectroscopic [35, 36] analyses; destructive techniques, such as thermogravimetric and Pyr-GC/MS [37, 38]; and advanced techniques, such as satellite and aerial observations, have been reported for the estimation of MPs. They are costly and labor-intensive, and the results vary greatly among observers, which is subject to error [39–41]. FTIR, ATR, NIR-based hyperspectral imaging, Raman spectroscopic procedures [42–45], and fluorescence staining with Nile red dye derivatives have also been overlooked in the characterization and quantification of MPs, which are relatively fast and easy sample preparations. In these techniques, defining the edge-driven interaction of an MP and the surrounding uncertainty affects the detection limit. These conventional methods present several limitations: they often result in inaccurate estimations, prolonged detection times, and high false-positive results. There is a critical need for more efficient, accurate, and cost-effective analytical techniques to enhance the environmental monitoring of MPs. The integration of automatic particle detection with advanced electrochemical and optical methods [46] and image analysis enhances the detection efficiency and accuracy of MPs in aquatic environments.

Detecting PFASs in aqueous environments is difficult because of their low concentrations in parts per trillion (ppt) and cocontaminants. Globally, numerous research groups are actively investigating the contamination and remediation of PFAS. The USEPA recommended LC-MS/MS methods integrated with polystyrene-divinylbenzene SPE for highly sensitive detection of 29 PFASs in a short analysis time [49–51]. Similarly, semiquantitative assays such as TOP and TOF assays offer a comprehensive analysis of PFASs, encompassing short-chain and oxidizable precursors [50–52]. In India, several research groups are working on PFAS chemical analysis. Binu et al. used HPLC to separate PFAS chemicals and extract and analyze PFBA, PFOA, and PFNA.55 Namasivayam Vasudevan et al. used LC-MS to screen PFASs and reported L-PFBS (up to 136.27 $ng.L^{-1}$) and PFOA (up to 77.61 $ng.L^{-1}$) in water sources of Chennai [54]. Rajiv Janardhanan et al. analyzed PFOA and PFOS in rivers, i.e., Noyyal and Cauvery, and lakes around Chennai, via ultrafast liquid chromatography [55]. Although these methods are well-established and validated, they require complicated sample preparation and preconcentration, expensive instrumentation, long analysis times, and high costs per sample and are limited by a lack of portability. The potential analytical techniques for the detection and quantification of MPs and PFAS are shown in Fig. 16.3 and Table 16.1.

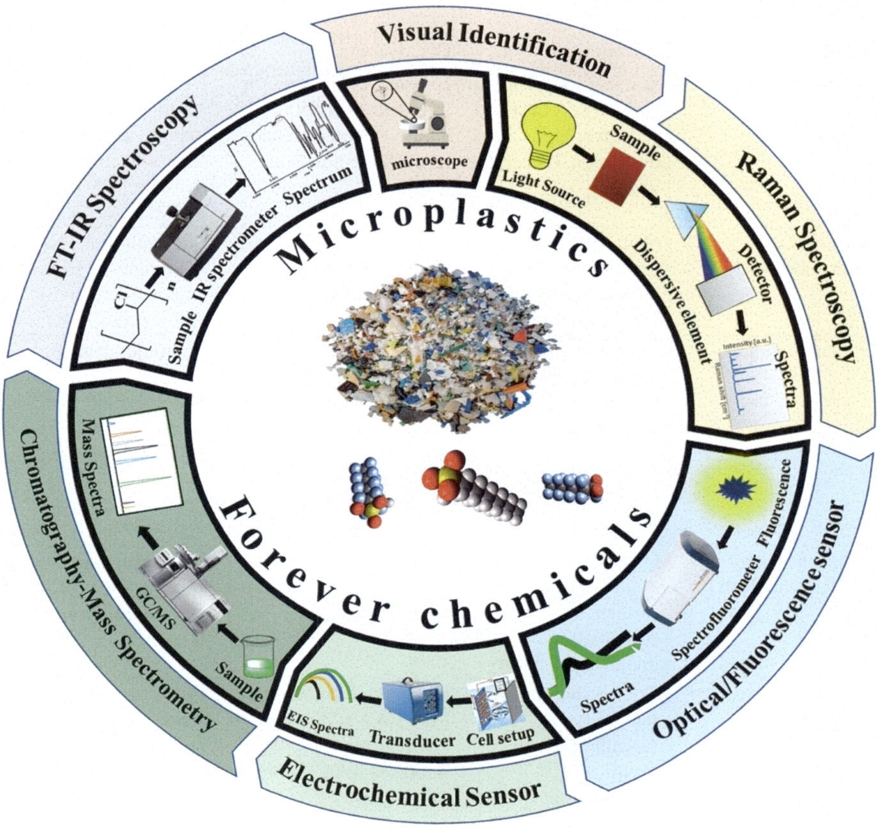

Fig. 16.3 Potential analytical techniques for the detection and quantification of MPs and PFAS

Table 16.1 Strengths and limitations of the analytical techniques used for the detection and quantification of MPs and PFAS [88–96]

Method	Advantages	Disadvantages	Future directions
MPs [88–93]			
Visual identification: Optical microscope	Suitable for macro/ meso size MPs, simple, on-site, easy to use, low-cost	Time-consuming, difficulty in identifying type, size, and distribution of MPs, high subjectivity, misclassification rates up to 70%	Can be applied as a preselection step before using advanced techniques
Stereo electron microscope (SEM/ TEM-EDAX), DLS	Suitable for μm size MPs, high-resolution morphology, and chemical composition analysis	Tedious sample preparation; MPs having almost same electron densities complicates in distinguishing types by TEM	Coupling with other methods (Pyr-GC/MS, μ-Raman, FIIR) for quantification as well as imaging of MPs

(continued)

Table 16.1 (continued)

Method	Advantages	Disadvantages	Future directions
Fluorescence microscopy Staining with dyes	Suitable for <200 μm size MPs, robust and sensitive, high-resolution 3D surface imaging	Expensive, Nonspecific interactions from additives, interferences result false-positive results	Require validation of staining protocol, explore selective and high-affinity staining dyes for MPs over NOMs
Atomic force microscopy (AFM)	Nanoscale-level topographical imaging of MPs with high resolution	Applicable only at controlled lab scale, expensive, and time-consuming	Integration of TERS with nano-FTIR will enhance sensitivity and resolution
Optical techniques:	Nondestructive, highly accurate, label-free, real-time, and robust	Time-consuming, costly, sophisticated techniques and lab-scale analysis, variable results due weathering, wavelength limitations, need consolidated library reference for validation	Photonics-based portable integrated in situ detection systems in aquatic environments are yet to be explored for real-time analysis
IR spectroscopy FT-IR, near-IR, ATR	Quantitative, suitable to identify and confirm the composition of MPs, imaging possibility	Limited accessibility of the instrument, NOMs as interference	Improving adequate spectral library toward weathered byproducts from MPs
Raman spectroscopy RS imaging, μ-Raman, ASPEx-μ-Raman	Compatible for most sample types, valuable insights on chemical composition, spatial distribution, and size of MPs, sensitive	Unable to differentiate MPs from additives and NOM pigments, fluorescent background interference	Focus on μ-Raman for spatial imaging, SERS for improved detection of weak signals, coupling deep learning with RS through neural network
Spectroscopic satellite and aerial observation, hyperspectral imaging	Screening enormous area from rural to remote, excellent tool for plastic debris and precursors of MPs, facilitate frequent monitoring and mapping studies	Low-resolution, unsuitable for small MPs, unable to provide quantitative results, require standardization of image processing software, continuous oceanographic monitoring	Extendable to assess marine litter and monitor the water cycle and ocean fluxes via IMDOS integrating ship, buoy, and aircraft with automated systems
Optical (bio)sensors	Sensitive, simple, inexpensive, easy-to-use and fast, low-cost	Indirect quantification through the use of redox dyes or QDs, LOD in the ppm or ppb range for optical assays (>regulatory permissible limits)	Designing specific receptor to enhance selectivity and sensitivity

(continued)

Table 16.1 (continued)

Method	Advantages	Disadvantages	Future directions
Thermal decomposition chromatography	Suitable for quantification of NPs and its OPAs via pyrolysis byproducts	Destructive, tedious, and time-consuming, access restricted to remote areas, needs well-trained experts	These techniques could be independently utilized for accurate mass determination, coupling with advanced PLE methods for quantifying plastic debris and
LC-MS	Size exclusion-based MP characterization	Low-resolution and high cost, no information regarding number, size, or shape	Sample preparation strategies
Pyrolysis-GC-MS	Identifies the chemical composition and type of polymer, no need for sample pretreatment	Complex process, lab-scale, unsuitable for mixed samples	Explores different working modes of Pyr-GC/MS for specific MPs
TGA/TED-GC/MS, DSC	Selective, sensitive, provide detailed chemical information and quantitative multicomponent analysis, have large sample capacity	Lack in distinguishing complex MPs, operator dependent, low LOD and LOQ	Sample preparation from complex NOMs
Electrochemical (bio) sensors	Quantitative and qualitative information of selected contaminants at ultratrace levels, flexibility and portability in measurements, potential of simultaneous and multiplexed assays	Lack of selectivity and sensitivity at environmental concentrations	Miniaturization of the device to reduce the energy requirements and prioritize validation against reference analyses and conducting demonstration studies to showcase platform capabilities. Design more sensitive and selective receptor materials to reach permissible limits
PFAS [94–96]			
Chromatography coupled with mass spectrometric methods (GC/LC/HPLC with MS/MS or Orbitrap/ToF-MS)	Quantitative and selective, sensitive in the range of ppt, well-established and validated	High cost per sample (>$100), off-site analysis, sophisticated instrumentation, tedious and time-consuming, requires skilled technicians	Focuses on enhancing size fractionation to identify size distribution, significant capital investment

(continued)

Table 16.1 (continued)

Method	Advantages	Disadvantages	Future directions
Optical techniques	Nondestructive, highly accurate, label-free, real-time, and robust	Costly and time-consuming, lab-scale analysis, variable results due to degradation, need consolidated library reference for validation	AI-integrated portable in situ detection systems in aquatic environments are yet to be explored
FT-IR	Quantitative	Limited accessibility of the instrument, NOMs as interference	Standard database formulation
Raman spectroscopy SERS surface (AuNPs/AgNPs)	Sensitive, characteristic peaks, chemical composition		Creating new SERS substrates, portable devices with real-world applications, increasing the range of PFAS compounds,
Optical (bio)sensors	Sensitive, simple, inexpensive, easy-to-use and fast, low-cost	Indirect quantification through the use of redox dyes or QDs, LOD in the ppm or ppb range for optical assays (>regulatory permissible limits)	Designing specific receptor to enhance selectivity and sensitivity
Semiquantitative assays: TOP and TOF	Used for prescreening tool, enable measurements of PFAS precursors and specific analysis of short-chain and oxidizable precursors	Limited selectivity and sensitivity require expertise	Nonselective oxidation process results uncertainty in the TOP assay products Further research should be conducted to minimize the variation. Confirmation and validation of the TOP assay protocol is required
Electrochemical (bio) sensors (impedimetric, chemiluminescence), nanomaterials, QDs, MIPs	Versatile, simple, fast, inexpensive, and portable, highly sensitive Could provide a total PFAS content	Indirect quantifications and selectivity issues (interference of cocontaminants and NOMs) Electrochemical assays need to meet regulatory permissible limits Need further QC/QA validation	Focus on developing smartphone device for on-field applications with multichannel platforms for real-time sampling and analysis

4.1 *Visual Inspection*

Visual inspection is a popular method for identifying and classifying MPs (>1 mm). It involves direct examination of macroparticles or the use of tools such as stereoscopes and optical and electron microscopes to identify morphological aspects. This method offers several advantages, including the ability to categorize particles on the basis of physical attributes (size, color, and shape). However, a significant drawback of visual inspection is its subjectivity and weak signals emitted by MPs under microscopic analysis [56]. AFM-based imaging via a sharp electric tip-cantilever setup provides high-quality topographical information of MPs [57].

To improve the accuracy and efficiency of visual inspection, researchers have explored various methods, such as the use of selective staining dyes (Oil Red EGN, eosin B, Hostasol Yellow 3G, and Rose Bengal), to increase visibility via fluorescence microscopy. Furthermore, techniques such as prodding particles with needles or using a heated needle tip to test for plasticity have been proposed to aid in differentiating MPs from NOMs. For a more comprehensive analysis, visual inspection should be combined with complementary techniques such as spectroscopy to identify and quantify MPs. Visual inspection-based analysis is not possible for fluoropolymer-based organic contaminants.

4.2 *Optical Techniques*

Optical techniques driven by light-plastic interactions can be utilized to analyze the intrinsic optical properties of MPs and PFAS. These methods require postsample preparation and lab analysis, making real-world applications difficult.

4.2.1 FTIR and NIR Spectroscopy

Infrared spectroscopy is widely employed for identifying and characterizing the functional groups of MPs and PFAS present in aquatic environments. The technique works by measuring the change in dipole moment (molecular vibrations) during interaction with infrared light, producing a vibrational spectrum that can be matched with the spectrum of known polymer materials. However, the application of FTIR to different water matrices presents unique challenges, i.e., organic and inorganic contaminants that can interfere with the FTIR spectra [58]. Pretreatment steps such as the National Oceanic and Atmospheric Administration (NOAA) method [35], filtration [59], centrifugation [60], and staining [61] to obtain clear and unambiguous spectra further complicate the analysis. The presence of additional spectral bands resulting from weathering or continuous degradation leads to misidentification. Moreover, an unavailable spectral library may not accurately represent the degraded state [62]. Several advanced techniques have been developed to improve the accuracy and detection of MPs and PFAS decomposition products; i.e., coupling FTIR and microscopy (narrow infrared beam, μ-FTIR-FPA) focuses on selected spots on the MPs and enables the detection of MPs up to 10 μm in size, with an accuracy of more than 98% [63]. Additionally, the choice of FTIR mode (transmission, reflection, or diffuse reflectance) is not the same, and further research is needed for standardized methods and comprehensive spectral databases [64].

4.2.2 Raman Spectroscopy

Raman spectroscopy (RS) works on the basis of the polarizability of chemical bonds and provides a fingerprint molecular spectrum, which is a suitable tool for detecting and identifying MPs and PFASs in aquatic environments. SERS with nanostructured surfaces and μ-RS coupled with microscopy are effective at detecting MPs, including PE, PVC, PP, PC, PTFE, PS, PUR, PA, and polyolefin elastomers [65]. PE, PS, and PP are the most common sources of marine MPs, with characteristic RS signals falling in the range of 2800–3000 cm^{-1}. Raman spectroscopic mapping enabled the successful distinction between PE and PP, even when they were in close proximity within mixed microplastic samples [66]. Notably, MPs and PFASs subjected to natural weathering processes result in changes in physical and chemical properties, which do not interfere with the RS results. Cole et al. proposed CARS microscopy integrated with confocal laser scanning and a synchronized dual-wavelength picosecond laser source for the identification of MPs of various sizes. To differentiate and quantify PFAS, modifying Au/Ag nanostructured surfaces with specific alkanethiol molecules is recommended. Additionally, integrating machine learning algorithms, such as LDA and SVM, can help improve discrimination capabilities and achieve low detection limits in SERS [67]. The development of automated systems will facilitate large-scale and continuous environmental monitoring, reducing the need for manual intervention.

4.2.3 Spectroscopic Imaging Via Satellite and Aerial Observation

ESA's Sentinel-2 satellites, which are equipped with multispectral instruments, measure floating macroplastics in coastal marine environments via FDI and spectral signature analysis. By employing methods such as Gaussian mixture and naive Bayes classification, these satellites can achieve up to 86% accuracy in detecting floating MPs.71 Spaceborne radars using optical, SAR, hyperspectral, and TIR sensors have also been employed for the oceanic imaging of MPs [69]. Service providers such as Sately help analyze large-area PFAS contamination via multispectral and hyperspectral imaging from satellites, UAVs, planes, and fixed cameras, providing both alerts and qualitative results [70].

4.2.4 Optical (Bio)Sensors

The current trend focuses on developing receptors using nanostructured materials, quantum dots, proteins, dyes, and polymers to capture MPs and PFAS. The interaction between the receptor and pollutants results in a conformational change or a biological/chemical/electrochemical reaction, which is transduced into a bio−/electrochemiluminescent, fluorescent, colorimetric, and/or current signal [71]. Small dye molecule-based optical complexation assays [72], such as the astkCARE kit, utilize probes including methylene blue, eosin Y, erythrosin B, Janus Green B, Nile blue A, toluidine blue, and fluorine-swaddled amplifying fluorescent polymers, squaraine dye, and PFOS-specific mesoester-BODIPY receptors for colorimetric/fluorescence-based detection and quantification of PFAS [73]. Most of the reported sensors focus on PFOA and PFOS. To date, only one sensor has been developed for next-generation GenX fluorochemicals [74]. These reports are based on indirect quantification employing redox dyes, nanoparticles, and QDs, with detection limits typically in the ppm or ppb range, often falling short of regulatory permissible limits. In some reports, researchers have attempted to introduce molecular memory (MIPs) by covering the polymer coating over the surface of indicators for selectivity. However, the LOD is compromised [75]. The lack of optically active functionalities and chromophores, lack of a specific receptor design, minimal to no analyte-receptor interaction at the molecular level, uncertainty in precursor sources, and slow diffusion rates lead to high LODs.

4.3 Thermal Degradation and Chromatographic Analysis

Thermal decomposition combined with mass spectral analysis of decomposed end products is considered efficient for detecting the types of MPs and PFAS, which includes Pyr-GC/MS, TED-GC/MS, TGA-DSC, and TGA-TD-GC-MS. Pyrolysis depolymerizes MPs in the absence of oxygen, followed by MS analysis of evolved gases. This method can identify the chemical composition of plastic polymers by analyzing their characteristic thermal degradation products and related organic additives. It accurately determines the polymer type by comparison with pyrolysis reference maps of known pure polymers [76]. Fries et al. used Pyr-GC/MS to analyze beach soil samples, and the results indicated the presence of chlorinated PE, PP, PS, and polyamide particles with additives such as diethylhexyl phthalate, dibutyl phthalate, and diethyl phthalate [77]. Measuring MP weight loss as a function of time (or temperature) in TGA/TED-GC/MS can handle complex environmental samples, as it combines TGA and TD-GC/MS to quickly analyze and quantify common polymers in the environment (PE, PP, PS, PA6, and PET) [13].

Conventionally, extraction uses SPE/IPE/LLE/SPME, followed by chromatographic techniques (LC/HPLC/UPLC) combined with mass spectral analysis (MS/MS), and semiquantitative assays (TOP and TOF) are utilized and fine-tuned for the analysis of PFAS in various contamination matrices.82 Tabar et al. proposed a

PFOA-sensitive MIP receptor using a heat-transfer method, which operates on the basis of changes in the thermal conductivity of the polymer matrix in a flow cell in relation to the PFOA concentration, and its detection limit is 22 pM. With the ability to improve selectivity toward different PFAS, this approach could be further explored for MPs and cocontaminants [79].

4.4 Electrochemical (Bio)Sensors

The development of active and specific electrocatalyst-functionalized microelectrodes for screen-printed electrodes and suitable electrochemical techniques for emerging contaminants is of particular interest because of their relatively short response time, easy handling, and low cost associated with portable device fabrication [80]. MIPs-based voltametric sensors [81–84], g-C_3N_4 and TiO_2 nanotube arrays [85]-based electrochemiluminescence sensors, potentiometric sensors [86], and MOF-based microfluidic impedance sensors [87] have been reported for the detection of major PFAS contaminants. The dearth of electrochemically active functionalities, along with the presence of nonfluorinated cocontaminants sharing similar structures and properties, presents significant challenges. Minimal to no analyte-receptor interaction at the molecular level, uncertainty in precursor sources, and slow diffusion rates lead to high LODs.

5 Treatment of MPs and PFASs from Aquatic Environments

Removing MPs and associated PFASs from the aquatic environment is a tedious and prohibitively expensive task because of the combination of PFAS persistence with MPs, widespread contamination, complex chemistry, expensive remediation technologies, and regulatory challenges. WWTPs and landfills serve as the primary linkage for the distribution of MPs and associated contaminants to aquatic ecosystems. Physical, chemical, and biological treatment approaches such as screening, skimming, sedimentation, adsorption[101–102], filtration, dissolved air flotation, electrocoagulation, sedimentation, activated sludge treatment, adsorbents, MBR, and advanced oxidative processes have been applied in WWTPs to either remove or degrade these contaminants. While current treatment technologies are not specifically tailored for MP removal, they manage to reduce the majority of MPs in wastewater.

Conventional methods such as air stripping, thermal treatment, and chemical oxidation are ineffective for PFAS. PFASs, owing to their persistent nature, unexplored physicochemical properties, and partitioning constants, result in low to negative removal efficiencies, particularly for PFAA. Breaking the carbon-fluorine bonds in PFASs may require various reagent combinations. Site-specific strategies may necessitate a combination of different approaches for cost-effective treatment.

The selection of appropriate remedial strategies for specific PFASs depends on varying PFAS concentrations, water qualities, treatment goals, and other factors [99].

The large daily effluent volume still leads to significant discharge of MPs and PFASs into aquatic environments. Most of these techniques only remove/transfer them from one environmental matrix to another. These techniques require in situ analysis, are time intensive, and require the routine use of matrix-matched calibration standards. Advanced methods such as adsorption and degradation are effective at the laboratory scale, but their effectiveness decreases in real water bodies [100]. The lack of long-term degradation data and comprehensive studies urges research and technological advancements for practical applications. There is a growing interest in removing MPs and PFASs from contaminated samples through biotic and abiotic degradation processes such as bacterial, photocatalytic, and thermal methods. Bioremediation and phytoremediation have low reaction rates but are suitable for natural water bodies [101]. Advanced treatment methods such as electrochemical and photoelectrochemical methods are promising because of their simplicity, effectiveness, independence from pH changes, and the presence of oxygen or NOMs [28]. Treatment methods for the removal and degradation of MPs and PFAS are presented in Fig. 16.4.

5.1 Removal Techniques

The separation and removal of MP and PFAS contaminants is a challenge because of their persistence and widespread distribution in aquatic environments across the globe. Advanced techniques, such as filtration, adsorption, and membrane filtration, offer potential solutions for mitigating these pollutants.

5.1.1 Adsorption Method

Adsorption is a cost-effective treatment for MPs and long-chain PFAS, but it necessitates nondestructive pre- and posttreatment for each cycle. The adsorption efficiency and diffusion rate depend on the pH, temperature, adsorbent type, NOM, and ions. While certain adsorbents have shown high adsorption capabilities in laboratory tests, their effectiveness has decreased in real samples, especially for mixed PFAS streams [102]. The inexpensive and most recognized sorption approaches, such as GAC and ion exchange resin composites [103], result in high removal rates at the laboratory scale but are time-consuming and generate unwanted solid waste in practical applications. Currently, NOM-derived sponge materials, i.e., bamboo-derived GAC, cross-linked chitosan beads, straw, willow, and maize-derived ash composites, MIPs, and CNTs, exhibit high adsorption capacities exceeding 700 mg. g^{-1} (PFOS) due to hemimicelle formation and hydrophobic interactions. The polymer chain length of PFAS negatively affects the performance of adsorbents [104].

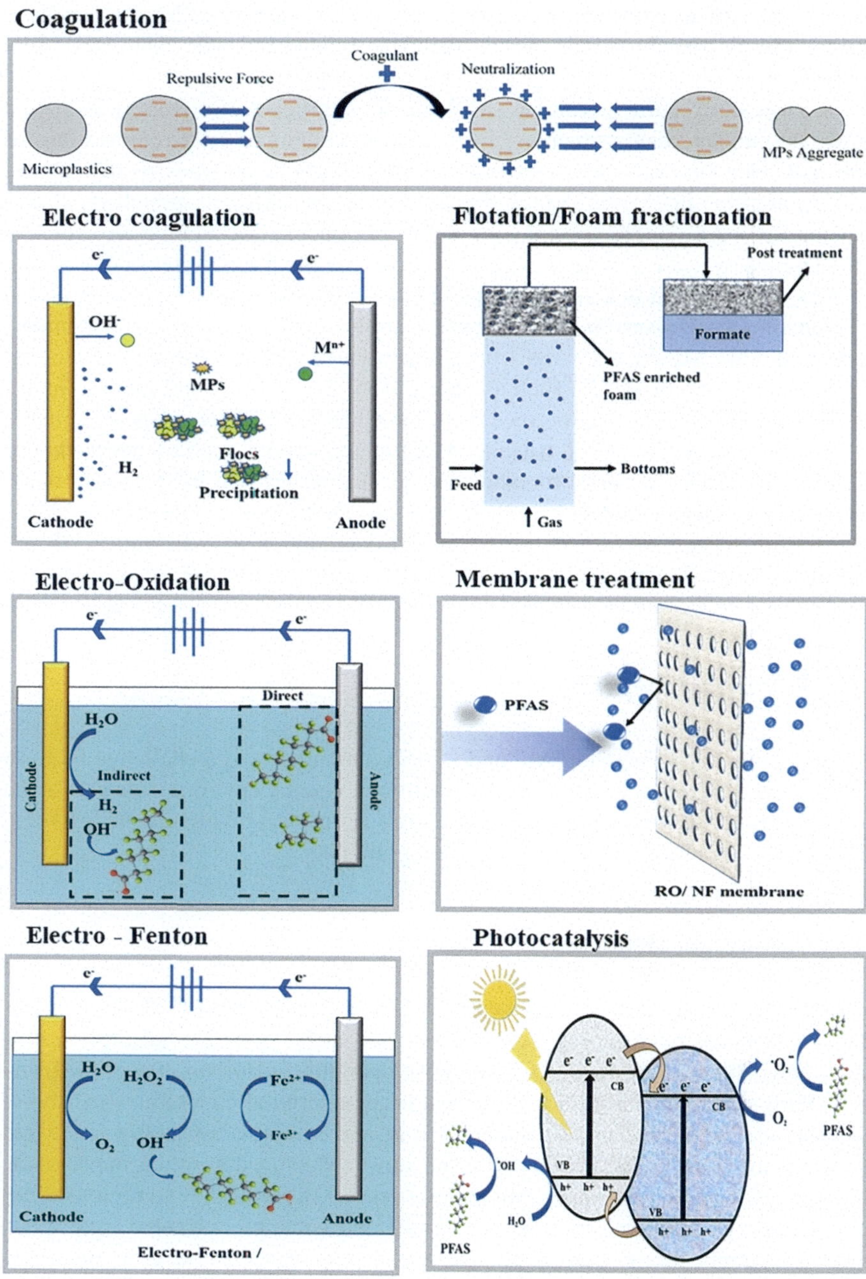

Fig. 16.4 Treatment methods for the removal and degradation of MPs and PFAS

However, the suitability of these adsorbents in pilot-scale treatment plants needs further verification depending on the contamination level and complex natural background matrices in real environments. This method is not suitable for removing PFASs, particularly in aquatic environments, because of their small amounts and hydrophilic characteristics.

Magnetic separation involves the grafting of magnetic seeds at the polymer surfaces of the MPs (magnetized MPs), which can then be efficiently extracted via external magnetic fields. This method typically involves two steps, (i) dispersion of magnetic seeds and (ii) agglomeration with the target MPs, with inherent magnetic properties, and these aggregates are easily separated from the solution via an external magnetic field. Its effectiveness, ease of operation, and potential for material reusability make it a valuable tool for mitigating MP and PFAS pollution in aquatic environments.

5.1.2 Ion Exchange Method (IX Resin)

Compared with blank adsorption media, IX resins are more efficient at removing PFASs from aquatic samples because of their complementary adsorption and ion exchange functionalities (Table 16.2). The PFAS removal efficiency of IX resins varies on the basis of multiple factors, similar to adsorbents. Short-chain PFASs more effectively access ion exchange sites and pair with counterions on resin surfaces, overcoming the limitations of GAC. However, the removal expenses range from \$300 to \$600 per sample, limiting the widespread and regular treatment of all MPs and PFAS [105].

5.1.3 Membrane Filtration

Membrane filtration and RO separation are viable methods for removing MPs from WWTPs. RSFs show superior performance in removing up to 98% of MPs, whereas microfiltration (pore size of 0.1–10 μm), ultrafiltration (0.01–0.1 μm), nanofiltration, and reverse osmosis (<0.01 μm) are more advanced filtration techniques that utilize membranes with progressively smaller pore sizes to capture progressively smaller MPs and are designed primarily for removing PFAS associated with MPs [106]. During this process, a dynamic membrane (DM), a secondary filtration layer, is formed on the main membrane from the particles and other flocculants within the wastewater itself. It also acts as an extended filtration layer, increasing the overall efficiency of the process. However, when the DM thickness increased, the removal efficiency decreased. Hence, DMs are cleaned via various methods, such as air scouring, water backwashing, and surface brushing. In the case of PFAS, nanofiltration membranes, particularly polyamide membranes, exhibit high removal rates for long-chain PFAS, whereas RO achieves up to 99% removal efficiency for both long- and short-chain PFAS [107]. However, the widespread adoption of these

Table 16.2 Various treatment approaches involve the removal and degradation of MPs and PFAS [25, 120–125]

Method	Matrices and efficiency	Advantages	Disadvantages	Future trends
Removal				
Microplastics				
Adsorption (GAC, ChGO, MOF, rGO)	Hydrophobic and π-π interactions, 89–95% removal efficiency	Reusable, no secondary pollution, suitable for wastewater recycling, low operational and maintenance cost, feasible, and energy-efficient	Clogging, demands multitreatment procedures for complete removal	Toward mechanistic understanding to increasing the efficiency and rotating continuous filter module to scale up
Magnetic separation (magnetic biochar, metal oxide composites)	Suitable for drinking water treatment, ~93% removal efficiency	Reusable and viable for tiny MPs	Not applicable for solid samples	Introducing specific functionalities in magnetic seeds to improve the removal of variety of MPs
Membrane filtration (RSFs; disk filters; micro-, ultra-, and nanofiltration; and reverse osmosis)	Electrostatic and H-bonding, > 85% removal efficiency	Complete removal, less secondary pollution, suitable for wastewater recycling	Demands high pressure, rerelease of captured MPs into effluent, membrane fouling	Mandating backwash to reduce fouling
Coagulation Chemical reagents Electrocoagulation (EC)	Functionalities driven agglomeration to produce solid sludges, >90% maximum removal efficiency, suitable for tiny MPs, electrode selectivity (Al > Fe)	Efficient, simple, cost-effective, requires low electric current, controllable operation conditions, no secondary contaminants, most of the natural weathering of MPs favors coagulation	Mineral settling, cathode passivation, high power consumption at large scale, toxic chemical additives, variable electrode requirements, and continuous replacement of anodes	Focus on novel reactor and electrode design, integration into real and sewage treatment plants, process optimization for scaling, understanding the EC process, electrofloating, and electrolytic redox processes taking place in coagulation unit
Flotation and foam fractionation (DAF, Posi-DAF, SAFF®, electroflotation)	Suitable for low- and medium-density MPs, > 90% saturation efficiency	Eco-friendly, inexpensive, sustainable, automatic control over gas input and output	Production of more sludge, energy consumption, insufficient data	Toward introducing new and selective functionalities in foaming agents to enhance the performance

PFAS				
Adsorbents (GAC, SPAC, AC-H_3PO_4, PAC-Fe_3O_4, Fe_3O_4@GAC, PDA-coated Fe_2O_3, PEI-CNC, GO-PEI)	Suitable for municipal sewage and drinking water, sorption via hydrophobic (long-chain) and electrostatic (short-chain) interactions and disposed via incineration or solidification, adsorption capacity range 53.0–876.9 $mg.g^{-1}$, >98% efficiency	Cost-effective, regeneration of spent GAC by simple alcohol wash, selective and specific for long-chain PFAS	Expensive operational and maintenance cost, require pre- and posttreatment before each cycle, low reusability. Nonspecific toward cocontaminants, less active for short-chain PFAS	Toward production of GAC composites with size-tunable pores for selective and increased adsorption surface, better disposal for used adsorbents
Membrane filtration Micro-, ultra-, and nanofiltration, thin-film nanocomposite, reverse osmosis	High-pressure driven electrostatic repulsion and size exclusion membrane technology	Sustainable, feasible, and highly effective, desirable pore size, suitable for treatment-train approach with GAC	Demand high pressure, membrane fouling	Developing simple and low-cost reactor design and automated timely cleaning of membrane materials
Ion exchange (IX) resins Polystyrene or polyacrylate-based resin beads (AER, A-714, A-244, A860, A592, IRA, Amberlite® IRA-400, Purolite® A600)	Exchange of cl^- and SO_4^{2-} counterions by ionic PFAS anchored on charged resin bead surface. Anion exchange resins remove >99% of PFOA, PFOS, and sulfonic acid PFAS via electrostatic interactions. β-CD polymers for zwitterionic PFAS	Cost-effective, tunable selectivity, amenable for both long-chain and short-chain PFAS, fixed polymer matrix with charged functional groups and movable counterions	Mixed results at complex PFAS composition in real conditions, aggregation/ micelle formation-induced fouling problem, resin regeneration, and reuse issues	To develop bench-scale column testing for resin specificity and short treatment cycle, focus on minimal salt consumption, and reduced pipeline corrosion
Coagulation Electrocoagulation (Al > Zn > Fe)	Addition/in situ generation of coagulating agents by sacrificial electrodes for floc formation, 100% and 81% removal efficiency for PFOS and PFOA and PFBS, respectively	Amenable to mobile and static configurations	High salt content restricts groundwater drainage	Focus on the subsequent drying, incineration, and disposal of PFAS flocs

(continued)

Table 16.2 (continued)

Method	Matrices and efficiency	Advantages	Disadvantages	Future trends
Flotation and foam fractionation (DAF, Posi-DAF, SAFF®, electroflotation)	Suitable for long-chain PFAS, 95% and 92% removal efficiencies for PFOS and PFOA	Safe, inexpensive, sustainable, automatic control over gas input and output, low volume, high-density PFAS concentrate	Rerelease of aerosolized PFAS in air, 10% removal efficiency of PFBS	Toward SAFF booster compounds for short-chain PFAS, tertiary fractionation container system
Degradation				
Microplastics				
Biodegradation Microbial, fungal, and enzymatic degradation	Microbes produce hydrolyzing enzyme to convert MPs into other useful products. Production of EPS enhances the bioflocculation property of MPs	Simple, eco-friendly, sustainable, avoids toxic additives, control over byproduct formation	Limited to biodegradable substances, slow for ex situ bioremediation, slow rate of degradation	Toward the selection of strong microorganisms over specific MPs, enzyme stability, and their efficiency against various MP types and environmental conditions
Membrane bioreactor (sMBR and iMBR)	FTIR analysis confirms degradation; 99.9% degradation efficiency	Simple, compact, efficient, high effluent quality with better disinfection	Energy-intensive, costly, seasonal fluctuations, and membrane fouling	Focus on adapting MBR to numerous unnamed MPs, specific filtration flux, fouling control via module/cassette/tank, efficient membrane cleaning strategies
Thermal degradation Incineration, gasification, and pyrolysis	Thermally destroying C–C or C–F bonds, 95% degradation efficiency	Effective for mineralization of MPs	Limited at the industrial scale; incomplete incineration could result in toxic and volatile byproducts; high energy requirements	Capture and recycle thermal energy during incineration
Advanced oxidation processes				
Chemical digestion, Fenton process, ozonalization	26% efficiency	Low cost, easy reagent preparation	Low efficiency, not applicable to all MPs	Toward integrating electro-Fenton process for high efficiency (75%)

Photocatalysis	Complete mineralization, > 90% degradation efficiency	Cheap, eco-friendly, mineralization of MPs with the help of solar energy	pH dependent, lack of selectivity in degrading MPs over NOMs, lab-scale, difficult in catalyst recovery	Developing cost-effective photocatalyst materials and reactor design
Electrochemical oxidation	Chemically breaking MP backbone via electrochemically produced ROS, 86.6% degradation efficiency	Cost-effective, amenable to mobile and static configurations	Toxic byproduct formation, nonselective interactions, electrode fouling	Integration with renewable energy sources
Functional material degradation	Chemically breaking MP backbone and converting into value-added end products	Efficient (better than other adsorption methods)	Toxic byproduct formation, catalyst recovery, pH dependent	Enhance surface-specific functionalities for selectivity. Implement high-performance vertical reaction columns using advanced materials for effective and in situ PFAS remediation
PFAS				
Biodegradation Microbial and fungal enzymatic degradation	Defluorination of short-chain PFCA via coenzyme A activation	Natural process, sustainable, eco-friendly, amenable to natural water bodies contaminated with AFFFs	Slow reaction rate compared to chemical and electrochemical methods, conflicting reports, lack of accessible hydrogen atom in the alkyl chain for primary microbial attack, unexplored experimental mechanism, incomplete defluorination	Focus on mixed cultures of specific microbes, specific delivery of fungal enzymes, coupling with other methods in treatment train to improve degradation efficiency and field implementation

(continued)

Table 16.2 (continued)

Method	Matrices and efficiency	Advantages	Disadvantages	Future trends
Constructed wetland and terrestrial phytoremediation	Combination of soil absorption and plant uptake, diffusion through small plasma membrane (passive) for short-chain and nonionic PFAS and specific transport receptors like carrier proteins, aquaporin channels, and anion channel (active) for long-chain PFAS	Sustainable, green method for large-scale treatment	Slow reaction rate compared to chemical and electrochemical methods	Study on role of transporter, root morphology, plant composition for improved activity
Membrane filtration and **MBR technology**	FTIR analysis confirms degradation; 99.9% removal efficiency	Simple, compact, efficient, high effluent quality with better disinfection	Energy-intensive, costly, seasonal fluctuations, and membrane fouling	Focus on adapting MBR to unknown PFAS and precursors, fouling control via module/cassette/tank, efficient membrane cleaning strategies, developing anaerobic membrane bioreactors (AnMBRs)
Thermal degradation incineration, gasification, and pyrolysis (600–1400 °C)	Thermally destroying C–C or C–F bonds; degradation temperature increases with increasing chain length	Effective, mineralization of MPs	Limited at the industrial scale; incomplete incineration could result in toxic and volatile byproducts; high energy requirements	Capture and treat incineration pollutants using scrubbers or filters
Hydrothermal alkaline treatment (HALT)	Alkaline addition enhances hydrothermal degradation via nucleophilic substitution of F- by OH- and C–C bond cleavage; 100% defluorination rate at 5 M NaOH system	Quick and complete mineralization, highly efficient for AFFF and sewage sludge, nondetectable level degradation	High energy consumption, not suitable for solid contaminations due to alkaline neutralization properties of soil	Suitable additives can greatly enhance the efficiency of PFAS degradation

Advanced oxidation processes				
Chemical digestion via $Na_2S_2O_8$, $KMnO_4$, H_2O_2, O_3, vitamin B12/Ti (III) citrate	Formation of shorter-chain PFAS obtained at active $S_2O_8^{2-}$ whereas F^- ions and CO_2 as final end productsobtainedatheat-treated$S_2O_8^{2-}$	Rapid, more effective, specific to long-chain PFAS; aqueous phase reaction suits aquatic systems	Mixed results, short-chain fluorocarbon byproducts, unexplored destruction mechanism	Further research at the pilot scale is needed
Photolysis, photocatalysis high-valent Fe, TNTs, In_2O_3, Ga_2O_3, MIP@TiO_2, BN/TiO_2, TiO_2-rGO	Suitable for long-chain PFAS, ROS (HO_2^- and $O_2^{\bullet}$)-driven reaction	Rapid, effective; aqueous phase reaction suits aquatic systems	Catalyst fouling and reduced activity, unexplored destruction mechanism, catalyst recovery	Explore the impact of UV light on degradation, PFAS-specific catalyst modification, optimize reactor design and portable installation of photocatalyst for continuous treatment
Electrochemical oxidation (BDDE, Ti_4O_7, Ag-coated Ti, Ce-PbO_2, SnO_2, 3DG-PbO_2, Fe electrodes)	Suitable for long-chain PFAS, electrocatalytic oxidation, degradation efficiency of 95% for PFOA in seawater	Cost-effective, amenable to mobile and static configurations	Toxic byproduct formation, nonselective interactions, mineral deposition on electrode, i.e., ClO_4^-, uncertain short-chain PFAS removal, insufficient data	Minimize electrode fouling with reagents like H_2O_2, integration with renewable energy sources
Functional material degradation Porous materials (coal cinder, hydrotalcite, wood pulp, h-BN, and polymer sponge), nanomaterials (biochar, LDHs, magnetic $MnFe_2O_4$@rGO), MOF, MXene composites	Hydrophobic, electrostatic, and acid-base interactions; calcium hydrotalcite has removal capacity of PFOS 1400 $mg.g^{-1}$; 98% removal efficiency for long-chain PFAS by Zr-MOF	Efficient (better than GAC) for long-chain PFAS	Inefficient for short-chain PFAS	Enhance surface functionalities for improved short-chain PFAS removal Implement high-performance vertical reaction columns using advanced materials for effective and in situ PFAS remediation

(continued)

Table 16.2 (continued)

Method	Matrices and efficiency	Advantages	Disadvantages	Future trends
Supercritical water oxidation (SCWO)	C–S bond cleavage-induced radical-driven C–F breakage, >99% PFAS destruction, and 63% defluorination efficiency	No intermediate fluorocarbon formation, highly effective	Not applicable to landfill and solid samples	More focus needed to understand the adaptability to other contaminants
Sonolysis/ultrasonication (200–1060 kHz)	Acoustic cavitation-driven breakdown of H_2O and O_2 to produce OH•, H•, and $O_2^{\bullet -}$ radicals, suitable for long-chain PFAS, i.e., PFOS, PFHpA, and PFHpS at 358 KHz, and short-chain PFAS, i.e., PFBA and PFBS at 610 KHz, >96% for PFOS at 400 KHz	Greener approach, no byproduct formation	Time-consuming, low degradation rates, limited to the laboratory scale	Focus on the correlation between threshold frequency and size of produced bubbles for degradation of individual PFAS and explore applicability in real conditions

technologies is hindered by issues such as high cost, membrane fouling, low water recovery, and saltwater disposal.

5.1.4 Coagulation

Coagulation, a charge neutralization process, remains a promising technique for removing MPs and PFASs from water. As the added ions attract and adhere to the contaminants, they cancel out the repulsive forces between them to clump together into larger flocs that can be easily separated from the water. Furthermore, research suggests that weathered polymeric materials with new functionalities are more susceptible to coagulation than pristine materials. These functional groups enhance the interaction with coagulants through both electrostatic attraction and hydrogen bonding, leading to more efficient removal [108]. Studies have shown that weathered MPs achieve a removal efficiency of up to 99% under optimized coagulation conditions, compared with only 82% removal for pristine particles.

Electrocoagulation (EC) is a well-established water treatment process that has promising potential for removing MPs and PFAS. EC utilizes electrical current to destabilize and remove contaminants. In an EC reactor, sacrificial metal electrodes (typically Al or Fe) dissolve to provide the necessary metal ions for coagulation. Al anodes have been demonstrated to eliminate the MPs of PE, PP, PMMA, and cellulose acetate with over 90% removal efficiency. The large-scale application of ECs faces challenges related to high power consumption and electrode passivation. Potential solutions include exploring renewable resources such as solar and wind energy to improve the power supply. Future research directions for the removal of MPs and fluorochemicals by ECs should focus on the effects of dissolved oxygen and NOM contents in water on removal efficiency [109].

5.1.5 Fractionation and Flotation

Adsorption-bubble flotation-based separation comprises a series of foam fractionation training stages. This method is suitable for fiber MPs and amphiphilic PFAS surfactants in aqueous solutions. The concentration is then destroyed via incineration. Compared with other foam-forming agents and other methods, such as adsorption and plasma treatment, SAF is a new foam-forming method that has high removal efficiency. Depending upon the nature of the MPs and PFASs, different strategies, such as foam and ozofractionation, dissolved air flotation (DAF), positive-modification DAF (Posi-DAF), and Surface Active Foam Fractionation (SAFF®), are used for specific contaminants, which must be extended to other MP and PFAS contaminants to increase their removal efficiency. [110]

5.2 Degradation Techniques

During degradation, MPs and PFASs are typically mineralized via physical and chemical processes, eradicating them from the environment. For example, incineration completely degrades MPs into CO_2 and H_2O.

5.2.1 Biodegradation

Microorganisms hold promise for expediting plastic biodegradation by secreting enzymes capable of degrading various plastic polymers and other pollutants. For example, cutinases and laccases have degraded PET and PA, PE, and PP, respectively. Studies on bacterial peroxidases are limited but hold promise for future exploration. Lipases, which are produced by various bacteria and fungi, play crucial roles in PET degradation. Despite its selectivity, several hurdles involved in bioremediation need to be addressed. Biodegradation rates are generally slow, necessitating further research to accelerate the process [111]. Additionally, the use of genetically modified organisms in bioremediation raises containment concerns due to unintended release into the environment [112]. Researchers have also focused on immobilizing enzymes on nanoparticles, which potentially enhances MP degradation efficiency. In another strategy, combining constructed wetland and terrestrial phytoremediation with biodegrading enzymes offers a natural treatment platform for emerging MP and PFAS contaminants [113].

5.2.2 Membrane Biodegradation

In a membrane bioreactor, a flow-type filtration unit integrated with biological reactors, such as enzymes/microbe-loaded beds, is effectively applied for municipal and industrial wastewater treatment. Integrating microbial enzyme suppliers with micro−/ultramembranes provides dual remediation: filtration followed by enzymatic degradation. Side-stream and immersed (sMBR and iMBR) membrane modules are currently used [106]. Conflicting experimental results have been reported in membrane-based bioreactors, which may be due to variations in MP samples and membrane pore sizes, highlighting the contamination of membranes and the complexity of MP removal.

5.2.3 Thermal Degradation

Thermal degradation is a process involving the breakdown of polymer structures through a series of chemical reactions initiated by heat. Understanding the mechanism of thermal degradation is crucial for predicting the stability and fate of MP and PFAS contaminants [114]. Incineration and plasma arc destruction (600–1400 °C) are recommended methods for the irreversible transformation of plastic and PFAS waste. Hydrothermal treatment (HALT) with additives quickly and completely mineralizes PFAS but demands high energy consumption [115].

5.2.4 Advanced Oxidation Degradation

In recent years, chemical, photocatalytic, electrochemical, and wet supercritical water oxidation methods have been used to remove MPs and PFASs from aquatic ecosystems via the use of reagents and functional materials. These methods are interesting approaches for the complete mineralization of MPs and associated PFAS pollutants.

5.2.4.1 Photocatalysis

Photocatalysis, a process employing light-active materials (photocatalysts), offers a promising approach for degradation. Upon exposure to light, these photocatalysts generate electron-hole pairs that drive redox reactions, facilitating the breakdown of polymeric materials. Studies have explored the use of various photocatalysts to achieve higher degradation efficiencies than earlier methods. Despite promising results, transitioning photocatalysis from lab-scale experiments to real-world applications requires further investigation [116]. Furthermore, compared with other NOMs, long-term exposure to photocatalysts results in the production of toxic byproducts, the leaching of toxic materials, and decreased selectivity in the degradation of MPs and PFASs, which needs to be addressed.

5.2.4.2 Electrochemical Oxidation

This is an effective method for the complete mineralization of highly stable, persistent, and difficult-to-oxidize MP and PFAS contaminants. Various high-valent metals, metal oxides, and water splitting electrocatalysts have been employed as catalysts to increase the selectivity and rate of degradation [117]. For example, MPs and PFASs, which are challenging to break down via other methods, are cleaved via the electro-Fenton process and photo−/electrocatalysis-based redox reactions via cathodically produced ROS. However, toxic byproduct formation, nonselective interactions, and electrode fouling due to mineral deposition are major hurdles that need to be overcome [118].

5.2.4.3 Functional Material Degradation

Advanced functional materials such as supramolecular and graphitic materials (MOFs, COFs, MXenes, Gr, GO, rGO, h-BN, g-C_3N_4), TMDs, LDHs, metal oxides, and their composites have been shown to be more efficient for the complete mineralization of PFASs and MPs [119]. This research also focuses on the conversion of plastic waste into value-added products and fuels through partial anaerobic degradation.

6 Challenges and Future Recommendations

A lack of awareness of the seriousness and unnoticed impact of irresponsible behavior and ignorance of the public are major failures that need to be addressed. Failure to address this issue will lead to escalating concentrations of these contaminants, resulting in harmful effects and making mitigation increasingly difficult. The wide range of end products and additives released by MPs and fluoropolymers creates uncertainty in data reporting, as some customized application-specific polymer-additive recipes. There is also insufficient disclosure of chemical content in consumer products, particularly those currently on the market, due to confidential business claims and the absence of strict regulatory frameworks. Additional research is needed to understand the toxicity of MPs and PFASs and their implications.

Alternatives to PFASs remain largely unknown due to their poorly characterized chemical structures, limited understanding of health impacts, and the lack of toxicological endpoint studies. Although significant efforts have been made to phase out many plastic and fluoropolymer-based products through alternative R&D initiatives, the risk of PFAS exposure, uptake, and accumulation in marine food chains persists due to the complexity of their sources and degradation pathways during environmental weathering. On the other hand, companies are moving toward the synthesis of less fluorinated polyfluoro (GenX) chemicals with similar required properties, which also pose a significant threat to the aquatic food web. Furthermore, short-chain PFASs generally transport faster, whereas long-chain PFASs are more hazardous. Long-chain PFASs are cleaved into short-chain PFASs through different pathways, such as C-chain breakage, $-CF_2$ elimination, the nucleophilic substitution of fluorine, and HF elimination. Hence, the prediction of intermediate and end products is very tedious.

Although significant research has focused on the identification, detection, quantification, removal, and degradation of MPs and PFASs, real field implementation of the above methods is in the early development stage due to the unmet performance of methods such as variable permissible and detection limits, contamination matrices, and efficiencies [126]. Therefore, identifying, quantifying, and treating MPs and associated cocontaminants via a single method is very difficult. Moreover, on the basis of our observations, each method has its own advantages, disadvantages, and inconsistencies in terms of standard QC/QA procedures for sample collection, detection, and treatment. In water bodies located near industrial and military bases, removal and detection methods, particularly for PFASs, do not work efficiently because of the coexistence of organic and inorganic pollutants and NOMs, increasing the difficulty of practical applications.

As mentioned in Table 16.1, the most effective techniques for detecting and quantifying MPs and PFAS include advanced spectroscopic techniques such as

Raman and FTIR spectroscopy (e.g., μ-FTIR and μ-Raman) and chromatographic techniques coupled with mass spectrometry (e.g., LC-MS-MS-MS). These techniques are currently used because they offer quantitative results with high sensitivity and specificity, especially for complex environmental samples. However, these techniques and protocols may not be suitable for all conditions, such as indifferent water matrices and changes in the type and size of MPs, leading to errors and irreproducibility. Additionally, challenges such as high costs, multistep sample preparation, sophisticated instrumentation, skilled technicians, and limited portability remain significant barriers. Recently, researchers have focused on on-site analysis methods that do not rely on sophisticated facilities, such as ISEs and smartphone app-based optical and electrochemical monitoring systems. However, these approaches still face challenges in achieving the desired detection limits and specificity. Accurate and efficient analytical methods for identifying MPs and PFASs, their precursors, and their transformation products need standardization for better research and documentation. Future developmental efforts should prioritize the standardization and optimization of analytical methods for sampling and analyzing MPs and PFAS in water and wastewater systems.

As shown in Table 16.2, highly efficient methods such as membrane filtration and adsorption are adapted for removal, and AOPs, MBRs, and incineration are used for degrading MPs and PFASs from aquatic environments, respectively. Adsorption using materials such as GAC and ion exchange resins is cost-effective and efficient, especially for long-chain PFASs, but further optimization is needed to increase selectivity for short-chain PFASs and reduce waste generation. Recent research has suggested that biochar filters outperform sand filters, with C500 filters achieving a remarkable 100% removal rate without releasing any MPs themselves [127]. Adsorbents and AIX beads can reduce PFCA and PFSA concentrations to below permissible limits (<1–2 ng.L^{-1}) in WWTPs but have a risk of pollution transfer rather than real depletion. Integrating RSFs with activated carbon enhances the adsorption capacity. Ion exchange resins are selective only for free ionic PFAS. However, it is not suitable for aquatic environments that contain nonionic and PFAS-adsorbed MP contaminants. Therefore, the chemical attributes of the resin should be carefully tailored to target and remove specific co-contaminants adsorbed onto the surfaces of PFAS and microplastics in aquatic environments. Membrane filtration, particularly nanofiltration and RO, achieves up to 99% removal efficiency, but challenges such as membrane fouling and high operational costs limit its widespread application. Recently, electrocoagulation processes using either Al or Fe electrodes have been used for the effective removal of MPs and PFAS in aquatic environments, particularly for fibers, PS and PP microparticles, and PFOS and PFOA from water. Currently, PFASs are preconcentrated with advanced functional material-modified electrodes via the electrosorption method and degraded via the electrochemical oxidation method. The merits of this method include removal and degradation occurring simultaneously; however, the efficiency and complete degradation of this method are shortfalls. Similarly, on the basis of reports on photo- and/or electrocatalytic oxidation, researchers are focusing on the complete mineralization of PFAS contaminants but face hurdles due to partial degradation, toxic byproduct formation, and nonselective interactions. Using the photochemical degradation method, complete mineralization could be achieved for partially chlorinated PFCA. Low-cost, nontoxic, and highly stable light-harvesting materials have yet to be explored for MPs and PFASs via photoelectrocatalytic degradation. Light-activated micromotors coated with photocatalysts hold promise for targeted MP removal and degradation. Plasma treatment and incineration, i.e., aerobic (CO_2, H_2O) and anaerobic (CH_4) methods, are the most promising degradation techniques for concentrated effluent. Future research should adopt a two-step approach for PFAS water remediation: first, PFAS pollutants should be concentrated with IX resin and/or GAC; second, PFAS should be degraded via advanced oxidation processes such as catalytic degradation, resulting in H_2O and CO_2.

In academia, there is a significant gap in the development of cost-effective, miniaturized, and easily transportable techniques focused on the pilot scale. Multidisciplinary attempts to combine various techniques to overcome such obstacles are recommended. The upcycling of plastic debris and MPs into value-added waste-derived materials, instead of completely breaking them down through mineralization, has recently attracted growing interest from environmental researchers [128]. For example, Zafar et al. transformed PE MPs into graphene via the APMP technique and applied it for the removal of PFOA [129]. The pyrolysis and AOP-based conversion of plastic debris into value-added chemicals via selectively breaking weak/hetero positions of the polymer chains to produce aliphatic and aromatic hydrocarbons, which can be utilized as fuels and energy byproducts. During degradation, MPs and PFAS may produce more prevalent and toxic by products, e.g., bisphenol A and n-fluoroheptane, further exacerbating the situation.

To overcome these situations, the adoption of bioremediation and MBR methods using enzyme-loaded membranes for converting PFAS and MPs into expected value-added byproducts is recommended. To achieve this goal, before selective enzyme-loaded MBR, a thorough structural analysis of specific types of MPs and PFAS should be performed using existing spectral techniques, and then, theoretical retrosynthesis predictions can be applied to predict the possible byproducts, which will be helpful for converting all the MPs and associated fluoropolymeric contaminants into value-added products. These advancements not only aid in preventing, reducing, and managing MP pollution but also serve as vital tools for assessing the risks posed by MPs and PFAS to human health and ecosystems.

7 Conclusion

The current management strategies fail to prevent plastics from polluting the environment. Several approaches require feasibility studies on the degradation and fate of emerging pollutants. Research should be expanded, and awareness of microplastic hazards must be increased. Emphasizing plastic recycling and promoting energy recovery from plastic waste can enhance the circular economy and reduce pollution. To conclude, this book chapter starts with the growth of global plastics and PFAS production and consumption and its consequences. The second and third parts address a research gap on the occurrence and distribution of MP and PFAS cocontaminants and the importance of addressing this issue. In the fourth part, an introduction to the identification and quantification methods followed by a detailed review of each method focusing on their merits and demerits with research gaps and future research trends are emphasized. In the fifth part, a detailed review covering the treatment of emerging contaminants via separation, removal, and degradation strategies is presented. The final part comprises an overall observation of the reported methods, challenges that require necessary attention, opportunities, and future scopes for monitoring/quantifying and removing plastic pollutants and associated PFAS cocontaminants.

References

1. Lim, X. (2023). Could the world go PFAS-free? Proposal to ban 'forever chemicals' fuels debate. *Nature, 620*(7972), 24–27. https://www.nature.com/articles/d41586-023-02444-5
2. Dai, Y., Zhao, J., Sun, C., et al. (2022). Interaction and combined toxicity of microplastics and per- and polyfluoroalkyl substances in aquatic environment. *Frontiers of Environmental Science & Engineering, 16*, 136. https://doi.org/10.1007/s11783-022-1571-2
3. Piringer, O. G., & Baner, A. L. (2000). *Plastic packaging materials for food: Barrier function, mass transport, quality assurance, and legislation*. Wiley-VCH.
4. Health and environmental impacts of single use plastics. Plastics Pollution Coalition. https://www.plasticpollutioncoalition.org/guides/singleuseplastics/healthimpacts
5. (a) https://www.statista.com/statistics/282732/global-production-of-plastics-since-1950/. (b) Evich, M. G., Davis, M. J., McCord, J. P., Acrey, B., Awkerman, J. A., Knappe, D. R., ... & Washington, J. W. (2022). Per-and polyfluoroalkyl substances in the environment. *Science, 375*(6580): eabg9065. https://doi.org/10.1126/science.abg9065.
6. Thompson, R. C., Olsen, Y., Mitchell, R. P., Davis, A., Rowland, S. J., John, A. W., & Russell, A. E. (2004). Lost at sea: Where is all the plastic? *Science, 304*(5672), 838–838.
7. https://www.oecd.org/environment/plastic-pollution-is-growing-relentlessly-as-waste-management-and-recycling-fall-short.htm
8. Borrelle, S. B., Ringma, J., Law, K. L., Monnahan, C. C., Lebreton, L., McGivern, A., et al. (2020). Predicted growth in plastic waste exceeds efforts to mitigate plastic pollution. *Science, 369*(6510), 1515–1518. https://doi.org/10.1126/science.aba3656
9. Plastic Atlas-2021 facts and figures about the world of synthetic polymers Asia edition. https://hk.boell.org/en/2021/04/22/plastic-atlas-asia-edition
10. MacArthur, D. E., Waughray, D., & Stuchtey, M. R. (2016). The new plastics economy. *Rethinking the Future of Plastics*. http://www.ellenmacarthurfoundation.org/publications
11. Jenkins, T., Persaud, B. D., Cowger, W., Szigeti, K., Roche, D. G., Clary, E., et al. (2022). Current state of microplastic pollution research data: Trends in availability and sources of open data. *Frontiers in Environmental Science, 10*, 912107. https://doi.org/10.3389/fenvs.2022.912107
12. Vaid, M., Mehra, K., & Gupta, A. (2021). Microplastics as contaminants in Indian environment: A review. *Environmental Science and Pollution Research, 1-28*, 68025. https://doi.org/10.1007/s11356-021-16827-6
13. Rodriguez, K. L., Hwang, J. H., Esfahani, A. R., Sadmani, A. A., & Lee, W. H. (2020). Recent developments of PFAS-detecting sensors and future direction: A review. *Micromachines, 11*(7), 667. https://doi.org/10.3390/mi11070667

14. Bilela, L. L., Matijošytė, I., Krutkevičius, J., Alexandrino, D. A., Safarik, I., Burlakovs, J., … & Carvalho, M. F. (2023). Impact of per-and polyfluorinated alkyl substances (PFAS) on the marine environment: Raising awareness, challenges, legislation, and mitigation approaches under the one health concept. Marine Pollution Bulletin, 194, 115309. doi:https://doi.org/10.1016/j.marpolbul.2023.115309.
15. Qian, N., Gao, X., Lang, X., Deng, H., Bratu, T. M., Chen, Q., et al. (2024). Rapid single-particle chemical imaging of nanoplastics by SRS microscopy. *Proceedings of the National Academy of Sciences, 121*(3), e2300582121. https://doi.org/10.1073/pnas.2300582121
16. Gellrich, V., Brunn, H., & Stahl, T. (2013). Perfluoroalkyl and polyfluoroalkyl substances (PFASs) in mineral water and tap water. *Journal of Environmental Science and Health, Part A, 48*(2), 129–135. https://doi.org/10.1080/10934529.2013.719431
17. (a) USEPA, Fact sheet: PFOA & PFOS drinking water health advisories, 2016. https://www.epa.gov/sites/default/files/2016-06/documents/drinkingwaterhealthadvisories_pfoa_pfos_updated_5.31.16.pdf. (b) USEPA, Per- and Polyfluoroalkyl Substances (PFAS), Proposed PFAS National Primary Drinking Water Regulation, 2023.https://www.epa.gov/sdwa/andpolyfluoroalkyl-substances-pfas.
18. Kang, P., Zhao, Y., Wei, T., Cai, Y., Ji, B., & Addo-Bankas, O. (2024). Interactions between MPs and PFASs in aquatic environments: A dual-character situation. *Journal of Environmental Management, 351*, 119907. https://doi.org/10.1016/j.jenvman.2023.119907
19. Microplastics: Sources, effects and solution. European Parliament. https://www.europarl.europa.eu/topics/en/article/20181116STO19217/microplastics-sources-effects-and-solutions
20. Osman, A. I., Hosny, M., Eltaweil, A. S., et al. (2023). Microplastic sources, formation, toxicity and remediation: A review. *Environmental Chemistry Letters, 21*, 2129–2169. https://doi.org/10.1007/s10311-023-01593-3
21. Duis, K., & Coors, A. (2016). Microplastics in the aquatic and terrestrial environment: Sources (with a specific focus on personal care products), fate and effects. *Environmental Sciences Europe, 28*, 1–25. https://doi.org/10.1186/s12302-015-0069-y
22. Corcoran, P. L. (2021). Degradation of microplastics in the environment. In T. Rocha-Santos, M. Costa, & C. Mouneyrac (Eds.), *Handbook of microplastics in the environment*. Springer. https://doi.org/10.1007/978-3-030-10618-8_10-1
23. Wu, W.-M., et al. (2017). Microplastics pollution and reduction strategies. *Frontiers of Environmental Science & Engineering, 11*, 1. https://doi.org/10.1007/s11783-017-897-7
24. Xu, Y., et al. (2021a). A critical review of microplastic pollution in urban freshwater environments and legislative progress in China: Recommendations and insights. *Critical Reviews in Environmental Science and Technology, 51*, 2637–2680. https://doi.org/10.1080/10643389.2020.1801308
25. Sutkar, P. R., Gadewar, R. D., & Dhulap, V. P. (2023). Recent trends in degradation of microplastics in the environment: A state-of-the-art review. *Journal of Hazardous Materials Advances*. https://doi.org/10.1016/j.hazadv.2023.100343
26. Sunderland, E. M., Hu, X. C., Dassuncao, C., Tokranov, A. K., Wagner, C. C., & Allen, J. G. (2019). A review of the pathways of human exposure to poly-and perfluoroalkyl substances (PFASs) and present understanding of health effects. *Journal of Exposure Science & Environmental Epidemiology, 29*(2), 131–147. https://doi.org/10.1038/s41370-018-0094-1
27. Junaid, M., Liu, S., Yue, Q., Wei, M., & Wang, J. (2023). Trophic transfer and interfacial impacts of micro (nano) plastics and per-and polyfluoroalkyl substances in the environment. *Journal of Hazardous Materials, 133243*. https://doi.org/10.1016/j.jhazmat.2023.133243
28. Kucharzyk, K. H., Darlington, R., Benotti, M., Deeb, R., & Hawley, E. (2017). Novel treatment technologies for PFAS compounds: A critical review. *Journal of Environmental Management, 204*, 757–764. https://doi.org/10.1016/j.jenvman.2017.08.016
29. Lohmann, R., Cousins, I. T., DeWitt, J. C., Gluge, J., Goldenman, G., Herzke, D., et al. (2020). Are fluoropolymers really of low concern for human and environmental health and separate from other PFAS? *Environmental Science & Technology, 54*(20), 12820–12828. https://doi.org/10.1021/acs.est.0c03244

30. Pelch, K. E., Reade, A., Wolffe, T. A., & Kwiatkowski, C. F. (2019). PFAS health effects database: Protocol for a systematic evidence map. *Environment International, 130*, 104851. https://doi.org/10.1016/j.envint.2019.05.045
31. Shim, W. J., Hong, S. H., & Eo, S. E. (2017). Identification methods in microplastic analysis: A review. *Analytical Methods, 9*(9), 1384–1391. https://doi.org/10.1039/C6AY02558G
32. Ruggero, F., Gori, R., & Lubello, C. (2020). Methodologies for microplastics recovery and identification in heterogeneous solid matrices: A review. *Journal of Polymers and the Environment, 28*, 739–748. https://doi.org/10.1007/s10924-019-01644-3
33. Sierra, I., Chialanza, M. R., Faccio, R., Carrizo, D., Fornaro, L., & Pérez-Parada, A. (2020). Identification of microplastics in wastewater samples by means of polarized light optical microscopy. *Environmental Science and Pollution Research, 27*, 7409–7419. https://doi.org/10.1007/s11356-019-07011-y
34. Hidalgo-Ruz, V., Gutow, L., Thompson, R. C., & Thiel, M. (2012). Microplastics in the marine environment: A review of the methods used for identification and quantification. *Environmental Science & Technology, 46*(6), 3060–3075. https://doi.org/10.1021/es2031505
35. Murphy, F., Ewins, C., Carbonnier, F., & Quinn, B. (2016). Wastewater treatment works (WwTW) as a source of microplastics in the aquatic environment. *Environmental Science & Technology, 50*(11), 5800–5808. https://doi.org/10.1021/acs.est.5b05416
36. Li, J., Liu, H., & Chen, J. P. (2018). Microplastics in freshwater systems: A review on occurrence, environmental effects, and methods for microplastics detection. *Water Research, 137*, 362–374. https://doi.org/10.1016/j.watres.2017.12.056
37. Mansa, R., & Zou, S. (2021). Thermogravimetric analysis of microplastics: A mini review. *Environmental Advances, 5*, 100117. https://doi.org/10.1016/j.envadv.2021.100117
38. Minor, E. C., Lin, R., Burrows, A., Cooney, E. M., Grosshuesch, S., & Lafrancois, B. (2020). An analysis of microlitter and microplastics from Lake Superior beach sand and surface-water. *Science of the Total Environment, 744*, 140824. https://doi.org/10.1016/j.scitotenv.2020.140824
39. Cooper, D. A., & Corcoran, P. L. (2010). Effects of mechanical and chemical processes on the degradation of plastic beach debris on the Island of Kauai, Hawaii. *Marine Pollution Bulletin, 60*(5), 650–654. https://doi.org/10.1016/j.marpolbul.2009.12.026
40. Dümichen, E., Barthel, A. K., Braun, U., Bannick, C. G., Brand, K., Jekel, M., & Senz, R. (2015). Analysis of polyethylene microplastics in environmental samples, using a thermal decomposition method. *Water Research, 85*, 451–457. https://doi.org/10.1016/j.watres.2015.09.002
41. Majewsky, M., Bitter, H., Eiche, E., & Horn, H. (2016). Determination of microplastic polyethylene (PE) and polypropylene (PP) in environmental samples using thermal analysis (TGA-DSC). *Science of the Total Environment, 568*, 507–511. https://doi.org/10.1016/j.scitotenv.2016.06.017
42. Meyers, N., Catarino, A. I., Declercq, A. M., Brenan, A., Devriese, L., Vandegehuchte, M., et al. (2022). Microplastic detection and identification by Nile red staining: Towards a semi-automated, cost-and time-effective technique. *Science of the Total Environment, 823*, 153441. https://doi.org/10.1016/j.scitotenv.2022.153441
43. da Silva, V. H., Murphy, F., Amigo, J. M., Stedmon, C., & Strand, J. (2020). Classification and quantification of microplastics (< 100 μm) using a focal plane array–Fourier transform infrared imaging system and machine learning. *Analytical Chemistry, 92*(20), 13724–13733. https://doi.org/10.1021/acs.analchem.0c01324
44. Vidal, C., & Pasquini, C. (2021). A comprehensive and fast microplastics identification based on near-infrared hyperspectral imaging (HSI-NIR) and chemometrics. *Environmental Pollution, 285*, 117251. https://doi.org/10.1016/j.envpol.2021.117251
45. Cabernard, L., Roscher, L., Lorenz, C., Gerdts, G., & Primpke, S. (2018). Comparison of Raman and Fourier transform infrared spectroscopy for the quantification of microplastics in the aquatic environment. *Environmental Science & Technology, 52*(22), 13279–13288. https://doi.org/10.1021/acs.est.8b03438

46. Kamel, A. H., Hefnawy, A., Hazeem, L. J., Rashdan, S. A., & Abd-Rabboh, H. S. (2024). Current perspectives, challenges, and future directions in the electrochemical detection of microplastics. *RSC Advances, 14*(3), 2134–2158. https://doi.org/10.1039/D3RA06755F
47. Shoemaker, J., & Tettenhorst, D. (2018). *Method 537.1: Determination of selected per-and polyfluorinated alkyl substances in drinking water by solid phase extraction and liquid chromatography/tandem mass spectrometry (LC/MS/MS)*. National Center for Environmental Assessment.
48. Rosenblum, L., & Wendelken, S. (2019). *Method 533: Determination of per-and polyfluoroalkyl substances in drinking water by isotope dilution anion exchange solid phase extraction and liquid chromatography/tandem mass spectrometry*. US Environmental Protection Agency.
49. Shoemaker, J., & Tettenhorst, D., O.D. US Environmental Protection Agency, N.C.f.E. Assessment, Determination of selected per-and polyfluorinated alkyl substances in drinking water by solid phase extraction and liquid chromatography/tandem mass spectrometry (LC/MS/MS) National Exposure Research Laboratory 2018.
50. Al Amin, M., Luo, Y., Nolan, A., Robinson, F., Niu, J., Warner, S., et al. (2021). Total oxidisable precursor assay towards selective detection of PFAS in AFFF. *Journal of Cleaner Production, 328*, 129568. https://doi.org/10.1016/j.jclepro.2021.129568
51. Schultes, L., Peaslee, G. F., Brockman, J. D., Majumdar, A., McGuinness, S. R., Wilkinson, J. T., et al. (2019). Total fluorine measurements in food packaging: How do current methods perform? *Environmental Science & Technology Letters, 6*(2), 73–78. https://doi.org/10.1021/acs.estlett.8b00700
52. Fang, C., Zhang, X., Dong, Z., Wang, L., Megharaj, M., & Naidu, R. (2018). Smartphone app-based/portable sensor for the detection of fluoro-surfactant PFOA. *Chemosphere, 191*, 381–388. https://doi.org/10.1016/j.chemosphere.2017.10.057
53. Binu, K. R., Yamashita, N., Prabhasankar, V. P., Praveenkumarreddy, Y., Shenoy, J. K., & Balakrishna, K. (2022). A first report of Perfluoroalkyl substances (PFAS) in a large west-Flowing River in southern India. In *Impact of COVID-19 on emerging contaminants: One health framework for risk assessment and remediation* (pp. 3–16). Springer. https://doi.org/10.1007/978-981-19-1847-6_1
54. Sunantha, G., & Vasudevan, N. (2016). Assessment of perfluorooctanoic acid and perfluorooctane sulfonate in surface water-Tamil Nadu, India. *Marine Pollution Bulletin, 109*(1), 612–618. https://doi.org/10.1016/j.marpolbul.2016.05.023
55. Koulini, G. V., & Nambi, I. M. Ankan Mukherjee Das, Rajiv Janardhanan Laboratory of Disease Dynamics and Molecular Epidemiology Amity Institute of Public Health Amity University Noida, India.
56. Lavers, J. L., Oppel, S., & Bond, A. L. (2016). Factors influencing the detection of beach plastic debris. *Marine Environmental Research, 119*, 245–251. https://doi.org/10.1016/j.marenvres.2016.06.009
57. Kurouski, D., Dazzi, A., Zenobi, R., & Centrone, A. (2020). Infrared and Raman chemical imaging and spectroscopy at the nanoscale. *Chemical Society Reviews, 49*(11), 3315–3347. https://doi.org/10.1039/C8CS00916C
58. Veerasingam, S., Ranjani, M., Venkatachalapathy, R., Bagaev, A., Mukhanov, V., Litvinyuk, D., et al. (2020). Microplastics in different environmental compartments in India: Analytical methods, distribution, associated contaminants and research needs. *TrAC Trends in Analytical Chemistry, 133*, 116071. https://doi.org/10.1016/j.trac.2020.116071
59. Leslie, H. A., Brandsma, S. H., van Velzen, M. J., & Vethaak, A. D. (2017). Microplastics en route: Field measurements in the Dutch river delta and Amsterdam canals, wastewater treatment plants, North Sea sediments and biota. *Environment International, 101*, 133. https://doi.org/10.1016/j.envint.2017.01.018
60. Carr, S. A., Liu, J., & Tesoro, A. G. (2016 Mar). Transport and fate of microplastic particles in wastewater treatment plants. *Water Research, 91*, 174. https://doi.org/10.1016/j.watres.2016.01.002

61. Ziajahromi, S., Neale, P. A., Rintoul, L., & Leusch, F. D. (2017). Wastewater treatment plants as a pathway for microplastics: Development of a new approach to sample wastewater-based microplastics. *Water Research, 112*, 93. https://doi.org/10.1016/j.watres.2017.01.042
62. Jun-Li, X., & Gowen, A. A. (2019). Investigation of plasticizer aggregation problem in casein based biopolymer using chemical imaging. *Talanta, 193*, 128–138. https://doi.org/10.1016/j.talanta.2018.09.094
63. Tagg, A. S., Sapp, M., Harrison, J. P., & Ojeda, J. J. (2015). Identification and quantification of microplastics in wastewater using focal plane Array-based reflectance micro-FT-IR imaging. *Analytical Chemistry, 87*, 6032. https://doi.org/10.1021/acs.analchem.5b00495
64. Jun-Li, X., Thomas, K. V., Luo, Z., & Gowen, A. A. (2019). FTIR and Raman imaging for microplastics analysis: State of the art, challenges and prospects. *TrAC Trends in Analytical Chemistry, 119*, 115629. https://doi.org/10.1016/j.trac.2019.115629
65. Ivleva, N. P. (2021). Chemical analysis of microplastics and nanoplastics: Challenges, advanced methods, and perspectives. *Chemical Reviews, 121*(19), 11886–11936. https://doi.org/10.1021/acs.chemrev.1c00178
66. Liu, K., Pang, X., Chen, H., & Jiang, L. (2024). Visual detection of microplastics using Raman spectroscopic imaging. *Analyst, 149*(1), 161–168. https://doi.org/10.1039/D3AN01270K
67. Rothstein, J. C., Cui, J., Yang, Y., Chen, X., & Zhao, Y. (2024). Ultra-sensitive detection of PFASs using surface enhanced Raman scattering and machine learning: A promising approach for environmental analysis. *Sensors & Diagnostics., 3*, 1272. https://doi.org/10.1039/D4SD00052H
68. Biermann, L., Clewley, D., Martinez-Vicente, V., & Topouzelis, K. (2020). Finding plastic patches in coastal waters using optical satellite data. *Scientific Reports, 10*(1), 5364. https://doi.org/10.1038/s41598-020-62298-z
69. Evans, M. C., & Ruf, C. S. (2021). Toward the detection and imaging of ocean microplastics with a spaceborne radar. *IEEE Transactions on Geoscience and Remote Sensing, 60*, 1–9. https://doi.org/10.1109/TGRS.2021.3081691
70. https://www.satelytics.com/resources/2020-pfas-detection-and-quantification-a-deep-dive-on-our-newest-algorithm
71. Tammina, S. K., Khan, A., & Rhim, J. W. (2023). Advances and prospects of carbon dots for microplastic analysis. *Chemosphere, 313*, 137433. https://doi.org/10.1016/j.chemosphere.2022.137433
72. Menger, R. F., Funk, E., Henry, C. S., & Borch, T. (2021). Sensors for detecting per-and polyfluoroalkyl substances (PFAS): A critical review of development challenges, current sensors, and commercialization obstacles. *Chemical Engineering Journal, 417*, 129133. https://doi.org/10.1016/j.cej.2021.129133
73. Glasscott, M. W., Vannoy, K. J., Kazemi, R., Verber, M. D., & Dick, J. E. (2020). M-MIP: Molecularly imprinted polymer-modified microelectrodes for the ultrasensitive quantification of GenX (HFPO-DA) in river water. *Environmental Science & Technology Letters, 7*(7), 489–495. https://doi.org/10.1021/acs.estlett.0c00341
74. Tang, Y., Hardy, T. J., & Yoon, J. Y. (2023). Receptor-based detection of microplastics and nanoplastics: Current and future. *Biosensors and Bioelectronics, 234*, 115361. https://doi.org/10.1016/j.bios.2023.115361
75. Fischer, M., & Scholz-Böttcher, B. M. (2017). Simultaneous trace identification and quantification of common types of microplastics in environmental samples by pyrolysis-gas chromatography-mass spectrometry. *Environmental Science & Technology, 51*(9), 5052–5060. https://doi.org/10.1021/acs.est.6b06362
76. Fries, E., Dekiff, J. H., Willmeyer, J., Nuelle, M. T., Ebert, M., & Remy, D. (2013). Identification of polymer types and additives in marine microplastic particles using pyrolysis-GC/MS and scanning electron microscopy. *Environmental Science. Processes & Impacts, 15*(10), 1949–1956. https://doi.org/10.1039/c3em00214d

77. Dümichen, E., Eisentraut, P., Bannick, C. G., Barthel, A.-K., Senz, R., & Braun, U. (2017). Fast identification of microplastics in complex environmental samples by a thermal degradation method. *Chemosphere, 174*, 572–584. https://doi.org/10.1016/j.chemosphere.2017.02.010
78. Tabar, F. A., Lowdon, J. W., Caldara, M., Cleij, T. J., Wagner, P., Diliën, H., et al. (2023). Thermal determination of perfluoroalkyl substances in environmental samples employing a molecularly imprinted polyacrylamide as a receptor layer. *Environmental Technology & Innovation, 29*, 103021. https://doi.org/10.1016/j.eti.2023.103021
79. Clark, R. B., Wagner, D. C., Holden, D. T., Roberts, J. J., Zumbro, E., Goodnight, L., et al. (2023). PFAS electroanalysis in low-oxygen river water using electrogenerated dioxygen. *Environmental Science & Technology, 57*(51), 21815–21822. https://doi.org/10.1021/acs.est.3c03967
80. Ananthappan, P., Thangarasu, M., & Vasantha, V. S. (2024). Effect of electrophoretic deposition for selective detection of Pb^{2+} ions using nitrogen and sulphur-doped reduced graphene oxide. *Microchemical Journal*, 110879. https://doi.org/10.1016/j.microc.2024.110879; (a) Priyasamy, A., Selvam, S., Chellakannu, A., Sivasamy, V.V., & Mariakuttikan, J. (2022). Electrochemical sensors for the detection of food adulterants in miniaturized settings. In: Chandra, P., & Panesar, P.S. (Eds), *Nanosensing and bioanalytical technologies in food quality control*. Springer. https://doi.org/10.1007/978-981-16-7029-9_7.
81. Karimian, N., Stortini, A. M., Moretto, L. M., Costantino, C., Bogialli, S., & Ugo, P. (2018). Electrochemosensor for trace analysis of perfluorooctanesulfonate in water based on a molecularly imprinted poly (o-phenylenediamine) polymer. *ACS Sensors, 3*(7), 1291–1298. https://doi.org/10.1021/acssensors.8b00154
82. Ugo, P., Karimian, N., Stortini, A. M., & Moretto, L. M. (2018). New molecularly-imprinted electrochemical sensors for perfluorooctansulfonate and analytical methods based thereon. https://hdl.handle.net/10278/3718107.
83. Gong, J., Fang, T., Peng, D., Li, A., & Zhang, L. (2015). A highly sensitive photoelectrochemical detection of perfluorooctanic acid with molecularly imprined polymer-functionalized nanoarchitectured hybrid of AgI–BiOI composite. *Biosensors and Bioelectronics, 73*, 256–263. https://doi.org/10.1016/j.bios.2015.06.008
84. Chen, S., Li, A., Zhang, L., & Gong, J. (2015). Molecularly imprinted ultrathin graphitic carbon nitride nanosheets–based electrochemiluminescence sensing probe for sensitive detection of perfluorooctanoic acid. *Analytica Chimica Acta, 896*, 68–77. https://doi.org/10.1016/j.aca.2015.09.022
85. Li, J., Feng, H., Cai, J., Yuan, L., Wang, N., & Cai, Q. (2014). Molecularly imprinted polymer modified TiO2 nanotube arrays for photoelectrochemical determination of perfluorooctane sulfonate (PFOS). *Sensors and Actuators B: Chemical, 190*, 745–751. https://doi.org/10.1016/j.snb.2013.09.048
86. Fang, C., Chen, Z., Megharaj, M., & Naidu, R. (2016). Potentiometric detection of AFFFs based on MIP. *Environmental Technology & Innovation, 5*, 52–59. https://doi.org/10.1016/j.eti.2015.12.003
87. Cheng, Y. H., Barpaga, D., Soltis, J. A., Shutthanandan, V., Kargupta, R., Han, K. S., et al. (2020). Metal–organic framework-based microfluidic impedance sensor platform for ultrasensitive detection of perfluorooctanesulfonate. *ACS Applied Materials & Interfaces, 12*(9), 10503–10514. https://doi.org/10.1021/acsami.9b22445
88. Adhikari, S., Kelkar, V., Kumar, R., & Halden, R. U. (2022). Methods and challenges in the detection of microplastics and nanoplastics: A mini-review. *Polymer International, 71*(5), 543–551. https://doi.org/10.1002/pi.6348
89. Özgenç, E. (2024). Advanced analytical techniques for assessing and detecting microplastic pollution in water and wastewater systems. *Environmental Quality Management*. https://doi.org/10.1002/tqem.22217
90. Ebere, E. (2023). Technology for Screening Microplastics in the environment. https://doi.org/10.13140/RG.2.2.21495.60324/1

91. Dąbrowska, A. (2021). Raman spectroscopy of marine microplastics-A short comprehensive compendium for the environmental scientists. *Marine Environmental Research, 168*, 105313. https://doi.org/10.1016/j.marenvres.2021.105313
92. Araujo, C. F., Nolasco, M. M., Ribeiro, A. M., & Ribeiro-Claro, P. J. (2018). Identification of microplastics using Raman spectroscopy: Latest developments and future prospects. *Water Research, 142*, 426–440. https://doi.org/10.1016/j.watres.2018.05.060
93. Singh, B., & Kumar, A. (2023). Advances in microplastics detection: A comprehensive review of methodologies and their effectiveness. *TrAC Trends in Analytical Chemistry, 117440*. https://doi.org/10.1016/j.trac.2023.117440
94. Rayne, S., & Forest, K. (2009). Perfluoroalkyl sulfonic and carboxylic acids: A critical review of physicochemical properties, levels and patterns in waters and wastewaters, and treatment methods. *Journal of Environmental Science and Health Part A, 44*(12), 1145–1199. https://doi.org/10.1080/10934520903139811
95. Rehman, A. U., Crimi, M., & Andreescu, S. (2023). Current and emerging analytical techniques for the determination of PFAS in environmental samples. *Trends in Environmental Analytical Chemistry, 37*, e00198. https://doi.org/10.1016/j.teac.2023.e00198
96. Thompson, D., Zolfigol, N., Xia, Z., & Lei, Y. (2024). Recent progress in per-and polyfluoroalkyl substances (PFAS) sensing: A critical mini-review. *Sensors and Actuators Reports, 100189*. https://doi.org/10.1016/j.snr.2024.100189
97. Fu, J., Liu, N., Peng, Y., Wang, G., Wang, X., Wang, Q., et al. (2023). An ultra-light sustainable sponge for elimination of microplastics and nanoplastics. *Journal of Hazardous Materials, 456*, 131685. https://doi.org/10.1016/j.jhazmat.2023.131685
98. Li, J., Shan, E., Zhao, J., Teng, J., & Wang, Q. (2023). The factors influencing the vertical transport of microplastics in marine environment: A review. *Science of the Total Environment, 870*, 161893. https://doi.org/10.1016/j.scitotenv.2023.161893
99. Greger, M., & Landberg, T. (2024). Removal of PFAS from water by aquatic plants. *Journal of Environmental Management, 351*, 119895. https://doi.org/10.1016/j.jenvman.2023.119895
100. Pan, Y., Gao, S. H., Ge, C., Gao, Q., Huang, S., Kang, Y., et al. (2023). Removing microplastics from aquatic environments: A critical review. *Environmental Science and Ecotechnology, 13*, 100222. https://doi.org/10.1016/j.ese.2022.100222
101. Xiang, P., Zhang, Y., Zhang, T., Wu, Q., Zhao, C., & Li, Q. (2023). A novel bacterial combination for efficient degradation of polystyrene microplastics. *Journal of Hazardous Materials, 458*, 131856. https://doi.org/10.1016/j.jhazmat.2023.131856
102. Boiteux, V., Dauchy, X., Bach, C., Colin, A., Hemard, J., Sagres, V., et al. (2017). Concentrations and patterns of perfluoroalkyl and polyfluoroalkyl substances in a river and three drinking water treatment plants near and far from a major production source. *Science of the Total Environment, 583*, 393–400. https://doi.org/10.1016/j.scitotenv.2017.01.079
103. Li, J., Li, Q., Li, L. S., & Xu, L. (2017). Removal of perfluorooctanoic acid from water with economical mesoporous melamine-formaldehyde resin microsphere. *Chemical Engineering Journal, 320*, 501–509. https://doi.org/10.1016/j.cej.2017.03.073
104. Woodard, S., Berry, J., & Newman, B. (2017). Ion exchange resin for PFAS removal and pilot test comparison to GAC. *Remediation Journal, 27*(3), 19–27. https://doi.org/10.1002/rem.21515
105. Intense need for PFAS innovative methods for degradation and detection of PFAS. https://scenarios-project.eu/intense-need-for-pfas-innovative-methods-for-degradation-and-detection-of-pfas/
106. Bodzek, M., & Pohl, A. (2023). Possibilities of removing microplastics from the aquatic environment using membrane processes. *Desalination and Water Treatment, 288*, 104–120. https://doi.org/10.5004/dwt.2023.29154
107. Tang, C. Y., Fu, Q. S., Criddle, C. S., & Leckie, J. O. (2007). Effect of flux (transmembrane pressure) and membrane properties on fouling and rejection of reverse osmosis and nanofiltration membranes treating perfluorooctane sulfonate containing wastewater. *Environmental Science & Technology, 41*(6), 2008–2014. https://doi.org/10.1021/es062052f

108. Liu, Y., Xiao-Min, H., Zhao, Y., Wang, J., Ming-Xin, L., Peng, F.-H., & Bao, J. (2018). Removal of perfluorooctanoic acid in simulated and natural waters with different electrode materials by electrocoagulation. *Chemosphere, 201*, 303–309. https://doi.org/10.1016/j.chemosphere.2018.02.129
109. Yang, B., Han, Y., Yu, G., Zhuo, Q., Deng, S., Wu, J., & Zhang, P. (2016). Efficient removal of perfluoroalkyl acids (PFAAs) from aqueous solution by electrocoagulation using iron electrode. *Chemical Engineering Journal, 303*, 384–390. https://doi.org/10.1016/j.cej.2016.06.011
110. We, A. C. E., Zamyadi, A., Stickland, A. D., Clarke, B. O., & Freguia, S. (2023). A review of foam fractionation for the removal of per-and polyfluoroalkyl substances (PFAS) from aqueous matrices. *Journal of Hazardous Materials, 133182*. https://doi.org/10.1016/j.jhazmat.2023.133182
111. Wang, W., Mi, X., Zhou, Z., Zhou, S., Li, C., Hu, X., et al. (2019). Novel insights into the competitive adsorption behavior and mechanism of per-and polyfluoroalkyl substances on the anion-exchange resin. *Journal of Colloid and Interface Science, 557*, 655–663. https://doi.org/10.1016/j.jcis.2019.09.066
112. Zhang, Z., Sarkar, D., Biswas, J. K., & Datta, R. (2022). Biodegradation of per-and polyfluoroalkyl substances (PFAS): A review. *Bioresource Technology, 344*, 126223. https://doi.org/10.1016/j.biortech.2021.126223
113. Kaur, R., Talan, A., Tiwari, B., Pilli, S., Sellamuthu, B., & Tyagi, R. D. (2020). Constructed wetlands for the removal of organic micro-pollutants. In *Current developments in biotechnology and bioengineering* (pp. 87–140). Elsevier. https://doi.org/10.1016/B978-0-12-819594-9.00005-X
114. Hori, H., Nagaoka, Y., Murayama, M., & Kutsuna, S. (2008). Efficient decomposition of perfluorocarboxylic acids and alternative fluorochemical surfactants in hot water. *Environmental Science & Technology, 42*(19), 7438–7443. https://doi.org/10.1021/es800832p
115. Hao, S., Choi, Y. J., Wu, B., Higgins, C. P., Deeb, R., & Strathmann, T. J. (2021). Hydrothermal alkaline treatment for destruction of per-and polyfluoroalkyl substances in aqueous film-forming foam. *Environmental Science & Technology, 55*(5), 3283–3295. https://doi.org/10.1021/acs.est.0c06906
116. Zhang, M., Wang, W., Gong, T., Wu, Y., & Chen, G. (2024). Cutting-edge technologies and relevant reaction mechanism difference in treatment of long-and short-chain per-and polyfluoroalkyl substances: A review. *Chemosphere, 141692*. https://doi.org/10.1016/j.chemosphere.2024.141692
117. Santiago, A. R., Medina, P. B., & Su, X. (2022). Electrochemical remediation of perfluoroalkyl substances from water. *Electrochimica Acta, 403*, 139635. https://doi.org/10.1016/j.electacta.2021.139635
118. Le, T. X. H., Haflich, H., Shah, A. D., & Chaplin, B. P. (2019). Energy-efficient electrochemical oxidation of perfluoroalkyl substances using a Ti4O7 reactive electrochemical membrane anode. *Environmental Science & Technology Letters, 6*(8), 504–510. https://doi.org/10.1021/acs.estlett.9b00397
119. Rekik, H., Arab, H., Pichon, L., El Khakani, M. A., & Drogui, P. (2024). Per-and polyfluoroalkyl (PFAS) eternal pollutants: Sources, environmental impacts and treatment processes. *Chemosphere, 142044*, 142044. https://doi.org/10.1016/j.chemosphere.2024.142044
120. Burns, D. J., Stevenson, P., & Murphy, P. J. (2021). PFAS removal from groundwaters using surface-active foam fractionation. *Remediation Journal, 31*(4), 19–33. https://doi.org/10.1002/rem.21694
121. Zhao, H., Helgason, A., Leng, R., Chowdhury, S., Clermont, N., Dinh, J., et al. (2024). Removal of microplastics/microfibers and detergents from laundry wastewater by microbubble flotation. *ACS ES&T Water, 4*(4), 1819–1833. https://doi.org/10.1021/acsestwater.3c00802
122. Wang, Y., Li, Y. N., Tian, L., Ju, L., & Liu, Y. (2021). The removal efficiency and mechanism of microplastic enhancement by positive modification dissolved air flotation. *Water Environment Research, 93*(5), 693–702. https://doi.org/10.1002/wer.1352

123. Martic, S., Tabobondung, M., Gao, S., & Lewis, T. (2022). Emerging electrochemical tools for microplastics remediation and sensing. *Frontiers in Sensors, 3*, 958633. https://doi.org/10.3389/fsens.2022.958633
124. Zhao, B., Richardson, R. E., & You, F. (2024). Microplastics monitoring in freshwater systems: A review of global efforts, knowledge gaps, and research priorities. *Journal of Hazardous Materials, 135329*. https://doi.org/10.1016/j.jhazmat.2024.135329
125. Chellakannu, A., Karuppathevan, R., Muniasamy, K., & Sivasamy, V. V. (2024). Synthesis and characterization of Mn-doped g-C3N4/ZnO nanocomposite for photocatalytic degradation and antibacterial activity applications. *Journal of Molecular Structure, 1314*, 138708. https://doi.org/10.1016/j.molstruc.2024.138708
126. astkCARE: anionic surfactant detection, Cooperative Research Centre for Contamination Assessment and Remediation of the Environment, 2020.
127. Wang, Z., & Sedighi, M. (2023). Transport and retardation of microplastic in biochar. In *9th International Congress on Environmental Geotechnics (ICEG2023)*. https://doi.org/10.53243/ICEG2023-250
128. Zafar, M. A., & Jacob, M. V. (2024). Instant upcycling of microplastics into graphene and its environmental application. *Small Science, 2400176*. https://doi.org/10.1002/smsc.202400176
129. Jie, X., Li, W., Slocombe, D., Gao, Y., Banerjee, I., Gonzalez-Cortes, S., et al. (2020). Microwave-initiated catalytic deconstruction of plastic waste into hydrogen and high-value carbons. *Nature Catalysis, 3*(11), 902–912. https://doi.org/10.1038/s41929-020-00518-5

Part VII
Mitigation and Recommendation

Chapter 17
Safe Level (Threshold) of Emerging Contaminants in Drinking Water

Jaspreet Kaur, Ina Thakur, and Surinder Kumar Mehta

1 Introduction

Emerging contaminants (ECs) have garnered much attention in the ever-changing ecological landscape, indicating a need for better knowledge of their complex and diverse effects. Earlier chapters delved into and explored in depth a variety of ECs, including per- and polyfluoroalkyl substances (PFASs), microplastics, pharmaceuticals, pesticides, and other new contaminants. Each of these ECs presents perilous issues due to their multifaceted chemical characteristics, natural persistence mechanisms, and possible health hazards [1]. PFASs, for example, are known to resist the process of degradation, which results in extensive contamination of water supplies and soil. Similarly, microplastics pose a potential health risk, as they can easily enter the food chain through the process of biomagnification and are responsible for inflammation and toxicity in the body [2]. Likewise, various organic pharmaceutical moieties, such as antibiotics *and antidepressants*, can contribute to antibiotic resistance, impair endocrine signaling, and lead to a multitude of neurological issues [3].

It is crucial to address and manage the effects of these ECs for several reasons. First, each contaminant interacts differently with the environment and living beings, which requires customizable monitoring and cleaning solutions [4]. Second, the long-term and continuing persistence of ECs in the ecosystem necessitates urgent creative measures and precise restrictions to reduce their distribution and impact.

J. Kaur (✉) · I. Thakur
School of Basic Sciences, Indian Institute of Information Technology, Una, Himachal Pradesh, India
e-mail: jas.sandhu3@iiitu.ac.in; inathakur@iiitu.ac.in

S. K. Mehta
Department of Chemistry, Panjab University, Chandigarh, India
e-mail: skmehta@pu.ac.in

M. K. Jindal et al. (eds.), *Emerging Contaminants in Water*, Emerging Contaminants and Associated Treatment Technologies,
https://doi.org/10.1007/978-3-031-89591-3_17

Microplastics, for example, are extremely difficult to eradicate in the environment because of their small size and extensive distribution [5]. Therefore, intensive methods for prevention, detection, and cleaning are indispensable. Pharmaceutical ECs entering aquatic environments can have significant environmental repercussions and may lead to the proliferation of drug-resistant bacteria, highlighting the need for better disposal techniques and regulations [6]. Given these issues, comprehending the intricate nature of ECs is vital for safeguarding human health and sustaining ecological balance.

Setting acceptable limits for ECs is a critical step toward managing these pollutants. The establishment of a safe limit will help in implementing specific regulatory interventions for safeguarding human and environmental health. Obviously, these safe limits are not static, and they are modified when new scientific knowledge comes to the forefront or if detection technology improves. Establishing these constraints requires meticulous investigation to explore the potential risks associated with multiple concentrations and then identify values that are acceptable for drinking purposes [7]. This dynamic approach provides the foundation for tailoring regulatory frameworks to new scientific information while keeping safety standards updated and efficiently implemented.

In the context of regulatory and safety regimes, thresholds play a cardinal role. Thresholds set levels at or above which the contamination becomes unacceptable. These values serve as standards for regulatory authorities and policymakers in developing and implementing rules to safeguard public health and the environment. It also increases the ease of monitoring and quantifying the pollution level, hence enabling prompt interventions and remediation. Setting and periodically modifying these threshold values allows regulatory frameworks to meet new and rising challenges posed by ECs, offering a proactive and responsive strategy for environmental protection [8].

In summary, regulating ECs necessitates an extensive understanding of their various health repercussions and the need to ascertain acceptable limits. Establishing limits and adhering to them is an imperative step in regulatory and safety frameworks, which ensures that adequate safety measures are in place that shield both individuals and their surroundings from the adverse impacts of persistent pollutants.

2 Framework for Establishing Safe Levels for Drinking Water Contaminants

Safeguarding drinking water is vital, and determining appropriate levels of pollutants in drinking water is a difficult process that is based on risk assessment approaches. This framework includes a variety of scientific, regulatory, and toxicological issues that contribute to the determination of safe thresholds for human consumption.

2.1 Risk Assessment Methodologies

Risk assessment is an organized approach to assess the potential health hazards associated with pollutants in drinking water. The fundamental elements of risk assessment include hazard identification, dose–response evaluation, and estimation of exposure [9]. The above steps are an essential component of the risk assessment approach and are essential for establishing permissible levels of drinking water contaminants.

Hazard Identification Before risk assessment begins, the first step is to assess the dangers associated with each pollutant. Hazard identification ascertains whether or not a hazard has the potential to affect human health. This finding was validated by laboratory investigations, epidemiological studies, and retrospective analyses of human exposures. Several contaminants, such as lead, arsenic, mercury, rare earth elements (REEs), and some pesticides, for example, have been designated as perilous owing to their potential adverse health impacts [3, 9].

Dose–Response Assessment After the initial assessment of hazard identification, the next step involves the analysis of pollutant dosage and the likelihood of unfavorable health consequences. This is referred to as a dose–response evaluation. This phase is critical because it helps to define the dosage point at which a pollutant becomes a matter of concern. The dose–response relationship is sometimes shown as a curve, demonstrating the link between rising dosages and the likelihood of undesirable effects. Even very low dosages of certain pollutants might be dangerous; therefore, extremely strict safe standards are needed [10].

Exposure Assessment An exposure assessment examines the degree to which people are exposed to a contaminant in drinking water. This includes estimating the chemical content of drinking water in addition to the quantity of water consumed by individuals. The exposure evaluation analyzes a number of variables, such as sex, age, lifestyle, and geographic location, all of which may influence the degree of exposure. Dose–response and exposure assessments work together to furnish a comprehensive picture of a hazard risk [9].

2.2 Toxicological Data

Toxicological data are critical for identifying acceptable doses of contaminants in drinking water. Toxicology is the study of the detrimental impacts of chemicals on living organisms, and it offers a scientific basis for risk assessment. The key toxicological data used in this context are investigations on the toxic effects, carcinogenicity, and reproductive implications of contaminants [3].

Key Toxicological Data The safe levels are calculated from data from acute, subchronic, and chronic toxicity investigations. Acute toxicity studies evaluate the immediate effects of a single high-dose exposure, whereas subchronic and chronic studies investigate the effects of repeated contaminant exposure over time. Carcinogenicity studies are very significant for pollutants implicated in causing cancer since they indicate whether a material may cause cancer in people. Reproductive and developmental toxicity studies are especially important because they evaluate the possible effects of pollutants on reproductive health and fetal growth [11].

Differences in Toxicity Profiles Among Contaminants Not all pollutants have similar toxicological profiles, and this diversity has a substantial impact on the determination of tolerable limits. For example, certain pollutants can trigger harm at extremely low doses, but others may only represent a concern at larger quantities. The toxicity profile of a contaminant is determined by its chemical structure, mechanism of action, and biological systems. For example, arsenic is known to have both carcinogenic and noncarcinogenic effects, whereas lead largely impairs brain development in infants [8]. Understanding these distinctions is critical for establishing acceptable safety levels.

2.3 *Regulatory Standards and Guidelines*

Various regulatory bodies across the globe have created rules and guidelines for acceptable levels of pollutants in drinking water. These criteria have been established on the basis of risk assessment procedures and toxicological data presented previously. A comparison of existing standards from various bodies reveals both parallels and differences in how safe levels are set and enforced.

The Environmental Protection Agency The Environmental Protection Agency (EPA) is in charge of protecting the quality and safety of water in the United States. This is accomplished via the creation and subsequent implementation of a pair of vital pieces of legislation, i.e., the Safe Drinking Water Act (SDWA) and the Clean Water Act. These laws aim to protect both the country's water supply and the greater water environment, which includes water bodies such as lakes and rivers and coastal areas (www.epa.gov/sdwa).

The Safe Drinking Water Act (SDWA) was passed in 1974 with the main aim of safeguarding the purity of drinking water from all public water systems. The SDWA directed the EPA to regulate standards for drinking water quality that would protect against both naturally occurring and man-made contaminants. The SDWA additionally allows the EPA to oversee water supply systems and enforce compliance with specific conditions. This includes conducting regular water quality testing, communicating the outcomes, and deploying action when norms are not fulfilled. Public water systems that fail to satisfy standards may face penalties, which may lead to the involvement of the EPA in ensuring that corrective measures are undertaken that restore water integrity [12].

The Clean Water Act (CWA) was passed in 1972 and enhanced the efficacy of the SDWA by addressing pollution in the country's waterways. The CWA seeks to improve and protect the biological, chemical, and physical health of the US water. The EPA established surface water quality regulations under the Clean Water Act, which include restrictions on pollutants discarded into lakes, rivers, and coastal waters from an array of sources, including industrial operations, wastewater treatment plants, and rainwater runoff [3].

European Union Management of the emerging pollutants was performed by implementing a robust system with two major regulations: (i) the Water Framework Directive (WFD) and (ii) Registration, Evaluation, Authorization, and Restriction of Chemicals (REACH). The main goal of these regulations is ensuring effective monitoring and management of water resources to protect humans and the surrounding environment throughout the European Union. In 2007, the Registration, Evaluation, Authorization, and Restriction of Chemicals (REACH) was established and is among the most extensive chemical laws worldwide. It has been made mandatory that any chemical that is imported or produced in the EU with a quantity of more than 1 ton per year must be registered with the European Chemicals Agency (ECHA). The requirements for the registration process include the submission of comprehensive details on the quality of the chemical, its properties, hazards, and potential risks to the environment and human health. The chemicals are thereafter evaluated and, if necessary, authorized or limited to mitigate any possible risks [13].

The Water Framework Directive (WFD), enacted in 2000, sets a comprehensive framework for evaluating the quality of water, analyzing risks, and implementing pollution mitigation strategies. Member nations are also required to provide a management plan for river basins that specifies strategies to attain and sustain excellent water quality [14]. By integrating the advantages of both regulations, such as REACH's focus on the safety of chemicals and WFD's focus on water quality, these regulations have the potential to create a more efficient and comprehensive framework to protect the environment from pollution.

United Nations Environment Program (UNEP) and World Health Organization (WHO) The UNEP and the WHO are formulating guidelines and recommendations to control emerging contaminants globally. The major focus of these organizations is to establish global standards to mitigate the impact of pollutants, including microplastics, pharmaceuticals, and other persistent pollutants. These regulations are essential for protecting humans and their surroundings. The UNEP and the WHO make suggestions that help different nations formulate and execute regulations to protect the environment and water bodies. They play a significant role in promoting a common global response to emerging contaminants by highlighting the importance of international collaboration and information exchange. This holistic strategy allows nations, particularly those with limited resources, to establish effective programs for monitoring, analyzing, and managing pollutants [15].

Bureau of Indian Standards (BIS) India's water regulation system is governed by BIS and IS 10500:2012, which provide extensive safe drinking water standards. These regulations outline two significant limits: (i) acceptable limits and (ii) permissible limits. The acceptable limits denote the ideal standard of water quality that is needed, whereas the permissible limits are applied when other water sources are not available (Drinking Water Quality Monitoring & Surveillance Framework, 2018).

When any contaminant or parameter exceeds permissible levels, an investigation is mandated. After the investigation, appropriate action needs to be taken; otherwise, the drinking water supply should be stopped until the quality of the water falls below the acceptable limit. When the water quality falls below the acceptable limit, the water supply needs to be stopped, and an alternative source of water needs to be provided to safeguard human health. To ensure compliance with the regulations, the drinking water sources must undergo regular testing, and if the water fails to meet the set standards, immediate corrective measures are needed.

China's National Standards for Drinking Water Quality China's national standards consider the unique environmental and industrial characteristics of the nation while aiming to protect public health and comply with international principles. On the basis of national circumstances and WHO principles, the Ministry of Health and the Ministry of Environmental Protection set the requirements. China's standards, which strike a compromise between international suggestions and regional circumstances, are comparable to those of the EU.

From the above discussion, it can be concluded that the WHO offers general recommendations, whereas the EU and EPA have strict, legally binding regulations. Because the EPA bases its criteria on both technological viability and health implications, there are some variances in the upper limits, such as a higher action level for lead. China and the EU set uniform norms that apply everywhere, but they also permit certain latitude in their enforcement. Similarly, India adjusts its laws to suit the needs of the region. The WHO guidelines are advisory, but the EPA and EU regulations are legally binding. There are large variations in enforcement, with the EPA being especially stringent. Globally, disparities in local environmental factors, public health concerns, and enforcement capacities influence how contaminants are regulated in drinking water.

Countries can choose from a set of frameworks for determining acceptable levels of drinking water pollutants, which is a collaborative endeavor that incorporates risk assessment approaches, toxicological data, and regulatory criteria. Risk assessors can establish acceptable levels of pollutants for human consumption by methodically identifying risks, calculating dose–response relationships, and evaluating exposure. The comparison of regulatory criteria from various agencies highlights the global effort to guarantee safe drinking water for everyone.

3 Evaluation of Safe Levels for Specific Contaminants

ECs are bioactive and often bioaccumulate. It is feasible for these contaminants to be widespread and persist over time. It is anticipated that as the world's population grows, more people will live in densely populated areas, leading to more contamination of ecosystems and increasing concentrations of ECs in the environment. Owing to the vast abundance of ECs and the lack of regulations, reporting their possible toxicity and regularly monitoring and examining EC levels in water supplies and wastewater discharges are imperative.

3.1 Per- and Polyfluoroalkyl Substances (PFASs)

PFASs are sometimes called "forever chemicals" since most of them are extremely unlikely to break down in the environment under normal circumstances. These substances are employed as anti-staining, waterproofing, and fireproofing agents in a variety of industrial and consumer goods. They end up building up in groundwater, surface water, and even drinking water supplies due to their tenacious nature. Long-chain PFAAs (≥8 carbons for carboxylates, ≥6 carbons for sulfonates) and other long-chain PFASs are extremely bioaccumulative in humans. Because long-chain PFAAs have a long human half-life (several years), the body levels of these compounds remain elevated for years after exposure ends [16]. Numerous studies have investigated the presence of PFASs and their destiny in water bodies because one of the main routes for human exposure is through water. Concerns regarding the possible health risks associated with perfluoroalkyl carboxylates (PFCAs) and perfluoroalkane sulfonic acids (PFSAs) have been raised by the consistent discovery of these chemicals in tap water samples at different sites. Drinking water samples from Europe, America, Asia, Australia, and other regions have also been reported to contain PFASs [17, 18].

Surprisingly, there are no drinking water standards for any PFAS, despite extensive pollution, broad exposure, and clinical proof of health impacts. The USEPA issued a nonenforceable lifelong health advisory of 70 ng/L for perfluorooctanoic acid (PFOA) and perfluorooctane sulfonate (PFOS), either alone or in combination, in 2016 instead of establishing a standard. There is no comprehensive assessment of PFAS in US drinking water since public water systems have no obligation to treat water that exceeds EPA HAs or to routinely test for PFAS in the absence of an enforceable standard [19].

The EPA states that the lifetime health advisory for PFOA and PFOS in drinking water has been set as a reference at 70 ng/L. This limit of tolerance applies to PFOA and PFOS separately; however, in the event of an unexpectedly high concentration, as per Australian guidelines, it applies to the total amount of both chemicals. Compared with other international rules, this figure appears to be less conservative. Since 2002, when more cautious levels were implemented due to increasing

Table 17.1 US PGAO/PFOS drinking water guideline levels [21]

Guideline	Advisory limit		Reference limit	
	PFAO (ng/L)	PFOS (ng/L)	PFAO (ng/kg-day)	PFOS (ng/kg-day)
US EPA, 2016 Health Advisory Level	70	70	20	20
Alaska DEC, 2016, Groundwater cleanup level	400	400	20	20
Maine DEP, 2016, Remedial action guideline	130	560	6	80
Minnesota DOH, 2017, Noncancer health-based level	35	27	18	5.1
New Jersey DEP, 2017, Maximum contaminant level	14	13	2	1.8
North Carolina DENR, 2012, interim maximum allowable concentration	1000	–	NA	NA
Texas CEQ, 2017, Protective concentration level	290	560	15	20
Vermont DEC/DOH, 2016, primary groundwater enforcement standard	20	20	20	20

concerns, the global focus on PFASs has grown quickly. Owing to its removal from the market, PFOS, a "long-chain" (C8) chemical, is subject to additional regulations.

However, the current regulation seems to be driven by many factors, including financial factors and detection limits [20]. In 2016, the EPA suggested that lifetime limit for PFOA and PFOS should be 70 ng/L, individually or in combination. However, some US states have argued that the EPA's guidelines are insufficient and do not address the potential associated health risks; hence, various PFAS threshold values have been developed differently from those of the EPA. The state's water guideline levels for PFOA and/or PFOS ranged from 13 to 1000 ng/L compared with 70 ng/L recommended by the EPA for both individual and combined compounds (Table 17.1).

3.1.1 Potential Effects of PFASs on Human Health Above the Admissible Limit

PFASs may accumulate in humans and cause serious health issues, such as metabolic syndrome (MetS), which is a precursor to heart disease, the leading cause of death worldwide. PFOSs and PFOA are two of the most well-known and researched PFASs because of their enduring presence in important environmental compartments, such as air, soil, and water. Long-chain polyfluoroalkyl substances (PFOA and PFOS) have been connected to a number of health problems, such as low birth weight, reduced sperm quality, thyroid abnormalities, kidney and testicular malignancies, hypertension, and immune system toxicity [22].

Exposure to PFASs can reduce an individual's immunological response. Research conducted in vitro has demonstrated that perfluorinated compounds affect immune cells by modifying the expression of cytokines. The primary mechanism of action for the prefrontal cortex (PFC)-induced modulation of cellular processes is the activation of receptors that activate peroxisome proliferators (PPARs). PFOA has been identified by the International Agency for Research on Cancer (IARC) as potentially carcinogenic to humans. Notably, there is a greater risk of cancer among community members whose drinking water is contaminated with PFOA [23]. Exposure to PFAS in the workplace and community has been connected to a number of cancers in humans, including testicular, liver, prostate, and kidney cancers. Numerous PFASs exhibit characteristics similar to those of carcinogens, including oxidative stress induction and receptor-mediated effect modulation.

3.2 *Microplastics*

Microplastics are solid particles of polymers that are ≤5 mm in size and are insoluble in water [24]. Although the lower size boundary has no official definition, particles smaller than 1 μm are commonly referred to as nanoplastics instead of microplastics. A significant portion of plastic contamination has gone unnoticed: microplastics. Owing to their natural hydrophobic nature, microplastics can enter freshwater environments through contaminated plastic waste, air deposition, surface runoff, treated and untreated sewage effluent, and air deposition. Microplastics from water distribution systems can enter bottled and tap water [25]. Different kinds of microplastics are found in freshwater, as shown in Table 17.2.

3.2.1 Challenges in Setting Safe Levels of Microplastics

1. One of the major hindrances in quantifying microplastics in drinking water is the lack of standardized detection and measurement procedures. The wide range of shapes and sizes makes the identification and quantification of microplastics more difficult. The results may differ among laboratories depending upon the tools and methodologies used. This makes it challenging to set a reliable baseline to set a permissible limit for microplastics in drinking water.
2. Microplastics originate from different sources, such as cosmetics, food packaging, the disintegration of large plastics, clothing, and industrial processes. Because of their different origins, microplastics have a wide range of characteristics, including varying chemical compositions, shapes, and sizes. This diversity of microplastics hinders the execution of definitive guidelines for their presence in drinking water. The consideration of these unique properties of microplastics, which may affect human health, is highly important for regulators.

Table 17.2. Different kinds of microplastics that appear in freshwater environments

Location	Detection method	Concentration	Type/color/size	Reference
River sediments, Shanghai, China	Density separation, microscopy, and micro-FTIR spectroscopy	802.0 items/kg dry weight	Spheres, fibers, and fragments; PP, PEST, rayon; red, white, transparent, and blue	[26]
Wei River Basin, China	FTIR spectroscopy	3.67 to 10.7 items/L	Fibers; PE, PVC, and PS	[27]
Yangtze River, China	Stereo microscopy, Raman spectroscopy	$492{,}000.0 \times 10^{-6}$ microplastic/m^2	PP, PE, PS, nylon, POM, cellulose, and EVA	[28]
Vistula River, Poland	Raman spectroscopy		PE, PP, PA, and PC; white, blue, and red	[29]
Seine River and Marne River, Paris, France	Stereo microscopy and FTIR spectroscopy	100.6 ± 99.9 fibers/m^{-3} in the Marne River and 48.5 ± 98.5, 27.9 ± 26.3, 27.9 ± 40.3, and 22.1 ± 25.3 fibers/m^{-3} in the Seine River	Fibers; PET, PP, PA, and PUR; blue	[30]
Charleston Harbor and Winyah Bay Estuaries, South Carolina, USA	SEM and FTIR spectroscopy	221.0 ± 25.6 particles/m^2 in sediment samples of Winyah Bay and 413.8 ± 76.7 particles/m^2 in sediment samples of Charleston Harbor	Fragments, fibers, foams, and spheres; white, green, gray, blue, black, and red	[31]
20 urban lakes, Hanjiang River, and Yangtze River, Wuhan, China	Stereoscopic microscopy, SEM and FTIR spectroscopy	1660.0 ± 639.1 to 8925.0 ± 1591.0 numbers/m^3	Fiber; PP and PET	[32]
Western Superior Lake, USA	Microscopy, ATR-FTIR spectroscopy And pyrolysis GC–MS	0.0 to 110000.0×10^{-6} particles/m^2	Fragments, fibers, and films; PVC, PP, PE, PET, PS, and PDMS	[33]

3. Another challenge is the lack of knowledge of the impact of microplastics on human health. While the literature on the impact of microplastics on humans is still very new, concerns regarding their toxicity are rising exponentially. These microplastics may contain harmful chemicals and can affect human health if consumed directly or indirectly. However, its long-term impact is still unknown. This uncertainty makes microplastics in drinking water thresholds difficult to determine.

3.2.2 Available Data and Emerging Standards

The literature indicates that the concentration of microplastics in drinking water is not regulated by any governing authority in the world. To address this issue, many countries and various international organizations are making collaborative efforts. A 2019 WHO report on microplastics in drinking water highlighted the need to perform a comprehensive study on potential health risks. The WHO has determined that the current concentration of microplastics in drinking water does not pose any health risks to humans. However, researchers still emphasize further research on standard techniques to quantify microplastics. The WHO is diligently looking into the conditions yet has not yet established any standards or guidelines for the same.

However, the EU has recognized these microplastics as potential threats to humans and the environment, particularly the contamination of water bodies. The EU's drinking water directives have made it mandatory for every state member to monitor the identified contaminants in drinking water. However, microplastics are not included in this directive. The EU is actively funding research to acquire extensive knowledge of the presence and impact of microplastics. Depending on the results of this research, rules or restrictions may be developed in the future.

Similarly, the USEPA has recognized the existence of microplastics in drinking water but has not set any guidelines or standards to regulate their concentration in the water. The EPA is currently carrying out research to assess the health impacts and potential hazards of microplastics. This research may guide future regulatory guidelines.

3.3 Pharmaceuticals and Personal Care Products (PPCPs)

Pharmaceuticals are prescription and nonprescription drugs used to treat and prevent conditions in humans and animals, whereas personal care products are primarily used to increase the quality of our daily lives [34]. One of the largest categories of ECs is covered by PPCPs. Over the last 10 years, there have been many concerns about the accidental presence of PPCPs in various marine ecosystems (biota, sediments, and water) at levels that have the potential to have detrimental effects on the immediate environment. Owing to the widespread and increasing use of PPCPs in both human and veterinary medicine, their constant release into the environment has become standard practice, which is a significant reason for concern.

3.3.1 Regulatory Standards for Controlling PPCPs

A toxicological study is necessary to define toxicity endpoints and evaluate the risk of ECs. Regulating criteria, however, are lacking for several groups of ECs containing PPCPs, mostly because toxicological data are inadequate. For this reason, appropriate EC laws are lacking in many nations. The Indian Standards (IS

10500) for drinking water in India include limits on the amount of PPCPs that can be used, however, they do not include the monitoring of ECs in wastewater or water treatment facilities. For a number of PPCP classes, the WHO has created thorough regulatory criteria. Forty-five priority compounds with environmental quality standards that need to be upheld in aquatic environments are included in the EU Water Framework Directive.

The Safe Drinking Water Act and the USEPA regulate the emergence of contaminants. It concentrates mainly on the substances that are listed on the Contaminant Candidate List (CCL) that are detected in public water supplies but are not yet regulated. Table 17.3 shows different PPCPs and their acceptable limits [35].

The presence of PPCPs in water bodies directly affects human health. The literature has shown the existence of various pharmaceutical residues in drinking water, consisting of analgesics, antidepressants, anticonvulsants, etc. As a consequence, concern over health impacts and long-term exposure has emerged. Even though the concentrations of these pollutants are very low, they can severely affect human health. Long-term exposure may lead to various health issues, such as increased risk of cancer, hormonal imbalances, and physical abnormalities. Furthermore, the disposal of pharmaceutical residue in water bodies has given rise to antibiotic-resistant bacteria in the environment. This may directly affect human health, as it has become much more difficult to treat common diseases. The complexity of these challenges becomes more complicated, especially in developing countries, because there is a lack of comprehensive guidelines and regulatory controls.

The increasing global utilization of cosmetics and pharmaceuticals is expected to adversely affect both humans and the environment. This highlights the pressing need for improved risk management techniques and highlights the urgent need for enhanced risk management strategies and more rigorous guidelines to safeguard the environment and human health.

Table 17.3 Regulation of PPCPs as per the WHO, USEPA, EU, and IS 10500 standards (μg/L)

PPCPs	WHO	USEPA	EU	IS 10500
Edetic acid	600	–	–	–
Disinfection byproducts	–	–	–	–
Chloroform	300	80		200
N-Nitrosodimethylamine	0.1			
Trichloroacetate	200	–	–	–
2,4,6-Trichlorophenol	200	–	–	–
Bromate	10	–	–	–
Bromodichloromethane	60	–	–	60
Bromoform	100	80	–	100
Chlorate	700	–	–	–
Dibromoacetonitrile	70			
Dibromochloromethane	100	80		100
Dichloroacetonitrile	20	–	–	–
Dichloroacetate	50	60	–	–

3.4 *Pesticides and Agrochemicals*

3.4.1 Existing Thresholds and Their Adequacy

Agrochemicals, or fertilizers and insecticides, are thought of as a way to advance crop production technology, even if they are costly inputs. The development of agriculture and pesticide use are strongly intertwined. The application of pesticides has increased crop productivity and helped to avoid losses from pest assault, but excessive use of these chemicals is contaminating groundwater. Water bodies next to agricultural areas typically receive pesticide residues, even though agricultural land is the principal consumer of these substances. Agrochemical residues may be widespread in surface water systems, particularly in irrigation drains, which can contaminate river and pond water and negatively impact aquatic ecosystems. Therefore, it is important to closely adhere to the standards for the use of pesticides to reduce their negative impacts on both the environment and human health [36]. The safe concentrations of these agrochemicals in drinking water are covered in India by IS 10500. Table 17.4 lists the extensive regulatory criteria that the EU, USEPA, and WHO have set for fertilizers and pesticides.

Table 17.4 Regulation of agrochemicals as per the WHO, USEPA, EU, and IS 10500 standards (μg/L)

Agrochemicals	WHO	USEPA	EU	IS 10500
DBCP	1	0.2	–	–
2,4-D	30	70	–	30
Chlordane	0.2	2	–	–
Lindane	2	0.2	–	2
Chlorpyrifos	30	10	0.1	30
DDT	1	–	0.025	1
Mecoprop	10	–	–	–
Pentachlorophenol	9	1	1	–
Endrin	0.6	2	0.01	–
Alachlor	20	2	0.7	–
Terbuthylazine	7	–	–	–
Other chemicals	–	–	–	–
Perchlorate	70	15	–	–
Toluene	700	1000	–	–
1,4-Dioxane	50		–	–
Benzo(a)pyrene	0.7	0.2	0.1	0.1
Styrene	20	100	–	–
Tetrachloroethene	40		–	–
Vinyl chloride	0.3	2	–	–
Xylenes	500	10,000	–	–
Trichloroethene	20		–	–
Polychlorinated biphenyls	–	0.5	–	0.5
PFOA/PFOS	–	0.07	36	–
Dichloromethane	20	–	20	–

3.4.2 Concerns Related to Agrochemical Contaminants in Water Above the Permissible Limit

Human health is greatly compromised by drinking water contaminated by agrochemicals. In addition to acute poisoning, neurotoxic pesticides such as organophosphates and carbamates can have long-term health impacts, including cancer, endocrine disruption, and reproductive problems. Fertilizer nitrate pollution is especially dangerous because it can cause methemoglobinemia, sometimes known as "blue baby syndrome," which lowers an infant's blood's ability to carry oxygen. Numerous cancers and other recurring health issues have also been connected to extended exposure to nitrates and other pesticide residues in water.

Agrochemicals also affect aquatic life, as they pass into water bodies through different processes, such as runoff, drift, and leaching via the soil, or they may be applied directly into surface water to control some aquatic pests or mosquitoes. Insecticides and herbicides can affect fish and invertebrates and alter the species composition of aquatic commodities by mixing in surface water through spray drift or runoff. Approximately 41% of all amphibian species are threatened (IUCN 2020). According to the IUCN (2009), one-third of 6000 amphibian species worldwide are threatened. In addition to being affected by habitat loss, overexploitation, or introduced species, amphibians are affected by the pollution of surface waters with fertilizers and pesticides [37]. Owing to the impact of agrochemical water becoming unfit for consumption, it promotes the growth of algae and leads to eutrophication due to the excess use of chemical fertilizers.

Agrochemicals can be sprayed directly into surface water to control some aquatic pests or mosquitoes, or they can enter water bodies through other mechanisms, such as runoff, drift, and leaching via the soil. In either case, they also have an impact on aquatic species. By combining with surface water through spray drift or runoff, insecticides and herbicides can have an impact on fish and invertebrates as well as change the species composition of aquatic commodities. According to the IUCN (2020), 41% of all amphibian species are vulnerable. One-third of the world's 6000 amphibian species are vulnerable, according to IUCN reports from 2009. In addition to overuse, invasive species, and habitat degradation, frogs are impacted by fertilizer and pesticide pollution in surface waters [37].

3.5 Industrial Chemicals and Metals

3.5.1 Safe Levels for Industrial Contaminants and Heavy Metals

Heavy metals and industrial chemicals are frequently utilized in a variety of manufacturing and industrial activities. Despite the vital functions these compounds play in various industries, exposure to hazardous amounts of these substances can have serious negative effects on one's health and the environment. To reduce these dangers and guarantee that exposure does not negatively impact the environment or human health, safe levels have been set.

Industrial Chemicals These include insecticides, polymers, dyes, and solvents. Permissible exposure limits (PELs) are established by regulatory agencies such as the Occupational Safety and Health Administration (OSHA) and the USEPA to guarantee worker safety. For example, during an 8-h shift, the PEL for benzene, a typical industrial solvent, is 1 part per million (ppm).

Heavy Metals Typical heavy metals include arsenic, cadmium, mercury, and lead. In biological processes, these metals can accumulate and lead to long-term health problems. Organizations such as the EPA and the WHO set safe levels. For example, lead is regulated by the EPA to 15 parts per billion (ppb) in safe drinking water.

The public's health depends on the regulation of industrial chemicals and heavy metals in drinking water because exposure to elevated amounts can cause major health issues. Many regulatory strategies have been devised by nations and organizations to reduce the amount of heavy metals in drinking water. Comparisons of different regulations for ECs are given in Tables 17.5 and 17.6 [15, 38].

Guidelines for rare earth elements (REEs) in drinking water are not well established because of the low concentrations of these elements. The WHO and USEPA have not established any guidelines or limits for REE. Similarly, BIS 10500:2012 and the EU lack standards for REE. Some studies have suggested that the tentative concentration of cerium is <0.07 mg/l and that of lanthanum is <0.02 mg/l. There are no international standards set for these elements, but certain regions, such as

Table 17.5 Comparison of drinking water standards for trace and toxic metals

	BIS 10500:2012				WHO		USEPA		EU	
Toxic metals	Requirement (acceptable limit)		Permissible limit in the absence of alternative source		Permissible limit		Maximum contaminant level		Drinking water directive	
	mg/L	mg/L	mg/L	mg/L	mg/L	mg/L	mg/L	mg/L	mg/L	mg/L
Arsenic	0.01	10	0.05	50	0.01	10	0.01	10	0.01	10
Cadmium	0.003	3	No relaxation		0.03	30	0.005	5	0.005	5
Chromium	0.05	50	No relaxation		0.05	50	0.1	100	0.05	50
Copper	0.05	50	1.5	1500	2	2000	1.3	1300	2	2000
Iron	0.30	300	No relaxation		–		0.3	300	0.2	200
Lead	0,01	10	No relaxation		0.01	10	0.015	15	0.005	5
Nickel	0.02	20	No relaxation		0.07	70	–	–	0.02	20
Zinc	5	5000	15	15,000	–		5	5000	–	–

Table 17.6 Comparison of drinking water standards for industrial chemicals

Toxic chemicals	BIS 10500:2012	WHO	USEPA	EU
	ppb	ppb	ppb	ppb
Benzene	1	10	5	1
Trichloroethylene	1	20	5	10
Vinyl chloride	0.02	0.3	2	0.5
Carbon tetrachloride	700	700	5	100

South Africa [39], have monitored these REEs under an emerging contaminant framework because of their potential human and environmental risks.

4 Integrating Thresholds into Drinking Water Standards

4.1 Integrating Threshold Limits into National and International Standards

Health-based targets, such as numerical guideline values and other targets outlined in the Guidelines for Drinking Water Quality, are meant to serve as the scientific foundation upon which national/international or regional numerical drinking water quality standards can be developed; they are not meant to be legally binding limits. There is not one strategy that works for everyone, and different nations and areas may have different requirements for drinking water in different ways [15]. A range of environmental, social, cultural, economic, nutritional, and other factors influencing possible exposure need to be considered when creating national drinking water guidelines. This could result in national standards that are significantly different from these guidelines in terms of their scope and risk objectives.

The communities that are served, as well as the main water users, should be involved in the standards-setting process to guarantee that the requirements are acceptable to consumers. The organizations in charge of the community's drinking water supply may not be as close to it as public health organizations. They also collaborate with other sectors (such as education) locally, and their joint efforts are crucial to guaranteeing engaged community participation. To integrate thresholds into national and international standards, the following steps are involved:

1. *Adapting guideline values to locally relevant standards:* The percentage of acceptable daily intake assigned to drinking water when establishing guideline values for numerous chemicals varies to account for variances in exposure from various sources (such as food and water) in different parts of the world. Authorities are urged to create context-specific guideline values that are suited to the unique circumstances and conditions of a given area where pertinent exposure data are available.
2. *Periodic review and revision of standards:* As our understanding increases, certain guideline values might be adjusted, or new risks to the safety of drinking water might be taken into account. Additionally, advances in drinking water treatment technologies and pollutant analysis techniques are anticipated. Therefore, national or subnational standards need to be reviewed on a regular basis and should be designed to allow easy modifications. Standards may need to be changed, parameters may need to be added or removed, or both, but no change should be made without a good reason on the basis of risk assessment and resource priority for public health protection. Notifying all parties involved in changes is crucial when they are warranted.

3. *Drinking water regulations and supporting policies and programs:* Regulations supporting the prioritization of drinking water quality parameters to be tested, rather than mandating the testing of every parameter in these guidelines, will be strengthened by the incorporation of preventive risk management and prioritization approaches. Additionally, appropriate sanitation measures should be implemented at the community and household levels, encouraging action to prevent or mitigate contamination at the source. Finally, the identification of drinking water supplies that pose the greatest risks to public health will be ensured, allowing for the appropriate allocation of resources.
4. *Regulations:*

The alignment of national drinking water quality regulations with the principles outlined in these guidelines will ensure the following:

- There is a direct connection between drinking water quality regulations and the protection of public health.
- Regulations are designed to ensure safe drinking water from the source to the consumer, utilizing multiple protective barriers.
- Regulations are grounded in proven best practices that have consistently been shown to be effective.
- A range of tools, including education and training programs, incentives for good practices, and penalties for noncompliance, are in place to ensure adherence to the regulations.
- Regulations are tailored to be practical and appropriate for national, subnational, and local contexts, with specific approaches for certain situations, such as small community water supplies.
- Stakeholder roles and responsibilities are clearly defined, including how they should collaborate.
- Clear communication of "what, when, and how" information is ensured among stakeholders—including consumers—with specific actions outlined for both normal operations and in response to incidents or emergencies.
- Regulations are adaptable to changes in context, understanding, and technological advancements, and they are periodically reviewed and updated.
- Regulations are supported by relevant policies and programs.

4.2 Monitoring and Enhancement

Water quality monitoring is a fundamental tool in the management of safe drinking water supplies. Water quality monitoring is defined by the International Organization for Standardization (ISO) as "the programmed process of sampling, measurement and subsequent recording or signaling, or both, of various water characteristics, often with the aim of accessing conformity to specified objectives." Water quality monitoring is one of the first steps required in the rational development and management of water resources. In the field of water quality management, there has been a steady evolution in procedures for designing systems to obtain information on the changes in water quality. "Monitoring" comprises all activities to obtain "information" with respect to the water system.

4.3 Monitoring Strategy

Owing to economic and practical considerations, monitoring network design, sampling frequencies, choice of variables, and frequency of laboratory analysis should be determined on the basis of the information requirements, hydraulic and hydrologic constraints, variability in water body characteristics, the end use of water that drains to and from the water body, the overall objectives of the monitoring program, and, finally, the costs involved and budgets allocated to the program. It is also important to optimize the number of efforts required and the information generated and its importance in fulfilling the set objectives.

The scoping and designing steps constitute the foundation of the entire water quality monitoring program. The main objective of the design should be to minimize the cost of monitoring without sacrificing the desired information to the level of precision. Scoping and designing water quality monitoring programs are based on a clear scientific understanding of the following:

- Issues.
- Relevant background information.
- Monitoring objectives.
- Desired outcomes.
- Appropriate methods.
- The dynamics and characteristics of water systems.

4.4 Guidelines Enforcement

Enforcement standards are not meant to shut down ineffective systems but rather to guarantee that consumers have access to a consistent supply of clean, drinkable water [38]. Nonetheless, the most commonly stated justification for regular water supply monitoring is to identify noncompliance with norms and laws, which can then be used to compel improvement. Thus, in rural areas, it is only worth considering planning the development of surveillance programs if there is political will to implement water management policies. Several water supply rules are used to evaluate many aspects of water quality, compare the results to a benchmark, and take appropriate action when shortcomings are found. However, standards and regulations have lost confidence in many nations because "reality" is far from the ideal conditions provided by many water supply standards and legislation. Therefore, it is appropriate to plan for and pursue clear enforcement implementation strategies to maximize the benefits of their utilization.

5 Addressing Challenges and Gaps

Detecting ECs is a difficult task owing to their complexity and diversity. A lower concentration in the environment is the major obstacle in identifying emerging contaminants. These lower concentrations are generally measured in parts per billion (ppb) or in part per trillion (ppt). Therefore, it becomes critical to use highly sensitive analytical techniques such as UPLC–MS and GC–MS to determine the concentrations of these pollutants. However, the sensitivity of these techniques must be combined with effective extraction methods to detect and quantify even tiny amounts of ECs precisely [1]. The complicated properties of pollutants in water make the detection of ECs more difficult. Water samples often contain different kinds of pollutants that react with each other, which interferes with the analysis. This complex nature hinders the process of accurate identification of pollutants. This makes it more difficult to analyze the impact of such ECs. Therefore, sophisticated analysis is required to monitor these complex combinations and provide accurate data regarding the concentration of specific ECs [40].

The variation in the concentration of EC with time and space makes their identification more complex. The concentration of such pollutants varies significantly with time, complicating the assessment of the pollutants and their impact globally. The comprehensive sampling procedures that identify these differences are critical for determining the concentration of ECs in the environment. The lack of standardized procedures and reference materials makes analysis and inconsistency in the data impossible and hinders the formulation of appropriate regulations. Therefore, developing standardized and extensive reference material procedures is essential for enhancing the accuracy of EC detection and monitoring [8].

The persistent emergence of new pollutants presents a continual challenge for detection techniques. As new ECs are found in various circumstances, traditional detection techniques are inadequate because of the undefined protocols for novel pollutants. Consequently, there is a lag between the identification and formulation of procedural guidelines. Regulators face significant challenges in formulating and providing guidelines for novel pollutants. Such concerns arise from the need for extensive research to identify the risks related to these drugs and to implement safety protocols. Moreover, increasing public involvement and awareness is significant in procuring support for strong regulatory actions [3, 4]. By combining people's views and scientific findings, policies can be implemented to mitigate the risks associated with ECs successfully. The effective mitigation of these toxins requires the cooperation of government policy, public engagement, and active participation from industries.

6 Conclusions

Owing to their complex properties and risk to humans, emerging contaminants present significant challenges for both environmental research and the public. Because of the inadequate regulation of these pollutants, including microplastics, pharmaceuticals, and nanomaterials, enhancing detection measures has become important. Addressing ECs necessitates a complete approach that integrates scientific findings with regulatory and administrative reforms. The constantly evolving variety of ECs makes identification, monitoring, and mitigation more complex. Moreover, the introduction of new contaminants at faster rates sometimes exceeds the ability of existing techniques to control them, increasing the associated risks.

To resolve these challenges, many vital remedies are necessary. To address these difficulties, many critical solutions are essential. The development of a sophisticated detection system is important for ensuring reliable monitoring and mitigation of pollutants. Moreover, establishing preventive strategies that may effectively manage ECs, even in the absence of scientific evidence, is essential. This limitation would prioritize the reduction of toxins in the environment. Public awareness is also necessary because, with an uninformed public, the regulation and mitigation of toxins from the environment may stall. Furthermore, for an extensive and successful EC management plan, stakeholders' collaboration with different sectors is also important. Integrating research with policymaking ensures that mitigation plans are in accordance with the latest scientific research, increasing their efficacy in addressing future risks.

In summary, the management of ECs is a complex yet essential task that requires combined effort across all domains. We can significantly enhance our capacity to protect the environment and public health by prioritizing advanced detection systems and promoting collaborative governance.

Competing Interests The authors have no conflicts of interest to declare that are relevant to the content of this chapter.

Author Contributions All the authors contributed to the chapter's conception and design. Material preparation and data collection were performed by Dr. Jaspreet Kaur and Dr. Ina Thakur. The first draft of the chapter was written by Dr. Jaspreet and Dr. Ina Thakur, and all the authors commented on previous versions of the manuscript. The conceptualization and final proof reading were performed by Dr. Surinder Kumar Mehta. All the authors read and approved the final manuscript.

Ethical Approval This study does not contain any studies with human or animal subjects performed by any of the authors.

References

1. Geissen, V., Mol, H., Klumpp, E., Umlauf, G., Nadal, M., van der Ploeg, M., van de Zee, S. E. A. T. M., & Ritsema, C. J. (2015). Emerging pollutants in the environment: A challenge for water resource management. *International Soil and Water Conservation Research, 3*, 57–65. https://doi.org/10.1016/j.iswcr.2015.03.002
2. Stapleton, M. J., Ansari, A. J., Ahmed, A., & Hai, F. I. (2023). Evaluating the generation of microplastics from an unlikely source: The unintentional consequence of the current plastic recycling process. *Science of the Total Environment, 902*, 166090. https://doi.org/10.1016/j.scitotenv.2023.166090
3. Li, X., Shen, X., Jiang, W., Xi, Y., & Li, S. (2024). Comprehensive review of emerging contaminants: Detection technologies, environmental impact, and management strategies. *Ecotoxicology and Environmental Safety, 278*, 116420. https://doi.org/10.1016/j.ecoenv.2024.116420
4. Kumar, R., Qureshi, M., Vishwakarma, D. K., Al-Ansari, N., Kuriqi, A., Elbeltagi, A., & Saraswat, A. (2022). A review on emerging water contaminants and the application of sustainable removal technologies. *Case Studies in Chemical and Environmental Engineering, 6*, 100219. https://doi.org/10.1016/j.cscee.2022.100219
5. Mancinelli, M., Stevanin, C., Ardit, M., Chenet, T., Pasti, L., & Martucci, A. (2022). PFAS as emerging pollutants in the environment: A challenge with FAU type and silver-FAU exchanged zeolites for their removal from water. *Journal of Environmental Chemical Engineering, 10*, 108026. https://doi.org/10.1016/j.jece.2022.108026
6. Osuoha, J. O., Anyanwu, B. O., & Ejileugha, C. (2023). Pharmaceuticals and personal care products as emerging contaminants: Need for combined treatment strategy. *Journal of Hazardous Materials Advances, 9*, 100206. https://doi.org/10.1016/j.hazadv.2022.100206
7. Naidu, R., Jit, J., Kennedy, B., & Arias, V. (2016). Emerging contaminant uncertainties and policy: The chicken or the egg conundrum. *Chemosphere, 154*, 385–390. https://doi.org/10.1016/j.chemosphere.2016.03.110
8. Wang, F., Xiang, L., Sze-Yin Leung, K., Elsner, M., Zhang, Y., Guo, Y., Pan, B., Sun, H., An, T., Ying, G., Brooks, B. W., Hou, D., Helbling, D. E., Sun, J., Qiu, H., Vogel, T. M., Zhang, W., Gao, Y., Simpson, M. J., Luo, Y., Chang, S. X., Su, G., Wong, B. M., Fu, T. M., Zhu, D., Jobst, K. J., Ge, C., Coulon, F., Harindintwali, J. D., Zeng, X., Wang, H., Fu, Y., Wei, Z., Lohmann, R., Chen, C., Song, Y., Sanchez-Cid, C., Wang, Y., El-Naggar, A., Yao, Y., Huang, Y., Cheuk-Fung Law, J., Gu, C., Shen, H., Gao, Y., Qin, C., Li, H., Zhang, T., Corcoll, N., Liu, M., Alessi, D. S., Li, H., Brandt, K. K., Pico, Y., Gu, C., Guo, J., Su, J., Corvini, P., Ye, M., Rocha-Santos, T., He, H., Yang, Y., Tong, M., Zhang, W., Suanon, F., Brahushi, F., Wang, Z., Hashsham, S. A., Virta, M., Yuan, Q., Jiang, G., Tremblay, L. A., Bu, Q., Wu, J., Peijnenburg, W., Topp, E., Cao, X., Jiang, X., Zheng, M., Zhang, T., Luo, Y., Zhu, L., Li, X., Barceló, D., Chen, J., Xing, B., Amelung, W., Cai, Z., Naidu, R., Shen, Q., Pawliszyn, J., Zhu, Y. G., Schaeffer, A., Rillig, M. C., Wu, F., Yu, G., & Tiedje, J. M. (2024). Emerging contaminants: A One Health perspective*, Innovation, 5*. https://doi.org/10.1016/j.xinn.2024.100612
9. Hartmann, J., van der Aa, M., Wuijts, S., de Roda Husman, A. M., & van der Hoek, J. P. (2018). Risk governance of potential emerging risks to drinking water quality: Analysing current practices. *Environmental Science & Policy, 84*, 97–104. https://doi.org/10.1016/j.envsci.2018.02.015
10. Dieter, H. H. (2014). Health related guide values for drinking-water since 1993 as guidance to assess presence of new analytes in drinking-water. *International Journal of Hygiene and Environmental Health, 217*, 117–132. https://doi.org/10.1016/j.ijheh.2013.05.001
11. Schriks, M., Heringa, M. B., van der Kooi, M. M. E., de Voogt, P., & van Wezel, A. P. (2010). Toxicological relevance of emerging contaminants for drinking water quality. *Water Research, 44*, 461–476. https://doi.org/10.1016/j.watres.2009.08.023
12. Levin, R., Villanueva, C. M., Beene, D., Cradock, A. L., Donat-Vargas, C., Lewis, J., Martinez-Morata, I., Minovi, D., Nigra, A. E., Olson, E. D., Schaider, L. A., Ward, M. H., & Deziel,

N. C. (2024). US drinking water quality: Exposure risk profiles for seven legacy and emerging contaminants. *Journal of Exposure Science & Environmental Epidemiology, 34*, 3–22. https://doi.org/10.1038/s41370-023-00597-z
13. Voulvoulis, N., Arpon, K. D., & Giakoumis, T. (2017). The EU water framework directive: From great expectations to problems with implementation. *Science of the Total Environment, 575*, 358–366. https://doi.org/10.1016/j.scitotenv.2016.09.228
14. T. Moss, G. Bouleau, J. Albiac, L. Slavíkova, The EU water framework directive twenty years on: Introducing the special issue, n.d. www.water-alternatives.org.
15. A global overview of national regulations and standards for drinking-water quality ii A global overview of national regulations and standards for drinking-water quality, 2018. http://apps.who.int/bookorders.
16. Domingo, J. L., & Nadal, M. (2019). Human exposure to per- and polyfluoroalkyl substances (PFAS) through drinking water: A review of the recent scientific literature. *Environmental Research, 177*, 108648. https://doi.org/10.1016/j.envres.2019.108648
17. Liu, B., Zhang, H., Li, J., Dong, W., & Xie, L. (2017). Perfluoroalkyl acids (PFAAs) in sediments from rivers of the Pearl River Delta, southern China. *Environmental Monitoring and Assessment, 189*, 213. https://doi.org/10.1007/s10661-017-5921-7
18. Eriksson, U., Kärrman, A., Rotander, A., Mikkelsen, B., & Dam, M. (2013). Perfluoroalkyl substances (PFASs) in food and water from Faroe Islands. *Environmental Science and Pollution Research, 20*, 7940–7948. https://doi.org/10.1007/s11356-013-1700-3
19. Post, G. B. (2021). Recent US State and Federal Drinking Water Guidelines for Per- and Polyfluoroalkyl substances. *Environmental Toxicology and Chemistry, 40*, 550–563. https://doi.org/10.1002/etc.4863
20. M.H. Snow, Too little, too late: Congress's attempt to regulate forever chemicals through military appropriations, 2021. https://perma.cc/N6WM-BV39.
21. Cordner, A., De La Rosa, V. Y., Schaider, L. A., Rudel, R. A., Richter, L., & Brown, P. (2019). Guideline levels for PFOA and PFOS in drinking water: The role of scientific uncertainty, risk assessment decisions, and social factors. *Journal of Exposure Science & Environmental Epidemiology, 29*, 157–171. https://doi.org/10.1038/s41370-018-0099-9
22. Carlson, L. M., Angrish, M., Shirke, A. V., Radke, E. G., Schulz, B., Kraft, A., Judson, R., Patlewicz, G., Blain, R., Lin, C., Vetter, N., Lemeris, C., Hartman, P., Hubbard, H., Arzuaga, X., Davis, A., Dishaw, L. V., Druwe, I. L., Hollinger, H., Jones, R., Kaiser, J. P., Lizarraga, L., Noyes, P. D., Taylor, M., Shapiro, A. J., Williams, A. J., & Thayer, K. A. (2022). Systematic evidence map for over one hundred and Fifty Per-and Polyfluoroalkyl Substances (PFAS). *Environmental Health Perspectives, 130*. https://doi.org/10.1289/EHP10343
23. Panikkar, B., Lemmond, B., Allen, L., Dipirro, C., & Kasper, S. (2019). Making the invisible visible: Results of a community-led health survey following PFAS contamination of drinking water in Merrimack, New Hampshire. *Environmental Health, 18*, 79. https://doi.org/10.1186/s12940-019-0513-3
24. Tirkey, A., & Upadhyay, L. S. B. (2021). Microplastics: An overview on separation, identification and characterization of microplastics. *Marine Pollution Bulletin, 170*, 112604. https://doi.org/10.1016/j.marpolbul.2021.112604
25. Zhang, Y., Kang, S., Allen, S., Allen, D., Gao, T., & Sillanpää, M. (2020). Atmospheric microplastics: A review on current status and perspectives. *Earth-Science Reviews, 203*, 103118. https://doi.org/10.1016/j.earscirev.2020.103118
26. Peng, G., Xu, P., Zhu, B., Bai, M., & Li, D. (2018). Microplastics in freshwater river sediments in Shanghai, China: A case study of risk assessment in mega-cities. *Environmental Pollution, 234*, 448–456. https://doi.org/10.1016/j.envpol.2017.11.034
27. Ding, L., Mao, R. F., Guo, X., Yang, X., Zhang, Q., & Yang, C. (2019). Microplastics in surface waters and sediments of the Wei River, in the northwest of China. *Science of the Total Environment, 667*, 427–434. https://doi.org/10.1016/j.scitotenv.2019.02.332
28. Xiong, X., Wu, C., Elser, J. J., Mei, Z., & Hao, Y. (2019). Occurrence and fate of microplastic debris in middle and lower reaches of the Yangtze River—From inland to the sea. *Science of the Total Environment, 659*, 66–73. https://doi.org/10.1016/j.scitotenv.2018.12.313

29. Rytelewska, S., & Dąbrowska, A. (2022). The Raman spectroscopy approach to different freshwater microplastics and quantitative characterization of polyethylene aged in the environment. *Microplastics, 1*, 263–282. https://doi.org/10.3390/microplastics1020019
30. Dris, R., Gasperi, J., Rocher, V., & Tassin, B. (2018). Synthetic and non-synthetic anthropogenic fibers in a river under the impact of Paris megacity: Sampling methodological aspects and flux estimations. *Science of the Total Environment, 618*, 157–164. https://doi.org/10.1016/j.scitotenv.2017.11.009
31. A.D. Gray, H. Wertz, R.R. Leads, J.E. Weinstein, Microplastic in two South Carolina estuaries: Occurrence, distribution, and composition, 2018.
32. Zhang, K., Xiong, X., Hu, H., Wu, C., Bi, Y., Wu, Y., Zhou, B., Lam, P. K. S., & Liu, J. (2017). Occurrence and characteristics of microplastic pollution in Xiangxi Bay of Three Gorges Reservoir, China. *Environmental Science & Technology, 51*, 3794–3801. https://doi.org/10.1021/acs.est.7b00369
33. Hendrickson, E., Minor, E. C., & Schreiner, K. (2018). Microplastic abundance and composition in Western Lake superior as determined via microscopy, Pyr-GC/MS, and FTIR. *Environmental Science & Technology, 52*, 1787–1796. https://doi.org/10.1021/acs.est.7b05829
34. Boxall, A. B. A., Rudd, M. A., Brooks, B. W., Caldwell, D. J., Choi, K., Hickmann, S., Innes, E., Ostapyk, K., Staveley, J. P., Verslycke, T., Ankley, G. T., Beazley, K. F., Belanger, S. E., Berninger, J. P., Carriquiriborde, P., Coors, A., DeLeo, P. C., Dyer, S. D., Ericson, J. F., Gagné, F., Giesy, J. P., Gouin, T., Hallstrom, L., Karlsson, M. V., Joakim Larsson, D. G., Lazorchak, J. M., Mastrocco, F., McLaughlin, A., McMaster, M. E., Meyerhoff, R. D., Moore, R., Parrott, J. L., Snape, J. R., Murray-Smith, R., Servos, M. R., Sibley, P. K., Straub, J. O., Szabo, N. D., Topp, E., Tetreault, G. R., Trudeau, V. L., & Van Der Kraak, G. (2012). Pharmaceuticals and personal care products in the environment: What are the big questions? *Environmental Health Perspectives, 120*, 1221–1229. https://doi.org/10.1289/ehp.1104477
35. Mathew, R. A., & Kanmani, S. (2020). A review on emerging contaminants in Indian waters and their treatment technologies. *Nature Environment and Pollution Technology, 19*, 549–562. https://doi.org/10.46488/NEPT.2020.V19I02.010
36. Syafrudin, M., Kristanti, R. A., Yuniarto, A., Hadibarata, T., Rhee, J., Al-Onazi, W. A., Algarni, T. S., Almarri, A. H., & Al-Mohaimeed, A. M. (2021). Pesticides in drinking water-a review. *International Journal of Environmental Research and Public Health, 18*, 1–15. https://doi.org/10.3390/ijerph18020468
37. R. Pandey, S.K. Mishra, M.K. Mishra, Agrochemicals and their impact on environment, n.d. https://www.researchgate.net/publication/375485756.
38. Guidelines for drinking-water quality. First addendum to third edition, volume 1, Recommendations, World Health Organization, 2006.
39. John, S. O. O., Olukotun, S. F., Kupi, T. G., & Mathuthu, M. (2024). Health risk assessment of heavy metals and physicochemical parameters in natural mineral bottled drinking water using ICP-MS in South Africa. *Applied Water Science, 14*. https://doi.org/10.1007/s13201-024-02267-3
40. Lin, X., Fu, S., Zhang, B., Li, S., Fang, H., Zhou, W., Ying, F., & Xu, J. (2023). Research progress on risk assessment of emerging contaminants in aquatic environment. In *E3S web of conferences*. EDP Sciences. https://doi.org/10.1051/e3sconf/202336901003

Chapter 18
A Toxic Beauty: The Environmental Impact of Cosmetic Contaminants in Water and Sustainable Solutions

Hansa Muvel, Manoj Kumar Jindal, Pradip Kumar Tewari, and Vikky Anand

1 Introduction

Cosmetic products are widely used across the globe, with an increasing variety of chemical compounds incorporated into their formulations to enhance qualities such as fragrance, stability, shelf life, and appearance. These chemicals include additives, preservatives, surfactants, dyes, and agents that improve shine. Owing to their widespread usage, residues from cosmetic products continuously enter the environment in various forms. Many of these chemicals are bioactive and persistent and have the potential to bioaccumulate, posing significant risks to both human health and ecosystems. As such, the excessive use of cosmetics could have serious consequences for public health. The toxic ingredients commonly found in cosmetic products include parabens, triclosan, benzalkonium chloride, plastic microbeads, formaldehyde, diazolidinyl urea, imidazolidinyl urea, sunscreen agents (organic and

H. Muvel · M. K. Jindal (✉)
Center for Sustainable Drinking Water, Indian Institute of Technology Jodhpur, Jodhpur, Rajasthan, India
e-mail: mjindal@iitj.ac.in

P. K. Tewari
Center for Sustainable Drinking Water, Indian Institute of Technology Jodhpur, Jodhpur, Rajasthan, India

Department of Chemical Engineering, Indian Institute of Technology Jodhpur, Jodhpur, Rajasthan, India
e-mail: pradiptewari@iitj.ac.in

V. Anand
Department of Chemical Engineering, Indian Institute of Technology Jodhpur, Jodhpur, Rajasthan, India
e-mail: vikky@iitj.ac.in

M. K. Jindal et al. (eds.), *Emerging Contaminants in Water*, Emerging Contaminants and Associated Treatment Technologies,
https://doi.org/10.1007/978-3-031-89591-3_18

inorganic UV filters), and trace metals. The potential biological risks of these substances include genotoxicity, cytotoxicity, neurotoxicity, mutagenicity, and estrogenicity. The global cosmetics market is expected to expand from USD 393.75 billion in 2024 to USD 758.05 billion by 2032, reflecting a compound annual growth rate (CAGR) of 9.8% throughout the forecast period. In the United States, the cosmetics market is anticipated to experience substantial growth, reaching an estimated value of USD 417.24 billion by 2030 [1]. From March 2023 to February 2024, the global import of raw materials for cosmetics increased 13%, with 36,831 shipments sourced from 4809 exporters and supplied to 1999 buyers. However, February 2024 experienced a 31% year-over-year decline in shipments, primarily from Germany, China, and France. The top three importers were Russia, India, and Vietnam, with Russia leading at 44,716 shipments [2]. There has been growing concern over the increasing use of pharmaceuticals, cosmetics, and personal care products, including insect repellents, disinfectants, and dietary supplements. These products are continuously released into aquatic environments, where their ecological impact is largely linked to their bioactivity, toxicity, and bioaccumulation potential [3]. Sewage treatment plants often fail to completely remove cosmetic ingredients, such as microplastics [4], UV filters, perfluoroalkyl compounds, and synthetic musks [5, 6]. Additionally, these chemicals accumulate in sewage sludge, which is sometimes used as fertilizer [7], leading to further environmental contamination. Unlike pharmaceuticals, which undergo metabolic transformations, cosmetic products are designed for external use and often remain unchanged as they enter the environment, especially during washing and showering [8]. This continuous release presents a persistent ecological risk throughout an individual's life. Toxic chemicals are commonly found in cosmetics and they pose biological risks to human health and aquatic systems. These findings emphasize the importance of understanding the potential dangers of these substances and suggest future research directions to mitigate these risks.

2 Global Import Trends of Raw Materials for Cosmetics: Import and Export Trends

Raw materials for cosmetics are imported by over 111 countries worldwide, with prominent global buyers such as Unique South Africa, Oriflame India Pvt. Ltd., and Asia Industrial Chemical Joint Stock Company featured in the Volza database, which tracks over 5190 active buyers. The top importing countries include Russia, with 44,716 shipments (30% market share), India, with 38,325 shipments (26%), and Vietnam, with 29,335 shipments (20%). Other key importers are Peru, the United States, Ukraine, Türkiye, Indonesia, Colombia, and Chile. In terms of global suppliers, Germany, China, and France dominate, with Germany exporting 16,872 units, China exporting 16,356 units, and France exporting 16,202 units, accounting for a significant share of the market. Other key exporters include the United States, Singapore, Japan, India, Spain, South Korea, and Poland. The Volza database also tracks over 12,696 suppliers, helping businesses find ideal partners by analyzing billions of shipping invoices across 20+ parameters [2]. These countries play crucial roles in the global supply chain for raw cosmetic materials, facilitating the development of beauty and skin care products worldwide.

3 Cosmetic Ingredients as New Environmental Contaminants

The beauty and skin care industries are major contributors to global carbon emissions, and consumers are increasingly scrutinizing the environmental impact of cosmetics. Cosmetics, which encompass any products used to cleanse or enhance the appearance of the body, include a wide range of commercial items. Makeup, in particular, serves as a popular form of self-expression, allowing individuals to highlight their unique style. However, the environmental impact of the beauty industry cannot be overlooked. The cosmetics sector faces significant sustainability challenges, including packaging waste, the depletion of natural resources, and the carbon emissions associated with individual products. Table 18.1 displays the environmental impact of common beauty products and their marketing in terms of their contribution to the carbon footprint, their impact on the environment, and their sustainable alternative solutions.

Table 18.1 Environmental impact of common beauty products and sustainable alternatives

Category	Carbon footprint	Environmental issues	Sustainable solutions	References
Packaging waste	Major contributor to environmental impact; 120 billion pieces produced annually, with 95% ending up in landfills	Substantial amounts of plastic, glass, and metal packaging waste Microbeads in cosmetics contribute to marine pollution, plastics (single use), excessive use of nonrecyclable materials, energy in production and transport	Eco-friendly packaging, Compostable packaging, biobased products	[10, 11]
Fragrances	One bottle may account for 16 kg of CO_2. Some industry leaders are now using recaptured CO_2 in perfume production	Use of animal-derived ingredients like ambergris, musk, and civet. Nonrecyclable glass bottles and volatile organic compounds (VOCs) from disposal	Vegan and cruelty-free-certified fragrances Improved packaging recycling and proper disposal practices Opt for refillable roll-on or rub-on perfumes, preferably made from recaptured CO_2	[9, 12, 13]

(continued)

Table 18.1 (continued)

Category	Carbon footprint	Environmental issues	Sustainable solutions	References
Makeup	Lipstick tubes are typically too small to be recycled, leading to landfill waste. A single tube of lipstick may produce over 2 kg of CO_2	Harmful chemicals (parabens, sulfates, silicones, heavy metals such as lead, cadmium, nickel, chromium) in products linked to health issues Resource-intensive production	Shift to natural, nontoxic ingredients Eco-friendly, recyclable packaging options Choose refillable lipsticks	[14–16]
Nail products	Manufacturing involves harmful chemicals like formaldehyde and toluene, contributing to approximately 30% of the product's environmental impact. Packaging, especially plastic, accounts for 35% of the carbon emissions in the beauty industry, with 120 billion packaging pieces produced annually	Harmful chemicals like toluene, formaldehyde, dibutyl phthalate Glitter is made from microplastics Nonrecyclable glass bottles High carbon emissions from raw material sourcing and manufacturing	Eco-friendly, water-based nail polish options Reduced use of microplastics in products	[9, 17, 18]
Sunscreen	Often packaged in small, hard-to-recycle plastics. Products with added sunscreen or UV protection have an even higher impact. Ingredients like palm oil further harm the environment. The carbon footprint varies based on the product's composition	Chemicals like oxybenzone and octinoxate damage coral reefs and marine ecosystems	"Reef-safe" sunscreen labels Choose refillable, organic moisturizers	[14, 19, 20]
Tattoos	The production and transportation of tattoo inks, which often contain synthetic chemicals and pigments	Traditional tattoo ink contains toxic heavy metals like lead and mercury Lack of eco-friendly practices in some studios Ink production and disposal contribute to carbon emissions; environmental impact of tattoo removal processes is also notable	Use of plant-based, biodegradable inks Adoption of eco-friendly tattooing practices and materials	[21–23]

(continued)

Table 18.1 (continued)

Category	Carbon footprint	Environmental issues	Sustainable solutions	References
Makeup removal wipes	Single-use makeup wipes contribute significantly to plastic pollution, with 3.4 billion kilograms disposed of each year. They are made from nonbiodegradable materials like polyester, taking hundreds of years to decompose. Their production also involves greenhouse gas emissions from manufacturing and transportation	Made of polypropylene; these wipes are nonrecyclable, can clog drains, and pose a threat to the environment and waterways due to single-use disposal	Use reusable makeup remover cloths or biodegradable wipes made from natural fibers	[24–26]
Exfoliating scrubs with microbeads	These plastic particles are washed down drains and enter waterways. The production of these scrubs also involves energy-intensive processes, contributing to their overall carbon footprint	Contain plastic microbeads that do not dissolve and end up in aquatic and marine environments. These microbeads are harmful as they are ingested by marine life, leading to environmental pollution	Choose natural exfoliants like sugar, salt, or oatmeal that are biodegradable and safe	[27–29]
Face sheet masks	The carbon footprint of face sheet masks arises from the materials used (often nonbiodegradable), the production processes, and packaging waste. Many masks are made from synthetic fibers that take a long time to break down in landfills	Produce significant waste (plastic pouches, masks, and wrap) that are not recyclable or compostable, contributing to more landfill waste	Use reusable cotton or bamboo fabric masks; look for compostable packaging options	[26, 30, 31]
Skin-whitening creams	The carbon footprint of skin-whitening creams includes emissions from the manufacturing processes and packaging. Many creams contain harmful chemicals that can leach into the environment, affecting ecosystems	Contain harmful silicone, which, like microbeads, washes into marine environments and harms marine life. Additionally, these products reflect an unhealthy societal obsession with fair skin	Opt for products with natural ingredients and eco-friendly packaging. Avoid skin-whitening claims	[32, 33]

(continued)

Table 18.1 (continued)

Category	Carbon footprint	Environmental issues	Sustainable solutions	References
Sprays and mists	Aerosol sprays emit higher emissions than roll-ons	Release harmful aerosols and hydrocarbons, which contribute to atmospheric pollution and ozone layer depletion. They also add to landfill waste	Use pump sprays or refillable mist bottles; choose products with eco-friendly aerosols Opt for deodorant creams with organic ingredients in compostable packaging	[34–36]
Shampoo	Major carbon emissions arise from manufacturing, plastic packaging, and the heated water used in rinsing. Manufacturing a bottle generates approximately 0.07 kg of CO_2, equivalent to charging eight smartphones. Consumer use and disposal further add to emissions	Contains chemicals that can harm aquatic life (e.g., parabens, sulfates). Plastic packaging contributes to plastic waste	Choose shampoo bars in recyclable or compostable packaging	[23, 37]
Hair care (dyes/spray)	Hair dyes are harmful due to ingredients, manufacturing processes, and the vapors produced during use. They also contribute to significant amounts of the carbon footprint which vary depending on the composition and application	Contains harmful chemicals (e.g., ammonia) that can pollute water and soil synthetic dyes are nonbiodegradable	Opt for plant-based dyes or colors	[38, 39]
	Hair spray releases toxic VOCs directly into the air, contributing to ozone depletion. In 2023, a firm has reported that the average cradle-to-shelf emissions for each product sold amounted to 434 g of CO_2 equivalent	Aerosol sprays release volatile organic compounds (VOCs), contributing to air pollution, and often contain plastic microbeads harmful to marine life when washed off	Choose natural pump hair sprays	

3.1 Harmful Chemicals in Cosmetic Ingredients

The cosmetics industry uses a variety of chemicals, some of which have been linked to health and environmental concerns. Although not every cosmetic chemical is harmful, it is important to be aware of product ingredients for both personal well-being and ecological sustainability [40]. Choosing products made with natural or organic ingredients and supporting cruelty-free brands can help reduce the negative impacts of the beauty industry. The beauty sector faces many challenges related to sustainability, including the use of toxic ingredients and excessive packaging waste. However, by opting for sustainable products and supporting companies that prioritize environmental responsibility, consumers can significantly contribute to reducing the industry's ecological footprint [41]. A few harmful chemicals and their uses, impacts, risks, and alternatives are listed in Table 18.2 and shown in Fig. 18.1.

Table 18.2 Overview of various common cosmetic and personal care product chemicals, their structure, applications, environmental impact, health and ecological risks, and possible alternatives

Chemical	Structure	Used in	Environmental impact	Health and ecological risks	Alternatives	References
Butylated hydroxytoluene (BHA) and Butylated hydroxyanisole (BHT)	$C_{17}H_{23}O$ and $C_{17}H_{22}O_2$	Moisturizers, cosmetics	Linked to potential environmental damage, bioaccumulation in marine species, harmful to amphibians	Suspected hormone disruptors, genetic mutations in amphibians	Organic, herbal, eco-friendly products	[41–45]
Triclosan	$C_{12}H_7C_{13}O_2$	Antibacterial items (hand sanitizers, detergents)	Accumulates in soil, toxic to aquatic life, forms dioxins when combined with other chemicals in rivers	Alters biochemistry of amphibians, fish, and aquatic plants, slow to degrade	Natural, biodegradable antibacterials	[46]
Oxybenzone	$C_{14}H_{12}O_3$	Sunscreens	Harmful to coral reefs, contributes to coral decline, alters coral DNA	Fatal to baby coral, environmental damage to marine life	Reef-safe, chemical-free sunscreens	[47–49]
Lead	Pb	Toothpaste, lipsticks	Does not degrade easily, contributes to environmental contamination	Heavy metal toxicity, harmful to human health	Lead-free, nontoxic cosmetics	[50–52]
Parabens	$C_8H_8O_3$, $C_9H_{10}O_3$, $C_{10}H_{12}O_3$, $C_{11}H_{14}O_3$	Cosmetics, skin care products	Potential contamination of water systems, disrupt ecological balance	Linked to hormonal disruption, may affect wildlife and human health	Paraben-free, natural preservatives	[53]
Phthalates	$C_{12}H_{14}O_4$, $C_{16}H_{22}O_4$, $C_{24}H_{38}O_4$, $C_9H_{10}O_4$, $C_{26}H_{42}O_4$	Fragrances, plastics	Bioaccumulation in the environment, harmful to wildlife	Associated with reproductive and developmental issues, hormone disruption	Phthalate-free, natural fragrances	[54, 55]

Sodium lauryl sulfate (SLS)	$C_{12}H_{25}NaO_4S$	Soaps, shampoos, toothpastes	Pollutes water sources, can be toxic to aquatic life	Can irritate sensitive skin, eyes, and mucous membranes	SLS-free, natural foaming agents	[56, 57]
Formaldehyde	CH_2O	Preservative in some beauty products	Contributes to environmental pollution, toxic to aquatic organisms	Known carcinogen, skin irritation, respiratory issues	Formaldehyde-free, natural preservatives	[58, 59]
Glycerin	$C_3H_8O_3$	Moisture to the skin, cleansers, lip balms, and facial masks	Biodegradable, minimal environmental impact, can be water-soluble	Considered safe for humans, no significant ecological risks	Hyaluronic acid, aloe vera extract (plant-derived, environmentally friendly)	[60]
Sodium benzoate	$C_7H_5NaO_2$	Shampoos, conditioners, lotions, and deodorants	Biodegradable but can be toxic to aquatic life at high concentrations; may degrade into benzene	Safe at low concentrations but linked to skin irritation, allergic reactions, and potential carcinogenic risks (benzene)	Rosemary extract, grapefruit seed extract (natural)	[61]
Cyclopentasiloxane	Silicone-based, cyclic compound	Hair care products, skin care, deodorants	Nonbiodegradable, accumulates in water, can harm aquatic life	Skin and eye irritation, endocrine disruption	Natural oils	[62]
Phenoxyethanol	Ethanol derivative with phenolic group	Preservative in cosmetics, personal care products	Toxic to aquatic organisms, can persist in water	Skin irritation, neurotoxic effects at high concentrations	Ethylhexylglycerin, caprylyl glycol	[63]
Magnesium stearate	Magnesium salt of stearic acid	Cosmetics, tablets, powders	Low environmental impact, biodegradable	Mild skin irritation in some people, nontoxic	Rice starch, kaolin clay, candelilla wax, beeswax	[64]
Titanium dioxide	Metal oxide, TiO_2	Sunscreens, makeup, toothpaste, paints	Bioaccumulation in aquatic organisms, potential damage to aquatic ecosystems	Respiratory issues if inhaled, skin irritation in sensitive individuals	Natural mineral sunscreens	[65]

(continued)

Table 18.2 (continued)

Chemical	Structure	Used in	Environmental impact	Health and ecological risks	Alternatives	References
Iron oxides	Oxides of iron (Fe_2O_3, Fe_3O_4)	Cosmetics (blush, eye shadow), pigments	Low environmental impact, but may affect water quality if washed off in large quantities	Minimal health risks, nontoxic, but may cause irritation for sensitive individuals	Natural mineral pigments	[66]
Lanolin	Animal-derived wax (fatty alcohols)	Skin care, hair care, lip balms	Biodegradable, low environmental impact	Allergic reactions in sensitive individuals, potential contamination of water sources	Plant-based oils, synthetic lanolin	[67]
Polybutene	Synthetic polymer (hydrocarbon)	Lip gloss, mascara, skin care products	Nonbiodegradable, can persist in water	Skin irritation, clogging pores	Natural waxes (e.g., candelilla wax, carnauba wax), hydroxyethylcellulose	[68]
Butyl acetate	Ester of butanol and acetic acid	Nail polish, perfumes, paints	Volatile organic compound (VOC), contributes to air pollution	Respiratory irritation, dizziness, skin irritation	Natural essential oils	[69]
Benzalkonium chloride	Quaternary ammonium compound	Antiseptic, preservatives, disinfectants	Toxic to aquatic life, persistence in water	Skin irritation, respiratory issues, endocrine disruption	Grapefruit seed extract	[70]
Octinoxate	Ester of octyl alcohol and cinnamic acid	Sunscreens, makeup	Harmful to marine life, disrupts coral reefs	Endocrine disruption, skin irritation	Reef-safe sunscreens, raspberry seed oil, carrot seed	[71]
Oxybenzone	Benzophenone derivative	Sunscreens, skin care products	Pollutes oceans, harms coral reefs and marine organisms	Hormonal disruption, skin irritation, allergic reactions	Carrot seed oil, red raspberry seed oil	[71]

Zinc oxide	Inorganic compound (ZnO)	Sunscreens, cosmetics, diaper creams	Safe for the environment, biodegrades naturally	Generally safe but can cause irritation for sensitive skin	Natural mineral-based sunscreens	[72]
Glycolic acid	Alpha-hydroxy acid (AHA)	Exfoliants, peels, skin care	Low environmental impact, biodegradable	Skin irritation, sun sensitivity at high concentrations	Natural exfoliants (fruit enzymes (papaya or pineapple) and jojoba beads or bamboo powder)	[73]
Hyaluronic acid	Polysaccharide (sugar compound)	Skin care, antiaging products	Biodegradable, safe for the environment	Mild skin irritation, nontoxic	Plant-based alternatives (e.g., cassia seed)	[74]
Finasteride	Synthetic steroidal compound	Hair loss treatments	Minimal environmental impact, poorly absorbed in water	Hormonal side effects, birth defects if exposed during pregnancy	Saw palmetto, pumpkin seed oil	[75]
Resorcinol	Benzene derivative ($C_6H_4(OH)_2$)	Hair colorants, acne treatments	Toxic to aquatic organisms, persistent in water	Skin irritation, potential endocrine disruption, carcinogenic properties	Natural henna, plant-based colorants	[76]
PPD (paraphenylenediamine)	Organic compound ($C_6H_4(NH_2)_2$)	Hair dye, cosmetics	Toxic to aquatic life, persists in water	Allergic reactions, carcinogenic potential, skin irritation	Henna, indigo, plant-based dyes	[77]

(continued)

Table 18.2 (continued)

Chemical	Structure	Used in	Environmental impact	Health and ecological risks	Alternatives	References
Toluene	Aromatic hydrocarbon ($C_6H_5CH_3$)	Nail polish, adhesives, paints	VOC, air and water pollutant, toxic to aquatic life	Central nervous system toxicity, skin irritation, reproductive toxicity	Water-based nail polish, nontoxic adhesives	[77, 78]
Petrolatum	Mixture of hydrocarbons (alkanes, cycloalkanes, and aromatics)	Lip balms, ointments, moisturizers, and body creams	Nonbiodegradable, can persist in water and soil, contributing to long-term pollution	Potential to clog pores and cause skin irritation, may contribute to aquatic pollution when washed off	Plant-based oils, beeswax	[71, 79]

Fig. 18.1 Harmful beauty chemicals used by both women and men and their effects on health

4 Environmental Impact of Cosmetic Contamination: Pathways, Persistence, and Ecological Risks

After being introduced into the environment, cosmetic components can remain for an extended period because of their low volatility, high polarity, and water-attracting properties. These substances can travel over extensive distances, frequently through water systems, and might ultimately become part of the food chain. The degree of this distribution relies on the physicochemical attributes of the cosmetics and the traits of the surrounding environment [80]. After use, washed-off cosmetics typically enter household wastewater, which is then directed to sewage treatment plants (STPs) or wastewater treatment plants (WWTPs). These plants serve as a major source of contaminants from personal care products (PCPs) to the environment [81]. While wastewater treatment methods aim to remove pollutants, they often do not completely eliminate cosmetics and their components, resulting in residual contamination in the treated discharge [81–83]. Effluent from wastewater treatment plants is typically released into surface freshwater bodies; however, the removal of personal care products is often unreliable, resulting in traces of these substances remaining in the water [83, 84]. In some cases, treated wastewater is used for irrigation, and treated sewage sludge is applied as fertilizer in agriculture, further increasing the likelihood of cosmetic constituents entering the environment. This pollution can lead to the accumulation of cosmetic substances in plants, animals, and aquatic organisms, including fish and crustaceans. As a result, cosmetic components may enter the human food supply through the consumption of contaminated food. In addition to wastewater discharge, manufacturing effluents containing cosmetic ingredients are often sent to sewage treatment plants (STPs), which contribute to the overall environmental burden of personal care products (PCPs) [86]. While wastewater treatment plants (WWTPs) are designed to process sewage for recycling into drinking water or irrigation, they frequently fail to remove all traces of cosmetic substances. Moreover, sewage sludge, which may contain residual pollutants, is often used as fertilizer, potentially causing further environmental contamination. Eventually, these substances can filter into groundwater, potentially serving as a source of drinking water, which creates hazards for both human health and aquatic ecosystems. Although the levels of cosmetics found in the environment (ranging from ng/L to mg/L) may seem insignificant, their persistent presence poses a long-term ecological risk. The continuous release of cosmetic substances, many of which are poorly biodegradable, has raised significant concerns about their growing impact on aquatic ecosystems. These substances are considered "pseudopersistent" because they remain in the environment for extended periods, even at low concentrations [87]. In recent years, the widespread use of cosmetics has emerged as a significant environmental threat. However, monitoring of cosmetic ingredients in the environment is rarely included in environmental regulations worldwide, which makes it challenging to assess the full extent of their impact [88]. The biogeochemical cycles of food chains can be disrupted by the release of these substances, leading to potential perturbations in aquatic ecosystems [89]. Both the hydrophilic and lipophilic components of cosmetics can contribute to metabolic breakdown and rapid elimination in organisms, but some chemicals may accumulate in the food chain, causing bioaccumulation and potentially biomagnification [89, 90]. Studies have investigated the interactions of hydrophobic cosmetic chemicals with fish [91–93]. These substances may be concentrated in organisms at relatively high trophic levels [94]. Environmental conditions, such as changes in temperature or altered degradation processes, can increase the availability of cosmetic pollutants in the environment, further exacerbating this problem [95]. During wastewater treatment, PCPs may undergo microbial biodegradation, which can reduce their toxicity. However, this biodegradation is not always effective for the wide range of cosmetic ingredients found in wastewater, as biodegradation processes are often substrate-specific and may be inhibited by high concentrations of pollutants [96]. The presence of these substances can also harm the microorganisms involved in biodegradation, reducing their ability to break down pollutants. Environmental changes resulting from cosmetic pollution can sometimes be identified by observing shifts in the behavior or health of species that act as bioindicators [97]. However, assessing these changes is challenging and requires more comprehensive research. Therefore, a thorough understanding of the impacts and potential risks of cosmetic contamination is crucial, highlighting the need for further research and enhanced monitoring to address the environmental and health threats posed by cosmetics [98] (Fig. 18.2).

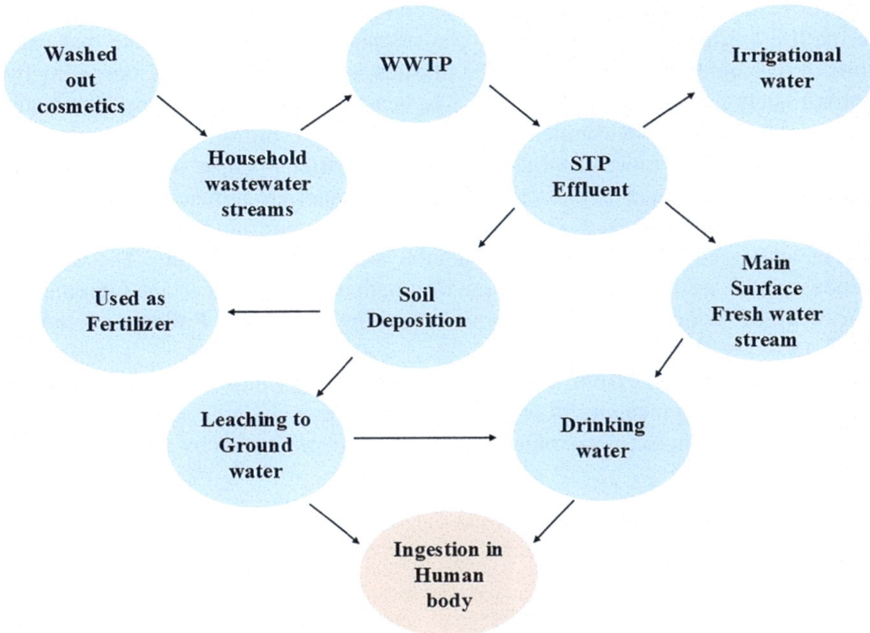

Fig. 18.2 The movement of cosmetic products into the environment

5 Environmental Challenges in the Global Cosmetics and Beauty Industry

The global cosmetics and beauty industry has a significant environmental footprint, ranging from the sourcing of raw materials to manufacturing processes and consumer waste [99]. This impact is not limited to landfills filled with disposable plastics; it also affects oceans, air quality, and ecosystems that have thrived for thousands of years [100]. The industry has contributed to the decline of various species due to its manufacturing practices and methods of product disposal [101]. Additionally, animal testing remains prevalent, affecting wildlife in 80% of the world's regions [102]. A major issue in the beauty industry is the presence of toxic substances, as cosmetics often contain ingredients such as parabens, plasticizers, formaldehyde, BHA, pesticides, and coal tar. Sunscreens exemplify the environmental damage caused by these synthetic components; they release approximately 14,000 tons of harmful chemicals into coral reefs annually [19]. Notably, substances such as octinoxate and oxybenzone identified by the Environmental Working Group are among the most destructive UV filters found in many personal care products. The Environmental Working Group's Skin Deep Cosmetics Database revealed that approximately 3600 products contain parabens and over 3500 products include polyethylene, a plastic used in various formulations [103]. Microplastics from cosmetics, such as microbeads and glitter, contribute to ocean pollution and harm marine life by disrupting their digestive systems [104]. Studies indicate that two-thirds of fish and 90% of seabirds ingest plastics, with ingestion rates increasing by approximately 2% each year [104, 105]. The beauty sector is undergoing significant scrutiny regarding its environmental practices. With 120 billion units of packaging produced each year, most of which are discarded after a single use, the industry must address its reliance on plastics and toxic substances to reduce its environmental impact. The industry faces significant ethical and environmental challenges that must be addressed to create a more sustainable and responsible future. These issues include animal testing, the harmful environmental effects of certain ingredients, poor working conditions, and the overwhelming problem of plastic pollution (Table 18.3, Fig. 18.3).

Collaborative efforts from informed consumers and innovative companies are essential for driving meaningful progress. Despite these challenges, the global cosmetics sector continues to experience substantial growth, driven by a growing focus

Table 18.3 Environmental impact of the beauty industry

Ethical issue	Description
Animal testing	Animal testing remains widespread in the beauty industry, with 88% of major brands still testing on animals. This leads to harm from chemicals and irritants, and the reversal of regulatory bans, such as the UK'S ban on animal testing, perpetuates the issue. Animal testing also contributes to environmental harm, including endangerment of species like rhesus macaques and ecological disruption [98, 100, 106, 107]
Water usage and transportation emissions	The beauty industry consumes large amounts of water, especially as a filler in cosmetics, which raises sustainability concerns amid global water stress. Additionally, the transportation of beauty products, averaging 4800 miles to the USA, results in high carbon emissions, though efforts to reduce emissions through different transport methods are underway [108]
Harmful chemicals and environmental impact	The beauty industry uses harmful chemicals, including parabens, formaldehyde, and pesticides, which pose risks to human health and the environment. Sunscreens contain chemicals like octinoxate and oxybenzone that harm coral reefs, while microplastics used in exfoliating products pollute oceans, causing long-term ecological harm
Plastic pollution and packaging waste	The beauty industry generates 120 billion units of plastic packaging annually, much of which is nonrecyclable and contributes to waste in landfills and incineration. Plastic packaging is responsible for up to 70% of the industry's carbon emissions, exacerbating global warming
Irresponsible sourcing of raw materials	Unsustainable sourcing practices, like palm oil and mica mining, contribute to deforestation, environmental degradation, and exploitation. Palm oil production leads to rainforest clearing and carbon release, threatening wildlife like orangutans, while mica mining often involves child labor and harm to ecosystems
The role of consumers and companies in driving change	Consumer demand for ethical products has led some beauty brands to adopt sustainable practices, including eco-friendly packaging, nontoxic ingredients, and ethical sourcing. Consumers can help drive change by supporting brands that prioritize cruelty-free, sustainable, and fair-trade practices, promoting a more responsible beauty industry

Fig. 18.3 Environmental impact of cosmetics

on health, hygiene, and self-care. The demand for beauty products, particularly skin care, hair care, and cosmetics, is increasing, as both younger and older generations place greater emphasis on wellness and appearance. This has led to increased sales of products such as sunscreen, antiaging creams, and hair dyes. Additionally, the influence of social media and the beauty industry has fueled the demand for makeup, especially among younger consumers. Manufacturers are expanding their offerings with personalized and digitalized products. Additionally, there has been a shift toward sustainable and organic cosmetics, with a growing demand for nontoxic, vegan, and eco-friendly products. However, concerns over synthetic chemicals, the rise of cosmetic surgery, high product costs, and the impact of the COVID-19 pandemic on retail outlets may slow growth.

6 Conclusion

In recent years, a broader variety of chemical substances have been used in the production of cosmetics, beauty products, and personal care products. These chemicals function as additives, fragrances, preservatives, stabilizers, surfactants, colorants, and shine enhancers to improve product quality, effectiveness, and longevity. Especially in terms of climate change, where cold climates are becoming too cold and hot climates are becoming very hot, using this cosmetic is even more essential for self-care. Considering this climate change factor and the essential utility of these products, monitoring the chemicals used in the cosmetic and beauty sectors is crucial. In this context, the results of this study indicate that chemicals that are biologically active, persistent in the environment, and bioaccumulative need to be monitored, and further research on alternatives is essential for commercialization to reduce their risks to human health and the environment and prevent long-term exposure. Prohibiting cosmetics that harm the environment is impractical, as these products are essential for daily use. Instead, innovative strategies and approaches are needed to address these challenges effectively. While international cosmetic regulatory agencies acknowledge these concerns, stricter regulations are necessary when new ingredients that may pose health risks are introduced. These measures must focus on protecting both the environment and public health by thoroughly assessing the ingredients in cosmetic products. Additional studies are needed to fully understand the impact of these substances, with an emphasis on both the immediate and long-term effects of major pollutants. This will facilitate more accurate monitoring and evaluation of their health and environmental consequences. Moreover, increasing consumer awareness about the environmental and health impacts of cosmetic products, particularly in relation to packaging, will promote more responsible usage and informed choices.

Acknowledgments This work was supported by the Jal Jeevan Mission, Ministry of Jal Shakti, Government of India.

Conflicts of Interest There are no conflicts to declare.

References

1. Fortune Business Insights. (2024). *Cosmetics market*. Accessed 18 Dec 2024. [Online]. Available: https://www.fortunebusinessinsights.com/amp/cosmetics-market-102614.
2. Volza Grow Global. (2024). Raw material for cosmetics imports in world—Market size & demand based on Import Trade Data [Online]. Source: https://www.volza.com/p/raw-material-for-cosmetics/import/.
3. Gao, P., & Lei, T. (2018). Bioaccessible trace metals in lip cosmetics and their health risks to female consumers. *Environmental Pollution, 238*, 554–561.
4. Conley, K., Clum, A., Deepe, J., Lane, H., & Beckingham, B. (2019). Wastewater treatment plants as a source of microplastics to an urban estuary: Removal efficiencies and loading per capita over one year. *Water Research X, 3*, 100030. https://doi.org/10.1016/j.wroa.2019.100030
5. Ramos, S., & Homem, V. (2016). A review of organic UV-filters in wastewater treatment plants. *Environment International, 86*, 24–44.
6. Díaz-Cruz, M. S., et al. (2009). Analysis of selected emerging contaminants in sewage sludge. *TrAC Trends in Analytical Chemistry, 28*(11), 1263–1275. https://doi.org/10.1016/j.trac.2009.09.003
7. Joss, A., & Siegrist, H. (2004). *Reviewed: Scrutinizing pharmaceuticals and personal care products in wastewater treatment*. ACS Publications.
8. 2025 Edition, Bologna, Italy, 2025 Edition, And Bologna, Italy Fair District. *The Global Beauty Industry: The Markets to Monitor in the Next Future*. 2025 Edition, Bologna, Italy Fair District, May 13, 2024." May 13, 2024.
9. Sophie Smith. (2024). *The Industry Beauty, Billions of beauty packaging goes unrecycled every year*. Available: https://theindustry.beauty/billions-of-beauty-packaging-goes-unrecycled-every-year. June 20, 2024.

10. Mendes, A. C., & Pedersen, G. A. (2021). Perspectives on sustainable food packaging:—Is biobased plastics a solution? *Trends in Food Science and Technology, 112*, 839–846. https://doi.org/10.1016/j.tifs.2021.03.049
11. Carter, P. (2024). The Beauty Industry's Boom. *Global Edge*. Available: https://globaledge.msu.edu/blog/post/57407/the-beauty-industry's-boom. Apr. 08, 2024.
12. Cook, L. (2016). The facts about Cruelty-Free Fragrances. *One Seed*. Available: https://oneseedperfumes.com/blogs/news/cruelty-free-perfume-101-the-facts-about-cruelty-free-fragrances. Jul. 20, 2016.
13. Gotsill, G. (2019). Animal Musks: The dark secret of perfume. *Purrfumery*. Available: https://purrfumery.com/pages/animal-musks-the-dark-secret-of-perfume.
14. *Fairglow*. (2023). Cosmetics carbon footprint. Available: https://www.fairglow.fr/blog-detail/breaking-down-cosmetics-carbon-footprint.
15. Martins, A. M., & Marto, J. M. (2023). A sustainable life cycle for cosmetics: From design and development to postuse phase. *Sustainable Chemistry and Pharmacy, 35*, 101178. https://doi.org/10.1016/j.scp.2023.101178
16. *Skin Kraft*. (2022). 16 toxic chemicals to avoid in cosmetics and skincare. Available: https://skinkraft.com/blogs/articles/toxic-chemicals-in-cosmetics.
17. Gutierrez, L. (2023). Beauty at what cost? The environmental impact of cosmetic ingredients and materials. *Sustainable Earth*. Available: https://sustainable-earth.org/cosmetic-ingredients.
18. *Amala Earth*. (2023). Sustainable nail care: Reducing our environmental footprint. Available: https://amala.earth/blogs/amala-earth-blog/sustainable-nail-care-reducing-our-environmental-footprint.
19. *National Oceanic and Atmospheric Administration*. (2024). Skincare Chemicals and Coral Reefs. Available: https://oceanservice.noaa.gov/news/sunscreen-corals.html.
20. *Down to Earth*. (2024). Sunscreens can hurt the marine environment—Here's how to choose one that's healthy for you and the sea. Available: https://www.downtoearth.org.in/environment/sunscreens-can-hurt-the-marine-environment-heres-how-to-choose-one-thats-healthy-for-you-and-the-sea.
21. Negi, S., & Bala, L. (2022). Tattoo inks are toxicological risks to human health: A systematic review of their ingredients, fate inside skin, toxicity due to polycyclic aromatic hydrocarbons, primary aromatic amines, metals, and overview of regulatory frameworks. *Toxicology and Industrial Health, 38*(7), 417–434.
22. Abhishek Gautam, "Health risk assessment: Heavy metals (antimony, arsenic, barium, cadmium, chromium, lead, mercury, and nickel) in tattoo ink." , 2023.
23. carbon offsets credits. (2024). Carbon footprint of cosmetics: Emissions from Makeup (By Item). Available: https://8billiontrees.com/carbon-offsets-credits/carbon-footprint-of-cosmetics/.
24. *The Good boutique*. (2023). The environmental impact of disposable makeup wipes. Available: https://www.thegoodboutique.com/inspiration/the-environmental-impact-of-disposable-makeup-wipes.
25. *Business Waste*. (2021). Millions of makeup wipes piled as high as the Eiffel Tower going to landfills every day. Available: https://www.businesswaste.co.uk/news/millions-of-makeup-wipes-piled-as-high-as-the-eiffel-tower-going-to-landfills-every-day/.
26. *MyKindaEarth*. (2024). The ultimate guide to reusable makeup wipes: Eco-friendly beauty at your fingertips. Available: https://mykindaearth.co.uk/blogs/news/the-ultimate-guide-to-reusable-makeup-wipes-eco-friendlybeauty-at-your-fingertips.
27. Alex, R. K. (2024). India's growing microplastic crisis spreads from skin to sea: Microbeads are hidden dangers in your self-care products. *Down to Earth*.
28. *Global Citizen*. (2017). The tiny microbeads in soap may pose a big environmental problem. Source: https://www.globalcitizen.org/en/content/why-you-shouldnt-use-microbead-soap.
29. Cheung, P. K., & Fok, L. (2017). Characterization of plastic microbeads in facial scrubs and their estimated emissions in Mainland China. *Water Research, 122*, 53–61.
30. Brucculieri, J. (2019). Are your beloved sheet masks killing the planet? *HuffPost*. Available: https://www.huffpost.com/entry/sheet-masks-not-environmentally-friendly.

31. Ajaj, R., Al Dweik, R., Syed Ali, S. A., & Stietiya, M. H. (2023). Understanding the environmental impacts of facemasks: A review on the facemask industry and existing life cycle assessment studies. *Sustainable Environment Research, 33*(1), 20. https://doi.org/10.1186/s42834-023-00182-x
32. United Nations Environment Programme by Chemicals & Pollution Action. (2023). *Gabon, Jamaica and Sri Lanka unite to fight hazardous skin lightening products*. Available: https://www.unep.org/news-and-stories/press-release/gabon-jamaica-and-sri-lanka-unite-fight-hazardous-skin-lightening.
33. *Skin Lightening Products Found to Contain Mercury, Hydroquinone and/or Steroids*. Department of Health, Spring 2021.
34. Jiang, X., & Hua, M. (2020). Impacts of mist spray on rice field micrometeorology and rice yield under heat stress condition. *Scientific Reports, 10*(1), 1579.
35. Oh, W., Ooka, R., Nakano, J., Kikumoto, H., & Ogawa, O. (2020). Evaluation of mist-spraying environment on thermal sensations, thermal environment, and skin temperature under different operation modes. *Building and Environment, 168*, 106484. https://doi.org/10.1016/j.buildenv.2019.106484
36. Knight, W. L., Fraser, M. P., & Herckes, P. (2021). Impact of misting systems on local particulate matter (PM) levels. *Aerosol and Air Quality Research, 21*(5), 200431. https://doi.org/10.4209/aaqr.200431
37. Anthony, J. (2024, May 2). The environmental impact of shampoo and conditioner packaging.
38. Farah. (2024, September 01). Why sustainable hair care is taking over: A look at the green beauty movement. *Jungle Hugger*.
39. *loreal*. L'Oréal—2023 universal registration document. Source: https://www.loreal-finance.com/eng/2023-universal-registration-document/en/article/223/.
40. Sheikh, A. (2017). Beauty care products—A chemistry deleterious to human chemistry. *International Journal of Medical Science and Public Health Research*, 1. https://doi.org/10.5455/ijmsph.2018.1130118112017
41. Chandra Pal, K. (2022). Environmental pain with human beauty.
42. Aerts, O., Goossens, A., Lambert, J., & Lepoittevin, J.-P. (2017). Contact allergy caused by isothiazolinone derivatives: An overview of noncosmetic and unusual cosmetic sources. *European Journal of Dermatology, 27*(2), 115–122. https://doi.org/10.1684/ejd.2016.2951
43. Zhang, X.-J. (2023). A review of the occurrence, metabolites and health risks of butylated hydroxyanisole (BHA). May 01, 2023. [Online]. Available: https://doi.org/10.1002/jsfa.12676.
44. Wasilewski, T., et al. (2022). Sustainable green processing of grape pomace using micellar extraction for the production of value-added hygiene cosmetics. *Molecules, 27*(8), 2444. https://doi.org/10.3390/molecules27082444
45. Chawla, S. (2022). Utilization of consumer nanoproducts for cosmetics and their impacts. In *Handbook of consumer nanoproducts* (pp. 1073–1094). Springer.
46. Glaser, Aviva. 2007, Triclosan—Alaska Community Action on Toxics
47. DiNardo, J. C., & Downs, C. A. (2018). Dermatological and environmental toxicological impact of the sunscreen ingredient oxybenzone/benzophenone-3. *Journal of Cosmetic Dermatology, 17*(1), 15–19. https://doi.org/10.1111/jocd.12449
48. Ruszkiewicz, J. A., Pinkas, A., Ferrer, B., Peres, T. V., Tsatsakis, A., & Aschner, M. (2017). Neurotoxic effect of active ingredients in sunscreen products, a contemporary review. *Toxicology Reports, 4*, 245–259. https://doi.org/10.1016/j.toxrep.2017.05.006
49. Agin, P. P. (2008). Rates of allergic sensitization and irritation to oxybenzone-containing sunscreen products: A quantitative meta-analysis of 64 exaggerated use studies. *Photodermatology, Photoimmunology & Photomedicine, 24*(4), 211–217.
50. Gunawardena, S. (2016). *Hazardous effects of cosmetics, perfumes and detergents*.
51. Ara, A., & Usmani, J. A. (2015). Lead toxicity: A review. *Interdisciplinary Toxicology, 8*(2), 55–64.

52. Burger, P., Landreau, A., Azoulay, S., Michel, T., & Fernandez, X. (2016). Skin whitening cosmetics: Feedback and challenges in the development of natural skin lighteners. *Cosmetics, 3*(4), 36. https://doi.org/10.3390/cosmetics3040036
53. Vinardell, M., & Mitjans, M. (2017). Alternative methods to animal testing for the safety evaluation of cosmetic ingredients: An overview. *Cosmetics, 4*(3), 30. https://doi.org/10.3390/cosmetics4030030
54. Guarnotta, V., Amodei, R., Frasca, F., Aversa, A., & Giordano, C. (2022). Impact of chemical endocrine disruptors and hormone modulators on the endocrine system. *International Journal of Molecular Sciences, 23*(10), 5710. https://doi.org/10.3390/ijms23105710
55. Wang, Y., & Qian, H. (2021). Phthalates and their impacts on human health. *Healthcare, 9*(5), 603. https://doi.org/10.3390/healthcare9050603
56. Dutta, S., & Haggerty. (2020). Phthalate exposure and long-term epigenomic consequences: A review. *Frontiers in Genetics, 11*, 405.
57. Bondi, C. A., & Marks, J. L. (2015). Human and environmental toxicity of sodium lauryl sulfate (SLS): Evidence for safe use in household cleaning products. *Environmental Health Insights, 9*, EHI-S31765.
58. Liang, W. (2020). Toxicity and effect of chemicals in skin care products on human health. *IOP Conference Series: Earth and Environmental Science, 512*(1), 012081. https://doi.org/10.1088/1755-1315/512/1/012081
59. Lin, R. R., & Lin, D. A. (2024). Toxic ingredients in personal care products: A dermatological perspective. *Dermatitis, 35*(2), 121–131.
60. Becker, L.C., 2019, "Safety assessment of glycerin as used in cosmetics. International Journal of Toxicology, 38(3_suppl), pp.6S–22S."
61. *Independent Chemical*. (2019). Is sodium benzoate safe in cosmetics? Available: https://independentchemical.com/blogs/is-sodium-benzoate-safe-in-cosmetics-20058.
62. Ko, Y., & Lim, D. (2024). Risk assessment of cyclohexasiloxane D6 in cosmetic products. *Toxicological Research*, 1–9.
63. Nonino, E. D. C. W. (2022). Aquatic environmental impact of cosmetics. In *Handbook of cosmetic science and technology* (pp. 398–407). CRC Press.
64. Mohiuddin, A. K. (2022). *International Journal of Pharmacy and Pharmaceutical Science*.
65. Khan, A. D., & Alam, M. N. (2019). Cosmetics and their associated adverse effects: A review. *Journal of Applied Pharmaceutical Sciences and Research, v2i1.1*, 1–6. https://doi.org/10.31069/japsr
66. Wargala, E., Sławska, M., Zalewska, A., & Toporowska, M. (2021). Health effects of dyes, minerals, and vitamins used in cosmetics. *Women, 1*(4), 223–237. https://doi.org/10.3390/women1040020
67. Williams, R. M. (2004). *Cosmetic chemicals and safer alternatives. Townsend letter for doctors and patients (247–248)* (pp. 32–35).
68. Burnett, C. L. (2020). Safety assessment of polyene group as used in cosmetics. *International Journal of Toxicology, 39*(2_suppl), 59S–90S.
69. Makuvara, Z., Marumure, J., Simbanegavi, T. T., Machingura, J., Chaukura, N., & Gwenzi, W. (2024). Unveiling the dark side of beauty? Beauty parlours as potential pollution and human exposure hotspots. *Process Safety and Environment Protection, 182*, 1153–1170. https://doi.org/10.1016/j.psep.2023.12.045
70. Wali, B., Othman, N., Rashid, S., Hamid, S., & Issa, H. (2024). A short overview of Benzalkonium chloride in pharmaceutical sector: Safety, environmental impact and alternatives. *ChemRxiv*. https://doi.org/10.26434/chemrxiv-2024-zfj3p
71. M. N. Mali, S. Shinde, and V. B. Wamane, "A Review on Toxic Chemicals in Cosmetics," vol. 7, no. 11, 2022.
72. Materna, J., & Nieradko-Iwanicka, B. (2020). Zinc detection in cosmetics. *Journal of Education, Health and Sport, 10*(6), 117–123.
73. Epstein, S. S. (2009). *Toxic beauty: How cosmetics and personal-care products endanger your health... and what you can do about it*. BenBella Books, Inc.

74. Juncan, A. M., Morgovan, C., Rus, L.-L., & Loghin, F. (2023). Development and evaluation of a novel anti-aging cream based on hyaluronic acid and other innovative cosmetic actives. *Polymers, 15*(20), 4134. https://doi.org/10.3390/polym15204134
75. Belay, M. H. (2022). *Evaluation of contaminants of emerging concern and their transformation products in water.*
76. Guryanov, I., Naumenko, E., & Fakhrullin, R. (2022). Hair surface engineering: Combining nanoarchitectonics with hair topical and beauty formulations. *Applied Surface Science Advances, 7*, 100188. https://doi.org/10.1016/j.apsadv.2021.100188
77. Sharma, A. K. (2022). *Cosmetics and their associated adverse effect.*
78. Kim, K. H., & Kabir, E. (2016). The use of personal hair dye and its implications for human health. *Environment International, 89*, 222–227.
79. Burnett, C. L., et al. (2010). Safety assessment of Toluene-2,5-Diamine, Toluene-2,5-Diamine sulfate, and Toluene-3,4-diamine as used in cosmetics. *International Journal of Toxicology, 29*(3_suppl), 61S–83S. https://doi.org/10.1177/1091581810361964
80. Caliman, F. A., & Gavrilescu, M. (2009). Pharmaceuticals, personal care products and endocrine disrupting agents in the environment–A review. *CLEAN–Soil, Air, Water, 37*(4–5), 277–303.
81. Daughton, C. G., & Ternes, T. A. (1999). Pharmaceuticals and personal care products in the environment: Agents of subtle change? *Environmental Health Perspectives, 107*(suppl 6), 907–938.
82. Chen, W., Xu, J., Lu, S., Jiao, W., Wu, L., & Chang, A. C. (2013). Fates and transport of PPCPs in soil receiving reclaimed water irrigation. *Chemosphere, 93*(10), 2621–2630. https://doi.org/10.1016/j.chemosphere.2013.09.088
83. Castiglioni, S., Bagnati, R., Fanelli, R., Pomati, F., Calamari, D., & Zuccato, E. (2006). Removal of Pharmaceuticals in Sewage Treatment Plants in Italy. *Environmental Science & Technology, 40*(1), 357–363. https://doi.org/10.1021/es050991m
84. Santos, J. L., Aparicio, I., & Alonso, E. (2007). Occurrence and risk assessment of pharmaceutically active compounds in wastewater treatment plants. A case study: Seville city (Spain). *Environment International, 33*(4), 596–601.
85. Yang, Y.-Y., & Toor, G. S. *Contaminants in the Urban Environment: Pharmaceuticals and Personal Care Products (PPCPs)— Part 1.*
86. Fick, J., Söderström, H., Lindberg, R. H., Phan, C., Tysklind, M., & Larsson, D. G. J. (2009). Contamination of surface, ground, and drinking water from pharmaceutical production. *Environmental Toxicology and Chemistry, 28*(12), 2522–2527. https://doi.org/10.1897/09-073.1
87. Daughton, C. G. (2003). Cradle-to-cradle stewardship of drugs for minimizing their environmental disposition while promoting human health. I. Rationale for and avenues toward a green pharmacy. *Environmental Health Perspectives, 111*(5), 757–774. https://doi.org/10.1289/ehp.5947
88. Molins-Delgado, D., Díaz-Cruz, M. S., & Barceló, D. (2015). *Introduction: personal care products in the aquatic environment. Personal care products in the aquatic environment* (pp. 1–34).
89. Zenker, A., Cicero, M. R., Prestinaci, F., Bottoni, P., & Carere, M. (2014). Bioaccumulation and biomagnification potential of pharmaceuticals with a focus to the aquatic environment. *Journal of Environmental Management, 133*, 378–387. https://doi.org/10.1016/j.jenvman.2013.12.017
90. Meador, J., 2006., "Rationale and procedures for using the tissue-residue approach for toxicity assessment and determination of tissue, water, and sediment quality guidelines for aquatic organisms Human and Ecological Risk Assessment: An International Journal, 12(6), pp.1018–1073."
91. Wennmalm, Å., & Gunnarsson, B. (2009). Pharmaceutical management through environmental product labeling in Sweden. *Environment International, 35*(5), 775–777.

92. Bruggeman, W. A., Opperhuizen, A., Wijbenga, A., & Hutzinger, O. (1984). Bioaccumulation of super-lipophilic chemicals in fish. *Toxicological & Environmental Chemistry, 7*(3), 173–189.
93. Thomann, R. V. (1989). Bioaccumulation model of organic chemical distribution in aquatic food chains. *Environmental Science & Technology, 23*(6), 699–707. https://doi.org/10.1021/es00064a008
94. Sweetman, A. J., Valle, M. D., Prevedouros, K., & Jones, K. C. (2005). The role of soil organic carbon in the global cycling of persistent organic pollutants (POPs): Interpreting and modeling field data. *Chemosphere, 60*(7), 959–972. https://doi.org/10.1016/j.chemosphere.2004.12.074
95. Helbling, D. E., Hollender, J., Kohler, H.-P. E., Singer, H., & Fenner, K. (2010). High-throughput identification of microbial transformation products of organic micropollutants. *Environmental Science & Technology, 44*(17), 6621–6627. https://doi.org/10.1021/es100970m
96. Onesios, K. M., Yu, J. T., & Bouwer, E. J. (2009). Biodegradation and removal of pharmaceuticals and personal care products in treatment systems: A review. *Biodegradation, 20*(4), 441–466. https://doi.org/10.1007/s10532-008-9237-8
97. Ebele, A. J., Abou-Elwafa Abdallah, M., & Harrad, S. (2017). Pharmaceuticals and personal care products (PPCPs) in the freshwater aquatic environment. *Emerging Contaminants, 3*(1), 1–16. https://doi.org/10.1016/j.emcon.2016.12.004
98. Van der Oost, R., Beyer, J., & Vermeulen, N. P. (2003). Fish bioaccumulation and biomarkers in environmental risk assessment: A review. *Environmental Toxicology and Pharmacology, 13*(2), 57–149.
99. Young, O. (2022). *The environmental impact of cosmetics is tremendous—Here's how they're harmful*.
100. Bilal, M., Mehmood, S., & Iqbal, H. M. N. (2020). The beast of beauty: Environmental and health concerns of toxic components in cosmetics. *Cosmetics, 7*(1), 13. https://doi.org/10.3390/cosmetics7010013
101. Groff, K., Bachli, E., Lansdowne, M., & Capaldo, T. (2014). Review of evidence of environmental impacts of animal research and testing. *Environments, 1*(1), 14–30. https://doi.org/10.3390/environments1010014
102. European Parliament. (2018). A*nimal testing of cosmetics: MEPs call for worldwide ban*. Available: https://www.europarl.europa.eu.
103. Martins, F. C., Videira, R. A., Oliveira, M. M., Silva-Maia, D., Ferreira, F. M., & Peixoto, F. P. (2021). Parabens enhance the calcium-dependent testicular mitochondrial permeability transition: Their relevance on the reproductive capacity in male animals. *Journal of Biochemical and Molecular Toxicology, 35*(3), e22661.
104. Bashir, S. M., Kimiko, S., Mak, C.-W., Fang, J. K.-H., & Gonçalves, D. (2021). Personal care and cosmetic products as a potential source of environmental contamination by microplastics in a densely populated Asian city. *Frontiers in Marine Science, 8*, 683482. https://doi.org/10.3389/fmars.2021.683482
105. Savoca, M. S., McInturf, A. G., & Hazen, E. L. (2021). Plastic ingestion by marine fish is widespread and increasing. *Global Change Biology, 27*(10), 2188–2199. https://doi.org/10.1111/gcb.15533
106. Wilcox, C., Van Sebille, E., & Hardesty, B. D. (2015). Threat of plastic pollution to seabirds is global, pervasive, and increasing. *Proceedings of the National Academy of Sciences, 112*(38), 11899–11904. https://doi.org/10.1073/pnas.1502108112
107. "Facts and Statistics About Animal Testing." People for the Ethical Treatment of Animals.
108. UN-Water. (2020). *UN World Water Development Report 2020—Water and Climate Change*.

Chapter 19
Mitigating Pharmaceuticals and Personal Care Products from Water Sources: A Review of Removal Strategies

Hajar Farzaneh, Jayaprakash Saththasivam, and Gordon McKay

1 Introduction

Due to water scarcity issues across the world, there has recently been a growing awareness of contamination in various water sources, such as surface water, groundwater, wastewater, and even drinking water. Pharmaceuticals, personal care products (PPCPs), and endocrine-disrupting compounds (EDCs) are among these contaminants, and due to their adverse effects on the environment as well as human health, extensive studies and investigations aiming at their mitigation have been conducted worldwide [28, 45, 46, 66, 68, 79, 87].

In this chapter, several published studies on the occurrence and removal of various types of PPCPs and EDCs are discussed. Their elimination by conventional water and wastewater methods is examined and compared with that of advanced technologies such as membrane filtration and activated carbon adsorption. In addition, this chapter explores removal by advanced oxidation processes (AOPs) and combined approaches that have shown effective PPCP and EDC removal at minimal cost with reduced production of harmful byproducts or wastes.

H. Farzaneh · J. Saththasivam · G. McKay (✉)
College of Science & Engineering, Hamad Bin Khalifa University, Qatar Foundation, Doha, Qatar
e-mail: jsaththasivam@hbku.edu.qa; gmckay@hbku.edu.qa

M. K. Jindal et al. (eds.), *Emerging Contaminants in Water*, Emerging Contaminants and Associated Treatment Technologies,
https://doi.org/10.1007/978-3-031-89591-3_19

2 Removal of PPCPs and EDCs by Conventional Processes

Most PPCPs and EDCs cannot be eliminated by conventional water and wastewater treatment processes. In the conventional water treatment process, only hydrophobic compounds that have high organic carbon content connected with particulate or colloidal substances are expected to be removed during the coagulation step. On the other hand, the removal of PPCPs and EDCs in conventional wastewater treatment plants is based on compound properties, which are essentially removed by biological processes such as activated sludge, biofiltration, and soil aquifer treatment. Most studies have shown that the removal of PPCPs and EDCs by conventional treatment methods is inadequate, and they suggest the use of advanced technologies for higher treatment quality, as follows [28, 33, 45, 66, 80, 87, 92]:

Gros et al. [33] studied the removal of 73 different pharmaceutical compounds in 7 conventional wastewater treatment plants along the Ebro River in Spain. They analyzed the influent and effluent samples of these seven plants and reported insufficient removal of all the pharmaceuticals. Assuming that the removal rate follows pseudo-first-order kinetics, they estimated the degradation rate. They reported comparatively high half-lives of most pharmaceuticals, which shows that higher hydraulic retention times are required for greater removal.

An investigation reported the presence of 56 PPCPs and EDCs in the rivers Taff and Ely in the United Kingdom, where most of the pharmaceuticals were antibacterial, anti-inflammatory, and antiepileptic compounds [47, 48]. The detected personal care products included sunscreen agents, preservatives, disinfectants, and antiseptics, with a total PPCP discharge amount of ~6 kg per day into the two rivers. Among the 56 PPCPs, the concentrations of paracetamol, tramadol, codeine, gabapentin, and atenolol exceeded 10 μg/L. The removal of these PPCPs was investigated in two plants: one with activated sludge and the other with trickling filter beds. Higher removal efficiencies were reported with the use of activated sludge reaching 85%, while the use of trickling filter bed could not obtain higher than 70% removal of the PPCPs and EDCs in their investigation. They concluded that the presence of PPCPs and EDCs in the rivers Taff and Ely is due to insufficient treatment processes.

In the industrial city of Ulsan, Korea, the removal of 20 PPCPs was investigated in 5 wastewater treatment plants by Behera et al. [9]. The compounds with the highest concentrations in the sewage influent were acetaminophen, atenolol, and lincomycin, which exceeded 10 μg/L. These conventional treatment plants include tertiary treatments with UV disinfection, dissolved air flotation, and fiber filtration. The average removal of acetaminophen, caffeine, estriol, and estradiol in all five plants was 99%. On the other hand, sulfamethazine, metoprolol, and carbamazepine had relatively low removals of less than 30%. According to their investigation, the antibiotic removal efficiencies were between 11 and 69%. They concluded that the main removal process takes place in the secondary treatment by biological processes, whereas the removal in the primary treatment (the physicochemical process) was less than 28%.

In China, the removal of 15 PPCPs and EDCs was investigated in four wastewater treatment plants in Beijing by Sui et al. [83]. They reported negligible removal by the primary treatment process, whereas during the secondary treatment process, the removal efficiency ranged from 12% to complete removal. The efficiency of PPCPS and EDC removal in secondary treatment is highly related to the nature and properties of the compound. The tertiary treatment showed different results using different treatment methods; for instance, sand filtration had inadequate removal, whereas the use of advanced technologies such as ozonation and microfiltration combined with reverse osmosis (MF/RO) had significantly high removal efficiencies.

The efficiency of PPCP and EDC removal by membranes is briefly reviewed in the following section, along with a brief review of removal by other advanced technologies.

3 Removal of PPCPs and EDCs by Advanced Technologies

Different studies have been conducted on advanced technologies for the removal of a variety of PPCPs and EDCs. The most studied technologies are membranes and adsorption by activated carbon. An example is the investigation by Snyder et al. [78] which had extensively studied the role of membranes and activated carbon in the removal of a variety of pharmaceuticals and EDCs. They had evaluated different membranes, including microfiltration (MF), ultrafiltration (UF), nanofiltration (NF), reverse osmosis (RO), membrane bioreactors (MBR), and combinations of these membranes in series. The membrane investigations were conducted at the pilot and/or full scale, whereas the activated carbon was studied at the bench scale columns and full scale. They had indicated that PPCP and EDC rejection by MF and UF was not significant, although some steroidal-type compounds were removed to high extents. NF and RO could remove almost all the studied PPCPs and EDCs with the detection of some trace levels of the compounds in the membrane effluents. The removal by granular activated carbon (GAC) was reported to be very efficient, with almost complete removal of all target compounds. A study of activated carbon breakthrough curves revealed that PPCPs and EDCs with higher hydrophilicity can be adsorbed to activated carbon faster than hydrophobic compounds, which was also reported by Kim et al. [52]. However, a full-scale study of removal by activated carbon revealed that non-regenerated carbon could not remove the targeted PPCPs and EDCs, which means that in full-scale plants, regular regeneration is important to avoid contaminant breakthrough.

A study by Kim et al. [52] on the removal of PPCPs and EDCs revealed inefficient removal by activated sludge, coagulation, and sand filtration in conventional wastewater and drinking water processes. They reported 99% removal of these contaminants by GAC. In their study on removal by MBR in wastewater treatment plants, effective removal of estriol, testosterone, androstenedione, acetaminophen, ibuprofen, and caffeine was achieved, whereas ineffective removal of erythromycin, tri(2-chloroethyl) phosphate (TCEP), trimethoprim, naproxen, diclofenac, and carbamazepine was reported. They had indicated very high removals by RO and NF of greater than 95% for all the studied PPCPs and EDCs, whereas the MF and UF membranes had selective contaminant removal and were highly dependent on the contaminant properties. The use of MBR followed by RO or NF for enhanced removal was suggested. In addition, they had indicated that although activated carbon can remove almost all organic contaminants, the removal capacity is limited by contact time, competition with natural organic matter (NOM), the solubility of the contaminant, and the type of carbon. This finding is also in good agreement with the study by Yu et al. [97] comparing two different GAC with different pore sizes for the removal of naproxen, carbamazepine, and nonylphenol. In general, the removal efficiency by adsorption is highly dependent on the contact time and the adsorbent dose. Water quality is also important, as NOM competes with targeted compounds to occupy adsorption sites. NOM blocks the pores of the adsorbent and thus adversely affects the removal efficiency of contaminants [31, 42, 81].

A review by Nasirabadi et al. [74] on the removal of PPCPs and EDCs from wastewater by membrane processes indicated that a combination of activated sludge with membrane filtration in MBR technologies has significant advantages over conventional activated sludge systems in terms of removal. This phenomenon has also been reported by Le-Minh et al. [55], who reported that the high sludge retention time and the concentration of biosolids in MBR systems are the reasons behind greater removal than conventional activated sludge. It has also been reported that the removal efficiency of RO systems is greater than 80% for almost all the studied compounds regardless of the membrane material. Removal by UF has been reported to be between no removal and 80%, depending on the properties of the compound. On the other hand, the NF membrane showed a high removal efficiency of 96% for complete removal and was the preferred membrane. Removal by membranes is highly dependent on the membrane pore size, the quality of the water, and the properties of the target compound [96]. Carbamazepine, which is a highly persistent pharmaceutical, has a low removal rate of less than 30% in MBR systems, and these membranes are effective for the removal of hormones and estrogens. Increasing the sludge retention time in MBR processes can result in higher removal efficiencies. Although MF and UF are not very efficient for the removal of PPCPs and EDCs, their combination with other technologies, such as powdered or granular activated carbon, can achieve greater removal [31, 72].

In a study in four different wastewater treatment plants in China, the influent of gemfibrozil to the plants ranged between 24 and 140 ng/L [83]. According to this study, the removal of gemfibrozil by conventional methods is very poor (5% in the primary treatment and 25% during the secondary treatment). Wastewater treatment plants with advanced treatment processes can remove gemfibrozil at relatively high rates. For example, removals of >90% were achieved via the MF/RO system, and removals of up to 50% were achieved via the UF system.

An extensive review on the removal of PPCPs and EDCs from water and wastewater by adsorption was conducted by Jung et al. [46]. They focused mainly on adsorption by carbon nanotubes (CNTs) but also reviewed removal by other high-binding adsorbents, such as activated carbon, biochar, and zeolite. They concluded that activated carbon is effective for the removal of a variety of pesticides, pharmaceuticals, and estrogenic compounds. Among the different adsorbents under review, CNTs have been reported to have relatively high adsorption capacities and high PPCP and EDC removal efficiencies owing to their structure, which has a high surface area and aspect ratio. The main issue associated with using CNTs in large-scale treatment plants is their cost, which is very high compared with that of activated carbon, in addition to their toxicity effects on the environment, which should be investigated before their utilization.

The use of carbide-derived carbon (CDC) as an adsorbent was reported to be efficient for the removal of ibuprofen from deionized water and treated sewage effluent by Almanassra et al. [2]. Guedidi et al. [34] reported that surface modification of commercial granular activated carbon by H_2O_2 formed carbonyl groups, which increased the adsorption of ibuprofen. Another study showed that the adsorption of ibuprofen is greater at lower pH values when activated carbon cloth is used [35]. Similarly, a study by Baccar et al. [7] on the removal of ibuprofen, ketoprofen, naproxen, and diclofenac via activated carbon produced from olive waste cake revealed increased adsorption with decreasing pH. The removal of ibuprofen was reported to be greater than that of paracetamol when activated carbon produced from local plant waste was studied by Mastre et al. [61]. The increased adsorption of ibuprofen was reported to be due to electrostatic interactions between ibuprofen molecules and the surface charge of activated carbon. Other waste materials have been used to produce activated carbon with a high surface area and have been tested for the removal of different PPCPs and EDCs; for example, the effective removal of atenolol, acebutolol, and carbamazepine was achieved in a study using palm kernel shells [86]. Other studies have investigated the adsorption of a variety of pesticides [37, 40], dyes [1, 13, 60], heavy metals [1, 14, 58], and pharmaceuticals [10, 21] via activated carbon produced from date stone. For example, Farzaneh et al. [25] reported high removal rates of ibuprofen using activated carbon produced from phosphoric acid-treated date stones, reported high adsorption capacities, and investigated adsorbent minimization in a two-stage batch process.

In addition to adsorption and membrane filtration, photodegradation by UV light is another advanced method that can degrade a range of PPCPs and EDCs. Kim and Tanaka [50] investigated the photodegradation of 30 PPCPs from pure water at the laboratory scale via two different UV lamps: one emits light only at 254 nm, and the other emits light at both 254 and 185 nm. A lamp that emits at two different wavelengths was reported to be more efficient. UV irradiation causes photolysis of water molecules, and then OH radicals, which are strong oxidizing agents, are produced. To enhance the oxidation process, the use of advanced oxidation processes by combining UV with H_2O_2 or O_3 to increase the production of OH radicals is suggested.

Although membranes and adsorbents are very effective at removing a wide range of organic contaminants, including PPCPs and EDCs, they contribute to several environmental issues. The main problem with adsorbents is their toxicological effects on the environment, which need to be studied prior to their application, whereas the main issue with membranes is brine disposal [31, 78]. The rejected compounds from the membranes are very concentrated in the brine streams and require proper disposal, which are more toxic than the influent water. Similarly, the used adsorbents need to be disposed or regenerated carefully. Furthermore, membrane processes require a significant amount of energy to maintain the required operating pressure [32, 72, 78]. The same issue applies for adsorption processes where thermal regeneration requires a significant amount of energy. Table 19.1 summarizes the PPCP removal efficiency and energy consumption of different types of membranes and adsorbents according to the literature [19, 51, 56, 64, 67, 72, 73, 77, 78, 89, 90]. The high energy consumption of these processes makes them expensive and can adversely affect the environment, as they produce wastes (concentrated brine in membrane filtration and used adsorbents in adsorption processes); thus, the focus should be on more cost-efficient and environmentally friendly processes, such as advanced oxidation processes (AOPs) [28, 32, 49, 63, 72, 76, 78]. The application of AOPs is easier with no waste production and easier in terms of maintenance and upgrades [49, 76]. Different types of AOPs are described in detail in the following section.

Table 19.1 Comparison between filtration and adsorption for the removal of PPCPs and EDCs

Method	Type	Removal efficiency (%)	Energy consumption (kWh/m^3)	Advantages	Disadvantages
Membrane filtration	MF	≤20	0.05–1	– Low energy cost – Operates at low pressures and does not require additional safety considerations – Compact modules	– Selective removal depending on the properties of PPCPs and EDCs – Concentrated reject water with the contaminants which requires proper disposal – Membrane fouling
	UF	≤20	0.1–4	– Low energy cost – Operates at low pressures – Compact modules	– Selective removal depending on the properties of PPCPs and EDCs – Concentrated reject water with the contaminants which requires proper disposal – Membrane fouling
	NF	≥90	3.2–10	– Efficient removal of a wide range of PPCPs and EDCs – Removal of hardness and salts – Compact modules	– Concentrated reject water with the contaminants which requires proper disposal – Membrane fouling – Operates at relatively high pressures
	RO	≥90	3.6–18.2	– Efficient removal of a wide range of PPCPs and EDCs – Removal of dissolved contaminants	– Concentrated reject water with the contaminants which requires proper disposal – High pressure requirement

(continued)

Table 19.1 (continued)

Method	Type	Removal efficiency (%)	Energy consumption (kWh/m^3)	Advantages	Disadvantages
Adsorption	PAC	≥90	0.05–0.45	– Efficient removal of a wide range of PPCPs and EDCs – Can be spiked through the flow of the water Fresh PAC is used within the treatment process	– Removal of used PAC requires additional step such as filtration – Proper mixing is required – Removal efficiency is limited by PAC dose, contact time, water quality, and properties of PPCPs and EDCs – Proper disposal of the used PAC is required – NOM and TOC competition with the targeted compounds.
	GAC	≥90	0.05–0.2	– Efficient removal of a wide range of PPCPs and EDCs – Recovery of the adsorbed components is possible	– Regular regeneration is required to avoid PPCPs and EDCs breakthrough which requires high energy consumption – Removal efficiency is limited by contact time, water quality, and properties of PPCPs and EDCs – NOM and TOC competition with the targeted compounds – Proper disposal of the used adsorbents is required
	CNT	≥90	0.05–0.19	– Efficient removal of a wide range of PPCPs and EDCs – High adsorption capacity due to the structure of the adsorbents with high aspect ratio – Recovery of the adsorbed components is possible	– Applications in large scales are very expensive – Their toxicological impacts on the environment should be studied – Proper disposal of the used adsorbents is required – Adsorbent regeneration requires high energy – Removal efficiency is limited by contact time and water quality – NOM and TOC competition with the targeted compounds

4 Advanced Oxidation Processes

As discussed in the previous sections, conventional water and wastewater treatment technologies are insufficient for the removal of PPCPs and EDCs. On the other hand, some advanced technologies, such as activated carbon and UF and RO, are efficient, but their main concerns are their toxicological effects as well as their cost-intensive processes [16, 70, 84]. Therefore, an advanced treatment technology called advanced oxidation processes (AOPs) has attracted increasing interest in recent decades [70, 93]. In AOPs, degradation occurs via reactions with free radicals, which are mainly hydroxyl radicals (•OH) that have very short lifetimes of microseconds [93]. •OH can be produced in different ways during different AOPs, such as Fenton, photoFenton, photocatalysis, ozone, O_3/H_2O_2, and O_3/UV [3, 70, 84, 93].

These radicals are highly reactive species with the ability to oxidize a wide range of organics at a rate constant of 10^8–10^{10} M^{-1} s^{-1} [84] and a redox potential of 2.85 eV, which is much greater than the redox potentials of ozone (2.07 eV) and H_2O_2 (1.77 eV) [93]. The limited selectivity of these radicals makes them superior oxidants for the removal of a variety of organics in water treatment processes [3, 93]. When organics are completely decomposed by •OH, harmless inorganic compounds, such as water and carbon dioxide, are produced. Another advantage of degradation by •OH is its fast decomposition under normal conditions without the need for an additional step to remove it from the water [84]. Among the different types of AOPs, O_3, H_2O_2, and titanium dioxide photolysis are the most popular methods [70].

4.1 Factors Affecting Oxidation Processes

Similar to other processes, AOPs can be affected by several factors that can reduce the treatment quality. These factors are explained below and should be considered before the utilization of AOPs for water and wastewater treatment.

4.1.1 Solution Temperature

Temperature is an important factor in ozone-related processes. Ozone solubility is greater at lower temperatures; thus, it is more stable in water and takes a longer time to decompose [20, 23]. The ozone solubility in pure water is zero at 60 °C and increases with decreasing temperature to reach 0.64 L ozone/L water at 0 °C [36]. The effect of temperature on ozone decay in treated wastewater has been investigated in several studies. An example is the study by Csefalvay et al. [20], where three different temperatures were studied. The pH of the wastewater was 8, and the dissolved organic carbon (DOC) concentration was approximately 9 mg/L, with an initial ozone dosage of 11.5 mg/L. They reported that ozone decays through first-order kinetics at all the studied temperatures (8, 18, and 27 °C) in 10 min. The rate constants were determined to be 0.0059, 0.0114, and 0.0241 s^{-1} for temperatures of 8, 18, and 27 °C, respectively, which showed that ozone decays faster at higher temperatures. They concluded that when the temperature was increased to 10 °C, the rate constantly doubled. Elovitz et al. [23] studied the effect of temperature on ozone decay in natural source water (Lake Zurich), which has an alkalinity of approximately 125 mg/L as $CaCO_3$ and a pH of 8. The temperature varied from 5 to 35 °C, and the ozone decay rate was increased by increasing the temperature with an initial ozone dosage of 1 mg/L. They reported that at temperatures of 5 and 15 °C, ozone decays in two phases, an initial fast decay phase and a slower decay phase, where both phases follow first-order kinetics. For the temperatures of 25 and 35 °C, the initial phase was negligible, and the ozone decay followed first-order kinetics throughout one entire phase. They also studied the exposure of •OH and reported a constant value for all the studied temperatures, which shows that the generation of hydroxyl radicals is independent of temperature variations.

4.1.2 Solution pH

The rate of radical formation is highly dependent on the pH of the solution, making it an important parameter to consider in oxidation processes [3]. At acidic pH values, molecular ozone is the major oxidant, whereas at pH values higher than 7, the major oxidant is the hydroxyl radical due to faster OH^--related ozone decomposition. As explained earlier, oxidation by less selective hydroxyl radicals is faster and more efficient than ozone oxidation is [119]. Thus, in most cases, the removal of organic compounds is faster and more efficient at higher pH values, as this is strongly dependent on the nature of the contaminant to be removed as well as the water quality. Since wastewater contains numerous reactive species in organic matter, at higher pH values, the decomposition of ozone in wastewater could be attributed to the protonation and/or deprotonation of reactive species [17, 23]. When the wastewater pH is changed, the speciation of HCO_3^- and CO_3^{2-} is altered, and in turn, the rate of •OH scavenging is affected [23, 82]. Typically, the solution pH decreases because of the formation of acidic species in the photooxidation process. Thus, adjusting the pH value is necessary to achieve high treatment efficiency [3].

In a study by Hewes and Davison [43], the decomposition of ozone in water was found to be of second order at pH 2 and 4 regardless of the temperature. At pH 6, the reaction order was determined to be 1.5–2, and when the pH was increased to 8, it was found to follow first-order kinetics. The rate of ozone decomposition increased with increasing pH, as reported in other studies, where at pH 8, the rate constant was determined to be 12.152 h^{-1} at 20 °C. Similarly, Buffle et al. [17] reported that increasing the pH caused an increase in the ozone decomposition rate and the production of •OH. They studied the effect of pH on wastewater treatment plant effluent. Slow ozone decomposition was observed at pH 2, and increasing the pH to 7.9 resulted in an increase in the decomposition rate. Ozone decomposition was fitted according to a second-order kinetic model. The production of •OH also increased with increasing pH, reaching a maximum concentration at pH 6.7, and increasing the pH further to 7.9 did not change the concentration of •OH. They reported an increase in the concentration of OH^- with increasing pH, but its rate of reaction with ozone was too small to explain the increase in ozone decomposition at higher pH values. Thus, they concluded that the effect of pH on ozone decomposition in wastewater could be due to the protonation and deprotonation of the reactive species in the organic matter. In Staehelin and Holgne's [82] investigation of ozone decay in pure water at 20 °C, pseudo-first-order kinetics were reported at pH values between 8 and 10. These findings agreed with other studies in which shorter half-lives were obtained at higher pH values. They reported that pH-dependent ozone decomposition is controlled by the concentration of OH^-, which is of first order for both ozone and OH^- concentrations. The increased production of OH^- at pH values higher than 7 was also reported by Bakcioglu and Otker [8], who observed an increased oxidation rate at pH >7 because oxidation by radicals is faster than direct oxidation.

4.1.3 Alkalinity and Scavengers

Water alkalinity has considerable effects on ozonation processes since carbonate and bicarbonate decrease the efficiency of water treatment. Carbonate and bicarbonate are OH radical scavengers, and the produced $CO_3^{\bullet-}$ is less reactive than the hydroxyl radical is [3, 62]. The following reactions illustrate the scavenging mechanism by carbonate and bicarbonate [3]:

$$HCO_3^- + \bullet OH \rightarrow CO_3^{\bullet-} + H_2O \tag{19.14.1}$$

$$CO_3^{2-} + \bullet OH \rightarrow CO_3^{\bullet-} + OH^- \tag{19.14.2}$$

The presence of some inorganic ions and salts, such as chloride, sulfate, and phosphate, can affect the efficiency of contaminant removal by AOPs. By increasing the water salinity, a longer time is required to achieve complete removal of a contaminant. Anions can scavenge hydroxyl radicals and form anion radicals, which are less reactive than hydroxyl radicals and may react with hydrogen peroxide [6, 39].

The •OH scavenging effect in water with high alkalinity has been studied by different researchers. Broceus et al. [16] reported better ozonation performance for a drinking water source with 9 mg/L $CaCO_3$ alkalinity than for source water with 80 mg/L $CaCO_3$. Elovits et al. [23] investigated the effect of alkalinity on the ozonation of drinking water from a lake. They decarbonated the water to obtain different alkalinities, and the results revealed greater •OH scavenging effects at higher alkalinities and thus slower ozone decomposition. Staehelin and Hoigne [82] analyzed ozone decomposition in pure water and natural water. Similarly, they reported faster ozone decomposition in pure water and slower decomposition in natural waters where the carbonate concentrations are high than those in pure water and can scavenge •OH radicals. They also reported that at low carbonate concentrations, the rate of ozone decay depends on the initial ozone dosage and time, whereas at higher carbonate concentrations, the rate of ozone decay becomes independent of the initial dosage. These results were valid at pH 8 to 10 at 20 °C and followed pseudo-first-order kinetics. They concluded that the reactions of OH^- and HO_2^- with ozone initiate radical chain reactions. The kinetic chain length of these reactions depends on the rate of the reaction of ozone with the formed radicals in competition with other solutes present in the water. As a result, these chain reactions are more important in pure water than in drinking or wastewater because of the scavenging effect of OH radicals by organic impurities and bicarbonate.

4.1.4 Natural Organic Matter

Natural organic matter (NOM) can affect the oxidation process of targeted contaminants and thus reduce their removal efficiency. NOM may act as an inhibitor and hinder the production of oxidizing radicals such as OH radicals through scavenging effects. In some cases, it has been reported that NOM acts as an oxidant and enhances the removal efficiency of target contaminants by producing more oxidizing radicals [18, 53]. This is highly dependent on the nature of the water matrix and the properties of NOM. In photooxidation and photocatalytic processes, NOM can reduce the efficiency of illumination as well as the catalyst adsorption capacity because of NOM adsorption and competition with target compounds [5, 15].

4.1.5 UV Light Absorbers and Blockers

In UV-related AOPs, different particles can affect the process in different ways. Some organics can absorb UV light, undergo photochemical reactions, and decompose; therefore, the portion of the UV light that is absorbed by these organics is not involved in the oxidation process and is wasted. In addition, light irradiation can block or scatter due to the presence of suspended particles, thus decreasing the light absorption efficiency [3]. The schematic in Fig. 19.1 shows the effects of different particles on the UV absorbance and efficiency.

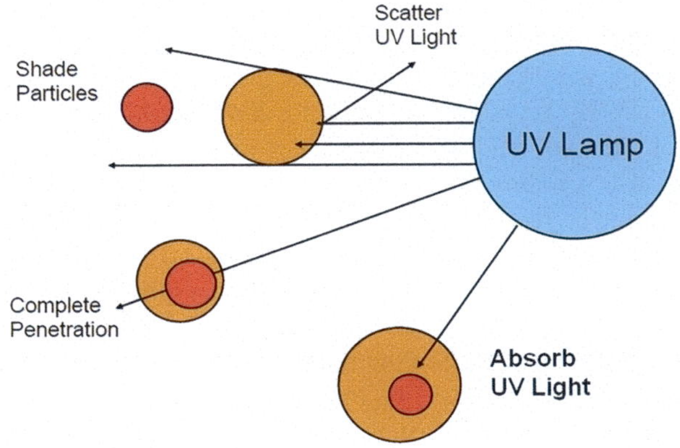

Fig. 19.1 Schematic of UV light absorbers and blockers

4.2 Removal of PPCPs and EDCs Via AOPs

The removal of a variety of microorganic and inorganic components using AOPs has been reported to be successful. This section provides a brief summary of different studies on the removal of PPCPs and EDCs by AOPs. As discussed earlier, PPCPs and EDCs are also detected in some drinking water sources, and attempts at their removal have been investigated by many researchers. For example, in one study [45], the presence of 55 different PPCPs and EDCs was investigated in raw water used as a drinking water source. The concentrations of these compounds were also analyzed after the treatment process. The analysis revealed that raw water is contaminated with 35 PPCPs and EDCs at concentrations of up to 1200 ng/L. The water treatment process consists of prechlorination, coagulation, sand filtration, ozonation, GAC filtration, and postchlorination. The treatment process resulted in the complete removal of most of the compounds except phenytoin, atenolol, hydrochlorothiazide, sotalol, and carbamazepine epoxide. Although these compounds were still detected in the water, their removal efficiencies exceeded 95%. They reported that the main removal step was chlorination and ozonation, with negligible removal via coagulation, flocculation, and sand filtration, whereas GAC was found to be effective for compounds with high hydrophobicity.

In a pilot plant study by Zhang et al. [98], the optimum combination of treatment processes was investigated for the removal of 15 PPCPs and EDCs as indicators from different chemical groups. The pilot plant processes included dissolved air flotation (DAF), pre- and intermediate ozonation with and without H_2O_2, intermediate chlorination, dual media filtration, GAC, and a UV/H_2O_2 system. They concluded that GAC and UV/H_2O_2 were efficient for the removal of more than 90% of most of the compounds. Intermediate ozonation was reported to be more efficient than preozonation. Since the ozone process is dependent on the ozone decomposition rate, the addition of H_2O_2 was suggested to accelerate its decomposition. The lowest removal rates were reported for fire-retardant tri(2-chloroethyl) phosphate (TCEP) and cotinine, which are nicotine metabolites. Although their removal rates by other processes were low, intermediate ozonation followed by dual media filtration and GAC resulted in removals higher than 90%. Cotinine removal by UV/H_2O_2 was also promising, with removal rates higher than 80%. The use of intermediate ozonation and a UV/H_2O_2 system for the removal of atrazine, DEET (N,N-diethyl-meta-toluamide), atenolol, and iopromide was found to be the most efficient process, with removals exceeding 90%, while removal by other processes was less than 75%. Acetaminophen, carbamazepine, sulfamethoxazole, trimethoprim, and 17β-estradiol were reported to be significantly removed by all the different AOP combinations. On the other hand, DAF and dual media filtration were insignificant, with removals of less than 50% for most of the compounds.

Tokumura et al. [87] investigated the removal of carbamazepine and diclofenac via Fe/H_2O_2/UV, TiO_2/UV, and O_3/H_2O_2 in model wastewater and pure water. In the Fe/H_2O_2/UV process, the model wastewater showed an improved Fenton reaction, as the water included extra iron ions; on the other hand, the photodegradation process was affected by the formation of complex reactions in the water media and the scattering of UV light. The degradation of carbamazepine and diclofenac during the TiO_2/UV process was affected by the scavengers and UV light scattering in the model wastewater in addition to the adsorption of other compounds onto the surface of TiO_2. In the O_3/H_2O_2 process, pharmaceutical removal was influenced by scavengers, which hindered the production of •OH as well as ozone dissolution in the model wastewater. The mass transfer of ozone from gas to the liquid phase can be disturbed by the compounds in the wastewater, causing ozone to separate from gas and hence lowering the removal efficiency. They concluded that O_3/H_2O_2 is more efficient for the removal of carbamazepine and diclofenac, where complete removal was achieved after 4 min of reaction time in pure water. Complete removal via the other two processes in pure water was obtained after 15 min and was increased to more than 2 h at different parameters. Using a process that degrades PPCPs and EDCs faster is more feasible for the treatment of wastewater in large amounts on a daily basis. In recent years, many studies have been conducted in this area worldwide, showing that the removal of PPCPs and EDCs is highly dependent on the properties of the compound and that different AOP methods yield different results accordingly. Table 19.2 shows the efficiency of different types of AOPs for the removal of a variety of PPCPs and EDCs from different groups.

Borikar et al. [12] reported only 1–4% removal of gemfibrozil and 20–26% removal of ibuprofen in pilot-scale tests using coagulation and lime softening treatments from a contaminated water source. The conventional process of the pilot scale was followed by the O_3/H_2O_2 AOP. The complete removal of gemfibrozil and

Table 19.2 Removal efficiencies of various PPCPs and EDCs by different AOPs

Group	Compound	Ozone	O_3/H_2O_2	H_2O_2/UV	O_3/H_2O_2/UV
Analgesic	Ibuprofen	Good	Good	Excellent	Excellent
	Acetaminophen	Excellent	Good	Excellent	Excellent
	Diclofenac	Excellent	Excellent	Excellent	Excellent
Antibiotic	Sulfamethoxazole	Excellent	Excellent	Excellent	Excellent
	Trimethoprim	Excellent	Excellent	Excellent	Excellent
Antiepileptic	Carbamazepine	Excellent	Excellent	Excellent	Excellent
Steroid	17β-estradiol	Excellent	Excellent	Excellent	Excellent
Beta-blocker	Atenolol	Good	Good	Excellent	Excellent
Blood lipid regulator	Gemfibrozil	Good	Moderate	Excellent	Excellent
Psychomotor stimulant	Caffeine	Good	Excellent	Excellent	Excellent
Nicotine metabolite	Cotinine	Poor	Poor	Good	Excellent
X-ray contrast agent	Iopromide	Poor	Poor	Excellent	Excellent
Pesticide	Atrazine	Good	Good	Excellent	Excellent
	DEET	Good	Excellent	Excellent	Excellent
Fire retardant	TCEP	Poor	Poor	Poor	Poor

Excellent ≥95%, good 75–94%, moderate 50–74%, poor <50%

88% removal of ibuprofen were achieved with an ozone dosage of 3.3 mg/L in combination with 0.08 mg/L H_2O_2. A lower ozone dose of 0.8 mg/L resulted in the removal of only 12% gemfibrozil. Water sources with higher alkalinity, turbidity, and dissolved organic carbon (DOC) had relatively lower removal rates because of the scavenging effects of carbonate particles. They also investigated the removal of gemfibrozil and ibuprofen by conventional treatment followed by UV/H_2O_2 in pilot-scale tests. The overall removal efficiency was reported to be approximately 83% when the UV intensity was increased to 2000 mJ/cm^2, and increasing the contact time resulted in complete removal with 5 mg/L H_2O_2. The removal of ibuprofen ranged from 80% for the case with the highest turbidity, DOC, and alkalinity to 98% for the water source with lower turbidity, alkalinity, and DOC. This finding shows that UV light absorbance is affected by the particles in the water, which may absorb the light and shade the targeted contaminant or scatter the UV light. They reported that the use of O_3/H_2O_2 is more efficient than the use of UV/H_2O_2 AOPs, which can result in greater removal in a shorter time.

The effects of alkalinity and DOC on the removal efficiency of different pharmaceuticals were also investigated by Huber et al. [44] via ozonation and AOP with O_3 combined with H_2O_2 in different natural waters. Depending on the water quality, ibuprofen removal varies from 31% to 77% with the ozonation process. This value increased to 78–90% removal with the addition of H_2O_2 at an initial concentration of ~0.102 mg/L. They conducted their experiments at 10 °C, pH 8, with a 2 mg/L ozone dose, and the contact time was fixed at 10 minutes.

Shu et al. [75] studied the removal of gemfibrozil with medium-pressure UV/H_2O_2. They concluded that by increasing the H_2O_2 dose, greater removal can be achieved. In addition, they reported that lower initial concentrations of gemfibrozil resulted in greater removal. Their investigation revealed a pH of 7 and high initial gemfibrozil concentrations between 20 and 40 mg/L. They also studied the removal of ibuprofen with medium-pressure UV/H_2O_2. Similar to their results with gemfibrozil, it was concluded that increasing the H_2O_2 dose results in greater removal, while it was also reported that when the initial concentration of ibuprofen is lower, greater removal can be obtained. A different study by Giri et al. [30] on the removal of 1 mg/L of gemfibrozil in combination with 15 other pharmaceuticals by 6 different AOPs revealed the complete removal of gemfibrozil in the first 5 min with O_3/TiO_2. A removal of ~80% was achieved via ozonation as well as the combination of O_3 and UV, whereas moderate removals were achieved via UV/H_2O_2, UV/TiO_2, and UV. According to their study, the best removal efficiency of ibuprofen was achieved via O_3/UV/TiO_2, with ~95% removal in 1 h. An O_3/UV system also yielded relatively good results, with approximately 75% removal in 1 h, whereas ~70% removal was achieved with UV/TiO_2. They also reported moderate removal with UV and O_3/TiO_2 and poor removal with UV/H_2O_2 and O_3. Madhavan et al. [57] achieved 98% removal of 18.5 mg/L ibuprofen using UV/TiO_2 (1 h contact time), whereas no removal occurred when the sample was exposed to UV alone. They also studied the effect of using UV/Fe^{3+} for the removal of high concentrations of ibuprofen, which resulted in 82% removal.

Ternes et al. [85] studied the removal of ibuprofen along with a variety of other PPCPs from the effluent of a sewage treatment plant in Germany at a concentration of 0.13 μg/L. They obtained 48% ibuprofen removal with 5 mg/L ozone, and the removal increased to >62% when the ozone dosage was increased to 15 mg/L.

A study investigating the removal of ibuprofen and gemfibrozil from treated sewage effluent by chlorination, ozonation, and O_3/H_2O_2 revealed insignificant removal by chlorination of both pharmaceuticals [24]. On the other hand, very fast removal of both compounds was achieved via ozone and O_3/H_2O_2. The study of the rate of degradation by second-order kinetics revealed that after 1.5 min, 80% of the ibuprofen was removed by O_3/H_2O_2, whereas after 5 min of reaction, 80% of the ibuprofen was removed by ozonation. The limited removal of ibuprofen was related to the complete consumption of radicals by other organics in the treated sewage effluent. Gemfibrozil was removed within the first 30 s for both the ozonation and O_3/H_2O_2 processes, which was attributed to its chemical structure having active sites. They investigated the effect of temperature in the range of 20–40 °C for both processes and reported negligible effects. The pH variation resulted in lower ibuprofen removal at pH 6 with ozonation. The formation of bromate as an oxidation byproduct was evaluated and was found to increase as the initial bromide concentration increased. By increasing the bromide concentration, ibuprofen removal decreased in the ozonation experiments, as a higher ratio of ozone to •OH was consumed by the increased bromide concentration. In the O_3/H_2O_2 system, the removal of ibuprofen was not affected by increasing bromide concentration because the faster production of •OH resulted in high removal of ibuprofen within 5 min regardless of the increased bromide concentration. To increase the removal efficiency of PPCPs and reduce possible oxidation byproducts, they suggested the use of a catalytic ozonation process in which ozone is combined with a catalyst such as activated carbon.

In addition to removal efficiency, evaluating the cost of different types of AOPs is also important prior to their application. Table 19.3 summarizes the cost efficiency related to energy consumption. The table is generated according to the literature [4, 26, 27, 29, 38, 51, 54, 59, 69, 88, 89, 94, 95].

Table 19.3 Advantages and disadvantages of different types of AOPs

Type	Energy consumption (kWh)	Advantages	Disadvantages
O_3	0.03–1.09	– Efficient removal of a wide range of PPCPs and EDCs – Has disinfection capability – Removal process is very fast – Ozone degradation is fast and there is no need for additional process for its removal – Dosage can easily be adjusted	– Possibility of formation of oxidation byproducts and bromate – Highly dependent on the quality of the water such as pH and alkalinity – Possibility of radical scavenging effects and thus reduction on the efficiency of oxidation process – Requires routine inspection – Adjustment of the influent water quality might be required

(continued)

Table 19.3 (continued)

Type	Energy consumption (kWh)	Advantages	Disadvantages
O_3/H_2O_2	0.035–0.39	– Enhanced PPCPs and EDCs removal due to the formation of highly reactive hydroxyl radicals – Has disinfection capability – Removal process is very fast – Ozone degradation is fast and there is no need for additional process for its removal – Dosages can easily be adjusted	– Possibility of formation of oxidation byproducts and bromate – Highly dependent on the quality of the water such as pH and alkalinity – Possibility of radical scavenging effects and thus reduction on the efficiency of oxidation process – Requires routine inspection – Adjustment of the influent water quality might be required
H_2O_2/UV	0.16–1.8	– No potential for bromate formation – Efficient removal of a wide range of PPCPs and EDCs – Removal process is very fast – UV lamp can provide disinfection – H_2O_2 dosage can easily be adjusted	– Possibility of formation of oxidation byproducts – Requires lamp replacement and routine inspection – Adjustment of the influent water quality might be required – Turbidity can affect the UV penetration – UV lamps are energy intensive – Water contamination with mercury in case of UV lamp failure
O_3/UV	0.37–1.19	– Efficient removal of a wide range of PPCPs and EDCs – Has disinfection capability – Removal process is very fast – O_3 dosage can easily be adjusted	– Possibility of formation of oxidation byproducts and bromate – Requires lamp replacement and routine inspection – Adjustment of the influent water quality might be required – Turbidity can affect the UV penetrat – UV lamps are energy intensive – Water contamination with mercury in case of UV lamp failure

(continued)

Table 19.3 (continued)

Type	Energy consumption (kWh)	Advantages	Disadvantages
TiO_2 photocatalysis	0.33–0.5	– Quick formation of hydroxyl radicals – Can remove a variety of PPCPs and EDCs	– Possibility of formation of oxidation byproducts – Requires high maintenance and pH adjustment – Proper mixing is required – Replacement of TiO_2 catalysts – Turbidity can affect the UV penetration – UV lamps are energy intensive – Expensive catalysts – Water contamination with mercury in case of UV lamp failure – Difficult for full-scale applications – Long contact time is required (usually more than 30 min)
Fenton	0.039–0.11	– The process is very simple with easy application – Operates with low energy consumption	– Possibility of formation of oxidation byproducts – Highly dependent on the pH and pH adjustments will increase the cost – Proper mixing is required – Formation of sludge which requires proper disposal – Difficult for full-scale applications

Although ozone-related processes are very fast and efficient for most contaminants, the cost and production of oxidation byproducts and bromate are major issues. For this reason, in addition to ozone and ozone/hydrogen peroxide processes, a combined integrated oxidation and adsorption process called catalytic ozonation is suggested. This combined method is expected to show high efficiency for the removal of different contaminants and is preferable over other advanced technologies, such as membranes. The catalytic ozonation process is briefly explained below.

5 Catalytic Ozonation Process

In the catalytic ozonation process, catalysts increase the removal rate by enhancing the oxidation process. This happens on the surface of the catalysts, which speeds up ozone decomposition and the formation of •OH. In addition, removal by the

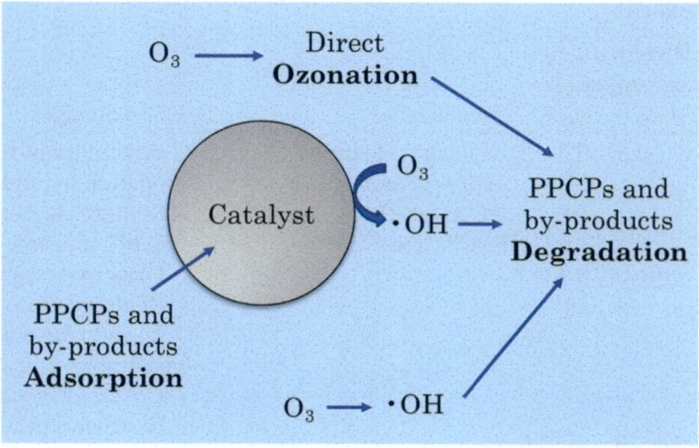

Fig. 19.2 Schematic of the catalytic ozonation process

adsorption process can occur in catalysts, which are usually activated carbon. Another advantage of the catalytic ozonation process is the possibility of adsorption of the produced oxidation byproducts. These adsorbed compounds can then be oxidized by ozone and/or •OH, hence creating more active sites for further adsorption. This in situ regeneration increases the lifetime of the adsorbents/catalysts. Compared with both the ozonation process and the adsorption process, the catalytic ozonation process has the benefit of cost reduction. In the catalytic ozonation process, a lower ozone dose might be sufficient for greater removal of PPCPs and EDCs than ozone alone; however, there is no need for periodic regeneration, as is required in the adsorption process, which is a disadvantage of these two processes. Figure 19.2 shows a schematic of the catalytic ozonation process for the degradation of PPCPs from water.

Catalytic ozonation is a new process for the removal of PPCPs and EDCs and has attracted increased interest in recent years. In comparison with degradation by ozone alone, enhanced efficiency has been reported. For example, Wang et al. [91] reported 53% ibuprofen removal by ozonation and 86% removal by a catalytic ozonation process using activated carbon produced from sewage sludge and corncob treated with $ZnCl_2$. Similarly, in an investigation by Bentancur-Corredor et al. [11], the ibuprofen removal efficiency increased from 27% by ozone alone to 55% when manganese copper alumina catalysts were used in the catalytic ozonation process. The degradation of ibuprofen and acetylsulfamethoxazole was reported to be faster than that of ozone alone when multiwalled carbon nanotubes were used in the catalytic ozonation process [22]. In very recent works, mesoporous manganese dioxide was used as a catalyst for the removal of ibuprofen and metoprolol by He et al. [41], whereas Ponnusamy et al. [65] investigated the removal of ibuprofen by MnO_2 nanosheets grown on woven carbon microfibers, and both studies reported faster removal by catalytic ozonation, with an increase in ibuprofen removal from 65 to 90% in the latest study.

In a recent work by Saeid et al. [71], the removal of ibuprofen from aqueous solution with ozonation and catalytic ozonation was investigated in a semibatch reactor. According to their study, for 95% removal of 10 mg/L ibuprofen, an ozone flow rate of 1000 ml/min in 4 h is needed. They also studied the effect of temperature on ibuprofen removal via ozonation by varying the temperature from 5 to 35 °C. In the first 2 h of the experiment, higher removal rates were achieved at lower temperatures, but after 2 h, the influence of temperature was minimal. A study of the effect of initial ibuprofen concentration revealed a higher removal rate at higher ibuprofen concentrations, whereas the final removal efficiency after 4 h was the same for the different initial concentrations. Finally, the addition of zeolite catalysts could enhance the removal of ibuprofen.

6 Conclusions and Recommendations

PPCPs and EDCs mainly find their way into water bodies via poorly treated effluent emanating from conventional water and wastewater treatment plants. To achieve higher removal efficiencies of these pollutants, advanced methods are suggested. Many water and wastewater treatment plants equipped with UF and RO membrane filtrations have achieved significant removal of a wide variety of PPCPs and EDCs. The high operation cost of these processes is a major concern because they are energy-intensive processes. This is in addition to the regular maintenance required for the membranes and the proper disposal of the brine stream, which is more concentrated than the feed water. Adsorption processes by different adsorbents, such as activated carbon, have also been widely used in treatment plants and are successful for the removal of different PPCPs and EDCs. Like in membrane systems, the high cost required for regular regeneration of adsorbents is a drawback of adsorption processes in addition to the cost needed for adsorbent production. In many cases, the toxicity of adsorbents is another issue that should be taken into account. High removal rates of PPCPs and EDCs are obtained in plants via ozonation and AOPs. The main disadvantage of oxidation processes is the possibility of the formation of oxidation byproducts, which mostly have greater toxicological effects than the parent chemical does.

Although these advanced technologies are currently used in many treatment plants, new methods that are more cost-efficient with fewer adverse effects on the environment are viable. An example is the utilization of the catalytic ozonation process in water and wastewater treatment plants. The catalytic ozonation process is a combination of ozonation and adsorption processes. Few studies have investigated the removal of some PPCPs and EDCs by catalytic ozonation, and all these investigations have concluded that, compared with the use of each process alone, each process has greater removal efficiency. In the combined process, a lower ozone dosage can be applied, and the need for frequent adsorbent regeneration will be eliminated because of the degradation of the adsorbed compound by ozone or •OH; hence, the operation cost will be lower. To reduce the cost of the catalytic ozonation process further, the use of adsorbents produced from waste materials, such as activated carbon produced from fruit stones, is advised. The activation method of adsorbents should also be investigated to find the most convenient method in terms of quality and cost.

To fill the gaps in this area, comprehensive assessments should be carried out to assess the performance of different catalysts under different treatment parameters and conditions by studying the formation of degradation byproducts and the rates of PPCP and EDC removal. It is also suggested that scale-up studies be conducted to evaluate the viability and practicality of employing a catalytic ozonation process at larger water and wastewater treatment plants. In conclusion, the successful use of a catalytic ozonation process can contribute to the development of more efficient and sustainable water treatment technologies.

References

1. Al-Ghouti, M. A., Li, J., Salamh, Y., et al. (2010a). Adsorption mechanisms of removing heavy metals and dyes from aqueous solution using date pits solid adsorbent. *Journal of Hazardous Materials, 176*, 510–520. https://doi.org/10.1016/j.jhazmat.2009.11.059
2. Almanassra, I. W., Kochkodan, V., Ponnusamy, G., et al. (2020). Carbide Derived Carbon (CDC) as novel adsorbent for ibuprofen removal from synthetic water and treated sewage effluent. *Journal of Environmental Health Science and Engineering, 36*, 1375. https://doi.org/10.1007/s40201-020-00554-0
3. Andreozzi, R. (1999). Advanced oxidation processes (AOP) for water purification and recovery. *Catalysis Today, 53*, 51–59. https://doi.org/10.1016/S0920-5861(99)00102-9
4. Asaithambi, P., Saravanathamizhan, R., & Matheswaran, M. (2015). Comparison of treatment and energy efficiency of advanced oxidation processes for the distillery wastewater. *International journal of Environmental Science and Technology, 12*, 2213–2220. https://doi.org/10.1007/s13762-014-0589-9
5. Autin, O., Hart, J., Jarvis, P., et al. (2013). The impact of background organic matter and alkalinity on the degradation of the pesticide metaldehyde by two advanced oxidation processes: UV/H2O2 and UV/TiO2. *Water Research, 47*, 2041–2049. https://doi.org/10.1016/j.watres.2013.01.022
6. Bacardit, J., Stötzner, J., Chamarro, E., & Esplugas, S. (2007). Effect of salinity on the photo-fenton process. *Industrial and Engineering Chemistry Research, 46*, 7615–7619. https://doi.org/10.1021/ie070154o
7. Baccar, R., Sarrà, M., Bouzid, J., et al. (2012). Removal of pharmaceutical compounds by activated carbon prepared from agricultural by-product. *Chemical Engineering Journal, 211–212*, 310–317. https://doi.org/10.1016/j.cej.2012.09.099
8. Balcioglu, I. A., & Otker, M. (2003). Treatment of pharmaceutical wastewater containing antibiotics by O3 and O3/H2O2 processes. *Chemosphere, 50*, 85–95.
9. Behera, S. K., Kim, H. W., Oh, J. E., & Park, H. S. (2011). Occurrence and removal of antibiotics, hormones and several other pharmaceuticals in wastewater treatment plants of the largest industrial city of Korea. *Science of the Total Environment, 409*, 4351–4360. https://doi.org/10.1016/j.scitotenv.2011.07.015
10. Belhamdi, B., Merzougui, Z., Trari, M., & Addoun, A. (2016). A kinetic, equilibrium and thermodynamic study of L-phenylalanine adsorption using activated carbon based on agricultural waste (date stones). *Journal of Applied Research and Technology, 14*, 354–366.
11. Betancur-Corredor, B., Soltan, J., & Peñuela, G. A. (2016). Mineralization of ibuprofen and humic acid through catalytic ozonation. *Ozone: Science and Engineering, 38*, 203–210. https://doi.org/10.1080/01919512.2015.1114880

12. Borikar, D., Mohseni, M., & Jasim, S. (2015). Evaluation and comparison of conventional and advanced oxidation processes for the removal of PPCPs and EDCs and their effect on THM-formation potentials. *Ozone Science & Engineering, 37*, 154–169. https://doi.org/10.1080/01919512.2014.940028
13. Bouchemal, N., Azoudj, Y., Merzougui, Z., & Addoun, F. (2012). Adsorption modeling of orange G dye on mesoporous activated carbon prepared from Algerian date pits using experimental designs. *Desalination and Water Treatment, 45*, 284–290. https://doi.org/10.1080/19443994.2012.692042
14. Bouhamed, F., Elouear, Z., & Bouzid, J. (2012). Adsorptive removal of copper(II) from aqueous solutions on activated carbon prepared from Tunisian date stones: Equilibrium, kinetics and thermodynamics. *Journal of the Taiwan Institute of Chemical Engineers, 43*, 741–749. https://doi.org/10.1016/j.jtice.2012.02.011
15. Brame, J., Long, M., Li, Q., & Alvarez, P. (2015). Inhibitory effect of natural organic matter or other background constituents on photocatalytic advanced oxidation processes: Mechanistic model development and validation. *Water Research, 83*, 362–371. https://doi.org/10.1016/j.watres.2015.07.044
16. Broseus, R., Vincent, S., Aboulfadl, K., et al. (2009). Ozone oxidation of pharmaceuticals, endocrine disruptors and pesticides during drinking water treatment. *Water Research, 43*, 4707–4717. https://doi.org/10.1016/j.watres.2009.07.031
17. Buffle, M., Schumacher, J., Meylan, S., et al. (2006). Ozonation and advanced oxidation of wastewater: Effect of O3 dose, pH, DOM and HO- scavengers on ozone decomposition and HO generation. *Ozone Science & Engineering, 28*, 247–259. https://doi.org/10.1080/01919510600718825
18. Camel, V., & Bermond, A. (1998). The use of ozone and associated oxidation processes in drinking water treatment: Review paper. *Water Research, 32*, 3208–3222.
19. Cornejo, P. K., Santana, M. V. E., Hokanson, D. R., et al. (2014). Carbon footprint of water reuse and desalination : a review of greenhouse gas emissions and estimation tools. *Water Reuse, 4*, 238–252. https://doi.org/10.2166/wrd.2014.058
20. Cséfalvay, E., Nöthe, T., & Mizsey, P. (2007). Modelling of wastewater ozonation—Determination of reaction kinetic constants and effect of temperature. *Periodica Polytechnica: Chemical Engineering, 51*, 13–17.
21. Darweesh, T. M., & Ahmed, M. J. (2017). Batch and fixed bed adsorption of levofloxacin on granular activated carbon from date (Phoenix dactylifera L.) stones by KOH chemical activation. *Environmental Toxicology and Pharmacology, 50*, 159–166. https://doi.org/10.1016/j.etap.2017.02.005
22. Du, M. S., Chen, K. P., & Lin, Y. P. (2019). Degradation of ibuprofen and acetylsulfamethoxazole by multi-walled carbon nanotube catalytic ozonation: Surface properties, kinetics and modeling. *Environmental Science: Water Research and Technology, 5*, 1758–1768. https://doi.org/10.1039/c9ew00492k
23. Elovitz, M. S., von Gunten, U., & Kaiser, H.-P. (2000). Hydroxyl radical/ozone ratios during ozonation processes, II. The effect of temperature, pH, alkalinity, and DOM properties. *Ozone Science & Engineering, 22*, 123–150. https://doi.org/10.1080/01919510008547216
24. Farzaneh, H., Loganathan, K., Saththasivam, J., & McKay, G. (2020). Ozone and ozone/hydrogen peroxide treatment to remove gemfibrozil and ibuprofen from treated sewage effluent: Factors influencing bromate formation. *Emerging Contaminants, 6*, 225–234. https://doi.org/10.1016/j.emcon.2020.06.002
25. Farzaneh, H., Saththasivam, J., McKay, G., & Parthasarathy, P. (2022). Adsorbent minimization for removal of ibuprofen from water in a two-stage batch process. *Processes, 10*, 453. https://doi.org/10.3390/pr10030453
26. Fischbacher, A., Lutze, H. V., & Schmidt, T. C. (2017). Ozone/H2O2 and ozone/UV processes. In M. Stefan (Ed.), *Advanced oxidation processes for water treatment: Fundamentals and applications* (pp. 163–194). IWA.

27. Freyria, F. S., Geobaldo, F., & Bonelli, B. (2018). Nanomaterials for the abatement of pharmaceuticals and personal care products from wastewater. *Applied Sciences, 8*, 1–16. https://doi.org/10.3390/app8020170
28. Ganiyu, S. O., Van Hullebusch, E. D., Cretin, M., et al. (2015). Coupling of membrane filtration and advanced oxidation processes for removal of pharmaceutical residues: A critical review. *Separation and Purification Technology, 156*, 891–914. https://doi.org/10.1016/j.seppur.2015.09.059
29. Gerrity, D., Ryu, H., Crittenden, J., & Abbaszadegan, M. (2008). Photocatalytic inactivation of viruses using titanium dioxide nanoparticles and low-pressure UV light. *Journal of Environmental Science and Health Part A, 43*, 1261–1270. https://doi.org/10.1080/10934520802177813
30. Giri, R. R., Ozaki, H., Ota, S., et al. (2010). Degradation of common pharmaceuticals and personal care products in mixed solutions by advanced oxidation techniques. *International journal of Environmental Science and Technology, 7*, 251–260. https://doi.org/10.1007/BF03326135
31. Grassi, M., Kaykioglu, G., Belgiorno, V., & Lofrano, G. (2012). Chapter 2: Removal of emerging contaminants from water and wastewater by adsorption process. In *Emerging compounds removal from wastewater* (pp. 15–38). Springer.
32. Grassi, M., Rizzo, L., & Farina, A. (2013). Endocrine disruptors compounds, pharmaceuticals and personal care products in urban wastewater: Implications for agricultural reuse and their removal by adsorption process. *Environmental Science and Pollution Research, 20*, 3616–3628. https://doi.org/10.1007/s11356-013-1636-7
33. Gros, M., Petrovic, M., Ginebreda, A., & Barcelo, D. (2010). Removal of pharmaceuticals during wastewater treatment and environmental risk assessment using hazard indexes. *Environment International, 36*, 15–26. https://doi.org/10.1016/j.envint.2009.09.002
34. Guedidi, H., Reinert, L., Lévêque, J. M., et al. (2013). The effects of the surface oxidation of activated carbon, the solution pH and the temperature on adsorption of ibuprofen. *Carbon, 54*, 432–443. https://doi.org/10.1016/j.carbon.2012.11.059
35. Guedidi, H., Reinert, L., Soneda, Y., et al. (2017). Adsorption of ibuprofen from aqueous solution on chemically surface-modified activated carbon cloths. *Arabian Journal of Chemistry, 10*, 3584–3594. https://doi.org/10.1016/j.arabjc.2014.03.007
36. Guzel-seydim, Z. B., Greene, A. K., & Seydim, A. C. (2004). Use of ozone in the food industry. *Lebensmittel-Wissenschaft & Technologie, 37*, 453–460. https://doi.org/10.1016/j.lwt.2003.10.014
37. Hameed, B. H., Salman, J. M., & Ahmad, A. L. (2009). Adsorption isotherm and kinetic modeling of 2, 4-D pesticide on activated carbon derived from date stones. *Journal of Hazardous Materials, 163*, 121–126. https://doi.org/10.1016/j.jhazmat.2008.06.069
38. Hansen, K. M. S., & Andersen, H. R. (2012). Energy effectiveness of direct UV and UV/H2O2 treatment of estrogenic chemicals in biologically treated sewage. *International Journal of Photoenergy ID, 270320*, 9. https://doi.org/10.1155/2012/270320
39. Hassaan, M. A., El Nemr, A., & Madkour, F. F. (2017). Testing the advanced oxidation processes on the degradation of Direct Blue 86 dye in wastewater. *Egyptian Journal of Aquatic Research, 43*, 11–19. https://doi.org/10.1016/j.ejar.2016.09.006
40. Hassan, S. S., Al-Ghouti, M. A., Abu-Dieyeh, M., & McKay, G. (2020). Novel bioadsorbents based on date pits for organophosphorus pesticide remediation from water. *Journal of Environmental Chemical Engineering, 8*, 103593. https://doi.org/10.1016/j.jece.2019.103593
41. He, Y., Wang, L., Chen, Z., et al. (2021). Catalytic ozonation for metoprolol and ibuprofen removal over different MnO2 nanocrystals: Efficiency, transformation and mechanism. *Science of the Total Environment, 785*, 147328. https://doi.org/10.1016/j.scitotenv.2021.147328
42. Hemidouche, S., Assoumani, A., Favier, L., et al. (2017). Removal of some endocrine disruptors via adsorption on activated carbon. In *E-health and bioengineering conference* (pp. 410–413). IEEE.
43. Hewes, C. G., & Davison, R. R. (1971). Kinetics of ozone decomposition and reaction with organics in water. *American Institute of Chemical Engineers, 17*, 141–147.

44. Huber, M. M., Canonica, S., Park, G. Y., & von Gunten, U. (2003). Oxidation of pharmaceuticals during ozonation and advanced oxidation processes. *Environmental Science and Technology, 37*, 1016–1024. https://doi.org/10.1021/es025896h
45. Huerta-Fontela, M., Galceran, M. T., & Ventura, F. (2011). Occurrence and removal of pharmaceuticals and hormones through drinking water treatment. *Water Research, 45*, 1432–1442. https://doi.org/10.1016/j.watres.2010.10.036
46. Jung, C., Son, A., Her, N., et al. (2015). Removal of endocrine disrupting compounds, pharmaceuticals, and personal care products in water using carbon nanotubes: A review. *Journal of Industrial and Engineering Chemistry, 27*, 1–11. https://doi.org/10.1016/j.jiec.2014.12.035
47. Kasprzyk-Hordern, B., Dinsdale, R. M., & Guwy, A. J. (2009). The removal of pharmaceuticals, personal care products, endocrine disruptors and illicit drugs during wastewater treatment and its impact on the quality of receiving waters. *Water Research, 43*, 363–380. https://doi.org/10.1016/j.watres.2008.10.047
48. Kasprzyk-Hordern, B., Dinsdale, R. M., & Guwy, A. J. (2008). The occurrence of pharmaceuticals, personal care products, endocrine disruptors and illicit drugs in surface water in South Wales, UK. *Water Research, 42*, 3498–3518. https://doi.org/10.1016/j.watres.2008.04.026
49. Khulbe, K. C., Feng, C. Y., Matsuura, T., & Ismail, A. F. (2012). Progresses in membrane and advanced oxidation processes for water treatment. *Membrane Water Treatment, 3*, 1–44. https://doi.org/10.12989/mwt.2012.3.3.181
50. Kim, I., & Tanaka, H. (2009). Photodegradation characteristics of PPCPs in water with UV treatment. *Environment International, 35*, 793–802. https://doi.org/10.1016/j.envint.2009.01.003
51. Kim, I., & Tanaka, H. (2011). Energy consumption for PPCPs removal by O3 and O3/UV. *Ozone: Science & Engineering, 33*, 150–157. https://doi.org/10.1080/01919512.2011.549427
52. Kim, S. D., Cho, J., Kim, I. S., et al. (2007). Occurrence and removal of pharmaceuticals and endocrine disruptors in South Korean surface, drinking, and waste waters. *Water Research, 41*, 1013–1021. https://doi.org/10.1016/j.watres.2006.06.034
53. Klavarioti, M., Mantzavinos, D., & Kassinos, D. (2009). Removal of residual pharmaceuticals from aqueous systems by advanced oxidation processes. *Environment International, 35*, 402–417. https://doi.org/10.1016/j.envint.2008.07.009
54. Lee, Y., Gerrity, D., Lee, M., et al. (2016). Organic contaminant abatement in reclaimed water by UV/HO and a combined process consisting of O/HO followed by UV/HO: Prediction of abatement efficiency, energy consumption, and by-product formation. *Environment Science and Technology, 50*, 3809–3819. https://doi.org/10.1021/acs.est.5b04904
55. Le-Minh, N., Coleman, H. M., Khan, S. J., et al. (2010). The application of membrane bioreactors as decentralised systems for removal of endocrine disrupting chemicals and pharmaceuticals. *Water Science and Technology, 61*, 1081–1088. https://doi.org/10.2166/wst.2010.884
56. Li, F., Xia, Q., Cheng, Q., et al. (2017). Conductive cotton filters for affordable and efficient water purification. *Catalysts, 7*, 1–12. https://doi.org/10.3390/catal7100291
57. Madhavan, J., Grieser, F., & Ashokkumar, M. (2010). Combined advanced oxidation processes for the synergistic degradation of ibuprofen in aqueous environments. *Journal of Hazardous Materials, 178*, 202–208. https://doi.org/10.1016/j.jhazmat.2010.01.064
58. Mangwandi, C., Kurniawan, T. A., & Albadarin, A. B. (2020). Comparative biosorption of chromium (VI) using chemically modified date pits (CM-DP) and olive stone (CM-OS): Kinetics, isotherms and influence of co-existing ions. *Chemical Engineering Research and Design, 156*, 251–262. https://doi.org/10.1016/j.cherd.2020.01.034
59. Melemeni, M., Stamatakis, D., Xekoukoulotakis, N. P., et al. (2009). Disinfection of municipal wastewater by TiO2 photocatalysis with UV-A, visible and solar irradiation and BDD electrolysis. *Global NEST Journal, 11*, 357–363.
60. Messaoudi, N. E., Khomri, M. E., Bentahar, S., et al. (2016). Removal of crystal violet by biosorption onto date stones. *Scientific Study & Research Chemistry & Chemical Engineering, 17*, 151–167(17).
61. Mestre, A. S., Bexiga, A. S., Proença, M., et al. (2011). Activated carbons from sisal waste by chemical activation with K 2CO 3: Kinetics of paracetamol and ibuprofen removal

from aqueous solution. *Bioresource Technology, 102*, 8253–8260. https://doi.org/10.1016/j.biortech.2011.06.024
62. Munter, R. (2001). Advanced oxidation processes- current status and prospect. *Proceedings of the Estonian Academy of Sciences. Chemistry, 50*, 59–80. https://doi.org/10.1002/9780470561331.ch18
63. Oturan, M. A., & Aaron, J. (2014). Advanced oxidation processes in water/wastewater treatment: Principles and applications. A review. *Critical Reviews in Environmental Science and Technology, 44*, 2577–2641. https://doi.org/10.1080/10643389.2013.829765
64. Pizzichini, M., & Russo, C. (2011). *Water purification-desalination with membrane technology supplied with renewable energy* (pp. 1–10). Italian National Agency for New Technology Energy and the Environment.
65. Ponnusamy, G., Farzaneh, H., Tong, Y., et al. (2021). Enhanced catalytic ozonation of ibuprofen using a 3D structured catalyst with MnO2 nanosheets on carbon microfibers. *Scientific Reports, 11*, 1–11. https://doi.org/10.1038/s41598-021-85651-2
66. Rahman, M. F., Yanful, E. K., & Jasim, S. Y. (2009). Endocrine disrupting compounds (EDCs) and pharmaceuticals and personal care products (PPCPs) in the aquatic environment: Implications for the drinking water industry and global environmental health. *Journal of Water and Health, 7*, 224–243. https://doi.org/10.2166/wh.2009.021
67. Rautenbach, R., Vossenkaul, K., Linn, T., & Katz, T. (1996). Waste water treatment by membrane processes—New development in ultrafiltration, nanofiltration and reverse osmosis. *Desalination, 108*, 247–253.
68. Richardson, B. J., Lam, P. K. S., & Martin, M. (2005). Emerging chemicals of concern: Pharmaceuticals and personal care products (PPCPs) in Asia, with particular reference to southern China. *Marine Pollution Bulletin, 50*, 913–920. https://doi.org/10.1016/j.marpolbul.2005.06.034
69. Roccaro, P., Sgroi, M., & Vagliasindi, F. G. A. (2013). Removal of xenobiotic compounds from wastewater for environment protection: Treatment processes and costs. *Chemical Engineering Transactions, 32*, 505–510.
70. Rueda-Márquez, J. J., Sillanpää, M., Pocostales, P., et al. (2015). Post-treatment of biologically treated wastewater containing organic contaminants using a sequence of H2O2 based advanced oxidation processes: Photolysis and catalytic wet oxidation. *Water Research, 71*, 85–96. https://doi.org/10.1016/j.watres.2014.12.054
71. Saeid, S., Tolvanen, P., Kumar, N., et al. (2018). Advanced oxidation process for the removal of ibuprofen from aqueous solution: A non-catalytic and catalytic ozonation study in a semi-batch reactor. *Applied Catalysis B: Environmental, 230*, 77–90. https://doi.org/10.1016/j.apcatb.2018.02.021
72. Schäfer, A. I., & Waite, T. D. (2002). *Removal of endocrine disrupters in advanced treatment—The Australian approach* (pp. 37–51). International Water Association.
73. Sewilam, H., & Nasr, P. (2015). Desalinated water for food production in the Arab region. In *Water security in a new world*. Springer.
74. Shojaee Nasirabadi, P., Saljoughi, E., & Mousavi, S. M. (2016). Membrane processes used for removal of pharmaceuticals, hormones, endocrine disruptors and their metabolites from wastewaters: A review. *Desalination and Water Treatment, 3994*, 1–30. https://doi.org/10.1080/19443994.2016.1140081
75. Shu, Z., Bolton, J. R., Belosevic, M., & Gamal El Din, M. (2013). Photodegradation of emerging micropollutants using the medium-pressure UV/H2O2 advanced oxidation process. *Water Research, 47*, 2881–2889. https://doi.org/10.1016/j.watres.2013.02.045
76. Silva, L. L. S., Moreira, C. G., Curzio, B. A., & Fonseca, F. V. (2017). Micropollutant removal from water by membrane and advanced oxidation processes—A review. *Journal of Water Resource and Protection, 9*, 411–431. https://doi.org/10.4236/jwarp.2017.95027
77. Singh, R. (2011). Analysis of energy usage at membrane water treatment plants. *Desalination and Water Treatment, 29*, 63–72. https://doi.org/10.5004/dwt.2011.2988

78. Snyder, S. A., Adham, S., Redding, A. M., et al. (2007). Role of membranes and activated carbon in the removal of endocrine disruptors and pharmaceuticals. *Desalination, 202*, 156–181. https://doi.org/10.1016/j.desal.2005.12.052
79. Snyder, S. A., & Benotti, M. J. (2010). Endocrine disruptors and pharmaceuticals: Implications for water sustainability. *Water Science and Technology, 61*, 145–154. https://doi.org/10.2166/wst.2010.791
80. Snyder, S. A., Westerhoff, P., Yoon, Y., & Sedlak, D. L. (2003). Pharmaceuticals, personal care products, and endocrine disruptors in water: Implications for the water industry. *Environmental Engineering Science, 20*, 449–469. https://doi.org/10.1089/109287503768335931
81. Sotelo, J. L., Rodríguez, A. R., Mateos, M. M., et al. (2012). Adsorption of pharmaceutical compounds and an endocrine disruptor from aqueous solutions by carbon materials. *Journal of Environmental Science and Health—Part B, 47*, 640–652. https://doi.org/10.1080/03601234.2012.668462
82. Staehelin, J., & Holgné, J. (1982). Decomposition of ozone in water: Rate of initiation by hydroxide ions and hydrogen peroxide. *Environmental Science and Technology, 16*, 676–681. https://doi.org/10.1021/es00104a009
83. Sui, Q., Huang, J., Deng, S., et al. (2010). Occurrence and removal of pharmaceuticals, caffeine and DEET in wastewater treatment plants of Beijing, China. *Water Research, 44*, 417–426. https://doi.org/10.1016/j.watres.2009.07.010
84. Suzuki, H., Araki, S., & Yamamoto, H. (2015). Evaluation of advanced oxidation processes (AOP) using O3, UV, and TiO2 for the degradation of phenol in water. *Journal of Water Process Engineering, 7*, 54–60. https://doi.org/10.1016/j.jwpe.2015.04.011
85. Ternes, T. A., Stüber, J., Herrmann, N., et al. (2003). Ozonation: A tool for removal of pharmaceuticals, contrast media and musk fragrances from wastewater? *Water Research, 37*, 1976–1982. https://doi.org/10.1016/S0043-1354(02)00570-5
86. To, M. H., Hadi, P., Hui, C. W., et al. (2017). Mechanistic study of atenolol, acebutolol and carbamazepine adsorption on waste biomass derived activated carbon. *Journal of Molecular Liquids, 241*, 386–398. https://doi.org/10.1016/j.molliq.2017.05.037
87. Tokumura, M., Sugawara, A., Raknuzzaman, M., et al. (2016). Comprehensive study on effects of water matrices on removal of pharmaceuticals by three different kinds of advanced oxidation processes. *Chemosphere, 159*, 317–325. https://doi.org/10.1016/j.chemosphere.2016.06.019
88. Tuerk, J., Sayder, B., Boergers, A., et al. (2010). Efficiency, costs and benefits of AOPs for removal of pharmaceuticals from the water cycle. *Water Science and Technology, 61*, 985–993. https://doi.org/10.2166/wst.2010.004
89. Venditti, S., Klepiszewski, K., & Christian, K. (2017). Lessons learned from European experiences and presentation of case studies. In *Hospital wastewaters* (pp. 153–169).
90. Vince, F., Aoustin, E., Bréant, P., & Marechal, F. (2008). LCA tool for the environmental evaluation of potable water production. *Desalination, 220*, 37–56. https://doi.org/10.1016/j.desal.0000.00.000
91. Wang, H., Zhang, L., Qi, F., et al. (2014). Removal performance and mechanism of Ibuprofen from water by catalytic ozonation using sludge-corncob activated carbon as catalyst. *Journal of Nanoscience and Nanotechnology, 14*, 7266–7271. https://doi.org/10.1166/jnn.2014.8942
92. Westerhoff, P., Yoon, Y., Snyder, S., & Wert, E. C. (2005). Fate of endocrine-disruptor, pharmaceutical, and personal care product chemicals during simulated drinking water treatment processes. *Environment Science and Technology, 39*, 6649–6663.
93. Yang, D., & Zhao, R. (2015). Advanced Oxidation Processes (AOPs) in wastewater treatment. *Water Pollution, 1*, 167–176. https://doi.org/10.1007/s40726-015-0015-z
94. Ying, M., Wu, J., Oakes, K. D., & Hu, A. (2015). Removal of pharmaceuticals and personal care products from wastewater. In L. M. Barrett (Ed.), *Wastewater treatment* (pp. 81–109). Nova Science Publishers.
95. Yonar, T. (2011). *Decolorisation of textile dyeing effluents using advanced oxidation processes* (pp. 1–26). Advances in Treating Textile Effluent.

96. Yoon, Y., Westerhoff, P., Snyder, S. A., et al. (2007). Removal of endocrine disrupting compounds and pharmaceuticals by nanofiltration and ultrafiltration membranes. *Desalination, 202*, 16–23. https://doi.org/10.1016/j.desal.2005.12.033
97. Yu, Z., Peldszus, S., & Huck, P. M. (2009). Adsorption of selected pharmaceuticals and an endocrine disrupting compound by granular activated carbon. 1. Adsorption capacity and kinetics. *Environmental Science and Technology, 43*, 1467–1473. https://doi.org/10.1021/es801961y
98. Zhang, S., Gitungo, S., Axe, L., et al. (2016). A pilot plant study using conventional and advanced water treatment processes: Evaluating removal efficiency of indicator compounds representative of pharmaceuticals and personal care products. *Water Research, 105*, 85–96. https://doi.org/10.1016/j.watres.2016.08.033

Index

M. K. Jindal et al. (eds.), *Emerging Contaminants in Water*, Emerging Contaminants and Associated Treatment Technologies,
https://doi.org/10.1007/978-3-031-89591-3

W

The manufacturer's authorised representative in the EU is Springer
Nature Customer Service Centre GmbH, Europaplatz 3, 69115 Heidelberg,
Germany. If you have any concerns regarding our products, please
contact ProductSafety@springernature.com

Printed and bound by CPI Group (UK) Ltd, Croydon, CR0 4YY
05/10/2026
02228127-0016